T0135335

Communications
in Computer and Information Science   1831

Rationale

The CCIS series is devoted to the publication of proceedings of computer science conferences. Its aim is to efficiently disseminate original research results in informatics in printed and electronic form. While the focus is on publication of peer-reviewed full papers presenting mature work, inclusion of reviewed short papers reporting on work in progress is welcome, too. Besides globally relevant meetings with internationally representative program committees guaranteeing a strict peer-reviewing and paper selection process, conferences run by societies or of high regional or national relevance are also considered for publication.

Topics

The topical scope of CCIS spans the entire spectrum of informatics ranging from foundational topics in the theory of computing to information and communications science and technology and a broad variety of interdisciplinary application fields.

Information for Volume Editors and Authors

Publication in CCIS is free of charge. No royalties are paid, however, we offer registered conference participants temporary free access to the online version of the conference proceedings on SpringerLink (http://link.springer.com) by means of an http referrer from the conference website and/or a number of complimentary printed copies, as specified in the official acceptance email of the event.

CCIS proceedings can be published in time for distribution at conferences or as post-proceedings, and delivered in the form of printed books and/or electronically as USBs and/or e-content licenses for accessing proceedings at SpringerLink. Furthermore, CCIS proceedings are included in the CCIS electronic book series hosted in the SpringerLink digital library at http://link.springer.com/bookseries/7899. Conferences publishing in CCIS are allowed to use Online Conference Service (OCS) for managing the whole proceedings lifecycle (from submission and reviewing to preparing for publication) free of charge.

Publication process

The language of publication is exclusively English. Authors publishing in CCIS have to sign the Springer CCIS copyright transfer form, however, they are free to use their material published in CCIS for substantially changed, more elaborate subsequent publications elsewhere. For the preparation of the camera-ready papers/files, authors have to strictly adhere to the Springer CCIS Authors' Instructions and are strongly encouraged to use the CCIS LaTeX style files or templates.

Abstracting/Indexing

CCIS is abstracted/indexed in DBLP, Google Scholar, EI-Compendex, Mathematical Reviews, SCImago, Scopus. CCIS volumes are also submitted for the inclusion in ISI Proceedings.

How to start

To start the evaluation of your proposal for inclusion in the CCIS series, please send an e-mail to ccis@springer.com.

Ning Wang · Genaro Rebolledo-Mendez ·
Vania Dimitrova · Noboru Matsuda ·
Olga C. Santos
Editors

Artificial Intelligence in Education

Posters and Late Breaking Results, Workshops
and Tutorials, Industry and Innovation Tracks,
Practitioners, Doctoral Consortium and Blue Sky

24th International Conference, AIED 2023
Tokyo, Japan, July 3–7, 2023
Proceedings

 Springer

Editors
Ning Wang ⓘ
University of Southern California
Los Angeles, CA, USA

Vania Dimitrova ⓘ
University of Leeds
Leeds, UK

Olga C. Santos ⓘ
UNED
Madrid, Spain

Genaro Rebolledo-Mendez
University of British Columbia
Vancouver, BC, Canada

Noboru Matsuda ⓘ
North Carolina State University
Raleigh, NC, USA

ISSN 1865-0929 ISSN 1865-0937 (electronic)
Communications in Computer and Information Science
ISBN 978-3-031-36335-1 ISBN 978-3-031-36336-8 (eBook)
https://doi.org/10.1007/978-3-031-36336-8

This Springer imprint is published by the registered company Springer Nature Switzerland AG
The registered company address is: Gewerbestrasse 11, 6330 Cham, Switzerland

Preface

The 24th International Conference on Artificial Intelligence in Education (AIED 2023) marked the 30th anniversary of the International Artificial Intelligence in Education Society and the 24th edition of its International Conference. This year, the conference's theme was AI in Education for a Sustainable Society and it was held in Tokyo, Japan. It was conceived as a hybrid conference allowing face-to-face and online contributions. AIED 2023 was the next in a series of annual international conferences for presenting high-quality research on intelligent systems and the cognitive sciences for the improvement and advancement of education. It was hosted by the prestigious International Artificial Intelligence in Education Society, a global association of researchers and academics specialising in the many fields that comprise AIED, including, but not limited to, computer science, learning sciences, educational data mining, game design, psychology, sociology, and linguistics among others.

The conference aimed to stimulate discussion on how Artificial Intelligence shapes and can shape education for all sectors, how to advance the science and engineering of intelligent interactive learning systems, and how to promote broad adoption. Engaging with the various stakeholders – researchers, educational practitioners, businesses, policy makers, as well as teachers and students – the conference set a wider agenda on how novel research ideas can meet practical needs to build effective intelligent human-technology ecosystems that support learning.

AIED 2023 attracted broad participation. For the main program, published in the main volume, we received 311 submissions, of which 251 were submitted as full papers, and 60 were submitted as short papers. Of the full-paper submissions, 53 were accepted as full papers (thus the full-paper acceptance rate was 21.11%) and another and 14 were accepted as short papers. Of the 60 short-paper submissions we received, 12 were accepted as short papers. These papers are included in Volume 1 of the AIED 2023 proceedings. In addition to the 53 full papers and 26 short papers, 65 submissions were selected as posters. This set of posters was complemented with the other poster contributions submitted for the Poster and Late Breaking Results track that are compiled in this volume of the AIED 2023 proceedings.

AIED 2023 offered other venues to present original contributions beyond the main paper presentations, including a Doctoral Consortium Track, an Industry and Innovation Track, Interactive Events, Posters/Late-Breaking Results, a Practitioner Track and a track where published researchers from the International Journal of AIED could present their work. Since this year marks the 30th anniversary of the International AIED Society, the conference had a BlueSky Track that included papers reflecting upon the progress of AIED in the last 30 years and envisioning what's to come in the next 30 years. All of these additional tracks defined their own peer-review process. The conference also had a Wide AIED track where participants from areas of the World not typically represented at the conference could present their work as oral presentations. All these contributions are compiled in the present volume of the AIED 2023 proceedings.

For making AIED 2023 possible, we thank the AIED 2023 Organizing Committee, the hundreds of Program Committee members, the Senior Program Committee members and the AIED Proceedings Chairs - Irene-Angelica Chounta (Volume 1) and Christothea Herodotou (Volume 2).

July 2023

Ning Wang
Genaro Rebolledo-Mendez
Vania Dimitrova
Noboru Matsuda
Olga C. Santos

Organization

General Conference Chairs

Rose Luckin   University College London, UK
Vania Dimitrova   University of Leeds, UK

Program Co-chairs

Ido Roll   Technion - Israel Institute of Technology, Israel
Danielle McNamara   Arizona State University, USA

Industry and Innovation Track Co-chairs

Steve Ritter   Carnegie Learning, USA
Inge Molenaar   Radboud University, The Netherlands

Workshop and Tutorials Co-chairs

Mingyu Feng   WestEd, USA
Alexandra Cristea   Durham University, UK
Zitao Liu   TAL Education Group, China

Interactive Events Co-chairs

Mutlu Cukurova   University College London, UK
Carmel Kent   Educate Ventures, UK
Bastiaan Heeren   Open University, The Netherlands

Local Co-chairs

Sergey Sosnovsky   Utrecht University, The Netherlands
Johan Jeuring   Utrecht University, The Netherlands

Proceedings Chair

Christothea Herodotou   Open University, UK

Publicity Chair

Elle Yuan Wang   Arizona State University, USA

Web Chair

Isaac Alpizar-Chacon   Utrecht University, The Netherlands

International Artificial Intelligence in Education Society

Management Board President

Rose Luckin   University College London, UK

President-Elect

Vania Dimitrova   University of Leeds, UK

Secretary/Treasurer

Bruce M. McLaren   Carnegie Mellon University, USA

Journal Editors

Vincent Aleven   Carnegie Mellon University, USA
Judy Kay   University of Sydney, Australia

Finance Chair

Ben du Boulay   University of Sussex, UK

Membership Chair

Benjamin D. Nye   University of Southern California, USA

Publicity Chair

Manolis Mavrikis   University College London, UK

Executive Committee

Ryan S. Baker	University of Pennsylvania, USA
Min Chi	North Carolina State University, USA
Cristina Conati	University of British Columbia, Canada
Jeanine A. DeFalco	CCDC-STTC, USA
Vania Dimitrova	University of Leeds, UK
Rawad Hammad	University of East London, UK
Neil Heffernan	Worcester Polytechnic Institute, USA
Christothea Herodotou	Open University, UK
Akihiro Kashihara	University of Electro-Communications, Japan
Amruth Kumar	Ramapo College of New Jersey, USA
Diane Litman	University of Pittsburgh, USA
Zitao Liu	TAL Education Group, China
Rose Luckin	University College London, UK
Judith Masthoff	Utrecht University, The Netherlands
Noboru Matsuda	Texas A&M University, USA
Tanja Mitrovic	University of Canterbury, New Zealand
Amy Ogan	Carnegie Mellon University, USA
Kaska Porayska-Pomsta	University College London, UK
Ma. Mercedes T. Rodrigo	Ateneo de Manila University, Philippines
Olga Santos	UNED, Spain
Ning Wang	University of Southern California, USA

Program Committee Members

Senior Programme Committee

Giora Alexandron	Weizmann Institute of Science, Israel
Ivon Arroyo	UMASS, USA
Roger Azevedo	University of Central Florida, USA
Ryan Baker	University of Pennsylvania, USA

Stephen B. Blessing   University of Tampa, USA
Min Chi   NCS, USA
Mutlu Cukurova   University College London, UK
Carrie Demmans Epp   University of Alberta, Canada
Vania Dimitrova   University of Leeds, UK
Dragan Gasevic   Monash University, Australia
Sébastien George   LIUM, Le Mans Université, France
Floriana Grasso   University of Liverpool, UK
Peter Hastings   DePaul University, USA
Neil Heffernan   Worcester Polytechnic Institute, USA
Seiji Isotani   Harvard University, USA
Irena Koprinska   The University of Sydney, Australia
Vitomir Kovanovic   The University of South Australia, Australia
Sébastien Lallé   Sorbonne University, France
H. Chad Lane   University of Illinois at Urbana-Champaign, USA
James Lester   North Carolina State University, USA
Shan Li   McGill University, Canada
Roberto Martinez-Maldonado   Monash University, Australia
Noboru Matsuda   North Carolina State University, USA
Manolis Mavrikis   UCL, UK
Gordon McCalla   University of Saskatchewan, Canada
Bruce McLaren   CMU, USA
Eva Millan   UMA, Spain
Tanja Mitrovic   Intelligent Computer Tutoring Group, University
                             of Canterbury, Christchurch, New Zealand
Riichiro Mizoguchi   Japan Advanced Institute of Science and
                             Technology, Japan
Kasia Muldner   CUNET, Canada
Roger Nkambou   Université du Québec à Montréal, Canada
Benjamin Nye   USC, USA
Andrew Olney   University of Memphis, USA
Jennifer Olsen   University of San Diego, USA
Luc Paquette   University of Illinois at Urbana-Champaign, USA
Kaska Porayska-Pomsta   UCL Knowledge Lab, UK
Thomas Price   North Carolina State University, USA
Kenneth R. Koedinger   CMU, USA
Genaro Rebolledo-Mendez   University of British Columbia, Mexico
Ido Roll   Technion - Israel Institute of Technology, Israel
Jonathan Rowe   North Carolina State University, USA
Nikol Rummel   RUB, Germany
Vasile Rus   The University of Memphis, USA
Olga Santos   aDeNu Research Group (UNED), Spain

Sergey Sosnovsky	Utrecht University, Netherlands
Mercedes T. Rodrigo	Department of Information Systems and Computer Science, Ateneo de Manila University, Philippines
Marco Temperini	Sapienza University of Rome, Italy
Vincent Wade	Trinity College Dublin, Ireland
Ning Wang	USA
Diego Zapata-Rivera	Educational Testing Service, USA

Program Committee

Seth Adjei	Northern Kentucky University, USA
Bita Akram	North Carolina State University, USA
Burak Aksar	Boston University, USA
Laia Albó	Universitat Pompeu Fabra, Spain
Azza Abdullah Alghamdi	KAU, Saudi Arabia
Samah Alkhuzaey	University of Liverpool, Saudi Arabia
Laura Allen	University of Minnesota, USA
Antonio R. Anaya	Universidad Nacional de Educacion a Distancia, Spain
Tracy Arner	Arizona State University, USA
Ayan Banerjee	Arizona State University, USA
Michelle Barrett	Edmentum, USA
Abhinava Barthakur	University of South Australia, Australia
Sarah Bichler	Ludwig Maximilian University of Munich, Germany
Gautam Biswas	Vanderbilt University, USA
Emmanuel Blanchard	IDÛ Interactive Inc., Canada
Nathaniel Blanchard	Colorado State University, USA
Geoffray Bonnin	Université de Lorraine - LORIA, France
Nigel Bosch	University of Illinois Urbana-Champaign, USA
Bert Bredeweg	University of Amsterdam, Netherlands
Julien Broisin	Université Toulouse 3 Paul Sabatier - IRIT, France
Christopher Brooks	University of Michigan, USA
Armelle Brun	LORIA - Université de Lorraine, France
Jie Cao	University of Colorado Boulder, USA
Dan Carpenter	North Carolina State University, USA
Alberto Casas-Ortiz	UNED, Spain
Wania Cavalcanti	Universidade Federal do Rio de Janeiro COPPEAD, Brazil
Li-Hsin Chang	University of Turku, Finland
Penghe Chen	Beijing Normal University, China

Zixi Chen   University of Minnesota, USA
Ruth Cobos   Universidad Autónoma de Madrid, Spain
Cesar A. Collazos   Full Professor, Colombia
Maria de Los Angeles Constantino   Tecnológico de Monterrey Campus Laguna,
  González   Mexico
Seth Corrigan   University of California, Irvine, USA
Maria Cutumisu   University of Alberta, Canada
Jeanine DeFalco   US Army Futures Command, USA
M Ali Akber Dewan   Athabasca University, Canada
Konomu Dobashi   Aichi University, Japan
Tenzin Doleck   Simon Fraser University, Canada
Mohsen Dorodchi   University of North Carolina Charlotte, USA
Fabiano Dorça   Universidade Federal de Uberlandia, Brazil
Cristina Dumdumaya   University of Southeastern Philippines,
                                              Philippines
Yo Ehara   Tokyo Gakugei University, Japan
Ralph Ewerth   L3S Research Center, Leibniz Universität
                                              Hannover, Germany
Stephen Fancsali   Carnegie Learning, Inc., USA
Alexandra Farazouli   Stockholm University, Sweden
Effat Farhana   Vanderbilt University, USA
Mingyu Feng   WestEd, USA
Márcia Fernandes   Federal University of Uberlândia, Brazil
Carol Forsyth   Educational Testing Service, USA
Reva Freedman   Northern Illinois University, USA
Selen Galiç   HACETTEPE UNIVERSITY, Turkey
Wenbin Gan   National Institute of Information and
                                              Communications Technology, Japan
Michael Glass   Valparaiso University, USA
Benjamin Goldberg   United States Army DEVCOM Soldier Center,
                                              USA
Alex Sandro Gomes   Universidade Federal de Pernambuco, Brazil
Aldo Gordillo   Universidad Politécnica de Madrid (UPM), Spain
Monique Grandbastien   LORIA, Universite de Lorraine, France
Beate Grawemeyer   University of Sussex, UK
Nathalie Guin   LIRIS - Université de Lyon, France
Jason Harley   McGill University, Canada
Bastiaan Heeren   Open University, The Netherlands, Netherlands
Laurent Heiser   Université Côte d'Azur, Inspé de Nice, France
Wayne Holmes   UCL, UK
Anett Hoppe   TIB Leibniz Information Centre for Science and
                                              Technology; L3S Research Centre, Leibniz
                                              Universität Hannover, Germany

Lingyun Huang	McGill University, Canada
Yun Huang	Carnegie Mellon University, USA
Ig Ibert-Bittencourt	Federal University of Alagoas, Brazil
Tomoo Inoue	University of Tsukuba, Japan
Paul Salvador Inventado	California State University Fullerton, USA
Mirjana Ivanovic	University of Novi Sad, Faculty of Sciences, Department of Mathematics and Informatics, Serbia
Stéphanie Jean-Daubias	Université de Lyon, LIRIS, France
Johan Jeuring	Utrecht University, Netherlands
Yang Jiang	Columbia University, USA
Srecko Joksimovic	Education Future, University of South Australia, Australia
David Joyner	Georgia Institute of Technology, USA
Akihiro Kashihara	The University of Electro-Communications, Japan
Mizue Kayama	Shinshu University, Japan
Hieke Keuning	Utrecht University, Netherlands
Yeojin Kim	North Carolina State University, USA
Kazuaki Kojima	Teikyo University, Japan
Amruth Kumar	Ramapo College of New Jersey, USA
Tanja Käser	EPFL, Switzerland
Andrew Lan	University of Massachusetts Amherst, USA
Mikel Larrañaga	University of the Basque Country, Spain
Hady Lauw	Singapore Management University, Singapore
Nguyen-Thinh Le	Humboldt Universität zu Berlin, Germany
Tai Le Quy	Leibniz University Hannover, Germany
Seiyon Lee	University of Pennsylvania, USA
Marie Lefevre	LIRIS - Université Lyon 1, France
Blair Lehman	Educational Testing Service, USA
Carla Limongelli	Università Roma Tre, Italy
Fuhua Lin	Athabasca University, Canada
Nikki Lobczowski	University of Pittsburgh, USA
Yu Lu	Beijing Normal University, China
Vanda Luengo	Sorbonne Université - LIP6, France
Collin Lynch	North Carolina State University, USA
Sonsoles López-Pernas	Universidad Politécnica de Madrid, Spain
Aditi Mallavarapu	University of Illinois at Chicago, USA
Mirko Marras	University of Cagliari, Italy
Jeffrey Matayoshi	McGraw Hill ALEKS, USA
Kathryn McCarthy	Georgia State University, USA
Guilherme Medeiros-Machado	ECE Paris, France
Angel Melendez-Armenta	Tecnologico Nacional de Mexico, Mexico

Wookhee Min	North Carolina State University, USA
Phaedra Mohammed	The University of the West Indies, Trinidad and Tobago
Daniel Moritz-Marutschke	College of Global Liberal Arts, Ritsumeikan University, Japan
Fahmid Morshed-Fahid	North Carolina State University, USA
Ana Mouta	Escuela de Doctorado, USAL, Portugal
Tomohiro Nagashima	Saarland University, Germany
Huy Nguyen	CMU, USA
Narges Norouzi	UC Berkeley, USA
Nasheen Nur	Florida Institute of Technology, USA
Marek Ogiela	AGH University of Science and Technology, Poland
Ranilson Paiva	Universidade Federal de Alagoas, Brazil
Radek Pelánek	Masaryk University Brno, Czechia
Eduard Pogorskiy	Open Files LTD, UK
Elvira Popescu	University of Craiova, Romania
Miguel Portaz	UNED, Spain
Ethan Prihar	Worcester Polytechnic Institute, USA
Shi Pu	Education Testing Service, USA
Mladen Rakovic	Monash University, Australia
Ilana Ram	Technion Israel Institute of Technology, Israel
Sowmya Ramachandran	Stottler Henke Associates Inc., USA
Martina Rau	University of Wisconsin - Madison, Department of Educational Psychology, USA
Traian Rebedea	University Politehnica of Bucharest, Romania
Carlos Felipe Rodriguez-Hernandez	Tecnologico de Monterrey, Mexico
Rinat B. Rosenberg-Kima	Technion - Israel Institute of Technology, Israel
Demetrios Sampson	Curtin University, Australia
Sreecharan Sankaranarayanan	Carnegie Mellon University, USA
Mohammed Saqr	University of Eastern Finland, Finland
Petra Sauer	beuth university of applied sciences, Germany
Robin Schmucker	Carnege Mellon University, USA
Filippo Sciarrone	Universitas Mercatorum, Italy
Kazuhisa Seta	Osaka Prefecture University, Japan
Lele Sha	Monash, Australia
Davood Shamsi	Ad.com, USA
Lei Shi	Newcastle University, UK
Aditi Singh	Kent State University, USA
Daevesh Singh	Indian Institute of Technology, India
Sean Siqueira	Federal University of the State of Rio de Janeiro (UNIRIO), Brazil

Abhijit Suresh	University of Colorado Boulder, USA
Vinitra Swamy	EPFL, Switzerland
Michelle Taub	University of Central Florida, USA
Maomi Ueno	The University of Electro-Communications, Japan
Josh Underwood	Independent, Spain
Maya Usher	Technion - Israel Institute of Technology, Israel
Masaki Uto	The University of Electro-Communications, Japan
Rosa Vicari	Universidade Federal do Rio Grande do Sul, Brazil
Maureen Villamor	University of Southeastern Philippines, Philippines
Alessandro Vivas	UFVJM, Brazil
Alistair Willis	The Open University, UK
Chris Wong	University of Technology Sydney, Australia
Peter Wulff	Heidelberg University of Education, Germany
Kalina Yacef	The University of Sydney, Australia
Nilay Yalcin	University of British Columbia, Canada
Sho Yamamoto	Kindai University, Japan
Andrew Zamecnik	University of South Australia, Australia
Stefano Pio Zingaro	Università di Bologna, Italy
Gustavo Zurita	Universidad de Chile, Chile

Additional Reviewers

Peter Ahn	Jorge Ghosh
Martin Ali	Alex Gomes
Christian Almeida	Vladimir Gómez
Marco Bauer	Laura Ha
Andrea Becker	Bruno Hall
Mark Brown	William Han
Richard Chen	Zhang Hou
Marc Cheng	Christophe Hsieh
Eric Cheung	Michel Hsu
Chris Chung	Victor Ito
Jonathan Cohen	Jens Jain
Nicolas Costa	Yang Jang
Florian De Souza	Igor Jin
Tobias Deng	Fabio Jing
Alberto Ding	Gabriel Johnson
Matthew Fernandes	Philipp Kim
Pedro Fernández	Sven King
Luca Ferreira	Jürgen Klein

Sara Lam
Hiroshi Lau
Joseph Le
Olga Lim
Guillaume Lima
Claudia Lin
Ralf Ludwig
Frédéric Luo
Charles Ma
Anton Miller
Alejandro Min
Ioannis Mishra
Andrzej Nguyen
Claudio Ni
Simone Nie
Petr Patel
Davide Paul
Johan Pei
Dmitry Qian
Silvia Qiao
Holger Qin
Yuki Ribeiro
Takeshi Roberts
Julia Robinson
Walter Sánchez
Marcel Santos
Jean Sato

Alexey Shao
Dimitris Sharma
Mikhail Shen
Nikolaos Singh
Edward Smith
Jonas Son
Catherine Tanaka
Masahiro Tang
Sylvain Tao
Timo Tsai
René Tseng
Omar Tu
Lorenzo Weiß
Gilles Wen
Mauro Weng
Christina Xiang
Jacques Xiao
Tony Xie
Grzegorz Yang
Ryan Yao
Elizabeth Ye
Guido Young
Antoine Yu
Andrés Yuan
Gregory Zheng
Aaron Zhong
Anders Zhou

Contents

Four Interactions Between AI and Education: Broadening Our Perspective on What AI Can Offer Education

Sina Rismanchian$^{(\boxtimes)}$ and Shayan Doroudi

University of California, Irvine, Irvine, CA 92697, USA
{srismanc,doroudis}@uci.edu

Abstract. In the 30th anniversary of the International Artificial Intelligence in Education Society, there is a need to look back to the past to envision the community's future. This paper presents a new framework (AI × Ed) to categorize different interactions between artificial intelligence (AI) and education. We use our framework to compare papers from two early proceedings of AIED (1985 & 1993) with AIED proceedings papers and IJAIED papers published in 2021. We find that two out of four kinds of interactions between AI and education were more common in the early stages of the field but have gained less attention from the community in recent years. We suggest that AI has more to offer education than a pragmatic toolkit to apply to educational problems; rather, AI also serves as an analogy for thinking about human intelligence and learning. We conclude by envisioning future research directions in AIED.

Keywords: AI and education · Applied epistemology · Cognitive modeling · Intelligent tutoring systems · AI literacy · Large language models

1 Introduction

In 1977—over 15 years prior to the establishment of the International Artificial Intelligence in Education Society—Kenneth Kahn published a paper entitled "Three Interactions between AI and Education" [13]. Motivated by his work as a graduate student in the Logo Group, Kahn suggested that there are (at least) three roles that artificial intelligence (AI) could play in education:

1. Children can write "simple AI programs"
2. Children can interact with "AI 'teaching programs,'" and
3. "AI theories of intelligence and learning" can guide the design of learning environments.

Forty-five years later, we contend that the Artificial Intelligence in Education (AIED) community has not given equal focus to these three interactions. Among these three, the AIED community has disproportionately focused on the second.

N. Wang et al. (Eds.): AIED 2023, CCIS 1831, pp. 1–12, 2023.
https://doi.org/10.1007/978-3-031-36336-8_1

The first has only recently gained popularity under the banners of "AI education" and "AI literacy." The third was a focus in the early days of AIED, but has since become unpopular. In this paper, we propose a new framework that can help categorize the diversity of roles that AI can play in education. Our framework was inspired by that of Kahn but also resonates with a recent framework proposed by Porayska-Pomsta and Holmes [23,24]. What all three of our frameworks share in common is the notion that the relationship between AI and education is more than a purely instrumental one which has become increasingly common. Rather, AI and machine learning (ML) can also act as analogies to human intelligence (HI) and learning, which can in turn influence how we think about and design educational environments.

Below we introduce our framework for thinking about the relationships between AI and education. We propose two axes, which give rise to four quadrants, each of which highlights a particular role that AI can play in education. We then locate every paper from three proceedings of AIED and one volume of the International Journal of Artificial Intelligence in Education (IJAIED) in our framework to compare the focus of the field before 1993 with the focus today (using the year 2021 as a representative example). We discuss examples of papers in the different quadrants of our framework and show that while the field used to have a mix of instrumental applications of AI in education (e.g., AI "teaching programs") and research focusing on the analogy between AI and HI, today it almost exclusively focuses on instrumental applications. In order to show that earlier approaches may still be relevant today, we then discuss several strands of recent work that exemplify the less-represented uses of AI in education. Our overall approach in this paper is to examine the past to reflect on what the future could be. Our overall message is that AI has more to offer education than we are currently taking advantage of.

2   The AI×Ed Framework

AI×Ed uses two main axes to interpret the position of any particular research project in AIED; it is depicted in Fig. 1. We conceptualize work in AIED as primarily being about some kind of interaction between AI and one or more people (which we refer to as end users). However, this interaction can look quite different depending on the role that AI plays and the kind of end user, giving rise to the two axes of AI×Ed. Below, we first describe the two axes and then we describe how the axes interact with one another, mentioning representative research in each of the four quadrants.

2.1   End User Axis

The horizontal axis considers the role of the end user who is interacting with AI. We put forward a dichotomy between whether the person *primarily* interacting with AI is a researcher or a learner. Of course, in a paper where a learner is the primary end user, a researcher can also benefit downstream (e.g., by learning

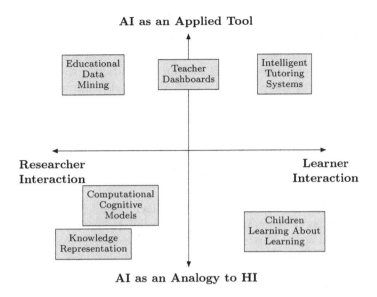

Fig. 1. AI×Ed framework (with examples of various subfields of AIED)

about learning) and vice versa. To clarify our perspective, we bring up instances representing each end of the spectrum.

Learner End: Most of the papers in areas such as ITS, adaptive assessment, and personalized learning, are categorized at this end as they serve as a means to teach *learners*, assess *learners*, and personalize *learners'* experience. In these papers, learners are the first-hand consumers of the contribution, which means that the results of the papers are intended to improve learners' education immediately. While the focus here is on the use of AI to advance education, we can extend the AI×Ed framework to also consider the use of education to advance AI. In this case, work on AI literacy and AI education would also be examples of learners interacting with AI.

Researcher End: Studies at this end typically involve researchers using AI to improve our understanding of learning and educational systems. Most research in the educational data mining (EDM) field is categorized at the researcher end. These contributions apply machine learning tools to enhance the community's understanding of teaching and learning [14]. Other examples include using computational models to study how children learn [2,29] or applying fundamental AI concepts to education to generate novel insights and research questions [8]. Although the way that researchers interact with AI is different in these two types, they are similar in that researchers are the primary end users of the AI, even if the insights generated are intended to benefit learners downstream. The implications of these contributions are beneficial for other researchers and practitioners to push forward the field.

We postulate that this axis is more of a spectrum rather than a binary, which means that we leave room in between the ends of the spectrum to include teachers, practitioners, and parents as the end users of AI. Thus, we position contributions such as using AI tools to improve teachers' experience (e.g., AI-based dashboards and at-risk detectors) or AI curricula designed for teachers in the middle of the axis.

2.2   Role of AI Axis

The vertical axis indicates the role that AI is playing. The two ends for this spectrum are AI as an applied tool vs. AI as an analogy to HI.

AI as an Applied Tool: This end of the axis includes research that emphasizes employing AI techniques toward pragmatic educational ends, analogous to how researchers would apply AI to other fields such as medicine, physics, business, or law. We contend this is what people typically mean today when they refer to "AI in education." In essence, the novelty of these works is in the way they can utilize AI as an applied tool to tackle educational problems. This includes developing new educational technologies (e.g., intelligent tutoring systems and personalized learning platforms) and applying data mining techniques to educational datasets.

AI as an Analogy to Human Intelligence: This end of the axis refers to a more principled interplay between AI and education: they are both about agents that think and learn; they both investigate epistemological questions. Using AI as an analogy to HI—and vice versa—is rooted in cognitive science and was much more popular in the early days of the AIED community (as we show below). Indeed, many early AI researchers were interested in exploring this interplay between AI and education [9]. The field of information-processing psychology, which has been very influential in AIED and education research more broadly, initially formed out of research that involved reasoning about thinking in both humans and machines.

Work in this area can fall under at least one of two varieties. First, researchers can build computational models of how people learn, such as ACT-R [2], Cascade [29], and SimStudent [19]. This involves improving our understanding of how people learn (and how we can better teach them) by building machines that can learn (and be taught). The development of such rule-based AI models has also influenced the design of intelligent tutoring systems like cognitive tutors [1].

Second, *theoretical* concepts in AI and machine learning might have analogs in HI and human learning. For example, knowledge representation is an important area of research in both AI and education [23]. Relatedly, Seymour Papert and Marvin Minsky developed the notion of a microworld in the context of programming AI and then applied that concept to how kids learn—giving rise to the notion of educational microworlds [9]. More recently, the field of machine teaching is interested in how insights from how to optimally teach machine learners may apply to teaching human learners [33]. Recent work has also shown how the bias-variance tradeoff in machine learning may help explain recent results on individual differences in how people learn [10] as well as debates around what instructional strategies are optimal [8].

2.3 Interaction of the Two Axes

Although we described the two axes separately, the nature of different kinds of research at the intersection of AI and education will depend on their position on both axes. Learners can learn by interacting with applied AI tools like intelligent tutoring systems or conversational agents; learners can also learn *about* applied AI (upper right quadrant). Researchers can also learn about learning by using AI tools to tackle educational questions (upper left quadrant). Alternatively, researchers can use AI as an analogy to human intelligence (lower left quadrant) to gain insights into how people learn and how to teach, for example by simulating computational models of learning or by applying results from machine teaching. Finally, researchers are not the only ones who can use AI as an analogy to human intelligence; learners can also do this in order to improve their own learning (lower left quadrant), for example by writing AI programs and reflecting on the nature of the program's cognitive capacities and their own.

This last use case might be the most foreign to AIED researchers and deserves further exposition. The idea was initially proposed by Papert and colleagues [21]:

The aim of AI is to give concrete form to ideas about thinking that previously might have seemed abstract, even metaphysical. It is this concretizing quality that has made ideas from AI so attractive to many contemporary psychologists. We propose to teach AI to children so that they, too, can think more concretely about mental processes. While psychologists use ideas from AI to build formal, scientific theories about mental processes, children use the same ideas in a more informal and personal way to think about themselves. (pp. 157-158)

Interestingly, Papert suggests this idea first came up (but was forgotten) in a conversation he had with Jean Piaget (likely in the early1960 s) when "engag[ing] in playful speculation about what would happen if children could play at building little artificial minds" [22]. However, as we see below, this idea has been rather underexplored in recent decades.

3 Comparison to Other Frameworks

AI×Ed was largely inspired by Kahn's early list of three interactions between AI and education but it was expanded to account for the range of work at the intersection of AI and education, both in the past and today. In Kahn's framework, the first interaction is having children writing simple AI programs. As discussed above, this could fall in either the top right or bottom right quadrants of AI×Ed. Having children learn about their own learning (bottom right) was indeed one of the key motivations to having children program AI at the time [13]. The second form of interaction refers largely to intelligent tutoring systems and would fall under the upper right quadrant of AI×Ed. Finally, the third form of interaction—using AI theories to inform our understanding of how children learn and in turn guide the design of learning environments—falls under the bottom right quadrant of AI×Ed (with Papert's microworlds being a clear example).

More recently, Porayska-Pomsta has also called on a categorization of the roles of AI in education that moves beyond seeing AI as a purely engineering discipline [23,24]. Porayska-Pomsta and Holmes have suggested three roles for AI in education: AI can act "as a form of *civil engineering* concerned with addressing immediate education-related challenges within society at large;" it can act as an *applied philosophy* that investigates foundational questions around how people learn and think; and it can act as a *research methodology* that concretely applies AI tools to investigate the nature of teaching and learning (e.g., through computational cognitive models) [24]. The civil engineering approach is the same as our notion of AI as an applied tool that interacts with learners and teachers (top right quadrant), though perhaps it also includes some applications that are researcher-focused (top left quadrant). The use of AI as applied philosophy and research methodology both fall under our bottom left quadrant. The distinction between applied philosophy and research methodology is interesting, but we believe the more salient feature is that they both rely on the analogy between AI and HI. We note that Porayska-Pomsta's categorization appears to be similar to an earlier tripartite categorization in AIED; Baker suggested that models (broadly conceived) can be applied to AIED in three ways: as a component in an AIED system, as a basis for design, and as a scientific tool, [4].

All of these frameworks are useful for thinking about the ways in which AI can interact with education, but AI×Ed tries to identify more general categories that capture the range of work in AIED over the past few decades. For example, the upper left quadrant of our framework has not been represented in previous frameworks (largely because the use of EDM to learn about learning is a recent phenomenon). However, we note that all of these frameworks agree with our own on the fact that AI is not merely an engineering discipline, but also a fundamental tool that can be used to investigate the nature of learning and teaching and influence the design of educational environments.

4   Locating AIED Research on AI×Ed

We position papers from the 2nd International Conference on Artificial Intelligence and Education (AIED 1985) proceedings [32] and the AIED 1993 proceedings [5] in AI×Ed to provide a big picture of the diversity of papers in the early stages of the community. Then, we locate papers in the AIED 2021 proceedings and the 2021 volume of IJAIED (Volume 31) to show how research foci, in terms of AI×Ed, shifted from the early years to recent years.

4.1   Proceedings of AIED 1985

As shown in Fig. 2a, we find a diversity of papers spanning three of the four quadrants. The top right quadrant includes papers focusing on ITSs (e.g., "Computers

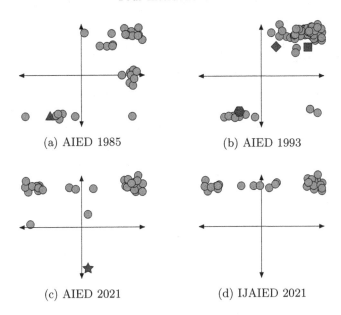

(a) AIED 1985   (b) AIED 1993

(c) AIED 2021   (d) IJAIED 2021

Fig. 2. Locating early and recent AIED articles in AI×Ed. The exact positions of points within clusters are arbitrary. The special shapes signify articles discussed in the paper.

Teaching Programming" and "Architectures for Intelligent Tutoring"). On the other hand, there are a substantial number of works that lie at the learner interaction end and in the middle of the AI role axis. Some of these articles focus on microworlds, Logo, and AI curricula. While microworlds are conceptually rooted in the Papert's work in AI, some of these works are more about the design of microworld learning environments (sometimes without using any applied AI). Other papers describe curricula and analyses of the impacts of teaching AI to non-researchers and children. We place these articles in middle of the y-axis because in the 1980s, teaching children AI was motivated by both pragmatic reasons and for students to learn about human learning. Finally, we find that there are several papers where researchers utilize the analogy between AI and HI (bottom left quadrant). For example, Carley describes an agent-based model of "knowledge acquisition as a social phenomenon" which combines AI-based knowledge representation formalisms with a socio-cultural theory of learning [6] (triangle).

4.2 Proceedings of AIED 1993

We find that 1993 has a similar distribution to the papers from 1985, except that there are no longer papers in the middle of the y-axis at the learner interaction end; see Fig. 2b. This is because the AIED community had moved away from emphasizing microworlds and AI curricula.

In the upper right, we find that AIED continued to emphasize work on ITSs, and moving beyond developing the technical abilities of such systems, researchers called for using these systems in school settings [16] (diamond). Another new trend that becomes visible in the proceedings is the endeavor to develop intelligent curricula and intelligent systems for teachers (e.g., [31]). These articles are mostly placed in the middle of the x-axis, as their first-hand users are teachers and practitioners. The development of new knowledge acquisition models and student models continued to be a research focus in 1993. VanLehn et al. developed Cascade, a program to simulate cognitive procedures in self-explanation based on previous experiments [29], which we classify as an example of AI as an analogy for HI (hexagon). However, VanLehn and colleagues then use this model as a component of OLAE, their Bayesian student modeler to assist human assessors [28], which we classify as an example of AI as an applied tool (square). In this proceedings, we also see the emergence of Bayesian models in two articles [26,28]. The models are early examples in AIED of data-driven models to be incorporated into intelligent tutoring systems. The use of Bayesian models became ubiquitous in developing ITSs in the following years [7].

4.3   Proceedings of AIED 2021 and IJAIED 2021

There is a dramatic shift from early AIED articles to articles published in the AIED proceedings and IJAIED in 2021, as shown in Figs. 2c and 2d. In 2021, we see a substantial number of papers at the intersection of AI as an applied tool and researcher interaction. With advances in the fields of ML, natural language processing, and network analysis leading to the prominence of data-driven techniques, researchers employed these techniques to investigate learners' behaviors using large fine-grained datasets. These papers largely belong to the recent fields of EDM and learning analytics. Additionally, Research on ITSs continues to garner the most significant attention in the field, though the AI techniques have transformed from rule-based models to data-driven ones [17]. As intelligent tutoring systems could be optimized via state-of-the-art optimization techniques, there is increasing emphasis on data-driven student models. This is perhaps one of the reasons why the community has shifted away from papers emphasizing AI as an analogy to HI. Finally, We found only one paper in 2021 that lies at the bottom of AI×Ed (star). Kent et al. [15] "explores the relationship between unsupervised machine learning models, and the mental models of those who develop or use them." They show that learners (e.g., managers of an organization) can engage in a co-learning process with unsupervised ML models to mitigate biases in their mental models.

5   Envisioning the Future

When examining the current state of the AIED community from the lens of AI×Ed, we notice that a large swath of potential research at the intersection of AI and education—namely research in the bottom half of AI×Ed—has been largely absent from the current discourse on AI in education. However, this kind of research was popular in the early days of AIED. We contend that AI has more to offer education than simply a tool that can be pragmatically applied like it is in any other field. Here we articulate several ways that the AIED community might do this, drawing on recent strands of work. We do not claim these are necessarily the most productive routes, but they may help set a research agenda AIED's future that is more expansive about the role that AI plays in education.

5.1   AI Literacy

Proposing AI curricula for educating students has been one of the categories of research articles in the early stages of the AIED community but only recently has it gained renewed attention; as shown by Ng et al., the number of published AI literacy articles has increased sharply during recent years [20]. We contend that recent work in AI literacy has almost exclusively focused on the top right quadrant (i.e., teaching AI toward pragmatic ends), whereas earlier work also extended to the bottom right quadrant (i.e., teaching AI to encourage students to reflect on how they learn). Reinvigorating this latter strand of work could push AI literacy research in new directions.

A key area of AI literacy is teaching students about AI ethics, including issues of fairness, transparency, safety, etc. Learning about the ethics of applying certain AI tools in society or learning techniques for improving the fairness of AI would fit under the top right quadrant. However, students can also use AI ethics as an opportunity to revisit fundamental questions about humanity that researchers have been asking for decades (bottom right quadrant). For instance, self-driving cars has led to a renewed interest in the trolley problem in philosophy [3]. The need to answer such questions in order to design ethical self-driving cars is now inevitable. Relatedly, a discussion of algorithmic bias could deal with the similarities between algorithmic bias and human cognitive biases [12] and what our understanding of unfair AI says about ourselves.

5.2   Teachable Agents

In contrast to intelligent tutoring systems, learners can also learn by teaching teachable agents, or what we might call "intelligent *tutorable* systems." Although this work might be classified as using AI as an applied tool, we note that it relies on the ability of AI to mimic student-like behaviors and the student's ability to reflect on their own learning as they help the AI learn—hence it draws upon the analogy between AI and HI. While there has been some recent work on developing teachable agents in AIED [18,19], it appears to be relatively sparse compared to work on other AI-based educational technologies. Perhaps by putting more emphasis on the analogy between AI and HI, the AIED community can pursue novel directions in developing teachable agents.

In the coming years, we also foresee more work on a different kind of teachable agent rooted in the rise of black-box ML models and the emphasis on AI literacy. Namely, we anticipate students can learn about AI *and* human intelligence by teaching ML agents. Vartiainen et al. [30] illustrate that when 3–9 year old children interact with Google's Teachable Machines, they can contrast their learning with the way the machines learn. For instance, an 8-year-old child describes the way the machine learns as: "Well, we showed that to it many times, a little over a hundred times ... and then it learned that". As discussed above, depending on the focus of the AI curriculum, such work could span the top and bottom of the learner interaction side of our framework.

5.3 Agent-Based Models and Complex Systems Methodology

We contend that two reasons why computational models of human learning have gone out of favor in AIED are that (1) AI techniques have moved away from rule-based systems to data-driven AI and (2) over the past few decades many education researchers have turned to learning theories that suggest learning is too complex to be described by simple rule-based cognitive systems. Recently, there has been a call toward complex systems methodology in education research [11]. Agent-based modeling is a particular complex systems method that could be used to simulate learning in a more bottom-up fashion. This may help bridge the divide between contemporary theories of learning and the insights that computational modeling might provide [4]; for example, we are currently investigating an agent-based model for the ICAP framework [25]. (Once again, Papert had long ago advocated for the use of agent-based models for theorizing about how kids learn [21].) Beyond benefiting researchers (bottom left quadrant), we contend that agent-based models of learning could also be used to teach learners about the complex nature of learning (bottom right quadrant), building on the importance of teaching children about complex systems [11].

5.4 The Curious Case of Large Language Models

The emergence of large language models (LLM) such as ChatGPT has recently brought up many opportunities and challenges for the field. Here, we describe different possibilities for leveraging these models in the future of AIED and where they fit in the AI×Ed framework. Perhaps the most widely-antcipated usage of these models is as a writing assistant [27] which is located at the top right quadrant of our framework since learners are using AI as an applied tool. At the upper left, researchers can benefit from the technical backbone of LLMs as an applied tool by fitting such models to datasets that are optimized for educationally relevant tasks. On the other hand, there are underexplored opportunities for using LLMs as an analogy to HI. At the bottom right, for instance, LLMs can serve as teachable agents (see Sect. 5.2) where a student may teach an LLM a procedure or concept in order to investigate their knowledge gaps in dialogue with the teachable agent. This relies on the capacity of the pre-trained model to learn through natural language like humans. Finally, at the bottom left of AI×Ed lies epistemological research where researchers could use thought experiments and simulations involving LLMs to tackle questions that may inform debates around the nature of knowledge, language, and learning. For example, what is the difference between human and machine language skills and how does that influence our understanding of how students can learn from dialogue?

6 Conclusion

As we look toward the future, it is tempting to think about how technological advances in AI (e.g., deep learning and large language models) and the increasing ubiquity and computational power of technology will change education. But as we argue here, it is worth taking a step back to better understand the kinds of relationships that AI can have and *has had* with education. The future of AIED may look different if we reinvigorate the emphasis on AI as an analogy to human intelligence. We hope the AI×Ed framework is a useful tool for thinking about how we can combine some of the early theoretical and conceptual advances of the AIED community with the technological advances of our times.

References

1. Anderson, J.R., Boyle, C.F., Corbett, A.T., Lewis, M.W.: Cognitive modeling and intelligent tutoring. Artif. Intell. **42**(1), 7–49 (1990)
2. Anderson, J.R., Lebiere, C.J.: The Atomic Components of Thought. Psychology Press (2014)
3. Awad, E., et al.: The moral machine experiment. Nature **563**(7729), 59–64 (2018)
4. Baker, M.J.: The roles of models in artificial intelligence and education research: a prospective view. J. Artif. Intell. Educ. **11**, 122–143 (2000)
5. Brna, P., Ohlsson, S., Pain, H. (eds.): Proceedings of AI-ED 93 World Conference on Artificial Intelligence in Education. AACE (1993)
6. Carley, K.: Knowledge acquisition as a social phenomenon. In: Yazdani, M., Lawler, R. (eds.) 2nd International Conference on Artificial Intelligence and Education, pp. 27–29. University of Exeter (1985)
7. Conati, C.: Bayesian student modeling. In: Advances in Intelligent Tutoring Systems, pp. 281–299 (2010)
8. Doroudi, S.: The bias-variance tradeoff: how data science can inform educational debates. AERA Open **6**(4), 2332858420977208 (2020)
9. Doroudi, S.: The intertwined histories of artificial intelligence and education. Int. Artif. Intell. Educ., 1–44 (2022)
10. Doroudi, S., Rastegar, S.A.: The bias-variance tradeoff in cognitive science. Cogn. Sci. **47**(1), e13241 (2023)
11. Jacobson, M.J., Wilensky, U.: Complex systems in education: scientific and educational importance and implications for the learning sciences. J. Learn. Sci. **15**(1), 11–34 (2006)
12. Johnson, G.M.: Algorithmic bias: on the implicit biases of social technology. Synthese **198**(10), 9941–9961 (2021)
13. Kahn, K.: Three interactions between AI and education. Mach. Intell. **8**, 422–429 (1977)
14. Käser, T., Schwartz, D.L.: Modeling and analyzing inquiry strategies in open-ended learning environments. Int. J. Artif. Intell. Educ. **30**(3), 504–535 (2020)

15. Kent, C., et al.: Machine learning models and their development process as learning affordances for humans. In: Roll, I., McNamara, D., Sosnovsky, S., Luckin, R., Dimitrova, V. (eds.) AIED 2021. LNCS (LNAI), vol. 12748, pp. 228–240. Springer, Cham (2021). https://doi.org/10.1007/978-3-030-78292-4_19
16. Koedinger, K.R., Anderson, J.R.: Effective use of intelligent software in high school math classrooms. In: Proceedings of AI-ED 93 World Conference on Artificial Intelligence in Education. AACE (1993)
17. Koedinger, K.R., Brunskill, E., Baker, R.S., McLaughlin, E.A., Stamper, J.: New potentials for data-driven intelligent tutoring system development and optimization. AI Mag. **34**(3), 27–41 (2013)
18. Leelawong, K., Biswas, G.: Designing learning by teaching agents: the Betty's Brain system. Int. J. Artif. Intell. Educ. **18**(3), 181–208 (2008)
19. Matsuda, N., et al.: Learning by teaching SimStudent – an initial classroom baseline study comparing with cognitive tutor. In: Biswas, G., Bull, S., Kay, J., Mitrovic, A. (eds.) AIED 2011. LNCS (LNAI), vol. 6738, pp. 213–221. Springer, Heidelberg (2011). https://doi.org/10.1007/978-3-642-21869-9_29
20. Ng, D.T.K., Leung, J.K.L., Chu, S.K.W., Qiao, M.S.: Conceptualizing AI literacy: an exploratory review. Comput. Educ. Artif. Intell. **2**, 100041 (2021)
21. Papert, S.: Mindstorms: Children, Computers, and Powerful Ideas. Basic Books, Inc. (1980)
22. Papert, S.: The Children's Machine: Rethinking School in the Age of the Computer. Basic Books, Inc. (1993)
23. Porayska-Pomsta, K.: AI as a methodology for supporting educational praxis and teacher metacognition. Int. J. Artif. Intell. Educ. **26**, 679–700 (2016)
24. Porayska-Pomsta, K., Holmes, W.: Toward ethical AIED. arXiv preprint arXiv:2203.07067 (2022)
25. Rismanchian, S., Doroudi, S.: A computational model for the ICAP framework: exploring agent-based modeling as an AIED methodology. In: 24th International Conference on Artificial Intelligence in Education (2023, in press)
26. Sime, J.A.: Modelling a learner's multiple models with Bayesian belief networks. In: Proceedings of AI-ED 93 World Conference on Artificial Intelligence in Education, pp. 426–432. AACE (1993)
27. Tate, T., Doroudi, S., Ritchie, D., Xu, Y., Warschauer, M.: Educational research and AI-generated writing: confronting the coming tsunami (2023)
28. VanLehn, K.: OLAE. progress toward a multi-activity, Bayesian student modeler. In: Proceedings of AI-ED 93 World Conference on Artificial Intelligence in Education. AACE (1993)
29. VanLehn, K.: Cascade: a simulation of human learning and its applications. In: Proceedings of AI-ED 93 World Conference on Artificial Intelligence in Education. AACE (1993)
30. Vartiainen, H., Tedre, M., Valtonen, T.: Learning machine learning with very young children: who is teaching whom? Int. J. Child-Comput. Interact. **25**, 100182 (2020)
31. Winkels, R., Breuker, J.: Automatic generation of optimal learning routes. In: Proceedings of AI-ED 93 World Conference on Artificial Intelligence in Education, pp. 330–337. AACE (1993)
32. Yazdani, M., Lawler, R. (eds.): 2nd International Conference on Artificial Intelligence and Education. University of Exeter (1985)
33. Zhu, X., Singla, A., Zilles, S., Rafferty, A.N.: An overview of machine teaching. arXiv preprint arXiv:1801.05927 (2018)

Computational Models of Learning: Deepening Care and Carefulness in AI in Education

Daniel Weitekamp[✉] and Kenneth Koedinger

Carnegie Mellon University, Pittsburgh, PA 15289, USA
weitekamp@cmu.edu

Abstract. The field of Artificial Intelligence in Education (AIED) *cares* by supporting the needs of learners with technology, and does so *carefully* by leveraging a broad set of methodologies to understand learners and instruction. Recent trends in AIED do not always live up to these values, for instance, projects that simply fit data-driven models without quantifying their real world impact. This work discusses opportunities to deepen careful and caring AIED research by developing theories of instructional design using computational models of learning. A narrow set of advances have furthered this effort with simulations of inductive and abductive learning that explain how knowledge can be acquired from experience, initially produce mistakes, and become refined to mastery. In addition to being theoretically grounded, explainable, and empirically aligned with patterns in human data, these systems show practical interactive task learning capabilities that can be leveraged in tools that interactively learn from natural tutoring interactions. These efforts present a dramatically different perspective on machine-learning in AIED than the current trends of data-driven prediction.

Keywords: Student Models · Simulated Learners · Computational Models of Learning · Interactive Task Learning · Caring AI

1 Introduction

These proceedings celebrate the 30th anniversary of the Artificial Intelligence in Education society. At its founding in 1993 John Self was the editor of the JAIED journal and the AIED society's president. Self's belief was that AIED is set apart from other fields of computing because—as Kay and Mcella have paraphrased nicely—AIED "cares" and is "careful" [7] (or in Self's words "cares, precisely" [23]). Self argued that AIED *cares* by adapting to and supporting the needs of human learners. Unlike theoretical computer science, AIED engages with human participants and embraces the reality that educational technology has an impact on people in the real world. AIED cares by supporting learners as they are, not as we imagine them to be. AIED is *careful* because it uses a variety of research tools to deepen its understanding of its subjects. It draws from and contributes back to a broad set of disciplines, including artificial intelligence, cognitive psychology, education, and sociology. AIED researchers use interdisciplinary tools as needed to understand and model learners, build theories of learning and instruction, and effectively help learners become what they aim to be.

Self coined the term "computational mathetics" to refer to a broad set of disciplines that study matters of learning, including many forms of student modeling and

N. Wang et al. (Eds.): AIED 2023, CCIS 1831, pp. 13–25, 2023.
https://doi.org/10.1007/978-3-031-36336-8_2

diagnostic agents familiar to us today [22]. However, in recounting the legacy of computational mathetics, he lamented that student modeling had failed to produce solid theoretical groundings to inform Intelligent Tutoring System (ITS) design, compared to for instance, other fields of engineering like aeronautics:

> "Aircraft design has progressed through many centuries of visions and a few decades of serious experimentation to largely depend on the theory of aeronautics and specialised test environments. Would ITSs ever be built by a blend of beautiful theory and empirical fine tuning?" (1998, pg. 354)

Self imagined that if student models were not just data-structures, but also executable programs then they could "be used not only descriptively but also predictively, to predict how a student would solve problems in the future (assuming the model were accurate)" (1998, pg. 351). In other words, if student models were simulations of learning they could be executed on instructional material to make relative predictions about instructional design choices, much like a wind tunnel guides aeronautic design. Self believed that the field was hesitant to approach the topic of student simulation because it was too "complex" (1995, pg. 92) for the theories of learning and AI of the time.

In 2023, it is worth reconsidering if Self's high fidelity computational models of learning are on the horizon. As we argue in the following pages, the answer is resoundingly yes!—but not necessarily because of the technologies driving the current AI hype. But, if we adopt an appropriate mindset of care and carefulness in choosing our foundations, there does seem to be an exciting path forward.

First, we should consider the state of *care* and *carefulness* in current AIED research. It's all too easy to claim those virtues by association without knowing how to, or caring to put them into practice, and there is a fine line between the ideal and debatable cases. Throughout its history AIED has thrived from the use of caring and careful interdisciplinary methods like user studies and A/B experiments that reach outside of our core computer science toolbelt, into the real world. But the consideration of care and carefulness is not characterized only by checking methodological boxes in the classroom. Care and carefulness are as much applicable in purely analytical or computational programs of research that begin and end behind a computer screen. The bare minimum requirement of *caring* in AIED research is that we quantify and build arguments for the value of our work in the real world. In advance of our consideration of how computational models of learning can help us deepen these values, let us digress to highlight a subfield of AIED in which this *caring* justification often falls short. In consideration of Self's sentiment that learning engineering had methodological holes [23], we will take an analytical estimation approach—an approach common among engineers but underutilized in AIED. With a simple back-of-the-napkin calculation we will estimate the potential real world impact of building better knowledge tracers and in doing so highlight some epistemic pitfalls to engaging in *caring* and *careful* student modeling.

2   Do Better Student Models Produce Better Knowledge Tracing?

Knowledge tracers [5] estimates students' knowledge of individual skills, concepts, or facts as the probability of answering future practice items correctly [21], or of achieving a latent "mastery" state [5]. Students continue practicing problems associated with particular skills, concepts, or facts until the tracer's estimate of mastery exceeds some threshold. Knowledge tracers are typically fit to student performance data and utilize time, number of practice opportunities, or other features to estimate mastery.

Let us raise and attempt to answer the question analytically: If we built a 5% better statistical model for knowledge tracing how much time would we expect the new tracer to save students? Consider the following calculation (with code[1]). Our analytic approach, confronts us immediately with an often overlooked consideration: that the model's performance in the neighborhood of the mastery threshold is the only region where performance matters since this is where the tracer decides if it should stop selecting items of each kind. Between an old model and a new one a total RMSE (Root-Mean-Square-Error) improvement of 5% is a fairly large one—comparable for instance to the difference between the best model in [6] (0.329 RMSE) and normal Performance-Factors-Analysis (0.343 RMSE) [21][2]. But model improvements are rarely reported only around the mastery threshold. So we'll assume, perhaps incorrectly, that a total model improvement of 5% implies a 5% improvement in this critical neighborhood.

Consider a logistic regression model like Cen. et al. [3] where number of opportunities is our only feature. We'll assume that the performance threshold point for mastery is 95% and that we have a new model with 5% better RMSE at the threshold point than an older one. Independent of our choice of model, we can compute analytically that the old will report 87.84% student accuracy instead of 95% (assuming it was underestimating). Let us also assume that our new model is actually the ground-truth model and set a few constants of this ground truth: a first opportunity intercept of 35% accuracy, and an average of 12 opportunities to achieve mastery. Assuming equal proportions of change in intercept and slope, we can expect our new model to save students an average of 4.19 extraneous practice attempts. If every attempt takes 15 s, and there are 500 knowledge components learned over a year, then over the course of that year, the old 5% worse model would have students practice 8.75 h more or about 35% of the total 25 h expected in ground-truth. Compared to similar model-driven calculations reported by Yudelson et al. [31,32], our analytically derived 35% improvement is an overestimate, but well within the right order of magnitude.

Should we then conclude that the project of building incrementally better knowledge tracers is a valuable one? Yet more careful considerations notwithstanding[3], it would seem that the answer is yes!—at least insofar as we can build yet better student models. But, it would seem that this project is only accidentally aligned with a justifiable real-world impact. If our calculation had found that a 5% RMSE improvement produced only a 0.1% time saving, or if a review of the last two decades of student modeling showed no improvements in the neighborhood of the mastery threshold (only in the region preceding it), then we would be forced to accept those efforts as time wasted. The typical measures of overall fit statistics simply do not address these concerns. They do not quantify improvement in terms of the real-world experiences of students.

---

[1] https://github.com/DannyWeitekamp/Quantifying-Knowledge-Tracing-Time-Saved.

[2] KDD Cup 2010 EDM Challenge: Algebra I 2005–2006 dataset.

[3] Knowledge tracers are fit to formative assessment data not delayed summative assessments. An improved fit to formative data is no guarantee of improved supports for long-term retention.

In this particular subfield, the *caring* consideration of connecting models to their real world impact is decidedly the exception and not the rule. This pattern should concern us. Neglecting to connect models to their potential real world impact is like building a car and never test driving it. The apparent source of this cognitive dissonance is an overemphasis on the logic of justification of big-data machine-learning which consider systematic benchmarks and improvements in fit statistics as valuable in their own right—hardly a caring or careful logic if it makes no direct connection to its impact on real human learners. Taking the final step of justifying the real world impact of models should be standard practice in AIED.

2.1   Examples of Fitting Models with Care

Today's data-driven machine-learning allows for an approach where problems are solved *just as prediction problems*. But we should consider what is lost in that simplicity. We can almost always better support learning by seeking to theoretically understand it. AIED researchers should always consider engaging with their research from a learning scientist's perspective—raising and attempting to answer questions regarding their subjects of inquiry: human learners and how to best support their learning with technology. Exemplary instances of this mindset include learning curve analysis and automatic domain model selection [3]. These tools also fit student performance models—the difference is how they are used. In these cases, fit statistics and patterns of performance are not ends in themselves, they are used diagnostically for finding real issues in learning materials. Negative signals like flat learning curves and poor relative domain model fits can help learning engineers identify issues in instructional materials [3]. Nevertheless, interpreting these signals and revising instruction in response is still more of an art than science, requiring the engineer to make educated guesses about how best to best modify instruction. While these tools deepen our understanding of our subjects of inquiry, there is a great deal more we can do to deepen our understanding of learners and instruction beyond what existing tools allow.

3   Going Deeper with Computational Models of Learning

A common mentality of AI researchers well into the 1980s was that AI research could be a means of building and testing theories of learning—a complementary approach to the experimental methods of cognitive psychology. AI research has shifted away from this perspective toward purely technical and practical concerns, but it is worth reflecting on whether we've lost elements of care and carefulness in that shift.

Unlike purely practical AI, computational models *care* by simulating elements of cognition *explicitly*, instead of treating them *implicitly* by reducing cognitive phenomena to numerical quantities or probabilities like in statistical models that fit parameters to data. A computational model embodies an executable theory which is often more detailed than an experimentally driven theory because the computational theory must be precise enough to be implemented in simulation, yet general enough to reproduce broad sets of human behaviors, and furthermore plausible enough to satisfy known constraints of human cognition, like for instance, having only particular prior knowledge or a finite working memory, to name just a few [12]. When computationally modeling *learning*, the modeler is confronted not only with these computational and cognitive constraints, but also with an empirical reality glimpsed through student data. The simulation must succeed at learning from the kinds of natural instruction that humans experience, [16] and reproduce patterns of correct performance and errors.

As a practical means of deepening our insights into human learning, computational models of learning can be used to make *a priori* predictions without need for prior collection of student data. For instance: Which of two forms of instruction will work best? What kinds of misinterpretations and mistakes will students make as they practice? How can we most optimally adapt to particular student behaviors or traits and produce the best learning outcomes? In answering any of these questions, a computational model has a key advantage over a statistical one: simulated cognitive reasons underlying its predictions. A computational model simulates the formation of knowledge in response to particular instructional events, and the execution of that acquired knowledge on new material. As a predictor, this goes well beyond simply fitting parameters to patterns in data, and holds the promise of enabling learning engineers to not only predict which instructional design decisions are optimal but also predict why. These explainable patterns of simulated cognition also allows researchers to debug and revise their theories intelligently when the computational model's predictions prove to be incorrect.

Developing AI under the many constraints of computational modeling is no doubt challenging. It is a cyclic project of refinement aimed at building general purpose executable simulations of learning that can be applied to a wide array of domains, learn from several forms of instruction, and match human learning behavior faithfully. It is a harder and more principled project than the development of purely practical or predictive AI where performance statistics are often the sole guiding objective. Nonetheless, it is likely the most direct and precise means of building a theory of instructional design.

3.1 The Baby in the Good Old-Fashion AI Bathwater

In setting foundations for computational models of learning, it is tempting to forgo early forms of AI and look to the trends of data-driven machine-learning. A dominant theme of the last decade of AI has been to demonstrate that new capacities can be acquired by fitting high-dimensional deep-learning [13] models to copious amounts of data. Deep-learning has been used to train many state-of-the-art AI like Alpha-Go and Chat-GPT, yet it offers us very little toward the purpose of modeling human learning. Despite the complexity of its high-dimensional multi-layered structure, deep-learning relies on a single over-simplified and data-hungry learning mechanism: regression via gradient descent. By comparison, humans are remarkably data-efficient learners able to achieve mastery in academically relevant tasks from only a handful of instructional and practice opportunities. Humans, no doubt, owe their learning efficiency to a variety of forms of reasoning that can help them rapidly construct and verify knowledge. Since our objective with computational modeling is to explain and intelligently support these remarkable learning capacities, the data-inefficiency and inexplicability of deep-learning's *blackbox* knowledge structures make for poor foundations.

If deep-learning has dispensed with the capabilities we care about in computational modeling, perhaps we should return to the hard-coded expert-systems of the

Good Old-Fashion AI (GOFAI) era? This too would be a poor choice. Expert-systems model the execution of human-like expertise using hard-coded rules, but do not model the acquisition of expertise from experience. One could be forgiven for thinking that we are necessarily stuck between two bad options: the choice between the flexibility of acquiring blackbox representations in a data-driven manner, and the rigidity of hard-coded but explainable symbolic knowledge structures. Even cognitive architectures like SOAR [10] and ACT-R [2], that aim to model learning, align heavily with the hard-coded inflexibility of GOFAI. They rely heavily on predefined domain-specific rules—modeling learning only by tuning the activation patterns of hard-coded rules or by recombining them into new specialized structures. The gaping hole in these theories is how those initial rules came to be acquired through experience, and how that acquired knowledge can produce human errors without the sources of errors being explicitly programmed.

To achieve the best of both options, there is a very clear, but seriously underutilized approach to machine-learning—a neoclassical approach where expert-system-like rules can be efficiently acquired in a highly flexible yet mostly symbolic bottom-up fashion. One place to look for such advances is in the emerging field of interactive tasking learning (ITL) [11]. ITL systems seek to acquire performance capabilities from just a few instances of human provided instruction. Some ITL systems like Rosie [19], are built on cognitive architectures, and share many of their assumptions. Many more ITL innovations may deviate entirely from feasible human cognition, so our choice of adopting these innovations should be made carefully.

In the near-term the more directly useful neoclassical approaches to machine-learning can be found in past AIED research on simulated students, which have been touted as foundations of ITL in their own right [11]. Three such systems, Sierra [24], SimStudent [18] and the Apprentice Learner (AL) [17], are able to learn to perform academically relevant procedural tasks (including but not limited to, learning math and science procedures) from at least two kinds of instruction: demonstrations of correct performance, and positive and negative correctness feedback—core forms of instruction that students experience in a one-on-one tutoring setting or while working in an intelligent tutoring system. Since these systems can learn in a bottom-up fashion from learning materials like intelligent tutoring systems (ITSs), they are a compelling basis for Self's proposed executable student models. They provide an answer to the question of how knowledge can be efficiently induced from first experiences, initially produce errors, and then become refined into expertise through practice.

3.2 Setting Our Foundations in Induction and Abduction

There are several broad categories of learning consistent with this neoclassical view of machine-learning that we might consider as foundations for a computational model of learning. I highlight Sierra, SimStudent and AL because they implement purely inductive and abductive learning mechanisms. Induction is the formation of knowledge by generalizing from examples. Abduction is the use of existing knowledge to find the most likely reason underlying an example. Both forms of reasoning can produce imperfect knowledge structures from instruction and reproduce human errors.

There are certainly alternative foundations. For instance, deduction: finding that which follows logically from what is already known, and—drawing from the suggestions of prior works [16,27]—learning from being told, learning by analogy [8], planning and debugging towards a goal [10], reinforcement learning [20], speed-up learning through practice [2,10], just to name a few. It is all too easy to fool ourselves into thinking that any one of these particular learning capabilities supersedes the rest, encompassing a vaguely characterized notion like *logical reasoning* or *general intelligence*. We should consider the scope and applicability of each of these proposed mechanisms individually and perhaps aim to unite these disparate methods under a unified toolset, and with it build a sort of model-human learner.

The reason to favor induction and abduction as foundations over other choices is they offer a means of generating potentially buggy knowledge from experiences. Many forms of learning require some initial prior domain-specific knowledge. Even learning from being told—which is the closest thing to directly programming a human, but not anywhere nearly as precise or unambiguous—requires prior language comprehension capabilities and vocabulary. If prior knowledge is always hard-coded as infallible expertise then we run the risk of creating toy-models that replicate the acquisition of capacities but fail to reproduce patterns of human error. Thus, the property of induction and abduction of generating buggy knowledge from experiences establishes a solid foundation for mechanistically modeling the real messiness of human learning. We shouldn't resign ourselves to the idea that learner errors are purely random, or that states of knowledge are simply scalars on a continuum from unknown to known. Principled computational modeling rooted in induction and abduction can simulate the reasons underlying human errors, and perhaps with this deeper theoretical understanding, allow us to make more intelligent instructional decisions.

3.3 Evidence of Executable Student Models

Results from prior work with Sierra, SimStudent and the Apprentice Learner (AL) show a compelling foundation for using inductive and abductive learning mechanisms in executable theories of learning. An algorithmic level discussion of these systems is beyond the scope of this work. Although, it is worth noting that they all rely on a combination of multiple, mostly symbolic machine-learning mechanisms, and share a common high-level structure—a testament to the fact that two computational modelers held to the same constraints tend to come to similar solutions.

Generators of Errors and Domain Models: VanLehn et al. found that when taught multi-column subtraction, Sierra could reproduce two-thirds of the types of errors students produced on quizzes—more than twice the errors explained by ad-hoc expert-system based theories characterized by hard-coded "repairs" [25]. Sierra's general theory of inductive and abductive learning, was simultaneously more parsimonious than the hard-coded theories yet better at generating a broad set of observed human errors. A similar result, by Li et al. showed that SimStudent could produce a better fitting domain model of a simple algebra ITS than those hand-built by learning engineers [14]. The domain model was constructed by using the execution of SimStudent's induced representations and production-rules to form knowledge-component to item mappings. These examples show evidence that general theories of bottom-up learning from experiences can generate empirically better predictions of overall student performance behaviors than ad hoc human-generated ones. An appropriate high-level theory can produce surprising and highly specific theoretical predictions.

Models of Individuals: To explain the behavior of a theoretical *average* student is one thing. It is another to explain the behavior of particular individuals. In this regard, simulated learner technologies are still very young, but the results to date offer a compelling picture of where they can take us. Maclellan et al. showed that when Apprentice Learner (AL) agents were taught on the same sequences of ITS problems as students, they could reproduce student error rates opportunity-by-opportunity in several different domains—producing population learning-curves that aligned well with student performance data without explicitly fitting to it [17]. As an interactive task learning system AL is remarkably data-efficient and can produce learning curves orders of magnitude closer to human curves than, for instance, deep reinforcement-learning.

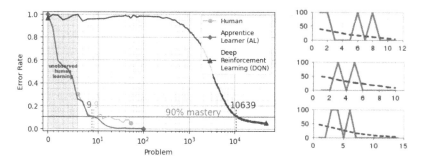

Fig. 1. (left) AL learning curves on fraction addition and multiplication. Learning-rate is far closer to humans than deep reinforcement-learning. (right) 1/0 pattern of correctness of single AL agents on denominator conversion steps.

Building learning curves by applying AL's theory of learning to an instructional environment (like an ITS) opens a profoundly different perspective on student modeling than the typical data-driven one. When AL works in an instructional environment, it learns to solve problems and produce actual step-by-step responses. By contrast, a typical statistical model reduces student attempts to a binary random variable where the probability of correct performance increases over time—an approach that seems appropriate because the patterns of correct (1) and incorrect (0) responses in student data often seem non-deterministic. For instance, a typical performance pattern might be "0010111", where the student answers correctly, but then incorrectly on later items before consistently answering correctly. AL can reproduce this seemingly random behavior (Fig. 1, right) without explicitly injecting randomness. AL agents use evidence from individual instances of instruction to refine induced skills with every practice opportunity, especially on ones where its current knowledge structures produce mistakes that receive negative feedback. So an AL agent's pattern of performance can oscillate between correct and incorrect as it solves new problems and converges toward mastery. As a theory, this envisions the particular step-by-step differences in responses between individual learners as arising from differences in prior experiences, creating particular knowledge that when applied to particular new problems, produces particular responses.

This is not to say that this prior work has achieved a sort of Laplace's demon of human learning—an infallible predictor of future responses from a known starting point. A student's starting point, the state of their prior knowledge before using an ITS, is always a matter of uncertainty. Students can encounter instructional opportunities in the classroom or at home that an ITS can't be expected to know about. In these cases we have to make guesses. For instance Weitekamp et al., found better fits in AL generated curves by pre-training agents with a number of random practice opportunities commensurate with student's predicted prior knowledge [29]—the content of prior experience was guessed while the quantity was tailored to the individual. Other cognitive considerations, not included in these works like mechanisms of forgetting and attention are harder to model purely mechanistically, and may require some probabilistic treatment. Beyond these considerations there is a litany of individual biological, social, and environmental factors—did the student sleep well? eat breakfast? do they have ADHD?—that one could imagine injecting theories of into such a system.

In the near-term getting the general theory of learning broadly correct, is more important than adding precise theories of, typically unobservable, individual factors. For instance, Weitekamp et al. have reported discrepancies between trends in the types of errors made by simulated and real students attempt-by-attempt [30]—so the cycle of theory refinement is well underway. Our larger takeaway should be that the path toward Self's proposed executable student model is very clear. These demonstrations of computational models of learning are surprisingly parsimonious, domain-general, and accurate *a priori* first-order approximations. And, since they are inspectable and debuggable there is a path forward of further incisive and deliberate model development constrained by computational, empirical, and cognitive constraints.

Testers of Instruction: So how closely do these simulated students need to align with human learning behavior before we begin to trust them to make decisions about instructional design? Maclellan et al. showed that AL agents could replicate broad differences in patterns of student performance between blocked and interleaved instruction [17]. Beyond this, several unpublished results from participants in an AL-base workshop series capture the notion that if a simulated student fails to learn from an ITS, then it may not be adaptive enough to support low achieving human students. A simulated student failing to learn from an ITS can be a signal that the target knowledge taught by the ITS does not follow logically from the provided instruction, even by induction over several problems. A high-achieving, high prior knowledge student might succeed despite this lack of support. But just like a human student with low-prior knowledge a simulated student can typically only learn from instruction that begins from first principles and breaks down problems into fine-grained substeps.

In addition to succeeding or failing outright a simulated learner can sometimes learn more efficiently from one form of instruction than another. For instance, AL agents can learn multi-column addition considerably faster when the ITS teaches a version of the procedure where 0's are explicitly carried for each partial sum instead of being omitted. This explicit-zero version of addition is easier for the agent because it provides a place in the problem interface for explicitly accounting for having calculated the carry value. So, the overall procedure follows a more consistent, easy to learn pattern. As a theory this result makes the novel prediction that the same would be true for students.

4   Additional Practical Uses of Simulated Learners

So far I've addressed how we can gain a deeper theoretical understanding of learning and instruction by building computational models of learning. Those of us accustomed to wielding AI as a sort of magic wand of prediction will surely see more direct data-driven or ready-made solutions (i.e. like chat GPT) to a variety of practical problems in education. In the near-term I'm certain that a great deal will be achieved this way, but in the long-term I'm less certain that this will be a consistently fruitful approach. It's hard to imagine deep-learning's fuzzy mimicry holding a monopoly over more cogni-tively principled approaches, especially when it comes to matters of supporting human learning.

If deep-learning is today's AI magic wand then simulated learners may well be tomorrow's—or at least find similarly wide applicability among the options in our AI toolbag. As purely practical AI, simulated learners are remarkably data-efficient and domain-general interactive tasks learners. They stand apart among ITL systems because of an emphasis on cheating less with hard-coded elements and specialized toy-environments, and not at all on mimicking patterns from large datasets. Instead they have sought domain-general mechanisms of learning that can efficiently construct knowledge bottom-up from natural tutoring interactions. For instance, Li [15] added a representation-learning mechanism to SimStudent to eliminate hard-coded represen-tational prior knowledge. Maclellan [17] with the creation of AL loosened require-ments of special supplementary instructional experiences accompanying demonstra-tions. Soon mechanisms for learning from natural language instruction will be incorpo-rated into AL. All of these efforts have manifested, in simulation, a human-like ability to learn in diverse ways from diverse experiences. This flexibility presents an opportunity to build experiences where non-programmers teach simulated learners interactively.

Authoring ITSs: For instance, prior work has shown that simulated learners can be used to author ITSs [18] faster than existing tools [28]. Programming production-rule based ITSs require about 200–300 hour per hour of instruction authored. GUI-based tools cut this time down by about half, but add restrictions on what can be built [1]. By instead authoring via interactively tutoring a simulated learner, the production rules of an ITS can be induced from natural instruction instead of being programmed. This could reduce authoring times to about the time taken to tutor a student: one hour per hour of instruction. One does not need to do a particularly complex calculation to see the potential impact here—the multiplicative effect of a wide array of people building highly adaptive ITSs faster. With such a tool ITSs might be able to scale in the way that MOOCs have, transitioning learning at scale away from passive content delivery toward the highly adaptive deliberate practice based instruction characteristic of ITSs [9].

Teacher Training: There is a vast difference between knowing how to do something and knowing how to teach it. A core element of that difficulty is that humans are rather poor at reflecting on, and articulating the content of their tacit knowledge [4]. Another is that it is hard to make *a priori* predictions about the misconceptions that novices may acquire, and so it is difficult to tailor instruction to address or avoid them. If teachers practiced by teaching simulated learners they may acquire a deeper sense of why misconceptions arise since they can inspect the agent's knowledge and inspect instances of how buggy knowledge is constructed. They could also test various forms of instruction to see which methods are most effective. VanLehn et al. have articulated this vision in much greater detail [26]. But, I might add to that vision the possibility of creating domain-general versions of such tools for teaching new or specialized material. Professional expertise is often acquired through experience, but not taught deliberately. Future simulated learners may aide everyday professional development by lending explicit cognitive support in workplace apprenticeship relationships.

5   Conclusion

This work has explored John Self's themes of care and carefulness in AIED, especially with respect to analytical and computational projects of research. We have argued that applications of data-driven machine-learning in AIED don't always live up to these virtues, as they often follow a misguided logic of justification based on improving performance statistics that rarely take the final step of quantifying real-world impact. We have proposed computational models of learning as means of engaging in more deeply theoretical, caring and careful computational AIED research. Computational models of learning care by seeking an explicit and detailed understanding of the mechanisms underlying learning, which are often treated only implicitly through statistical modeling.

In any applied field of research there is a balance between theoretical and practical advances. But we shouldn't forget that in many fields of engineering theoretical advances have streamlined the development of the greatest practical advances. Self's view was that instructional design could be improved significantly if we had high fidelity theories of learning that could inform instructional design. He imagined executing such theories on instructional material, much like a wind tunnel tests aeronautic designs. I've demonstrated here that prior simulated learner systems have begun this project in earnest. The results to date show a promising path toward realizing Self's vision of executable theories of learning. These systems additionally show several practical interactive task learning use cases like ITS authoring, and teacher training. Perhaps the most exciting element of these advances is their origins in AIED research. The unique caring and careful perspective of AIED can contribute as much to the field of machine-learning as it does to the classroom.

References

1. Aleven, V., et al.: Example-tracing tutors: intelligent tutor development for non-programmers. Int. J. Artif. Intell. Educ. **26**(1), 224–269 (2016)
2. Anderson, J.R., Matessa, M., Lebiere, C.: ACT-R: a theory of higher level cognition and its relation to visual attention. Hum. Comput. Interact. **12**(4), 439–462 (1997)
3. Cen, H., Koedinger, K., Junker, B.: Learning factors analysis – a general method for cognitive model evaluation and improvement. In: Ikeda, M., Ashley, K.D., Chan, T.-W. (eds.) ITS 2006. LNCS, vol. 4053, pp. 164–175. Springer, Heidelberg (2006). https://doi.org/10.1007/11774303_17
4. Clark, R.E., Feldon, D.F., Van Merrienboer, J.J., Yates, K.A., Early, S.: Cognitive task analysis. In: Handbook of Research on Educational Communications and Technology, pp. 577–593. Routledge (2008)
5. Corbett, A.T., Anderson, J.R.: Knowledge tracing: modeling the acquisition of procedural knowledge. User Model. User-Adap. Inter. **4**(4), 253–278 (1994)
6. Gervet, T., Koedinger, K., Schneider, J., Mitchell, T., et al.: When is deep learning the best approach to knowledge tracing? J. Educ. Data Min. **12**(3), 31–54 (2020)
7. Kay, J., McCalla, G.I.: The careful double vision of self. Int. J. Artif. Intell. Educ. **13**(1), 11–18 (2003)
8. Klenk, M., Forbus, K.: Analogical model formulation for transfer learning in AP physics. Artif. Intell. **173**(18), 1615–1638 (2009)
9. Koedinger, K.R., Kim, J., Jia, J.Z., McLaughlin, E.A., Bier, N.L.: Learning is not a spectator sport: doing is better than watching for learning from a MOOC. In: Proceedings of the Second ACM Conference on Learning@ Scale, pp. 111–120 (2015)
10. Laird, J.E.: The Soar Cognitive Architecture. MIT Press (2019)
11. Laird, J.E., et al.: Interactive task learning. IEEE Intell. Syst. **32**(4), 6–21 (2017)
12. Langley, P.: The computational gauntlet of human-like learning. In: Proceedings of the AAAI Conference on Artificial Intelligence, vol. 36, pp. 12268–12273 (2022)
13. LeCun, Y., Bengio, Y., Hinton, G.: Deep learning. Nature **521**(7553), 436–444 (2015)
14. Li, N., Cohen, W.W., Koedinger, K.R., Matsuda, N.: A machine learning approach for automatic student model discovery. In: EDM, pp. 31–40. ERIC (2011)
15. Li, N., Matsuda, N., Cohen, W.W., Koedinger, K.R.: Integrating representation learning and skill learning in a human-like intelligent agent. Artif. Intell. **219**, 67–91 (2015)
16. MacLellan, C.J., Harpstead, E., Marinier, R.P., III, Koedinger, K.R.: A framework for natural cognitive system training interactions. Adv. Cogn. Syst. **6**, 1–16 (2018)
17. Maclellan, C.J., Harpstead, E., Patel, R., Koedinger, K.R.: The apprentice learner architecture: closing the loop between learning theory and educational data. In: International Educational Data Mining Society (2016)
18. Matsuda, N., Cohen, W.W., Koedinger, K.R.: Teaching the teacher: tutoring SimStudent leads to more effective cognitive tutor authoring. Int. J. Artif. Intell. Educ. **25**(1), 1–34 (2015)
19. Mininger, A., Laird, J.: Interactively learning a blend of goal-based and procedural tasks. In: Proceedings of the AAAI Conference on Artificial Intelligence, vol. 32 (2018)
20. Nason, S., Laird, J.E.: Soar-RL: integrating reinforcement learning with Soar. Cogn. Syst. Res. **6**(1), 51–59 (2005)
21. Pavlik, P.I., Jr., Cen, H., Koedinger, K.R.: Performance factors analysis-a new alternative to knowledge tracing. Online Submission (2009)
22. Self, J.: Computational mathetics: towards a science of learning systems design (1995)
23. Self, J.: The defining characteristics of intelligent tutoring systems research: ITSs care, precisely. Int. J. Artif. Intell. Educ. (IJAIED) **10**, 350–364 (1998)
24. VanLehn, K.: Learning one subprocedure per lesson. Artif. Intell. **31**(1), 1–40 (1987)

25. VanLehn, K.: Mind Bugs: The Origins of Procedural Misconceptions. MIT Press (1990)
26. VanLehn, K., Ohlsson, S., Nason, R.: Applications of simulated students: an exploration. J. Artif. Intell. Educ. **5**, 135–135 (1994)
27. VanLehn, K.A.: Arithmetic procedures are induced from examples. Technical report (1985)
28. Weitekamp, D., Harpstead, E., Koedinger, K.: An interaction design for machine teaching to develop AI tutors. In: CHI (2020)
29. Weitekamp, D., Harpstead, E., MacLellan, C.J., Rachatasumrit, N., Koedinger, K.R.: Toward near zero-parameter prediction using a computational model of student learning. In: International Educational Data Mining Society (2019)
30. Weitekamp, D., Ye, Z., Rachatasumrit, N., Harpstead, E., Koedinger, K.: Investigating differential error types between human and simulated learners. In: Bittencourt, I.I., Cukurova, M., Muldner, K., Luckin, R., Millán, E. (eds.) AIED 2020. LNCS (LNAI), vol. 12163, pp. 586–597. Springer, Cham (2020). https://doi.org/10.1007/978-3-030-52237-7_47
31. Yudelson, M., Ritter, S.: Small improvement for the model accuracy–big difference for the students. In: Industry Track Proceedings of 17th International Conference on Artificial Intelligence in Education (AIED 2015), Madrid, Spain (2015)
32. Yudelson, M., Koedinger, K.: Estimating the benefits of student model improvements on a substantive scale. In: Educational Data Mining 2013 (2013)

Towards the Future of AI-Augmented Human Tutoring in Math Learning

Vincent Aleven[1], Richard Baraniuk[2], Emma Brunskill[3], Scott Crossley[4],
Dora Demszky[3], Stephen Fancsali[5], Shivang Gupta[1], Kenneth Koedinger[1],
Chris Piech[3], Steve Ritter[5], Danielle R. Thomas[1(✉)], Simon Woodhead[6],
and Wanli Xing[7]

[1] Carnegie Mellon University, Pittsburgh, PA 15213, USA
shivangg@andrew.cmu.edu, {koedinger,drthomas}@cmu.edu
[2] Rice University, Houston, TX 77005, USA
[3] Stanford University, Stanford, CA 94305, USA
[4] Vanderbilt University, Nashville, TN 37203, USA
[5] Carnegie Learning, Pittsburgh, PA 15219, USA
[6] Eedi, London, England, UK
[7] University of Florida, Gainesville, FL 32611, USA

Abstract. One of the primary obstacles to improving middle school math achievement is lack of equitable access to high-quality learning opportunities. Human delivery of high-dosage tutoring can bring significant learning gains, but students, particularly economically disadvantaged students, have limited access to well-trained tutors. Augmenting human tutor abilities through the use of artificial intelligence (AI) technology is one way to scale up access to tutors without compromising learning quality. This workshop aims to highlight the challenges and opportunities of AI-in-the-loop math tutoring and encourage discourse in the AIED community to develop human-AI hybrid tutoring and teaching systems. We invite papers that provide clearer understanding and support the progress of human and AI-assisted personalized learning technologies. The structure of this full-day hybrid workshop will include presentations of accepted papers, small or whole group discussion, and a panel discussion focusing on common themes related to research and application, key takeaways, and findings imperative to increasing middle school math learning.

Keywords: Tutoring · Personalized learning · AI-assisted tutoring

1 Leveraging AI and Human Tutoring

The primary challenge to improving middle school math achievement is providing all students equitable access to the existing high-quality learning opportunities that we know to be effective. Students from economically disadvantaged and historically underserved backgrounds can learn just as well as their peers when given the same opportunities, but they are more likely to experience learning gaps due to a lack of access to these learning opportunities [2]. High-dosage human tutoring can produce dramatic learning gains, particularly if tutors are well-trained in providing students social-motivational support [4]. However, low-income students lack access to well-trained tutors, evidenced by the 16 million low-income children on the waitlist for high-quality afterschool programs [1]. In addition, the estimated costs of $2500+ per student for individualized tutoring prohibits student access [3]. Human tutoring alone cannot meet present students need. Sustainable and scalable tutoring infrastructures are possible through the combined synergy of artificial intelligence (AI)-assisted and human technologies that can be achieved through novel and well-engineered AI-supported tutoring models.

AI-assisted tutoring shows promise and can potentially double learning outcomes [5], but analytics show that many students, especially from low-income backgrounds, are not getting sufficient learning opportunities. Student inaccessibility can be attributed to a variety of factors, including: not having sufficient access to the medium of using AI, such as digital devices and internet; issues facing inclusion with inadequate support of diverse student needs, such as English language learners and students with disabilities; and a lack of understanding of AI capabilities and limitations [7]. The challenges facing math learning related to access, equity, fairness, and inclusion have fostered collaborative and focused efforts on AI-assisted human-technology ecosystems that increase learning opportunities for all students.

This workshop aims to facilitate discussion and engagement among the Artificial Intelligence in Education (AIED) community regarding AI-assisted individualized learning tools to improve middle school teaching and tutoring. In particular, the workshop hosts updates on progress, findings, and challenges to AI-supported, personalized instruction. We invite empirical and theoretical papers aligned with the theme particularly (but not exclusively) within the following areas of research and application:

- **AI-assisted and Human Tutoring Systems:** Insight into better understanding and supporting human, AI-assisted, and interactive learning technologies related to individualized instruction
- **Delivery and Scale:** Efficacy of different human tutoring delivery systems (e.g., video, audio, chat) and the corresponding needed differentiated support; Different models for scaling including peer tutoring, computer tutoring, etc.
- **Training Development:** Tutor/teacher training development that recognizes diverse experiences and backgrounds, in relation to AI-assisted tutoring support structures
- **Equity and Inclusion:** Issues facing equity and inclusion, with focus on intelligent techniques to support students from under resourced communities
- **Ethics:** Privacy and transparency of intelligent techniques, such as using federated machine learning and explainable AI to examine data ownership and human-AI collaboration; Transferability and fairness of predictive models across contexts
- **Evaluation:** Program evaluation, such as applications using large-language models or dataset development for reinforcement learning of models; Methods of measuring student growth, with possible insights into dosage; Evidence of learning outcomes
- **Key Challenges:** Barriers, considerations, and challenges to providing human and AI-based tutoring and individualized instruction at scale
- **Interoperability:** How do AI and human tutoring systems interact with existing technological and social systems?

2   Relevance and Call for Papers from the AIED Community

There is a concerted effort within the AIED community to increase learning opportunities among economically disadvantaged and historically underrepresented students. The COVID-19 pandemic had a severe impact on education globally. The U.S. has lost nearly twenty years of math progress among middle school students [6], with racial and economic learning gaps preventing millions of students from realizing their potential. By leveraging the power of AI, the AIED community is working to provide equitable learning opportunities and helping bridge the persistent opportunity gap in action.

The workshop will include presentations of accepted papers and facilitated discussion sessions. We will solicit papers relevant to the themes using the short- or long-paper format described in the conference proceedings guidelines. Papers will go through a single-blind review process, with reviewers anonymous and authors known. Reviewers will be required to make a recommendation of either acceptance or rejection of the paper and explain their reasoning behind their decision. They will assess the paper based on three criteria, using a scoring system of $-1, 0$, or 1; alignment with the workshop's theme, level of interest to AIED, and overall quality. Authors of accepted papers will provide presentations at the conference. The paper submission deadline is May 26, 2023, with schedule details on the workshop website: https://sites.google.com/andrew.cmu.edu/aied2023workshop/.

3   Target Audience and Participation

The target audience consists of researchers, educational practitioners, businesses, policy-makers, and anyone among the AIED community interested in enriching their knowledge of AI-assisted and human tutoring working in synergy to provide individualized learning experiences. The workshop intends to set a wider agenda related to AI and human tutoring to grasp novel research ideas, interesting findings, and key insights to meet the practical needs of those involved in the research and development of AI-assisted and human tutoring ecosystems. This workshop is hybrid with no limit on the number of remote participants.

4   Workshop Format

We propose a full-day hybrid workshop with following activities: 1) presentations of accepted papers with Q&A, 2) small-group discussions on the conference themes, 3) reports of small-group discussions, 4) a moderated panel with audience participation focused on next steps, and 5) a closing summary and discussion. These activities will be oriented toward the workshop goals to develop shared understanding of the current state of AI-augmented human math tutoring and to pose key research questions and challenges for future research. The length of presentations will be determined by the organizing committee based on the maturity of the work, level of interest among AIED, and significance to the themes. Small group discussions will follow, and these will be aligned with the eight research areas described. A subsequent whole-group discussion will contain a question-and-answer period with in-person panelists. A summary of the key issues and responses from the panel discussion, along with commonalities among accepted papers, will be published in the workshop proceedings.

5 Organizing Committee (Alphabetical Order)

Vincent Aleven, Ph.D., Carnegie Mellon University, aleven@cs.cmu.edu
Vincent is a Professor of Human-Computer Interaction at Carnegie Mellon University, with 30 years of experience in research of AI-based learning. His lab created Mathtutor, an AI-based tutoring software for middle school math and the tools for AI-based software, CTAT and Tutorshop. Vincent has written over 250 publications, with he and his team winning 11 best paper awards at international conferences and has acted as PI or co-PI on 20 major research grants. Currently, Vincent is co-editor-in-chief of the *International Journal of Artificial Intelligence in Education* (IJAIED).

Richard Baraniuk, Ph.D., OpenStax, Rice University, richb@rice.edu
Richard is the C. Sidney Burrus Professor of Electrical and Computer Engineering at Rice University and the Founding Director of OpenStax. He is a Member of the National Academy of Engineering and American Academy of Arts and Sciences and a Fellow of the National Academy of Inventors, American Association for the Advancement of Science, and IEEE. For his work in open education, he has received the C. Holmes MacDonald National Outstanding Teaching Award, the Tech Museum of Innovation Laureate Award, the Internet Pioneer Award from the Berkman Center for Internet and Society at Harvard Law School, and many other prestigious awards.

Emma Brunskill, Ph.D., Stanford University, ebrun@cs.stanford.edu
Emma is an Associate Professor in the Computer Science Department at Stanford University where she aims to create AI systems that learn from a few samples to robustly make good decisions. Her work is inspired by the positive impact AI may have in education and healthcare, with interests in large language models to advance AI-assisted human tutoring. Emma is part of the Stanford AI Lab, the Stanford Statistical ML group, and AI Safety @Stanford. She has received an NSF CAREER award, Office of Naval Research Young Investigator Award, and many other awards. Emma and her lab have received multiple best paper nominations for their AI and machine learning work.

Scott Crossley, Ph.D., Vanderbilt University, scott.crossley@vanderbilt.edu
Scott is a Professor of Special Education at Vanderbilt University. His primary research focus is on natural language processing and the application of computational tools and machine learning algorithms in language learning, writing, and text comprehensibility. His main interest area is the development and use of natural language processing tools in assessing writing quality and text difficulty. He is also interested in the development of second language learner lexicons and the potential to examine lexical growth and lexical proficiency using computational algorithms.

Dora Demszky, Ph.D., Stanford University, ddemszky@stanford.edu
Dora is an Assistant Professor in Education Data Science at Stanford University. Her research focuses on measuring equity, representation, and student-centeredness in educational texts, with the goal of providing insights to educators to improve instruction. She develops measures based on natural language processing that work well for high-dimensional, unstructured data, and she applies these measures to provide feedback to educators. Dr. Demszky has received her PhD in Linguistics at Stanford.

Stephen Fancsali, Ph.D., Carnegie Learning, sfancsali@carnegielearning.com
Stephen is Director of Advanced Analytics at Carnegie Learning. With over a decade of experience in educational data science, he specializes in statistical and causal modeling of data produced by learners as they interact with AI-driven instructional software. He works on innovative learning analytics and models of student learning underlying MATHia, LiveLab, MATHstream, and other products. Stephen has published in the *Journal of Learning Analytics* and many conference proceedings. He received a Ph.D. in Logic, Computation, and Methodology from Carnegie Mellon University.

Shivang Gupta, Carnegie Mellon University, shivangg@andrew.cmu.edu
Shiv is the Head of Product at PLUS - Personalized Learning Squared at Carnegie Mellon University. A graduate of the Masters in Educational Technology and Applied Learning Science (METALS) program at CMU, Shiv was the lead curriculum developer at First Code Academy in Hong Kong and previously worked on corporate training in the metaverse.

Kenneth Koedinger, Ph.D., Carnegie Mellon University, koedinger@cmu.edu
Ken is the Hillman professor of Computer Science and Psychology at Carnegie Mellon University and founder of PLUS tutoring. He is a co-founder of CarnegieLearning, Inc. that has brought Cognitive Tutor based courses to millions of students since it was formed in 1998, and leads LearnLab, the scientific arm of CMU's Simon Initiative. Through extensive research and development in human-computer tutoring, Ken has demonstrated a doubling of math learning among middle school students, with future aims at bringing similar high-quality tutoring at scale. He has authored over 300 research papers and over 60 grant proposals.

Chris Piech, Ph.D., Stanford University, piech@cs.stanford.edu
Chris is an Assistant Professor in Computer Science at Stanford University. His research is in AI (and other computational methods) for education. He teaches introduction to Computer Science, CS106A and the online offering, Code in Place. The secret ingredient to both courses is high-quality human tutoring at scale.

Steve Ritter, Ph.D., Carnegie Learning, sritter@carnegielearning.com
Steve Ritter is Founder and Chief Scientist at Carnegie Learning. Dr. Ritter earned a doctorate in cognitive psychology at Carnegie Mellon University. He was instrumental in the development of the Cognitive Tutors for math, which led to Carnegie Learning, where it forms the basis of the MATHia intelligent tutoring system. Dr. Ritter is the author of many papers on the design and evaluation of adaptive instructional systems

and is recognized as an expert in the field. Dr. Ritter leads a research team devoted to using learning engineering to improve the efficacy of the company's products.

Danielle R. Thomas, Ed.D., Carnegie Mellon University, drthomas@cmu.edu
Danielle is a systems scientist at Carnegie Mellon University and research lead on the PLUS - Personalized Learning Squared tutoring project. She is a former middle school math teacher and school administrator, founding several mentoring programs supporting young women and youth in STEM. Danielle leverages her past experiences to advance research and development of tutor training and the creation of AI-assisted tutor feedback. She has first-authored over a dozen peer-reviewed papers since 2021.

Simon Woodhead, Ph.D., Eedi, simon.woodhead@eedi.co.uk
Simon is a co-founder of Eedi and also host of the Data Science in Education meetup. He coordinates Eedi's machine learning research, which has been conducted in collaboration with Microsoft Research, and turns this into new product features. With experience leading both product development and research, he has created award-winning edtech solutions with strong data science foundations.

Wanli Xing, Ph.D., University of Florida, wanli.xing@coe.ufl.edu
Wanli is an assistant professor of educational technology at the College of Education. His research themes are: (1) explore and leverage educational big data in various forms and modalities to advance the understanding of learning processes; (2) design and develop fair, accountable and transparent learning analytics, and AI-powered learning environments; (3) create innovative strategies, frameworks, and technologies for AI, Data Science, and STEM education.

References

1. Afterschool Alliance. America After 3PM: Demand Grows, Opportunity Shrinks (2020)
2. Chine, D., et al.: Educational equity through combined human-AI personalization: a propensity matching evaluation. In: Rodrigo, M.M., Matsuda, N., Cristea, A.I., Dimitrova, V. (eds.) AIED 2022. LNCS, vol. 13355, pp. 366–377. Springer, Cham (2022). https://doi.org/10.1007/978-3-031-11644-5_30
3. Kraft, M., Falken, G.: A Blueprint for Scaling Tutoring Across Public Schools (EdWorkingPaper: 20-335). Annenberg Institute at Brown University (2021)
4. Nickow, A., Oreopoulus, P., Quan, V.: The impressive effects of tutoring on prek-12 learning: a systematic review and meta-analysis of the experimental evidence. National Bureau of Economic Research (NBER), Working paper # 27476 (2020)
5. Pane, J., Griffin, B., McCaffrey, D., Karam, R.: Effectiveness of cognitive tutor algebra I at scale. Educ. Eval. Policy Anal. **36**, 127–144 (2014)
6. U.S. Department of Education. Institute of Education Sciences, National Center for Education Statistics, National Assessment of Educational Progress (NAEP) (2022)
7. Vincent-Lancrin, S., van der Vlies, R.: Trustworthy artificial intelligence (AI) in education: promises and challenges, OECD Education Working Papers, No. 218 (2020)

Empowering Education with LLMs - The Next-Gen Interface and Content Generation

Steven Moore[1](✉), Richard Tong[2], Anjali Singh[3], Zitao Liu[4], Xiangen Hu[5], Yu Lu[6], Joleen Liang[7], Chen Cao[8], Hassan Khosravi[9], Paul Denny[10], Chris Brooks[3], and John Stamper[1]

[1] Carnegie Mellon University, Pittsburgh, PA, USA
StevenJamesMoore@gmail.com
[2] Carnegie Learning, Pittsburgh, PA, USA
[3] University of Michigan, Ann Arbor, MI, USA
[4] Jinan University, Guangzhou 510632, Guangdong, China
[5] The University of Memphis, Memphis, TN, USA
[6] Beijing Normal University, Beijing 100875, China
[7] Squirrel AI Learning, Shanghai 200030, China
[8] University of Sheffield, Sheffield, UK
[9] University of Queensland, St Lucia, QLD 4072, Australia
[10] University of Auckland, Auckland 1010, New Zealand

Abstract. We propose the first annual workshop on Empowering Education with LLMs - the Next-Gen Interface and Content Generation. This full-day workshop explores ample opportunities in leveraging humans, AI, and learning analytics to generate content, particularly appealing to instructors, researchers, learning engineers, and many other roles. The process of humans and AI cocreating educational content involves many stakeholders (students, instructors, researchers, instructional designers, etc.), thus multiple viewpoints can help to inform what future generated content might be useful, new and better ways to assess the quality of such content and to spark potential collaborative efforts between attendees. Ultimately, we want to illustrate how everyone can leverage recent advancements in AI, making use of the latest machine learning methods and large language models (LLMs), and engage all participants in shaping the landscape of challenges and opportunities in this space. We wish to attract attendees interested in scaling the generation of instructional and assessment content and those interested in the use of online learning platforms.

Keywords: Large language models · Educational content creation · human-AI partnerships · learnersourcing · robosourcing

1 Motivation and Theme

Language models that leverage Generated Pre-trained Transformers (GPT) have significantly advanced the field of natural language processing (NLP) and led to the development of various language-based applications [7]. One potential application of language models is as communication interfaces in human-in-the-loop systems, where the model serves as a mediator among the teacher, students and the machine capabilities including its own. This approach has several benefits, including the ability to personalize interactions, allow unprecedented flexibility and adaptivity for human-AI collaboration and improve the user experience. However, several challenges still exist in implementing this approach, including the need for more robust models, designing effective user interfaces, and ensuring ethical considerations are addressed. Additionally, instructors often find themselves having to generate large banks of resources such as practice and assessment questions to accommodate the growing shift towards online learning formats. The continual creation and improvement of assessment items allows for a greater breadth of topic coverage, helps to identify well-constructed and valid assessments, and as a result, enables improved learning analytics [6]. However, instructors and teaching staff rarely have the time or incentive to develop large numbers of high-quality questions for formative assessments that are often used for personalization and adaptive learning; instead their efforts tend to be focused on creating high-stakes assessments such as quiz or exam questions. To address this challenge, there is great potential in exploring educational content creation via partnerships that involve pairings of instructors, students, and AI. Partnerships for cocreating educational content often involve four distinct and iterative phases: creation, evaluation, utilization, and instructor/expert oversight. These four phases are also utilized by *learnersourcing*, which involves students generating their own educational resources and content that can be leveraged by future learners [4, 8]. This offers a domain agnostic way to help scale the creation of high-quality assessments, while also helping students learn the course content. Leveraging advances in large language models (LLMs), learnersourcing can be combined with AI to provide students with near instant feedback on the educational content they create, yielding higher quality contributions [2].

Partnerships between student-AI and instructor-AI present ample opportunity for content creation and evaluation [11]. Advances in natural language processing (NLP) and generative models provide space for AI to play a fundamental role in the co-creation of content with humans or to assist with the automated evaluation of its quality. The quality evaluation of this content can be further supported by learning analytics related to how students perform on these human-AI cocreated questions, compared to traditional assessments. Related work has explored the quality and novelty of AI-generated learning resources [10], and leveraged NLP [5], trust-based networks [1], and deep learning methods [9] to assist students in the evaluation of both student- and AI-generated content. While human input remains critical in this creation and evaluation process, more work needs to look at leveraging AI to further support students and instructors as they create educational content.

2 Objectives

This workshop aims to bring together researchers and practitioners from academia and industry to explore the potential of LLMs as the communication and collaboration interfaces in human-in- the-loop systems. The objectives of the workshop are to:

- The application of Large Language Models (LLMs) in educational settings

- Generation and evaluation of educational content with the help of LLMs
- Co-creation of educational partnerships, where the human or AI might benefit the most
- Ethical considerations in the use of LLMs as communication interfaces in educational settings
- Designing effective and standardized user interfaces for LLM-based educational systems
- Crowdsourcing & Learnersourcing in conjunction with LLMs

3   Call for Submission

While no submission is required to participate in the workshop, we encourage 4–6 page submissions of work-in-progress or position papers that are related to partnerships for co-creating educational content. The call for papers will open on March 27, 2023, and the submission deadline will be May 27, 2023. The submitted papers will be reviewed by three members of the program committee and the authors will be notified of their paper acceptance on June 03, 2023.

Papers can target a range of topics related to human-AI partnerships for educational content creation. When it comes to the evaluation process of having students or AI review and revise student-generated content, there is a challenge regarding how we can assist students in optimally acting on the suggestions and revisions. Further work remains to investigate how we might leverage AI to assist students in making consistently high-quality learnersourced contributions [3]. Among these challenges with humans and AI cocreating educational content lie many opportunities to explore ways of making it more accessible and beneficial to student learning. A clear opportunity regarding the creation of student-generated content is the different ways we can encourage students to make high-quality contributions, such as leveraging learnersourcing interventions [12]. There are limitless activity types that can be created and evaluated using a plethora of techniques. For instance, students could work in conjunction with a large language model, like ChatGPT or GPT-3, to develop and refine assessment questions or explanations of learning content [10]. This can help them quickly improve the content they generate, while also engaging them in critical thinking as they review the model's suggestions, such as recommended distractors, and explore the limitations of the models. LLMs are trained on broad data produced by humans, and thus are known to suffer from biases like humans. Using automatically generated content without human oversight for educational content generation runs the risk of perpetuating some of these biases. We see a human-in-the-loop approach, involving both students and instructors, as essential for moderating such biases and for improving and tailoring the performance of the underlying generative models for suitability in educational contexts.

4   Workshop Format and Activities

The workshop will run as an interactive full-day session with mini-presentations and round-table discussions on the theme. The provisional schedule is given below:

- **Introductions**: Introductions of workshop organizers and participants, and a background to the focus of the workshop.

- **Invited Guest Speaker**: A leading researcher not on the organizing committee will give a presentation regarding the use of AI in educational content creation
- **Short Presentations**: Authors of accepted submissions present their work which would be followed by a Q&A session
- **Round-table discussion**: Participants will move around specific topics of interest related to various types of human-AI partnerships for educational content including creating, evaluating, utilizing and overseeing.
- **Mini-Hackathon**: Participants will work alongside the organizers to build out prototypes of educational applications that leverage LLMs.
- **Concluding remarks and community engagement**: Closing remarks on the workshop will be made with future steps and opportunities for collaboration between participants.

The main goal of this workshop is to explore how partnerships between students, instructors, and AI can be leveraged towards educational content. We believe participants from a wide range of backgrounds and prior knowledge on learning sciences, machine learning, natural language processing, and learning analytics can both benefit and contribute to this workshop. As the creation of educational content involves many stakeholders, multiple viewpoints can help to inform what future student and AI-generated educational content might be useful, better ways to assess the quality of the content and spark collaboration efforts between attendees. The accepted submissions will be published as part of a CEUR proceedings. We will also provide participants with a Slack channel and mailing list dedicated to sharing related work, along with updating our workshop website.

5 Organizers

- **Steven Moore** is a PhD student at Carnegie Mellon University and is advised by Dr. John Stamper. His research is focused on leveraging NLP with learnersourcing, finding ways to assess the quality of student-generated content.
- **Richard Tong** is an experienced technologist, executive, entrepreneur and one of the leading evangelists for global standardization efforts for learning technologies and artificial intelligence. He serves as the current chair of IEEE Artificial Intelligence Standards Committee.
- **Anjali Singh** is a PhD candidate at the University of Michigan and is advised by Dr. Christopher Brooks. Her research uses learnersourcing and AI to improve Data Science education, in formal and informal learning settings like MOOCs.
- **Zitao Liu** is the Dean of Guangdong Institute of Smart Education, Jinan University, Guangzhou, China. His research is in the area of machine learning, and includes contributions in the areas of artificial intelligence in education and educational data mining.
- **Xiangen Hu** is a professor in the Department of Psychology, Department of Electrical and Computer Engineering and Computer Science Department at The University of Memphis (UofM) and senior researcher at the Institute for Intelligent Systems (IIS) at the UofM and is professor and Dean of the School of Psychology at Central China Normal University (CCNU).

- **Yu Lu** is currently an Associate Professor with the School of Educational Technology, Faculty of Education, Beijing Normal University (BNU), where he also serves as the director of the artificial intelligence lab at the advanced innovation center for future education (AICFE) and the director of ICT center at Sino- Finnish Joint Learning Innovation Institute (JoLii).
- **Joleen Liang** is a Ph.D candidate in the Intelligent Science System at Macao University of Science and Technology. She is the executive director of the AI and Robotics Education Committee of the China Education Development Strategy Society, the deputy head of the Technology and Standards Working Group, and the Smart Education Working Committee of the Internet Society of China.
- **Chen Cao** is a PhD candidate at the Information School at the University of Sheffield and is advised by Professor Frank Hopfgartner, Dr Laura Sbaffi and Dr Xin (Skye) Zhao. Her research leverages LLMs to create intelligent tutoring systems with generative content, creative analogies, and adaptive feedback.
- **Hassan Khosravi** is an Associate Professor in the Institute for Teaching and Learning Innovation and an Affiliate Academic in the School of Information Technology and Electrical Engineering at the University of Queensland. He has conducted extensive learnersourcing research and leads the development and dissemination efforts of the RiPPLE system.
- **Christopher Brooks** is an Assistant Professor at the University of Michigan and is an applied Computer Scientist who builds and studies the effects of educational technologies in higher education and informal learning environments. He has led learnersourcing efforts on the Coursera platform, where he investigated student choice in the generation of multiple-choice questions.
- **Paul Denny** is an Associate Professor in Computer Science at the University of Auckland, New Zealand. He leads the PeerWise project, which hosts educational content created by students from 90 countries, and has pioneered work evaluating the implications of LLMs for computing education..
- **John Stamper** is an Associate Professor at the Human-Computer Interaction Institute at Carnegie Mellon University and the Technical Director of the Pittsburgh Science of Learning Center DataShop. His work involves leveraging educational data mining techniques and the creation of data tools.

References

1. Darvishi, A., Khosravi, H., Sadiq, S.: Employing peer review to evaluate the quality of student generated content at scale: a trust propagation approach. Proceedings of the Eighth ACM Conference on Learning@Scale, pp. 139–150 (2021)
2. Denny, P., Sarsa, S., Hellas, A., Leinonen, J.: Robosourcing Educational Resources--Leveraging Large Language Models for Learnersourcing (2022). arXiv preprint arXiv:2211.04715
3. Huang, A., et al.: Selecting student-authored questions for summative assessments. Research in Learning Technology 29 (2021)
4. Khosravi, H., Demartini, G., Sadiq, S., Gasevic, D.: Charting the design and analytics agenda of learnersourcing systems. LAK21: 11th International Learning Analytics and Knowledge Conference, pp. 32–42 (2021)

5. Moore, S., Nguyen, H.A., Stamper, J.: Evaluating crowdsourcing and topic modeling in generating knowledge components from explanations. International Conference on Artificial Intelligence in Education, pp. 398–410 (2020)

6. Moore, S., Nguyen, H.A., Stamper, J.: Leveraging students to generate skill tags that inform learning analytics. In: Proceedings of the 16th International Conference of the Learning Sciences-ICLS 2022, pp. 791–798 (2022)

7. Moore, S., Nguyen, H.A., Bier, N., Domadia, T., Stamper, J.: Assessing the quality of student-generated short answer questions using GPT-3. In: Educating for a New Future: Making Sense of Technology-Enhanced Learning Adoption: 17th European Conference on Technology Enhanced Learning, EC-TEL 2022, Toulouse, France, September 12–16, 2022, Proceedings, pp. 243–257. Springer International Publishing, Cham (2022, September)

8. Moore, S., Stamper, J., Brooks, C., Denny, P., Khosravi, H.: Learnersourcing: Student-generated Content@ Scale. In: Proceedings of the Ninth ACM Conference on Learning@ Scale, pp. 259–262 (2022, June)

9. Ni, L., et al.: Deepqr: Neuralbased quality ratings for learnersourced multiple-choice questions. Proceedings of the AAAI Conference on Artificial Intelligence **36**(11), 12826–12834 (2022)

10. Sarsa, S., Denny, P., Hellas, A., Leinonen, J.: Automatic generation of programming exercises and code explanations using large language models. In: Proceedings of the 2022 ACM Conference on International Computing Education Research-Volume 1, pp. 27–43 (2022)

11. Singh, A., Brooks, C., Doroudi, S.: Learnersourcing in theory and practice: synthesizing the literature and charting the future. In: Proceedings of the Ninth ACM Conference on Learning@ Scale, pp. 234–245 (2022)

12. Singh, A., Brooks, C., Lin, Y., Li, W.: What's in it for the learners? Evidence from a randomized field experiment on learnersourcing questions in a MOOC. In: Proceedings of the Eighth ACM Conference on Learning@ Scale, pp. 221–233 (2021, June)

Conducting Rapid Experimentation with an Open-Source Adaptive Tutoring System

Zachary A. Pardos[✉], Ioannis Anastasopoulos, and Shreya K. Sheel

School of Education, University of California, Berkeley, Berkeley, USA
{pardos,ioannisa,shreya_sheel}@berkeley.edu

Abstract. Intelligent Tutoring Systems have been an area of particular relevance and importance to AIED. In this tutorial, we introduce uses of a new tool to accelerate the speed at which the community can innovate and experiment in the general area of computer tutoring systems. We showcase the field's first fully fledged and open-source adaptive tutoring system (OATutor) with a completely creative commons licensed problem library based on popular open-licensed algebra textbooks. We demonstrate, with hands-on tutorials, how the system can be deployed on github.io and used to help AIED researchers rapidly run A/B experiments, including how to add or modify content in the system, analyze its log data, and link OATutor content to learning management system with LTI. Our open-sourcing of three textbooks worth of questions and tutoring content in a structured data format (JSON) also opens up avenues for AIED researchers to apply new and legacy educational data mining and NLP techniques.

Keywords: Intelligent Tutoring Systems · Authoring · Creative Commons · A/B Testing · HCI

1 Background

Adaptive learning systems based on Intelligent Tutoring System (ITS) principles have long been established as effective tools for improving learning gains as compared to conventional classroom instruction [3]. Most notably, a study by the RAND corporation demonstrated a clear benefit when introducing a successful commercial ITS called Cognitive Tutor into a middle school classroom. The median student reference score moved up from the 50th to the 58th percentile, when compared to a classroom of the same year that did not utilize the tutor, which is equivalent to making up for the learning loss that occurs during the transition from 8th to 9th grade [6]. Other mastery-based computer tutoring programs have shown positive results in a variety of randomized controlled trials, including an ASSISTments study with seventh graders across 43 schools in Maine [9], a SimCalc intervention in Texas for seventh and eighth graders [10], and an ITS reading comprehension program study with middle school students [12,13].

© The Author(s), under exclusive license to Springer Nature Switzerland AG 2023
N. Wang et al. (Eds.): AIED 2023, CCIS 1831, pp. 38–43, 2023.
https://doi.org/10.1007/978-3-031-36336-8_5

OATutor serves as the first open-source[1] adaptive tutor based on Intelligent Tutoring System (ITS) principles [3]. Opening up resources, courses, datasets, and algorithms has led to the adoption of digital resources at scale and accelerated the advancement of learning sciences research. Open Educational Resources (OERs), resources that are openly provisioned for use under a creative commons license for non-commercial purposes [5], allow for broader access and dissemination of learning materials for researchers to pull content from.

Massive Open Online Courses (MOOCs), open access instructional content released through an online platform, gained traction as digital versions of courses from well-known universities, made available to the world [2]. While MOOCs are open access, their content is propriety and not creative commons like an OER.

A variety of educational technology platforms have opened up their data, releasing anonymized data logs, promoting offline experimentation replication and generalization of methods across datasets. Carnegie Learning has released dozens of datasets, which have contributed considerably to educational data mining [14]. Their 2010 dataset for the Association for Computing Machinery (ACM) Special Interest Group on Knowledge Discovery and Data-mining's (KDD) annual competition with over 9 GB of student data became the most cited from an educational technology platform [11] and the largest the competition had seen at the time. ASSISTments, a free web-based tutoring platform has also been generous with the release of its datasets that have been used in over 100 papers.

Open algorithms remove barriers for the educational data mining community to expand and replicate research. For example, pyBKT [1] provides the first python implementation of a popular algorithm for cognitive mastery estimation, which is the same algorithm used in the Cognitive Tutor/MATHia.

A combination of open resources, datasets, and algorithms in the form of an open platform has been rare. Cognitive Tutor's Geometry 1997 dataset is public, but the system itself is not open and not free to use. ASSISTments, an open platform, encourages outside researcher collaboration to run A/B experiments on different content-based approaches [4], but does not release the content of its platform, nor its codebase.

OATutor was built with researchers in mind, supporting rapid experimentation and continuous improvement. The system's open-source code and creative common content can be forked, modified, and experimented with, and then linked to in a publication, allowing for better replication of research.

2 OATutor System

Open Adaptive Tutor (OATutor) is an open source adaptive tutoring system inspired by Intelligent Tutoring System (ITS) principles [7]. It features knowledge tracing-based mastery learning assessment, A/B testing capabilities, LTI support, and ability to be quickly deployed on GitHub Pages. OATutor's initial

---

[1] https://github.com/CAHLR/OATutor.

release includes three Creative Commons adaptive algebra textbooks, with questions derived from OpenStax[2], with source code made available under an MIT license, allowing for commercial use and modification. OATutor was designed with the intent to help facilitate and accelerate tutoring research in the learning sciences community. The system has been piloted in several algebra classrooms over three years and is showing early signs of reliable learning gains in nascent experiments [8].

3   Rapid Experimentation

OATutor's logging provides all the information needed to evaluate various outcomes of experimental learning gain designs. The system's in-build A/B testing facilities can be used, whereby any arbitrary can be wrapped in an if statement, checking the randomly assigned condition of the user. Alternatively, the system can be embedded in external survey systems, like Qualtrics, that can handle the experimental design and random assignment. OATutor provides system toggles that can be easily turned on or off, such as displaying or not displaying correctness feedback, providing hints, and tracking mastery. These toggles can be defined at the lesson level, allowing for pre-test and post-test lessons to be setup without feedback and acquisition phase lessons to have feedback and hints turned on. The ability to disable hints also enables researchers to ask general introductory questions (such as questions regarding a user's age, gender, etc.) within OATutor itself, rather than having to utilize an external third party platform as well.

In OATutor, questions are typically formatted as follows: problem header, problem body, step header, step body, allowing the question or parts of the question to be displayed on four distinct areas on the platform. This gives researchers the flexibility to experiment with different ways to ask questions and break up problems.

OATutor provides for easy implementation of multiple experimental conditions. This is best demonstrated by a recent experiment conducted. OATutor was utilized for an experiment between the efficacy of OATutor hints and hints generated by LLMs (specifically ChatGPT). The experiment required three conditions: hints from OATutor, hints from ChatGPT, and no hints (to control for prior knowledge). Firstly, a simple duplication of the OATutor questions/hints was made. Then, the hints from these were substituted with the ChatGPT version of the hints in the raw json files. This became the ChatGPT experimental condition. For the control condition with no hints, the hint and feedback features were simply disabled. Finally, OATutor's log data allows researchers to track actions such as a user asking for a hint, time the user took each action for, whether the user answered the question correctly or not. This data could then be easily analyzed in order to summarize A/B testing learning gain results. The entire study was setup and run in just a few weeks.

---

[2] https://openstax.org/subjects/math.

4   Pre-tutorial Plans

Prior to the tutorial we will promote access to following:

- OATutor codebase, including the 3 algebra textbooks from its 1.5 release
- The OATutor Website, featuring information geared for both teachers and researchers.
  - The teacher page of the website will include information on how to curate content as well as how to integrate OATutor into a Canvas Course
  - The researcher page of the website will include Google Collaboratory notebooks regarding deploying the system and examples of using the platform for rapid experimentation.
- Initial tutorial resources
  - A video demonstrating the spreadsheet format for content authoring
  - The OATutor formatting guide
- A pre-tutorial survey asking participants to rate their interest in specific areas of focus of the workshop, allowing us to tailor the tutorial sessions to participant needs and interests.

5   Tutorial Structure

The first half of this full-day tutorial will begin with examining three main components of OATutor (research capabilities, content creation framework, and LMS integration) followed by a hands-on working session where participants will have the opportunity to apply what they have learned into their own research in the second half.

Introduction: We will begin with a brief introduction of the system, its three-year development process, and related efforts. Participants will then introduce themselves and briefly describe their research interests in order to enrich discussion and questions throughout the tutorial, as well as facilitate potential collaboration across similar research domains.

Part 1: The first key section of the tutorial will be focused on research capabilities of the system. We will demonstrate OATutor's A/B testing capabilities and discuss how we can analyze experiment data. Furthermore, the tutorial will cover re-tagging content with different skills through the use of machine learning, as well as giving real-time personalized hints using a backend server.

Part 2: The second key section will focus on the content environment of OATutor, with a short introduction and demonstration of the system's content tools, followed by an introduction to LTI tools with a demonstration integrating the system into a Canvas Course. The Canvas teacher dashboard will also be showcased.

Part 3: The final key section will feature a hands-on activity, with participants creating their own content in the spreadsheet format, and displaying it within the tutor. There will be further discussion regarding data analysis of log data from an OATutor experiment, as well as a tutorial on creating new skill models.

Hands-on Working Session: During the second half of the tutorial, participants will have the opportunity to apply OATutor into their own research domains. Participants can begin integrating OATutor's research capabilities with their own research questions, applying skills from the tutorial and asking any standing questions to tutorial facilitators along the way. Our goal for this working session is for it to act as a jumpstart for researchers to enable them in asking novel questions and exploring design decisions with OATutor.

6   Post Workshop Plans

We plan to study the ways in which researchers could see themselves using OATutor to conduct research, initiating rapid research and development in order to continue suiting the needs of the research community. Through interacting with researchers in conference workshops from different domains under learning sciences, we aim to understand researchers' points of view and how our system can support their needs in asking questions in their domains and explore design decisions. We plan on collecting more open-ended feedback responses from researchers, where workshop attendees will have the opportunity to fill out a survey if they are interested in further using the system in their research. The survey will be used as a pseudo application form to gather interest from researchers to understand their fields and research interests with the goal of partnering with their research labs to continue using OATutor and evaluate the system's extensibility in their respective fields.

7   Organizers

- **Zach Pardos** is an associate professor in the Berkeley School of Education.
- **Shreya Sheel** is a second year doctoral student in Learning Sciences and Human Development at University of California, Berkeley. Her research interests are in leveraging human computer interaction principles for the development and implementation of AI-assisted technologies.
- **Ioannis Anastasopoulos** is an incoming first year doctoral student in Learning Sciences and Human Development at University of California, Berkeley. After having completed his masters, also in Learning Sciences at Berkeley, he plans to focus his future research on adaptive tutoring system usability in classrooms. Further research interests include learnersourcing and facilitation of replicable content creation environments.
- **Shreya Bhandari** is a first year undergraduate in EECS and has been working with the system to run ChatGPT-related experiments.
- **Matthew Tang** is a software engineer at Google and former undergraduate in EECS who led the initial implementation of the system. Matthew will be holding remote Zoom office hours during the second half of the tutorial.

Acknowledgments. We would like to thank the Office of Undergraduate Research & Scholarships at UC Berkeley for facilitating the Undergraduate Research Apprenticeship Program that supported our content team members. This work was partially supported by an educational data science fellowship grant from Schmidt Futures.

References

1. Badrinath, A., Wang, F., Pardos, Z.: pyBKT: an accessible python library of Bayesian knowledge tracing models. arXiv preprint arXiv:2105.00385 (2021)
2. Baturay, M.H.: An overview of the world of MOOCs. Procedia. Soc. Behav. Sci. **174**(2015), 427–433 (2015)
3. Corbett, A.T., Koedinger, A.T., Anderson, J.R.: Intelligent tutoring systems. In: Handbook of Human-Computer Interaction, pp. 849–874. Elsevier (1997)
4. Heffernan, N.T., Heffernan, C.L.: The ASSISTments ecosystem: building a platform that brings scientists and teachers together for minimally invasive research on human learning and teaching. Int. J. Artif. Intell. Educ. **24**(2014), 470–497 (2014)
5. Johnstone, S.: Forum on the impact of open courseware for higher education in developing countries-final report. Educ. Q. **3**(2005), 15–18 (2005)
6. Pane, J.F., Griffin, B.A., McCaffrey, D.F., Karam, R.: Effectiveness of cognitive tutor algebra I at scale. Educ. Eval. Policy Anal. **36**(2), 127–144 (2014)
7. Pardos, Z., Tang, M., Anastasopoulos, I., Sheel, S.K., Zhang, E.: OATutor: an opensource adaptive tutoring system and curated content library for learning sciences research. In: Proceedings of the 2023 CHI Conference on Human Factors in Computing Systems. ACM, Hamberg, Germany (2023)
8. Pardos, Z.A., Bhandari, S.: Learning gain differences between ChatGPT and human tutor generated algebra hints. arXiv preprint arXiv:2302.06871 (2023)
9. Roschelle, J., Feng, M., Murphy, R.F., Mason, C.A.: Online mathematics homework increases student achievement. AERA Open **2**(4) (2016). https://doi.org/10.1177/2332858416673968
10. Roschelle, J.: Integration of technology, curriculum, and professional development for advancing middle school mathematics: three large-scale studies. Am. Educ. Res. J. **47**(4), 833–878 (2010)
11. Stamper, J., Pardos, Z.A.: The 2010 KDD cup competition dataset: engaging the machine learning community in predictive learning analytics. J. Learn. Anal. **3**(2), 312–316 (2016)
12. Wijekumar, K.: Multisite randomized controlled trial examining intelligent tutoring of structure strategy for fifth-grade readers. J. Res. Educ. Effectiveness **7**(4), 331–357 (2014)
13. Wijekumar, K.K., Meyer, B.J.F., Lei, P.: Large-scale randomized controlled trial with 4th graders using intelligent tutoring of the structure strategy to improve nonfiction reading comprehension. Educ. Tech. Res. Dev. **60**(2012), 987–1013 (2012)
14. Fischer, C., et al.: Mining big data in education: affordances and challenges. Rev. Res. Educ. **44**(1), 130–160 (2020)

AI Education for K-12: A Survey

Ning Wang[1(✉)] and James Lester[2]

[1] University of Southern California, Los Angeles, CA, USA
nwang@ict.usc.edu
[2] North Carolina State University, Raleigh, NC, USA

Abstract. Artificial Intelligence (AI) is a foundational technology that is permeating every aspect of our daily lives. It is also profoundly transforming our workforce around the globe. It is critical to prepare future generations with basic knowledge of AI, not just through higher education, but beginning with childhood learning. There is a growing body of research into how students, especially K-12 students, construct an understanding of and engage with core ideas in the field, that can ground the design of learning experiences in evidence-based accounts of how youth learn AI concepts, how understanding progresses across concepts, or what concepts are most appropriate for what age-levels. On July 7, 2023, we held a workshop on *AI Education for K-12* at the 24th International Conference on Artificial Intelligence in Education, in Tokyo, Japan, in a synchronous hybrid format, allowing both in-person and virtual contributions. The workshop brought together researchers in K-12 AI education to discuss the state-of-the-art and identify research gaps and future directions to bring evidence-based AI education in and out of the classrooms.

Keywords: K-12 AI education · youth AI education · workshop

Artificial Intelligence (AI) is a foundational technology that is permeating every aspect of our daily lives. It has become the driving force for social, economic, and scientific advancement, and rapidly transforms the workforce around the globe. As AI becomes more widely used across a variety of disciplines, it is increasingly important to teach future generations basic knowledge of AI, not just through higher education, but beginning with childhood learning. Driven by the demand for youth AI education, there has been a recent surge of research on AI education programs that engage K-12 students in hands-on experience to learn AI, and prepare teachers to teach in classrooms and informal settings.

To integrate AI education into K-12 schools, defining AI literacy [36,61] and developing curricula and guidelines [15,28,45,50] are one of the first and most important steps. Around the world, countries race to promote AI education in K-12 [20], including the United States [15], Canada [38], Europe [19,26], China [63], Singapore [20], Thailand [46], Israel [47], Australia [22], and many more. Given the pervasiveness of AI in youth's daily life and future workplace, AI ethics education has become the focus for several research programs [13,27,29,42].

N. Wang et al. (Eds.): AIED 2023, CCIS 1831, pp. 44–49, 2023.
https://doi.org/10.1007/978-3-031-36336-8_6

Most youth AI education programs integrate ethics education with learning of AI algorithms and applications.

Since K-12 schools often have packed schedules, AI education often finds its place in informal settings. Many AI curriculum are designed for youth summer camps [1,4,25,54], including those in museums [5], with some designed as exhibits at museums [57]. Other informal AI education programs, such as Technovation, partner with national and international media outlets such as Youth Radio, National Public Radio (NPR) and Teen Vogue to reach youth far and wide [52]. Informal learning contexts can serve as rich environments to conduct research about how children build understanding about concepts outside of standards to which formal learning contexts are generally bound. Even with recent investigations on AI learning experiences for museum settings [35], there is still a dearth of research being conducted about how to promote AI literacy and engagement in informal learning spaces [37,65].

Given that students are front and center of most of the youth AI education program, some researchers have begun to investigate student's perspective on AI Literacy, and their intention and interest in learning AI [8,48,51]. Recognizing the important role teachers play in youth AI literacy development, many AI education programs include teacher professional development (PD) to prepare teachers with AI content knowledge, ethical use of AI, and methods to integrate AI learning into existing STEM classrooms or after-school activities [14,30,62]. Many emerging initiatives involve teachers in the process of co-creating a curriculum associated with their context [10,11,64]. However, the literature reports on teacher training in AI [34,41] indicate that, at present, there is a shortage of teachers with adequate training to teach AI in schools [62].

While AI includes many topics, from formal logic to probabilistic reasoning [18], machine learning is by far the most popular topics covered in many youth AI education programs [3,21,23,31,40,44,55,56,58,59], with the exception of a few programs begins with fundamental concepts such as classical search [2,32]. As AI permeates our lives through various tools and services, there is an increasing need to consider how to teach young learners about AI in a relevant and engaging way. Regarding the context through which AI education takes place, robotics have been a natural topic even before the recent surge in youth AI education movement [7,12,17,23,60]. Others support exploration of AI utilized to realize human-like intelligence [22,24,57], with natural language processing (NLP) AIs emerged as one of the popular vehicles [9,33,39,49,53]. Given the complexity of AI behind NLP, uncovering how such AI processes unfold has been a challenge, with a recent interactive tool for visually exploring word embeddings in NLP offers a promising approach [6]. Other research programs have built self-paced game-based learning environments for formal and informal settings [16,19,31,32] or leverage innovative wearable computing to engage learners in culturally responsive ways [43].

This short survey was prepared in advance for a workshop on *AI education in K-12* at the 24th International Conference on Artificial Intelligence in Education. The workshop brought together researchers in K-12 AI education to discuss the state-of-the-art and identify research gaps and future directions to bring evidence-based AI education to K-12 in and out of the classrooms. The workshop was held in Tokyo, Japan, in a synchronous hybrid format, allowing both in-person and virtual contributions, on July 7, 2023. This survey on the current research in K-12 AI education is an attempt to serve as a starting point to the fruitful discussions at the workshop.

References

1. Learn and design artificial intelligence, with ethics in mind: one-week workshop taught in collaboration with MIT media lab, April 2023. https://empow.me/ai-ethics-summer-program-mit
2. Akram, B., Yoder, S., Tatar, C., Boorugu, S., Aderemi, I., Jiang, S.: Towards an AI-infused interdisciplinary curriculum for middle-grade classrooms. In: Proceedings of the AAAI Conference on Artificial Intelligence, vol. 36, pp. 12681–12688 (2022)
3. Ali, S., DiPaola, D., Lee, I., Hong, J., Breazeal, C.: Exploring generative models with middle school students. In: Proceedings of the 2021 CHI Conference on Human Factors in Computing Systems, pp. 1–13 (2021)
4. Alvarez, L., Gransbury, I., Cateté, V., Barnes, T., Ledéczi, Á., Grover, S.: A socially relevant focused AI curriculum designed for female high school students. In: Proceedings of the AAAI Conference on Artificial Intelligence, vol. 36, pp. 12698–12705 (2022)
5. AMNH: Science research mentoring program: the official unofficial site for SRMP (2023). https://www.srmp4life.com/
6. Bandyopadhyay, S., Xu, J., Pawar, N., Touretzky, D.: Interactive visualizations of word embeddings for K-12 students. In: Proceedings of the AAAI Conference on Artificial Intelligence, vol. 36, pp. 12713–12720 (2022)
7. Burgsteiner, H., Kandlhofer, M., Steinbauer, G.: IRobot: teaching the basics of artificial intelligence in high schools. In: Proceedings of the AAAI Conference on Artificial Intelligence, vol. 30 (2016)
8. Chai, C.S., Lin, P.Y., Jong, M.S.Y., Dai, Y., Chiu, T.K., Qin, J.: Perceptions of and behavioral intentions towards learning artificial intelligence in primary school students. Educ. Technol. Soc. **24**(3), 89–101 (2021)
9. Chao, J., et al.: StoryQ: a web-based machine learning and text mining tool for K-12 students. In: Proceedings of the 53rd ACM Technical Symposium on Computer Science Education, vol. 2, p. 1178 (2022)
10. Chiu, T.K.: A holistic approach to the design of artificial intelligence (AI) education for K-12 schools. TechTrends **65**(5), 796–807 (2021)
11. Dai, Y., et al.: Collaborative construction of artificial intelligence curriculum in primary schools. J. Eng. Educ. **112**(1), 23–42 (2023)
12. Eguchi, A.: AI-robotics and AI literacy. In: Malvezzi, M., Alimisis, D., Moro, M. (eds.) EDUROBOTICS 2021. SCI, vol. 982, pp. 75–85. Springer, Cham (2021). https://doi.org/10.1007/978-3-030-77022-8_7
13. Forsyth, S., Dalton, B., Foster, E.H., Walsh, B., Smilack, J., Yeh, T.: Imagine a more ethical AI: using stories to develop teens' awareness and understanding of AI and its societal impacts. In: Conference on Research in Equitable and Sustained Participation in Engineering, Computing, and Technology, pp. 1–2 (2021)
14. Gardner-McCune, C., et al.: Co-designing an AI curriculum with university researchers and middle school teachers. In: Proceedings of the 54th ACM Technical Symposium on Computer Science Education, vol. 2, pp. 1306–1306 (2022)

15. Gardner-McCune, C., Touretzky, D., Martin, F., Seehorn, D.: AI for K-12: making room for AI in K-12 CS curricula. In: ACM Technical Symposium on Computer Science Education, pp. 1244–1244 (2019)
16. Geldhauser, C., Matt, A.D., Stussak, C.: I am AI gradient descent-an open-source digital game for inquiry-based CLIL learning. In: Proceedings of the AAAI Conference on Artificial Intelligence, vol. 36, pp. 12751–12757 (2022)
17. Gonzalez, A.J., et al.: AI in informal science education: bringing turing back to life to perform the turing test. Int. J. Artif. Intell. Educ. **27**, 353–384 (2017)
18. Greenwald, E., Leitner, M., Wang, N.: Learning artificial intelligence: insights into how youth encounter and build understanding of AI concepts. In: Proceedings of the AAAI Conference on Artificial Intelligence, vol. 35, pp. 15526–15533 (2021)
19. Guerreiro-Santalla, S., Mallo, A., Baamonde, T., Bellas, F.: Smartphone-based game development to introduce K12 students in applied artificial intelligence. In: Proceedings of the AAAI Conference on Artificial Intelligence, vol. 36, pp. 12758–12765 (2022)
20. Heintz, F.: Three interviews about K-12 AI education in America, Europe, and Singapore. KI-Künstliche Intelligenz **35**(2), 233–237 (2021)
21. Hitron, T., Orlev, Y., Wald, I., Shamir, A., Erel, H., Zuckerman, O.: Can children understand machine learning concepts? The effect of uncovering black boxes. In: Proceedings of the 2019 CHI Conference on Human Factors in Computing Systems, pp. 1–11 (2019)
22. Ho, J., et al.: Classroom activities for teaching artificial intelligence to primary school students. In: Proceedings of International Conference on Computational Thinking Education, pp. 157–159. The Education University of Hong Kong (2019)
23. Holowka, P.: Teaching robotics during covid-19: machine learning, simulation, and AWS DeepRacer. In: 17th International Conference on Cognition and Exploratory Learning in Digital Age, CELDA, vol. 2020 (2020)
24. Jordan, B., Devasia, N., Hong, J., Williams, R., Breazeal, C.: PoseBlocks: a toolkit for creating (and dancing) with AI. In: Proceedings of the AAAI Conference on Artificial Intelligence, vol. 35, pp. 15551–15559 (2021)
25. Judd, S.: Activities for building understanding: how AI4ALL teaches AI to diverse high school students. In: Proceedings of the 51st ACM Technical Symposium on Computer Science Education, pp. 633–634 (2020)
26. Kandlhofer, M., et al.: Enabling the creation of intelligent things: bringing artificial intelligence and robotics to schools. In: IEEE Frontiers in Education Conference, pp. 1–5 (2019)
27. Kaspersen, M.H., Bilstrup, K.E.K., Van Mechelen, M., Hjort, A., Bouvin, N.O., Petersen, M.G.: High school students exploring machine learning and its societal implications: opportunities and challenges. Int. J. Child-Comput. Interact., 100539 (2022)
28. Kim, S., et al.: Why and what to teach: AI curriculum for elementary school. In: Proceedings of the AAAI Conference on Artificial Intelligence, vol. 35, pp. 15569–15576 (2021)
29. Krakowski, A., Greenwald, E., Hurt, T., Nonnecke, B., Cannady, M.: Authentic integration of ethics and AI through sociotechnical, problem-based learning. In: Proceedings of the EAAI, vol. 36, pp. 12774–12782 (2022)
30. Lee, I., Perret, B.: Preparing high school teachers to integrate AI methods into stem classrooms. In: Proceedings of the AAAI Conference on Artificial Intelligence, vol. 36, pp. 12783–12791 (2022)

31. Lee, S., et al.: AI-infused collaborative inquiry in upper elementary school: a game-based learning approach. In: Symposium on Education Advances in Artificial Intelligence, vol. 35, pp. 15591–15599 (2021)
32. Leitner, M., Greenwald, E., Montgomery, R., Wang, N.: Design and evaluation of ARIN-561: an educational game for youth artificial intelligence education. In: Proceedings of The 30th International Conference on Computers in Education (2022)
33. Lin, P., Van Brummelen, J., Lukin, G., Williams, R., Breazeal, C.: Zhorai: designing a conversational agent for children to explore machine learning concepts. In: AAAI Conference on Artificial Intelligence, vol. 34, pp. 13381–13388 (2020)
34. Lindner, A., Berges, M.: Can you explain AI to me? Teachers' pre-concepts about artificial intelligence. In: Frontiers in Education Conference, pp. 1–9. IEEE (2020)
35. Long, D., Jacob, M., Magerko, B.: Designing co-creative AI for public spaces. In: Proceedings of the 2019 on Creativity and Cognition, pp. 271–284 (2019)
36. Long, D., Magerko, B.: What is AI literacy? Competencies and design considerations. In: CHI Conference on Human Factors in Computing Systems, pp. 1–16 (2020)
37. Moore, R.L., Jiang, S., Abramowitz, B.: What would the matrix do?: A systematic review of K-12 AI learning contexts and learner-interface interactions. J. Res. Technol. Educ. **55**(1), 7–20 (2023)
38. Nisheva-Pavlova, M.: AI courses for secondary and high school-comparative analysis and conclusions. In: ERIS, pp. 9–16 (2021)
39. Norouzi, N., Chaturvedi, S., Rutledge, M.: Lessons learned from teaching machine learning and natural language processing to high school students. In: Proceedings of the AAAI Conference on Artificial Intelligence, vol. 34, pp. 13397–13403 (2020)
40. Olari, V., Cvejoski, K., Eide, Ø.: Introduction to machine learning with robots and playful learning. In: Proceedings of the AAAI Conference on Artificial Intelligence, vol. 35, pp. 15630–15639 (2021)
41. Olari, V., Romeike, R.: Addressing AI and data literacy in teacher education: a review of existing educational frameworks. In: The 16th Workshop in Primary and Secondary Computing Education, pp. 1–2 (2021)
42. Payne, B.H.: AI+ ethics curriculum for middle school (2019)
43. Payne, W.C., et al.: danceON: culturally responsive creative computing. In: CHI Conference on Human Factors in Computing Systems, pp. 1–16 (2021)
44. Rodríguez-García, J.D., Moreno-León, J., Román-González, M., Robles, G.: Evaluation of an online intervention to teach AI with LearningML to 10–16 year old students. In: ACM Symposium on Computer Science Education, pp. 177–183 (2021)
45. Sabuncuoglu, A.: Designing one year curriculum to teach artificial intelligence for middle school. In: Proceedings of the 2020 ACM Conference on Innovation and Technology in Computer Science Education, pp. 96–102 (2020)
46. Sakulkueakulsuk, B., et al.: Kids making AI: integrating machine learning, gamification, and social context in stem education. In: 2018 IEEE International Conference on Teaching, Assessment, and Learning for Engineering (TALE), pp. 1005–1010. IEEE (2018)
47. Shamir, G., Levin, I.: Transformations of computational thinking practices in elementary school on the base of artificial intelligence technologies. In: Proceedings of EDULEARN20, vol. 6, pp. 1596–1605 (2020)
48. Sing, C.C., Teo, T., Huang, F., Chiu, T.K., Xing Wei, W.: Secondary school students' intentions to learn AI: testing moderation effects of readiness, social good and optimism. Educ. Technol. Res. Dev. **70**(3), 765–782 (2022)

49. Song, Y., et al.: AI made by youth: a conversational AI curriculum for middle school summer camps. In: Proceedings of the EAAI 2023 (2023)
50. Steinbauer, G., Kandlhofer, M., Chklovski, T., Heintz, F., Koenig, S.: A differentiated discussion about AI education K-12. KI-Künstliche Intelligenz **35**(2), 131–137 (2021)
51. Suh, W., Ahn, S.: Development and validation of a scale measuring student attitudes toward artificial intelligence. Sage Open **12**(2), 21582440221100464 (2022)
52. Technovation: Technovation: AI family challenge (2023). https://www.curiositymachine.org/lessons/lesson/
53. Van Brummelen, J., Heng, T., Tabunshchyk, V.: Teaching tech to talk: K-12 conversational artificial intelligence literacy curriculum and development tools. In: Proceedings of the EAAI, vol. 35, pp. 15655–15663 (2021)
54. Vandenberg, J., Min, W., Cateté, V., Boulden, D., Mott, B.: Promoting AI education for rural middle grades students with digital game design. In: Proceedings of the SIGSCE, vol. 2, pp. 1388–1388 (2022)
55. Vartiainen, H., Tedre, M., Valtonen, T.: Learning machine learning with very young children: who is teaching whom? Int. J. Child-Comput. Interact. **25**, 100182 (2020)
56. Verner, I.M., Cuperman, D., Reitman, M.: Exploring robot connectivity and collaborative sensing in a high-school enrichment program. Robotics **10**(1), 13 (2021)
57. VHX: VHX: AI behind virtual humans (2023). https://www.k12aied.org/projects/vhx/
58. Virtue, P.: GANs unplugged. In: Proceedings of the AAAI Conference on Artificial Intelligence, vol. 35, pp. 15664–15668 (2021)
59. Wan, X., Zhou, X., Ye, Z., Mortensen, C.K., Bai, Z.: SmileyCluster: supporting accessible machine learning in K-12 scientific discovery. In: Proceedings of the Interaction Design and Children Conference, pp. 23–35 (2020)
60. Williams, R., Kaputsos, S.P., Breazeal, C.: Teacher perspectives on how to train your robot: a middle school AI and ethics curriculum. In: Proceedings of the AAAI Conference on Artificial Intelligence, vol. 35, pp. 15678–15686 (2021)
61. Wong, G.K., Ma, X., Dillenbourg, P., Huan, J.: Broadening artificial intelligence education in K-12: where to start? ACM Inroads **11**(1), 20–29 (2020)
62. Xia, L., Zheng, G.: To meet the trend of AI: the ecology of developing AI talents for pre-service teachers in China. Int. J. Learn. Teach. **6**(3), 186–190 (2020)
63. Xiao, W., Song, T.: Current situation of artificial intelligence education in primary and secondary schools in china. In: The Sixth International Conference on Information Management and Technology, pp. 1–4 (2021)
64. Yau, K.W., et al.: Co-designing artificial intelligence curriculum for secondary schools: a grounded theory of teachers' experience. In: Proceedings of ISET, pp. 58–62. IEEE (2022)
65. Zhang, K., Aslan, A.B.: AI technologies for education: recent research & future directions. Comput. Educ. Artif. Intell. **2**, 100025 (2021)

Tutorial: Educational Recommender Systems

Yong Zheng$^{(\boxtimes)}$ (iD)

Department of Information Technology and Management, Illinois Institute of
Technology, Chicago, IL 60616, USA
yzheng66@iit.edu

Abstract. Recommender systems (RecSys) have found widespread use
in a variety of applications, including e-commerce platforms like Ama-
zon.com and eBay, online streaming services such as YouTube, Netflix,
and Spotify, and social media sites like Facebook and Twitter. The
success of these applications in improving user experience and decision
making by providing personalized recommendations highlights the effectiveness of RecSys. Over the past few decades, RecSys has also made
its way into the field of education, which results in the development
of educational recommender systems (EdRec). Its applications in this
field include personalized learning experiences, recommending appropriate formal or informal learning materials, suggesting learning peers,
and adapting learning to context-aware or mobile environments, and
so forth. Recently, the development of RecSys has been advanced by a
series of interesting and promising topics, such as multi-task learning,
multi-objective optimization, multi-stakeholder considerations, concerns
of fairness, accountability, and transparency, etc. However, this progress
made in the field of recommender systems was not adequately dissemi-
nated to the education community or the development of EdRec. In this
tutorial, we will introduce the background, motivations, knowledge &
skills associated with the current development of EdRec, and discuss a
set of emerging topics and open challenges, along with case studies in
EdRec.

Keywords: education · recommender system · educational
recommender system · EdRec

1 Introduction and Motivations

The emergence of data science and artificial intelligence technologies has made
technology-enhanced learning (TEL) [11,20] a promising area of research and
practice in education. TEL is a domain that generally covers technologies to
support teaching and learning activities, including recommendation technologies
that facilitate retrieval of relevant learning resource [20]. One of the significant
contributions of TEL has been the development of personalized learning and
educational recommender systems (EdRec).

N. Wang et al. (Eds.): AIED 2023, CCIS 1831, pp. 50–56, 2023.
https://doi.org/10.1007/978-3-031-36336-8_7

Recommender systems (RecSys) are a class of recommendation models or algorithms designed to make personalized recommendations to users based on their past behaviors and preferences. These systems are used in various applications, including e-commerce, social media, online streaming services, and so forth. The recommendation models can be used to predict which items the user is likely to be interested in and recommend them accordingly. RecSys have become increasingly popular due to their ability to improve user experience, increase engagement, and drive sales.

RecSys are also used in the field of education to provide personalized learning experiences, help students discover appropriate learning materials, and suggest learning peers. Relevance and rankings are usually the major targets in most of the recommendation applications, such as online streaming services. In EdRec, researchers usually seek more functions by going beyond recommending relevant items only. For example, EdRec can also be used or assessed to improve student learning performance [7], detect misunderstandings or find students with difficulties [7], enhance social learning by bookmarks or tags [7,19], recommend learning pathways [10,14], assist peer-matching [7,10], and so forth.

RecSys have been evolved towards the development of different types of recommendation systems or models, rather than dealing with traditional information (e.g., user preferences and item features) only. These various types of RecSys may include but not limited to, group RecSys [13], context-aware RecSys [28], multi-criteria RecSys [2], cross-domain RecSys [4], multi-stakeholder RecSys [1], multi-task RecSys [23], multi-objective RecSys [32], RecSys with fairness considerations [22], and so forth.

Despite the significant progress made in these fields, the advancements have not been adequately shared with the education community or incorporated into the development of EdRec. This tutorial aims to address this gap by providing an introduction to the background, motivations, knowledge and skills associated with the current development of EdRec. Additionally, the tutorial will discuss a range of emerging topics along with case studies, as well as a discussion of open challenges in EdRec.

2 Educational Recommender Systems

In this section, we deliver a sketch of outlines for our tutorial talk.

2.1 EdRec: Characteristics

First of all, it is necessary to distinguish EdRec from RecSys applied in other domains. In a summary, we are able to identify a list of the following characteristics in EdRec.

- *Heterogeneous data.* A set of user preferences (e.g., explicit ratings or implicit feedback) are required to build RecSys. In the educational domain, this information may be from heterogeneous resources [7], rather than a single source.

For example, user preferences can be collected from explicit ratings through questionnaire, click-through data from learning management system, interaction data from intelligent tutoring systems, and so forth.

- *Special features.* Traditional RecSys usually rely on user preferences, item features or user demographic information. By contrast, EdRec may need extra features in the education domain, i.e., pedagogical features [5,7,17], such as instructional rules, pre/post requisites, learning history, knowledge levels, educational standards, assessment methods, and so forth.
- *Dynamic preferences.* User preferences may vary from contexts to contexts (e.g., time, location, companion, etc.). In EdRec, user preferences may also be easily changed or altered due to pedagogical features [7], such as teaching or learning objectives, learning styles, dynamic knowledge levels, educational standards, etc.
- *Degree of personalization.* Most of RecSys aim to deliver personalized recommendations to individuals. In EdRec, recommended items may be better to be personalized at the level of group learners, rather than individuals. For example, recommendations may be distributed to a group of students due to their knowledge levels (e.g., beginner, intermediate, advanced).
- *Multiple contextual factors.* Context-aware RecSys can adapt the recommendations to different contextual situations, while scope of contextual variables in traditional RecSys may be limited to time, location, companion, and so forth. In EdRec, there are multiple and more contextual factors [20], such as device, bandwidth, quality of Internet connections, and pedagogical features (e.g., learning style, learning history, knowledge levels, etc.).
- *Going beyond relevance.* As mentioned before, EdRec can recommend relevant or user-preferred items, but we also expect EdRec to improve learning experiences or qualities. In this case, EdRec may work together with other educational data mining or artificial intelligence techniques, e.g., course design, peer-matching, predicting student performance, identifying students with difficulties, etc.
- *Real-life evaluations.* Traditional RecSys may have three common ways for evaluations – offline evaluations by using offline metrics (e.g., relevance or ranking metrics), user satisfaction through user studies or questionnaire, online A/B tests. In EdRec, we may additionally have real-life evaluations, where we can measure the actual effects on teaching and learning by EdRec from a long run [6,12,20].

2.2   EdRec: Classifications

According to *the recommendation opportunities,* EdRec can be classified into learning objects recommendations, learning pathway recommendations, peer recommendations, though most of the efforts were focused on learning objects recommendations. In addition, the recommendation models can also be revised or adapted to other educational tasks, such as predicting student performance.

By considering *the receiver of the recommendations*, EdRec can be built for students/learners, instructors, course developers, parents, publishers or other stakeholders.

Based on *the number of stakeholders considered* in the recommendation model, EdRec can be put into two categories – EdRec by considering the receiver of recommendations only, and EdRec by considering multiple educational stakeholders (e.g., learners, instructors, parents, publishers, etc.).

According to *the type of RecSys*, EdRec could be traditional RecSys, group RecSys, multi-criteria RecSys, cross-domain RecSys, and so forth. Based on the *the type of recommendation models or algorithms*, EdRec can be built by using collaborative filtering, content-based approaches, demographic information-based models, utility-based models, knowledge-based approaches or hybrid recommendation models.

2.3   EdRec: Current Development

As mentioned previously, most of the efforts in EdRec were contributed to learning objects recommendations, e.g., course [9] or MOOC [18] recommendations, book recommendations [21], after-school program recommendations [3], and so forth. Limited research was focused on learning pathway recommendations [14]. Fewer researchers attempted to take pedagogical features [14,17] into consideration. One of the major reasons is probably the limitation in data availability in EdRec. In terms of different types of RecSys, researchers have incorporated extra information and made attempts to develop diverse EdRec. The corresponding case studies in the tutorial include but not limited to the practice in context-aware EdRec [20], group EdRec [24,25], multi-criteria EdRec [31], multi-stakeholder EdRec [27], EdRec with human factors [16,29], EdRec with consideration of fairness [8], transparency [30] and explainabilities [15].

2.4   EdRec: Open Challenges

EdRec still have the challenges inherited from traditional RecSys, such as the sparsity issue, cold-start problems, novelty and diversity concerns, etc. There are also several challenges specifically to be alleviated or solved in EdRec.

- *Limited open data sets.* There is limited data available for EdRec, especially large-scale data with diverse features. We released the ITM-Rec data [26] recently and it can be considered as a test bed to build and evaluate multiple types of EdRec.
- *Pedagogical challenges.* Pedagogical features were usually treated as either constraints or contexts in EdRec. It is necessary to exploit other usages of pedagogical features.
- *Adaptation to dynamic user preferences.* Learners' preferences may be easily changed to according to their evolved knowledge levels, learning history, learning style or objectives. It is still challenging to identify and adapt to interest drifts.

- *Multi-objective or Multi-task learning.* E-commerce has made the attempts to incorporate recommendation performance and operation goals (e.g., number of subscriptions, revenue) into a multi-objective or multi-task learning process. In EdRec, it is promising to combine recommendation quality and educational objectives (e.g., student performance, learning interests, etc.) in similar processes.
- *Visualization, transparency and explainability.* Better visualizations and explanations are required to enhance transparency, so that users or learners can trust the recommendations and be willing to share more data with us.
- *Responsible recommendations.* Recommendations should also be produced and delivered by considering social and educational responsibilities.

References

1. Abdollahpouri, H., et al.: Multistakeholder recommendation: survey and research directions. User Model. User-Adap. Inter. **30**(1), 127–158 (2020). https://doi.org/10.1007/s11257-019-09256-1
2. Adomavicius, G., Manouselis, N., Kwon, Y.O.: Multi-criteria recommender systems. In: Ricci, F., Rokach, L., Shapira, B., Kantor, P.B. (eds.) Recommender Systems Handbook, pp. 769–803. Springer, Boston, MA (2011). https://doi.org/10.1007/978-0-387-85820-3_24
3. Burke, R., Zheng, Y., Riley, S.: Experience discovery: hybrid recommendation of student activities using social network data. In: Proceedings of the 2nd International Workshop on Information Heterogeneity and Fusion in Recommender Systems, pp. 49–52 (2011)
4. Cantador, I., Fernández-Tobías, I., Berkovsky, S., Cremonesi, P.: Cross-domain recommender systems. In: Ricci, F., Rokach, L., Shapira, B. (eds.) Recommender Systems Handbook, pp. 919–959. Springer, Boston, MA (2015). https://doi.org/10.1007/978-1-4899-7637-6_27
5. Drachsler, H., Verbert, K., Santos, O.C., Manouselis, N.: Panorama of recommender systems to support learning. In: Ricci, F., Rokach, L., Shapira, B. (eds.) Recommender Systems Handbook, pp. 421–451. Springer, Boston, MA (2015). https://doi.org/10.1007/978-1-4899-7637-6_12
6. Erdt, M., Fernandez, A., Rensing, C.: Evaluating recommender systems for technology enhanced learning: a quantitative survey. IEEE Trans. Learn. Technol. **8**(4), 326–344 (2015)
7. Garcia-Martinez, S., Hamou-Lhadj, A.: Educational recommender systems: a pedagogical-focused perspective. In: Tsihrintzis, G., Virvou, M., Jain, L. (eds.) Multimedia Services in Intelligent Environments: Recommendation Services, pp. 113–124. Springer, Heidelberg (2013). https://doi.org/10.1007/978-3-319-00375-7_8
8. Gómez, E., Shui Zhang, C., Boratto, L., Salamó, M., Marras, M.: The winner takes it all: geographic imbalance and provider (un) fairness in educational recommender systems. In: Proceedings of the 44th International ACM SIGIR Conference on Research and Development in Information Retrieval, pp. 1808–1812 (2021)
9. Guruge, D.B., Kadel, R., Halder, S.J.: The state of the art in methodologies of course recommender systems-a review of recent research. Data **6**(2), 18 (2021)

10. Khribi, M.K., Jemni, M., Nasraoui, O.: Recommendation systems for personalized technology-enhanced learning. In: Kinshuk, Huang, R. (eds.) Ubiquitous Learning Environments and Technologies, pp. 159–180. Springer, Heidelberg (2015)
11. Kirkwood, A., Price, L.: Technology-enhanced learning and teaching in higher education: what is 'enhanced' and how do we know? A critical literature review. Learn. Media Technol. **39**(1), 6–36 (2014)
12. Klašnja-Milićević, A., Ivanović, M., Nanopoulos, A.: Recommender systems in e-learning environments: a survey of the state-of-the-art and possible extensions. Artif. Intell. Rev. **44**(4), 571–604 (2015). https://doi.org/10.1007/s10462-015-9440-z
13. Masthoff, J.: Group recommender systems: combining individual models. In: Ricci, F., Rokach, L., Shapira, B., Kantor, P.B. (eds.) Recommender Systems Handbook, pp. 677–702. Springer, Boston, MA (2011). https://doi.org/10.1007/978-0-387-85820-3_21
14. Nabizadeh, A.H., Leal, J.P., Rafsanjani, H.N., Shah, R.R.: Learning path personalization and recommendation methods: a survey of the state-of-the-art. Expert Syst. Appl. **159**, 113596 (2020)
15. Pesovski, I., Bogdanova, A.M., Trajkovik, V.: Systematic review of the published explainable educational recommendation systems. In: 2022 20th International Conference on Information Technology Based Higher Education and Training (ITHET), pp. 1–8. IEEE (2022)
16. Salazar, C., Aguilar, J., Monsalve-Pulido, J., Montoya, E.: Affective recommender systems in the educational field. a systematic literature review. Comput. Sci. Rev. **40**, 100377 (2021)
17. Thongchotchat, V., Kudo, Y., Okada, Y., Sato, K.: Educational recommendation system utilizing learning styles: a systematic literature review. IEEE Access (2023)
18. Uddin, I., Imran, A.S., Muhammad, K., Fayyaz, N., Sajjad, M.: A systematic mapping review on MOOC recommender systems. IEEE Access **9**, 118379–118405 (2021)
19. Vassileva, J.: Toward social learning environments. IEEE Trans. Learn. Technol. **1**(4), 199–214 (2008)
20. Verbert, K., et al.: Context-aware recommender systems for learning: a survey and future challenges. IEEE Trans. Learn. Technol. **5**(4), 318–335 (2012)
21. Wang, F., Zhang, L., Xu, X.: A literature review and classification of book recommendation research. J. Inf. Syst. Technol. Manag. **5**(16), 15–34 (2020)
22. Wang, Y., Ma, W., Zhang, M., Liu, Y., Ma, S.: A survey on the fairness of recommender systems. ACM Trans. Inf. Syst. **41**(3), 1–43 (2023)
23. Wang, Y., et al.: Multi-task deep recommender systems: a survey. arXiv preprint arXiv:2302.03525 (2023)
24. Zheng, Y.: Exploring user roles in group recommendations: a learning approach. In: Adjunct Publication of the 26th Conference on User Modeling, Adaptation and Personalization, pp. 49–52 (2018)
25. Zheng, Y.: Identifying dominators and followers in group decision making based on the personality traits. In: The HUMANIZE Workshop at ACM Conference on Intelligent User Interfaces (2018)
26. Zheng, Y.: ITM-Rec: an open data set for educational recommender systems. In: Companion Proceedings of the 13th International Conference on Learning Analytics & Knowledge (LAK) (2023)
27. Zheng, Y., Ghane, N., Sabouri, M.: Personalized educational learning with multi-stakeholder optimizations. In: Adjunct Publication of the 27th Conference on User Modeling, Adaptation and Personalization, pp. 283–289 (2019)

28. Zheng, Y., Mobasher, B.: Context-Aware Recommendations, pp. 173–202. World Scientific Publishing (2018)
29. Zheng, Y., Subramaniyan, A.: Personality-aware recommendations: an empirical study in education. Int. J. Grid Util. Comput. **12**(5–6), 524–533 (2021)
30. Zheng, Y., Toribio, J.R.: The role of transparency in multi-stakeholder educational recommendations. User Model. User-Adap. Inter. **31**(3), 513–540 (2021). https://doi.org/10.1007/s11257-021-09291-x
31. Zheng, Y., Wang, D.: Multi-criteria ranking: next generation of multi-criteria recommendation framework. IEEE Access **10**, 90715–90725 (2022)
32. Zheng, Y., Wang, D.X.: A survey of recommender systems with multi-objective optimization. Neurocomputing **474**, 141–153 (2022)

Equity, Diversity, and Inclusion in Educational Technology Research and Development

Adele Smolansky[1]([✉]) [ID], Huy A. Nguyen[2] [ID], Rene F. Kizilcec[1] [ID], and Bruce M. McLaren[2]

[1] Cornell University, Ithaca, NY 14850, USA
as2953@cornell.edu
[2] Carnegie Mellon University, Pittsburgh, PA 15213, USA

Abstract. Modern education stands to greatly benefit from technological advances, especially in Artificial Intelligence (AI), that aim to enable effective and personalized learning for all students. However, to improve learning for the majority of students, AI solutions may exclude those who are under-represented due to unique differences in their demographic background or cognitive abilities. Towards combating this issue, we propose a workshop that will initiate conversations about equity, diversity, and inclusion in educational technology research and development. The workshop invites papers from the AIED community about equitable and inclusive educational technology that supports diverse populations, with selected authors being invited to present their work. The workshop is structured around three stages of learning engineering - system design, experimental study, and data analysis - with informational presentations, guest speakers, paper presentations, and group discussions relevant to each stage. Through the participation of community members from multiple disciplines, we seek to formulate a framework for developing and assessing equitable and inclusive educational technology.

Keywords: Equity and Inclusion in AIED · Educational Technology · Cognitive Abilities and Disabilities · Demographic Factors

1 Relevance and Importance to the AIED Community

Modern technologies have transformed teaching and learning by increasing access to information, enabling virtual or remote learning, and diversifying learning activities through multimedia channels [2]. However, technological advancements may not be uniform across learner populations. For instance, students from low-income or rural areas may not have the technology access required for digital learning; students with disabilities may struggle to use technologies that were not designed with accessibility in mind, and developers may not be critically evaluating AI fairness [8]. Advances in AI allow for a greater degree of personalization but may amplify existing inequities and negatively impact student learning [5].

N. Wang et al. (Eds.): AIED 2023, CCIS 1831, pp. 57–62, 2023.
https://doi.org/10.1007/978-3-031-36336-8_8

Thus, to fully realize the potential of educational technology, it is crucial to ensure the commitment to *equity*, whereby all students are treated fairly with equal access to the resources necessary for their success, and *inclusion*, whereby the *diversity* of students in terms of race, social class, ethnicity, religion, gender, and ability are recognized and incorporated into the design of educational technology [1]. Recognizing this need, we propose a workshop that explores equity, diversity, and inclusion (EDI) in educational technology research and development, focusing on understanding the existing barriers and identifying solutions to overcome them.

This workshop will bring together researchers from different disciplines to collectively formulate a framework for developing and assessing equitable and inclusive educational technology that supports diverse populations. The workshop structure is based on the three stages of learning engineering – system design, experimental study, and data analysis – with each stage featuring brief presentations to build common ground, a guest speaker, and paper presentations. Throughout the workshop, participants will have the opportunity to learn from the field experience of researchers, practitioners, and developers, critically reflect on the inclusiveness of existing educational technology, and contribute to refining best practices in EDI. Participants will also engage in discussions about emerging topics in this area, including the bi-directional relationship between AI and EDI [10], the proper usage of demographics data in AIED research [6], and the influence of new generative AI technologies, such as ChatGPT[1] and GPT-4 [9], on EDI in education.

This workshop is expected to foster rich discussions and connections between junior and senior researchers across different disciplines. Through their participation, the audience will be able to (1) develop an understanding of where inequity or exclusion may manifest in educational technology, (2) share and learn from others' experiences in promoting equity and inclusion, and (3) gather suggestions for making their own learning systems more inclusive. Upon completion of the workshop, we will invite participants to join a Discord space and mailing list dedicated to sharing advances in this area. Additionally, the synthesis of lessons learned in AIED-based equity and inclusion research will be summarized by the workshop organizers and shared with the broader community through the publicly available workshop proceedings. In a broader sense, we view this workshop as a follow-up to previous conversations on ethics in AIED [3,4] and an opportunity to connect participants for future collaborations, such as on a literature review paper in this area of research.

2   Call for Papers

We will solicit papers from the AIED community about equity, diversity, and inclusion in educational learning tools research and development for K-12 and

---

[1] https://openai.com/blog/chatgpt.

higher education. Educational technology is a broad term encompassing intelligent tutoring systems, educational games, learning systems and tools, and other technology that help people in education [7].

Authors of accepted papers will be invited to give a paper presentation of 10–20 minutes (depending on the number of papers selected). We will also publish accepted papers in the workshop proceedings.

2.1 Paper Topics

Paper topics include, but are not limited to, the following topics:

- Designing educational technology for underrepresented demographic populations (e.g., females in STEM, students with disabilities, racial minorities).
- Developing AI algorithms that adapt to gender, cognitive and physical ability, learning preferences, socio-economic status, etc.
- Improving and evaluating the accessibility of educational technology.
- Analyzing data from educational technology with demographic elements.
- Developing inclusive learner models considering demographics.
- Exploring the bi-directional relationship between AI and EDI.
- Reflecting on the proper usage of demographics data in AIED research.
- Discussing the influence of emerging large language models, such as ChatGPT and Bard, on EDI in education.

2.2 Submission and Review Process

Papers may be extended abstracts (2–4 pages, not including references) or short papers (4–6 pages, not including references). The event and the call-for-papers will be advertised through mailing lists, research communities, and personal connections. Information on the workshop and the call for papers is also available on a website created by the workshop organizers: https://adelesmolansky.com/aied23-edi-edtech/.

The review process will be single-blind (reviewers are anonymous, and authors are not), in which papers are assigned a score of -1, 0, or 1 based on the following criteria: relevance to the workshop, importance to the AIED community, and paper quality. Workshop organizers will tabulate scores and select the top 4–6 papers. Authors will be notified of the reviews and decision via email.

2.3 Important Dates

The schedule for the call for paper and review is as follows:

- Call for papers opens: April 1, 2023
- Paper submission deadline: May 19, 2023
- Paper review period: May 20–June 7, 2023
- Final paper decisions: June 8–11, 2023
- Notification of acceptance: June 11, 2023
- Camera-ready deadline: June 30, 2023
- Workshop day: July 7, 2023

3   Workshop Format and Activities

We will organize a full-day workshop with presentations from organizers, guest speakers, paper presentations, and structured discussions. The workshop will be offered in person (Toyko, Japan) and online (Zoom).

3.1   Workshop Activities

The workshop will begin with a presentation about different demographic factors and the importance of considering demographics in the research and development of digital learning tools. Subsequently, we will focus on best practices to promote equity, diversity, and inclusion in each stage of the learning engineering cycle. Paper presentations, guest speakers, and discussions will be organized around each stage of the learning engineering cycle: (1) system design, (2) experimental study, and (3) data analysis. The workshop will feature a keynote speech from Professor Ryan Baker (University of Pennsylvania), who has extensively studied all three steps of the learning engineering cycle. The workshop will conclude with a discussion about how researchers can implement equitable and inclusive practices in their work.

For system design (stage 1), we will present prior evidence of how students' interaction with digital learning platforms differs by demographic factors, discuss how children with physical, cognitive, and behavioral disabilities can be considered in the initial design process, and feature presentations from researchers who developed learning tools for children with disabilities. To evaluate the effectiveness of learning tools through experimental study (stage 2), we will introduce different types of experimental studies (e.g., lab studies, classroom studies, observational studies), the principles of a good experimental study, and examples of rigorous studies from prior works. Lastly, towards iteratively improving the learning tools and better understanding the student's learning process with data analysis (stage 3), we will present several analytical techniques, discuss the insights gained from these techniques, and hear from authors who have submitted papers on this topic. We will conclude with a discussion about future research directions to promote equity and inclusion with educational technology, taking into account recent advances in social studies, technological accessibility, and AI methodologies. Through the participation of community members from multiple disciplines, we seek to formulate a framework for developing and assessing equitable and inclusive educational technology.

3.2   Workshop Schedule

The specific schedule for the workshop is as follows:

- **10:00–10:45**: Introductions from the workshop members; presentation from workshop organizers about background information on the workshop topic.
- **10:45–11:30**: Guest speaker from organizing committee member Ryan Baker.

- **11:30–12:30**: Paper presentations for all accepted papers on designing and evaluating new learning tools *(learning engineering cycle steps 1 and 2)*.
- **12:30–13:30**: BREAK
- **13:30–14:15**: Paper presentations for all accepted papers on the design and evaluation of new learning tools *(learning engineering cycle step 3)*.
- **14:15–14:30**: Introduce the final discussion.
- **14:30–15:30**: Small group discussions organized by learning engineering cycle steps.
- **15:30–16:00**: Full group discussion.

4   Organizers

- **Adele Smolansky** will be a first-year Ph.D. student at Stanford University starting Autumn 2023. She recently graduated from Cornell University with a BS in Computer Science. Adele's research is on applications of AI to educational technology and creating accessible and inclusive learning tools for all students.
- **Huy Nguyen** is a fifth-year Ph.D. student in Human-Computer Interaction at Carnegie Mellon University. His research employs experimental studies and educational data mining to examine how digital learning games can bridge the gender gap in middle-school math education.
- **Rene Kizilcec** is an Assistant Professor of Information Science at Cornell University. He studies the impact of technology in formal and informal learning environments and scalable interventions to broaden participation and reduce achievement gaps. His recent work is on academic progress, algorithmic transparency, and fairness in predictive analytics in higher education.
- **Bruce M. McLaren** is an Associate Research Professor in the Human-Computer Interaction Institute at Carnegie Mellon University and was the President of AIED from 2017 to 2019. His research focuses on digital learning games, intelligent tutoring systems, and collaborative learning. He has around 200 publications, spanning journals, conferences, and book chapters.

5   Program Committee

The program committee will include researchers from various research areas such as accessibility, educational technology, AI fairness, cultural studies, games, gender, racial equity, diversity, and educational neuroscience. Members include Professors Jessica Hammer (Carnegie Mellon University), Richard Ladner (University of Washinton), Shima Salehi (Stanford University), Nick Haber (Stanford University), Ryan Baker (University of Pennsylvania), Ivon Arroyo (University of Massachusetts Amherst), and Rod Roscoe (Arizona State University); research scientists Hao-Fei Cheng (Amazon) and Benjamin Shapiro (Apple); and Ph.D. students Na Li (Penn State) and Kimberly Williamson (Cornell University).

References

1. Ainscow, M.: Promoting inclusion and equity in education: lessons from international experiences. Nord. J. Stud. Educ. Policy **6**(1), 7–16 (2020)
2. Bull, G., et al.: Evaluating the impact of educational technology (2016)
3. Holmes, W., Bektik, D., Denise, W., Woolf, B.P.: Ethics in aied: who cares? In: Artificial Intelligence in Education: 19th International Conference, AIED 2018, London, UK, 27–30 June 2018, Proceedings, Part II, pp. 551–553. Springer, Heidelberg (2018)
4. Holmes, W., Bektik, D., Di Gennaro, M., Woolf, B.P., Luckin, R.: Ethics in aied: Who cares? In: Artificial Intelligence in Education: 20th International Conference, AIED 2019, Chicago, USA, June 25–29, 2019, Proceedings, Part II. pp. 424–425. Springer, Heidelberg (2019)
5. Holstein, K., Doroudi, S.: Equity and artificial intelligence in education. In: The Ethics of Artificial Intelligence in Education, pp. 151–173. Routledge (2022)
6. Karumbaiah, S., Ocumpaugh, J., Baker, R.S.: Context matters: differing implications of motivation and help-seeking in educational technology. Int. J. Artif. Intell. Educ., 1–40 (2021)
7. Long, Y., Aleven, V.: Educational game and intelligent tutoring system: a classroom study and comparative design analysis. ACM Trans. Comput.-Hum. Interact. (TOCHI) **24**(3), 1–27 (2017)
8. Lynch, P., Singal, N., Francis, G.A.: Educational technology for learners with disabilities in primary school settings in low-and middle-income countries: a systematic literature review. Educ. Rev., 1–27 (2022)
9. OpenAI: Gpt-4 technical report (2023)
10. Roscoe, R.D., Salehi, S., Nixon, N., Worsley, M., Piech, C., Luckin, R.: Inclusion and equity as a paradigm shift for artificial intelligence in education. In: Artificial Intelligence in STEM Education, pp. 359–374. CRC Press (2022)

Artificial Intelligence and Educational Policy: Bridging Research and Practice

Seiji Isotani[1,2](✉) ⓘ, Ig Ibert Bittencourt[1,2] ⓘ, and Erin Walker[3] ⓘ

[1] Harvard Graduate School of Education, Cambridge, MA 02138, USA
{seiji_isotani,ig_bittencourt}@gse.harvard.edu
[2] NEES: Center for Excellence in Social Technologies, Federal University of Alagoas, Maceio, AL 57072-970, Brazil
[3] University of Pittsburgh, Pittsburgh, PA 15260, USA
eawalker@pitt.edu

Abstract. The use of artificial intelligence (AI) in education has been on the rise, and government and non-government organizations around the world are establishing policies and guidelines to support its safe implementation. However, there is a need to bridge the gap between AI research practices and their potential applications to design and implement educational policies. To help the community to address this challenge, we propose a workshop on AI and Educational Policy with the theme "Opportunities at the Intersection between AI and Education Policy." The workshop aimed to identify global challenges related to education and the adoption of AI, discuss ways in which AI might support learning scientists in addressing those challenges, learn about AI and education policy initiatives already in place, and identify opportunities for new policies to be established. We intend to develop action plans grounded in the learning sciences that identify opportunities and guidelines for specific AI policies in education.

Keywords: Policy design · Policy implementation · Educational technology

1 Introduction

Since the United Nations Declaration of Human Rights, the expansion of educational opportunity globally has been spectacular [5], on which humanity moved from 45% of access to Education in 1948 to 95% in 2022. Although the Global Educational Movement has progressed to promote the expansion of educational opportunities, the challenge we face nowadays is also manifold, such as quality of education [26], the well-being of students and teachers [1], availability of technological and educational resourcesc [19], lack of digital capabilities of the teachers [28], and so on. We are not only living in an unequal educational system, but the learning poverty [3] and inequality are increasing in the last decade and have deepened during the Covid-19 pandemic [20, 21].

The situation is even more dramatic when we talk about the Global South, which has most of the low-middle-income countries in the world. The challenges in Global South Education involve [5]:

- a high number of nonliterate people;
- a high number of students still in primary education;
- a high number of adolescents and youths out of secondary school;
- a high gender gap, and so on.

Furthermore, the digital divide is one of the challenges that is deepening even more the inequality between the Global South and Global North. Digital Divide implies worldwide explosive growth of the Internet, but data has shown it is an uneven, multidimensional phenomenon [4]. Indeed, technological innovations have transformed different sectors of the economy, promoting more development and embedding value in the chain worldwide by facilitating collective action in the direction of peace, justice, and sustainability.

Some would also argue that Artificial Intelligence (AI) and other digital technologies can support the global movement to achieve the civilizational goals of the 2030 agenda [33]. However, a recent study on Nature Communications about the role of Artificial Intelligence in Education (AIED) to achieve the Sustainable Development Goals varies according to each goal. The bad news is that SDG4: Ensuring inclusive and equitable quality education and promote lifelong learning opportunities for all, is one of the goals with more negative impacts on the use of AI, indicating that 70% of the targets were affected/inhibited by AI [33]. Such challenges are not exclusive to AIED but to digital technologies in education as a whole. Although ICTs helped to reshape the ways we live and learn, due to the digital divide, it has been just another layer of promoting inequality in education. Therefore, digital divides persist in internet access and the skills and competencies needed to leverage technology for collective and personal aims [5].

We are in an Educational Crisis. Nevertheless, there are several advances promoted by digital technologies and AIED. Digital technologies helped humanity respond quickly to the Covid-19 pandemic and provide remote learning to billions of students worldwide [18]. For several months, 1.7 billion students had no access to education, and digital technologies were used as the main strategy in all countries, even in the Global South. Additionally, there are recent studies and reports discussing and presenting the benefits of Artificial Intelligence to promote education in the Global South, such as AI Literacy [11, 13], AI for public policies and policymakers [12], Ethics of AIED [9, 25], AI applied to effective learning [10, 16, 31], AI to support teachers [2, 17], Improving management of schools with AI [12], and so on.

As a result, interest and investment in AI for Education are accelerating. And with it, concerns about the issues that arise when AI is widely implemented in educational technologies, such as bias, fairness, data privacy, and data security have also increased [23, 24]. UNESCO's Guide to Artificial Intelligence in Education [21] emphasizes that public policies will likely not be able to cope with the pace of AI innovation, and calls for more participation from public institutions and researchers, like those in the AIED community, to address issues of ethics, sustainability, and equity of using AI in education.

In an ecosystem of rapid AI innovation, government and non-government organizations around the world are beginning to establish policies and guidelines to support a safer implementation of AI in education. Japan's minister of education is preparing for the next wave of educational technology by considering what needs technology can and cannot address, suggesting that curriculums must focus on human skills, strengthening liberal arts education. Both China and Singapore have placed focus on encouraging leadership in AI innovation and personalized learning, with China's national plans for artificial intelligence (see also [22]) and Singapore's National AI Strategy. The policy think tank of the Government of India has released a National AI strategy with a dual focus on AI's ability to transform India's economy and the need to develop AI for all in a safe and inclusive manner, and states have followed up by creating AI policy roadmaps [14, 15].

Non-government organizations have begun to establish a policy think tank for AI in Africa [34] as a whole, while specific countries develop their own game plan such as South Africa's new data privacy regulations. Meanwhile, the U.S. recently released the AI Bill of Rights [30] and AI curriculum planning is beginning to occur on a state level through organizations like AI for K-12, while policies like COPPA and FERPA address privacy and security of student data nationally. E.U. Member countries are taking steps to implement AI and computing curricula in schools and as of April 2021 have released sweeping regulations on the use and development of AI.

On a global scale, some organizations are sticking together recommendations across regions; the Institute for Ethical AI in Education's Ethical Framework for AI in Education, based in the U.K., sets international ethical standards, and the Beijing Consensus on Artificial Intelligence and Education lays out recommendations developed as part of a workshop in 2019's AIED Conference.

In this context, we strongly believe that it is necessary to create a channel to bridge the gap between AIED research practices and their potential applications to design and implement educational policies. We also need to discuss global policy initiatives that leverage AI in education and embrace contributions that address the challenges of designing solutions to disseminate AI-enhanced education to underserved populations.

Thus, to address this challenge, we are proposing a workshop on AIED and Policy with the theme "Opportunities at the Intersection between AI and Education Policy." We are interested in exploring real-world applications and practical experiences of AI in Education, considering broader objectives such as increasing equity and access and responding to the digital divide, and important needs such as the intersection between AI technologies, curriculum integration, and teacher training. More specifically, our goals are to:

- Identify global challenges related to education including barriers and opportunities for the adoption of AI, with a focus on incorporating perspectives from the Global South;
- Discuss the ways in which AI might support learning scientists in addressing those challenges;
- Learn about AI and education policy initiatives already in place;
- Identify opportunities for new policies to be established;
- Develop action plans, grounded in the learning sciences, that identify opportunities and guidelines for specific AI policies. As we identify opportunities, we will ensure we discuss related risks in the areas of Fairness, Accountability, Transparency, and Ethics of AI in Education.

Acknowledgments. This workshop initiative is supported by the Center for Integrative Research on Computer and Learning Sciences (CIRCLS - https://circls.org), a center that connects learning sciences projects in the United States, and where AI and Education Policy is a key topic in the intersection between research and practice.
Other Organizing Committee Members:
• Deblina Pakhira, Digital Promise;
• Dalila Dragnic-Cindric, Digital Promise;
• Cassandra Kelley, University of Pittsburgh;
• Judi Fusco, Digital Promise;
• Jeremy Roschelle, Digital Promise.
The authors have used Grammarly and ChatGPT to improve the text.

References

1. Bai, W., et al.: Anxiety and depressive symptoms in college students during the late stage of the COVID-19 outbreak: a network approach. Transl. Psychiatry **11**(1), 638 (2021)
2. Baker, R.S.: Stupid tutoring systems, intelligent humans. Int. J. Artif. Intell. Educ. **26**(2), 600–614 (2016)
3. World Bank: The State of Global Learning Poverty: 2022 Update. Technical Report. A new joint publication of the World Bank, UNICEF, FCDO, USAID, the Bill & Melinda Gates Foundation (2022)
4. Calzada, I., Cobo, C.: Unplugging: deconstructing the smart city. J. Urban Technol. **22**(1), 23–43 (2015). https://doi.org/10.1080/10630732.2014.971535
5. Carney, S.: Reimagining our futures together: a new social contract for education. Comparative Education, 1–2 (2022). https://doi.org/10.1080/03050068.2022.2102326
6. Department of International Cooperation, Ministry of Science and Technology (MOST), P.R.China. (2017, September 15). Next generation artificial intelligence development plan issued by State Council. China Science and Technology Newsletter. https://www.mfa.gov.cn/ce/cefi//eng/kxjs/P020171025789108009001.pdf
7. European Union: New rules for artificial intelligence – questions and answers (2021, April 26). https://ec.europa.eu/commission/presscorner/api/files/document/print/en/qanda_21_1683/QANDA_21_1683_EN.pdf
8. Friedman, L., Black, N.B., Walker, E., Roschelle, J.: Safe AI in Education Needs You. ACM (2021, November 8). https://cacm.acm.org/blogs/blog-cacm/256657-safe-ai-in-education-needs-you/fulltext
9. Holmes, W., et al.: Ethics of AI in education: towards a community-wide framework. Int. J. Artif. Intell. Educ. **32**(3), 504–526 (2021)
10. Joaquim, S., Bittencourt, I.I., de Amorim Silva, R., Espinheira, P.L., Reis, M.: What to do and what to avoid on the use of gamified intelligent tutor system for low-income students. Educ. Inf. Technol. **27**(2), 2677–2694 (2021). https://doi.org/10.1007/s10639-021-10728-4
11. Madaio, M.A., et al.: Collective support and independent learning with a voice-based literacy technology in rural communities. In: Proceedings of the 2020 CHI Conference on Human Factors in Computing Systems, pp. 1–14 (2020)
12. Miao, F., Holmes, W., Huang, R., Zhang, H.: AI and Education: A Guidance for Policymakers. UNESCO Publishing (2021). https://unesdoc.unesco.org/ark:/48223/pf0000376709
13. Miao, F.: K-12 AI Curricula: A Mapping of Government-Endorsed AI Curricula. UNESCO Publishing (2022). https://unesdoc.unesco.org/ark:/48223/pf0000380602

14. NITI Aayog: Tamil Nadu Safe & Ethical Artificial Intelligence Policy 2020 (2022, November). https://indiaai.gov.in/research-reports/tamil-nadu-safe-ethical-artificial-intelligence-pol icy-2020
15. NITI Aayog: National Strategy for Artificial Intelligence (2018). https://indiaai.gov.in/res earch-reports/national-strategy-for-artificial-intelligence
16. Oyelere, S.S., et al. Artificial intelligence in african schools: towards a contextualized approach. In: 2022 IEEE Global Engineering Education Conference (EDUCON), pp. 1–7. IEEE (2022). https://doi.org/10.1109/educon52537.2022.9766550
17. Paiva, R., Bittencourt, I.I.: Helping Teachers Help Their Students: A Human-AI Hybrid Approach. In: Bittencourt, I.I., Cukurova, M., Muldner, K., Luckin, R., Millán, E. (eds.) AIED 2020. LNCS (LNAI), vol. 12163, pp. 448–459. Springer, Cham (2020). https://doi.org/ 10.1007/978-3-030-52237-7_36
18. Reimers, F.: Education and Covid-19: Recovering from the Shock Created by the Pandemic and Building Back Better. International Academy of Education and UNESCO, Geneva (2021). http://www.ibe.unesco.org/en/news/education-and-covid-19-rec overing-shock-created-pandemic-and-building-back-better-educational
19. Reimers, F., Amaechi, U., Banerji, A., Wang, M.: An Educational Calamity: Learning and Teaching During the Covid-19 Pandemic. Independently published (2021). https://www.ama zon.com/educational-calamity-Learning-teaching-Covid-19/dp/B091DYRDPV
20. Reimers, F.M. (ed.): Primary and Secondary Education During Covid-19. Springer, Cham (2022). https://doi.org/10.1007/978-3-030-81500-4
21. Reimers, F.M., Amaechi, U., Banerji, A., Wang, M.: Education in crisis. transforming schools for a post-covid-19 renaissance. In: Education to Build Back Better. Springer International Publishing, pp. 1–20 (2022). https://doi.org/10.1007/978-3-030-93951-9_1
22. Roberts, H., Cowls, J., Morley, J., Taddeo, M., Wang, V., Floridi, L.: The Chinese approach to artificial intelligence: an analysis of policy, ethics, and regulation. AI & Soc. **36**(1), 59–77 (2020). https://doi.org/10.1007/s00146-020-00992-2
23. Roschelle, J., Lester, J., Fusco, J., (eds.): AI and the future of learning: Expert panel report, Digital Promise (2020). https://circls.org/reports/ai-report
24. Santos, J., Bittencourt, I., Reis, M., Chalco, G., Isotani, S.: Two billion registered students affected by stereotyped educational environments: an analysis of gender-based color bias. Humanities and Social Sciences Communications **9**(1), 1–16 (2022). https://www.nature. com/articles/s41599-022-01220-6
25. Schiff, D.: Education for AI, not AI for education: the role of education and ethics in national AI policy strategies. Int. J. Artif. Intelli. Edu. **32**(3), 527–563 (2021). https://doi.org/10.1007/ s40593-021-00270-2
26. Schleicher, A.: World Class. How to Build a 21st-Century School System. OECD (2018). https://doi.org/10.1787/9789264300002-en
27. Smart Nation Digital Government Office: National AI Strategy (2019). https://www.smartn ation.gov.sg/files/publications/national-ai-strategy.pdf
28. TALIS: TALIS 2018 Results: Teachers and School Leaders as Lifelong Learners. OECD Report. (2019). https://doi.org/10.1787/23129638
29. The Institute for Ethical AI in Education: Developing an ethical framework for ai in education. University of Buckingham (2021, March 30). https://www.buckingham.ac.uk/research-the-institute-for-ethical-ai-in-education/
30. The White House: Blueprint for an ai bill of rights: Making automated systems work for the American people. The United States Government (2022, October 4). https://www.whiteh ouse.gov/ostp/ai-bill-of-rights/
31. Uchidiuno, J.O., Ogan, A., Yarzebinski, E., Hammer, J.: Going global: Understanding english language learners' student motivation in english language MOOCs. Int. J. Artif. Intelli. Edu. **28**(4), 528–552 (2017). https://doi.org/10.1007/s40593-017-0159-7

32. UNESCO: Beijing Consensus on Artificial Intelligence and Education (2019). https://une sdoc.unesco.org/ark:/48223/pf0000368303
33. Vinuesa, R., et al.: The role of artificial intelligence in achieving the sustainable development goals. Nature Communications 11(1) (2020). https://doi.org/10.1038/s41467-019-14108-y
34. RIA: Research ICT Africa Launches New AI Policy Think Tank for Africa (2021). https://researchictafrica.net/2021/01/29/ria-launches-ai-policy-research-centre/

Automated Assessment and Guidance of Project Work

Victoria Abou-Khalil[1](✉) ⓘD, Andrew Vargo[2] ⓘD, Rwitajit Manjoumdar[3] ⓘD,
Michele Magno[1] ⓘD, and Manu Kapur[4] ⓘD

[1] Center for Project-Based Learning, ETH Zurich, Zurich, Switzerland
[2] Osaka Metropolitan University, Osaka, Japan
[3] Graduate School of Informatics, Kyoto University, Kyoto, Japan
[4] Department of Humanities, Social and Political Sciences, ETH Zurich, Zurich, Switzerland

Abstract. Project-Based Learning improves student achievement and prepares students for a sustainable society, but the multimodality and unstructured nature of project work makes it difficult to comprehensively assess and support it. The recent availability of high-frequency sensors and the accessibility of machine learning techniques provide opportunities to create technologies that capture and analyze data in real time during projects. The workshop aims to bring together experts from the fields of AI, learning analytics, sensing technologies, and learning sciences to discuss technologies for detecting, understanding, and predicting learning during projects, as well as to discuss what is pedagogically desirable, technically feasible, and ethically sound. This workshop is important to the AIED community, particularly those working on developing technologies to capture project work or more generally, complex multimodal student interactions.

Keywords: Project-Based Learning · MultiModal Learning Analytics · Assessment · Automatic Guidance

1 Background

Project work has been demonstrated to be an effective method for improving student achievement (Chen and Yang, 2019; Kingston, 2018). It also offers the promise of preparing students to be designers and members of a sustainable society by fostering skills such as problem-solving, decision-making, teamwork, communication, and self-directed learning (Berbegal Mirabent et al., 2017; Cörvers et al., 2016; Guo et al., 2020; Moliner et al., 2015; Vogler et al., 2018).

Even though projects are an essential part of today's education, gathering real-time data during projects, and analyzing it to predict success or design personalized support remain challenging due to the multimodal and unstructured nature of project work. For example, students move freely between individual tasks and collaboration, they work on hardware and software, conduct discussions, and build prototypes. This fluidity makes it difficult to capture and assess the project comprehensively and hinders efforts to make projects more widespread in education. The best practices for guiding and structuring

N. Wang et al. (Eds.): AIED 2023, CCIS 1831, pp. 69–73, 2023.
https://doi.org/10.1007/978-3-031-36336-8_10

project activities are still unclear (Jonassen and Hung, 2008). Since we teach what we can measure, it is very important to measure the complexity of projects and use these measurements to predict success and support students.

The growing field of MultiModal Learning Analytics (MMLA) makes use of the recent availability of high-frequency sensors and advancements in machine learning and offers potential solutions to these problems. These technologies make it possible to capture and analyze complex interactions during projects, leading to a deeper understanding of the learning process (Ochoa and Worsley, 2016). Text, speech, gestures, movements, sketches, facial expressions, and handwriting analysis have been used to analyze student learning (Blikstein and Worsley, 2016; Raca, 2015; Raca et al., 2014, 2013; Spikol et al., 2018; Worsley and Blikstein, 2015). Recent work in AI in Education has taken a predictive approach and aimed at detecting learning outcomes from students' behaviors (e.g.: from videos by (Goldberg et al., 2021). Despite its importance, research on the topic is still nascent.

The majority of MMLA innovations are learning-design agnostic, which could be threatening to the field as learning analytics that is designed considering the particularities of the learning task design have more chances to impact students' learning (Knight et al., 2020; Yan et al., 2022). During projects, learning takes place across the digital and physical space and the different project structures. Projects can have different structures and can involve different phases (e.g.: problem definition, solution design, implementation, and prototyping) (Honglin and Yifan, 2022; Skillen, 2019). During each stage, students interact differently and studies on project-based learning need to account for the resulting differences in learning (English and Kitsantas, 2013). Finally, we cannot talk about capturing student behaviors without mentioning cultural differences. Even though the use of MMLA is increasing, the majority of models are developed and evaluated with participants from Western populations (Chua et al., 2019). Predictive features that apply to a certain population might not apply to another and can cause unfitting conclusions and biases. Consequently, this workshop aims to discuss the monitoring and support of diverse project types, phases, and student cultures both online and offline.

2   Workshop Description

2.1   Objectives

The development of tools that measure and support projects is a difficult task that includes technological, instructional, and design-based challenges. This workshop aims to activate an interdisciplinary community focused on studying complex interactions during projects. Experts from the fields of AI, learning analytics, sensing technologies, and learning sciences will present and discuss technologies for detecting, understanding, and predicting learning during projects, realistic obstacles for designing and using these technologies in practical settings, as well as future pathways for research. Additionally, we aim to foster interactions between researchers, psychologists, and teachers to discuss what is pedagogically desirable, technically feasible, and ethically sound when capturing and analyzing data during projects.

2.2 Relevance and Importance to the AIED Community

This workshop is important to the AIED community because it addresses the challenge of capturing and assessing complex, multimodal student interactions that are not easily measured through log data. Projects are increasingly being used as part of educational instruction, but there is limited support for assessing and supporting them, leaving the whole responsibility to the teacher. This workshop aims to provide guidance and discussions that can be beneficial to the AIED community, particularly those working on developing technologies to capture project work or more generally, complex and unstructured student interactions.

2.3 Content and Themes

The call for participation welcomes theoretical or empirical submissions of position papers, case studies, or ongoing research on the following topics:

Data Collection: Papers exploring the types of data that can be collected both offline and online during project work and the methods for collecting this data.

Modeling behaviors during project work: papers that identify patterns of behavior among students and discuss how they can be used to provide feedback, coaching, or recommendations.

Predicting learning outcomes: Studies detecting learning outcomes from real-time interactions during projects.

Assessment approaches and ontologies: Research on how to assess students' project work, as well as discussions on setting criteria for the assessment of projects and fairness in evaluation.

Intelligent systems: Papers examining the development and use of intelligent systems to support project work and the challenges faced in implementing these systems.

Cultural considerations and other characteristics for personalization: Studies on the role of culture or individual characteristics in the development, testing, and implementation of technologies to support project work.

Translation to practice: Research on the barriers and challenges faced in translating research to practice and the strategies for overcoming these obstacles.

Ethics, regulation, and privacy: Analyses of the ethical considerations for collecting student data during projects and the steps that need to be taken in order to safeguard student privacy. Discussions on the steps that need to be taken to ensure that these technologies are beneficial to students as well as mechanisms to mitigate or exploit the observer effect.

3 Workshop Organizers

Victoria Abou-Khalil is a postdoctoral researcher at the Center for Project-Based Learning at ETH Zurich. She studies the effects of technologies on learning, learning in low-resource settings, and the effectiveness of project-based learning.

Andrew Vargo is a research assistant professor at Osaka Metropolitan University. His research is focused on ubiquitous sensing technologies and learning augmentation.

Rwitajit Majumdar is a senior lecturer at the Graduate School of Informatics at Kyoto University. His research interests include technology-enhanced learning environment design, Learning Analytics, and studying human-data interactions in educational contexts.

Michele Magno is a senior scientist at ETH Zurich and the head of the Center for Project-Based Learning. He received his master's and Ph.D. degrees in electronic engineering from the University of Bologna, Italy in 2004 and 2010, respectively. Additionally, he is a guest full professor at Mid University in Sweden. He has published more than 250 articles in international journals and conferences, in which he received multiple best papers and best poster awards. The key topics of his research are wireless sensor networks, wearable devices, machine learning at the edge, energy harvesting, power management techniques, and the extended lifetime of battery-operated devices.

Manu Kapur holds the Professorship for Learning Sciences and Higher Education at ETH Zurich, Switzerland, and directs The Future Learning Initiative (FLI) at ETH Zurich to advance research on the science of teaching and learning in higher education contexts. Prior to this, Manu was a Professor of Psychological Studies at the Education University of Hong Kong. Manu also worked at the National Institute of Education (NIE/NTU) of Singapore as the Head of the Curriculum, Teaching, and Learning Department, as well as the Head of the Learning Sciences Lab (LSL). Manu conceptualized and developed the theory of Productive Failure to design for and bootstrap failure for learning mathematics better. He has done extensive work in real-field ecologies of STEM classrooms to transform teaching and learning using his theory of productive failure across a range of schools and universities around the world.

References

Berbegal Mirabent, J., Gil Doménech, M.D., Alegre, I.: Where to locate? A project-based learning activity for a graduate-level course on operations management. Int. J. Eng. Educ. **33**, 1586–1597 (2017)

Blikstein, P., Worsley, M.: Multimodal learning analytics and education data mining: using computational technologies to measure complex learning tasks. J. Learn. Anal. **3**, 220–238 (2016)

Chen, C.-H., Yang, Y.-C.: Revisiting the effects of project-based learning on students' academic achievement: a meta-analysis investigating moderators. Educ. Res. Rev. **26**, 71–81 (2019)

Chua, Y.H.V., Dauwels, J., Tan, S.C.: Technologies for automated analysis of co-located, real-life, physical learning spaces: where are we now?. In: Proceedings of the 9th International Conference on Learning Analytics & Knowledge, pp. 11–20 (2019)

Cörvers, R., Wiek, A., Kraker, J. de, Lang, D.J., Martens, P.: Problem-based and project-based learning for sustainable development. In: Sustainability Science, pp. 349–358. Springer (2016)

English, M.C., Kitsantas, A.: Supporting student self-regulated learning in problem-and project-based learning. Interdiscip. J. Probl.-Based Learn. **7**, 6 (2013)

Goldberg, P., et al.: Attentive or not? Toward a machine learning approach to assessing students' visible engagement in classroom instruction. Educ. Psychol. Rev. **33**, 27–49 (2021)

Guo, P., Saab, N., Post, L.S., Admiraal, W.: A review of project-based learning in higher education: student outcomes and measures. Int. J. Educ. Res. **102**, 101586 (2020)

Honglin, L., Yifan, N.: The construction of project-based learning model based on design thinking. In: 2022 4th International Conference on Computer Science and Technologies in Education (CSTE). IEEE, pp. 173–177 (2022)

Kingston, S.: Project Based Learning & Student Achievement: what Does the Research Tell Us?. PBL Evidence Matters, Vol. 1, No. 1. Buck Inst. Educ (2018)

Knight, S., Gibson, A., Shibani, A.: Implementing learning analytics for learning impact: taking tools to task. Internet High. Educ. **45**, 100729 (2020)

Moliner, M.L., et al.: Acquisition of transversal skills through PBL: a study of the perceptions of the students and teachers in materials science courses in engineering. Multidiscip. J. Educ. Soc. Technol. Sci. **2**, 121–138 (2015)

Ochoa, X., Worsley, M.: Augmenting learning analytics with multimodal sensory data. J. Learn. Anal. **3**, 213–219 (2016)

Raca, M.: Camera-based estimation of student's attention in class. EPFL (2015)

Raca, M., Tormey, R., Dillenbourg, P.: Sleepers' lag-study on motion and attention. In: Proceedings of the Fourth International Conference on Learning Analytics and Knowledge, pp. 36–43 (2014)

Raca, M., Tormey, R., Dillenbourg, P.: Student motion and it's potential as a classroom performance metric. In: 3rd International Workshop on Teaching Analytics (IWTA) (2013)

Skillen, P.: Project Based Learning (PBL): am I Doing it Right? [WWW Document]. Powerful Learn. Pract. (2019). https://plpnetwork.com/2019/02/14/pbl-right/, accessed 30 June 2022

Spikol, D., Ruffaldi, E., Dabisias, G., Cukurova, M.: Supervised machine learning in multimodal learning analytics for estimating success in project-based learning. J. Comput. Assist. Learn. **34**, 366–377 (2018)

Vogler, J.S., Thompson, P., Davis, D.W., Mayfield, B.E., Finley, P.M., Yasseri, D.: The hard work of soft skills: augmenting the project-based learning experience with interdisciplinary teamwork. Instr. Sci. **46**, 457–488 (2018). https://doi.org/10.1007/s11251-017-9438-9

Worsley, M., Blikstein, P.: Using learning analytics to study cognitive disequilibrium in a complex learning environment. In: Proceedings of the Fifth International Conference on Learning Analytics and Knowledge, pp. 426–427 (2015)

Yan, L., Zhao, L., Gasevic, D., Martinez-Maldonado, R.: Scalability, sustainability, and ethicality of multimodal learning analytics. In: LAK22: 12th International Learning Analytics and Knowledge Conference, pp. 13–23 (2022)

How to Open Science: Promoting Principles and Reproducibility Practices Within the Artificial Intelligence in Education Community

Aaron Haim$^{(\boxtimes)}$ⓘ, Stacy T. Shawⓘ, and Neil T. Heffernanⓘ

Worcester Polytechnic Institute, Worcester, MA 01609, USA
ahaim@wpi.edu

Abstract. Across the past decade, open science has increased in momentum, making research more openly available and reproducible. Artificial Intelligence (AI), especially within education, has produced effective models to better predict student outcomes, generate content, and provide a greater number of observable features for teachers. While completed, generalized AI models take advantage of available open science practices, models used during the actual research process are not made available. In this tutorial, we will provide an overview of open science practices and their benefits and mitigation within AI education research. In the second part of this tutorial, we will use the Open Science Framework to make, collaborate, and share projects - demonstrating how to make materials, code, and data open. The final part of this tutorial will go over some mitigation strategies when releasing datasets and materials so other researchers may easily reproduce them. Participants in this tutorial will learn what the practices of open science are, how to use them in their own research, and how to use the Open Science Framework.

The website (https://aied2023-tutorial.howtoopenscience.com/) and associated resources can be found on an Open Science Framework project (https://doi.org/10.17605/osf.io/yd9kr).

Keywords: Open Science · Reproducibility · Preregistration

1 Background

Open Science is a term used to encompass making methodologies, datasets, analyses, and results of research publicly accessible for anyone to use freely [6,16]. This term gained popularity in the early 2010 s s when researchers noticed that they were unable to replicate or reproduce prior work done within a discipline [15]. There was also a large amount of ambiguity when trying to understand what processes were used in a study. Open science, as a result, started to gain more traction to provide greater context, robustness, and reproducibility metrics. From there, many disciplines of research created their own formal definition

N. Wang et al. (Eds.): AIED 2023, CCIS 1831, pp. 74–78, 2023.
https://doi.org/10.1007/978-3-031-36336-8_11

and recommended practices. The widespread adoption of open science increased exponentially when large scale studies conducted in the mid 2010 s s found that numerous works were difficult or impossible to reproduce and replicate in psychology [2] and other disciplines [1].

There are numerous processes that open science can be broken down into such as open data, open materials, open methodology, and preregistration. **Open Data** specifically targets datasets and their documentation for public use without restriction, typically under a permissive license or in the public domain [8]. Not all data can be openly released (such as with personally identifiable information); but there are specifications for protected access that allow anonymized datasets to be released or a method to obtain the raw dataset itself. **Open Materials** is similar in regard except it concerns tools, source code, and their documentation [5]. This tends to be synonymous with **Open Source** in the context of software development, but materials are used to encompass the source in addition to available, free-to-use technologies. **Open Methodology** defines the full workflow and processes used to conduct the research, including how the participants were recruited and the procedure they went through [6]. The methodologies typically expand upon the original paper, such as technicalities that would not fit in the paper format, or survey items or test questions administered to participants. Finally, **Preregistration** acts as a time-stamped copy of the initial methodology and analysis plan before the start of a study, defining the process of research without knowledge of the outcomes [10, 11]. Preregistrations can additionally be updated to preserve the initial experiment conducted and the development as more context is generated and changes must be made (but all changes are documented).

A study is **reproducible** when the results reported can be obtained from the static input (e.g., dataset, configuration settings) and methodology (e.g., source code, software program) provided in the paper [12–14]. When source code is used to determine the results, there are numerous pitfalls that will make reproducibility difficult or impossible. If the libraries and versions of used languages or software are not reported, then it cannot be guaranteed that the source code will run. Libraries across different major or beta versions may not have the same structure or functions as defined in the source code. As such, a workspace needs to be recoverable from the provided source code. Another issue relates to non-seeded random values producing a different result than reported in the paper. However, when the random value is not seeded, the variance of the results should not differ majorly between executions. Therefore, while random values should be seeded for reproduction, the random values should have low variability on the results.

2   Tutorial Goals

Open science practices and reproducibility metrics are becoming more commonplace within numerous scientific disciplines. Within subfields of educational technology, however, the adoption and review of these practices and metrics are neglected or sparsely considered [9]. There are some subfields of education technology that have taken the initiative to introduce open science practices (special education [3]; gamification [4], education research [7]); however, other subfields have seen little to no adoption. Concerns and inexperience in what can be made publicly available to how to reproduce another's work are some of the few reasons why education researchers may choose to avoid or postpone discussion on open science and reproducibility. On the other hand, lack of discussion can lead to tediousness and repetitive communication for datasets and materials or cause a reproducibility crisis [1] within the field of study. As such, there is a need for accessible resources and understanding on open science, how it can be used, and how to mitigate any potential issues that may arise within one's work at a later date.

In this tutorial, we will cover some of the basic practices of open science and some of the challenges and mitigation strategies associated with education technology specifically. Next, we will provide a step-by-step explanation on using the Open Science Framework to create a project, collaborate with other researchers, post content, and preregister a study. Using examples from the field of educational technology, we will showcase how to incorporate common open science practices, in addition to practices that, when implemented, are designed to improve reproducibility.

3　Tutorial Organization

The tutorial focuses on introducing some common open science practices and their usage within education technology, providing an example on using the Open Science Framework to create a project, post content, and preregister studies, and using previous papers to apply the learned practices and any additional reproduction mitigation strategies. More specifically, the tutorial will cover the following:

- First, we will provide a presentation on an overview of a few problems when conducting research. Using this as a baseline, we will introduce open science and its practices and how they can be used to nullify some of these issues and mitigate others. In addition, we will attempt to dispel some of the misconceptions of these practices.
- Second, we will provide a live example of using the Open Science Framework (OSF)[1] website to make an account, create a project, add contributors, add content and licensing, and publicize the project for all to see. Afterwards, we will provide a guide to creating a preregistration, explaining best practices, and identifying how to create an embargo. Additional features and concerns, such as anonymizing projects for review and steps required to properly do so, will be shown.
- Third, we will discuss reproducibility metrics within work when providing datasets and materials. This will review commonly used software and

---

[1] https://osf.io/.

languages (e.g., Python[2], R[3] and RStudio[4], Visual Studio Code[5]) and how, without any steps taken, most work tends to be extremely tedious to reproduce or are not reproducible in general. Afterwards, we will provide some mitigation strategies needed to remove these concerns.
- Finally, we will use a few papers, each containing different issues, and apply the necessary steps needed to reproduce the results within the paper.

3.1 Organizers

Aaron Haim[6] is a Ph.D. student in Computer Science at Worcester Polytechnic Institute. His initial research focuses on developing software and running experiments on crowdsourced, on-demand assistance in the form of hints and explanations. His secondary research includes reviewing, surveying, and compiling information related to open science and reproducibility across papers published at education technology and learning science conferences.

Stacy T. Shaw[7] is an Assistant Professor of Psychology and Learning Sciences at Worcester Polytechnic Institute. She is an ambassador for the Center for Open Science, a catalyst for the Berkeley Initiative in Transparency in Social Sciences, and serves on the EdArXiv Preprint steering committee. Her research focuses on mathematics education, student experiences, creativity, and rest.

Neil T. Heffernan[8] is the William Smith Dean's Professor of Computer Science and Director of the Learning Sciences & Technology Program at Worcester Polytechnic Institute. He co-founded ASSISTments, a web-based learning platform, which he developed not only to help teachers be more effective in the classroom, but also so that he could use the platform to conduct studies to improve the quality of education. He has been involved in research papers containing some of the largest openly accessible data and materials in addition to convincing the Educational Data Mining conference to use the Open Science badges when researchers are submitting papers.

References

1. Baker, M.: 1,500 scientists lift the lid on reproducibility. Nature **533**(7604), 452–454 (2016)
2. Collaboration, O.S.: Estimating the reproducibility of psychological science. Science **349**(6251), aac4716 (2015). https://doi.org/10.1126/science.aac4716, https://www.science.org/doi/abs/10.1126/science.aac4716

---

[2] https://www.python.org/.
[3] https://www.r-project.org/.
[4] https://posit.co/products/open-source/rstudio/.
[5] https://code.visualstudio.com/.
[6] https://ahaim.ashwork.net/.
[7] http://stacytshaw.com/.
[8] https://www.neilheffernan.net/.

3. Cook, B.G., Collins, L.W., Cook, S.C., Cook, L.: A replication by any other name: a systematic review of replicative intervention studies. Remedial Special Educ. **37**(4), 223–234 (2016). https://doi.org/10.1177/0741932516637198

4. García-Holgado, A., et al.: Promoting open education through gamification in higher education: the opengame project. In: Eighth International Conference on Technological Ecosystems for Enhancing Multiculturality,. TEEM 2020, p. 399–404. Association for Computing Machinery, New York (2021). https://doi.org/10.1145/3434780.3436688

5. Johnson-Eilola, J.: Open source basics: definitions, models, and questions. In: Proceedings of the 20th Annual International Conference on Computer Documentation, SIGDOC 2002, p. 79–83. Association for Computing Machinery, New York (2002). https://doi.org/10.1145/584955.584967

6. Kraker, P., Leony, D., Reinhardt, W., Beham, G.: the case for an open science in technology enhanced learning. Int. J. Technol. Enhanced Learn. **3**(6), 643–654 (2011). https://doi.org/10.1504/IJTEL.2011.045454, https://www.inderscienceonline.com/doi/abs/10.1504/IJTEL.2011.045454

7. Makel, M.C., Smith, K.N., McBee, M.T., Peters, S.J., Miller, E.M.: A path to greater credibility: large-scale collaborative education research. AERA Open **5**(4), 2332858419891963 (2019). https://doi.org/10.1177/2332858419891963

8. Murray-Rust, P.: Open data in science. Nature Precedings **1**(1), 1 (2008). https://doi.org/10.1038/npre.2008.1526.1

9. Nosek, B.: Making the most of the unconference (2022). https://osf.io/9k6pd

10. Nosek, B.A., et al.: Preregistration is hard, and worthwhile. Trends Cognit. Sci. **23**(10), 815–818 (2019). https://doi.org/10.1016/j.tics.2019.07.009

11. Nosek, B.A., Ebersole, C.R., DeHaven, A.C., Mellor, D.T.: The preregistration revolution. Proceed. National Acad. Sci. 115(11), 2600–2606 (2018). https://doi.org/10.1073/pnas.1708274114, https://www.pnas.org/doi/abs/10.1073/pnas.1708274114

12. Nosek, B.A., et al.: Replicability, robustness, and reproducibility in psychological science. Ann. Rev. Psychol. **73**(1), 719–748 (2022). https://doi.org/10.1146/annurev-psych-020821-114157, pMID: 34665669

13. Patil, P., Peng, R.D., Leek, J.T.: A statistical definition for reproducibility and replicability. bioRxiv **1**(1), (2016). https://doi.org/10.1101/066803, https://www.biorxiv.org/content/early/2016/07/29/066803

14. National Academies of Sciences, E. Affairs, P. Committee on Science, E. Information, B. Sciences, D. Statistics, C. Analytics, B. Studies, D. Board, N. Education, D. et al: Reproducibility and Replicability in Science. National Academies Press, Washington, D.C., USA (2019), https://books.google.com/books?id=6T-3DwAAQBAJ

15. Spellman, B.A.: A short (personal) future history of revolution 2.0. Perspect. Psychol. Sci. **10**(6), 886–899 (2015). https://doi.org/10.1177/1745691615609918, pMID: 26581743

16. Vicente-Saez, R., Martinez-Fuentes, C.: Open science now: a systematic literature review for an integrated definition. J. Bus. Res. **88**, 428–436 (2018). https://doi.org/10.1016/j.jbusres.2017.12.043, https://www.sciencedirect.com/science/article/pii/S0148296317305441

AI and Education. A View Through the Lens of Human Rights, Democracy and the Rule of Law. Legal and Organizational Requirements

Wayne Holmes[1] , Christian M. Stracke[2] , Irene-Angelica Chounta[3](✉) ,
Dale Allen[4], Duuk Baten[5] , Vania Dimitrova[6] , Beth Havinga[7] ,
Juliette Norrmen-Smith[8], and Barbara Wasson[9]

[1] University College London, London, UK
wayne.holmes@ucl.ac.uk
[2] University of Bonn, Bonn, Germany
stracke@uni-bonn.de
[3] University of Duisburg-Essen, Duisburg, Germany
irene-angelica.chounta@uni-due.de
[4] Dxtera Institute & EdSAFE AI Alliance, Boston, USA
[5] SURF, Utrecht, The Netherlands
[6] University of Leeds, Leeds, UK
[7] European EdTech Alliance e.V., Bielefeld, Germany
[8] University of Oxford, Oxford, UK
[9] University of Bergen, Bergen, Norway

Abstract. How can we develop and implement legal and organisational requirements for ethical AI introduction and usage? How can we safeguard and maybe even strengthen human rights, democracy, digital equity and rules of law through AI-supported education and learning opportunities? And how can we involve all educational levels (micro, meso, macro) and stakeholders to guarantee a trustworthy and ethical AI usage? These open and urgent questions will be discussed during a full-day workshop that aims to identify and document next steps towards establishing ethical and legal requirements for Artificial Intelligence and Education.

Keywords: Ethics · Artificial Intelligence · human rights · democracy · rule of law

1 Introduction

It is evident and required, as demonstrated in the latest publications, studies, and events, that Artificial Intelligence – in the context of education and beyond – requires legal (that is, laws) and organisational requirements (for example, regulatory actions such as data protection regulations) for enabling, monitoring, and governance. This is widely reflected in the contributions of the Artificial Intelligence in Education (AIED) community where we come across rich collections of research works about the Ethics of

N. Wang et al. (Eds.): AIED 2023, CCIS 1831, pp. 79–84, 2023.
https://doi.org/10.1007/978-3-031-36336-8_12

AI in Education [3–5], Fairness, Accountability, Transparency, and Ethics in AIED [9], educational stakeholders' perceptions of AI [1] and Scrutable AI [6]. A similar picture is depicted by the public announcements and actions of national (e.g., Norwegian governmental task force report[1]) and international organisations such as the EU (AI Act[2]), UNESCO (International Forum on Artificial Intelligence and Education 2022[3], AI and Education: Guidance for policy -makers [8]) and the CoE (AI, Human rights, Democracy and the Rule of Law: A Primer Prepared for the Council of Europe [7]).

In particular, in 2021, the Council of Europe (CoE) – an international organisation founded in 1949 to uphold human rights, democracy and the rule of law in Europe[4] – established an Expert Group to investigate, and propose a legal instrument for the application of AI in education (AIED) and a recommendation for the teaching of AI (AL Literacy). The project is working towards developing an actionable set of recommendations for Member States, helping ensure that the application and teaching of AI in education (collectively referred to as "AI&ED") is for the common good.

To that end, the CoE Expert Group published the report "AI and Education. A Critical View Through the Lens of Human Rights, Democracy and the Rule of Law" [2] that discusses AI&ED from the perspective of the fundamental values of the CoE, aims to bring forward challenges, opportunities and implications of AI&ED, and is designed to stimulate and inform further critical discussion. In addition, the Expert group has undertaken a survey of Member States, which has been designed to inform the project and to stimulate further critical debate. The report and preliminary findings of the survey were presented in the two-day working conference "Artificial Intelligence and Education"[5] held by the CoE on 18th and 19th of October 2022 in Strasbourg. The conference attracted participation of experts from various domains (such as policy makers, academia, non-profit organisations and industry).

All workshop organisers are members of the CoE Expert Group and/or were invited by the CoE and contributed to the CoE Expert Conference. Drawing on their collective experience, the workshop organisers aimed to further the discussion about ethical AI and Education; that is, learning with AI, learning about AI, and preparing for AI. We envision that these discussions are necessary to set solid foundations regarding the development, deployment, and use of AI systems in educational contexts in ways that respect human rights and principles, ensure fair and ethical use and promote human agency [10].

---

[1] NOU Report from the Norwegian Ministry of Education Expert Task Force on Learning Analytics. 6 June 2023 https://laringsanalyse.no/.

[2] https://artificialintelligenceact.eu/.

[3] https://www.unesco.org/en/articles/international-forum-artificial-intelligence-and-education-2022.

[4] Currently, the CoE has 46 member states (19 more than the European Union), with a population of approximately 675 million.

[5] https://rm.coe.int/0900001680aa58ac.

2 Objectives, Relevance, and Importance

2.1 Workshop Objectives

The workshop intends to achieve the following objectives:

- To explore potential guidelines for trustworthy and ethical AI usage and to critically assess their structure and usage scenarios
- To develop ideas around personal future ramifications (including trustworthiness and ethical implications) and AI usage in education, and to collate, present and discuss these
- To explore in group work and in the plenary how legal guidelines for trustworthy and ethical AI usage can be developed and designed, in particular related to legal and organisational requirements
- To identify methods for dealing with the quick changing developments in AI, providing a legal framework that is flexible but also appropriately effective

To achieve these objectives, we will address the following questions to guide and contextualise the discussions:

- How can we develop and implement legal and organisational requirements for ethical AI introduction and usage?
- How can we keep and maybe even strengthen human rights, democracy, digital equity and rules of law through AI-supported education and learning opportunities?
- And how can we involve all educational levels (micro, meso, macro) and stakeholders to guarantee a trustworthy and ethical AI usage?

The main focus will be on the legal and organisational requirements to achieve a regulation in the fields of AI and Education (AI&ED). As a basis we will use the work and activities of the Council of Europe (CoE) and its appointed AI&ED Expert Group, such as the report "AI and Education. A Critical View Through the Lens of Human Rights, Democracy and the Rule of Law". Finally, we will seek to define what next is necessary in relation to the established requirements and goals.

2.2 Workshop Relevance and Importance to the AIED Community

To prepare for the Council of Europe Council of Ministers meeting in 2024, when the proposed text of the legal instrument on AI in Education will be debated and (hopefully) agreed, and when a final report from the CoE Expert Group will be published, the CoE Expert Group is currently eliciting expert input from key stakeholder groups: students, teachers, commercial organisations, NGOs, and academia.

The members of the AIED community are key stakeholders, such that it is critical that their views are properly addressed by the Council of Europe's in their work (particularly the proposed legal instrument) on AI and education. The workshop is also closely aligned with the AIED 2023 conference theme - "AI in Education for Sustainable Society". It will provide an arena for the AIED community to familiarise themselves with the CoE initiative and to engage in follow on activities. Therefore, the results from CoE Expert Conference "Artificial Intelligence and Education" (held on 18th and 19th of October

2022 in Strasbourg) will be presented and the initial report of the CoE Expert Group: *Artificial Intelligence and Education. A critical view through the lens of human rights, democracy and the rule of law* (2022)[6], will be introduced. The workshop will also facilitate the forming of a wider community of practice within the AIED community to consider the wider societal implication of AIED.

3   Workshop Format

The workshop will take place as a full-day interactive event with the intention to trigger discussions between participants. To do so, the workshop consists of two sessions – the morning and the afternoon sessions - that are interrelated and complement each other. The whole workshop will be hybrid with small working groups that are on-site as well as online and always followed by plenary sessions where the working groups are given the opportunity to inform each other on the discussions and results.

3.1   Program

The workshop will begin with an introduction to the program, the objectives, and a summary of the key messages from the initial report on AI&ED by the CoE Expert Group. The core part of the morning session will centre on four key questions:

- **Q1.** What issues does the report misunderstand or misrepresent, and how should this be corrected?
- **Q2.** What issues does the report ignore, miss or forget, and what other information should be included?
- **Q3.** What regulations, if any, should be put in place to protect human rights, democracy and the rule of law whenever AI is applied (AIED) or taught (AI Literacy) in educational contexts?
- **Q4.** What case studies can the participants suggest to enhance the public understanding of the issues raised in the report?

We will introduce each question with a 5-min presentation. Then, participants will form small groups to discuss and to respond to the issues introduced, and to reflect on potential theoretical and practical implications, before reporting back their views and ideas to the workshop. This session will conclude with a discussant (one of the workshop organisers) summarising the participants' input and offering a wrap-up.

In the afternoon, the workshop will follow the World Café format and will consist of three parts. In the first, introductory part (15 min), we will briefly present the World Café method and the topic "Legal and organisational requirements for ethical AI introduction and usage". In the second core part, we will discuss different and pre-prepared themes and aspects in groups of ideally 4 to 10 participants. The procedure will be identical on all group tables:

1. The theme will be read that always consists of an open question and clarified.

---

[6] https://rm.coe.int/artificial-intelligence-and-education-a-critical-view-through-the-lens/168 0a886bd.

2. Afterwards, the contributions (= answers on the open question of the theme) will be collected as post-it at a flipchart next to the group table.
3. Finally, all collected contributions will be sorted in open discussion and be clustered according to defined categories or criteria so that they can be presented. That can be done independently or using the categories or criteria from the other rounds before.

In the third part (30 to 45 min), the results from the group tables will be presented in the plenary and critically discussed and reflected. These can then be provided as summaries for publication after the event.

Table 1. The proposed agenda of the workshop "AI and Education. A view through the lens of human rights, democracy and the rule of law. Legal and organisational requirements."

Sessions	Agenda Item	Duration (minutes)
Morning session	Welcome / Introduction to the workshop	15'
	Group discussions on the CoE AI&ED report	105'
	Summary and reflections of group discussion	15'
	Wrap-up of morning session	10'
Afternoon session	Introduction to the World Café	15'
	Thematic group work about "Legal and organizational requirements for ethical AI introduction and usage"	60'
	Plenary talks / presentations	45'
	Open discussion with CoE Expert group	15'
Closing	Envisioning future steps and directions	10'

To conclude the workshop, we will collect all issues that the workshop participants wish to discuss and bring to the attention of the CoE Expert Group (15 min). Finally, we will define next steps and activities and conclude with a wrap up of the whole workshop (10 min). The proposed agenda is presented in Table 1.

3.2 Target Audience

We aim to reach out and involve all AIED community members regardless their expertise and experience. The interactive workshop is dedicated to interested beginners without any pre-knowledge as well as to experts. To promote inclusiveness, we chose a hybrid participation format (on-site and online) that we envision to foster discussions with all AIED community members worldwide.

3.3 Expected Outcomes

We envision that this workshop will results in three main outcomes. First, the presentation, discussion and collection of legal and organizational criteria, dimensions and requirements that are needed for the development and design of strategies and guidelines defining and regulating trustworthy and ethical AI and its introduction and usage (in education and beyond). Second, the workshop discussions will directly inform the Council of Europe's Expert Group on AI&ED's work towards the final report and the potential for a legal instrument to be debated by Council Ministers in 2024. Third, we intended to collaborate on a common publication based on the workshop discussions and results.

References

1. Chounta, I.-A., Bardone, E., Pedaste, M., Raudsep, A.: Exploring teachers' perceptions of Artificial Intelligence as a tool to support their practice in Estonian K-12 education. IJAIED (2021). https://doi.org/10.1007/s40593-021-00243-5
2. Holmes, W., Persson, J., Chounta, I.-A., Wasson, B., Dimitrova, V.: Artificial intelligence and education: a critical view through the lens of human rights, democracy and the rule of law. Council of Europe (2022)
3. Holmes, W., et al.: Ethics of AI in education: towards a community-wide framework. International Journal of Artificial Intelligence in Education 1–23 (2021)
4. Holmes, W., Porayska-Pomsta, K. (eds.): The Ethics of AI in Education. Practices, Challenges, and Debates. Routledge, New York (2023)
5. Holmes, W., Tuomi, I.: 'State of the Art and Practice in AI in Education'. European Journal of Education: research, Development and Policies n/a (n/a) (2022). https://doi.org/10.1111/ejed.12533
6. Kay, J., Kummerfeld, B., Conati, C., Porayska-Pomsta, K., Holstein, K.: Scrutable AIED. In: Handbook of Artificial Intelligence in Education, pp. 101–125. Edward Elgar Publishing (2023)
7. Leslie, D., Burr, C., Aitken, M., Cowls, J., Katell, M., Briggs, M.: Artificial intelligence, human rights, democracy, and the rule of law: a primer. Council of Europe (2021). https://www.turing.ac.uk/news/publications/ai-human-rights-democracy-and-rule-law-primer-prepared-council-europe
8. Miao, F., Holmes, W., Huang, R., Zhang, H.: AI and education: a guidance for policymakers. UNESCO Publishing (2021)
9. Woolf, B.: Introduction to IJAIED Special Issue, FATE in AIED. Int. J. Artifi. Intelli. Edu. **32**, 501–503 (2022)
10. Bozkurt, A., et al.: Speculative Futures on ChatGPT and Generative Artificial Intelligence (AI): A Collective Reflection from the Educational Landscape. Asian Journal of Distance Education **18**(1), 53–130 (2023). https://www.asianjde.com/ojs/index.php/AsianJDE/article/view/709/394

AI in Education. Coming of Age? The Community Voice

Wayne Holmes[1]([⊠]) [iD] and Judy Kay[2] [iD]

[1] University College London, London, UK
wayne.holmes@ucl.ac.uk
[2] University of Sydney, Camperdown, Australia
judy.kay@sydney.edu.au

Abstract. At AIED 2022 (Durham), a panel discussion took place in response to a provocation that acknowledged that the work of the AIED community is increasingly being commercialised and that, partly as a consequence, AIED is increasingly being criticised (e.g., for perpetuating poor pedagogic practices, datafication, and introducing classroom surveillance). The provocation went on to ask whether the AIED community should carry on regardless, continuing with its traditional focus, or whether it should seek a new role. As the panel discussion raised many important issues, the panellists were subsequently invited to publish their thoughts in opinion pieces in the IJAIED. To extend the discussion, several leading researchers from outside the community were also invited to contribute. The result was a Special Issue of the IJAIED ("AIED. Coming of age?", 2023). However, now that some of AIED's leading community members and some leading researchers from outside the community had had their say, there remained an important gap. What were the thoughts and opinions of the wider membership of the AIED community? This is particularly timely given the recent explosion of AIED tools, particularly due to the now widely available tools based on Large Language Models. This half-day workshop was dedicated to opening up the conversation, so that everyone in the community could have their say, and could contribute to the future of AIED.

Keywords: ethics · artificial intelligence · human rights · democracy · rule of law

1 Introduction

The AIED 2022 conference held in Durham, UK, included a panel discussion entitled *"AIED: Coming of Age?"* The panel began with the following provocation:

The AIED community has researched the application of AI in educational settings for more than forty years. Today, many AIED successes have been commercialised – such that, around the world, there are as many as thirty multi-million-dollar-funded AIED corporations, and a market expected to be worth $6 billion within two years. At the same time, AIED has been criticised for perpetuating poor pedagogic practices, datafication, and introducing classroom surveillance.

N. Wang et al. (Eds.): AIED 2023, CCIS 1831, pp. 85–90, 2023.
https://doi.org/10.1007/978-3-031-36336-8_13

The commercialisation and critique of AIED presents the AIED academic community with a conundrum. Does it carry on regardless, continue its traditional focus, researching AI applications to support students, in ever more fine detail? Or does it seek a new role? Should the AIED community reposition itself, building on past successes but opening new avenues of research and innovation that address pedagogy, cognition, human rights, and social justice?

Because of the level of interest generated by the multitude of issues raised by the discussion in Durham, challenges centred on the futures of AIED and the AIED research community [1], the panellists were each invited to publish their thoughts in opinion pieces in a Special Issue of the International Journal of Artificial Intelligence in Education, entitled "AIED. Coming of Age?" The contributors to that Special Issue (2023) included the original panellists, all of whom have been leading members of the AIED community for many years (see, e.g., [2–6]). In alphabetical order:

- Ben du Boulay (University of Sussex),
- Art Graesser (University of Memphis),
- Ken Koedinger (Carnegie Mellon University),
- Danielle McNamara (Arizona State University), and
- Maria Mercedes (Didith) T. Rodrigo (Ateneo de Manila University).

In addition, because of their notable contributions as members of the Durham panel's audience, the following leading AIED researchers were also invited to contribute opinion pieces:

- Peter Brusilovsky (University of Pittsburgh),
- René Kizilcek (Cornell University), and
- Kaśka Porayska-Pomsta (University College London).

Finally, in order to further open the discussion and to introduce some alternative voices into the debate – in other words, to prevent the Special Issue becoming an AIED echo chamber – several leading researchers from outside the community who had published in related areas (see, e.g., [7–9]) were also invited to contribute opinion pieces:

- Rebecca Eynon (University of Oxford),
- Caroline Pelletier (University College London),
- Jen Persson (DefendDigitalMe),
- Neil Selwyn (Monash University),
- Ilkka Tuomi (Meaning Processing),
- Ben Williamson (University of Edinburgh), and
- Li Yuan (Beijing Normal University).

All told, there were fifteen opinion pieces representing a spectrum of views, some complementary, others not, both from within and from outside the AIED community (see Appendix I for the full list of authors and papers).

The Special Issue was introduced by Wayne Holmes [10], who raised six challenges with which he argued the AIED community should seriously engage, his aim being to further provoke the discussion. Importantly, rather than being a criticism of the work of the AIED community, it was a recognition of the successes of the AIED community alongside the fact that the world had changed and it appears to be a time of seismic shift, due at least in part to the tsunami of tools based on Large Language Models being used in education.

However, given that some of AIED's leading community members and some leading researchers from outside the community had had their say, there still remained an important gap. What were the thoughts and opinions of the wider membership of the AIED community? Accordingly, this half-day workshop was dedicated to opening up the conversation, so that everyone in the community could have their say, and could contribute to the future development of AIED.

2 Objectives, Relevance, and Importance

2.1 Workshop Objectives

The workshop intended to achieve the following objectives:

- To give all members of the AIED community (novices and experienced researchers alike) the opportunity to reflect on the opinions expressed by the opinion writers – both from within the AIED community and from outside – published in the IJAIED Special Issue ("AIED. Coming of Age?", 2023).
- To give all members of the AIED community the opportunity to discuss some key issues centred on the future of AIED and AIED research.
- To give all members of the AIED community the opportunity to have their voice heard by the AIED community and the IAIED (International Artificial Intelligence in Education Society) leadership.

To achieve these objectives, we discussed eight questions (e.g., where should AIED go now, should the AIED community be extended to include more members with expertise in other disciplines, and how can the AIED community better address criticisms) to guide and contextualise the discussions (for the full list of questions, see Sect. 3.1).

2.2 Workshop Relevance and Importance to the AIED Community

The workshop was relevant to the AIED community for two key reasons:

1. It was essential that AIED community members, as key stakeholders, had the opportunity to express and explore their collective views about the future of AIED and AIED research, so that those views could be taken into account by the AIED and IAIED leadership.
2. The workshop also facilitated the forming of a wider community of practice within the AIED community to consider the future of AIED and AIED research.

3   Workshop Format

The workshop was a half-day interactive event that aimed to trigger and facilitate discussions between the participants. It was dedicated to opening up the conversation, so that everyone in the community could have their say, and could contribute to the future development of AIED. To prepare for the workshop, participants were encouraged to read as many of the opinion pieces from the IJAIED Special Issue ("AIED. Coming of Age?", 2023) as possible, prioritising those by authors whose work they did not already know. At the workshop, three participants gave short presentations each summarising and critiquing one opinion piece by an author with whom they were previously unfamiliar. The remainder of the workshop was centred on eight questions (see Sect. 3.1), with participants working in small working groups before reporting back to the whole workshop (i.e., plenary sessions in which the working groups were given the opportunity to inform each other on their discussions and outcomes). The workshop concluded with a verbal report by the discussant and a wrap up.

3.1   Program

The face-to-face workshop began with an introduction, led by Judy Kay and Wayne Holmes. It continued with three short presentations by participants summarising and critiquing one of the IJAIED Special Issue opinion pieces ("AIED. Coming of Age?", 2023) by an author with whom they were previously unfamiliar. The core part of the workshop centred on eight questions:

Q1. Where should AIED go now? What should it continue to research and what (if any) new areas should it start to research?

Q2. Should the AIED community be extended to include more members with expertise in other disciplines (e.g., humanities and social sciences). If so, what other disciplines and how might this be best achieved?

Q3. How can the AIED community best address criticisms (e.g., that AIED perpetuates poor pedagogic practices, datafication, and surveillance)?

Q4. What responsibilities does the AIED community have for when their research is commercialised, and what can be done to ensure that the commercial AIED still focuses on enhancing education?

Q5. What direction should the AIED community take for collaboration between academic and commercial AIED research?

Q6. Should we build the areas of research into human-centred interface aspects, where we focus on the interface that students experience?

Q7. What role should the AIED community have in researching the teaching of AI (i.e., AI Literacy), in recognition that AI is seeping into all aspects of our lives, creating new challenges for education?

Q8. What other issues do the AIED members wish to discuss and bring to the attention of the IAIED?

For each question, the participants were divided into small groups to discuss the question and its implications. Afterwards, the groups reconvened into a plenary, and each group reported back their discussion and key outcomes. All participants were then invited to discuss and comment. The workshop concluded with a verbal report by the discussant, Vania Dimitrova, and a wrap up by Judy Kay and Wayne Holmes.

3.2 Outcomes

The key outcome of the workshop is an academic paper (in preparation) that summarises and critically evaluates the discussion. The paper is being led by Vania Dimitrova, Judy Kay and Wayne Holmes, and it will acknowledge by name (with permission) the contribution of the workshop participants. The paper will be presented to the IAIED, to inform future developments, and will also be submitted to the IJAIED for consideration through the usual review process.

Appendix I.

IJAIED Special Issue "AIED. Coming of Age?" (2023) opinion piece authors.

Ben du Boulay	Pedagogy, Cognition, Human Rights, and Social Justice
Peter Brusilovsky	AI in Education, Learner Control, and Human-AI Collaboration
Rebecca Eynon	The future trajectory of the AIED community: defining the 'knowledge tradition' in critical times
Art Graesser, Xiangen Hu, John Sabatini, and Colin Carmon	Where is Ethics in the Evolution of AIED's Future?
Rene Kizilcec	To Advance AI Use in Education, Focus on Understanding Educators
Ken Koedinger	Ken Koedinger (in conversation with Wayne Holmes)
Danielle McNamara	AIED: Coming of Age. Opinion Piece by Danielle S. McNamara
Caroline Pelletier	Against personalised learning
Jen Persson	Artificial Intelligence and Education: Research, the Redistribution of Authority, and Rights
Kaśka Porayska-Pomsta	Towards a manifesto for the pro-active role of the AIED Community in agenda setting for human-centred AI and educational innovation
Maria Mercedes T. Rodrigo	Is the AIED Conundrum a First-World Problem?
Neil Selwyn	Constructive criticism? Working with (rather than against) the AIED back-lash
Ilkka Tuomi	Beyond mastery: Toward a broader understanding of AI in education
Ben Williamson	The social life of AI in education
Li Yuan	Where does AI-driven education in the Chinese context and beyond go next?

References

1. Holmes, W., Tuomi, I.: State of the art and practice in AI in education. European Journal of Education: research, development and policies **57**(4), 542–570 (2022). https://doi.org/10.1111/ejed.12533

2. Koedinger, K.R., Anderson, J.R., Hadley, W.H., Mark, M.A.: Intelligent tutoring goes to school in the big city. Int. J. Artif. Intell. Educ. **8**(1), 30–43 (1997)

3. du Boulay, B.: Escape from the Skinner Box: The case for contemporary intelligent learning environments. Br. J. Edu. Technol. **50**(6), 2902–2919 (2019). https://doi.org/10.1111/bjet.12860

4. Graesser, A.C., Person, N.K.: Question asking during tutoring. Am. Educ. Res. J. **31**(1), 104–137 (1994)

5. McNamara, D.S., Graesser, A.C., McCarthy, P.M., Cai, Z.: Automated evaluation of text and discourse with Coh-Metrix. Cambridge University Press (2014)

6. Rodrigo, M.M.T., Baker, R.S.J.: Coarse-grained detection of student frustration in an introductory programming course. In: Proceedings of the Fifth International Workshop on Computing Education Research, pp. 75–80 (2009)

7. Selwyn, N.: Should Robots Replace Teachers?: AI and the Future of Education (Digital Futures). Polity (2019)

8. Tuomi, I.: 'The Impact of Artificial Intelligence on Learning, Teaching, and Education', European Union: JRC, Seville (2018). Online. Available: https://publications.jrc.ec.eur opa.eu/repository/bitstream/JRC113226/jrc113226_jrcb4_the_impact_of_artificial_intelli gence_on_learning_final_2.pdf

9. Williamson, B., Eynon, R.: Historical threads, missing links, and future directions in AI in education. Taylor & Francis (2020)

10. Holmes, W., Persson, J., Chounta, I.-A., Wasson, B., Dimitrova, V.: Artificial Intelligence and Education a Critical View Through the Lens of Human Rights, Democracy and the Rule of Law', Council of Europe, Strasbourg, France (2022). Accessed: Oct. 20, 2022. Online. Available: https://rm.coe.int/artificial-intelligence-and-education-a-critical-view-thr ough-the-lens/1680a886bd

Designing, Building and Evaluating Intelligent Psychomotor AIED Systems (IPAIEDS@AIED2023)

Olga C. Santos$^{(\boxtimes)}$ (ID), Miguel Portaz (ID), Alberto Casas-Ortiz (ID),
Jon Echeverria (ID), and Luis F. Perez-Villegas

Artificial Intelligence Department, Computer Science School, UNED,
Calle Juan del Rosal, 16, 28040 Madrid, Spain
{ocsantos,mportaz,alberto.casasortiz,jecheverria,lfpvillegas}@dia.uned.es

Abstract. Psychomotor learning is an emerging research direction in the AIED (Artificial Intelligence in Education) field. This topic was introduced in the AIED research agenda back in 2016 in a contribution at the International Journal of AIED, where the SMDD (Sensing-Modelling-Designing-Delivering) process model to develop AIED psychomotor systems was introduced. Recently, a systematic review of the state of the art on this topic has also been published in the novel Handbook of AIED. In this context, the aim of the IPAIEDS tutorial is to motivate the AIED community to research on intelligent psychomotor systems and give tools to design, build and evaluate this kind of systems.

Keywords: psychomotor learning · motor skills · human movement computing

1 Background

Psychomotor learning involves the integration of mental and muscular activity with the purpose of learning a motor skill. Many types of psychomotor skills can be learned, such as playing musical instruments, dancing, driving, practicing martial arts, performing a medical surgery, or communicating with sign language. Each one has a different set of unique characteristics that can make the learning process even more complex. To define and categorize the learning process of psychomotor activities, several psychomotor taxonomies have been proposed (Dave (1970); Ferris and Aziz (2005); Harrow (1972); Simpson (1972); Thomas (2004)). These taxonomies are defined in terms of progressive levels of performance during the learning process, going from observation to the mastery of motor skills, as discussed elsewhere (Santos (2016a)). In this context, it is expected that AIED systems can be useful to enhance the performance of motor skills in a faster and safer way for learners and instructors.

O. C. Santos, M. Portaz, A. Casas-Ortiz, J. Echeverria, L. F. Perez-Villegas—Contributing authors.

To build procedural learning environments for personalized learning of motor skills the process model SMDD (Sensing-Modelling-Designing-Delivering) has been proposed (Santos (2016b)). This process model guides the flow of information about the movements performed when using an intelligent psychomotor AIED system along four interconnected phases: 1) **Sensing** the learner's corporal movement as specific skills are acquired within the context in which this movement takes place; 2) **Modelling** the physical interactions, which allows comparing the learner's movements against pre-existing templates of an accurate movement (e.g., a template of how an expert would carry out the movement); 3) **Designing** the feedback to be provided to the learner (i.e., what kind of support and corrections are needed, and when and how to provide them); and 4) **Delivering** the feedback in an effective non-intrusive way to advise the learner on how the body and limbs should move to achieve the motor learning goal.

To identify the situation of the research of psychomotor learning in the AIED field, we have carried out two systematic reviews (Santos (2019), Casas-Ortiz et al. (2023)). The first one show that the field was still very unmature, but most recent one resulted in the identification of 12 AIED psychomotor systems. Nonetheless, we have just ran the queries of that systematic review again obtaining a couple of new results. This shows the field is in continuous evolution and the tutorial can be a good opportunity to present this research and updates.

In our contribution to the Handbook of AIED we present KSAS and KUMITRON psychomotor systems as case studies. These two systems are in fact included as showcases in the AIED website (https://iaied.org/showcase). The purpose of KSAS is to assist during the learning of the order in which a set of movements (known in martial arts as katas, forms, or sets) is performed. KSAS was developed as a mobile application and presented as interactive event at AIED 2021 (Casas-Ortiz and Santos (2021)). Based on this experience, a new system called KLS that uses Extended Reality (a term that includes Virtual Reality, Augmented Reality and Mixed Reality) is being developed to make the psychomotor learning experience more engaging due to the immersive capabilities of these technologies.

In turn, the purpose of KUMITRON is to teach learners how to anticipate the opponent's movement during a martial arts combat (known as kumite) to train the combat strategy. It has been presented at AIED 2021 (Echeverria and Santos (2021b)) and also at MAIED workshop (Echeverria and Santos (2021a)). Since kumites involve two participants, collaborative learning in psychomotor activities needs to be further explored, and this is precisely addressed in (Echeverria and Santos (2023)).

Initially, we have selected martial arts for our research because it encompasses many of the characteristics common to other psychomotor activities like the management of strength and speed while executing the movements, visuomotor coordination of different parts of the body to respond to stimuli, participation of different agents during the learning like opponents or instructor, improvisation and anticipation against stimuli or even the use of tools accompanying the movement. Moreover, martial arts can also support the learning of other subjects as Physics following a kinestesic approach also called embodied learning, as in the Phy+Aik framework (Santos and Corbí (2019)).

However, we are also exploring other psychomotor domains, in particular, we have started to build a psychomotor system to recommend the physical activities and movements to perform when training in basketball, either to improve the technique, to recover from an injury or even to keep active when getting older. This system is called iBAID (intelligent Basket AID). It is being developed following a human-centric approach, hybrid artificial intelligence and the ideas discussed in (Portaz and Santos (2022)). Within this context, Portaz et al. (2023) disclose the iBAID system architecture taking into account the principles of ethics and governance of artificial intelligence according to the aforementioned paradigms, including i) privacy, ii) accountability, iii) security, iv) transparency and explainability, v) fairness and non-discrimination, vi) human control of technology, professional, vii) responsibility, and viii) promotion of human values.

Moreover, sign language communication is also being explored, where hand movements are involved. In particular, so far we have focused on segmenting each of the signs (Perez-Villegas et al. (2022)), which can be used to build datasets that can feed the development of a psychomotor system to learn sign language.

2   Relevance and Importance to the AIED Community

AIED research has traditionally focused on cognitive/meta-cognitive and affective skills. However, in the last years some AIED researchers have suggested the need to take into account the psychomotor domain in the AIED field, thus addressing the three domains defined in Bloom's taxonomy of learning objectives (Bloom et al. (1956)).

The IJAED paper in the Special Issue of the 25th anniversary (Santos (2016b)) was the first contribution that explicitly called for an expansion of AIED research into motor skill learning and asked to revisit AIED's roots and to highlight the powerful potential marriage between AIED and existing (non-adaptive) systems for motor skill learning.

This opened the path to other AIED resarchers who have also called the attention to carry out research in the psychomotor domain, such as the workshop on Authoring and Tutoring Methods for Diverse Task Domains: Psychomotor, Mobile, and Medical at AIED 2018 (https://easychair.org/cfp/AIED_WKSP_Psychol). Another related workshop is the workshop on Multimodal Artificial Intelligence in Education (MAIEd) that was run in AIED 2021 (https://maied.edutec.science/) addressing two complementary approaches: i) extending computer-based learning activities with physical sensors for tracking learners' behaviour, and ii) tracing learning activities that require levels of physical coordination.

Furthermore, it should also be noted that for the first time in 2023, the Call for Papers for AIED (https://www.aied2023.org/cfp.html) includes explicitly the following items in the lines of research and application proposed for the contributions: "representing and modelling psychomotor learning" and "learning of motor skills".

Finally, and as aforementioned, the proposal of this tutorial coincides also with the publication of the Handbook of Artificial Intelligence in Education by professors Ben du Boulay, Tanja Mitrovic and Kalina Yacef (https://www.elgaronline.com/edcollbook/book/9781800375413/9781800375413.xml). This Handbook is aimed to present the most important topics and future trends of the AIED field, and whose Chap. 18 (authored by some of the organizers of this tutorial) focuses on intelligent systems for psychomotor learning, presenting, as mentioned before, a systematic review and two cases of study.

3   Format and Activities

The tutorial will start with a theoretical introduction of the field. In particular, we will present the motivation for the psychomotor research, the state of the art of the field and the SMDD process model to design intelligent psychomotor AIED systems.

After that, we will provide practical examples and exercises to the participants showing different proceesing approaches along the SMDD cycle. We will prepare these practical activities in Google Colab notebooks so that participants can try them during the tutorial to learn how to build a psychomotor system following the SMDD phases. We will cover both data gathering from inertial signals and video signals for the sensing stage. These activities could be done individually or in pairs/groups.

We also plan to focus on evaluation approaches for psychomotor systems. For this, we will show participants some of the intelligent AIED psychomotor systems that we are developing to learn some motor skills (e.g., KSAS, KLS, KUMITRON and iBAID). We then will show the information collected by the systems and how it can be used to measure to the learning progress.

Further details of the tutorial as well as the materials offered will be available from the tutorial website: (https://blogs.uned.es/phyum/publications/ipaieds23/).

4   Expected Target Audience and Participation

We think the tutorial can attract two types of participants. On the one hand, junior AIED researchers that want to take advantage of the new sensing and processing capabilities available in many devices to develop intelligent systems for their doctoral research.

On the other hand, we also expect some senior AIED researchers that want to get ideas to supervise new research projects on psychomotor learning.

In order to promote and support research in the psychomotor field we plan to make a proposal on psychomotor AIED systems for a special issue in the International Journal of AIED. Although the call for papers of the special issue will be open, all participants of the tutorial will be invited to contribute to it.

Acknowledgments. This tutorial is part of the project "HUMANAID-Sens: HUman-centered Assisted Intelligent Dynamic systems with SENSing technologies (TED2021-129485B-C41)" funded by MCIN/AEI/10.13039/501100011033 and the European Union "NextGenerationEU"/PRTR.

References

Bloom, B., Engelhart, M., Furst, E., et al.: Taxonomy of educational objectives: the classification of educational goals, in Handbook I: Cognitive Domain, vol. 3. Longman, New York (1956)

Casas-Ortiz, A., Santos, O.C.: KSAS: An AI application to learn martial arts movements in on-line settings. In: Interactive Events at AIED21 (Artificial Intelligence in Education), pp. 1–4 (2021). https://aied2021.science.uu.nl/file/KSAS-An-AI-Application-to-learn-Martial-Arts-Movements-in-on-line-Settings.pdf

Casas-Ortiz, A., Echeverria, J., Santos, O.C.: Chapter 18. Intelligent systems for psychomotor learning: a systematic review and two cases of study. In: du Boulay, B., Mitrovic, A., Yacef, K., (eds.) Handbook of Artificial Intelligence in Education, chap 18, p 390–418. Edward Elgar Publishing (2023)

Dave, R.H.: Psychomotor levels. In: Armstrong, R.J., (ed.) Developing and Writing Educational Objectives, pp. 33–34. Educational Innovators Press, Tucson, AZ (1970)

Echeverria, J., Santos, O.C.: KUMITRON: A Multimodal Psychomotor Intelligent Learning System to Provide Personalized Support when Training Karate Combats (2021a). https://aied2021.science.uu.nl/file/Kumitron.pdf

Echeverria, J., Santos, O.C.: KUMITRON: learning in Pairs Karate related skills with artificial intelligence support. In: Interactive Events at AIED21 (Artificial Intelligence in Education), Feb 2023 (2021b). https://aied2021.science.uu.nl/file/Kumitron.pdf

Echeverria, J., Santos, O.C.: Towards analyzing psychomotor group activity for collaborative teaching using neural networks. In: Proceedings of the International Conference on Artificial Intelligence in Education (AIED 2023) (2023)

Ferris, T., Aziz, S.: A psychomotor skills extension to bloom's taxonomy of education objectives for engineering education. In: Exploring Innovation in Education and Research (ICEER- 2005), pp. 1–5 (2005)

Harrow, A.: A Taxonomy of the Psychomotor Domain: A Guide for Developing Behavioral Objectives. David McKay Company Inc., New York (1972)

Perez-Villegas, L.F., Valladares, S. (supervisor), Santos, O.C. (supervisor): Sign Language Segmentation Using a Transformer-based Approach. Master Thesis. Master in Research in Artificial Intelligence, UNED (2022)

Portaz, M., Santos, O.C.: Towards personalised learning of psychomotor skills with data mining. In: Doctoral Consortium at EDM22 (Educational Data Mining) (2022). https://educationaldatamining.org/edm2022/proceedings/2022.EDM-doctoral-consortium.108/index.html

Portaz, M., Manjarrés, A., Santos, O.C.: Towards human-centric psychomotor recommender systems. In: Adjunct Proceedings of the 31st ACM Conference on User Modeling, Adaptation and Personalization (UMAP 2023 Adjunct) (2023). https://doi.org/10.1145/3563359.3596993

Santos, O.C.: Beyond Cognitive and Affective Issues: Designing Smart Learning Environments for Psychomotor Personalized Learning, Springer International Publishing, pp 1–24 (2016a). https://doi.org/10.1007/978-3-319-17727-4_8-1

Santos, O.C.: Training the body: the potential of aied to support personalized motor skills learning. Int. J. Artif. Intell. Educ. **26**(2), 730–755 (2016). https://doi.org/10.1007/s40593-016-0103-2

Santos, O.C.: Artificial intelligence in psychomotor learning: modeling human motion from inertial sensor data. Int. J. Artifi. Intell. Tools **28**(04), 1940,006

Santos, O.C., Corbí, A.: Can Aikido help with the comprehension of physics? a first step towards the design of intelligent psychomotor systems for steam kinesthetic learning scenarios. IEEE Access **7**, 176,458–176,469 (2019)

Simpson, E.: The classification of educational objectives in the psychomotor domain: The psychomotor domain, vol. 3. Gryphon House, Washington, DC (1972)

Thomas, K.: Learning Taxonomies in the Cognitive, Affective, and Psychomotor Domains (2004). http://www.rockymountainalchemy.com/whitePapers/rma-wp-learning-taxonomies.pdf

Intelligent Textbooks: The Fifth International Workshop

Sergey Sosnovsky[1]([✉]) [iD], Peter Brusilovsky[2] [iD], and Andrew Lan[3] [iD]

[1] Utrecht University, Princetonplein 5, 3584 CC Utrecht, The Netherlands
s.a.sosnovsky@uu.nl
[2] University of Pittsburgh, 135 North Bellefield Avenue, Pittsburgh, PA 15260, USA
peterb@pitt.edu
[3] University of Massachusetts Amherst, Amherst, MA 01003, USA
andrewlan@cs.umass.edu

Abstract. Textbooks have evolved over the last several decades in many aspects. Most textbooks can be accessed online, many of them freely. They often come with libraries of supplementary educational resources or online educational services built on top of them. As a result of these enrichments, new research challenges and opportunities emerge that call for the application of AIED methods to enhance digital textbooks and learners' interaction with them. Therefore, we ask: How can we use intelligent and adaptive technologies to facilitate the access to digital textbooks and improve the learning process? What new insights about knowledge and learning can be extracted from textbook content and data-mined from the logs of students interacting with it? How can these insights be leveraged to develop improved intelligent texts? How can we leverage new language technology to manage and augment textbooks? The Fifth International Workshop on Intelligent Textbooks features research contributions addressing these and other research questions related to intelligent textbooks. It brings together researchers working on different aspects of learning technologies to establish intelligent textbooks as a new, interdisciplinary research field.

Keywords: intelligent textbooks · digital and online textbooks · open educational resources (OER) · modelling and representation of textbook content · assessment generation · adaptive presentation and navigation · content curation and enrichment

1 Introduction

Textbooks remain one of the main methods of instruction, but – just like other educational tools – they have been evolving over the last several decades in many aspects (how they are created, published, formatted, accessed, and maintained). Most textbooks these days have digital versions and can be accessed online. Plenty of textbooks (and similar instructional texts, such as tutorials) are freely available as open educational resources (OERs). Many commercial textbooks come with libraries of supplementary educational resources or even distributed as parts of online educational services built on top of them. The transition of textbooks from printed copies to digital and online formats has facilitated numerous attempts to enrich them with various kinds of interactive functionalities including search and annotation, interactive content modules, automated assessments, chatbots and question answering, etc.

As a result of these enrichments, new research challenges and opportunities emerge that call for the application of artificial intelligence (AI) methods to enhance digital textbooks and learners' interaction with them. There are many research questions associated with this new area of research; examples include:

- How can one facilitate the access to textbooks and improve the reading process?
- How can one process textbook content to infer knowledge underlying the text and use it to improve learning support?
- How can one process increasingly more detailed logs of students interacting with digital textbooks and extract insights on learning?
- How can one find and retrieve relevant content "in the wild", i.e., on the web, that can enrich the textbooks?
- How can one leverage advanced language technology, such as chatbots, to make textbooks more interactive?
- How can one better understand both textbooks and student behaviors as they learn within the textbook and create personalized learner experiences?

The Fifth International Workshop on Intelligent Textbooks invited research contributions addressing these and other research questions related to the idea of intelligent textbooks. While the pioneer work on various kinds of intelligent textbook technologies has already begun, research in this area is still rare and spread over several different fields, including AI, human-computer interaction, natural language processing, information retrieval, intelligent tutoring systems, educational data mining, and user modeling. This workshop brings together researchers working on different aspects of intelligent textbook technologies in these fields and beyond to establish intelligent textbooks as a new, interdisciplinary research field.

2   Workshop Background and Topics

This workshop will build upon the success of the four previous workshops on Intelligent Textbooks that we organized in conjunction with AIED'2019 [1], AIED'2020 [2], AIED'2021 [3], and AIED'2022 [4]. Altogether, the four previous workshops featured 45 papers and brought together more than a hundred researchers exploring various aspects of intelligent textbooks. An analysis of topics presented at these workshops [5] revealed seven prominent areas (Table 1), which to some extent represent prospects of researchers coming from different fields and applying knowledge and approaches from these fields to the research on intelligent textbooks [6].

Table 1. Categories of intelligent textbook papers over the years.

Topic/Year	2019	2020	2021	2022	Special Issue
Intelligent interfaces	5	2	1	0	0
Smart content	1	1	2	2	0
Knowledge extraction	1	2	0	1	1
Learning content construction	0	1	3	3	3
Intelligent textbook generation	1	2	0	1	1
Interaction mining and crowdsourcing	1	1	3	1	4
Domain-focused textbooks and prototypes	1	0	2	0	0
Miscellaneous	4	0	2	1	0

Following success of the first four workshops, we proposed a special issue of the Journal of Artificial Intelligence in Education on Intelligent Textbooks. The proposal has been approved. The Call for Papers brought 19 submission proposals out of which guest editors approved 9 for further consideration. Currently, the submissions are undergoing a peer review. The categorisation of submissions according to the topics identified in our earlier analysis is presented in Table 1 as well. It is important to notice that (1) these numbers are projections, as the submissions are still under review; (2) papers submitted to a journal typically present more comprehensive projects that can cover multiple topics and belong to several categories at once. Nevertheless, the data shows, that the key research directions keep attracting considerable interest of researchers year by year, however, the focus of attention gradually shifts following the development of the field. We could acknowledge the gradual increase of interest in research on learner interaction mining, which is promoted by the increased number of available digital textbooks and rapid accumulation of data [7]. This topic is explored in grater details by a related series of workshops on the Analysis of Reading Behavior organized at Learning Analytics and Knowledge Conference series[1]. Similarly, we can observe a gradually increasing research in learning content construction based on the material of intelligent textbooks [8,9], which corresponds to the rapid progress in the natural language processing techniques. We expect the shift of focus will continue as new AI technologies become available. In particular, it is natural to expect the rise of research on interactive question-answering and other forms of dialogue in context of digital textbooks. While pioneer research on this topic started more than 10 years ago [10], the capabilities of large language models like ChatGPT remarkably expanded research opportunities in this area.

To reflect the recurrent and emergent research directions, the topics discussed at the 5th workshop include but are not limited to:

1. Modelling and representation of textbooks: examining the prerequisite and semantic structure of textbooks to enhance their readability; b) Analysis and

---

[1] https://sites.google.com/view/lak23datachallenge/home.

mining of textbook usage logs: analyzing the patterns of learners' use of textbooks to obtain insights on learning and the pedagogical value of textbook content;

2. Collaborative technologies: building and deploying social components of digital textbooks that enable learners to interact with not only content but other learners;

3. Generation, manipulation, and presentation: exploring and testing different formats and forms of textbook content to find the most effective means of presenting different knowledge;

4. Assessment and personalization: developing methods that can generate assessments and enhance textbooks with adaptive support to meet the needs of every learner using the textbook;

5. Content curation and enrichment: sorting through external resources on the web and finding the relevant resources to augment the textbook and provide additional information for learners.

6. Dialog-assisted textbooks: leveraging chatbot technology to support and facilitate learner-textbook dialogues.

3 Workhop Organizers

3.1 Chairs

Sergey Sosnovsky is an Associate Professor of Software Technology for Learning and Teaching at the Department of Information and Computing Sciences, Utrecht University. His research interests include various aspects of designing, developing and evaluating adaptive educational systems and personalized information systems in general. Dr. Sosnovsky holds a PhD degree in Information Sciences from University of Pittsburgh (Pittsburgh, PA, USA). Before joining Utrecht University, Dr. Sosnovsky worked as the head of the e-Learning lab at German Center for Artificial Intelligence (DFKI) and as a senior researcher at Saarland University (Saarbrücken, Germany).

Peter Brusilovsky is a Professor of Information Science and Intelligent Systems at the University of Pittsburgh, where he directs Personalized Adaptive Web Systems (PAWS) lab. Peter Brusilovsky has been working in the field of adaptive educational systems, user modeling, and intelligent user interfaces for more than 30 years. He published numerous papers and edited several books on adaptive hypermedia, adaptive educational systems, user modeling, and the adaptive Web. Peter is the past Editor-in-Chief of IEEE Transactions on Learning Technologies and a board member of several journals including User Modeling and User Adapted Interaction and ACM Transactions on Interactive Intelligent Systems. Peter has been exploring the topic of intelligent textbooks for over 20 years. Together with G. Weber he developed one of the first online intelligent textbooks ELM-ART [11], which received the 1998 European Academic Software award.

Andrew S. Lan is an Assistant Professor in the College of Information and Computer Sciences, University of Massachusetts Amherst. His research focuses on the development of human-in-the-loop machine learning methods to enable scalable, effective, and fail-safe personalized learning in education, by collecting and analyzing massive and multi-modal learner and content data. Prior to joining UMass, Andrew was a postdoctoral research associate in the EDGE Lab at Princeton University. He received his M.S. and Ph.D. degrees in Electrical and Computer Engineering from Rice University in 2014 and 2016, respectively. He has also co-organized a series of workshops on machine learning for education; see http://ml4ed.cc/ for details.

3.2   Program Committee

- Isaac Alpizar Chacon, Utrecht University
- Debshila Basu Mallick OpenStax, Rice University
- Paulo Carvalho, Carnegie Mellon University
- Vinay Chaudhri, SRI International
- Brendan Flanagan, Kyoto University
- Reva Freedman, Northern Illinois University
- Benny Johnson, VitalSource Technologies
- Roger Nkambou, Université du Québec à Montréal
- Noboru Matsuda, North Carolina State University
- Andrew Olney, University of Memphis
- Philip Pavlic Jr., University of Memphis
- Cliff Shaffer, Virginia Tech
- Khushboo Thaker, University of Pittsburgh
- Ilaria Torre, University of Genoa

4   Workshop Format and Materials

We plan a full-day hybrid workshop that will combine regular presentations with more interactive formats such as a demonstration of working prototypes. The workshop program and other information is available on the workshop website (https://intextbooks.science.uu.nl/workshop2023/) where we publish the call for contributions, the final program, and the first version of the proceedings. The final version of the proceedings will be published at http://ceur-ws.org/. Slides of workshop presentations will be made available at https://www.slideshare.net/.

References

1. Sosnovsky, S., et al., (eds.) Proceedings of the First Workshop on Intelligent Textbooks at 20th International Conference on Artificial Intelligence in Education (AIED 2019). CEUR Workshop Proceedings, vol. 2384, p, 145. CEUR, Chicago, USA (2019)

2. Sosnovsky, S., et al., (eds.) Proceedings of the 2nd Workshop on Intelligent Textbooks at 21st International Conference on Artificial Intelligence in Education (AIED 2020). CEUR Workshop Proceedings, vol. 2674. CEUR, Chicago, USA (2020)
3. Sosnovsky, S., et al., (eds.) Proceedings of the 3rd Workshop on Intelligent Textbooks at 22nd International Conference on Artificial Intelligence in Education (AIED 2021). CEUR Workshop Proceedings, vol. 2895. CEUR (2021)
4. Sosnovsky, S., Brusilovsky, P., Lan, A., (eds.) Proceedings of the 4th Workshop on Intelligent Textbooks at 23rd International Conference on Artificial Intelligence in Education (AIED 2022). CEUR Workshop Proceedings, vol. 3192. CEUR (2022)
5. Sosnovsky, S., Brusilovsky, P., Lan, A.: Intelligent textbooks: themes and topics. In: 23rd International Conference on Artificial Intelligence in Education, AIED 2022, Part 2, Durham, UK. Springer (2022). https://doi.org/10.1007/978-3-031-11647-6_19
6. Brusilovsky, P., Sosnovsky, S., Thaker, K.: The return of intelligent textbooks. AI Mag. 43(3) (2022)
7. Lu, O.H.T., et al.: A quality data set for data challenge: featuring 160 students' learning behaviors and learning strategies in a programming course. In: the 30th International Conference on Computers in Education. Asia-Pacific Society for Computers in Education (2022)
8. Pavlik Jr., P.I., et al.: The mobile fact and concept textbook system (MoFaCTS). In: Second Workshop on Intelligent Textbooks at 21st International Conference on Artificial Intelligence in Education (AIED 2020). CEUR (2020)
9. Van Campenhout, R., Hubertz, M., Johnson, B.G.: Evaluating AI-generated questions: a mixed-methods analysis using question data and student perceptions. In: 23rd International Conference on Artificial Intelligence in Education, AIED 2022, Part 1, Durham, UK. Springer (2022). https://doi.org/10.1007/978-3-031-11644-5_28
10. Chaudhri, V.K., et al.: Inquire biology: a textbook that answers questions. AI Mag. 34(3), 55–72 (2013)
11. Weber, G., Brusilovsky, P.: ELM-ART: An adaptive versatile system for Web-based instruction. Int. J. Artif. Intell. Educ. 12(4), 351–384 (2001)

AI to Support Guided Experiential Learning

Benjamin Goldberg[1](✉) ⓘ and Robby Robson[2] ⓘ

[1] U.S. Army DEVCOM Soldier Center, 12423 Research Parkway, Orlando, FL 32826, USA
benjamin.s.goldberg.civ@army.mil
[2] Eduworks Corporation, 400 SW 4th Street STE 110, Corvallis, OR 97333, USA
robby.robson@eduworks.com

Abstract. Kolb defined experiential learning as "the process whereby knowledge is created through the transformation of experience" [1], often expressed as learning-by-doing. In the context of this workshop, Guided Experiential Learning (GEL) is a pedagogical framework for learning-by-doing that emphasizes longitudinal skill development and proficiency gained through focused, repetitive practice under real world-like conditions [2]. GEL often requires scaffolding psychomotor, affective, and cognitive skill acquisition across multi-modal learning experiences, including games and simulations delivered using virtual, augmented, and mixed reality based applications. In addition, the skills targeted using a GEL framework are generally developed over time via episodic events with controlled conditions dictated by learner states and learning theory [1, 3].

The complexity of designing and assessing experiential learning in the technology-enabled, data-rich environments in which GEL takes place make GEL an ideal candidate for using AI. AI can assist in the design, delivery, and evaluation of experiential events that contribute to longer-term skill and proficiency objectives and to optimize learning. This workshop addresses the research challenges involved in applying AI to GEL. These include multi-modal data strategies, in which data comes from physical, virtual, and mixed-reality training systems; AI models for estimating and predicting skill acquisition and competency; designing learning experiences and assessments to produce optimal data for AI-based models and that provide data under variations in condition and complexity; and applications of AI to instructional support, feedback and coaching [4].

Keywords: Guided Experiential Learning · Competencies · GIFT · AI

1 Introduction

AI is revolutionizing the way we learn, work, and acquire new skills. With its ability to process and analyze vast amounts of data, automatically generate content, and provide intelligent tutoring support, AI is helping educators and trainers develop and deliver personalized, effective, and engaging learning experiences. This workshop explores how AI can be used to influence and optimize Guided Experiential Learning (GEL), with a focus on connecting researchers across our community and establishing a forum to monitor progress and to drive collaboration and information sharing.

© The Author(s), under exclusive license to Springer Nature Switzerland AG 2023
N. Wang et al. (Eds.): AIED 2023, CCIS 1831, pp. 103–108, 2023.
https://doi.org/10.1007/978-3-031-36336-8_16

Specifically, we aim to explore how AI can be applied to GEL that involves acquiring knowledge, skill, and behaviors (i.e., competencies) across a combination of psychomotor, affective, and cognitive learning domains. This requires tools, methods and standards that apply to classroom instruction, serious games, simulators, mixed reality, and live exercises. The goal is to drive a "learn by doing" acquisition curve, which is the defining characteristic of experiential learning, and to do in a way that guides learners to practice targeted skills and sub-skills under a variety of conditions with objective assessments and intelligent coaching functions. This approach can establish the automated performance responses exhibited by experts and aligns to principles of deliberate practice and skill acquisition theory [3].

In music, for example, GEL might involve practicing scales, practicing a piece written for the sole purpose of applying those scales, and performing that piece in front of an audience of fellow students. In our context, training might take place in mixed reality where the output of the instrument and movements of the student are tracked, and a live audience is replaced by a virtual one in an immersive environment. This setup produces data that can be analyzed to determine performance and enables the virtual audience to be manipulated to simulate conditions such as noisy distractions, a disengaged or intensely engaged audience, or a medical crisis that requires the musician to stop completely and start the performance over again.

2 Six Steps in GEL

GEL follows the tenets of intelligent tutoring and competency-based learning. It requires operationalized models at different levels of learning interaction, where tools and methods are combined to (1) optimize each learning event based on a targeted set of objectives, (2) combine learning events to optimize longitudinal skill progression across competency development goals, and (3) maintain competency over time by modeling skill decay and personalizing training schedules. While each learning interaction takes place in a single environment, GEL assumes that an ecosystem of resources will be combined to meet overall learning and development objectives.

From our perspective, an important question for the AIED community is how technology-mediated GEL is best applied to support competency development and enable a group of learners to progress from novice to expert. Many AI tools and applications are relevant and can be used to implement data-driven processes with explicit consideration given to the design, authoring, and configuration requirements of a GEL event. With these objectives in mind, it is necessary to:

1. Collect data from a heterogenous collection of systems,
2. Interpret the data in terms of performance,
3. Model and track skill acquisition,
4. Identify which skills and conditions to target next,
5. Design scenarios that practice those skills and conditions, and
6. Provide coaching and feedback to learners.

In the traditional domains addressed by AI in Education, such as secondary school mathematics or language learning, these steps are relatively well understood. Instruction is usually delivered in classrooms or via traditional eLearning. The domains have been broken down into knowledge components (KCs) by many researchers, and these KCs are assessed via traditional quizzes and tests. Intelligent tutors implement a variety of methods for assessing and tracking knowledge, recommending the next KC or activity, and providing including coaching and feedback [5].

Complexities quickly arise when we move to GEL. The data are no longer straightforward to collect or interpret. In the music example, for instance, videos of a student and digital data from an instrument must be translated into data about how a student interacted with their instrument in the context of a specified task. Different systems may produce different types of data, all of which must be translated into meaningful performance evaluations that are interpretable with other performance measurement sources. Skill states cannot be modeled without first constructing an appropriate skills framework. Such a framework might exist already, but it may be necessary to engage subject matter experts and practitioners to create one. The models used to estimate skills may need to take relationships among skills and performance under different conditions into account, and once the models are in place, they must be trained to identify the optimal set of skills to address given a learner's current state and desired state. Finally, GEL scenarios must be designed, which could involve constructing scenarios in games, mixed reality, and field environments.

2.1 An Archetypical Example

An archetypical example of GEL is given by US Army training and the Synthetic Training Environment Experiential Learning for Readiness (STEEL-R) project [6]. STEEL-R supports the Army's investment in modernizing its core simulation-based training assets by leveraging advancements in gaming technology and immersive mixed reality to provide soldiers and civilians meaningful learning opportunities in realistic synthetic scenarios. STEEL-R involves soldiers, teams of soldiers, and teams of teams (e.g., squads and platoons) engaged in training drills that are defined by Army doctrine. The drills involve several interdependent tasks, each of which has its own performance requirements relative to cognitive, psychomotor and affective competencies. Training sessions are scheduled to provide hands-on learning opportunities and can take place across game-based environments, mixed reality, and eventually in live field exercises. These resources and environments are combined across a training calendar with an established culminating live event used assess readiness at the individual and team level.

Data from all training environments are collected by the Generalized Intelligent Framework for Tutoring (GIFT) [7] and used by GIFT to evaluate performance on tasks. Results and the conditions under which the tasks were evaluated are reported in a standardized format to a system that translates performance evaluations into assertions about skills in a skills framework. A model that accumulates these assertions over time is used to estimate and report skills proficiency. A human-in-the-loop examines these reports to determine what skills should be addressed next, under what conditions they should be trained, and what level of competence is targeted. This is fed into a component that looks for existing suitable scenarios based on current skill levels of the soldiers and teams and the skills and conditions that have been specified. If no such scenario can be found, a tool is available to create new ones.

2.2 Roles of Standards

STEEL-R's implementation is grounded in data standards developed for more general types of learning technologies. The underlying vision is using common learning engineering practices and standards [8] to create interoperable learning ecosystems [9]. GEL is dependent on system interoperability, with the eXperience Application Programming Interface (xAPI) [10] serving as an initial adopted standard, used to establish a controlled syntax for reporting performance outcomes. From an experiential learner modeling view, xAPI statements must be carefully crafted to capture relevant context to help drive competency modeling techniques. This involves establishing metadata standards for defining performance, with careful attention paid to difficulty, stress and other variables used to build assertions about an individual's or team's ability.

3 Uses of AI in GEL

Recent years have seen an increased interest in applications of AI to adult learning and workforce development, spurred by rapid changes in technology, the workplace environment, and an interest in increasing diversity and inclusion. A workshop on the role of AI in GEL is relevant to the AIED community because it brings together two crucial components of modern education: AI and hands-on, practical learning experiences. GEL involves active participation, problem-solving, and reflection. These are essential for developing real-world skills and for preparing the workforce of the future, which will require workers to adapt to new technologies, collaborate with others, and solve complex problems. AI can help optimize this form of learning and longitudinal competency development, and the AIED community can directly influence this area. In the following subsections we briefly introduce the high-level themes covered by contributors to the workshop. These highlight research and development topics that drive a foundational element of GEL. For a full breakdown of the research landscape, we highly recommend reading the full proceedings.

3.1 Data Collection

GEL takes place in live environments where learners interact with physical objects, and increasingly in games and simulations supported by virtual, augmented, and mixed reality. Before any AI or machine learning (ML) techniques can be applied, strategies are needed to collect and interpret data across a set of disparate technologies and environments. Multimodal learning analytics is an emerging area in the AIED community [11]. Exciting advancements are being made in the use of video, audio, and sensor data to deliver insights into student learning processes and support learner modeling. To get to AI-enabled GEL, researchers must mature a data strategy to define how sources can be collected and applied across interactive multi-modal environments.

Beyond collection, data management processes are required to support innovative assessment techniques. Based on the subjectivity of assessing competency over time, and the abundance of data available during a GEL event, AI will be necessary to objectively evaluate performance across complex task environments. Evaluating whether a task was completed, or an action taken, can be reduced to a formula. Evaluating whether a musical piece was played well or if a team performed cohesively involves "judgment" that is currently done by experts but is within the reach of modern AI.

3.2 Modeling Skill Acquisition and Competency

Predictive learner analytics is another maturing area of interest in the AIED community. Models that turn performance data into skills estimates have been used in intelligent tutoring systems (ITS) for many years. However, the structure of skills frameworks used in GEL – together with the need to incorporate repetition, skill decay, and conditions – creates a requirement for new types of models and for methods that can train these models with limited data. By observing learner behaviors and outcomes over time and across multiple environments, and by using sources of data that provide evidence of learning, AI can be applied to develop predictive models of learner and team competency state [12], which will help inform next-step pedagogical decisions and provide insights for talent management. Evaluating the quality of a learning event on overall competency acquisition is also a critical function to enable self-optimizing learning environments designed for GEL.

3.3 Experience Design

Assessing the proficiency of an individual or team on a real-world task is no longer a matter of simply giving them a test and scoring it. Instead, it is necessary to observe performance across multiple trials, delivered under multiple conditions. From this perspective, a learning experience must not only be designed to support task execution; it must also be designed to allow for context-specific monitoring and assessment of foundational behaviors, processes, and procedures across a set of tasks, conditions, and standards. In addition, to support GEL, a learner requires several experiential opportunities under variations in condition and complexity, thus creating a content creation and curation challenge. There is some exciting work in this area examining standardized approaches for configuring measurable experience events within a larger experiential learning scenario, providing a foundational data schema to drive adoption of a GEL informed experience design workflow [13].

3.4 Coaching, Feedback and Support

Instructional support, feedback, and coaching are critical for GEL. Determining when, how, and what forms of feedback and support to deliver to facilitate effective learning remain open areas of research [6]. In addition, when considering GEL and longitudinal skill development, assisting learners in defining what they should practice next, and how best to practice it is a useful question that AI can help through analysis and decision support functions. In real-world live GEL, instructors determine what skills and conditions should be targeted. In many ITS, this is done algorithmically by identifying which skills are in the zone of proximal development. The same can be done in instrumented GEL environments. In STEEL-R, for example, targeted skills and conditions are input by an instructor, but the system checks whether a learner is ready to acquire those skills.

4 Promising Areas of Research

As indicated by the breadth of topics in this Workshop, GEL touches on many areas of research, including learner models and state estimation, design and optimization of learner experiences, and standards and technology. All of these are promising areas of research, and the recent generative AI is likely to have a profound impact on all types of learning, including GEL. Generative AI could be used to aid scenario design (one can imagine the prompt *"design five scenarios to practice persuasive two-minute arguments in which a learner presents to a moderately hostile audience of over 100 people"*). Such semi-automation of scenario design has the potential to vastly reduce the cost of implementing GEL but will also require work on data strategies, learner modeling, AI-based optimization, and other areas that are already target areas and will make research in these areas both more valuable and more urgent.

References

1. Kolb, D.A.: The Process of Experiential Learning. Prentice-Hall, Englewood Cliffs, NJ (1984)
2. Hernandez, M., et al.: Enhancing the total learning architecture for experiential learning. In: I/ITSEC (2022)
3. Ericsson, K.A.: Deliberate practice and acquisition of expert performance: a general overview. Acad. Emerg. Med. **15**, 988–994 (2018)
4. Lester, J.C., Spain, R.D., Rowe, J.P., Mott, B.W.: Instructional support, feedback, and coaching in game-based learning. In: Handbook of Game-based Learning, pp. 209–237. MIT Press, Cambridge, MA (2020)
5. Graesser, A.C., Conley, M.W., Olney, A.: Intelligent Tutoring Systems. In: Harris, K.R., Graham, S., Urdan, T., Bus, A.G., Major, S., Lee Swanson, H. (eds.) APA Educational Psychology Handbook, Vol 3: Application to Learning and Teaching., pp. 451–473. American Psychological Association, Washington (2012). https://doi.org/10.1037/13275-018
6. Goldberg, B., Owens, K., Gupton, K., Hellman, K., Robson, R., et al.: Forging Competency and Proficiency through the Synthetic Training Environment with an Experiential Learning for Readiness Strategy. In: I/ITSEC (2021)
7. GIFT Portal: https://www.gifttutoring.org. Last accessed 7 May 2023
8. Goodell, J., Kolodner, J.: Learning Engineering Toolkit: Evidence-Based Practices from the Learning Sciences, Instructional Design, and Beyond. Routledge, New York (2022)
9. Walcutt, J. J., Schatz, S.: Modernizing learning: building the future learning eco-system. (2019). Government Publishing Office, Washington, DC. License: Creative Commons Attribution CC BY 4.0 IGO
10. Experience API: https://adlnet.gov/projects/xapi/. Last accessed 7 May 2023
11. Emerson, A., Cloude, E.B., Azevedo, R., Lester, J.: Multimodal learning analytics for game-based learning. Br. J. Edu. Technol. **51**, 1505–1526 (2020)
12. Robson, R., Hu, X., Robson, W., Graesser, A.C.: Mathematical models to determine competencies. In: Design Recommendations for ITS, vol. 9, pp. 107–112. US Army (2022)
13. Blake-Plock, S., Owens, K., Goodell, J.: The value proposition of GIFT for the field of learning engineering. In: Proceedings of the 11th Annual GIFT Users Symposium (2023)

An Automated Approach to Assist Teachers in Recommending Groups of Students Associated with Collaborative Learning Techniques Using Learning Paths in Virtual Learning Environments

Ilmara Monteverde Martins Ramos[1,2](✉) ⓘ, David Brito Ramos[2] ⓘ,
Bruno Freitas Gadelha[1] ⓘ, and Elaine Harada Teixeira de Oliveira[1] ⓘ

[1] Institute of Computing, Federal University of Amazonas (UFAM), Manaus, AM, Brazil
{ilmaramonteverde,bruno,elaine}@icomp.ufam.edu.br
[2] Federal Institute of Education, Science, and Technology of Amazonas (IFAM), Campus
Parintins, Parintins, AM, Brazil
david.brito@ifam.edu.br

Abstract. This paper presents a doctoral research proposal for an Automated approach to assist teachers in recommending groups of students associated with Collaborative Learning Techniques using Learning Paths in Virtual Learning Environments. The main objective is to create and validate an automated approach to assist professors in recommending groups of students associated with collaborative learning techniques in introductory courses on programming in Virtual Learning Environments. In particular, the approach will be based on student interaction actions with a VLE called CodeBench, in blended classes, through Learning Paths in graph format, suggestions of students clusters resulting from the groups formation, techniques of collaborative learning, with the carrying out of case studies, in basic and higher education classes in teaching programming carried out in Brazil at the Federal University of Amazonas and the Federal Institute of Amazonas. The doctorate began in March 2021, therefore, it will have 28 months of development until the date of the AIED 2023, with completion expected in March 2025.

Keywords: Formation Groups · Collaborative Learning Techniques · Programming Teaching

1 Motivation for Research

In the scholar context, in some cases it is observed that the use of technological resources has some resistance, after all, the proposal requires changes in the teachers' pedagogical practices. It is still a challenge for educators and institutions to develop an evidence-based culture of using data to make instructional decisions and improve instruction [Romero and Ventura 2020]. Universities have increasingly used online environments, given their flexibility, especially for professors and students, to build a collaborative and efficient learning environment [Wella and Tjhin 2017]. Virtual Learning Environments (VLEs) are online classrooms that provide communication, collaboration, administration, and reporting tools [Romero and Ventura 2010]. These systems store a large amount of data from students during their interactions with the environment. Thus, tools are needed to automatically analyze entirely the wealth of educational data that can be used to understand how students learn [Romero and Ventura 2020]. One of the possibilities for using VLEs is to conduct group activities. Educational platforms widely use group work, as it encourages the student's active participation in the learning process and the pursuit of the collective construction of knowledge. In VLEs, there is a frequent need to use Collaborative Learning Techniques (CoLTs). CoLTs are guidelines to be applied in activities that require collaboration, and one or more techniques can be combined to be adopted in activities, making class discussions more collaborative [Barkley, Major and Cross 2014]. However, forming groups in a face-to-face or distance learning course can be an unnatural process [Barkley, Major, and Cross 2014]. In addition, for collaborative learning to be successful, it is important to form groups that can satisfactorily fulfill the objectives of the activity. There are three types of approaches to forming groups [Ounnas, Davis, and Millard 2007]. Random selection: groups are created at random, with no specific criteria defined, generally used to form informal and temporary groups; Self-selection: where students are allowed to choose the group they want to belong to and negotiate with whom they want to work; Teacher-selection: groups are initiated by the teacher, allowing them to create or direct the formation of groups, considering shared interests and social ties. In addition, the work of [Felder and Brent 2001] describes a fourth approach called automatic selection, in which the system automatically creates groups using a based-algorithm method. The first three alternatives have several flaws, such as: random selection can generate disproportionate groups, self-selection requires affinity between people, and selection by teachers presents difficulty in dealing with many students and with complex grouping criteria [Borges et al. 2018]. Therefore, given these approaches, the automatic selection is a viable solution, widely used in Computer-Supported Collaborative Learning (CSCL) environments. To use an automatic grouping technique, it is necessary to characterize the students' behavior [Ratnapala, Ragel and Deegalla 2014] and group them based on similar behavioral characteristics to promote collaborative learning during the use of the VLE [Tang and McCalla 2005]. Such information can be obtained by analyzing the Learning Paths (LP). To suggest groups in VLEs, it is relevant to know what students do in it. Educational data mining allows you to build analytical models that find patterns, behaviors and trends that allow the automatic discovery of hidden information in the data [Romero, Ventura and García 2008]. In this sense, this research will use LP as a resource to suggest groups of learners to teachers to carry out collaborative activities. Based on the work [Ramos I. et al. 2021], this proposal intends to improve it with a focus on group recommendations integrated to CoLTs for introductory programming disciplines. The LP concept adopted will be the one proposed by [Ramos D. et al. 2021], to designate the path taken by the student during their interaction with the VLE, represented in a graph format. Each vertex represents a resource/activity, weighted by the number of student's interactions. The edges are of 3 types: the standard edge represents the LP structure defined by the teacher, the advance edge represents the route that the student took from one resource/activity to another further ahead than the established by the teacher, and finally, the return edges represent the returns made by students to previous resources/activities. This proposal will also select which techniques will be worked on from the CoLTs concepts defined by [Barkley, Major and Cross 2014]. In this way, it is clear how challenging it is for teachers to form groups for collaborative activities, both in distance education, face-to-face or blended, which is the focus of this research. It is necessary to help them, especially when it is essential to establish groups to carry out collaborative activities within the VLE, for this, it is important to establish criteria to carry out the groupings. In this context, teachers must have tools that can help them in this task.

2   Research Question

The problem addressed by this study consists of allowing the teachers, who uses a VLE, to have an alternative approach that can help them in the process of forming groups of students. In general, clusters consider some clustering techniques to generate clusters, such as those described in the works of [Jagadish 2014] and [Yathongchai et al. 2013] and some works, such as [Zakrzewska 2008] and [Ramos I. et al. 2021], provide the approach of groups selected by the teacher. However, it is not enough just to group according to this information. The grouping should also consider the CoLTs that the teacher wants to apply in each activity. However, with the existence of several techniques, teachers may need a lot of time to choose which of the techniques they can use in each collaborative activity. Therefore, the purpose of this research is to create and validate an automated approach to assist teachers in recommending groups of students associated with CoLTs for activities in VLEs. In particular, the approach will be based on the students' LPs using a VLE called CodeBench and on CoLTs. CodeBench is an online judge that supports the process of teaching and learning in programming. In addition to creating a recommendation approach for groups of students integrated into the CoLTs, it will be possible to: (i) choose one or combine the CoLTs, according to the activity to be carried out; (ii) associate the of groupings to the CoLTs; and (iii) identify which student attributes are better related to each CoLTs worked on in this research project. The proposed research aims to answer the following research questions: Is it possible to identify better group formation strategies for the techniques applied to students of introductory programming disciplines? What different combinations of techniques could be used in blended teaching? Does the approach help the teacher in forming groups? Did the formation of groups help to improve the performance (grade) of the students? Despite the focus on blended teaching, it is expected that the approach can contribute to any teaching modality that uses a VLE as an environment for teaching, learning, communication, and interaction between teacher, student, and content.

3   Research Methodology

This section provides a description of the methodology to be applied during the execution of this proposal. The research methodology is based on case study methodologies and design science research. It will be exploratory and applied research, with quantitative and qualitative analyses, with experiments carried out in the form of case studies and data collected from CodeBench and with information collected through questionnaires and interviews. First, it points out the need to obtain knowledge of the state of the art in the literature, and then, a survey of what data will be needed to develop the approach. Next, an observational study will be carried out, and, finally, the application of the approach in a case study. To obtain state-of-the-art knowledge, it is necessary to search the current literature for works that seek to solve the problem addressed by this proposal, or that are similar, that can contribute to an updated theoretical basis and well aligned with this work proposal. In this first moment, bibliographic research will be carried out through a systematic review of the literature on suggestions/recommendations of groups applied in programming teaching and, mainly, to identify if there are already works, in this context, that use CoLTs in their approach. This phase will allow a better understanding of how this proposal can contribute to other works, which gaps can be filled, and where progress is being made in the state of the art. Once identified as the best contribution within the group formation area, there is a need to obtain the input data, select them and process them to compose the implementation of the approach. Thus, it will also be necessary to carry out a field survey, where data from CodeBench will be collected. The data will be explored, analyzed, and organized to find an efficient way to use them. At first, the data will be analyzed to generate a model based on graphs, to identify the students' LP based on previous studies [Ramos D. et. al. 2021, Ramos I. et. al. 2021]. The LP will be analyzed using data mining and machine learning techniques to generate student models based on the information obtained. In the next step, with the LP formed, the necessary attributes will be extracted for the development of the tool that will carry out the formation suggestion of student groups. Then, the groups that best correspond to each type of CoLTs will be selected. The method of grouping the learners will be chosen according to the characteristics identified in the previous step. At the end of this phase, the proposed approach will be able to suggest groups of students for collaborative activities according to the CoLTs chosen by the teacher. The proof of concept, that is, the application of the approach in a real context, will be done through two case studies that will be carried out in basic and higher education classes. The first will be a pilot study, to carry out the necessary refinements, and the second will be more comprehensive. For this, a tool will be developed that will apply the previous process and present the recommendations to the teacher. The tool will be applied in CodeBench. Validation will be carried out with the help of data collection through questionnaires and interviews with teachers and questionnaires with students participating in the research. This last phase will only start after being submitted and approved by the Research Ethics Council. Finally, all research and results will be presented in thesis format. Completed Activities: Bibliographic Research – a survey of the state of the art of group formation in programming teaching, data extraction of LP, implementation of Visualization of LP. Ongoing activities: Field Research - a collection of real data to conduct implementation tests of the approach; Modeling LP – organize student data into attributes to run tests. Planned activities, but not started yet: Bibliographic Research – CoLTs in programming teaching; Selection of CoLTs that will be part of the research; Selection of the apprentices grouping method; Selection of characteristics, of students and groups, better correlated to CoLTs; Development of the tool for practical application of the approach in CodeBench; Pilot case study: testing the tool to make possible adjustments; Case Study: Testing the Tool;

Test data collection; Validation of Results – interviews with professors and application of questionnaires with students and professors; Organization and presentation of results.

4 Expected Contributions

This research intends to recommend effective groups in which the students are satisfied, learning, and getting good grades, that the teacher has one more automatic tool that can help him in the formation of groups in VLEs. Also, no works were found in the literature that recommend the formation of groups of learners integrated to CoLTs in teaching programming. Below we present the expected contributions.

The grouping of students according to their LP in the VLE. The LP provides subsidies to extract the attributes to be used in the clustering algorithms; The identification of the most relevant attributes to help recommend groups; The third contribution desired is the conceptual framework for the automatic detection of students' groups that explores the identified attributes, serving as a basis for future research that wants to improve the investigation on group recommendation in VLEs applied to the CoLTs; The fourth contribution consists of the implementation of three similarity metrics to be used by clustering algorithm, allowing to the professor more than one option to choose the most adequate clusters to be applied for the collaborative activity; Analysis of results acquired with the approach in basic and higher education classes, since, in the state of the art, there are still no approaches used in basic education with this theme; Identify which CoLTs can be used in teaching programming, according to the formation of groups for collaborative activities to be applied by teachers; Developing a tool from the framework, so that the teacher can have an automated alternative to form groups of students according to their interactions within the environment. The initiative of this research aims to improve the teaching-learning process within any modality that uses a VLE and needs groups recommendations for collaborative activities integrated with CoLTs, in introductory programming disciplines. This research intends not only to benefit EaD but any teaching modality that uses VLEs as an environment for teaching, learning, communication, and interaction between teacher, student, and content. These are ours expected contributions to the CSCL community.

Acknowledgment. Federal Institute of Education, Science, and Technology of Amazonas (IFAM) – Campus Parintins. The present work is the result of the Research and Development (R&D) project 001/2020, signed with Federal University of Amazonas and FAEPI, Brazil, which has funding from Samsung, using resources from the Informatics Law for the Western Amazon (Federal Law n° 8.387/1991), and its disclosure is in accordance with article 39 of Decree No. 10.521/ 2020. This study was financed in part by Conselho Nacional de Desenvolvimento Científico e Tecnológico - Brasil - CNPq (Process 308513/2020–7) and Fundação de Amparo à Pesquisa do Estado do Amazonas - FAPEAM (Process 01.02.016301.02770/2021–63).

References

Barkley, E.F., Major, C.H., Cross, K.P.: Collaborative Learning Techniques: A Handbook for College Faculty, 2nd edn. Jossey-Bass (2014)

Borges, S., Mizoguchi, R., Bittencourt, I.I., Isotani, S.: Group formation in cscl: a review of the state of the art. In: Cristea, A.I., Bittencourt, I.I., Lima, F. (eds.) HEFA 2017. CCIS, vol. 832, pp. 71–88. Springer, Cham (2018). https://doi.org/10.1007/978-3-319-97934-2_5

Felder, R.M., Brent, R.: Effective strategies for cooperative learning. J. Cooperation Collab. Coll. Teach. **10**, 69–75 (2001)

Jagadish, D.: Grouping in collaborative e-learning environment based on interaction among students. In: International Conference on Recent Trends in Information Technology. IEEE, Chennai, India (2014)

Ounnas, A., Davis, H. C., Millard, D. E.: Towards semantic group formation. In: IEEE International Conference on Advanced Learning Technologies (ICALT). IEEE, Niigata, Japan (2007)

Ramos, D.B., Ramos, I.M.M., Gasparini, I., Oliveira, E.H.T.: A new learning path model for e-learning systems. Int. J. Distance Educ. Technol, (IJDET) **19**(2), 34–54 (2021). https://doi.org/10.4018/IJDET.20210401.oa2

Ramos, I., Ramos, D., Gadelha, B., de Oliveira, E.H.T.: An approach to group formation in collaborative learning using learning paths in learning management systems. IEEE Trans. Learn. Technol. **14**(5), 555–567 (2021). https://doi.org/10.1109/TLT.2021.3117916

Ratnapala, I.P., Ragel, R.G., Deegalla, S.: Students behavioural analysis in an online learning environment using data mining. In: International Conference on Information and Automation for Sustainability, Colombo (2014)

Romero, C., Ventura, S.: Educational data mining and learning analytics: an updated survey. WIREs Data Mining Knowl. Discov. **10**, 1–21 (2020)

Romero, C., Ventura, S.: Educational data mining: a review of the state of the art. In: IEEE Transactions on Systems, Man, and Cybernetics, Part C (Applications and Reviews), pp. 601–618. IEEE (2010)

Romero, C., Ventura, S., García, E.: Data mining in course management systems: Moodle case study and tutorial. Comput. Educ. **51**, 368–384 (2008)

Tang, T., McCalla, G.: Smart recommendation for an evolving e-learning system: architecture and experiment. Int. J. E-Learn. **4**, 105–129 (2005)

Wella, W., Tjhin, V.U.: Exploring effective learning resources affecting student behavior on distance education. In: International Conference on Human System Interactions (HSI). IEEE, Ulsan, South Korea (2017)

Zakrzewska, D.: Cluster analysis for user's modeling in intelligent e-learning systems, New Frontiers in Applied Artificial Intelligence. In: Nguyen, N.T., Borzemski, L., Grzech, A., Ali, M. (eds.) IEA/AIE 2008. Lecture Notes in Computer Science, pp. 209–214. Poland, Springer (2008)

Yathongchai, C., Angskkun, T., Yathongchai, W., Angskun, J.: Learner classification based on learning behavior and performance. In: IEEE Conference on Open Systems (ICOS). IEEE, Kuching, Malaysia (2013)

Structures in Online Discussion Forums: Promoting Inclusion or Exclusion?

Kimberly Williamson$^{(\boxtimes)}$ and René F. Kizilcec

Cornell University, Ithaca, USA
{khw44,kizilcec}@cornell.edu

Abstract. While race and gender academic disparities have often been categorized via differences in final grade performance, the day-to-day experiences of minoritized student populations may not be accounted for when only concentrating on final grade outcomes. However, more fine-grained information on student behavior analyzed using AI and machine learning techniques may help to highlight the differences in day-to-day experiences. This study explores how linguistic features related to exclusion and social dynamics vary across discussion forum structures and how the variation depends on race and gender. We applied linear mixed-effect analysis to discussion posts across six semesters to investigate the effect of discussion forum structure, race, and gender on linguistic features. These results can be used to suggest design changes to instructors' online discussion forums that will support students in feeling included.

Keywords: Social Network Analysis · LIWC · Inclusion

1 Introduction

Higher education institutions worldwide seek to improve equity and inclusion for their students. In the United States, colleges and universities have attempted to improve the underrepresentation of minoritized students and eliminate disparities in academic persistence and performance. These disparities have led to careful examinations of academic policies and instructional design practices for potential disparate impacts on minoritized groups of students [7]. Considering the increased use of technology for teaching and learning in higher education, it is important to examine how these technologies, especially those that incorporate artificial intelligence and machine learning, affect student experiences and outcomes in ways that impact academic disparities.

Academic disparities are commonly assessed in terms of achievement gaps, computed as the difference in the average final grade for two groups of students. However, this measure is coarse and reductive and poorly represents the student experience in the course. Milner [13] argues that the discourse surrounding the racial achievement gap prioritizes comparing minoritized students to white students without nuanced answers explaining the reason for the differences. However, the use of data mining and machine learning in higher education has shown

© The Author(s), under exclusive license to Springer Nature Switzerland AG 2023
N. Wang et al. (Eds.): AIED 2023, CCIS 1831, pp. 115–120, 2023.
https://doi.org/10.1007/978-3-031-36336-8_18

some optimism for using more fine-grained information to better represent student behavior in a course [3,11]. Moreover, this fine-grained information regarding students' engagement in a course can provide a more complete description of course activities and behaviors and a more holistic measure of academic disparities. In turn, holistic measurements of academic disparities can yield more insights for potential interventions in academic policy and instructional design to eliminate these disparities.

Research into online discussion forums has shown promising opportunities to yield rich data for assessing the student experience in the course. Many course instructors use forums to encourage the asynchronous exchange of ideas and question-answering [6]. These exchanges in online discussions have been analyzed using natural language processing (NLP) methods to measure students' demonstration of essential learning processes, like critical thinking [15], cognitive presence [9], and collaborative problem-solving [4], to name a few. However, despite discussion forums being communal spaces to communicate and demonstrate learning, discussion forums may also feel like isolating spaces for marginalized students [12]. Atkinson & DePalma [2] reported that, like face-to-face environments, microaggressions and bias occur in online environments, leading to students sometimes feeling excluded from forum discussions.

This study explored how linguistic features associated with exclusion and social dynamics vary in different discussion forums. The results of this work contribute to the recent calls from the AIED community to improve equity and inclusion in education for AI-assisted and interactive technologies by extending the methods that can be used to identify academic disparities.

2   Related Work

NLP methods are used to analyze written discourse in academic courses. The simplest technique is bag-of-words, which calculates frequencies from words and phrases. While this technique is simple to conduct and interpret, including every word/phrase can lead to context-specific representations, making this an overly-specific technique when summarizing across multiple courses [1]. On the other hand, dictionary methods only count words in selected dictionaries rather than every word. To have confidence that the specific words in a dictionary match a construct, such as positive words, the dictionary must be validated, like The Linguistic Inquiry and Word Count (LIWC), a popular dictionary [18]. Prior studies explored how exclusion may be detected in online environments. For example, Klauke & Kauffeld [8] experimentally investigated the linguistic features of participants whom other participants in online chats ostracized. They found ostracized participants had higher uses of first-person singular pronouns (i.e., I or me) and lower uses of first-person plural pronouns (i.e., we or our) and articles (i.e., a or the). These results are consistent with other studies on social status and hierarchy, finding more interpersonal distress with individuals using more first-person singular pronouns [19]. In course discussions, this pattern could manifest by students who feel included using significantly more language, such as, "We did this" or "Our class," as opposed to students who feel excluded. Additionally, other studies have explored LIWC summary categories, analytical thinking, clout, and authenticity to detect social dynamics in written discourse. Given this research, LIWC summary categories can represent psychological constructs related to power, status, and expertise.

Along with written content, the course discussion forums structure can also be characterized differently. Social Network Analysis (SNA) is used to understand course discussion structure. For example, SNA has been used in education to measure students' discussion forum participation [5] and predict graduation [14]. The combination of multiple SNA measures (i.e. density, centralization, and homophily) can be used to explore the structure further. Grouping course discussion networks into meaningful structures allows for exploring how structures may facilitate learning processes and power dynamics differently.

RQ1. How does the prevalence of linguistic features of exclusion and social dynamics (LIWC categories: word count, analytical thinking, clout, authenticity, first-person singular pronouns, and first-person plural pronouns) vary across different discussion structures?

The abundance of research and real-world applications for using linguistic features to evaluate or predict educational performance highlights the need for research that investigates how these patterns may differ across race and gender. Prior research has already found gender differences in the linguistic features that predict students' grade outcomes [10]. For example, Lin et al. [10] found a positive relationship between cognitive language use and academic performance for female students when the analysis was disaggregated by gender. This type of analysis is important because different races and genders may display varying linguistic patterns in a given discussion structure. We argue that these findings, combined with the experiences provided by different discussion structures, necessitate research into whether the variation in linguistic features across discussion structures is the same when the data is disaggregated by gender and race.

RQ2. How does the prevalence of linguistic features of exclusion and social dynamics (LIWC categories: word count, analytical thinking, clout, authenticity, first-person singular pronouns, and first-person plural pronouns) varying across different discussion structures depend on student gender and race?

3 Methodology

3.1 Dataset

The study sample consists of Canvas discussion post entries for courses offered in a three-year period (2019–2022) at a selective research university in the United States. This period was chosen because the Fall 2019 semester was the first semester in which Canvas was the official LMS, and Spring 2022 was the dataset's last entire semester of data. In addition, we only analyzed Fall and Spring courses with at least 50 posts that replied to other posts in a course over the semester to ensure a critical mass of students and communication to construct a network. After applying this criterion, our dataset resulted in 891 courses, with

17,075 students making 430,531 posts. Finally, we obtained student enrollment and sociodemographic data from the university's Student Information System (SIS). The student demographics are 56.4% women, 43.6% men, 35.5% White, 21.0% Asian, 17.1% International, 13.8% Hispanic, 7.1% Black, 5.1% Two or More Races, 0.3% American Indian, 0.1% Hawaiian/Pacific Islander. While we acknowledge the differences between race and ethnicity, we had no control over the data being provided to us as a single field.

3.2 Clustering Analysis

A directed edge network was created for each course using the iGraph R package (https://igraph.org/r/), where edges are created from direct replies between course participants (students and instructional staff). Fifteen SNA metrics were then computed for each course network. The fifteen metrics included measures for centralization, density, diameter, homophily, reciprocity, and transitivity. We then applied a principal component analysis to reduce the number of dimensions inputted into a k-means clustering algorithm. Next, we used k-Means clustering over the selected principal components to identify discussion network clusters. We determined the optimal number of clusters using the within-cluster sums of squares, average silhouette, and gap statistics for values of k between 2 and 10. After k was selected, each course was labeled with a cluster using the k-means clustering results. Both PCA and k-Means were run in R using the stats package. To further validate the cluster analysis, syllabi from ten courses were randomly selected from each of the three cluster groups. Finally, the instructions and practices of the online discussion in the syllabi were reviewed to see if the syllabi instructions matched the cluster label.

3.3 LIWC Processing

We extracted the discussion-related Canvas LMS data and cleaned the post's text by removing the HTML tags. Next, the LIWC-22 analysis for all categories was run on the cleaned text. After the LIWC analysis, the observational unit was each student's post in a course. In order to conduct further analysis, the dataset was pared down to be a single-student course observation by averaging the LIWC values for each student in the course. This data reduction is consistent with other LIWC studies, such as [16], that averaged the LIWC values for multiple essays to create a single value for each student. Lastly, the records for course grades, course clusters, and student demographics were added to the dataset.

3.4 Statistical Analysis

For this analysis, specific LIWC categories were selected and analyzed independently to understand the effects of SNA structure, race, and gender on language use. The LIWC categories selected were: Word Count, Analytical

Thinking, Clout, Authenticity, First Person Singular, and First Person Plural. We used a linear mixed-effects model with crossed random factors approach, which is a recommended method for analyzing such datasets [17], allowing for a more accurate effect of discussion structure on language use by controlling for the variance associated with differences in individual students and courses. For each LIWC category, three linear mixed-effects models were constructed: Structure, Structure:Race, and Structure:Gender. The Structure model only included a fixed effect for the discussion structure. The Structure:Race model included fixed effects for discussion structure and race and an interaction term of structure and race. The Structure:Gender model included fixed effects for discussion structure and gender and an interaction term of structure and gender. Courses and students were modeled as random intercepts in all models to account for variance due to individual differences and course differences. Statistical analyses were conducted using R (v.3.6.3) software for statistical analysis with package *lme4* (https://cran.r-project.org/web/packages/lme4/vignettes/lmer.pdf) for fitting linear mixed-effects models.

References

1. Ahuja, R., et al.: Machine learning and student performance in teams. In: Bittencourt, I.I., Cukurova, M., Muldner, K., Luckin, R., Millán, E. (eds.) AIED 2020. LNCS (LNAI), vol. 12164, pp. 301–305. Springer, Cham (2020). https://doi.org/10.1007/978-3-030-52240-7_55
2. Atkinson, E., DePalma, R.: Dangerous spaces: constructing and contesting sexual identities in an online discussion forum. Gend. Educ. 20(2), 183–194 (2008). https://doi.org/10.1080/09540250701797192
3. Baker, R.S., Martin, T., Rossi, L.M.: Educational data mining and learning analytics. In: Rupp, A.A., Leighton, J.P. (eds.) The Handbook of Cognition and Assessment, pp. 379–396. John Wiley & Sons Inc, Hoboken, NJ, USA (Nov 2016). https://doi.org/10.1002/9781118956588.ch16
4. Dowell, N.M., Cade, W.L., Tausczik, Y., Pennebaker, J., Graesser, A.C.: What works: creating adaptive and intelligent systems for collaborative learning support. In: Trausan-Matu, S., Boyer, K.E., Crosby, M., Panourgia, K. (eds.) ITS 2014. LNCS, vol. 8474, pp. 124–133. Springer, Cham (2014). https://doi.org/10.1007/978-3-319-07221-0_15
5. Dowell, N.Met al.: modeling learners' social centrality and performance through language and discourse. In: International Educational Data Mining Society (2015)
6. Fehrman, S., Watson, S.L.: A systematic review of asynchronous online discussions in online higher education. Am. J. Distance Educ. 35(3), 200–213 (2021). https://doi.org/10.1080/08923647.2020.1858705
7. Hurtado, S., Alvarez, C.L., Guillermo-Wann, C., Cuellar, M., Arellano, L.: A Model for Diverse Learning Environments: The Scholarship on Creating and Assessing Conditions for Student Success. In: Smart, J.C., Paulsen, M.B. (eds.) Higher Education: Handbook of Theory and Research, vol. 27, pp. 41–122. Springer, Netherlands, Dordrecht (2012). https://doi.org/10.1007/978-94-007-2950-6_2, higher Education: Handbook of Theory and Research
8. Klauke, F., Kauffeld, S.: Does it matter what i say? using language to examine reactions to ostracism as it occurs. Front. Psychol. 11, 558069 (2020). https://doi.org/10.3389/fpsyg.2020.558069

9. Kovanović, Vet al.: Towards automated content analysis of discussion transcripts: a cognitive presence case. In: Proceedings of the Sixth International Conference on Learning Analytics & Knowledge - LAK 2016, pp. 15–24. ACM Press, Edinburgh, United Kingdom (2016). https://doi.org/10.1145/2883851.2883950

10. Lin, Y., Yu, R., Dowell, N.: LIWCs the same, not the same: gendered linguistic signals of performance and experience in online STEM courses. In: Bittencourt, I.I., Cukurova, M., Muldner, K., Luckin, R., Millán, E. (eds.) AIED 2020. LNCS (LNAI), vol. 12163, pp. 333–345. Springer, Cham (2020). https://doi.org/10.1007/978-3-030-52237-7_27

11. Macgilchrist, F.: Cruel optimism in edtech: when the digital data practices of educational technology providers inadvertently hinder educational equity. Learn. Media Technol. **44**(1), 77–86 (2019). https://doi.org/10.1080/17439884.2018.1556217

12. McKee, H.: "Your views showed true ignorance!!!": (Mis)Communication in an online interracial discussion forum. Comput. Compos. **19**(4), 411–434 (2002). https://doi.org/10.1016/S8755-4615(02)00143-3

13. Milner, H.R.: Rethinking achievement gap talk in urban education. Urban Educ. **48**(1), 3–8 (2013). https://doi.org/10.1177/0042085912470417

14. Nur, N., et al.: Student network analysis: a novel way to predict delayed graduation in higher education. In: Isotani, S., Millán, E., Ogan, A., Hastings, P., McLaren, B., Luckin, R. (eds.) AIED 2019. LNCS (LNAI), vol. 11625, pp. 370–382. Springer, Cham (2019). https://doi.org/10.1007/978-3-030-23204-7_31

15. O'Riordan, T., Millard, D.E., Schulz, J.: Is critical thinking happening? Testing content analysis schemes applied to MOOC discussion forums. Comput. Appl. Eng. Educ. **29**(4), 690–709 (2021). https://doi.org/10.1002/cae.22314

16. Pennebaker, J.W., Chung, C.K., Frazee, J., Lavergne, G.M., Beaver, D.I.: When small words foretell academic success: the case of college admissions essays. PLoS ONE **9**(12) (2014). https://doi.org/10.1371/journal.pone.0115844

17. Raudenbush, S.W.: A crossed random effects model for unbalanced data with applications in cross-sectional and longitudinal research. J. Educ. Stat. **18**(4), 321–349 (1993). https://doi.org/10.3102/10769986018004321

18. Tausczik, Y.R., Pennebaker, J.W.: The psychological meaning of words: LIWC and computerized text analysis methods. J. Lang. Soc. Psychol. **29**(1), 24–54 (2010). https://doi.org/10.1177/0261927X09351676

19. Zimmermann, J., Wolf, M., Bock, A., Peham, D., Benecke, C.: The way we refer to ourselves reflects how we relate to others: Associations between first-person pronoun use and interpersonal problems. J. Res. Pers. **47**(3), 218–225 (2013). https://doi.org/10.1016/j.jrp.2013.01.008

Assessment in Conversational Intelligent Tutoring Systems: Are Contextual Embeddings Really Better?

Colin M. Carmon[1,2(✉)], Xiangen Hu[1,2], and Arthur C. Graesser[1,2]

[1] University of Memphis Psychology, Memphis, TN 38152, USA
cmcarmon@memphis.edu
[2] Institute for Intelligent Systems, University of Memphis, Memphis, TN 38152, USA

Abstract. This research investigates the ability of semantic text models to assess student responses during tutoring compared with expert human judges. Recent interest in text similarity has led to a proliferation of models that can potentially be used for assessing student responses; however, whether these models perform as well as traditional distributional semantic models like Latent Semantic Analysis for student response assessment in automatic short answer grading is unclear. We assessed 5166 response pairings of 219 participants across 118 electronics questions and scored each with 13 different computational text models, including models that use regular expressions, distributional semantics, word embeddings, contextual embeddings, and combinations of these features. We show a few semantic text models performing comparably to Latent Semantic Analysis, and in some cases outperforming the model. Furthermore, combination models outperformed individual models in agreement with human judges. Choosing appropriate computational techniques and optimizing the text model may continue to improve the accuracy, recall, weighted agreement and therefore, the effectiveness of conversational ITSs.

Keywords: Agents · Automatic short answer grading · AutoTutor · Computational linguistics · Dialogue · Distributional semantics · Embeddings · Context embeddings · Intelligent tutoring systems · Natural language processing

1 Introduction

Distributional semantics models develop methods for quantifying and categorizing semantic similarities between linguistic text segments based on their distributional properties in large samples of language data. There are many contexts where distributional semantics techniques can be applied for assessing text similarity, such as automatic grading of students' text productions. Automatic grading of natural language can be broken down according to question types: fill in the blank, short answer, and essay. Fill in the blank answers generally consider one key term or phrase. Automatic Short Answer Grading (*ASAG*; [3]) refers to assessment of short natural language responses to objective questions using computational methods. We can consider many modern conversational intelligent tutoring systems (*ITS*) to be interactive forms of ASAG.

© The Author(s), under exclusive license to Springer Nature Switzerland AG 2023
N. Wang et al. (Eds.): AIED 2023, CCIS 1831, pp. 121–129, 2023.
https://doi.org/10.1007/978-3-031-36336-8_19

1.1 Conversational ITSs

ITSs that incorporate natural language communication with students are called conversational ITSs. Examples of early, notable conversational ITSs include *CIRCSIM-Tutor* [10], *Why2Atlas* [28], and *AutoTutor* [15]. Modern conversational ITSs provide instruction, evaluative feedback (positive, negative, neutral) and qualitative feedback to improve answers (hints, prompts, tutoring questions, etc.) to the student with similarities to human tutoring [12, 24, 25, 27]. In this paper, we focus on student verbal input assessment in a conversational ITS that teaches an electronics curriculum through conversation.

ElectronixTutor [14, 23] teaches electronics by holding a conversation in natural language with students [12, 13] in addition to linking them to other ITS and learning resources. The AutoTutor component asks students questions and guides them to an expected answer through a series of dialogue moves in conversation with the goal of probing the students for concepts and ideas that they may know, but do not initially articulate in their answers to questions. The dialogue moves include pumps (e.g., tell me more), hints, assertions, corrections of errors, and proper branching between dialogue turns in order to conduct seemingly human-like conversation. The *initial answer* of the student represents the student's first attempt to fully respond to a question and has a large impact on how the multiturn dialogue evolves, so it is important to optimize the accuracy of semantic matches between the content of student answers and the expectations. The system's dialogue and assessment of student input is currently driven by a combination of Regular Expressions (*RegEx*; [17]) and Latent Semantic Analysis (*LSA*; [18]).

1.2 Models for Assessing Conversational Input

Systems like ElectronixTutor and AutoTutor have traditionally relied on LSA and Regular Expressions for semantic models of answer grading. LSA is a robust tool for information retrieval, information extraction, and computation of semantic similarity between text segments; LSA and other similar distributional semantics techniques have been around since the 1980's and 1990's [9, 11]. Over the past few decades, and especially since the 2010's, new computational semantics techniques have been developed including word embedding models, such as Word2Vec (*W2V*; [19, 20]) and transformer models (i.e., contextual embeddings) such as Bidirectional Encoder Representations from Transformers (*BERT*; [8]), Sentence Bidirectional Encoder Representations from Transformers (*SBERT*; [26]), and Sentence Generative Pretrained Transformer (*SGPT*; [21]).

1.3 Latent Semantic Analysis

Latent Semantic Analysis [18] is a distributional semantics model for assessing the similarity of pairs of texts. "Chair", "table", and "eat", for example, often appear in the same documents and, as such, have high semantic similarity to each other. A statistical method called Singular Value Decomposition (SVD) reduces the corpus. There are several metrics on individual words that can be derived from the word by document matrix and the SVD. The metric of similarity is a cosine match score from -1 to 1, with 0 representing no semantic similarity and 1 representing a perfect similarity between the student response and the designated expectation by virtue of the 110–500 dimensions constructed in the LSA space. Given two vectors, a cosine similarity score is computed.

1.4 Word2Vec and Variants

LSA is a count-based method that computes the co-occurrence of words in documents. In contrast, W2V is a prediction-based method that accounts for local predictive statistics on words that appear in a corpus of documents. Bigram2Vec (*W2VB*), is another variation of W2V observed in the study. This is a variation of the same W2V model that focuses on bigram embeddings rather than single word embeddings. Doc2vec is also a variant of the Word2Vec model. W2V and variants, like LSA, have the ability to compute cosine similarity scores between two vectorized text inputs. This is the basis for our comparison of models in terms of agreement with the human judges. In this study, we use a Skip-gram W2V model. Additionally, we use the softmax method for training the models. A softmax, or normalized exponential function, converts a vector of real numbers into a probability distribution of possible outcomes that sum to 1.

1.5 Transformer Model Variants

In addition to substituting Word2Vec models for LSA in automatic semantic assessment tasks, we also considered two new models, SBERT and SGPT. The original BERT model is pre-trained on a large corpus of 3.3 billion words. The SBERT model observed in this paper is also trained on the SNLI and Multi-Genre NLI corpora [1, 29]. SGPT is a small variant of *GPT-3* by OpenAI [2] created specifically for sentence embeddings. At base, BERT uses 110 million parameters with 12 hidden layers, whereas GPT-3 uses 175 billion parameters and 96 hidden layers. It is important to point out the training corpora of BERT, SBERT, and SGPT are many orders of magnitude larger than those for LSA and Word2Vec and variants.

SBERT produced favorable performance over BERT in a number of tasks including semantic text similarity for sentence embeddings [26]. However, this was achieved through a considerably high degree of supervision. Accordingly, SBERT and other large, pre-trained transformers may remain ineffective and impractical in contexts such as ASAG for a domain-specific topic such as electronics where large, labeled data may be scarce [30] or where relevant terms are distinct to the subject matter of electronics rather than general English language. The literature favors unsupervised approaches in these tasks [16].

2 Method

We evaluate the performance of various language models, including LSA, Word2Vec and its variants, SBERT, and SGPT, using standard metrics in computational linguistics and social sciences (i.e., precision, recall, F1, accuracy, and Cohen's κ). The models are also combined with RegEx. The paper primarily focuses on F1 to compare the agreement between the computer models and human judges, as well as between human judges.

This paper assesses the system's ability to identify initial answers rated as 6 by human experts. Comparing initial answers of students to good answers to electronics questions is essential for ElectronixTutor to select subsequent dialogue moves to accurately adapt to the individual student.

We analyze student response data from ElectronixTutor [4–6] where student short-answer responses were scored by the computer and then compared to humans. We collected the dataset on 219 unique Amazon Mechanical Turk workers (MTurkers) who answered 118 questions asked by AutoTutor in ElectronixTutor. MTurkers attempted a full answer to these questions in an open-ended fashion. Each response was paired with the ideal answer and each expectation to the main question, resulting in 5166 ($n = 5166$) total response pairings. The following is an example of a question in ElectronixTutor, the ideal answer, and a breakdown of the ideal answer into three expectations (Fig. 1):

Main Question: How does the way the Zener diode is connected help protect the bulb in a circuit?

Ideal Answer: The Zener diode is connected as a reverse biased diode. The break-down voltage is less than the burn out voltage of the bulb. Extra current flows through the diode instead of the bulb.

Expectation 1: The Zener diode is connected as a reverse biased diode.

Expectation 2: The breakdown voltage is less than the burn out voltage of the bulb.

Expectation 3: Extra current flows through the diode instead of the bulb.

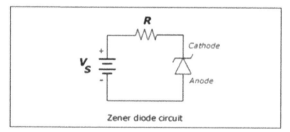

Fig. 1. Associated figure for main question: "How does the way the Zener diode is connected help protect the bulb in a circuit?"

2.1 Human Judges

Two subject-matter experts on electronics independently rated the user responses on a scale ranging from 1–6. (1) There was no attempt to answer the question, (2) The answer is not on topic or contains metacognitive language, (3) The answer is on topic, but completely incorrect, (4) The answer is mostly incorrect but contains a small degree of truth value, (5) The answer is mostly correct, and (6) The answer is ideal. Subject-matter experts with graduate degrees in electronics received a training session on evaluating semantic similarity between main questions, expectations, and participant responses.

3   Results and Discussion

We compared models trained on the same electronics-specific corpus to a pre-trained SBERT model which was pretrained on a large domain-general corpus [7]. Performance on these models needs to be interpreted in comparison to agreement between experts. The agreement between two human experts was as accuracy of 0.867, with a precision score of 0.467, recall score of 0.618, F1 measure of 0.532, and a kappa score of 0.456.

3.1   Agreement Between Humans and Stand-Alone Computer Models

This next section contains tables showing results of model performances compared with the two expert judges. *LSAet* is the original LSA space from a 2019 study [5, 6]; *LSAc* is the model trained on the new combined electronics and Newtonian physics corpus in 2021; *RE* is Regular Expressions; *W2V* is Word2Vec; *W2VB* is Bigram2Vec; *D2V* is Doc2Vec; *SBERT* is SBERT; *SGPT* is SGPT. Additionally, we combine each of the corpus-based computational models with Regular Expressions (Tables 1 and 2).

Table 1. Agreement between Judge 1 and 2 versus standalone computer models (Physics and Electronics) Plus Pretrained SBERT

Performance Metrics	Judge 1							
	LSAc	LSAet	W2V	W2VB	D2V	RE	SBERT	SGPT
Precision	.583	.350	.596	.560	.539	.448	.571	.475
Recall	.272	.215	.292	.250	.297	.440	.277	.284
Accuracy	.680	.750	.707	.660	.720	.794	.690	.722
F1	.370	.260	.386	.347	.383	.444	.373	.356
κ	.194	.160	.220	.160	.221	.366	.200	.193
Judge 2								
Precision	.577	.350	.521	.526	.481	.450	.510	.429
Recall	.201	.215	.204	.176	.203	.585	.187	.195
Accuracy	.670	.750	.690	.643	.706	.794	.670	.713
F1	.398	.260	.293	.264	.286	.509	.274	.268
κ	.144	.160	.142	.100	.138	.428	.116	.120

Table 2. Agreement between Judge 1 and 2 versus combination computer models (Physics and Electronics) plus a pretrained SBERT combination

Performance Metrics	Judge 1					
	RE + LSAc	RE + W2V	RE + W2VB	RE + D2V	RE + SBERT	RE + SGPT
Precision	.479	.502	.450	.462	.450	.473
Recall	.584	.562	.585	.587	.584	.577
Accuracy	.852	.856	.859	.860	.859	.859
F1	.526	.530	.508	.517	.517	.519
κ	.446	.446	.428	.437	.366	.438
Judge 2						
Precision	.482	.529	.496	.458	.448	.468
Recall	.440	.405	.349	.440	.440	.431
Accuracy	.851	.847	.825	.862	.862	.860
F1	.458	.460	.410	.449	.444	.448
κ	.371	.372	.312	.312	.366	.368

4 General Discussion

The scores in the stand-alone corpus-based semantic models are not impressive. However, a combination of the corpus-based semantics models with Regular Expressions had performance in some models almost as high as agreement between human experts. RegEx alone, which was the best of the stand-alone models, had an average agreement with the two judges on the F1 measure of 0.477, which is respectable compared with the agreement between human judges of 0.532. When we consider the best model combined with RexEx, namely RE + W2V, the comparison to humans has an F1 measure of 0.495. Consequently, there is some added value of the semantic models in conjunction with RegEx and the resulting combined models are almost comparable to human experts (0.495 versus 0.532). Interestingly, the domain-general SBERT model was not as high, but impressive at 0.481.

Bigram2Vec had substantially lower κ values than each of the other models observed in study. Further, models trained on the physics + electronics corpus consistently performed better than those trained only on the physics corpus in the study.

4.1 Conclusion and Implications

The RegEx and W2V combination model had the highest overall performance with substantially higher precision than other models observed. The highest F1 score observed in combination models came from the RegEx and W2V combination reaching .530. This is perhaps of the most notable findings in the paper considering agreement between humans measured an uncannily similar F1 score of .532, while accuracy between humans was .867, and accuracy reached .856 in the RegEx and W2V combination model. W2V performed slightly better on the task than state of the art transformer models pre-trained on a large corpus of general language which indicates the importance of domain-specific training corpora for such a task. It is unclear how fine-tuning pre-trained transformer models will affect performance in the task. Additionally, combination models featuring RegEx show clear benefit over stand-alone models.

4.2 Future Research

A critical step for follow-up research remains running current, newer, and adjusted-training models (e.g., fine-tuned pre-trained transformers) on new data that have been collected from Navy electrical engineers in 2022 [22]. There were semantic judgments between student responses and expectations from 3 human experts in this recent study. There were also many student contributions in the multi-turn follow-up conversations after the initial response of the human from the main question. How will the different models compare in evaluating student contributions across many turns in the AutoTutor conversations?

Acknowledgments. This research was supported by the Office of Naval Research (N00014-00-1-0600, N00014-15-P-1184; N00014-12-C-0643; N00014-16-C-3027) and the National Science Foundation Data Infrastructure Building Blocks program (ACI-1443068). Any opinions, findings, and conclusions or recommendations expressed in this material are those of the authors and do not necessarily reflect the views of ONR or NSF.

References

1. Bowman, S.R., Angeli, G., Potts, C., Manning, C.D.: A large annotated corpus for learning natural language inference. arXiv preprint arXiv:1508.05326 (2015)
2. Brown, T., et al.: Language models are few-shot learners. In: Advances in Neural Information Processing Systems, vol. 33, pp. 1877–1901 (2020)
3. Burrows, S., Gurevych, I., Stein, B.: The eras and trends of automatic short answer grading. Int. J. Artif. Intell. Educ. 25(1), 60–117 (2014). https://doi.org/10.1007/s40593-014-0026-8
4. Carmon, C.M.: Semantic matching evaluation: optimizing models for agreement between humans and AutoTutor. Electronic Theses and Dissertations. 2148 (2021). https://digitalcommons.memphis.edu/etd/2148
5. Carmon, C.M., Hampton, A.J., Morgan, B., Cai, Z., Wang, L., Graesser, A.C.: Semantic matching evaluation of user responses to electronics questions in AutoTutor. In: Sixth (2019) ACM Conference on Learning @ Scale. ACM, Chicago (2019). https://doi.org/10.1145/3330430.3333649
6. Carmon, C., Morgan, B., Hampton, A.J., Cai, Z., Graesser, A.C.: Semantic matching evaluation in ElectronixTutor. In: Proceedings of the 11th International Conference on Educational Data Mining, Buffalo, NY, pp. 580–583 (2018)
7. Condor, A., Litster, M., Pardos, Z.A.: Automatic short answer grading with SBERT on out-of-sample questions. In: Hsiao, S., Sahebi, S. (eds.) Proceedings of the 14th International Conference on Educational Data Mining, pp. 345–352. EDM (2021)
8. Devlin, J., Chang, M.W., Lee, K., Toutanova, K.: BERT: pre-training of deep bidirectional transformers for language understanding. arXiv preprint arXiv:1810.04805 (2018)

9. Dumais, S.T., Furnas, G.W., Landauer, T.K., Deerwester, S., Harshman, R.: Using latent semantic analysis to improve access to textual information. In: Proceedings of the SIGCHI Conference on Human Factors in Computing Systems, CHI 1988, pp. 281–285. ACM Inc., New York, NY, USA. https://doi.org/10.1145/57167.57214 (1988)
10. Evens, M.W., et al.: CIRCSIM-tutor: an intelligent tutoring system using natural language dialogue. In: Proceedings of the 12th Midwest Artificial Intelligence and Cognitive Science Conference, Oxford, pp. 16–23 (2001)
11. Furnas, G.W., Landauer, T.K., Gomez, U.M., Dumais, S.T.: Statistical semantics: analysis of the potential performance of key-word information systems. Bell Syst. Tech. J. **62**(6), 1753–1806 (1983)
12. Graesser, A.C.: Conversations with AutoTutor help students learn. Int. J. Artif. Intell. Educ. **26**, 124–132 (2016)
13. Graesser, A.C.: Learning science principles and technologies with agents that promote deep learning. In: Learning Science: Theory, Research, and Practice, pp. 2–33. McGraw-Hill, New York (2020)
14. Graesser, A.C., et al.: ElectronixTutor: an intelligent tutoring system with multiple learning resources for electronics. Int. J. STEM Educ. Innov. Res. **5**(1) (2018). https://doi.org/10.1186/s40594-017-0072-5
15. Graesser, A.C., et al.: AutoTutor: a tutor with dialogue in natural language. Behav. Res. Methods Instrum. Comput. **36**, 180–193 (2004)
16. Hill, F., Cho, K., Korhonen, A.: Learning distributed representations of sentences from unlabelled data. arXiv preprint arXiv:1602.03483 (2016)
17. Jurafsky, D., Martin, J.: Speech and Language Processing. Prentice Hall, Englewood (2008)
18. Landauer, T.K., McNamara, D.S., Dennis, S., Kintsch, W.: Handbook of Latent Semantic Analysis. Erlbaum, Mahwah, NJ (2007)
19. Mikolov, T., Chen, K., Corrado, G.S., Dean, J.: Efficient estimation of word representations in vector space. In: Proceedings of Workshop at ICLR (2013)
20. Mikolov, T., Sutskever, I., Chen, K., Corrado, G.S., Dean, J.: Distributed representations of words and phrases and their compositionality. In: Advances in Neural Information Processing Systems, vol. 26 (2013)
21. Muennighoff, N.: SGPT: GPT sentence embeddings for semantic search. arXiv preprint arXiv: 2202.08904 (2022)
22. Nazaretsky, T., Hershkovitz, S., Alexandron, G.: Kappa learning: a new item-similarity method for clustering educational items from response data. In: Proceedings of the 12th International Conference on Educational Data Mining, EDM 2019, pp. 129–138 (2019)
23. Nye, B.D., Core, M., Swartout, B., Hu, X., Morgan, B., Graesser, A.: ElectronixTutor content and system testing to support adaptive learning for nuclear field electronics. Final report (2022)
24. Nye, B.D., Graesser, A.C., Hu, X.: AutoTutor and family: a review of 17 years of natural language tutoring. Int. J. Artif. Intell. Educ. **24**(4), 427–469 (2014)
25. Olney, A.M., et al.: Guru: a computer tutor that models expert human tutors. In: Cerri, S.A., Clancey, W.J., Papadourakis, G., Panourgia, K. (eds.) ITS 2012. LNCS, vol. 7315, pp. 256–261. Springer, Heidelberg (2012). https://doi.org/10.1007/978-3-642-30950-2_32
26. Reimers, N., Gurevych, I.: Sentence-BERT: sentence embeddings using siamese BERT-networks. arXiv preprint arXiv:1908.10084 (2019)
27. VanLehn, K., Graesser, A.C., Jackson, G.T., Jordan, P., Olney, A., Rose, C.P.: When are tutorial dialogues more effective than reading? Cogn. Sci. **31**, 3–62 (2007)
28. VanLehn, K., et al.: The architecture of Why2-Atlas: a coach for qualitative physics essay writing. In: Cerri, S.A., Gouardères, G., Paraguaçu, F. (eds.) Intelligent Tutoring Systems, vol. 2363, pp. 158–167. Springer, Heidelberg (2002). https://doi.org/10.1007/3-540-47987-2_20

29. Williams, A., Nangia, N., Bowman, S.R.: A broad-coverage challenge corpus for sentence understanding through inference. arXiv preprint arXiv:1704.05426 (2017)

30. Zhang, Y., He, R., Liu, Z., Lim, K.H., Bing, L.: An unsupervised sentence embedding method by mutual information maximization. arXiv preprint arXiv:2009.12061 (2020)

A Recommendation System for Nurturing Students' Sense of Belonging

Aileen Benedict[1]([✉]) [iD], Sandra Wiktor[1], Mohammadali Fallahian[1][iD],
Mohsen Dorodchi[1]([✉]) [iD], Filipe Dwan Pereira[2][iD], and David Gary[1][iD]

[1] University of North Carolina at Charlotte, Charlotte, NC, USA
abenedi3@uncc.edu
[2] Universidade Federal de Roraima, Boa Vista, Brazil

Abstract. Nurturing a sense of belonging in a classroom can positively impact student attrition, especially for underrepresented groups. In this work, we develop and study the effectiveness of a recommendation system that aims to foster belonging by sharing challenges and possible solutions that other students have had while taking the same course. Our system uses sentence transformers and calculates the similarity between students' reflections. We measure participating students' sense of belonging before and after they view the top 5 challenges and solutions received from other students. Using a significance level of 0.05, we found a p-value of 0.0145, indicating that there is a significant increase in overall belonging values. Students also rated 61% of the solutions as useful. This work allows future students to also benefit from the experiences of those before them. By showing students that those before them also had similar challenges and overcame them, we can show students that they do, in fact, belong among their peers.

Keywords: sense of belonging · educational recommender system · semantic similarity

1 Problem and Motivation

"We can never get a recreation of community and heal our society without giving our citizens a sense of belonging." - Patch Adams

To have a sense of belonging is to view oneself as a valued and accepted member of the community [2]. Students who feel a sense of belonging are more likely to have higher levels of academic engagement, motivation, and self-esteem [5]. This sense of belonging is crucial for students' overall well-being and ability to thrive academically [5]. Research has also shown that underrepresented minorities and first-generation college students often face challenges in belonging, tending to have a lower sense of belonging than majority students, and is usually due to a lack of representation and bias [2]. It is necessary to take a sense of belonging into account when developing strategies and interventions for student success.

N. Wang et al. (Eds.): AIED 2023, CCIS 1831, pp. 130–135, 2023.
https://doi.org/10.1007/978-3-031-36336-8_20

The goal of this work is to develop a recommendation system that can (1) serve as an intervention in increasing students' sense of belonging and (2) recommend possible solutions to students' current challenges in a course. This system was heavily inspired by research in student belonging interventions. For example, one study has found that brief interventions targeting students' sense of belonging have yielded promising results, as in the case of one study aiming to improve feelings of belonging of African American students, who are notably susceptible to experiencing feelings of uncertainty with regards to belonging [7]. A brief, hour-long intervention and subsequent follow-up assignments aiming to express a narrative that social adversity is "shared and short-lived" was utilized to address these uncertainties. As a result, this study found a GPA increase in the participating African-American students compared to control groups and improvements in self-reported health [7].

We build upon our previous work described in [1]. A previous limitation was a small sample size. For this iteration, we performed the study across two semesters and seven total classes, obtaining a larger sample of 47 students. Another limitation was the inability to know whether a student implemented the solution or not before rating them. We modified the post-survey in hopes of fixing this by providing an option for students to select if they did not attempt the recommended solution, instead of rating them.

2   Methodology

2.1   Study Method

This study aims to evaluate a recommendation system that anonymously shares challenge and solution reflections written by other students. The study aims to determine if (1) accessing similar challenges written by other students can help increase sense of belonging and (2) the recommended solutions provided are useful to their learning experiences. The research uses the quasi-experimental pretest-posttest design to measure the effectiveness of the recommendation system as an intervention for student belonging.

Participating Courses. To participate in our study, we looked for courses that already included student reflections as a part of their curriculum or were willing to do so. Among the undergraduate courses that participated were a software engineering course and an operating systems and networking course during both the Summer and Fall 2022 semesters. Moreover, Introduction to Programming, Introduction to Data Mining, and Data Structures courses took part in the Fall 2022 semester.

Student Self-Reflections. Student reflections serve as the foundation of our dataset and the capability of our system to identify similarities between students and create recommendations. In the context of education, reflection refers to the process of analyzing what enabled learning to occur, as well as developing links between the learned material and the learner's thoughts and feelings towards it [3]. Prompts for students' reflections were given through Canvas, the university's Learning Management System. To generate recommendations, this study required the following prompts for data collection:

1. What was a challenge you overcame?
2. What was a solution that helped you overcome that challenge?
3. Do you consent to anonymously share your above responses with other students who may experience similar challenges?
4. Do you have any current challenges?

Questions 1 through 3 are used to build a dataset of challenge-solutions pairs to recommend to future students. Questions could also be reworded to fit their course material and schedule. For example, Question 1 may be worded as "What was a challenge you overcame these past few weeks?" or "What was a challenge you overcame during the last project?" Question 4 is to help generate a recommendation for the participating student responding to that prompt.

Recruitment and Participants. This study was announced as an extra credit opportunity in participating courses to recruit students. The requirement to participate was to be at least age 18+ and to complete the reflection(s) as a part of their course. All student participants were made aware of the content and goals of the research. Our analysis only included consenting students who responded to both surveys. We had 47 participants across all seven courses who completed all study parts. Out of the 47 participants, 31 (66%) are male, 15 (32%) are female, and one is unlisted (2%). Furthermore, 19 (40%) of our respondents are White, 13 (28%) are Asian, 7 (15%) are Hispanic, five (13%) are African American, one (2%) is listed as international, and one (2%) is listed as multiple. It is important to note that this demographic information is extremely limited, such as gender being binary and racial options lacking. It would be beneficial for a more nuanced demographic survey to be added.

Study Procedure. We evaluated our system by having participating students take a survey before and after reviewing the system's results, which are the top five most similar challenges and the paired solutions written by other students. These surveys are referred to as the pre-survey and post-survey, respectively. Before participating, students were provided with a consent form explaining the study's background and purpose. All consent forms and surveys were delivered electronically through Google Forms.

The pre-survey consisted of a questionnaire designed to measure sense of belonging in CS courses, which had been validated by Moudgalya et al. [4]. The questionnaire consists of 26 prompts that students rate on a 7-point Likert scale ranging from *Strongly Disagree* to *Strongly Agree*. Each prompt pertained to one of six factors associated with a student's overall sense of belonging, including membership, affect, acceptance, trust, and the desire to fade [4]. Membership assesses the student's perception of themselves as a member of the computer science community. Acceptance measures the student's perception of how accepted they are by their community. Affect evaluates the student's comfort level. Trust assess how much the student trusts the instruction team to create a fair learning environment. Finally, the desire to fade measured the student's desire to withdraw from the community.

After the participants completed the pre-survey, they were sent an email with their top five personalized results from our system. The system utilized their most recent reflected-upon challenge to find (1) challenges faced by other students that were most similar and (2) the paired solutions written by those same students for that particular challenge. Participants were then asked to complete the post-survey after reviewing their results and possibly attempting any of the suggested solutions.

The post-survey contained two parts. The first part was identical to the pre-survey, with questions measuring student belonging. The second part then asked participants to (1) rate how similar they thought each returned challenge was to their own, and to (2) evaluate the usefulness of the solutions received. Possible ratings ranged from 1 (Not Similar or Not Useful) to 7 (Very Similar or Very Useful). Students could also enter a 0 when rating solution usefulness if they chose not to use the solution provided. The post-survey also asked students to provide qualitative feedback to explain the reasoning behind their ratings, though optional. Since students rated five challenge-solution pairs each, we had a total of 235 ratings.

2.2   System Design

The system being discussed here is designed to identify similarities among students' challenges, and subsequently recommend relevant challenges and solutions from students who have similar experiences. We follow the same process as described in Benedict et. al [1]. The code for this system is publicly available on GitHub[1].

Dataset. This research employed reflections from prior semesters to investigate the types of issues students encountered in their courses and the potential solutions that could aid them. Any reflections that were missing a response to either the challenge or solution prompts were removed.

Students participating in the Summer 2022 software engineering course had recommendations generated from a dataset of 1154 historic student reflections collected since the Fall 2020 semester. The software engineering dataset for the Fall 2022 students then had a total of 1470 reflections. The data structures dataset had a total of 311 reflections, collected from the Spring 2022 and current Fall 2022 semester. The introduction to data mining course had reflections collected since Spring 2022 and had a total of 241 reflections. The networking

---

[1] https://github.com/samfallahian/reflection_rec_sys.

and operating systems course had a total of 38 reflections for the Summer 2022 participants, only starting data collection since that summer, and a total of 97 reflections for the Fall 2022 participants. The introduction to programming course had a total of 241 reflections collected that semester.

Data Privacy Considerations. In order to ensure the protection of student privacy and compliance with the Federal Educational Privacy Rights Act (FERPA), we took measures to anonymize all previous student reflection responses. This made it impossible to identify the authoring student based solely on the reflection text. For reflections starting in the Summer 2022 semester or later, an additional prompt was added asking students if they consent to anonymously share their responses with other students who may have similar challenges. If a student did not give consent, their reflection was removed from the dataset entirely. This approach was approved by the Institutional Review Board (IRB) under the numbers 21-0415 and 21-0211.

Generating Recommendations. The raw text from the dataset of student reflection challenge responses, as well as the challenge of the current student who will be receiving recommendations, are converted to numerical representations in a fixed-length vector space for each sentence. In this study, we use the pretrained stsb-roberta-large sentence transformer model [6] to perform this conversion. We then calculate cosine similarity using the embeddings from the sentence transformer model to find student challenges that are semantically similar to those of a current student. We filter down to the top five most similar challenges. From these most similar challenges, we then find the matching solutions provided. We also filter out any responses written by the current student so that they do not receive a solution they had written previously.

3   Contributions

3.1   Results

A one-tailed (greater) paired t-test was performed to compare the difference in belonging scores before and after the intervention to determine if belonging increased. The overall sense of belonging was found to have a t-test value of 2.25, a p-value of 0.01 and a mean difference of 0.03 with a standard deviation of 0.10. The test used an alpha level of 0.05, suggesting a statistically significant increase between the before and after belonging measures. Across a sample size of 47 students, test values also suggest a significant increase in *membership* (t-test: 1.88, p-value: 0.03, mean difference: 0.06, with standard deviation: 0.21) and *affect* (t-test: 2.22, p-value: 0.02, mean difference: 0.05, with standard deviation: 0.15). In comparison, our previous work had a small sample size and was therefore unable to find a significant difference [1]. Here, we also see a significant increase in overall belonging for Male students, as well as those in the White, Asian, and African American demographics, with a slightly larger effect size for African

American students. There was also a relatively large effect size for Hispanic students' affect ($p = 0.05$ and $d = 0.79$). We also performed a two-tailed paired t-test and a one-tailed (lesser) paired t-test, confirming that none of the belonging factors for any demographic had a significant *decrease* using an alpha level of p<0.01, p<0.05, and p<0.10.

3.2 Impact and Future Work

Our results suggest potential for further exploring this system as a method for promoting student belonging. However, a limitation of this study design is that it cannot control for other factors in the classroom that may also impact student belonging. Expanding the study to a wider range of courses could account for these variables. Currently, the system only utilizes textual data similarity for generating recommendations. There is ample opportunity to incorporate additional features, such as student and instructor ratings of solutions, and to incorporate these ratings back into the system. Additionally, incorporating factors such as race, gender, age, or other aspects of identity could shed light on how intersectionality affects different types of challenges. Education is a complex and individualized process, and AI can play a crucial role in personalizing interventions for each person. In the future, we aim to investigate the possibility of creating a conversational agent for students to interact with. Currently, we send results to students via automated emails. We believe that a deeper interaction with the system would be beneficial.

References

1. Benedict, A., Al-Hossami, E., Dorodchi, M., Benedict, A., Wiktor, S.: Pilot recommender system enabling students to indirectly help each other and foster belonging through reflections. In: LAK22: 12th International Learning Analytics and Knowledge Conference, pp. 521–527 (2022)
2. Fan, X., Luchok, K., Dozier, J.: College students' satisfaction and sense of belonging: differences between underrepresented groups and the majority groups. SN Soc. Sci. **1**(1), 1–22 (2021)
3. Lew, M.D.N., Schmidt, H.G.: Self-reflection and academic performance: is there a relationship? Adv. Health Sci. Educ. **16**, 529–545 (2011)
4. Moudgalya, S.K., Mayfield, C., Yadav, A., Hu, H.H., Kussmaul, C.: Measuring students' sense of belonging in introductory cs courses. In: Proceedings of the 52nd ACM Technical Symposium on Computer Science Education, pp. 445–451 (2021)
5. Osterman, K.F.: Teacher Practice and Students' Sense of Belonging, pp. 239–260. Springer, Netherlands, Dordrecht (2010). https://doi.org/10.1007/978-90-481-8675-4_15
6. Reimers, N., Gurevych, I.: Sentence-bert: Sentence embeddings using siamese bert-networks. arXiv preprint arXiv:1908.10084 (2019)
7. Walton, G.M., Cohen, G.L.: A brief social-belonging intervention improves academic and health outcomes of minority students. Science **331**(6023), 1447–1451 (2011)

Desirable Difficulties? The Effects of Spaced and Interleaved Practice in an Educational Game

Jonathan Ben-David[✉] and Ido Roll

Faculty of Education in Science and Technology, Technion-Israel Institute of Technology, Haifa, Israel
jonathanbd@campus.technion.ac.il

Abstract. Educational games benefit from incorporating evidence-based learning principles. Most of these principles have been studied in conventional learning settings, and their applicability and possible benefits to games is unclear. Two such principles within the general framework of Desirable Difficulties are Spacing and Interleaving. Both refer to the idea of distributed practice over time (as opposed to blocked), with interleaving having the added benefit of presenting opportunities for comparison across stimuli. We designed a digital game on the topic of multiplication facts that implemented three possible sequencing of problems – blocked, spaced, and interleaved. One hundred and fifty elementary school students were randomly assigned to one of the conditions. In-game learning curve analysis of logs found that blocked presentation improves in-game performance. However, out-of-game tests found that interleaved presentation improves out-of-game performance efficiency. This work demonstrates the applicability and benefits of incorporating evidence-based principles into digital games.

Keywords: Educational games · Desirable difficulties · Learning rates · Classroom experiment · Multiplication facts

1 Introduction and Background

There is a substantial body of research showing that games can be effective tools for learning and assessment [1, 2]. Still, studies indicate that educational games can foster better learning by incorporating more evidence-based learning principles, grounded in cognitive theories of learning [3, 4]. Such principles have mainly been researched in the context of conventional learning environments (e.g., textbooks, lectures), and more research is needed to assess their applicability and inform their successful integration into game environments.

To contribute to this effort, we evaluated the effects of incorporating two robust learning principles into a digital game environment. The principles we focused on are Spacing and Interleaving. Both principles are considered 'desirable difficulties'[5], and deal with the sequential arrangement of the learning material – a common problem in

© The Author(s), under exclusive license to Springer Nature Switzerland AG 2023
N. Wang et al. (Eds.): AIED 2023, CCIS 1831, pp. 136–141, 2023.
https://doi.org/10.1007/978-3-031-36336-8_21

the design of educational games and tutors [6]. These difficulties may hinder immediate performance but lead to better long-term retention and possibly transfer ability [7].

We designed a single-digit multiplication practice game with three possible arrangements of problems – blocked, spaced, and interleaved. We evaluated the effects of these conditions by a combined approach of in- and out-of-game assessments. The in-game assessments are based on performance indicators and learning curve analysis of game logs. Out-of-game assessment is based on statistical analysis of pre- and post-test performance. We discuss lessons learned and implications for evidence based educational games design and evaluation.

Our research question is the following: What is the impact of spacing and interleaving of practice on in-game performance as well as transfer to out-of-game environment? We hypothesize that the game environment will show similar patterns to those in other settings: Blocked practice will facilitate immediate performance, while interleaved practice will support improved recall ability (retention) and speed.

2 Methods

2.1 Research Design

We used a pre-treatment-post randomized experiment to evaluate the effects of implementing Spacing and Interleaving in an educational game context. Additionally, in a conscious effort to generate inclusive and representative sample, we performed the experiment in 3 ethnically and religiously varied schools. A total of 154 third grade students participated in the study as part of their regular algebra curriculum. The study procedure was: pretest→game play session→posttest→demographic questionnaire.

2.2 Materials

The Game. We designed and implemented a web-based drill-and-practice game focused on single-digit multiplication. The game offers a fantasy setting, where players take on the role of dinosaur museum manager. As managers, they responsible on recruiting fossil diggers, feeding them and overseeing the diggings. Each of these actions is achieved by solving relevant representations of multiplication problems, with clear goals, feedback and progression systems as is common in games. The game included 9 multiplication problems, each repeating three times (for a total of 27 problems). There are three game versions corresponding to the three study conditions – Blocked, Interleaved and Spaced. Conditions differ *only* in the sequence of the multiplication problems in the game. In the Blocked version – each problem appeared 3 times in a row AND all of the problems of the same multiplicator (e.g., multiplications of 5s) were blocked together (Table 1). The Blocked condition is the baseline for this study. In the Interleaved version – problems were blocked by the same main multiplicator, but any two adjacent problems had a different second multiplicator (e.g., 5 * 6 followed by 5 * 4). Having the same main multiplicator, adjacent problems had something substantial in common, which offered opportunity for comparison and inferring from one problem to the other. In the Spaced version – semi-random sequence of problems, with the single limitation

Table 1. Sequencing of multiplication problems under the 3 conditions in the game. The table is color-coded so that all the problems with the same main multiplicator (e.g., 5) have the same overall color, and the second multiplicand determines the hue within the given color (e.g., light/medium/dark blue, in the case of multiplications of 5). This table only covers the first 6 questions, as an example.

Blocked	Interleaved	Spaced
5*5	5*5	5*5
5*5	5*6	7*6
5*5	5*4	8*9
5*6	5*5	5*6
5*6	5*6	7*9
5*6	5*4	8*8

that any two adjacent problems were NOT of the same main multiplicator (e.g., 5 * 4 followed by 7 * 7). Thus, spacing was created between same and similar multiplications.

The Data. We collected in-game data by recording every action taken within the game in game-logs (e.g., solution attempts, hints). Additionally, we administered pre- and post-tests online, and collected students answers along with time stamps.

2.3 Analysis

We focused our analysis on students' change in accuracy and time-on-task when solving problems in-game and in conventional out-of-game tests [8]. Thus, to evaluate the effect of condition on learning *in-game*, we conducted learning curve analysis (LCA) to assess the learning rate as well as statistical comparisons of *total* game time.

LCA is based on the classical theory of cognitive skill acquisition and the law of practice [9], which states that the more practice opportunities students get, the better their performance is. The analysis looks at students' performance on repeated opportunities of the same skills and can estimate the learning rate [10].

To assess students' transfer ability *out-of-game*, we statistically compared score and duration on pre- and post-test.

3   Results

3.1   In-Game Performance

Table 2 provides a summary of the means and standard deviations of average total game time and learning curve analysis. Results show that students on the Blocked group solved the game significantly faster (ANOVA, $F[2, 129] = 5.03, p = .008$, and $\eta^2 = .048$) and demonstrated the steepest learning curve – suggesting they learned faster than the other conditions. Spaced and Interleaved did not differ significantly from each other on either measure.

Learning curve analysis demonstrated negative slopes for all conditions, indicating a decrease in error rate, which suggests that learning occurs. The regression coefficients,

Table 2. In-game performance measures by conditions. Means and standard deviations (in parenthesis) of Total game time, and logistics regression coefficients with standard deviations (in parenthesis) of the learning curve analysis for the three conditions. * P-value < 0.05, ** P-value < 0.01

Condition	N	Total game time [min]	LCA learning estimate
Blocked	47	31.68 (9.62)	0.083 (0.017)**
Spaced	55	**36.67 (11.21)***	0.053 (0.002)**
Interleaved	52	**37.81 (8.88)***	0.044 (0.002)**

which are a measure of the learning rate between opportunities, were all significantly different from 0 ($p < .001$). Students in the blocked condition showed about double the learning rate when compared to the Interleaved group, which had the slowest rate (Fig. 1).

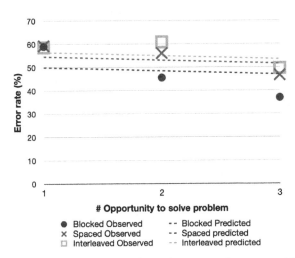

Fig. 1. Learning curve analysis – students' change in error rate by opportunities and condition.

3.2 Out-of-Game Learning Gains

Overall, when comparing Pre to Posttest performance, students did not significantly improve their test score: t (153) = 0.94, p = .35). However, students did improve significantly in test duration, solving the Posttest 30% faster on average across conditions: t (153) = 9.01 p = < 0.001, with effect size of 0.83 Cohen's D, which is considered a large effect. Table 3 provides a summary of the means and standard deviations of test scores and test duration, as a function of the three game conditions.

All three conditions showed a significant reduction in time duration from pre- to post-test, with students in the Interleaved group improving significantly more (almost double) than those on the Blocked group.

Table 3. Means and standard deviations (in parenthesis) for tests scores, normalized score change, test duration, and efficiency (see text for details). *P-value <= 0.01.

Condition	N	Test Score		Test Duration (Sec)	
		Pre	Post	Pre	Post
Blocked	47	6.72(1.6)	6.68(1.9)	233.24(118.2)	*159.80(92.7)**
Spaced	55	6.29(2.0)	6.51(2.1)	256.70(125.0)	*179.02(126.8)**
Interleaved	52	5.77(1.9)	6.02(2.2)	292.06(142.5)	*149.86(82.5)**

4   Discussion and Contribution

Our results demonstrate that in-game indicators often differ from out-of-game indicators. We thus used a combination of in-game performance indicators and learning curve analysis with out-of-game transfer and learning gains assessment with conventional multiplication tests. Others have combined in-game performance data with close expert monitoring and interpretation [11], or in-game data and verified tests [12].

The literature on desirable difficulties has repeatedly demonstrated the possible benefits of spaced and interleaved practice on the learning process in non-game environments. Our study confirms that these principles can have similar positive effects on learning in a game environment. Spacing, like interleaving, hindered immediate progress, but did not have the same positive impact on test duration. Since test duration was not limited in our experiment, we can regard it as an objective measure for fluency and cognitive difficulty [13]. Thus, this condition significantly reduced cognitive difficulty on an out-of-game task. This suggests that opportunities to compare and contrast may contribute more to learning than just spacing in a varied game environment.

These findings provide insights into the process of rigorously evaluating the effects of implementing learning principles in a digital game environment. In addition, this work demonstrates the applicability of desirable difficulties to game environments and problem sequencing in drill-and-practice math games.

References

1. Zhonggen, Y.: A meta-analysis of use of serious games in education over a decade. Int. J. Comput. Games Technol. **2019**, 4797032 (2019). https://doi.org/10.1155/2019/4797032
2. Lei, H., Chiu, M.M., Wang, D., et al.: Effects of game-based learning on students' achievement in science: a meta-analysis. J. Educ. Comput. Res. **60**, 1373–1398 (2022). https://doi.org/10.1177/07356331211064543
3. Mayer, R.E.: Computer games in education. Annu. Rev. Psychol. **70**, 531–549 (2019). https://doi.org/10.1146/annurev-psych-010418-102744
4. McLaren, B.M., Nguyen, H.: AIED digital learning games: a review. In: AIED Digital Learning Games (2020)
5. Bjork, R.A.: Institutional impediments to effective training. In: Learning, Remembering, Believing: Enhancing Individual and Team Performance, pp. 295–306. National Academy Press, Washington, DC (1994)

6. Pavlik, P., Bolster, T., Wu, S.-M., Koedinger, K., MacWhinney, B.: Using optimally selected drill practice to train basic facts. In: Woolf, B.P., Aïmeur, E., Nkambou, R., Lajoie, S. (eds.) ITS 2008. LNCS, vol. 5091, pp. 593–602. Springer, Heidelberg (2008). https://doi.org/10.1007/978-3-540-69132-7_62

7. Clark, C.M., Bjork, R.A.: When and why introducing difficulties and errors can enhance instruction. In: Benassi, V., Overson, C.E., Hakala, C.M. (eds.) Applying Science Learning Education Infusing Psychological Science in to the Curriculum, pp. 20–30. Society for the Teaching of Psychology (2014)

8. Rau, M.A., Aleven, V., Rummel, N.: Interleaved practice in multi-dimensional learning tasks: which dimension should we interleave? Learn. Instr. **23**, 98–114 (2013). https://doi.org/10.1016/j.learninstruc.2012.07.003

9. Newell, A., Rosenbloom, P.S.: Mechanisms of Skill Acquisition and the Law of Practice. In: Cognitive skills and their aquisition, pp. 1–56. Lawrence Erlbaum Associates Inc. (1981)

10. Martin, B., Koedinger, K.R., Mitrovic, A., Mathan, S.: On Using Learning Curves to Evaluate ITS. In: AIED, pp 419–426 (2005)

11. Harpstead, E., Aleven, V.: Using empirical learning curve analysis to inform design in an educational game. pp, 197–208 (2015). https://doi.org/10.1145/2793107.2793128

12. Shute, V., Rahimi, S., Smith, G., et al.: Maximizing learning without sacrificing the fun: Stealth assessment, adaptivity and learning supports in educational games. J. Comput. Assist. Learn. **37**, 127–141 (2021). https://doi.org/10.1111/JCAL.12473

13. Van Gog, T., Paas, F.: Instructional efficiency: revisiting the original construct in educational research. Taylor Francis **43**, 16–26 (2008). https://doi.org/10.1080/00461520701756248

Evaluating a Conversational Agent for Second Language Learning Aligned with the School Curriculum

Elizabeth Bear[(✉)] and Xiaobin Chen

University of Tübingen, Walter-Simon-Str. 12, 72072 Tübingen, Germany
elizabeth.bear@uni-tuebingen.de

Abstract. Interaction and its underlying processes have long been considered foundational to second language learning. Language learners, however, are often presented with few opportunities for meaningful, interactive practice. The rapid advancements in speech and language technologies in recent years have given rise to a number of conversational agents seeking to address this need within the interdisciplinary research area of dialogue-based computer-assisted language learning (CALL). While dialogue-based CALL has been shown to foster language proficiency development, it is crucial that a system is evaluated from multiple dimensions, from the technological sufficiency of the underlying components to learning and intercultural outcomes. In this paper, we present the plans and status of three studies targeting different aspects of a conversational agent, Aisla, developed for use in German secondary schools and aligned with the linguistic and intercultural goals of the English as a foreign language curriculum.

Keywords: Conversational agents · Second language acquisition · Computer assisted language learning

1 Introduction

Second language acquisition (SLA) research has been largely informed by the Interaction Hypothesis [22], which posits that, through the process of *negotiation for meaning*, interaction not only serves as a means of practice but is also a mechanism in which language is learned [12,22]. Language learners, however, are often presented with few opportunities for meaningful, interactive practice, especially in foreign language learning environments [25]. In the context of the German education system, less than 30% of German adults recently claimed that they were proficient speakers of a foreign language [11], suggesting a particular need for more speaking practice.

Supported by the German Federal Ministry of Education and Research and the LEAD Graduate School & Research Network, which is funded by the Ministry of Science, Research and the Arts of the state of Baden-Württemberg within the framework of the sustainability funding for the projects of the Excellence Initiative II.

The rapid advancements in speech and language technologies in recent years have given rise to a number of conversational agents, e.g., [17,29], seeking to address this need for more meaningful speaking practice. Conversational agents, also known as spoken dialogue systems for those targeting spoken language, constitute a major research strand within the broader interdisciplinary area of dialogue-based computer-assisted language learning (CALL) [1]. A recent meta-analysis has affirmed the effectiveness of dialogue-based CALL on language proficiency development and identified effective system features [2].

Conversational agents developed for second/foreign language learning differ from more general-purpose chatbots in their approach and goals [8]. In [6], we argued for the design of a conversational agent based on SLA theories, namely the interaction approach [12], and task-based language teaching (TBLT) [9]. We further noted our use of commercial systems for most of the AI services, which has accelerated our system development and "allows us to focus on designing learning contents and managing the learning process" [6, p. 584].

The research agenda presented in this paper is situated within a three-year project funded by the German Federal Ministry of Education and Research titled, *Aisla: An intelligent agent for second language English learning in real-life contexts*. To date, we have designed and implemented a range of tasks targeting the linguistic and intercultural goals of the English as a foreign language (EFL) curriculum in the state of Baden-Württemberg. This paper presents the plans and status of three studies targeting different aspects of our conversational agent, Aisla. The first explores the influence of learner and linguistic variables on the performance of major commercial automatic speech recognition (ASR) systems. The second evaluates the feasibility and effectiveness of automatic corrective feedback during interaction with a conversational agent. Given Aisla's naturalistic contexts within a TBLT framework, the third study examines if practice with a task-based conversational agent increases learners' intercultural communicative competence (ICC) [5], an overarching goal of the EFL curriculum.

2 Background

2.1 Automatic Speech Recognition

It is well-established that ASR systems may have difficulty with learner speech due to the acoustic properties, e.g., accents, and language, e.g., grammar and non-word errors [27,32]. Much of this work has targeted non-native accents and an ASR's acoustic model [32]. Within CALL, successful systems such as [31] have employed their own speech recognizers trained on learner language and targeted a limited number of language structures. Due to the accelerating technological developments, increased attention has been paid to commercial ASR performance on non-native speech [15], albeit with a continued focus on accents. The recently-developed conversational agents [29] and [17] used a commercial ASR, Google's, with positive outcomes but did not report detailed ASR metrics. We reexamine commercial ASR use in CALL from the perspective of *learner* factors, e.g., language background and proficiency level, and *language* factors, e.g., grammatical structure and error types at the word and sentence level.

2.2   Corrective Feedback

During interaction, opportunities arise for feedback on meaning, e.g., clarification requests, and on form, referred to as corrective feedback within SLA [12]. Providing corrective feedback automatically involves Grammatical Error Correction (GEC), ranging from rule-based to neural machine translation approaches [3]. While rule-based CALL systems have enhanced proficiency development [23], it is less clear whether automatic feedback can be adequately provided in less constrained conditions such as a spoken dialogue system, especially in conjunction with the ASR difficulties described in 2.1. Engwall et al. recently asserted, "Technologically, it is currently beyond the state of the art to robustly detect participant errors in free L2 conversations and provide adequate, explicit feedback" [10, p. 13]. Other work has suggested, however, that simple heuristics and interactivity can overcome technological limitations [28]. We will assess if automatic feedback based on currently available technologies, such as the open-source LanguageTool [19], is robust enough to enhance learning gains, at least for certain language structures, during interaction with a conversational agent.

2.3   Intercultural Communicative Competence

Specific targets for ICC are found within Common European Framework of Reference for Languages (CEFR) descriptors [7] and within the curriculum plan of Baden-Württemberg, with its EFL goals centered around ICC [18]. Despite the curricular embedding, the teaching of culture can be challenging for teachers [4], is often underexplored in school textbooks [20], and from an AIED perspective, represents an ill-defined domain [24]. ICC is also underrepresented in CALL research despite the affordances that technology can offer toward its development [14]. While ICC has been shown to be enhanced through, among other technologies, immersive simulations [16], open educational resources [21], and virtual exchange [26], as far as the authors are aware, ICC development has not been explored in the context of interaction with a task-based conversational agent in a mobile application, particularly one designed for use in secondary schools.

3   Research Questions

We summarize our research questions below:

1. How readily usable are major commercial ASRs for CALL targeting spoken language? How is ASR accuracy affected by *learner factors*—namely native language, proficiency level, and gender? How is ASR accuracy affected by *language factors*—grammatical correctness and structure type?
2. Does the provision of automatic corrective feedback increase learning gains during interaction with a conversational agent? What is the interplay between feedback effectiveness and technological limitations?
3. Does interaction with a task-based conversational agent foster the development of learners' ICC?

4 Methodology

RQ1 The evaluation of our first research question is underway in an online study aiming to recruit a heterogeneous group of speakers. At the time of submission, we have collected data from over 150 participants of 20 language backgrounds from A1 to C2 proficiency levels. Participants complete two task types: (1) reading aloud grammatical and ungrammatical sentences and (2) freer production dialogues mirroring those in Aisla. Our tasks are based on prompts from recent SLA research covering a range language structures occurring at different acquisition stages [13]. The recordings will be automatically processed through major commercial Asrs. The generated transcriptions will then be evaluated against the ground truth, i.e., transcriptions by human raters. We plan to analyze our data with mixed-effects models examining predictors related to the learner—native language, proficiency, and gender—and related to language—grammatical correctness and structure type at the word and sentence level.

RQ2 The second research question will take place this year in 7th grade English classes in Baden-Württemberg. For homework over a two-week period, participants will complete eight tasks (e.g., paying at a restaurant), four of which target comparatives, and four target modal verbs. In the feedback condition, we provide corrective feedback at each conversation turn using a rule-based approach. Based on the methodology of [23], we randomize within-class by target structure. In other words, half of each class will receive feedback on the four comparative tasks, and the other half will receive feedback on the modals tasks. Learning gains will be measured by pre- and post-tests, in which we expect larger improvements for the structure on which participants received corrective feedback. For a subset of the data, we also plan to examine the precision and recall of the feedback mechanism and ASR accuracy during real use in EFL classes.

RQ3 This follow-up study to RQ2 will take place over a period of at least four-weeks in 7th grade English classes in Baden-Württemberg. The homework tasks will follow a similar structure to those in RQ2, with the addition of elements for prediction and reflection based on [24]. Following [30]'s framework, we plan to measure gains in ICC through an enactment, e.g., a role play, which will be evaluated via a coding scheme, and a questionnaire administered at the start and end of the intervention. As with other AIED systems, in addition to learning outcomes, insights into the learning process can be viewed through student actions in Aisla during the tasks, predictions, and reflections, which we will also examine from a mixed-methods lens.

5 Outlook

Since our first two studies are underway or will take place shortly, we hope to share preliminary results. The results will guide AIED researchers on the feasibility and capabilities of using current ASR and feedback technologies in language learning applications and inform system and task design decisions for enhancing learning outcomes. Our third study, projected to occur during the following school year, will add an interdisciplinary perspective and further insights into the teaching of an ill-defined domain in AIED. Future directions include examining the characteristics, e.g., interactional style, of our agent, the use of avatars during interaction, and connections to learner affect.

References

1. Bibauw, S., François, T., Desmet, P.: Discussing with a computer to practice a foreign language: Research synthesis and conceptual framework of dialogue-based call. Comput. Assist. Lang. Learn. **32**(8), 827–877 (2019)
2. Bibauw, S., Van Den Noortgate, W., François, T., Desmet, P.: Dialogue systems for language learning: a meta-analysis. Language Learn. Tech. **26**(1) (2022)
3. Bryant, C., Yuan, Z., Qorib, M.R., Cao, H., Ng, H.T., Briscoe, T.: Grammatical Error Correction: A Survey of the State of the Art. arXiv preprint arXiv:2211.05166 (2022)
4. Byram, K., Kramsch, C.: Why is it so difficult to teach language as culture? Ger. Q. **81**(1), 20–34 (2008)
5. Byram, M.: Teaching and assessing intercultural communicative competence: Revisited, 2nd edn. Multilingual Matters, Bristol (2020)
6. Chen, X., Bear, E., Hui, B., Santhi-Ponnusamy, H., Meurers, D.: Education theories and AI affordances: design and implementation of an intelligent computer assisted language learning system. In: AIED 2022, pp. 582–585. Springer, Cham (2022). https://doi.org/10.1007/978-3-031-11647-6_120
7. Council of Europe: Common European Framework of Reference for Languages: Learning, teaching, assessment - Companion volume. Council of Europe Publishing, Strasbourg (2020)
8. Divekar, R.R., Lepp, H., Chopade, P., Albin, A., Brenner, D., Ramanarayanan, V.: Conversational agents in language education: where they fit and their research challenges. In: Stephanidis, C., Antona, M., Ntoa, S. (eds.) HCII 2021. CCIS, vol. 1499, pp. 272–279. Springer, Cham (2021). https://doi.org/10.1007/978-3-030-90179-0_35
9. Ellis, R.: Task-based language learning and teaching. Oxford University Press, Oxford (2003)
10. Engwall, O., et al.: Learner and teacher perspectives on robot-led L2 conversation practice. ReCALL, 1–16 (2022)
11. Eurostat: Foreign language skills statistics. https://ec.europa.eu/eurostat/statistics-explained/index.php/Foreign_language_skills_statistics, (Accessed 14 Jan 2023)
12. Gass, S., Mackey, A.: Input, interaction and output in second language acquisition. In: VanPatten, B., Keating, G.D., Wulff, S. (eds.) Theories in Second Language Acquisition: An Introduction, pp. 192–222, 3rd edn. Routledge, New York (2020)
13. Godfroid, A., Kim, K.M.: The contributions of implicit-statistical learning aptitude to implicit second-language knowledge. Stud. Second. Lang. Acquis. **43**(3), 606–634 (2021)
14. González-Lloret, M.: Technology and L2 pragmatics learning. Annu. Rev. Appl. Linguist. **39**, 113–127 (2019)

15. Hollands, S., Blackburn, D., Christensen, H.: Evaluating the performance of state-of-the-art ASR systems on non-native English using corpora with extensive language background variation. In: Proceedings of Interspeech 2022, pp. 3958–3962 (2022)
16. Johnson, W.L.: Using immersive simulations to develop intercultural competence. In: Ishida, T. (ed.) Culture and Computing. LNCS, vol. 6259, pp. 1–15. Springer, Heidelberg (2010). https://doi.org/10.1007/978-3-642-17184-0_1
17. Kim, H., Yang, H., Shin, D., Lee, J.H.: Design principles and architecture of a second language learning chatbot. Lang. Learn. Tech. **26**(1), 1–18 (2022)
18. Kultusministerium: Englisch als erste fremdsprache. In: Bildungsplan des Gymnasiums 2016 (2016), http://www.bildungsplaene-bw.de/, Lde/LS/BP2016BW/ALLG/GYM/E1 (Accessed 14 Jan 2023)
19. LanguageTool. https://languagetool.org
20. Limberg, H.: Teaching how to apologize: EFL textbooks and pragmatic input. Lang. Teach. Res. **20**(6), 700–718 (2016)
21. Lin, Y.J., Wang, H.C.: Using enhanced OER videos to facilitate English L2 learners' multicultural competence. Comput. Educ. **125**, 74–85 (2018)
22. Long, M.: The role of the linguistic environment in second language acquisition. In: Ritchie, W., Bhatia, T. (eds.) Handbook of Language Acquisition. Second Language Acquisition, vol. 2, pp. 413–468. Academic Press, San Diego (1996)
23. Meurers, D., De Kuthy, K., Nuxoll, F., Rudzewitz, B., Ziai, R.: Scaling up intervention studies to investigate real-life foreign language learning in school. Annu. Rev. Appl. Linguist. **39**, 161–188 (2019)
24. Ogan, A., Aleven, V., Jones, C.: Advancing development of intercultural competence through supporting predictions in narrative video. Int. J. Artif. Intell. Educ. **19**(3), 267–288 (2009)
25. Ortega, L.: Meaningful L2 practice in foreign language classrooms: A cognitive-interactionist SLA perspective. In: DeKeyser, R.M. (ed.) Practice in a second language: Perspectives from applied linguistics and cognitive psychology, pp. 180–207. Cambridge University Press, New York (2007)
26. O'Dowd, R.: Virtual exchange: moving forward into the next decade. Comput. Assist. Lang. Learn. **34**(3), 209–224 (2021)
27. Park, S., Culnan, J.: A comparison between native and non-native speech for automatic speech recognition. J. Acoust. Soc. Am. **145**(3), 1827–1827 (2019)
28. Robertson, S., Munteanu, C., Penn, G.: Designing pronunciation learning tools: The case for interactivity against over-engineering. In: Proceedings of the 2018 CHI Conference on Human Factors in Computing Systems, pp. 1–13 (2018)
29. Ruan, S., et al.: EnglishBot: an AI-powered conversational system for second language learning. In: 26th International Conference on Intelligent User Interfaces, pp. 434–444 (2021)
30. Sercu, L.: Assessing intercultural competence: A framework for systematic test development in foreign language education and beyond. Intercult. Educ. **15**(1), 73–89 (2004)
31. de Vries, B.P., Cucchiarini, C., Bodnar, S., Strik, H., van Hout, R.: Spoken grammar practice and feedback in an ASR-based CALL system. Comput. Assist. Lang. Learn. **28**(6), 550–576 (2015)
32. Wei, X., Cucchiarini, C., van Hout, R., Strik, H.: Automatic Speech Recognition and Pronunciation Error Detection of Dutch Non-native Speech: cumulating speech resources in a pluricentric language. Speech Commun. **144**, 1–9 (2022)

EngageMe: Assessing Student Engagement in Online Learning Environment Using Neuropsychological Tests

Saumya Yadav[1]([✉]) [ID], Momin Naushad Siddiqui[2] [ID], and Jainendra Shukla[1] [ID]

[1] HMI Lab, IIIT-Delhi, New Delhi, India
{saumya,jainendra}@iiitd.ac.in
[2] Jamia Millia Islamia, New Delhi, India

Abstract. In the proposed research, we investigated whether the standardized neuropsychological tests commonly used to assess attention can be used to measure students' engagement in online learning settings. Accordingly, we employed 73 students in three clinically relevant neuropsychological tests to assess three types of attention. Students' engagement performance, as evidenced by their facial video, was also annotated by three independent annotators. The manual annotations observed a high level of inter-annotator reliability (Krippendorffs' Alpha of 0.864). Further, by obtaining a correlation value of 0.673 (Spearmans' Rank Correlation) between manual annotation and neuropsychological tests score, our results show construct validity to prove neuropsychological test scores' significance as a latent variable for measuring students' engagement. Finally, using non-intrusive behavioral cues, including facial action unit and eye gaze data collected via webcam, we propose a machine learning method for engagement analysis in online learning settings, achieving a low mean squared error value (0.022). The findings suggest a neuropsychological test-based machine learning technique could effectively assess students' engagement in online education.

Keywords: Human-Computer Interaction · Affective Computing · Online Learning · Cognitive Engagement Assessment · Attention

1 Introduction and Motivation

The online education system offers many benefits, such as cost-effectiveness and accessibility to students, but the shift to online education during the COVID-19 pandemic has made it challenging for instructors to assess student engagement. Assessing engagement is important for both instructors and students. Researchers have used different modalities like audio and video to detect engagement in digital learning environments. The use of webcams is a common focus in recent research on assessing student engagement. Engagement is a multidimensional construct composed of three dimensions: emotional, behavioural, and

N. Wang et al. (Eds.): AIED 2023, CCIS 1831, pp. 148–154, 2023.
https://doi.org/10.1007/978-3-031-36336-8_23

cognitive. These dimensions are dynamically interrelated factors for every student. Emotional engagement focuses on effective reactions, while behavioural engagement focuses on the action of students [10]. Cognitive engagement is a psychological process involving attention, and investment [7].

In this study, attention is considered a fundamental aspect of cognitive engagement, and is assessed as a latent variable. Attention levels are determined using neuroanatomic theories, psychometric tests, cognitive processing, and clinical-based models. The clinical-based model comprises five components for measuring attention levels: Alternating, Divided, Focused, Selective, and Sustained attention. Alternating, Divided, and Focused attention are critical for students, as they facilitate shifting focus [8]. Selective attention is also relevant for academic foundations and can be improved through training [12]. Finally, Sustained attention refers to the ability to pay attention for a longer period [6].

Prior research is not consistent in considering the relationship between attention and engagement. Engagement is a multi-faceted concept that plays a vital role in learning and academic achievement. It is crucial to acknowledge that engagement and attention are separate cognitive processes, with attention being frequently mistaken as engagement in [4,13]. Further, prior literature has mostly utilized explicit methods of collecting the ground truth [7,9] which induces distractions, adds workload on the users, and does not provide objective information to presenters. To accurately measure attention among students, it is necessary to consider all aspects of attention. Currently, no dataset is available that computes the ground-truth labels for comprehensive attention. We propose EnagageMe, a novel technique that uses standardized neuropsychological assessments to create coherent ground truth for comprehensive attention assessment and utilizes non-intrusive behavioral cues through a webcam to measure student attention. We assess the validity of neuropsychological tests in recognizing comprehensive attention among students in an online environment. We evaluate the usefulness of non-intrusive features behaviourial cues such as facial action units (FAUs) and gaze points as indicators of comprehensive attention [5,16].

2   Methodology for EngageMe

After obtaining ethics approval from the Institutional Review Board of Indraprastha Institute of Information Technology, Delhi, India (IIITD/IEC/08/ 2021-4), data collection for training purposes was carried out from 73 participants. This included obtaining webcam recordings and cognitive task responses in CSV format. Both the CSV data and video were divided into 60-second segments, using timestamps to ensure no data loss during splitting. The CSV data provided labels for the participants' performance, while features were extracted from each video segment. The experiment was created using jsPsych [2]. No specialized hardware or software was required for data collection, participants were just allowed access to their webcams, resulting in a normal study environment experience. It took approximately 20 min for each participant.

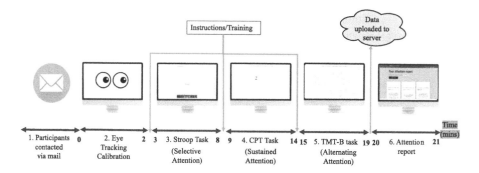

Fig. 1. Experiment timeline for EnagageMe data collection.

A webcam-based eye tracker (webgazer.js) was used in the experiment. The eye tracker needed to be calibrated before starting the cognitive tasks. Eye-tracker validation was also added to ensure that the webgazers' accuracy was greater than 60%. The experiment timeline is shown in Fig 1.

- Stroop Test: The Stroop test is a widely used measure of selective attention [15]. We used a computerized version of the test in which participants used arrow keys to report the font color, with a fixed stimulus time of 4000 ms and variable post-trial gap of 100–500 ms. The test was designed to match webgazer response coordinates in post-processing, and no feedback was provided during the actual experiment.
- Continuous Performance Test (CPT): The CPT test is a widely used measure for sustained attention [3]. In our experiment, participants had to respond to a sequence of letters appearing on the screen, and press "M" if the current letter was the same as the one presented two items back (N-back CPT). The stimulus duration was 1500 ms, and there was a 500 ms post-trial gap.
- Trail Making Test (TMT): The TMT test is a frequently used neuropsychological test for assessing cognitive performance, and different types of attention [11]. We used TMT-B task for analyzing the participants' alternating attention for our study. Participants were supposed to select a sequence of alphanumeric bubbles that turned green once selected. Three trials for the TMT-B task were conducted.

2.1 Label Generation from Generated CSVs

Data Exploration and Labelling:

- Stroop: In the Stroop test, a significant difference was found in reaction time between congruent and incongruent stimuli ($t(2,73)=3.058$, $p < 0.05$). The mean reaction time for congruent stimuli was 892.952 ms, and for incongruent stimuli was 937.497 ms. Correctness of response was used as the scoring metric and accuracy scores, ranging from 0–1, were calculated. 1 indicating the highest level of attention and 0 indicating the lowest level of attention.

- CPT: In the CPT test, attention is measured by the number of correct hits, and inattention/impulsivity by errors (omission/commission). A weak negative correlation was found between false alarms and median reaction time (-0.196, $p < 0.05$). The CPT score is based on accuracy and non-matching cases only, scaled to a range of 0–1, where 0 represents low attention and 1 represents high attention.
- TMT: In TMT, response time decreases as the test progresses due to fewer competing stimuli. The mean response time for three trials was 3148 ms, 2580 ms, and 3110 ms. The performance metric used was total time taken in TMT trials, which was scaled to a score range of 0–1. A higher score implies better performance, indicating less time taken.

2.2 Manual Annotation

We established construct validity for our new attention test. We compared neuropsychological or automated test scores with manual annotation scores. For this, we hired a team of 3 annotators to annotate the recorded clips of the participants. Annotators were instructed to label the clips based on attention, using a scale of 0–3. Attention categories were inspired by a previous study [14], and the scale ranges from highly attentive (3), moderately attentive (2), low attentive (1), and inattention (0). In the neuropsychological test, the continuous spectrum was discretized into four classes by dividing the distribution into segments within 0.5 of the standard deviation. Labels were decided based on performance scores to facilitate a multi-labeled classification model of attention.

2.3 Feature Extraction

We extracted head pose, and facial features from OpenFace [1]. OpenFace generated 712 features for each frame in a clip. The features are related to Gaze, Pose, FAUS, Landmarks locations in 2D and in 3D. During data pre-processing, we discarded the data for users where the output for success (successful tracking of the face in the frame) was zero.

3 Results and Discussion

In this section, we present the results of two experiments. The first experiment establishes the validity of the attention assessment tool by measuring the correlation between neuropsychological test scores and manual annotations. Krippendorff's Alpha was used to measure inter-annotator agreement, resulting in a high agreement score of 0.869. Spearman's correlation was used to measure the agreement between annotators, and a minimum correlation of over 81% was found, indicating strong agreement. The construct validity was further supported by a significant relationship (0.673) found between automated test scores and manual annotation using Spearman's Rank correlation. Our second experiment aimed to determine whether students' behavioral cues could be used to measure their attention levels during cognitive tasks. We used a Bi-LSTM model to generate embeddings from the extracted features of each clip, with the size of the embedding being $100 \times N$, where N is the number of clips for different tests. These embeddings were then input into Machine Learning (ML) algorithms to generate different regression models. The dataset was partitioned into training and testing sets with a ratio of 80:20, respectively, at a user level. The training set was trained for 5-fold cross-validation to evaluate the performance of each model, which were trained for 100 iterations with a learning rate of 0.1. Table 1 shows the mean squared error (MSE) for each model.

Table 1. MSE results on EngageMe

Model	MSE
AdaBoost	0.0279
Gradient Boosting (GB)	0.0541
Lightweight Gradient-Boosting Model (LGBM)	**0.0227**
Support Vector Regression (SVR)	0.546

The ensemble algorithms (GB, AdaBoost, and LGBM) outperformed the linear models (SVR). Specifically, the LGBM algorithm achieved the best results with a validation MSE of 0.022. This supports the hypothesis that facial behavioral cues can be reliable indicators of attention levels during cognitive tasks.

4   Conclusion and Future Work

The proposed research has demonstrated the potential for using standardized neuropsychological tests, commonly used to assess attention, as a measure of students' engagement in online learning environments. We found a high level of inter-annotator reliability and construct validity between the manual annotations of students' engagement and the neuropsychological test scores. Additionally, a ML method was proposed using non-intrusive behavioral cues, to analyze engagement in online learning settings, achieving a low mean squared error value. The results of this study suggest that the proposed neuropsychological tests-based ML technique could be a useful tool for instructors to assess students' engagement in online education contexts. Future work may also correlate it to the grasping capacity of a student, linking neuropsychological tests to attention and Intelligence Quotient (IQ).

Acknowledgements. This research work is funded by a research grant (Ref. ID.: IHUB Anubhuti/Project Grant/03) of IHUB Anubhuti-IIITD Foundation and is partly supported by the Infosys Center for AI and the Center for Design and New Media (A TCS Foundation Initiative supported by Tata Consultancy Services) at IIIT-Delhi, India.

References

1. Amos, B., Ludwiczuk, B., Satyanarayanan, M.: Openface: A general-purpose face recognition library with mobile applications. Tech. Rep. CMU-CS-16-118, CMU School of Computer Science (2016)
2. De Leeuw, J.: Jspsych: a javascript library for creating behavioral experiments in a web browser. Behav. Res. Methods **47**, 1–12 (2014)
3. Ghassemi, F., Moradi, M.H., Doust, M.T., Abootalebi, V.: Classification of sustained attention level based on morphological features of eeg's independent components. In: 2009 ICME International Conference on Complex Medical Engineering (2009)
4. Goldberg, P., et al.: Attentive or not? toward a machine learning approach to assessing students' visible engagement in classroom instruction. Educ. Psychol. Rev. **33**, 27–49 (2019)
5. Hutt, S., Krasich, K., R. Brockmole, J., K. D'Mello, S.: Breaking out of the lab: Mitigating mind wandering with gaze-based attention-aware technology in classrooms. In: Proceedings of the 2021 CHI Conference on Human Factors in Computing Systems, pp. 1–14 (2021)
6. Ko, L.W., Komarov, O., Hairston, W.D., Jung, T.P., Lin, C.T.: Sustained attention in real classroom settings: an eeg study. Front. Hum. Neurosci. **11**, 388 (2017)
7. Lackmann, S., Léger, P.M., Charland, P., Aubé, C., Talbot, J.: The influence of video format on engagement and performance in online learning. Brain Sci. **11**(2), 128 (2021)
8. Lai, Y.J., Chang, K.M.: Improvement of attention in elementary school students through fixation focus training activity. Int. J. Environ. Res. Public Health **17**(13), 4780 (2020)
9. Linson, A., Xu, Y., English, A.R., Fisher, R.B.: Identifying student struggle by analyzing facial movement during asynchronous video lecture viewing: Towards an automated tool to support instructors. In: Lecture Notes in Computer Science, pp. 53–65 (2022)
10. Renninger, K.A., Bachrach, J.E.: Studying triggers for interest and engagement using observational methods. Educ. Psychol. **50**, 58–69 (2015)
11. Sohlberg, M.K.M., Mateer, C.A.: Improving attention and managing attentional problems. Ann. N. Y. Acad. Sci. **931**, 359–375 (2006)
12. Stevens, C., Bavelier, D.: The role of selective attention on academic foundations: a cognitive neuroscience perspective. Dev. Cogn. Neurosci. **2**, S30–S48 (2012)
13. Szafir, D., Mutlu, B.: Pay attention! In: Proceedings of the SIGCHI Conference on Human Factors in Computing Systems (2012)
14. Whitehill, J., Serpell, Z., Lin, Y.C., Foster, A., Movellan, J.R.: The faces of engagement: automatic recognition of student engagement from facial expressions. IEEE Trans. Affect. Comput. **5**, 86–98 (2014)

15. Wright, B.C.: What stroop tasks can tell us about selective attention from child-hood to adulthood. Br. J. Psychol. **108**(3), 583–607 (2017)
16. Zagermann, J., Pfeil, U., Reiterer, H.: Studying eye movements as a basis for measuring cognitive load. In: Extended Abstracts of the 2018 CHI Conference On Human Factors In Computing Systems, pp. 1–6 (2018)

Exploring the Effects of "AI-Generated" Discussion Summaries on Learners' Engagement in Online Discussions

Xinyuan Hao[(✉)] [iD] and Mutlu Cukurova [iD]

University College London, London, UK
`xinyuan.hao.21@ucl.ac.uk`

Abstract. Discussion summaries have been considered important for promoting peer interactions in online discussion forums. However, most previous research only focused on student-generated summaries and their impact on learners in the summarizing roles. Even though recent artificial intelligence (AI) progress has demonstrated the possibility of generating decent forum post summaries, little attention has been paid to the educational implications of such auto-generated forum post summaries. To further understand the perceptions and reactions of students to AI-generated forum summaries, this research used a Wizard of Oz approach to collect students' online discussion forum log data with and without the "AI-generated" summaries. The results indicated that making an auto-generated summary available for students might not necessarily boost their interactions and engagements, especially for inactive students. However, the summary could serve as a reminder for students to participate in the discussion forum in a timely manner. Future research is needed to investigate whether the timing of providing an auto-generated summary might influence its impact.

Keywords: Auto-generated text summarization · Online discussion forum · Online engagement

1 Introduction

Online discussion forum has been seen as a valuable tool for promoting computer-supported collaborative learning (CSCL). However, because of the large scale of discussion and great diversity of interactions on the online discussion forums, it is difficult for learners to get an overview of the discussion content or find the specific information they are looking for [1]. Moreover, the current standard design of an online discussion forum might lead learners to focus more on the newest posts, which will result in the neglect of previous vital posts that are worth revisiting. Attending to others' posts in depth and breadth is the premise of deepening learners' understanding and creating interactive online dialogues. As summary has long been seen as an effective way to improve comprehension [2] and potentially initiate further discussion [3], having a summary of all the critical posts might be beneficial to online learners.

© The Author(s), under exclusive license to Springer Nature Switzerland AG 2023
N. Wang et al. (Eds.): AIED 2023, CCIS 1831, pp. 155–161, 2023.
https://doi.org/10.1007/978-3-031-36336-8_24

2 Related Work

2.1 Previous Studies on Manual Forum Post Summarization

A manual summarization can be provided by instructors or by students individually or collaboratively. Many scholars [4] claimed that learners with a summarizing role will "listen" to others' posts in breadth, and in return, they tend to outperform their peers in academic achievements [5]. Moreover, the student-generated summary also allows other learners to continue further discussion based on the summary [3], which indicated that it might not be necessary for students to participate in generating a summary to benefit from it. It is crucial to note that in the long term, asking students to generate discussion summaries either individually or collaboratively might demotivate them [6], as learners reported frustration and reluctance when being asked to summarize the ideas that they disagree with [7]. As writing a summary can be a time-consuming and laborious task, which instructors and students, especially those who do not even read the whole discussion, are unwilling to do so, an auto-generated summary might be an alternative to support online learners while reducing the adverse side effects.

2.2 Previous Studies on Auto-generated Forum Post Summarization

Despite the fact that auto-generated text summarization has gained numerous scholars' attention, only a few previous studies have explicitly focused on forum summarization [8]. Plus, most of them have focused on extractive summarization rather than abstractive summarization. Even though these researchers proposed that extractive summaries could represent the true meaning of the original text and have excellent readability [9], it is worth noticing that an extractive summary is incapable of synthesizing redundant posts with a different point of view [7]. This issue is particularly problematic for educational contexts in which one of the learning goals is to improve students' awareness and understanding of different views on the same topic discussed.

Nevertheless, the success in auto-generated extractive summary motivated scholars to challenge the more complicated abstractive summarization. A recent study using Bidirectional Encoder Representation from Transformer (BERT) successfully generated abstractive summaries of online news [10]. Although the development in the AI field has demonstrated the possibility of auto-generated forum summaries, there is a gap in research investigating the educational value of such summaries. Existing research on forum summarization mainly covers the technical issues of improving the accuracy of capturing contextual information [8] and the ease with which the summaries can be understood by humans [9] and does not discuss the impact of the summary on learners' online behaviors and outcomes. Furthermore, Zhang et al.[7] observed that learners might have trust issues and interact differently with summaries formed with the help of automatic summarization. Therefore, although the potential is highlighted in various narratives, little is known about how an auto-generated forum post summary impacts learners' engagement in online discussions and outcomes. In this paper, we investigate the influence of an AI-generated abstractive forum summary on learners' posting behaviors. More specifically, we aim to address the research question: To what extent do the AI-generated summaries have impacts on the number of students' posts, the length of the posts and the timing of posting?

3 Methodology

A Wizard of Oz approach was used in this research. Specifically, the first researcher manually summarized the posts following the same procedure as a transformer-based summarizer and posted the summaries online as an AI-agent. Existing transformer-based summarizers mainly deal with long, non-academic texts [11]. Few empirical studies demonstrated AI transformers' ability to produce summaries based on multiple authors' short, academic discussion posts. Thus, using the Wizard of Oz method, this study aims to get more information about potential engagement patterns [12] to help further fine-tune the AI-summarizers to be used in educational settings.

3.1 Participants

The research participants were postgraduate students enrolled in a postgraduate module related to education in the 2021–2022 and 2022–2023 academic years. It was not ethically and logistically possible to randomly allocate students from the same academic year to the control and experimental groups while ensuring the control group students remain blind to the provided summaries since students in the experimental group might cite things from the summaries or mention them in private conversations. Hence, the researchers used the data collected in the academic year of 2021–2022 without the experimental intervention with summaries as the control group data.

There were 50 participants in the control group and 42 participants in the experiment group. The module lecturer requested the participants to write one post and reply to at least two posts on the forums every week from week two to week nine. No summary was provided for the control group participants. The experiment group participants were informed that forum summaries provided by an AI transformer would be posted online using the first researcher's account on Tuesday and Friday evenings.

3.2 Research Procedure

The first researcher scrutinized the original posts to fix the spelling and grammar errors and deleted the hyperlinks to reduce the risk of misinterpretation [13], as an AI summarizer would do. Then, the individual summaries of every post were created manually. Each summary was 10% of the length of the original text at the shortest, and 30% at the longest. The range was decided with reference to the ratio between human written summary length and source text length provided in prior research [14].

The summaries of each post were used as input to produce weekly forum post summaries, which were topic-based and presented in bullet points. This categorized design was selected because it ranked the highest by students among the four mainstream designs [11]. All posts and replies created before every Tuesday at 12 pm were included in the first version of the weekly summary. Then it was posted on Tuesday evening. The timing choice was made due to practical reasons and the potential benefits of reading "mid-discussion" summaries on initiating a new level of discussion [3].

After each week's in-class session, the first researcher continued collecting the late student posts. The same procedure was repeated to produce the updated version of the weekly forum summary. The number of topics in the updated summary started from 10, as suggested in Xiao et al.'s [15] findings. All posts and replies created before every Friday at 3 pm were included in the updated summary. Then it was posted online by the end of Friday. The timing choice was made since the following week's tasks were accessible to students on Friday. Thus, Friday was considered the end of the week, and the updated summary was seen as the "end-of-discussion" summary.

The learning management system (LMS) hosting the discussion forum would automatically log data on participants' online activities, such as posting, replying, and viewing. In this publication, the researchers primarily concentrate on posting activities. Three variables were compared between the control and experiment groups: the number of posts, the average length of each post and the timing of posting. Replying and viewing activities as indicators of engagement were not included due to the length restriction of the paper. Additionally, participants could check others' posts and the summaries through emails. Since these viewing activities were not visible in the log data, interviews with participants are needed to understand their viewing experience.

Table 1. Comparison of the average length of each post between the control group and experiment group from week two to week nine

Average Post Length (Words)	Control Group	Experiment Group
Week 2	423.83	496.76
Week 3	429.21	365.37
Week 4	443.62	327.24
Week 5	448.63	334.37
Week 6	404.03	322.67
Week 7	418.69	300.93
Week 8	446.76	228.37
Week 9	399.28	307.41

Fig. 1. Comparison of the number of posts between the control group and experiment group from week two to week nine

4   Results and Discussion

From Fig. 1 and Table 1, it can be observed that for most of weeks, experiment group participants generated fewer posts than control group students and their posts were generally shorter. Even though the control group participants also experienced a drop in engagement, significantly more students persisted in writing lengthy posts until the end of the course than experiment group students. Interestingly, there was a significant fluctuation in the number of students posting on Tuesday in the experimental group on week six and eight. It is essential to note that these changes might be related to uncontrolled contextual variables. For instance, week six was the first week after the reading week, and the participants had a draft submission deadline for another significant module of their program in week eight. This might indicate that the experiment group students were more likely to be influenced by other pressures.

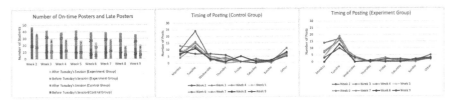

Fig. 2. Comparison of the number of on-time posters and late posters and the timing of posting between the control group and experiment group from week two to week nine

Judging from the number and length of posts, experiment group learners seemed less enthusiastic about engaging online, but some factors need to be considered when interpreting the phenomenon. It is worth noticing that whereas 13 people in the control group took the module entirely online, only five participants in the experiment group did the same. Moreover, since the impacts of COVID-19 on face-to-face learning were stronger in the 2021–2022 academic year, the control group students might be more inclined to interact online than the experiment group participants. Additionally, the data about posting time might show a different story. The suggested posting time for students was Tuesday before the beginning of the in-person session. Therefore, students who posted before Tuesday's session were classified as "on-time posters," and the participants who created their posts later than when the class started were seen as "late posters". Figure 2 presents that despite the drop in the total number of participants posting online in the experiment group, a higher percentage of the students who posted online were on-time posters than in the control group. Moreover, the late posters in the experiment group seemed to be more frequently "on time". Compared the control group, fewer people did the catch-up activities after the week ended in the experiment group. Another point worth mentioning is that instead of catching up on Thursday as the control group learners did, the experiment group late posters were more likely to finish this activity on Friday when the updated version of summaries was provided. Even though, from the log data comparison, making an auto-generated summary available to students might not necessarily lead to more online engagement, the summary might have a reminder effect on students to help them participate in an online discussion on time. Another hypothesis is that the timing of posting the summaries might have an impact on their functions. Previous research identified that offering an "end-of-discussion" summary might have a wrapping-up effect and sometimes might accidentally interrupt the discussion [3]. Even though the content of the first version of the weekly summary was collected in the middle of the discussion, by the time the first researcher finished the summarizing task and posted it online, the ongoing discussion was closer to its end than the middle. Hence, the Tuesday version of the summary was intended to serve as a "mid-discussion" summary, which has been proven to facilitate further discussion, but it ended up being similar to the "end-of-discussion" summary, which might interrupt the flow of the interaction.

5 Conclusion

In this study, we investigated the impact of "AI-generated" forum post summaries on students' online engagement, especially posting behaviors. As the results show, making an AI-generated forum post summary available to students might not necessarily boost their online engagement. However, the summary might be considered a reminder for students to participate in the discussion in a timely manner. Further conclusions can only be made with further research with participants to deepen our understanding of students' experience with the proposed summaries. Future research using real AI-generated summaries with tighter control of the contextual factors, such as reading week interruption, will be conducted, and the results will be compared with this study's to explore the impact of AI-generated summaries on learners' online engagement and any potential influence caused by the timing of posting the summaries.

References

1. Zhang, A.X.: Building systems to improve online discussion. In: Companion of the 2018 ACM Conference on Computer Supported Cooperative Work and Social Computing, pp. 65–68. Association for Computing Machinery, New York (2018)
2. King, A.: Comparison of self-questioning, summarizing, and notetaking-review as strategies for learning from lectures. Am. Educ. Res. J. **29**(2), 303–323 (1992)
3. Wise, A., Chiu, M.M.: Analyzing temporal patterns of knowledge construction in a role-based online discussion. Int. J. Comput.-Support. Collab. Learn. **6**(3), 445–470 (2011)
4. Schellens, T., Van Keer, H., De Wever, B., Valcke, M.: Scripting by assigning roles: does it improve knowledge construction in asynchronous discussion groups? Int. J. Comput.-Support. Collab. Learn. **2**(2), 225–246 (2007)
5. Peterson, A.T., Roseth, C.J.: Effects of four CSCL strategies for enhancing online discussion forums: Social interdependence, summarizing, scripts, and synchronicity. Int. J. Educ. Res. **76**, 147–161 (2016)
6. Wise, A.F., Chiu, M.M.: The impact of rotating summarizing roles in online discussions: effects on learners' listening behaviors during and subsequent to role assignment. Comput. Hum. Behav. **38**, 261–271 (2014)
7. Zhang, A., Verou, L., Karger, D.: Wikum: bridging discussion forums and wikis using recursive summarization. In: Proceedings of the 2017 ACM Conference on Computer Supported Cooperative Work and Social Computing (CSCW'17), pp. 2082–2096. Association for Computing Machinery, New York (2017)
8. Tarnpradab, S., Liu, F., Hua, K.: Toward extractive summarization of online forum discussions via hierarchical attention networks. In: Rus, V., Markov, Z. (eds.) The Thirtieth International Florida Artificial Intelligence Research Society Conference, pp. 288–292. The AAAI Press, Florida (2018)
9. Gerhana, Y.A., et al.: Text summarization using Textrank for knowledge externalization from Indonesian online discussion forums. In: 2021 7th International Conference on Wireless and Telematics (ICWT), pp. 1–7. IEEE, Danbung (2021)
10. Ramina, M., Darnay, N., Ludbe, C., Dhruv, A.: Topic level summary generation using BERT induced abstractive summarization model. In: 2020 4th International Conference on Intelligent Computing and Control Systems (ICICCS), pp. 747–752. IEEE, Madurai (2020)
11. Uddin, G., Baysal, O., Guerrouj, L., Khomh, F.: Understanding how and why developers seek and analyze API-related opinions. IEEE Trans. Softw. Eng. **47**(4), 694–735 (2021)

12. Riek, L.D.: Wizard of Oz studies in HRI: a systematic review and new reporting guidelines. J. Hum. Robot Interact. **1**(1), 119–136 (2012)
13. Gottipati, S., Shankararaman, V., Ramesh, R.: TopicSummary: a tool for analyzing class discussion forums using topic based summarizations. In: 2019 IEEE Frontiers in Education Conference (FIE), pp. 1–9. IEEE, Uppsala (2019)
14. Botarleanu, R.M., Dascalu, M., Allen, L.K., Crossley, S.A., McNamara, D.S.: Multitask summary scoring with longformers. In: Rodrigo, M.M., Matsuda, N., Cristea, A.I., Dimitrova, V. (eds.) Artificial Intelligence in Education, pp. 756–761. Springer, Durham (2022)
15. Xiao, C., Shi, L., Cristea, A., Li, Z., Pan, Z.: Fine-grained main ideas extraction and clustering of online course reviews. In: Rodrigo, M.M., Matsuda, N., Cristea, A.I., Dimitrova, V. (eds.) Artificial Intelligence in Education, pp. 294–306. Springer, Durham (2022)

Building Educational Technology Quickly and Robustly with an Interactively Teachable AI

Daniel Weitekamp[✉]

Carnegie Mellon University, Pittsburgh, PA 15213, USA
weitekamp@cmu.edu

Abstract. Intelligent tutoring systems (ITSs) aim to support student learning through comprehensive adaptive features, making for costly development times—about 200–300 hours of development time per hour of instruction. This proposal outlines plans to overcome several technical challenges toward building authoring tools whereby a non-programmer can build ITSs by interactively teaching simulated students. I propose both interaction design considerations and machine-learning innovations. These include a multi-modal natural language processing mechanism that mimics student learning from narrated tutorial instruction, and an active-learning mechanism that identifies training examples likely to eliminate inaccuracies in the simulated student's induced production rules. I propose to evaluate these features over 3 user studies and evaluate the generality of this authoring method in a final open-ended authoring study. This work aims to democratize ITS authoring by opening new authoring opportunities to non-programmers by making authoring as time-efficient and natural as human-to-human tutoring.

Keywords: Simulated Students · ITS Authoring · Interactive Task Learning

1 Introduction

This work proposes several advances on existing simulated student-based authoring tools [10,14]. The proposed work falls within the criteria of an interactive task learning (ITL) system [5]—it succeeds at inducing whole domain-specific programs via interactively teachable and domain-general mechanisms of learning. Beyond domains covered by prior work on programming by demonstration [2,6] learning ITS behavior for academic tasks has a number of unique challenges. For instance steps in STEM tasks often utilize mathematical formulae—not just simple atomic interface manipulations. Another challenge of these sorts of tasks is that they very often have multiple solution paths [1] and sequences of steps that vary depending on some context specific decision process. The proposed work introduces solutions to both of these challenges, and introduces a means of automatically verifying user programs.

© The Author(s), under exclusive license to Springer Nature Switzerland AG 2023
N. Wang et al. (Eds.): AIED 2023, CCIS 1831, pp. 162–168, 2023.
https://doi.org/10.1007/978-3-031-36336-8_25

1.1 Prior Work on ITS Authoring

Although powerful at supporting student learning [4], production rule based ITSs are notoriously difficult and time consuming to build. One estimate places the time cost of building one hour of Cognitive Tutor based instruction at 200–300 hours of highly specialized developer time [1]. Subsequent efforts to build GUI based authoring tools [1,12] have aimed to speed up ITS authoring with non-programmer friendly interactions. But these tools only support authoring of a subset of the adaptive functionalities possible with programming. There is a need within the ITS community for tools that are easier and faster to use than existing authoring systems while supporting the broader sets of behaviors achievable with rule-based authoring. This will hasten the availability of highly adaptive learning technologies like ITSs to classrooms world-wide. Furthermore, ITSs often need to be re-rebuilt due to changes in content delivery methods, differences in school curricula around the world, and for the purposes of experimenting with variations in instructional designs.

1.2 ITS Authoring with Machine Learning

Simulated student based ITS authoring research has been driven in response to this dilemma—trying to bridge the chasm between what is known to work in edTech and our ability to implement it efficiently. Prior attempts have been made to build ITSs with interactively trainable agents. This authoring modality starts like any other—by building an HTML interface, and setting a start state for a first problem. However, to author adaptive grading behavior, the author interactively trains a simulated student by demonstrating solution paths and providing correctness feedback on the simulation's attempts on subsequent problems. The simulated student generalizes a set of production rules from the author's training, building the general adaptive grading and bottom-out hint-generation behavior of an ITS from being tutored on just a handful of problems.

SimStudent for instance has been interactively trained to master simple algebra equation solving problems up to a precision and recall of about 95% [9]. SIERRA is an early example of a simulated student that has learned multi-digit subtraction for modeling student errors [13]. Prior work with the Apprentice Learner framework [7] has demonstrated agents that can be trained on a wide variety of domains, although typically falling short of perfect accuracy [8].

Our recent work published at CHI with the Apprentice Learner (AL) framework demonstrates a promising initial result that 100% model-tracing completeness can be achieved for the domain of multi-column addition [14]. Model-tracing completeness is a stronger metric than accuracy indicating the proportion of a holdout set of intermediate problem states in which the set of next actions the agent believes are correct consists only of the set of correct actions—of which multiple may exist in a multi-solution problem domain. This 100% completeness mark was only achievable by us (the authors) having some insight into how the agents learned and of known edge-case problems. Our 10 users however showed median levels of completeness of about 92%. Overall, however we found that compared to CTAT example-tracing authoring by training AL was about 7 times faster in terms of median authoring efficiency (i.e. 13% to 92% over 45 min). And users generally enjoyed training AL over using example-tracing.

1.3   The Structure and Objectives of This Proposal

By having a simulated student learn from an author's instructional interactions it can induce general capabilities as a human would, sparing the author a great deal of repetitive, specialized, or otherwise cognitively demanding work. Authoring problems for multi-step ITSs involves implementing two kinds of functionality. 1) The grading behavior and adaptive supports like hint generation for each step (i.e. then-parts of rules), and 2) the control structure of the steps (i.e. if-parts of rules). Conventional authoring tools like CTAT example-tracing [1], have means of allowing authors to write limited code to generalize grading behavior to multiple problems, but in cases where problems have complex control structures, its often necessary to fall back on hand-writing whole production rules—a challenging and time-consuming process. Toward improving authoring with simulated students I propose two general goals:

Goal 1 : Spare the author from the difficulties of programming formulas for generating and tracing next-step actions.

Goal 2 : Widen the scope of domains that authors can quickly author ITSs for by broadening the scope of control structures they can efficiently build with non-programming based interactions. I outline 3 proposals toward these goals in the following sections.

2   Proposal 1: Designing for Completeness and Navigability

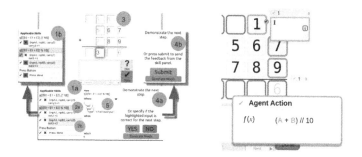

Fig. 1. (left) The training interface for our CHI2020 study. (right) A proposed overlay menu showing an agent's proposed actions.

Our prior prototype authoring tool presented at CHI 2020, used agents from the Apprentice Leaner (AL) framework [14]. Our qualitative observations reveal design issues, that could be fixed to streamline the process of tutoring the simulated learner. For instance, the spatial separation between the window showing the agent's proposed actions (Fig. 1 left-1a), and the tutoring interface (left-3) hampered authors' ability to quickly gauge if the agent was only proposing all the next correct actions, or if it needed extra feedback or demonstrations—a key signal the author must attend to across problem steps by clicking through the agent's proposed actions to determine if it exhibits model-tracing completeness on each step. To revise our prior interface, I propose super-imposing the agent's proposed actions over their associated tutoring system interface elements (Fig. 1 right) and add some additional ease of use features for quickly giving feedback to the agent.

We also encountered issues with participants having trouble tracking what they had taught the simulated learner. AL agent's learning, like human learning, is sensitive to the quality of instruction. If, for instance, the author forgets to show AL some important edge-case then this may reflect unpredictably—usually negatively—on the fidelity of the agent's eventual grading behavior. To help authors track training progress, and navigate easily through problem states of multi-solution problems, I also propose adding a behavior-graph like navigational interface for visualizing and navigating between problem states.

3 Proposal 2: Interactive Programming with Natural Language

A major barrier toward using simulated learners to succeeding at goal 1, is that prior work has relied on an often intractable form of brute-force search used to drive a learning mechanism called *how-learning* [10,14]. *How-learning* composes domain general functions to explain an author's demonstrated actions from values visible in a tutoring interface. The trouble is that the set of functions made available to *how-learning* must be made very large for a general purpose authoring tool. And with this large corpus of functions—which might include many kinds of functions beyond just arithmetic operations—*how-learning* may search through an intractably large space of function compositions and can be prone to stopping short on incorrect formulae that reproduce the demonstrated action but are incorrect in general.

Arguably this view of learning *how* to do individual steps in procedures via reverse engineering an instructor's demonstrations, deviates somewhat from human-to-human tutoring, where we might additionally expect accompanying natural language instruction. I propose *how-learning* can be made more robust by adding natural language processing into the mix: enabling an input modality whereby authors can verbally instruct simulated learners with operational language. By interpreting the natural language and example together, we can overcome the intractability of performing *how-learning* on demonstrated examples alone.

4   Proposal 3: Helping Authors to 100% Completeness

The interaction design improvements specified in proposal 1 should resolve prior issues concerning users' ability to quickly assess the model-tracing completeness of individual problem states. However, model-tracing completeness is a global objective—it is the certainty that when a student uses the resulting tutoring system, that for all states along all conceivable solution paths it only accepts and/or suggests the correct next actions. The question of when we are finished training is a matter of how we can come to train the AI to a point where we trust it with absolute certainty. Or more importantly, estimate the particular ways we mistrust the AI's generalizations, so we can provide precise instruction that helps it refine or replaces those generalizations to the point that they become trustworthy.

My proposed solution to this issue involves using an active-learning mechanism [11]: a mechanism for determining which training instances will help the agent learn most. The proposed mechanism could point the author to individual problem states that deviate from previous training experiences. This mechanism could locate edge-cases overlooked by the author's instructional strategy. When the set of problem states in the predicted solution paths of a corpus of problems are no longer considered uncertain by the agent, the author can be certain that they are finished training with confidence.

The typical active-learning paradigm involves a pool of unlabelled data from which to choose next optimal training [11]. A key differences in this case is that the pool of candidate training instances is, in effect, constructed by the agent. After some training, problems it has been shown, and produce unlabelled state-action pairs, for which it might query the author for correctness labels. The more important difference is that the agent in our case should ideally report when it is absolutely certain about its predictions of the correctness label. If the agent is uncertain of that label, it should estimate how uncertain it is.

The method I've derived, which is too complex to describe here, induces rule pre-conditions by maintaining a version-space of all pre-conditions consistent with training so far. A measure is produced of how much disagreement there is among the candidate pre-conditions in this space, and if no disagreement exists then the predicted label is considered certain. The technical challenges of this include producing this measurement without enumerating all enclosed pre-conditions, and building this version-space-like structure over arbitrary disjunctive normal logical statements—something previous work has argued is intractable [3], but that we build efficiently through approximation.

5   User Studies and Final Authoring Study

Proposals 1–3 will be evaluated with accompanying user studies similar to our 2020 CHI study. Studies 1 and 3 will evaluate if each proposed feature produces better median authoring efficiency and authoring completeness on multi-column arithmetic for another 8–10 participants each. Study 2 will co-occur with study

1 and evaluate the usability and robustness of our natural language processing extension. Study 2 does not evaluate efficiency because when *how-learning* operates as intended, it is only engaged in the early phases of training, meaning it doesn't contribute much to authoring time. User study participants will be primarily undergraduate and graduate students—a convenience population earlier in their careers than say the teachers or learning engineers that may be our primary end-users. We will select participants with varying programming expertise, to evaluate whether this is a factor in authoring success.

Of course we need to demonstrate that the proposed authoring tool works well for a wide variety of target domains. After these user studies are complete I will engage in a phase of internal authoring experiments—building a gauntlet of tasks in ITSs that this authoring method may succeed at building or not—to determine the limits of this authoring method. Finally, after some time spent building out the final features of the authoring tool I will conduct a final authoring study.

In this final culminating authoring study I plan to use the participants from the computational models of learning (CML) track at the LearnLab summer school to evaluate the broader goal of using AL as a general purpose authoring tool. I plan to 1) have participants author one new domain: stoichiometry, with both CTAT example-tracing and AL, and 2) have participants author one final domain of their choosing using AL only.

In the first part I would like to have participants build a stoichiometry ITS with elements of complex control structure that differ from multi-column addition. For comparison with CTAT-example tracing I will use a cross-over design where I randomly split participants into two groups that both use AL and CTAT-example tracing. One group will use CTAT-example tracing first and the other will use AL first. Participants will be given the tutor interface and will be allotted 30 min in each condition to author its grading and next-step hint functionality. To allow for data collection in a hybrid format I will have participants in each condition screen-share to a breakout room. Participants will be given a follow-up survey asking them to compare their experiences between conditions.

In part 2) participants will have 1 h to build an interface and train AL on a domain of their choice. This may go well for participants or they may fail to build the system they want to build. We will have participants screen record this session, and take a followup survey. In the survey participants will be explicitly asked to summarize their experience, indicate things that went wrong in the authoring process, and share how long it took them to fix or identify whether any issues encountered were unsolvable or required some programming to overcome. For instance, one way that AL can fail to learn a domain is if it does not have a sufficient set of features or functions to induce the target domain. This open-ended portion of the authoring process is an opportunity to evaluate whether the AL tutoring interface is able to point this out to users.

For both parts we will have participants upload their agents and tutoring system assets for evaluation. The AL agents from part 1) will be automatically evaluated for completeness on a holdout set of problem instances. For the AL agents trained in part 2) we will try to author these domains ourselves to get a rough sense of how complete the user's tutoring systems ultimately are.

References

1. Aleven, V., et al.: Example-tracing tutors: intelligent tutor development for non-programmers. Int. J. Artif. Intell. Educ. **26**(1), 224–269 (2016)
2. Gulwani, S., Harris, W.R., Singh, R.: Spreadsheet data manipulation using examples. Commun. ACM **55**(8), 97–105 (2012)
3. Hirsh, H.: Polynomial-time learning with version spaces. In: AAAI, pp. 117–122 (1992)
4. Kulik, J.A., Fletcher, J.: Effectiveness of intelligent tutoring systems: a meta-analytic review. Rev. Educ. Res. **86**(1), 42–78 (2016)
5. Laird, J.E., et al.: Interactive task learning. IEEE Intell. Syst. **32**(4), 6–21 (2017)
6. Li, T.J.J., Azaria, A., Myers, B.A.: SUGILITE: Creating multimodal smartphone automation by demonstration. In: Proceedings of the 2017 CHI Conference on Human Factors in Computing Systems, pp. 6038–6049. ACM (2017)
7. Maclellan, C.J., Harpstead, E., Patel, R., Koedinger, K.R.: The Apprentice Learner Architecture: Closing the loop between learning theory and educational data. International Educational Data Mining Society (2016)
8. MacLellan, C.J., Koedinger, K.R.: Domain-general tutor authoring with apprentice learner models. Int. J. Artif. Intell. Educ. **32**(1), 76–117 (2020). https://doi.org/10.1007/s40593-020-00214-2
9. Matsuda, N.: Teachable agent as an interactive tool for cognitive task analysis: a case study for authoring an expert model. Int. J. Artif. Intell. Educ. **32**(1), 48–75 (2022)
10. Matsuda, N., Cohen, W.W., Koedinger, K.R.: Teaching the teacher: tutoring sim-student leads to more effective cognitive tutor authoring. Int. J. Artif. Intell. Educ. **25**(1), 1–34 (2015)
11. Settles, B.: Active learning literature survey (2009)
12. Sottilare, R.A., Brawner, K.W., Goldberg, B.S., Holden, H.K.: The generalized intelligent framework for tutoring (gift). Orlando, FL: US Army Research Laboratory-Human Research & Engineering Directorate (ARL-HRED) (2012)
13. VanLehn, K.: Learning one subprocedure per lesson. Artif. Intell. **31**(1), 1–40 (1987)
14. Weitekamp, D., Harpstead, E., Koedinger, K.: An interaction design for machine teaching to develop ai tutors. CHI (2020 in press)

Investigating the Impact of the Mindset of the Learners on Their Behaviour in a Computer-Based Learning Environment

Indrayani Nishane$^{(\boxtimes)}$ ⓘ, Ramkumar Rajendran ⓘ, and Sridhar Iyer

Indian Institute of Technology Bombay, Mumbai, India
{indrayani.nishane,ramkumar.rajendran,sri}@iitb.ac.in

Abstract. Computer-based learning environments (CBLEs) are often used to provide customized learning experiences for students. To enhance their effectiveness, we propose analyzing learners' actions within CBLEs. In this study, we focus on the influence of learner mindset (fixed vs growth) on interaction patterns in a CBLE designed for teaching Python programming. Learner mindset refers to their beliefs about the malleability of their abilities. Individuals with a fixed mindset believe that their abilities are fixed, while those with a growth mindset believe abilities can be developed through learning. Using log data and pattern-mining techniques, we will identify learners' interaction patterns and behaviour while also assessing their mindset through a questionnaire. We will compare the interaction patterns of fixed and growth mindset learners using task models to support our findings.

Keywords: Learner Mindset · Learner Behaviour · Process Modelling for learners

1 Introduction

A Computer-based learning environment (CBLE) is an instructional method where learners use computer technology to acquire new knowledge. The learning process in the CBLE typically involves performing various tasks to understand concepts, apply the information to solve problems and receive feedback through assessments within the system. The log data generated from these actions can be analyzed to enhance the CBLE, making it more effective for remote education and promoting equity in learning opportunities for all.

Many factors, like learners' performance in various assessment activities, learner engagement, and learner-centric emotions, are reported to be used both in the classrooms and CBLEs to decide the pedagogical actions [1–3]. However, the learner mindset is not yet examined to analyse learner behaviour within CBLEs.

© The Author(s), under exclusive license to Springer Nature Switzerland AG 2023
N. Wang et al. (Eds.): AIED 2023, CCIS 1831, pp. 169–174, 2023.
https://doi.org/10.1007/978-3-031-36336-8_26

In cognitive psychology, mindset refers to activating different cognitive processes in response to a task, which influences how information is interpreted later on [4]. Mindsets are people's beliefs about human attributes [5] and shape the thoughts and perspectives people hold [6]. Mindset can also impact motivation and shape subjective and objective aspects of a task or a situation, especially in challenging conditions. During the learning process, learners may face difficulties such as unfamiliarity with the subject or difficulty answering assessment questions. Their mindset can play a crucial role in determining their actions while learning. All the learner's actions can be recorded in the CBLE as log data. Analysis of these actions will give insights into cognition and help improve the designs of CBLEs.

A "fixed mindset" sees abilities and qualities as unchanging, while a "growth mindset" believes they can be improved through effort, strategy, and sometimes assistance from others. These mindsets are based on beliefs about the nature of abilities, with a fixed mindset viewing them as unchangeable and requiring constant proof, while a growth mindset sees them as malleable and able to be developed through learning [7].

Mindset gets affected by various factors such as the subject, peer influence, and teachers. Numerous studies have shown that learners with a growth mindset tend to be more engaged and achieve better grades in traditional classroom settings [9–15]. However, the impact of mindset in computer-based learning environments (CBLEs) has not been studied yet. This makes a compelling case to consider mindset in the context of CBLEs.

2   Research Proposal

This section will discuss the research goal, research questions, proposed research plan, and expected outcomes.

2.1   Research Questions

The main goal of the research, based on the literature review and research gap, is *"to assess the effect of learners' mindset on their behaviour in a computer-based learning environment."* The research plan will address the following research questions related to this research goal.

1. What are the different interaction strategies or learning paths in a CBLE for cohorts of learners with different mindsets?
2. How do learners with different mindsets handle achievement and failure while learning in a CBLE?
3. What is the impact of mindset on the learners' emotions while interacting with the CBLE?
4. Can mindset be used to influence the design of CBLEs to provide more suited scaffolding?

2.2   Research Plan

We plan to use process models and other pattern-mining techniques to analyse the learner's behaviour from their log data. Process models depict the sequence of actions taken by learners pictorially and have been used in some studies to understand learning strategies [2,17,18]. The process models will show differences in learner behaviour based on their mindset. In the process models, each node represents an action taken by the learner, and each edge represents the transition from one action to another. Each node has a significance value (between 0 and 1), and each edge has a thickness indicating its significance, with darkness indicating correlation [16].

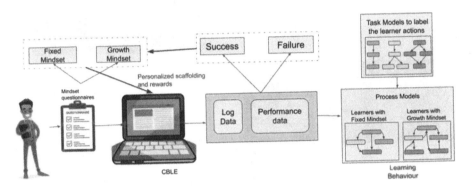

Fig. 1. Our proposed research approach and expected outcomes.

In a learning environment, learners acquire new knowledge to solve problems by performing tasks requiring cognitive skills. These tasks are organized into sub-tasks and mapped to actions in the CBLEs, known as the "Hierarchical Task Model" [19]. We plan on using such task models to identify the categories of actions performed by learners and compare the process models of learners with a fixed mindset versus those with a growth mindset. The prior research suggests that individuals with different mindsets react differently to failures and setbacks during learning [5,7,14]. The study aims to observe how learners' mindset affects their behaviour and emotions in a computer-based learning environment, especially with regard to achievement and failure within the system. Our proposed research approach and expected outcomes are shown in Fig. 1.

3   Methodology

The research goal is to understand whether the learner mindset impacts learners' behaviour while interacting with the CBLE. We are using the CBLE for capturing the learner's actions, the Mindset survey developed by Dr Dweck.

3.1   About PyGuru

We have developed a computer-based learning environment called PyGuru for teaching and learning Python programming skills [3]. It has components like a book reader where learners can read the content, a video player to watch the video content, Integrated Development Environment (IDE), and a discussion forum, as shown in Fig. 2.

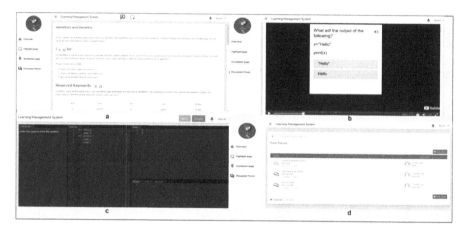

Fig. 2. PyGuru Environment (a) Book Reader (b) Video Player (c) IDE (d) Discussion Forum

The book reader component allows users to navigate text-based content and highlight text with colour. It also has an annotate feature for making notes, adding comments, and tagging selected text. The interactive video player has basic functions like play/pause/seek. It also has embedded questions, and it pauses until the learner answers. There are two IDEs for executing Python programs: an in-page IDE within the book reader and another that assesses code against test cases. All activities are logged.

3.2   Study Design

The study will be conducted with undergraduate students who preferably do not know Python programming. A pre-test is to be conducted before students start the interaction with the learning environment, along with a few demographic questions and the 3 item growth mindset scale questions developed by Dweck [6,8]. The same survey would also include consent for data collection for this research study. The post-test will be conducted after their interaction with PyGuru. Students will interact with the CBLE for 2–4 hours. We plan to conduct a semi-structured interview with the students to better understand their mindset, which will complement their self-reported mindset.

The log data will be collected in JSON format for the mouse clicks done and actions taken by the students while interacting with the CBLE. Data will be cleaned and analyzed using Python for the patterns, frequency, etc., of learner actions. The log data will be used with the ProM Tool, an open-source process mining tool (https://www.promtools.org/), to create the process models.

4   Current Status of Work

We conducted a study with 38 first-year undergraduate students enrolled for the B.Sc. (Information Technology) course between 18–19 years. The sample included 18 female and 20 male students. All the students interacted with the CBLE for 2 h daily during the two days in the workshop-like mode. Out of the 38 students, 15 have a Fixed Mindset (39.5%), and 23 have a Growth Mindset (60.5%). The scores in the pre-test and post-test were not statistically significantly different for both cohorts - learners with a growth mindset and learners with a fixed mindset. The study found variations in the process models for individuals with a growth mindset and those with a fixed mindset, as seen in the common sequences of actions. These differences align with the typical traits of individuals with a growth or fixed mindset.

We created a task model for our CBLE - PyGuru and used it to label the actions taken by students based on their log data. We then compared the sequences of actions from both groups and the task model to do a comparative analysis.

5   Proposed Plan

Our objective is to explore how learners' mindsets affect their behaviour and cognitive processes in CBLEs and to identify nuances in their actions. To achieve this, we plan to follow the action plan outlined below. -

1. Conducting another study to collect more data to establish the differences in the behaviour of students with Fixed and Growth mindsets.
2. Collecting multimodal data to understand the influence of mindset on learners' affect state.

References

1. Biswas, G., Segedy, J.R., Bunchongchit, K.: From design to implementation to practice a learning by teaching system: Betty's Brain. Int. J. Artif. Intell. Educ. **26**, 350–364 (2016)
2. Saint, J., Gašević, D., Pardo, A.: Detecting learning strategies through process mining. In: Pammer-Schindler, V., Pérez-Sanagustín, M., Drachsler, H., Elferink, R., Scheffel, M. (eds.) EC-TEL 2018. LNCS, vol. 11082, pp. 385–398. Springer, Cham (2018). https://doi.org/10.1007/978-3-319-98572-5_29

3. Singh, D., Rajendran, R.: Investigating learners' Cognitive Engagement in Python Programming using ICAP framework. In: Proceedings of the 15th International Conference on Educational Data Mining, p. 789 (2022)
4. French II, R.P.: The fuzziness of mindsets: Divergent conceptualizations and characterizations of mindset theory and praxis. Int. J. Org. Anal. **24**(4), 673–691 (2016)
5. Dweck, C.S.: Mindsets and human nature: promoting change in the Middle East, the schoolyard, the racial divide, and willpower. Am. Psychol. **67**(8), 614–622 (2012)
6. Dweck, C.S.: Mindset: The New Psychology of Success. Random House, New York, NY (2006)
7. Dweck, C.S.: Mindset - Changing the way you think to fulfil your potential. Little, Brown Book Group (2017)
8. Dweck, C.S.: Self-theories: Their role in motivation, personality, and development. Psychology Press, Philadelphia (1999)
9. Reid, K.J., Ferguson, D.M.: Work in progress-Enhancing the entrepreneurial mindset of freshman engineers. In: 2011 Frontiers in Education Conference (FIE), pp. F2D–1. IEEE (2011)
10. O'Brien, K., Lomas, T.: Developing a Growth Mindset through outdoor personal development: can an intervention underpinned by psychology increase the impact of an outdoor learning course for young people? J. Adv. Educ. Outdoor Learn. **17**(2), 133–147 (2017)
11. Stephens, J.M., Rubie-Davies, C., Peterson, E.R.: Do preservice teacher education candidates' implicit biases of ethnic differences and mindset toward academic ability change over time? Learn. Instr. **78**, 101480 (2022)
12. Limeri, L.B., Carter, N.T., Choe, J., Harper, H.G., Martin, H.R., Benton, A., Dolan, E.L.: Growing a growth mindset: characterizing how and why undergraduate students' mindsets change. Int. J. STEM Educ. **7**(1), 1–19 (2020)
13. Altunel, İ: Bridging the gap: a study on the relationship between mindset and foreign language anxiety. Int. Online J. Educ. Teach. **6**(3), 690–705 (2019)
14. Lou, N.M., Chaffee, K.E., Noels, K.A.: Growth, Fixed, nd Mixed Mindsets: Mindset System Profiles in Foreign Language Learners and their role in Engagement and Achievement. Studies in Second Language Acquisition, pp. 1–26 (2021)
15. Vongkulluksn, V.W., Matewos, A.M., Sinatra, G.M.: Growth mindset development in design-based makerspace: a longitudinal study. J. Educ. Res. **114**(2), 139–154 (2021)
16. Van Der Aalst, W.: Process mining: discovery, conformance and enhancement of business processes, vol. 2. Springer, Heidelberg (2011)
17. Nishane, I., Sabanwar, V., Lakshmi, T. G., Singh, D., Rajendran, R.: Learning about learners: Understanding learner behaviours in software conceptual design TELE. In: 2021 International Conference on Advanced Learning Technologies (ICALT), (pp. 297–301). IEEE (2021)
18. Rajendran, R., Munshi, A., Emara, M., Biswas, G.: A temporal model of learner behaviors in oeles using process mining. Proc. ICCE **2018**, 276–285 (2018)
19. Biswas, G., et al.: Multilevel learner modeling in training environments for complex decision making. IEEE Trans. Learn. Technol. **13**(1), 172–185 (2019)

Leave No One Behind - A Massive Online Learning Platform Free for Everyone

Alejandra Holguin Giraldo[1], Andrea Lozano Gutiérrez[1],
Gustavo Álvarez Leyton[1], Juan Camilo Sanguino[2],
and Rubén Manrique[2(✉)]

[1] GCFGlobalLearning LLC, North Carolina, USA
{alejandraholguin,andrealozano,gustavoalvarez}@gcfglobal.org
[2] Universidad de los Andes, Bogotá, Colombia
{jc.sanguino10,rf.manrique}@uniandes.edu.co

Abstract. GCFGlobalLearning is a non-profit organization committed to creating life-changing opportunities through an innovative virtual education project that incorporates technological tools. Our platform offers free courses to learners worldwide, available in English, Spanish, and Portuguese. Since our launch, we have welcomed millions of learners. Our current focus is on incorporating artificial intelligence tools that will enhance our learners' experience. To achieve this, we have developed a recommender system and a learning content search and organization tool that personalizes our learners' learning journey. Even with limited information about our learners, we can enhance their experience through the use of these AI tools. In this paper, we introduce the platform's primary components, detail how we overcome our limited learner information scenario, and share our vision of incorporating more artificial intelligence innovations in the future.

Keywords: Recommender systems · MOOC platform · Limited learner information scenarios

1 Introduction

GCFGlobalLearning is a non-profit organization that utilizes technological tools to create innovative and accessible virtual educational environments, with the aim of providing opportunities for a better life to those who lack the resources or opportunities to access basic education. Their primary objective is to help individuals acquire new knowledge to enhance their ability to navigate the modern world of work, ultimately improving their quality of life. In response to the COVID-19 pandemic in 2020, GCFGlobalLearning expanded their courses and learning content to cater not only to those without basic education, but also to mothers who head households, job-seekers, high school students, homeschooling parents, teachers, new entrepreneurs, and older adults who are not familiar with digital technology.

N. Wang et al. (Eds.): AIED 2023, CCIS 1831, pp. 175–186, 2023.
https://doi.org/10.1007/978-3-031-36336-8_27

Currently, GCFGlobalLearning offers 361 courses, counting more than 6,480 lessons, more than 2,590 videos, and more than 50 interactive activities and games". All courses are online, self-paced, and self-directed (without a tutor). The reach of our content is global with learners from a wide range of countries (see Fig. 1).

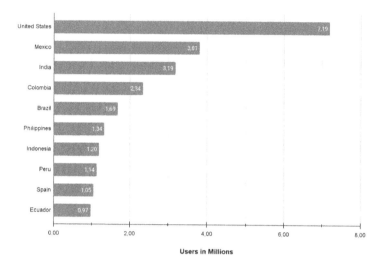

Fig. 1. 2022 top countries by number of learners

As a next step in the continuous improvement of our platform, we have incorporated a new set of tools that allow us to attack two problems: (i) improve the learner retention rate and (ii) offer a personalized experience. We define the retention rate as the number of learners who, after completing one of our learning resources or course, return to the platform to take a new course or learning resource. Historically, as shown in Fig. 2, our annual retention rate is around 25%. A diagnostic study conducted in 2021 reveals possible causes for this low retention. Firstly, due to the large volume of resources available on the platform and the absence of tutor guidance, for a self-directed learner is challenging to make an effective selection of learning resources. In several analyzed cases we found that the choice made by the learners does not follow a pedagogically logical organization (for example in terms of prerequisites or resources topologically related to previously seen), which leads to disconnected learning paths and directly impacts user interest. Secondly, and related to the previous one, the platform organization and search system are based on a human-made topic tree that does not reflect all the existing relationships between the courses.

In the above context, a recommender system (RS) and a personalized learning resources search and organization tool (Symphony) were developed in order to be integrated into the platform. In educational contexts, recommender systems have proven effective in retrieving learning resources that fit the interests of a learner to increase student retention and improve the learning experience [1]. Symphony, on the other hand, allows learners to create pages/boards with resources from the platform and from open external sources using a simple search interface. Symphony allows having related resources from multiple sources in one place, thus avoiding the so-called "split-attention effect" [4].

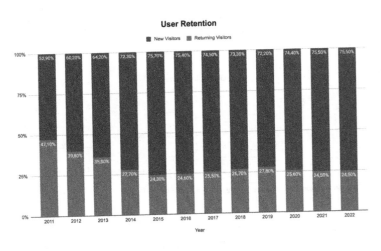

Fig. 2. Learner retention problem

A challenge in developing these tools was the lack of explicit information about the learners. As an organizational principle, the only requirement for learners to access courses is an Internet connection. For people inexperienced in technology domains registering or logging is considered an access barrier. Therefore registration and authentication on the platform are not mandatory. Only 2% of all learners are registered and logging in to the platform, so, only for them, we have explicit information related to their interests and we are able to uniquely identify their interaction with the platform. According to different authors, this is called "limited user/learner information scenarios" and poses a great challenge when it comes to personalizing the learning experience [3].

In this paper we present the main components of the platform (Sect. 2), with special emphasis on Symphony (Sect. 3) and RS (Sect. 4). In each of the components we present the strategies implemented to deal with the limited information scenario and the limitations that it entails. We end with a discussion on the challenges and opportunities of using AI tools in the context of our platform and in the process of building learning content (Sect. 6).

2   Learning Platform Overview

In pursuit of delivering an optimal learning experience, GCFGlobalLearning has developed an ecosystem of applications with diverse strategies for knowledge transfer to its users. Each of these applications operates independently, does not require mandatory login, and is available to all users. The applications within this ecosystem include "GCFGlobal LMS", "Symphony" and "Let's Play Stop!". A general overview of this ecosystem is depicted in Fig. 3.

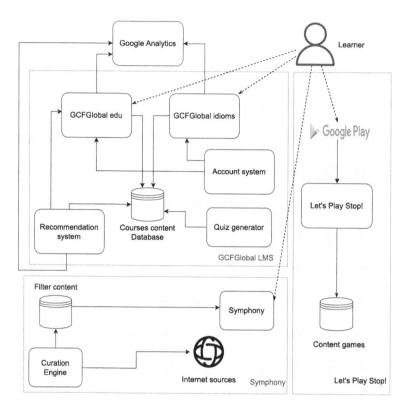

Fig. 3. Learning platform components.

2.1 GCFGlobal LMS

The GCFGlobal LMS (Learning Management System) serves as the cornerstone of our platform, facilitating the management of Massive Open Online Courses (MOOCs). Developed in-house, this application provides content designers with a suite of user-friendly tools for constructing learning content within our courses.

Furthermore, given that our courses are both self-directed and available on a massive scale, a significant portion of our development effort has been focused on ensuring ease of access and availability for the millions of users who engage with our platform. For those who opt to register with the platform, the LMS offers a range of follow-up functions, including course tracking, application to certified courses, certification management, and the opportunity to receive communications and feedback. The information pertaining to registered users' interactions is maintained within a local database that can be readily accessed by various system functionalities. For non-registered users, we store a range of interaction events using Google Analytics. We will subsequently elaborate on the intricacies associated with processing this data sourced from Google Analytics to develop our recommendation system.

The GCFGlobal LMS project encompasses two distinct domains. The first domain, "idiomas.GCFGlobal.org", is dedicated to teaching English to Spanish and Portuguese speakers. The second domain, "edu.GCFGlobal.org", offers courses designed to teach essential skills necessary for success in the 21st century. Both domains include different sets of courses that utilize a variety of learning materials, including videos, activities, quizzes, and text content. Moreover, the content is available in three languages: Spanish, Portuguese, and English. To generate quizzes, GCFGlobal LMS includes a quiz generator that creates multi-answer questions with feedback functionality. The answers are not saved because it is not possible in most cases to relate them univocally to a learner.

2.2   Let's Play Stop!

The game "Stop" is a popular Latin American pastime wherein participants select a letter from the alphabet and use it to find a set of words that start with this letter within the shortest possible time. The word must pertain to various predetermined categories, such as names, surnames, cities, objects, colors, animals, and foods. This game represents one of the most innovative techniques for acquiring proficiency in spelling and discovering new vocabulary across different languages. GCFGlobalLearning brought this game to mobile and smart devices as a support tool for language courses and it is available at no cost in Google Play for Android users.

3   Symphony

Due to the increasing efforts of multiple organizations and individuals, there is now an immense amount of freely accessible learning resources available on the Web in various domains and languages. However, the unorganized nature of the Web makes the search and selection process non-trivial. To address this issue, Symphony was created as a search tool to help people find valuable free learning content. The main motivation behind Symphony's development is to eliminate the friction between search and consumption. By simply entering a search term, the system generates a page of educational content, including definitions, a list of short and long videos, articles, free courses, and podcasts, all in one convenient location (see Fig. 4). The results not only include learning resources developed by the GCFGlobalLearning team, but also open learning resources on the internet.

The "Curation Engine" is an essential component of the project and consists of a list of pre-selected learning resources curated by the GCFGlobalLearning team. The team has carefully selected these resources based on their educational

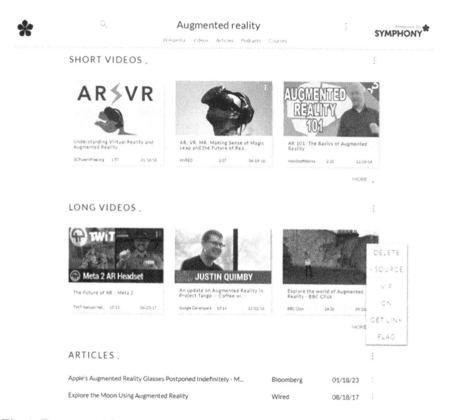

Fig. 4. Fragment of the page built by Symphony for the query "Augmented Reality"

value, reliability, and filtered inappropriate content. The curated list currently includes over 350,000 results that match more than 3,000 of the most popular learning terms. To further enhance the learning experience, each learner can customize the resulting page in various ways:

- Remove a particular learning resource or a complete source.
- Learning content subfilters that allow a selection by keywords.
- Retrieve additional related learning resources from a particular source.
- Flag a resource with one of these tags: Inappropriate, Unrelated, Broken Link. These annotations allow the GCFGlobalLearning team to conduct a process of reviewing and refining the curated list.
- Reorganize the results page and import learning resource links from other sources.

The tool has the potential to enable learners to create personalized learning spaces using carefully curated content. These spaces can be easily filtered in accordance with the learner's preferences. By consolidating all of the resources within a single, well-organized space, the tool facilitates learner concentration and helps to mitigate the impact of the "split-attention effect". However, the scaling of Symphony to cater to millions of users presents a range of limitations, particularly in terms of content curation. The diversity of learning needs amongst users is extensive, making it infeasible to curate resources that are relevant to all individuals. In such cases, Symphony retrieves learning resources that are ranked by traditional search engines, which may lack a teaching focus or fail to address the user's specific learning requirements. This notion is substantiated by user feedback received by the platform. The potential for Artificial Intelligence tools to facilitate the automatic curation of content is a promising avenue for making the process scalable.

4 Recommender System

Collaborative filtering (CF) is a widely-used recommendation technique in e-learning [6]. It works by analyzing users' behavior and course ratings to identify a group of users who share similar interests. Recommendations are then made based on the courses viewed by this group. Ratings can be explicit, such as when a learner assigns a score to a course, or implicit, as when ratings are inferred from the learner's interactions with the course content. These ratings are used to determine the similarity between users. Although CF has proven successful, it has limitations when the training data contains a limited number of user ratings. For instance, a CF-based system cannot group users with similar interests if the number of ratings is too low, leading to imprecise recommendations. This issue is known as the cold start problem. To overcome the cold start, recommendation systems often incorporate auxiliary content-based (CB) systems that utilize additional information - such as lesson and course features or textual information - to identify related courses. CB systems attempt to identify courses with similar or common characteristics to those previously viewed by the learner.

The components of our recommendation system are illustrated in Fig. 5, with the goal of recommending courses that align with a user's needs and interests. While the literature suggests that a CF strategy is the best recommendation model for educational platforms [5], the learning platform's web traffic primarily comprises new users (see Fig. 2). Consequently, relying solely on a CF model would be significantly affected by the cold start problem. To address this, we propose a hybrid model that combines both CF and CB models, allowing us to provide recommendations to both new and returning users. Our proposed architecture is based on a set of interconnected independent components, allowing for future updates or changes to be implemented with minimal impact. The content-based recommendation and collaborative filter-based recommendation components operate independently and are fed with two distinct information sources. First, the user interactions with the courses are stored in Google Analytics, which is used to build the CF model. Second, the contents of the lessons and courses are stored in a Mongo database, which serves as the data source for the CB recommendation model.

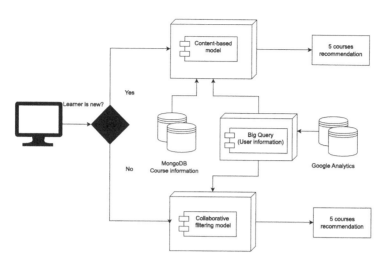

Fig. 5. Recommendation system components.

Google Analytics, a widely-used tool in e-commerce, has also become increasingly useful in learning platforms. Its popularity can be attributed to its ease of implementation and its ability to filter and analyze large volumes of logs. While log analysis may have limitations in identifying individual users, it is often the most commonly available type of data in existing web-based platforms. As logs contain data about the web pages visited by users, we can process them to extract information about the courses and lessons accessed by learners. In order to use these logs to build a recommendation system, we must make several assumptions, as outlined below: (i) learners access the platform using a unique device. This is because the device identifier serves as the user identifier in Google Analytics, and it is not possible to correlate the same learner accessing the platform using different devices; (ii) learners fully engage with the lessons they access. This second statement is quite strong since the existence of an access log no guarantee that a user will cover all or portions of the learning content.

Assuming that each learner accesses the learning platform from a single device can pose a threat to the validity of the recommendation in certain scenarios. The first scenario involves a single learner accessing the platform from two different devices, which could result in the learner being analyzed as two different users. This scenario, referred to as "data fragmentation," can lead to learners receiving different recommendations based on the device they use. The second scenario occurs when multiple learners access the platform from the same device, such as in shared computer rooms at educational institutions. In this case, the recommendation system operates for the collective, and individual user experiences may suffer due to the inability to differentiate between learners' interactions.

These limitations could easily be addressed by requiring users to identify themselves on the platform. However, as stated earlier, this conflicts with our principles of providing unrestricted access. Therefore, we accept these limitations to maintain our commitment that the sole requirement for accessing our courses is an Internet connection. A comprehensive description of the technical aspects underlying the implementation of the CF and CB systems, including the data enlistment and processing steps for each data source, can be found in the recent works [2,3]. The proposed recommendation methodology, utilizing logs from Google Analytics, has demonstrated the capability to suggest two out of five courses that are deemed relevant to the user, as per our evaluation. The achievement of such a result in the absence of explicit user information is considered noteworthy. Figure 6 shows a recommendation example for a particular learner.

Fig. 6. Recommendation example.

5   Challenges and Opportunities

Artificial intelligence (AI) has the potential to revolutionize the field of free open courses by enabling personalized and adaptive learning experiences. With the ability to process large amounts of data and identify patterns, AI can provide recommendations for individual learners, helping them to find courses that match their interests and abilities. As a result, AI can help to make free open courses more accessible and effective for a wider range of learners, ultimately contributing to a more inclusive and equitable education system.

Despite its potential, the gap between research and practical innovation with AI in online learning platforms remains a significant challenge that needs to be addressed. For example, we found after comprehensive research that most recommendation strategies can not operate in our limited user information scenario. The existence of an explicit profile of the learner is assumed, which is not our case. Some commercial solutions, indeed, suggested a change in the open way that we offer our courses in order to use their tools. In other words, the problem fits the tool instead of the tool fitting the problem.

We establish a partnership with an academic research group to design our recommendation system that aligns with the characteristics and operating principles of our system. This approach enabled us to remain at the forefront of research while delivering a product that can be implemented in a practical setting. This experience in developing the recommender system revealed a notable disparity between research priorities in the field of recommenders systems and real practical requirements. Research studies often concentrate on highly specialized areas that are not pertinent to the practical needs of actual learning content providers. This misalignment can impede the applicability of research findings in practical situations.

Our impetus for the development of Artificial Intelligence (AI) tools stemmed from our desire to provide learners with more meaningful and fulfilling learning experiences. Our collaboration with academia enabled us to recognize the potential of these tools for the internal construction processes of courses, which, in turn, has a profound impact on the quality of our educational offerings. As illustrated in Fig. 7, the life cycle of a course is intricately linked to the processes involved, the team responsible for executing these processes, and the time invested in each step. Clearly, the development stage requires the greatest effort, and it is in this stage that we have identified a larger set of opportunities for implementing AI tools. Below, we mention some of these opportunities that we will explore in the near future:

- **Automatic voice generation** could lead to cost savings in the production of course videos. Typically, a course consists of approximately 10 lessons and it takes approximately 5 d to record, edit and deliver clean audio for these lessons. If we were able to automate this process by cloning voices and generating audio from textual scripts we could significantly reduce the production time required.
- **Automatic Q&A generator**. In order for learners to obtain a course certificate, they are required to pass an evaluation that is manually constructed by the content creators. The process of building question banks for these evaluations is often time-consuming and can be assisted by tools that generate an initial set of questions and answers from the textual content of the course in an automatic way.
- **Assistance in script development** can be provided through the use of language models such as ChatGPT or BLOOM. These tools can assist in

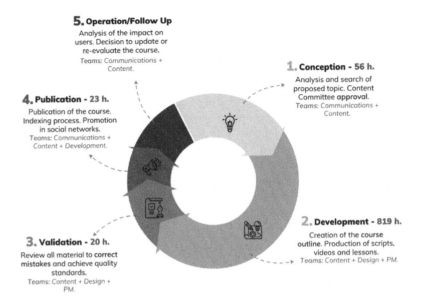

Fig. 7. Course lifecycle. Each stage involves the participation of a set of teams (content, design, communications). The hours reported in each stage were calculated by averaging over a sample of courses.

the creation of compelling and multilingual text, thereby aiding in the development of video scripts. Given the challenge of developing scripts in three different languages while remaining faithful to the original course design and using language that is easily understood by learners, such tools can be valuable in the creation of effective course content.

6   Conclusion

Open and free online education has become an increasingly important component of the modern educational landscape. This is due to the fact that online courses and educational resources are accessible to a much larger and more diverse audience than traditional classroom-based learning. Online education enables people from all over the world to access high-quality learning materials without the barriers of time, location, and cost. This democratization of education is particularly significant in countries and regions where access to traditional educational institutions and resources is limited. Additionally, open and free online education allows individuals to pursue learning at their own pace and according to their own interests, leading to lifelong learning and personal growth. By making education more accessible and inclusive, open and free online education has the potential to create a more knowledgeable and empowered society.

In favor of the above, GCFGlobalLearning is a non-profit organization that seeks to contribute to society by offering fully open and quality courses to the world. In this paper, we provide an overview of the key elements of the GCF-GlobalLearning platform designed to manage and deliver learning content, as well as two innovative initiatives involving the development of artificial intelligence tools aimed at enhancing the learner experience. Firstly, Symphony is introduced as a powerful tool that facilitates the discovery and consumption of learning resources, while also providing learners with the option of creating personalized learning spaces. Secondly, a recommendation system is presented, which has been designed to suggest relevant learning content to users, even in cases where user data is limited. The ultimate goal of these tools is to improve the learning experience and increase user engagement with high-quality, freely available educational resources. Ongoing efforts will focus on addressing scalability issues related to Symphony, and on exploring the potential of artificial intelligence models for generating automatic evaluations.

References

1. Aggarwal, C.C.: An Introduction to Recommender Systems, pp. 1–28. Springer International Publishing, Cham (2016). https://doi.org/10.1007/978-3-319-29659-3-1, https://doi.org/10.1007/978-3-319-29659-3-1
2. Sanguino, J., Manrique, R., Mariño, O., Linares, M., Cardozo, N.: Log mining for course recommendation in limited information scenarios. In: Mitrovic, A., Bosch, N. (eds.) Proceedings of the International Conference on Educational Data Mining, pp. 430–437. EDM'22, International Educational Data Mining Society (July 2022). https://doi.org/10.5281/zenodo.6853183
3. Sanguino Perez, J.C., Manrique, R.F., Mariño, O., Linares Vásquez, M., Cardozo, N.: A course hybrid recommender system for limited information scenarios. J. Educ. Data Mining 14(3), 162–188 (Dec 2022). https://doi.org/10.5281/zenodo.7304829, https://jedm.educationaldatamining.org/index.php/JEDM/article/view/608
4. Schmeck, A., Opfermann, M., van Gog, T., Paas, F., Leutner, D.: Measuring cognitive load with subjective rating scales during problem solving: differences between immediate and delayed ratings. Instr. Sci. 43(1), 93–114 (2014). https://doi.org/10.1007/s11251-014-9328-3
5. Shanshan, S., Mingjin, G., Lijuan, L.: An improved hybrid ontology-based approach for online learning resource recommendations. Education Tech. Research Dev. 69(5), 2637–2661 (2021). https://doi.org/10.1007/s11423-021-10029-0
6. Uddin, I., Imran, A.S., Muhammad, K., Fayyaz, N., Sajjad, M.: A systematic mapping review on mooc recommender systems. IEEE Access 9, 118379–118405 (2021). https://doi.org/10.1109/ACCESS.2021.3101039

Innovative Software to Efficiently Learn English Through Extensive Reading and Personalized Vocabulary Acquisition

Yo Ehara[✉]

Tokyo Gakugei University, Koganei, Tokyo 1848501, Japan
`ehara@u-gakugei.ac.jp`

Abstract. This paper introduces innovative software for efficient English language learning that incorporates machine learning and natural language processing techniques to personalize the vocabulary acquisition process for English language learners. The software was designed to enhance extensive reading by generating English language learning materials for learners based on their interests and proficiency levels. The software begins by administering a brief and straightforward vocabulary test to learners. The test is used to identify words that may be unknown to the learners from 12,000 words. The machine-learning algorithm identifies words that are most likely to be unknown to the learner based on their test performance. The software then prompts the learner to select a topic of interest such as science or music. Thereafter, the software generates personalized English language learning materials for the learner, which contain texts with specific vocabulary related to the selected topic. The material is generated using ChatGPT. The software highlights unknown words in the text, which the learner can check using a dictionary. The software then generates new material incorporating the unknown words that the learner has checked, ensuring that the learner is exposed to a wide vocabulary in his area of interest. This process is repeated multiple times with the software generating new materials and incorporating new words that the learner has checked, thereby facilitating the efficient acquisition of new vocabulary. Through this process, the learners can engage in extensive reading, enabling them to read more in English and develop their reading skills, while simultaneously acquiring new vocabulary related to their interests. The innovative approach of the software in English language learning offers a personalized, adaptive, and efficient approach to extensive reading. This can help learners improve their English proficiency by reading in a manner tailored to their interests and proficiency levels.

Keywords: Extensive Reading · ChatGPT · Generation

1 Introduction

English language learning has become increasingly essential in today's globalized world, with English serving as the primary language of communication in various professional and academic settings. While traditional language learning methods such as grammar drills and memorization are prevalent, they are often tedious, and boring, and may not cater to the individual needs and interests of learners. However, extensive reading, which involves reading large amounts of text, has been identified as a highly effective approach to language learning, particularly for vocabulary acquisition [5].

In this paper, we introduce innovative software for English language learning that incorporates extensive reading and personalized vocabulary acquisition. This software aims to address the challenges faced by learners using traditional language-learning methods by providing a personalized and adaptive approach to language learning. The software is designed to help learners acquire new vocabulary through extensive reading while ensuring that the materials are tailored to their interests and proficiency levels.

The software leverages machine learning and natural language processing (NLP) techniques to generate personalized English language learning materials for learners based on their interests and vocabulary levels. Specifically, the software uses a machine learning algorithm to identify words that the learners are likely to be unfamiliar with, based on their performance on a brief vocabulary test. The software then prompts the learners to select a topic of their interest and generates personalized English language learning materials using the service with the state-of-the-art natural language generation model, ChatGPT[1]. After generating the text, the software highlights unknown words in the text that the learners can check using a dictionary. The software then generates new materials, incorporating new words that the learner needs to learn based on the words that he has checked, ensuring that he is exposed to a variety of vocabulary within his area of interest.

The innovative approach to English language learning introduced in this study offers a personalized, adaptive, and efficient way to acquire new vocabulary through extensive reading. By following this approach, learners can engage in extensive reading, allowing them to read more in English and develop their reading skills, while simultaneously acquiring new vocabulary related to their interests. The paper concludes by discussing the potential impact of this innovative approach on English language learning and its implications for future research on language learning.

Relation to the Conference Theme "AI in Education for Sustainable Society": The use of our innovative technologies can contribute to the creation of a sustainable society, as education plays a critical role in promoting sustainable development. By enabling learners to acquire new knowledge and skills in English, they can contribute to building a more sustainable future by improving their communication with people in other countries. Therefore, it is crucial to continue to explore and develop innovative technologies that can enhance education and promote sustainable development. AI in Education for Sustainable Society is strongly related to our innovative software.

---

[1] https://openai.com/blog/chatgpt/.

2   Proposed Method

The proposed method for English language learning incorporates machine learning and natural language processing techniques to generate personalized English language learning materials for learners based on their interests and proficiency levels. This method consists of the following steps:

Vocabulary Test: The method begins with administering a brief and straightforward vocabulary test to the learner, which is used to identify words that are not included in the test and may not be known by the learner. The test is designed to be easily administered and graded, taking approximately 30 min to complete. The words were then ranked based on the learners' performance on the test.

Vocabulary Prediction: The machine learning algorithm identifies the words that are most likely unknown to the learner based on their performance on the test. The algorithm selects the top-ranked words, ensuring that they are neither too difficult nor too easy for the learners.

Topic Selection: The learner is prompted to select a topic of his interest, such as science or music, from a list of options. The selected topics are then used to generate personalized English language learning materials.

Text Generation: The software generates personalized English language learning material for the learner, which contains texts with specific vocabulary related to their chosen topic. The material is generated using a pre-trained natural language generation model called ChatGPT. The model generates coherent and natural-sounding text by predicting the most probable sequence of words given the input prompt. The software ensures that the text is grammatically correct and coherent, and highlights unknown words in the text. The ChatGPT model that we used in this system was the "Default" model for ChatGPT Plus at the time of Mar. 10, 2023.

Vocabulary Acquisition: The learner checks the unknown words highlighted in the text using a dictionary or other resources. The software generates new material that incorporates the words checked by the learner. The software repeats this process multiple times, incorporating new words each time to facilitate the efficient acquisition of new vocabulary.

Extensive Reading: By following this process, learners engage in extensive reading, allowing them to read more in English and develop their reading skills while simultaneously acquiring new vocabulary related to their interests. The software ensures that learners are exposed to a variety of vocabulary within their areas of interest, enhancing their motivation and engagement in the learning process.

The proposed method offers a personalized and adaptive approach to English language learning that can help learners acquire new vocabulary efficiently and

Generated: Science is a fascinating field of study, filled with nuance and intricacies that can be both fascinating and bizzarre. One of the most interesting aspects of science is the way in which it allows us to explore the world around us, from the smallest particles to the largest canary in the coal mine.

One of the most important principles in science is the idea of nuance. Scientists must be extremely precise in their observations and measurements, paying close attention to even the slightest details in order to gain a comprehensive understanding of the phenomena they are studying. For example, when studying the behavior of subatomic particles, scientists must account for the nuance of quantum mechanics, which describes the bizarre and often counterintuitive behavior of these tiny particles.

Another area of science that can be quite bizarre is the study of genetics. By examining the intricate details of DNA and the ways in which it is expressed in living organisms, scientists have been able to unlock a wealth of information about the natural world. From the canary in the coal mine to the tallest trees in the forest, genetics plays a crucial role in shaping the diversity of life on Earth.

Despite the many challenges that come with studying science, it is a field that is rich with opportunity for discovery and exploration. Whether you are interested in the nuance of subatomic particles or the bizarre intricacies of genetics, there is always something new to learn and

Chosen Topic: Science, Must-included word: bizzarre, nuance, canary

Fig. 1. The screenshot of the proposed system.

effectively. The method is innovative in its use of machine learning and natural language processing techniques to generate personalized learning materials for learners and has the potential to revolutionize English language learning approaches.

3   Software

Figure 1 shows the screen capture of the proposed system. In the text area, the text generated by ChatGPT is shown. Below the text area, "Science" indicates that the learner chose the topic "Science". The learner can change the topic by choosing another topic and clicking on the word.

Below the text area, the three words predicted to be unreadable to the learner are listed: "nuance", "bizarre", and "canary". The ChatGPT generates the text for the learner to read. The ChatGPT is instructed to include all three words in the text. It is also instructed to make all the words other than these three words as easy as possible, as long as the text is on the topic of "Science".

By moving the focus over the text area, words predicted to be unknown to the learner are automatically highlighted. By clicking on words in the text area, the learner can indicate which words are unknown to him. This click information is further used to predict words that are unknown to the learners.

4   Experiments

4.1   Setting

We used a vocabulary knowledge dataset [4]. The dataset consists of 12,000 English words. Fifteen English language learners rated their familiarity with each word on a 5-point scale. This dataset provides an accurate representation of the learners' knowledge of words typically found in newspaper articles.

From the vocabulary knowledge dataset, we randomly selected 43 words as the words also appear in the EVKD1 dataset, another vocabulary knowledge dataset distributed by us. We then created a classifier to determine whether the learners knew each word based on their performance on a vocabulary test. To

Table 1. The experiment result over the 10 trials (text generations).

Number of Trials	1	3	5	7	10
Percentage of Known Words	0.937	0.930	0.924	0.899	0.888
Accuracy of Prediction Vocabulary	0.939	0.939	0.929	0.928	0.913
Percentage of Readable Text	0.466	0.600	0.307	0.363	0.090

estimate vocabulary knowledge, we use a logistic regression model that uses the log frequency of words in both the British National Corpus [1] and Contemporary Corpus of American English [2] as features. Both corpora cover various topics and are used to assess language proficiency. This approach allowed us to achieve high accuracy in determining if each learner knows each word. To build the classifier, we followed [3], which provides hints to our system. However, [3] proposed a method for selecting text-to-read from many texts but did not propose text generation for extensive reading.

Thereafter, among the words that were classified to be unfamiliar to the learner, the words with the probability closest to 0.5 were selected as the words to be learned. These words were chosen because they were the most likely to occur most frequently among the words unknown to the learners and were important. The ChatGPT is requested to generate learning materials containing these words. Once ChatGPT generated a text, we again choose the words unknown to the learner but important words within the generated text. To this end, we again select the three words on the same criterion except that the previously chosen words should not be chosen as the three words to be contained in the generated text.

We repeated the generation for 10 times for each learner. After the generation, we measured their readability based on simulations using 12,000-word vocabulary data. The study [5] showed that learners need to know at least 95% of the words in a text to read it naturally and learn from it. The 12,000-word vocabulary covered most words that appeared in the generated text. Still, words not included in the vocabulary were assumed to be unknown to the learners. As for the topics, we simulated the case in which each learner choose a topic from three topics, namely science, sports, and culture, uniformly randomly.

4.2 Results

Table 1 shows the experimental results using the reading simulation. The first row shows the number of extensive reading trials, that is, the number of times the system generated text tailored to the learner. All values represent the average of the 15 learners. The second row shows the percentage of words known to the 15 learners on average. We can see that the percentage gradually decreases. This is rational because we ask ChatGPT to generate a text containing different words from the previous ones over trials: it is meaningless to generate a text containing the same set of three words repeatedly. The second row shows the accuracy of predicting whether each learner knows a word. We can see that the accuracy gradually decreases. However, the decrease is slight. Finally, The third row shows the percentage of readable texts over 15 texts tailored to each learner in each trial. The value decreases as the number of trials increases. This result is rational because, in this setting, ChatGPT generates texts containing different words from the previous ones over trials. Hence, if the learner memorizes the words to be contained in the generated texts, the decrease is expected to disappear.

This implies that the more text learners read, the more information about each learner's vocabulary knowledge is passed on to the system, and the system's further prediction is expected to become more accurate.

5   Conclusion

In this paper, we propose an innovative software for English language learning that incorporates extensive reading and personalized vocabulary acquisition. The software utilizes machine learning and natural language processing techniques to generate personalized English language learning materials for learners, based on their interests and proficiency levels.

Our experiments demonstrated that the proposed method is effective in facilitating vocabulary acquisition and reading comprehension skills. A personalized and adaptive approach to language learning exhibits the potential to provide learners with a more engaging and effective way to learn English.

Future studies should expand the scope of the learning materials to cover a wider range of topics and domains. In addition, the proposed method can be further refined to incorporate additional features such as speech recognition and pronunciation assessment.

Acknowledgements. This work was supported by JST ACT-X, Grant Number JPM-JAX2006, Japan.

References

1. BNC Consortium: The British National Corpus (2007)
2. Davies, M.: The corpus of contemporary american english (coca). https://www.english-corpora.org/coca/ (2008)
3. Ehara, Y.: Selecting reading texts suitable for incidental vocabulary learning by considering the estimated distribution of acquired vocabulary. In: Proceedings of Educational Data Mining (poster paper) (2022)
4. Ehara, Y., Shimizu, N., Ninomiya, T., Nakagawa, H.: Personalized Reading Support for Second-language Web Documents. ACM Trans. Intell. Syst. Technol. **4**(2), 31:1–31:19 (2013). https://doi.org/10.1145/2438653.2438666
5. Nation, I.: How Large a Vocabulary is Needed For Reading and Listening? Canadian Modern Language Review **63**(1), 59–82 (Oct 2006)

A Student-Teacher Multimodal Interaction Analysis System for Classroom Observation

Jinglei Yu[1,2], Zhihan Li[2], Zitao Liu[3], Mi Tian[4], and Yu Lu[1,2(✉)]

[1] School of Educational Technology, Faculty of Education,
Beijing Normal University, Beijing, China
luyu@bnu.edu.cn
[2] Advanced Innovation Center for Future Education, Beijing Normal University,
Beijing, China
[3] Guangdong Institute of Smart Education, Jinan University, Guangzhou, China
[4] TAL Education Group, Beijing, China

Abstract. Classroom observation is an effective way for teachers to improve professional development, and the analysis of student-teacher interactions is critical and significant to classroom observation. However, the traditional methods of the classroom observation are mainly based on manual coding by domain experts. Although several studies have been conducted to automate the coding and analyzing process, they are either based on audio information or video information collected from the classroom, which fails to jointly utilize multimodal information like domain experts. We thus propose a student-teacher multimodal interaction analysis system that conducts the analysis using both video and audio information and accordingly generates the informative reports based on the analysis results. A preliminary evaluation of the system validates the effectiveness of the built system and the analysis results could be further used for the evidence-based teaching behavior evaluation. The current limitations and possible optimization on the built system are discussed as well. We are planning to keep improving the system and deploy it to 1000 schools located at the rural areas in three years.

Keywords: Student-Teacher Interaction · Multimodal Interaction Analysis · Automatic Classroom Observation · Teaching Behavior

1 Introduction

Classroom observation provides an effective approach for teachers to have an explicit view on their classroom teaching and pedagogical practice [7,9]. During the observation, the comprehensive analysis on student-teacher interactions is critical, since it is directly related to the quality of teaching and learning [8]. In classroom observation, the classroom activities should be labeled through coding process. Traditionally, the coding process of the student-teacher interactions heavily relies on manual coding, which is time-consuming and hard to be

N. Wang et al. (Eds.): AIED 2023, CCIS 1831, pp. 193–199, 2023.
https://doi.org/10.1007/978-3-031-36336-8_29

deployed in large scale. Number of studies on automatic analysis have been conducted to facilitate the coding efficiency. For example, automatic speech recognition can be utilized on teacher's speech to segment and classify utterances for further analysis, such as distinguishing instructional activities and evaluating dialogic instructions [4]. Besides, computer vision techniques are also adopted to analyze classroom proxemics and students' states such as head orientation, upper body pose, raising hand, smiling, etc. [1]. However, classroom observation is a dual-coding process with both verbal and visual information input, and few existing systems jointly utilize such multimodal information for student-teacher interactions analysis. In fact, verbal information could provide visual analysis with semantic evidence to help understand course implementation. Meanwhile, visual information could assist to fill in details of student-teacher interaction with non-verbal behavior, such as nodding head.

We thus design and implement a student-teacher multimodal interaction analysis system. The system adopts a simple but informative observation method known as S-T analysis (i.e., student-teacher analysis) [3,5]. The S-T analysis is featured with instructive analysis results and intuitive visualization. In S-T analysis, the classroom activities are divided into teacher's and students' behaviors. The analysis result contains ratio of teacher's behavior (Rt), ratio of student-teacher interactions (Ch) and the predicted teaching mode. The visualization of the teaching mode is provided in the summary report. The system has been preliminary evaluated on 21 diverse classroom video recordings from different schools, covering multiple subjects ranging from Chinese, Math, English to Chemistry and Biology.

2   System Description

The system consists of three layers, namely data collection, multimodal processing and teaching analysis, as shown in Fig. 1. Firstly, in the data collection layer, video and audio streams are captured and data pre-processing tasks are fulfilled. Secondly, in the multimodal processing layer, teacher's actions and facial expressions from the video stream are detected, and meanwhile teacher's verbal information from the audio stream is recognized. Thirdly, teaching analysis layer conducts S-T analysis and teaching emotion analysis based on the statistics delivered through verbal information, actions and facial expressions. Finally, the system visualizes the analysis results by generating a summary report as feedback for teachers.

2.1   Data Collection

During the classroom teaching, video streams are recorded by a webcam towards the teacher, and audio streams are generally recorded through a microphone placed close to the teacher. In the data collection layer, image frames are extracted from the video streams and the audio streams are converted from raw waves into binary streams.

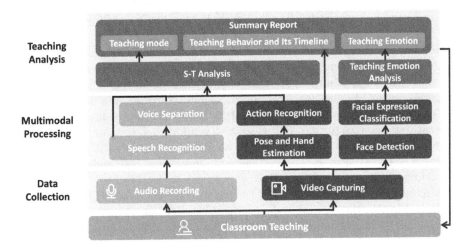

Fig. 1. Architecture of the system

2.2 Multimodal Processing

Multimodal processing layer conducts action recognition, facial expression classification and verbal information processing. For action recognition, the system supports six categories of actions, namely nodding head, shaking head, tilting head, as well as clapping hands, thumbing up and pointing. Since multiple teaching behaviors are dynamic actions, single frame could not present the action completely. We adopt sliding window with size of five frames and step of one frame to recognize the actions. For each sliding window, actions are detected in a two-step strategy. Firstly, the system locates 25 key points of the teacher's body and 21 key points of both left and right hands with the help of OpenPose models [2]. Secondly, with the pose estimation results, actions are recognized through pre-defined rules. An exemplary rule of shaking head is shown in Fig. 2B. We define the horizontal distance between left and right eye as D, and define the vertical difference between nose and neck as x. The rule of labeling each frame is that: when x >0.5D, the frame is labeled as r, and when x <-0.5D, the frame is labeled as l. In a sliding window, when an "l...r...l" or "r...l...r" pattern is

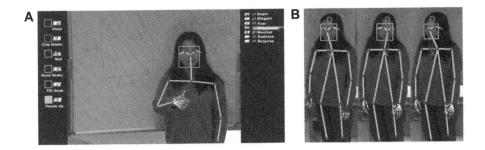

Fig. 2. Demonstration of action recognition and facial expression classification

observed, the shaking head action is detected. The timestamps from the beginning to the end of each action are recorded.

Besides, the facial expression is momentary and could be recognized through single frame. The system detects teacher's face and recognizes its facial expression via the facial attributes API. The facial expressions are classified into seven categories: anger, disgust, fear, happiness, sadness, surprise, and neutral.

Concurrently, for audio streams, system recognizes text content using the automatic speech recognition API which also supports voice separation. Besides, to provide semantic evidence for visual analysis, sentences with relevant keywords are recorded, such as "good", "correct", "excellent" and "right" for actions of nodding heads and clapping hands.

A monitoring panel is developed to demonstrate the processing results. As shown in Fig. 2A, the "thumb up" button is highlight on the left-side bar, and "happiness" with probability of 96% (i.e., the result of facial expression recognition) is shown on the right-side bar. Meanwhile, body and hands skeleton, the bounding box of face detection and captions are also shown on the panel.

2.3   Teaching Analysis

The teaching analysis layer conducts S-T analysis based on both verbal and visual information. The first step is to classify teacher's and students' behaviors during the class. In the built system, we define the teacher's behaviors refer to teacher's speech and teacher's six actions recognized in multimodal processing layer. The rest of time is classified as students' behaviors. To guarantee the recognized actions corresponding to the target semantics, we take advantage of semantic evidence from verbal information to filter appropriate action recognition results. Meanwhile, we utilize the recognized teaching actions to enrich verbal information with non-verbal teaching behaviors. For example, when a student is answering the question, teacher may nod head as non-verbal feedback to interact with the student.

Specifically, in S-T analysis, Rt refers to the duration ratio of teacher's behavior to the whole class, which reveals the dominant role of the class. Ch means the ratio of interaction times to the whole class duration, which to some extent reflects the degree of student-teacher interaction. We define the class duration as N, teacher's behavior duration as c, and the times of the same role appearing continuous behavior as g. Then, Rt and Ch could be calculated in (1) and (2), and teaching mode could be classified based on the criteria as shown in Table 1.

$$Rt = c/N \tag{1}$$

$$Ch = (g-1)/N \tag{2}$$

Finally, the system would generate a summary report that illustrates the teaching mode, teaching action and its timeline, as well as teaching emotion, as designed in Fig. 3. Note that the teaching emotions are summarized into three main categories, namely positive emotion (happiness), negative emotion (anger, disgust, fear, sadness) and neutral emotion (surprise and neutral).

Table 1. Teaching mode and criteria

Teaching Mode	Criteria
Lecturing	Rt ≤ 0.3
Practicing	Rt ≥ 0.7
Conversational	Ch ≥ 0.4
Mixed	0.3 < Rt < 0.7, Ch < 0.4

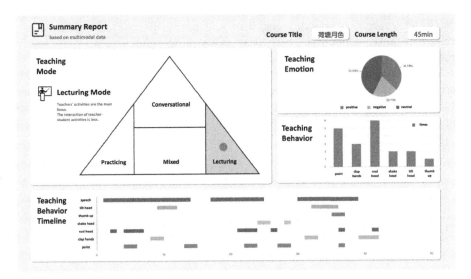

Fig. 3. An exemplar summary report showing the analysis results

3   System Evaluation

A preliminary evaluation has been conducted to test the S-T analysis results of the built system. We collected 21 classroom video recordings from different schools on the TKben website [6], where K-12 teachers teaching various subjects could share their videos. Each video is around 45 min and the subjects of tested videos cover Chinese, Math, English, Chemistry and Biology. To evaluate the system, each video has been annotated manually by domain experts as ground truth to classify the clips as teacher's or students' behavior. The evaluation result is shown in Table 2. To further analyze the results, we find the visual recognition could partially help supplement verbal information with non-verbal teaching behaviors. For example, when teacher tutoring students in group discussion, "pointing" action could be recognized and marked as teacher's behavior in between students' behaviors. Taking advantage of semantic evidence, several falsely recognized actions have been eliminated.

4 Discussion and Future Plan

The preliminary evaluation result shows the feasibility of using multimodal student-teacher information for automating the classroom observation. Meanwhile, we find both visual and verbal recognition require further improvement. For the visual modal, the role identification function should be developed. Moreover, the robustness of rule-based action recognition is limited. The system would further adopt the state-of-the-art deep learning models to increase the accuracy of action recognition. For the verbal modal, the quality of microphone directly affects the accuracy of speech recognition and voice separation. In addition, we would constantly expand the categories of teaching behaviors, and eventually provide more reliable feedback and comprehensive reports for teachers.

We are currently working on updating the system and equipping it with pedagogical knowledge to better evaluate teaching behaviors. We are planning to deploy the improved system to 1000 schools in rural areas in China to help improve the teaching qualities and teachers' professional development in three years.

Table 2. The preliminary evaluation result of S-T analysis

Number of Videos	Accuracy of Rt	Accuracy of Ch	Accuracy of Teaching Mode
21	59.7%	71.1%	47.6%

Acknowledgements. This work was supported in part by National Key R&D Program of China, under Grant No. 2020AAA0104500; in part by National Natural Science Foundation of China (Grant No. 62077006).

References

1. Ahuja, K., et al.: Edusense: practical classroom sensing at scale. Proc. ACM on Interact., Mobile, Wearable Ubiquit. Technol. **3**(3), 1–26 (2019)
2. Cao, Z., Simon, T., Wei, S.E., Sheikh, Y.: Realtime multi-person 2d pose estimation using part affinity fields. In: Proceedings of the IEEE Conference on Computer Vision and Pattern Recognition, pp. 7291–7299 (2017)
3. Chen, Y., Liu, Q., Wang, F., Wang, Y.: Research on the application of an improved, video-based s-t analysis method. e-Educ. Res. **37**(6), 90–96 (2016)
4. D'Mello, S.K., Olney, A.M., Blanchard, N., Samei, B., Sun, X., Ward, B., Kelly, S.: Multimodal capture of teacher-student interactions for automated dialogic analysis in live classrooms. In: Proceedings of the 2015 ACM on International Conference On Multimodal Interaction, pp. 557–566 (2015)
5. Fu, D., Zhang, H., Liu, Q.: Educational Information Processing. Beijing Normal University Press, Beijing, China (2021)
6. Advanced Innovation Center for Future Education, B.N.U.: Tkben. http://tkben. cn// Accessed March 9 (2023)

7. Millman, J., Darling-Hammond, L.: The new handbook of teacher evaluation: Assessing elementary and secondary school teachers. Corwin Press (1990)
8. Pianta, R.C.: Teacher-student interactions: measurement, impacts, improvement, and policy. Policy Insights Behav. Brain Sci. **3**(1), 98–105 (2016)
9. Zhang, H., Yu, L., Cui, Y., Ji, M., Wang, Y.: Mining classroom observation data for understanding teacher's teaching modes. Interact. Learn. Environ. **30**(8), 1498–1514 (2022)

Rewriting Math Word Problems to Improve Learning Outcomes for Emerging Readers: A Randomized Field Trial in Carnegie Learning's MATHia

Husni Almoubayyed[1]([✉]), Rae Bastoni[2], Susan R. Berman[1], Sarah Galasso[1], Megan Jensen[1], Leila Lester[2], April Murphy[1], Mark Swartz[1], Kyle Weldon[1], Stephen E. Fancsali[1], Jess Gropen[2], and Steve Ritter[1]

[1] Carnegie Learning, Inc., Pittsburgh, PA 15219, USA
halmoubayyed@carnegielearning.com
[2] CAST, Lynnfield, MA 01940, USA

Abstract. We present a randomized field trial delivered in Carnegie Learning's MATHia's intelligent tutoring system to 12,374 learners intended to test whether rewriting content in "word problems" improves student mathematics performance within this content, especially among students who are emerging as English language readers. In addition to describing facets of word problems targeted for rewriting and the design of the experiment, we present an artificial intelligence-driven approach to evaluating the effectiveness of the rewrite intervention for emerging readers. Data about students' reading ability is generally neither collected nor available to MATHia's developers. Instead, we rely on a recently developed neural network predictive model that infers whether students will likely be in this target sub-population. We present the results of the intervention on a variety of performance metrics in MATHia and compare performance of the intervention group to the entire user base of MATHia, as well as by comparing likely emerging readers to those who are not inferred to be emerging readers. We conclude with areas for future work using more comprehensive models of learners.

Keywords: machine learning · A/B testing · intelligent tutoring systems · reading ability · middle school mathematics

1 Introduction

A growing body of research has found connections between math learning outcomes and reading comprehension (See, e.g., [4,5]). Recent work seeks to develop more comprehensive models of learners as they use adaptive learning software, including efforts to build models that incorporate reading ability while students use adaptive software for mathematics instruction (e.g., [1,6]). Such work recognizes that students draw on skills outside of the target domain as they receive

N. Wang et al. (Eds.): AIED 2023, CCIS 1831, pp. 200–205, 2023.
https://doi.org/10.1007/978-3-031-36336-8_30

instruction and practice skills. With more comprehensive learner models, adaptive learning software like intelligent tutoring systems (ITSs) could go beyond existing approaches that adapt based on skills within the target domain (math) to adapt to factors like students' reading ability. These observations raise questions about the nature of supports that might be used in an ITS for mathematics to adapt to student reading ability.

2   MATHia and UpGrade

MATHia (formerly Cognitive Tutor [7]) is an ITS for middle- and high-school mathematics. Currently used by over 600,000 students across the US, MATHia content is presented to students organized by topic in "workspaces," which either take the form of "Concept Builders" or "Mastery Workspaces." Concept Builders present instructional content and interactive, exploratory tools along with a fixed sequence of multi-step problems that introduce students to new materials and develop students' conceptual understanding of target material. In Mastery Workspaces, students work through complex, multi-step problems towards mastery of each of a set of knowledge components (KCs; [3]) or skills associated with the workspace. Mastery is determined using Bayesian Knowledge Tracing [2]. We refer to students who successfully achieve mastery of all the KCs in a workspace as "graduated." Students get "promoted" when they are moved on to the next workspace if they encounter a pre-set maximum number of problems but still fail to achieve mastery of all KCs.

UpGrade is a free and open source platform for conducting randomized field trials (or "A/B tests") in educational software applications [8]. UpGrade enables large-scale randomized field trials in real classroom settings by integrating with EdTech software applications like MATHia and allowing researchers to manage experimental design and logistics through a simple, web-based user interface. UpGrade communicates with the EdTech application to randomly assign appropriate experimental conditions to learners. By integrating with MATHia, we are able to deliver instructional interventions across Carnegie Learning's sizeable customer base, within multiple workspaces.

3   Predicting Reading Ability

The student's first interaction with MATHia is a Concept Builder known as the Pre-Launch Protocol. This introductory activity prepares the student for working with MATHia, and is not particularly related to mathematics. Almoubayyed et al., 2023 [1] developed a neural-network based model to predict the end-of-year English Language Arts (ELA) scores of students based on student performance in the Pre-Launch Protocol, by training it on a sample of end-of-year ELA scores. We use this model to predict which students are emerging readers.

We use a version of the predictive model that is trained on predicting the probability that a student would pass the end-of-year ELA exam (the model has an AUC of around 0.8, please see [1] for model details). In this study, we define emerging readers as those whose probability of passing their ELA test is in the bottom quartile. We do not retrain the model as a binary classifier of being an emerging reader, because the training set in [1] comes from one school district, and it can be unclear what thresholds must be used to represent the bottom quartile of the student population that uses MATHia nationally. We define the bottom quartile for each workspace independently and for the sample in each condition independently.

4 Rewriting Word Problems

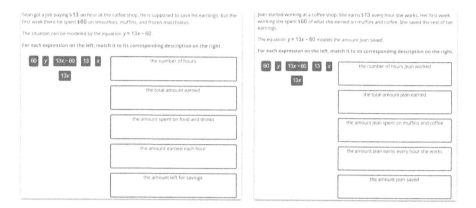

Fig. 1. An example of an unmodified word problem (left) in MATHia and the rewritten version (right).

Two workspaces were chosen to be rewritten from MATHia's Mastery Workspaces: "Analyzing Models of Two Step Linear Relationships" in Middle School Course 2 (Carnegie Learning's Grade 7 math sequence), which we refer to as 'Integers'; and "Analyzing Models of Linear Relationships" in Middle School Course 3 (Carnegie Learning's Grade 8 math sequence), which we refer to as 'Rationals.' These two workspaces were chosen due to the high correlation of student performance in these workspaces with students' ELA end-of-year state test scores, compared to the correlation with students' end-of-year state test math scores. Correlational relationships were based on a historical dataset for which ELA test scores were made available to researchers (via a data sharing agreement with the district). The content was rewritten based on internally developed principles by an instructional design team with two specific goals: (a) using only recognizable content (e.g., everyday objects or phenomena that appear in word problems) that is easily understandable and relevant to students across the country, and (b) using clear and precise language, that supports students in visualizing and connecting meaning across sentences. Explicitly, the rewrites aimed to preserve the underlying mathematical content difficulty. Rewriting was carried out,

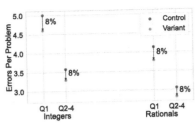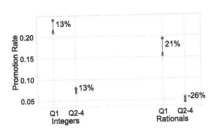

Fig. 2. The results of the metrics of the average Errors Per Problem and Promotion Rate. Q1 are students predicted to be emerging readers, with Q2-4 indicating the rest of the students. Q1 students show improvement in all cases, with the percent change annotated.

following the guidelines, by staff from Carnegie Learning and CAST, with every rewrite reviewed for quality assurance by a different person than the writer. An example of a problem before and after the rewrite process is provided in Fig 1.

5   Evaluation Across Reading Abilities

The study was randomized with equal probability of a student receiving the control or the variant condition as they encounter one of the target workspaces. It was deployed to the entire user base for a time period of 7 weeks, with a substantial number of students (14,767) enrolling in the experiment. Out of the students that enrolled in the experiment, we only use data for students that completed their assigned workspace and the Pre-Launch Protocol – a total of 12,374 students. In the Integers workspace, there were 4,113 in the variant and 3,894 in the control condition. In the Rationals workspace, there were 2,230 students in the variant and 2,137 in the control sample.

We use a variety of metrics to compare the results between the control and variant sample for each of the two workspaces. For all of these metrics, a smaller value indicates a better outcome. These metrics are:

- *Promotion Rate*: The number of students who were promoted in a sample (failed to master any skill in a workspace) over the number of all students in that sample.
- *Time Spent*: The total time spent by each student in a sample is computed, then the median is taken over all the students in that sample. We use the median here due to outliers with very large time values (e.g., when a student has MATHia open but is not paying attention). We report this metric independently for all students, and also for students who were 'graduated', or mastered all skills in the relevant workspace.
- *Total Errors*: The total number of errors that a student makes in the relevant workspace. The average is taken over all students in a sample.
- *Total Problems Completed*: The average of the total problems completed over all students in a sample. We report this metric independently for all students and for graduated students. Promoted students completed 25 problems each.

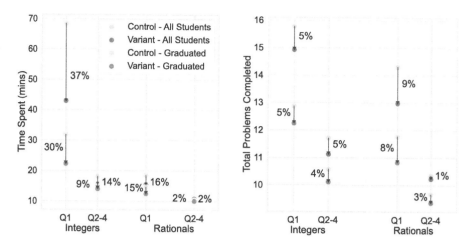

Fig. 3. The results of the Time Spent and Total Problems Completed metrics for all students (with percent change annotated to the right of each arrow) and graduated students (with percent change annotated to the left) independently. In all cases, there is an improvement for students receiving the rewritten problems, and this improvement is larger for Q1 students.

The results of the evaluation metrics are presented in Figs. 2 and 3. In all Quartile 1 (Q1, predicted to be emerging readers) and almost all Q2-Q4 (predicted not to be emerging readers) cases, the variant has better metric outcomes than the control. The improvement for Q1 is always better (sometimes dramatically) or at least equal to the improvement for Q2-4. In particular, the results show that for the population of students that were predicted by the model from [1] to be emerging readers, rewriting word problems following the guideline described in Sect. 4 results in making 8% fewer errors. Additionally 13–21% more emerging readers were able to master all the skills in the targeted workspaces. Out of the Q1 students who master all skills in the targeted workspaces, they do so in 15–30% less time, completing 5–8% fewer problems, which means they can spend more time on other material. We hope that these outcomes will result in higher end-of-year exam scores, but we leave that to future studies.

6   Conclusions

Using comprehensive learner models has the potential to help adaptively target and deliver supports to students who need them. We applied a machine learning model to classify emerging readers in MATHia, a math ITS, based on an introductory activity. We found that rewriting word problems with simple guidelines for added clarity and relevance led to large improvements in several performance metrics for students predicted to be emerging readers. More emerging readers were able to master all skills in the workspaces, do so in fewer problems, spend much less time in the workspaces, and make fewer errors. These improvements were not as high, and occasionally of potentially negative impact for other students, highlighting the importance of exploring the adaptive delivery of these kinds of supports. In future studies, we will explore whether these and similar improvements also positively affect end-of-year exam outcomes for these students. However, even at the level of specific workspaces, these improvements save emerging readers a significant amount of valuable time.

Acknowledgements. Research reported here was supported by Institute of Education Sciences, U.S. Department of Education, grant R324A210289 to CAST. Opinions expressed do not represent views of the IES or U.S. Department of Education.

References

1. Almoubayyed, H., Fancsali, S.E., Ritter, S.: Instruction-embedded assessment for reading ability in adaptive mathematics software. In: Proceedings of the 13th International Conference on Learning Analytics and Knowledge. LAK '23, Association for Computing Machinery, New York, NY, USA (2023)
2. Anderson, J.R., Corbett, A.T.: Knowledge tracing: modeling the acquisition of procedural knowledge. User Model. User-Adap. Inter. **4**, 253–278 (1995)
3. Koedinger, K.R., Corbett, A.T., Perfetti, C.: The knowledge-learning-instruction framework: Bridging the science-practice chasm to enhance robust student learning. Cogn. Sci. **36**(5), 757–798. https://doi.org/10.1111/j.1551-6709.2012.01245.x
4. Koedinger, K.R., Nathan, M.J.: The real story behind story problems: effects of representations on quantitative reasoning. J. Learn. Sci. **13**(2), 129–164 (2004). https://doi.org/10.1207/s15327809jls1302_1
5. Krawitz, J., Chang, Y.P., Yang, K.L., Schukajlow, S.: The role of reading comprehension in mathematical modelling: improving the construction of a real-world model and interest in germany and taiwan. Educ. Stud. Math. **109**, 337–359 (2022)
6. Richey, J.E., Lobczowski, N.G., Carvalho, P.F., Koedinger, K.: Comprehensive views of math learners: A case for modeling and supporting non-math factors in adaptive math software. In: Bittencourt, I.I., Cukurova, M., Muldner, K., Luckin, R., Millán, E. (eds.) Artificial Intelligence in Education, pp. 460–471. Springer International Publishing, Cham (2020)
7. Ritter, S., Anderson, J.R., Koedinger, K., Corbett, A.T.: Cognitive tutor: applied research in mathematics education. Psychonom. Bull. Rev. **14**, 249–255 (2007)
8. Ritter, S., Murphy, A., Fancsali, S.E., Fitkariwala, V., Lomas, J.D.: Upgrade: An open source tool to support a/b testing in educational software. In: Proceedings of the First Workshop on Educational A/B Testing at Scale. EdTech Books (2020)

Automated Essay Scoring Incorporating Multi-level Semantic Features

Jianwei Li[1,2] and Jiahui Wu[1(✉)]

[1] College of Network Education, Beijing University of Posts and Telecommunications,
Beijing 100088, China
amy_wjh@bupt.edu.cn
[2] Beijing Key Laboratory of Network System and Network Culture, Beijing University of Posts
and Telecommunications, Beijing 100876, China

Abstract. Essay writing might reveal the language proficiency of a student. Utilizing intelligent technology to automatically grade essays is an effective method of saving significant manpower and time resources, and improving the accuracy of score. The present models typically rely on shallow semantic features, deep semantic features and multi-level semantic features. Existing models, however, struggle to be superior in both scoring accuracy and generalization performance. As a result, we propose a model that incorporates multi-level semantic features. Specifically, we manually define and automatically extracted the shallow semantic features; we use the BERT pre-training model, convolutional neural networks and recurrent neural networks to extract the deep semantic features; and last, feature fusion is used to score essay automatically. The proposed model outperforms three state-of-the-art baseline methods, according to experimental results on two datasets. Additionally, the generalization of the model has been greatly enhanced. The study has a significant impact on automated essay scoring theoretical investigations and practical applications.

Keywords: Essay writing · Automated scoring · Feature fusion · Semantic features · Deep neural network

1 Introduction

The writing reflects the thinking and logical expression of the author and further reflects the language level. English writing has become a compulsory question in major language exams at home and abroad, including TOEFL, IELTS, GRE, and others. In the post-exam evaluation, objective questions are easy to mark because of their fixed answers. Nevertheless, English essays without standard answers sometimes call for several teachers to correct the same essay and eventually grade it in accordance with predetermined guidelines, which is labor intensive and time consuming. Furthermore, it is difficult to guarantee the accuracy and consistency of the rating results due to the reviewers' subjective factors, which have a significant impact on the results. With the development of artificial intelligence (AI), more and more researchers tried to use intelligent technologies to address the issue of evaluating English essays, leading to the creation of Automated Essay Scoring (AES).

N. Wang et al. (Eds.): AIED 2023, CCIS 1831, pp. 206–211, 2023.
https://doi.org/10.1007/978-3-031-36336-8_31

2   Conventional AES Models

The present AES models for English writing can be divided into three categories based on the depth of their semantic features: shallow semantic features, deep semantic features, and multi-level semantic feature fusion.

Early AES obtained the predicted scores of essays by performing linear regression operations or clustering the essays based on the shallow semantic features. The models, which take into account word and sentence length, etc., are highly interpretable but less accurate. With the rapid development of a new generation of AI technologies, represented by deep neural networks (DNN), researchers are exploring the use of DNN and NLP technologies to extract deep semantic features from essays. For instance, use Bi-LSTM [1], CNN and LSTM [2], Bi-LSTM [3], BERT and XLNet [4], and propose a meta-trained BERT with in-context tuning approach [5]. Although they achieve high accuracy, their interpretable is limited. Thus, researchers begin to explore the construction of multi-level semantic feature fusion models. For example, Qe-CLSTM uses hierarchical CNN and Bi-LSTM [6], TSLF-ALL employs Bi-LSTM and XGBoost [7], and MLSN merges CNN, CNN-LSTM, and hand-crafted features [8]. In summary, the performance of the models that incorporate multi-level semantic features is optimal. Nonetheless, these models are hard to generalize and use since their assessment effectiveness varies greatly across different datasets, and the generalizability of the current incorporated models still needs to be enhanced.

3   Design and Implementation of MLSF

Based on the aforementioned issues and considering the strengths and weaknesses of different types of models, we propose a deep neural network with multi-level semantic features (DNN-MLSF) model. Figure 1 depicts the architecture of the model.

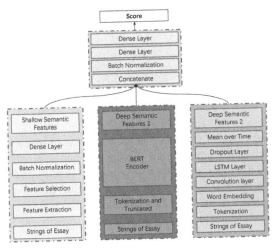

Fig. 1. The DNN-MLSF architecture.

Shallow Semantic Feature Extraction. We define and extract 20 different kinds of shallow semantic features from both the word and sentence dimensions. To calculate the correlation between the features and the scores, we used the Pearson correlation coefficient. Ten shallow features with the highest relevance to the scores were chosen for feature fusion. Data with various dimensions is input into the Batch Normalization (BN) layer for batch normalization. Finally, Dense, a fully connected layer, is added to extract the correlation factors between the shallow semantic features and map them to the output space via a nonlinear transformation. The number of neurons in the experiment is 10, and the activation function is Relu.

Deep Semantic Feature Extraction Using BERT. The BERT pre-training model "uncased L-12 H-768 A-12" that Google provided is used. We tokenize the input essay strings into words using FullTokenizer, and the result is truncated with a maximum length of 199. [CLS] is then added to the head of the truncated results to produce the final word array with a length of 200. The array is then introduced into the structure of BERT, which successively processes it using multi-head attention, residual connection, normalization, and a Dense layer over two layers. The output consists of two feature matrices, $'sequence\_output, cls\_output'$, one for each word and one for the full essay. To examine the integrity of the essay, the latter cls_output is chosen as one of the inputs of the feature fusion, namely the deep semantic feature 1.

Deep Semantic Feature Extraction Using CNN-LSTM. To begin, use NLTK to process the essay string. Secondly, the results are fed into the word embedding layer, and the words are encoded into a feature array of length 50. Thirdly, input the processed array into a one-dimensional convolution layer, which processes it and outputs a four-dimensional array. The parameters of this layer are set as follows: cnn_dim is 1, cnn_window_size is 3, cnn_border_mode is 'same', and $stride$ is 1. The fourth stage is inputting the four-dimensional array into the LSTM layer with the following settings: 300 for rnn_dim, Ture for $return\_sequence$, 0.5 for $dropout$, and 0.1 for $recurrent\_dropout$. Add a Dropout layer to make the model "drop" some nodes at each training session, thus preventing the model from overfitting. Finally, a Mean Over Time (MoT) layer is added, and the MoT layer processes the output vector H of the LSTM layer to produce the deep semantic features 2.

$$\text{MoT(H)} = \frac{1}{M} \sum_{t=1}^{M} h_t \qquad (1)$$

where M is the number of input vectors of the MoT layer and h_t is the element in the input vector.

Feature Fusion. Each type of semantic feature is combined into a tensor by the Concatenate layer, including the $input\_feature$, cls_output, and rnn_output. The processed tensor is input to the BN layer for batch normalization. Finally, add two Dense layers. The neuron count of the first layer is 100. The activation function employs the Sigmoid activation function and the formula is as follows. The neuron count of the second Dense layer is 1, and the activation function continues to employ the Sigmoid activation function.

$$\sigma(x) = \frac{1}{1 + e^{-x}} \qquad (2)$$

where x stands for the neuronal output features of the Dense layer.

4 Experiments

4.1 Datasets and Setup

In this work, we conducted experiments using two datasets: dataset 1 from the ASAP competition held by Kaggle and dataset 2 from an English exam data of an online unified examination in a Chinese institution. Every dataset is split into training set, validation set, and test set according to the ratio of 80:10:10. We employ a five-fold cross validation method and take the average of the results of five experiments as the final prediction score of the model.

We employ Keras, a widely utilized deep learning framework. In order to hasten the convergence of the procedure, the optimizer uses the RMSProp optimization algorithm. The loss function uses the mean square error (MSE). The best model among the training times is chosen as the output model, and the epoch is set to 50. For the selection of the hyperparameters, the initial settings were adopted from similar models available on GitHub [2]. In addition, throughout the procedure, we tuned one parameter at a time while holding the others constant.

4.2 Metrics and Baselines

In this study, we employ the Quadratic Weighted Kappa (QWK) as the experimental metric to evaluate the coherence of model predicted results with actual scores. The higher the QWK value, the better the evaluation performance. The generalizability of the models is evaluated by comparing the standard deviation (σ) and variance (σ^2), which may reflect the degree of dataset dispersion. Smaller values of them imply that there is little performance variance across various datasets and that the models are highly generalizable. Three models with superior assessment results from previous studies—BERT, CNN-LSTM, and BERT-Shallow semantic features were also explored as baseline approaches at the same time.

4.3 Analysis of Experimental Results

Table 1 displays the individual experimental outcomes of dataset 1. (1) It is obvious that the manually defined shallow semantic features have improved the performance of the model. For six prompts, the QWK values of the BERT-Shallow semantic features model are better than the BERT, and the average QWK is 5.09% higher than the single model. (2) The CNN-LSTM model outperformed other models in terms of evaluation performance on five prompts, but it had the lowest generalizability overall. (3) The DNN-MLSF model can effectively improve the average performance of the AES model, whose average QWK value for the eight prompts is 8.17%, 1.01%, and 2.92% greater than three baseline methods, respectively. (4) Utilizing multi-level semantic features may effectively improve the generalizability of the model. In comparison to the three baseline techniques, the DNN-MLSF model has the lowest standard deviation and variance values of the QWK. The variance is reduced by, for instance, 59.68%, 66.67%, and 30.56%, respectively. (5) The eighth prompt contains the largest range of scores (10–60), which is a significant element affecting the performance of the AES. In this prompt, all four models have the lowest QWK scores.

Table 1. Experimental results of dataset 1.

prompts	BERT	CNN + LSTM	BERT-Shallow semantic features	DNN-MLSF
1qwk	0.8091	0.7102	**0.8463**	0.786
2qwk	0.6528	0.6864	**0.6976**	0.6712
3qwk	0.6188	**0.7100**	0.6837	0.6964
4qwk	0.6869	**0.7948**	0.6746	0.7668
5qwk	0.7662	**0.8304**	0.7954	0.7928
6qwk	0.6948	**0.8162**	0.7044	0.7844
7qwk	0.7302	**0.7928**	0.7114	0.7778
8qwk	0.5425	0.5505	0.6684	**0.6754**
Avg qwk	0.6877	0.7364	0.7227	**0.7439**
σ qwk	0.0789	0.0868	0.0596	**0.0496**
σ^2 qwk	0.0062	0.0075	0.0036	**0.0025**

In addition to QWK, MSE, RMSE, MAE, and MAPE were introduced as evaluation metrics in the experiments of dataset 2, and the results are shown in Table 2. (1) The generalization of the DNN-MLSF model remains optimal after integrating the experimental results from two datasets. In comparison to the three baseline methods, the standard deviation of QWK from DNN-MLSF is 0.0547, the variance is 0.003, and the standard deviation is reduced by 46.27%, 35.95%, and 10.33%, respectively. The variance is also reduced by 71.13%, 58.9%, and 18.92%. (2) The DNN-MLSF model provides the most comprehensive performance. Although the CNN-LSTM model has the largest QWK value in dataset 2, the DNN-MLSF model has the best overall performance when all metrics are considered because it has the lowest MSE, RMSE, MAE, and MAPE values. (3) The performance of the AES is significantly impacted by the quality of the manual

Table 2. Experimental results of dataset 2.

model	QWK	σ	σ^2	MSE	RMSE	MAE	MAPE
BERT	0.4661	0.1019	0.0104	0.0781	0.2795	0.2297	0.8241
CNN + LSTM	**0.6582**	0.0854	0.0073	0.0593	0.2434	0.1924	0.4370
BERT- Shallow semantic features	0.6477	0.0610	0.0037	0.0615	0.2479	0.1962	0.8721
DNN-MLSF	0.6534	**0.0547**	**0.0030**	**0.0592**	**0.2430**	**0.1919**	**0.4340**

annotation of the dataset. It was discovered that dataset 1 had at least two teachers rating each essay. However, each essay in dataset 2 is graded by only one teacher and the entire dataset is graded by a range of teachers. This leads to low-quality manual annotation, which is another crucial factor affecting performance.

5 Summary and Prospect

This article focuses on methods to improve the generalization and accuracy of the AES using multi-level semantic features, upon which the following pioneering work has been done: suggest a DNN-MLSF model that incorporates multi-level semantic features. Specially, we extract deep semantic features from essays by using BERT to mine the in-context information, using CNN to extract local semantic features, and using LSTM to extract global semantic features, and incorporates them with manually defined and automatically extracted shallow semantic features. The model has the best scoring performance, and the generalization of the model is significantly improved compared to three baseline models, according to the outcomes of offline experiments on two datasets. In the future, we will continue to optimize the models though ablation experiment to select shallow semantic features, and migrate the model to the AES for Chinese essays to further evaluate the generalizability of the model among languages.

Acknowledgements. This paper was supported by Graduate Education Reform Project of Beijing University of Posts and Telecommunications (2022Y004), High-performance Computing Platform of BUPT.

References

1. Alikaniotis, D., Yannakoudakis, H., Rei, M.: Automated text scoring using neural networks. In: Proceedings of the 54th Annual Meeting of the Association for Computational Linguistics. Association for Computational Linguistics, pp. 715–725(2016)
2. Taghipour, K., Ng, H.T.: A neural approach to automated essay scoring. In: Proceedings of the 2016 Conference on Empirical Methods in Natural Language Processing, EMNLP, pp. 1882–1891(2016)
3. Jin, C., He, B., Hui, K., Sun, L.: TDNN: a two-stage deep neural network for prompt-independent automated essay scoring. In: Proceeding of the 56th Annual Meeting of the Association for Computational Linguistics, pp. 1088–1097. Association for Computational Linguistics, Melbourne (2018)
4. Rodriguez, P U., Jafari, A., Ormerod, C.M.: Language models and Automated Essay Scoring. ArXiv, 1909.09482 (2019)
5. Fernandez, N., Ghosh, A., Liu, N., Wang, Z., Choffin, B., Baraniuk, R., Lan, A.: Automated scoring for reading comprehension via in-context BERT tuning. In: 23rd International Conference, pp. 691–697. Springer International Publishing (2022)
6. Dasgupta, T., Naskar, A., Dey, L., Saha, R.: Augmenting textual qualitative features in deep convolution recurrent neural network for automated essay scoring. In: Proceedings of the 5th Workshop on Natural Language Processing Techniques for Educational Applications, pp. 93–102. Association for Computational Linguistics (2018)
7. Liu, J., Xu, Y., Zhu, Y.: Automated Essay Scoring based on Two-Stage Learning. ArXiv, 1901.0774 (2019)
8. Zhou, X., Fan, X., Ren, G., Yang, Y.: Automated English essay scoring method based on muti-level semantic features. J. Comput. Appl. **41**(08), 2205–2211 (2021). (in Chinese)

Promising Long Term Effects of ASSISTments Online Math Homework Support

Mingyu Feng[(⊠)] [ID], Chunwei Huang [ID], and Kelly Collins

WestEd, San Francisco, CA 94107, USA

{mfeng,chuan,kelly.collins}@wested.org

Abstract. Math performance continues to be an important focus for improvement. Many districts adopted educational technology programs to support student learning and teacher instruction. The ASSISTments program provides feedback to students as they solve homework problems and automatically prepares reports for teachers about student performance on daily assignments. During the 2018–19 and 2019–20 school years, WestEd led a large-scale randomized controlled trial to replicate the effects of ASSISTments in 63 schools in North Carolina in the US. 32 treatment schools implemented ASSISTments in 7th-grade math classrooms. Recently, we conducted a follow-up analysis to measure the long-term effects of ASSISTments on student performance one year after the intervention, when the students were in 8th grade. The initial results suggested that implementing ASSISTments in 7th grade improved students' performance in 8th grade and minority students benefited more from the intervention.

Keywords: ASSISTments · math learning · long-term effects · effective teaching · AI-based program

1 Introduction

Mathematics education continues to be an important focus for national improvement. Achievement gaps between demographic groups continue to be an important national and state-based concern. Due to the promise of technology as a tool for improving mathematics education and closing the achievement gap, the use of educational technology in K-12 education has expanded dramatically in recent years, accelerated by the COVID-19 pandemic. The AIED and intelligent tutoring systems researchers and developers have built numerous technology-based learning platforms and programs, but few of these products have been implemented at a large scale in authentic school classroom settings. ASSISTments [1] is one of the few digital platforms that have been used widely in the U.S. Over the past two years, ASSISTments use in schools increased significantly, going from supporting 800 teachers to supporting 20,000 teachers and their 500,000 students. At the beginning of the COVID pandemic, the U.S. Department funded a rapid review to synthesize existing evidence in online programs that promoted learning, and ASSISTments was one of the few digital learning programs recommended for use in response to the COVID pandemic [2].

© The Author(s), under exclusive license to Springer Nature Switzerland AG 2023
N. Wang et al. (Eds.): AIED 2023, CCIS 1831, pp. 212–217, 2023.
https://doi.org/10.1007/978-3-031-36336-8_32

The ASSISTments platform is a technology-based, formative assessment platform for improving teacher practices and student math learning outcomes. As students work through problems and enter their answers into ASSISTments, the system provides immediate feedback on the correctness of answers and offers additional assistance in the form of hints or scaffolds. ASSISTments provides teachers with real-time, easily accessible reports that summarize student work for a particular assignment, which teachers can use to target their homework review in class and tailor instruction to their students' needs. ASSISTments was identified as effective at improving 7th grade student's learning and changing teacher's homework review practices during an efficacy study in Maine ([3, 4], meeting What Works Clearinghouse standards without reservation, $g = .22$, $p < .01$). Escueta et al. [5] indicated that out of 29 studies they reviewed that met rigorous standards of randomization, ASSISTments was one of only "Two interventions in the United States [that] stand out as being particularly promising" (page 88).

Supported by the U.S. Department of Education, we conducted a large scale randomized controlled trial[1] to replicate the Maine study and see whether the found effects replicate in a heterogeneous population that more closely matches national demographics. In addition, we designed a follow-up study to measure the long-term effects of ASSISTments on student performance one year after the intervention was over. This late-breaking results paper reports on the findings from the recently completed preliminary analysis of the follow-up study. The follow-up study addresses two research questions (RQ): 1) *What is the impact of ASSISTments on student math outcomes at the end of Grade 8?* 2) *Do the effects of ASSISTments vary for students of different demographic characteristics?*

2   Background on ASSISTments

ASSISTments uses technology to give teachers new capabilities for assigning and reviewing homework and to give students additional support for learning as they do homework. Content in ASSISTments consists of mathematics problems with answers and hint messages. These mathematics problems are bundled into problem sets which teachers can use ASSISTments to assign to students in class or as homework. Students first do their assigned problems on paper and then enter their answers into ASSISTments to receive immediate feedback about the correctness of their answers, and/or hints on how to improve their answers or help separate multi-step problems into parts. One type of problem set is mastery-oriented "Skill Builders". Each skill builder provides opportunities for students to practice solving problems that focus on a targeted skill, until they reach a teacher-defined "mastery" threshold of proficiency (e.g., a streak of three correct answers on similar math problems). ASSISTments also automatically re-assessing students on skills and concepts that were "mastered" earlier at regular intervals and providing further opportunities for students to hone those skills.

---

[1] The study has been approved by the Institutional Review Board at Worcester Polytechnic Institute and WestEd. Participating teachers all signed consent forms. Parents received a notification letter and an opt-out form for their children in the study. The study was pre-registered (https://sreereg.icpsr.umich.edu/framework/pdf/index.php?id=2064).

ASSISTments provides teachers with real-time, easily accessible reports that summarize student work for a particular assignment in a grid format. These reports inform teachers about the average percent correct on each question, skills covered in the assignment, common wrong answers, and each student's answer to every question. Teachers use information about individual students to form instructional groups and address common difficulties.

The design and development of ASSISTments is built upon the theoretical foundations and empirical research. *Feedback* has been identified as a powerful way to increase student learning [6]. An extensive set of studies (e.g., [7–9]) has found significant learning gains in response to feedback within computer-based instruction. *Formative assessment* [10, 11] informs teachers of student learning and progress and students of their own performance in relation to learning goals. Several decades of research have shown formative assessment to be an effective way of improving both teaching and student learning [12, 13]. *Mastery learning* have been shown to lead to higher student achievement than more traditional forms of teaching [14, 15]. An U.S. Department Practice Guide [16] recommends *spacing practice* over time. Research has demonstrated that retention increases when learners are repeatedly exposed to content ([17]).

3   Research Design

The replication study used a school-level, clustered randomized experimental design. Schools within each district were paired based on their demographic characteristics and student prior performance on state math and English language arts (ELA) tests and then randomly assigned to a treatment or business-as-usual control condition. All 7th math teachers in a school had the same assignment of condition. Teachers in the treatment schools used ASSISTments to support math homework. In the control condition, teachers continued their existing homework practices, including any use of online tools, but not had access to ASSISTments. The intervention was implemented by all Grade 7 teachers in treatment schools over two consecutive years—teachers learned how to use ASSISTments for a year (2018–19) and then we measured the immediate impact for students in teachers' second year of experience with the system (2019–20). Students who were in Grade 7 during the 2019–20 school year comprised the analytic sample for the research questions. The students maintained their conditions and were followed longitudinally for another year to Grade 8 when their Grade 8 performance (long-term impact) was measured at the end of Grade 8 (see Fig. 1 for an overview). During the follow-up year in Grade 8, no interventions were be provided by the team to 8th grade teachers or students.

Sixty-three schools from 41 different districts were recruited into the sample and randomly assigned to condition (32 treatment schools and 31 control schools). The schools served several different grade levels (6–8, 8–12, K-8) and were distributed across rural, town, suburban, and city communities (33 rural, 11 town, 8 suburban, and 11 city). Of the 63 schools, 18 were charter schools, 45 were public schools, and 48 received Title 1 funding. One hundred and two 7th grade math teachers and their classrooms enrolled in the study. Teachers in the 32 treatment schools implemented ASSISTments in their classrooms for homework support for two years. They received two days of training during each summer in 2018 and 2019, and additional coaching and technical assistance were distributed across the school years of 2018–19 and 2019–20 via webinars, video conferencing, and in-person visits.

	Study Teachers	Study Students	Measures of Student Learning
2018-19: Warm-up	Learn to use ASSISTments with a different cohort of students	In 6th grade, no intervention	6th Grade State End of Grade Test (EoG) (Baseline)
2019-20: Measurement year	Use ASSISTments with 7th grade classrooms	In 7th grade, Use ASSISTments to complete homework assignments	7th Grade EoG (Immediate Outcome, missing due to COVID)
2020-21: Follow-up	No intervention	In 8th grade, no intervention	8th Grade EoG (Long Term Outcome)

Fig. 1. Study Overview

The onset of COVID-19 forced the closure of study schools by the middle of March 2020 and the move to remote instruction. The state End-of-Grade test (EOG), which was to be the immediate student learning outcome, was canceled for spring 2020. Majority of schools in North Carolina remained in remote instruction during the 2020–21 school year and the EOG test resumed in spring 2021, when participating students of the study were in 8th grade.

4   Data, Analysis, and Findings

Student's long-term learning outcome was measured by state standardized Grade 8 EOG math assessment. Student's 6th grade EOG scale scores served as the baseline measure, administered in 2018–19. Student demographic data accessed included gender, race/ethnicity, and economically disadvantaged status (eds). School-level covariates included average 6th grade EOG scale score, 7th grade enrollment size, Title 1 eligibility, percentage of students with eds, and percentage of ethnic groups. We obtained demographic, enrollment, prior performance data, and long-term outcome data from the state-wide database for all students of the 63 schools. The sample included 5,991[2] students with both 6th and 8th grade EOG test scores (2,961 treatment students, 3,030 control students)[3]. We examined student baseline equivalence on their 6th grade EOG test scores, gender, and ethnicity and found no significant difference between the two conditions.

To evaluate the efficacy of ASSISTments for improving students' long-term mathematics achievement in 8th grade (RQ 1), we conduct an intent-to-treat (ITT) analysis using two-level hierarchical linear regression models (HLM). The HLM was used to account for the clustering effect of the data (students were nested within schools). We modeled the mean differences in Grade 8 EOG scores (the long-term outcome) between students in treatment and control schools, controlling for the student's Grade 6 scores and other student- and school-level covariates to improve the precision of the impact

---

[2] About 1/3 students took an alternative math course and did not take the EOG test in 8th grade. Their scores were analyzed separately.

[3] School enrollment, and student demographic and state test data were acquired from the state longitudinal database and the agreement doesn't permit sharing of those data with any third parties. Other data collected during the study has been deposited to the Open ICPSR data repository (https://www.openicpsr.org/openicpsr/project/183645/version/V1/view).

estimate. To address RQ 2, we added the analytical model with a cross-level interaction term of the school-level treatment variable and indicator of student subgroup and conducted a series of moderator analysis examining whether the ASSISTments had a differential impact on minority (e.g., Hispanic, Black) versus nonminority (White) students. Similar models were also used to estimate the effects of the intervention on the students with other policy-relevant background variables.

The results showed that there was a statistically significant difference between students in the treatment and control conditions in their 8th grade EOG test scores ($p = 0.011$, effect size Hedges' $g = 0.10$) with students from the treatment schools performing significantly better than those in the control group (RQ1). The moderator analyses showed that the intervention benefited minority (non-White) students, who started with significantly lower 6th grade EOG scores, significantly more than majority (white) students ($p = 0.003$, $g = 0.14$) and that the impact was stronger for Hispanic students than non-Hispanic students ($p = 0.014$, $g = 0.13$) (RQ2). Overall, the results suggested that ASSISTments had a sustained long-term impact on students' math learning, even after one year of intervention implementation. The program also helped close the achievement gap among students of different ethnicity.

5　Conclusion and Future Work

Enhancing mathematics education is an imperative challenge, especially given the significant learning loss that has transpired during the pandemic in the past three years. In this paper, we presented the promising long-term effects from a rigorous efficacy study of 63 schools that evaluated the impact of the ASSISTments homework support program in diverse settings and implementation circumstances. Historically, when the policy relevant state standardized tests were used as an outcome measure, few studies demonstrated impact on student learning [5] and long-term effects were rarely detected. Therefore, the findings have strong implications for math interventions. The findings demonstrate the potency of the ASSISTments program and its value in promoting math learning in middle school, as well as the promise of leveraging ASSISTments, and similar AI-supported programs, to help overcome the academic setbacks caused by the pandemic. The team is continuing with benchmarking the effect sizes and further analysis of the data to explore the variation of effects for different populations and school settings and relationships between ASSISTments usage in Grade 7 and performance in Grade 8.

Acknowledgement. This material is based on work supported by Arnold Ventures and the Institute of Education Sciences of the U.S. Department of Education under Grant R305A170641. Any opinions, findings, and conclusions or recommendations expressed in this material are those of the authors and do not necessarily reflect the funders.

References

1. Heffernan, N.T., Heffernan, C.L.: The ASSISTments ecosystem: building a platform that brings scientists and teachers together for minimally invasive research on human learning and teaching. Int. J. Artif. Intell. Educ. **24**(4), 470–497 (2014). https://doi.org/10.1007/s40 593-014-0024-x

2. Sahni, S. D., et al.: A What Works Clearinghouse Rapid Evidence Review of Distance Learning Programs. U.S. Department of Education (2021)
3. Roschelle, J., Feng, M., Murphy, R.F., Mason, C.A.: Online mathematics homework increases student achievement. AERA Open 2(4), 233285841667396 (2016). https://doi.org/10.1177/2332858416673968
4. Murphy, R., Roschelle, J., Feng, M., Mason, C.: Investigating efficacy, moderators and mediators for an online mathematics homework intervention. J. Res. Educ. Effect. 13, 1–36 (2020). https://doi.org/10.1080/19345747.2019.1710885
5. Escueta, M., Quan, V., Nickow, A.J., Oreopoulos, P.: Education Technology: An Evidence-Based Review. NBER Working Paper No. 23744. August 2017 JEL No. I20,I29,J24 (2017). Retrieved from http://tiny.cc/NBER
6. Hattie, J., Timperley, H.: The power of feedback. Rev. Educ. Res. 77(1), 81–112 (2007)
7. DiBattista, D., Gosse, L., Sinnige-Egger, J.A., Candale, B., Sargeson, K.: Grading scheme, test difficulty, and the immediate feedback assessment technique. J. Exp. Educ. 77(4), 311–338 (2009)
8. Fyfe, E.R., Rittle-Johnson, B.: The benefits of computer-generated feedback for mathematics problem solving. J. Exp. Child Psychol. 147, 140–151 (2016)
9. Kehrer, P., Kelly, K., Heffernan, N.: Does immediate feedback while doing homework improve learning? In: FLAIRS 2013 – Proceedings of the 26th International Florida Artificial Intelligence Research Society Conference, pp. 542–545 (2013)
10. Black, P., Wiliam, D.: Classroom assessment and pedagogy. Assess. Educ.: Principles, Policy Practice 25(6), 551–575 (2018)
11. Bennett, R.E.: Formative assessment: a critical review. Assess. Educ.: Principles, Policy and Practice 18(1), 5–25 (2011)
12. OECD: Formative Assessment: Improving Learning in Secondary Classrooms. OECD Publishing, Paris, (2005). https://doi.org/10.1787/9789264007413-en
13. Faber, J.M., Luyten, H., Visscher, A.J.: The effects of a digital formative assessment tool on mathematics achievement and student motivation: Results of a randomized experiment. Comput. Educ. 106, 83–96 (2017)
14. Anderson, J.R.: Learning and Memory: An Integrated Approach, 2nd edn. John Wiley and Sons Inc., New York (2000)
15. Koedinger, K., Aleven, V.: Exploring the assistance dilemma in experiments with Cognitive Tutors. Educ. Psychol. Rev. 19, 239–264 (2007)
16. Pashler, H., Bain, P. M., Bottge, B. A., Graesser, A., Koedinger, K., McDaniel, M., et al.: Organizing instruction and study to improve student learning. Institute of Education Sciences (IES) Practice Guide (NCER 2007-2004). National Center for Education Research, Washington, D.C. (2007)
17. Rohrer, D.: The effects of spacing and mixing practice problems. J. Res. Math. Educ. 40, 4–17 (2009)

Decomposed Prompting to Answer Questions on a Course Discussion Board

Brandon Jaipersaud[3](✉) [ID], Paul Zhang[2] [ID], Jimmy Ba[1,3] [ID],
Andrew Petersen[2] [ID], Lisa Zhang[2] [ID], and Michael R. Zhang[1,3] [ID]

[1] University of Toronto, Toronto, ON, Canada
{jba,michael}@cs.toronto.edu
[2] University of Toronto Mississauga, Mississauga, ON, Canada
{pol.zhang,andrew.petersen,lc.zhang}@utoronto.ca
[3] Vector Institute, Toronto, ON, Canada
brandon.jaipersaud@mail.utoronto.ca

Abstract. We propose and evaluate a question-answering system that uses decomposed prompting to classify and answer student questions on a course discussion board. Our system uses a large language model (LLM) to classify questions into one of four types: *conceptual, homework, logistics*, and *not answerable*. This enables us to employ a different strategy for answering questions that fall under different types. Using a variant of GPT-3, we achieve 81% classification accuracy. We discuss our system's performance on answering conceptual questions from a machine learning course and various failure modes.

Keywords: Course Discussion Board · GPT-3 · Large-Language Models · Mixture of Experts · Prompting

1 Introduction

Course discussion boards are an important avenue for students to ask course-related questions. However, instructors and teaching assistants spend much time and effort responding to discussion board questions, especially for courses with large enrollment [3,8]. With recent advances in natural language processing, there is an opportunity to leverage general-purpose large language models (LLMs) to assist in answering these questions. While GPT-3 has been used in education to adapt assignments [6,7] and explanations [1], we are not aware of work using LLMs for course Q&A. Furthermore, non-LLM automated methods for answering student questions have been developed in prior work [2,3,8]. However, the cost of answering a question incorrectly is high: providing an incorrect answer can be detrimental for students and *increase* course staff workload.

To mitigate this risk, we propose a LLM-based question-answering system that uses decomposed prompting [4] to answer student questions. We first

Supported by Vector Institute, NSERC, Fujitsu, Amazon Research Award, and the CIFAR AI Chairs Program.

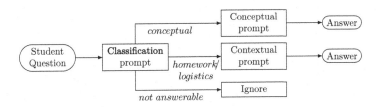

Fig. 1. Mixture of Experts for answering questions using decomposed prompting.

prompt a LLM to classify a student question into one of four types: *conceptual, homework, logistics,* and *not answerable.* Then, depending on the question type, we either ignore the question, or we further prompt a LLM to answer the question.

This Mixture of Experts approach can mitigate risk in a few ways. First, the risk and failure modes of an incorrect answer differ depending on the question type (e.g. a question about a deadline vs. explaining subtle concepts). Since each expert model can be deployed independently, educators have more control over the system. Second, we demonstrate that while LLMs can effectively differentiate between types of questions, different answering approaches and contextual cues may be effective for each question type. This modular approach suggests a framework for managing risk in real deployments and allows future work to focus on answering specific types of questions effectively.

Section 2 describes our approach. In Sect. 3.1, we show that *our classification system achieves an accuracy of 81%,* and analyze how the number of few-shot examples, the task description, and the question type labels affect classification accuracy. In Sect. 3.2, we use various metrics to show that our conceptual prompt works best for answering conceptual questions from an Intro to ML course. We use human evaluation to validate our automatic metrics and discuss the failure modes of our system on conceptual questions.

2   Mixture of Experts with Decomposed Prompting

Decomposed prompting [4] is an approach that uses LLM prompting to decompose a complex problem into simpler subtasks where each subtask can be solved by further prompting a LLM. Our mixture of experts design (Fig. 1) decomposes the task of answering student questions into two subtasks: **Question Classification** and **Question Answering**. We first classify questions using the prompt in Fig. 2 (left). Based on the classification, we further prompt a LLM to answer the question using Fig. 2 (right) or ignore the question.

The **Question Classification** task classifies student questions by type. As in prior work [8], our question types are based on the course-specific context required to answer the question. The four question types we use are:

1. **Conceptual Questions** that can be answered without course-specific context, e.g., *How do we choose the learning rate?*

Table 1. Precision, recall, and F-score for question classification type. The overall accuracy is 81%.

Type	Count	#Correct	Precision	Recall	F-Score
conceptual	13	11	0.79	0.85	0.81
homework	34	27	0.96	0.79	0.87
logistics	8	5	0.63	0.63	0.63
not answerable	14	13	0.68	0.93	0.79

2. **Homework Questions** that require the corresponding homework instructions to be answerable, e.g., *What does z refer to in Lab 1?*
3. **Logistics Questions** that require the course syllabus to be answerable, e.g., *Which room is the midterm in?*
4. **Not Answerable Questions** that require human intervention, e.g., *The instructor isn't here. Have office hours been cancelled?*

The **Question Answering** task proposes answers to a question. Our intention is for each question type to have a different prompting strategy: for example, *conceptual* questions do not require specialized context in the prompt. However, *homework* questions require that we identify and provide relevant sections of the assignment handout. Likewise, *logistics* questions require context from the course syllabus. Finally, *not answerable* questions are left to the instructors.

3   Results and Analysis

3.1   Question Classification

We evaluate our classification system on 72 historical student questions randomly sampled from the Fall 2022 instance of an upper-level Intro to ML course at a research-intensive institution in North America. We choose upper-year courses since, in our experience, there are more conceptual questions in these courses. To form the ground-truth type labels for the 72 questions, three course staff manually classified each question into the four types described in Sect. 2. 27 of the 72 annotated questions had a disagreement between two of the annotators, and we take the majority label as the ground truth. For 3 questions, all annotators disagreed on the ground truth label; we discard these questions from our analysis. Of the remaining 69 questions with ground truth labels, there are 13 *conceptual*, 34 *homework*, 8 *logistics*, and 14 *not answerable* questions.

The LLM we use for this task is *text-davinci-003*, an InstructGPT [5] variant of OpenAI's GPT-3 model fine-tuned on human instructions. Our prompt, which includes 31 in-context examples, is shown in Fig. 2 (left); we justify our choice of prompt below. In Table 1, we give a breakdown of the F-score by type; the total classification accuracy of our system is 81%.

Task Description. Our prompt begins with a task description outlining when a question belongs to each type (Fig. 2 (left)). To test the importance of this description, we evaluate three alternatives. With our descriptive prompt, the

Table 2. Classification accuracy by number of few-shot examples.

Examples	0	2	4	8	16	24	31	42
Accuracy	42%	61%	58%	67%	65%	77%	81%	70%

classification accuracy is 81%. Using no task description reduces the accuracy to 74%. Using only the first sentence of the description gives the lowest performance at 72%. Adding a sentence stating *Questions that point out corrections or typos should be classified as "homework"* gives an accuracy of 77%.

Number of Few-shot Examples. We use few-shot prompting and provide example questions and classifications (Fig. 2 (left)). To test the importance of these in-context examples, we vary the number of examples in our prompt. Table 2 shows that using 31 examples produces the highest classification accuracy.

Question Type Labels. The classification performance is sensitive to the choice of type labels used in the prompt. Renaming the types to *a, b, c, d* reduces accuracy to 70%. Renaming the types to *directly answerable, needs course material, needs administrative material, not answerable* reduces accuracy to 74%. Renaming the types to *conceptual, needs course material, needs administrative material, not answerable* reduces accuracy to 75%.

3.2   Question-Answering System

In this section, we present an initial effort to answer *conceptual* questions. In addition to the 69 questions from earlier, we include 63 additional questions from the Winter 2023 offering of the same Intro to ML course. Of these 63 questions, there are 20 *conceptual*, 31 *homework*, 8 *logistics* and 4 *not answerable* questions.

As before, we use *text-davinci-003*, now with the conceptual prompt shown in Fig. 2 (right). We use a temperature of 0.7 for answer generation since we have observed that this temperature produces more descriptive answers.

Difficulty of each Question Type. To justify focusing our attention on conceptual questions, we start by generating model answers to all 132 questions regardless of question type, using the conceptual prompt in Fig. 2 (right). We then evaluate the model's answers using the following metrics:

– The *Cosine similarity* between the embeddings of the model answer and the instructor answer, generated with Cohere's Embedding API.
– The *ROUGE score* between model and instructor answer to measure textual similarity between the answers.
– The *Perplexity of the instructor answer*, which measures how likely the model is to generate the instructor answer.

Table 3. Similarity between instructor and model answer by question type.

Question Type	Count	Cosine Similarity	ROUGE1/2/L	Perplexity
conceptual	34	0.62	0.30/0.07/0.18	7.61
homework	59	0.48	0.23/0.06/0.16	12.73
logistics	16	0.43	0.17/0.04/0.14	13.01
not answerable	23	0.52	0.19/0.03/0.13	34.32

Table 4. Human and automatic evaluation of conceptual answers.

Feedback	Count	Cosine Similarity	ROUGE1/2/L	Perplexity
good answer	8	0.83	0.65/0.51/0.57	3.05
bad answer	20	0.61	0.31/0.07/0.18	6.35

The result of applying these metrics across all 132 questions is shown in Table 3. The LLM performs best on conceptual questions across all three metrics. This agrees with our intuition that conceptual questions are easier for a general-purpose LLM to answer since they do not require course-specific details.

Human Evaluation on Conceptual Questions. The Intro to ML instructors labeled the 28 model answers to conceptual questions with *good answer* or *bad answer*, depending on whether the answer correctly and appropriately addresses the student's question. Table 4 shows the result of applying the automated metrics to each feedback category. The *good* answers are semantically and textually closer to their corresponding instructor answers. This validates our use of the metrics in Table 3.

Furthermore, we see that 8/28 (29%) of annotated answers were labeled with *good answer*. Of the 20 bad answers, we observe the following distribution of failure modes: 3 answers address questions that are misclassified as *conceptual* (rather

```
Task Description: Classify each question posted on an undergraduate course discussion board
into one of the following 4 types: conceptual", "homework", "logistics" or "not answerable".
A question that requires instructor intervention should be classified as "not answerable".
Questions that point out contradictions in assignment instructions and deadlines should be
classified as "not answerable". Conceptual questions should be classified as "conceptual".
Homework and lab questions that provide enough information to be answered should be
classified as "conceptual". Questions that need course content related context to be
answered such as an assignment handout should be classified as "homework". Questions that
need logistical context to be answered such as a course syllabus should be classified as "
logistics".

Question: Is the A1 Q2 code for debugging a neural network correct? It says we should debug
using a large dataset. However, using a small dataset seems to make more sense here.
Classification: homework

... (30 more question/classification pairs)

Question: <student question>
Classification:
```

```
Task: Answer the
following
question that was
posted by a
student on the
class discussion
board for an
introductory
machine learning
course. Your
answer should be
truthful, concise
and helpful to
the student.

Question: <
student question>

Answer:
```

Fig. 2. Prompt used for classification (left) and conceptual Q&A (right).

than *homework*). 5 answers are factually incorrect. For instance: *"The training set can be used to tune hyperparameters"*. 3 answers are correct but inappropriate for the student's level of knowledge. For example: *"Gradient descent can be used to optimize hyperparameters, although this is less common"* when this form of optimization was not discussed in the course. The remaining bad answers suffer from issues such as misunderstanding the question, incoherence, making incorrect assumptions, or adding irrelevant information to the answer.

4   Conclusion

We demonstrated that decomposed prompting is a useful strategy for classifying and answering student questions due to their context-dependent or unanswerable nature. Furthermore, our results have shown that LLMs can classify student questions with an accuracy of 81%, but yield poor performance when answering conceptual questions. Many of the poor answers arose due to misalignment with the preferences of a course instructor. Future work could aim to fine-tune the model on course discussion board questions. An interesting direction for future work is to combine LLMs with semantic processing or text extraction techniques [3,8] which should enable prompted LLMs to handle homework and logistics questions.

References

1. Drori, I., et al.: A neural network solves, explains, and generates university math problems by program synthesis and few-shot learning at human level. Proc. Natl. Acad. Sci. **119**(32), e2123433119 (2022)
2. Feng, D., Shaw, E., Kim, J., Hovy, E.: An intelligent discussion-bot for answering student queries in threaded discussions. In: Proceedings of the 11th international conference on Intelligent user interfaces, pp. 171–177 (2006)
3. Goel, A.K., Polepeddi, L.: Jill watson: A virtual teaching assistant for online education. In: Learning Engineering For Online Education, pp. 120–143, Routledge (2018)
4. Khot, T., et al.: Decomposed prompting: A modular approach for solving complex tasks (2022). https://doi.org/10.48550/ARXIV.2210.02406
5. Ouyang, L., et al.: Training language models to follow instructions with human feedback. arXiv:2203.02155 (2022)
6. Sarsa, S., Denny, P., Hellas, A., Leinonen, J.: Automatic generation of programming exercises and code explanations using large language models. In: Proceedings of the 2022 ACM Conference on International Computing Education Research-Volume 1, pp. 27–43 (2022)
7. Wang, Z., Valdez, J., Basu Mallick, D., Baraniuk, R.G.: Towards human-like educational question generation with large language models. In: Artificial Intelligence in Education: 23rd International Conference, AIED 2022, Durham, UK, July 27–31, 2022, Proceedings, Part I, pp. 153–166, Springer (2022). https://doi.org/10.1007/978-3-031-11644-5_13
8. Zylich, B., Viola, A., Toggerson, B., Al-Hariri, L., Lan, A.: Exploring automated question answering methods for teaching assistance. In: Bittencourt, I.I., Cukurova, M., Muldner, K., Luckin, R., Millán, E. (eds.) AIED 2020. LNCS (LNAI), vol. 12163, pp. 610–622. Springer, Cham (2020). https://doi.org/10.1007/978-3-030-52237-7_49

Consistency of Inquiry Strategies Across Subsequent Activities in Different Domains

Jade Mai Cock[1]([✉]) [ID], Ido Roll[2] [ID], and Tanja Käser[1] [ID]

[1] EPFL, Lausanne, Switzerland
{JadeMai.Cock,Tanja.Kaser}@epfl.ch
[2] Technion, Haifa, Israel
roll@technion.ac.il

Abstract. Interactive simulations encourage students to practice skills essential to understanding and learning sciences. Alas, inquiry learning with interactive simulations is challenging. In this paper, we seek to identify inquiry patterns across topics and evaluate their stability with regard to common behaviors and student membership. Applying a clustering approach, we propose an encoding through which we can model students' strategies in diverse environments. Specifically, we encode each sequence with three different levels of granularity which range from simulation-specific characteristics to simulation-agnostic features. Using this generalizable encoding, we find two clusters for each of two simulations. The formed groups exhibit similar learning patterns across environments. One systematically cycles through exploring and recording systematically over all variables. The other group explores the simulation more freely. This suggests that our feature encoding captures inherent quality of inquiry with simulations and can be used to characterize learners knowledge of productive exploration.

Keywords: interactive simulation · inquiry skills · spectral clustering · inquiry strategies · log data · open ended learning environments

1 Introduction

Open ended learning environments (OELE) have become omnipresent in scientific curricula [10,14]. Their assets are multiple: they foster inquiry learning by emulating complex phenomena in a safe, cheap, and accessible way, allowing students to explore and uncover the relationship among the various variables. Alas, navigating those environments is often difficult for learners [1,2,6,8]. Analyzing OELE interaction logs should help to identify the students who struggle, as well as suggest explanations for the sources of their difficulty. However, given their open-ended nature, identifying productive behaviors in an OELE is challenging.

A large body of work uses log data from interactive simulations to identify productive behaviors, create student profiles, and predict students' affiliation to

© The Author(s), under exclusive license to Springer Nature Switzerland AG 2023
N. Wang et al. (Eds.): AIED 2023, CCIS 1831, pp. 224–229, 2023.
https://doi.org/10.1007/978-3-031-36336-8_34

the profiles or their performances [13]. In doing so, strategies common to several environments have emerged [12].

Thus, most work so far has focused on investigating complex behaviors in one specific environment. Additionally, studies investigating multiple simulations were limited by their search for simpler behaviors such as *Control Variable Strategy* [3–5,11,12]. Indeed, because CVS is so well established, feature engineering is sometimes guided by the expectation of finding it [9] preventing us from extracting different types of strategies even though we do not know yet what effective inquiry learning looks like.

In this paper we are interested in understanding students' unguided interactions on two inquiry labs of unrelated topics without making any assumptions of what productive inquiry might look like. We propose a generalizable feature encoding and use it to engineer all students' interactions into transition matrices, which we then group through spectral clustering. We apply our clustering pipeline to interaction data from 144 students working on two sequential inquiry activities with different interactive simulations. We then investigate the resulting clusters to understand what patterns students adopt in the two environments and whether these patterns are the same across lab. With our analysis, we aim to address the following research questions: 1) What patterns do students adopt across inquiry labs, and how consistent are those strategies across topics (**RQ1**)? and 2) How consistent is students' use of these patterns across simulations (**RQ2**)? We find two clusters with similar patterns across topics, suggesting that they capture topic-independent inquiry patterns. Furthermore, cluster membership tends to be stable across lab, with "unstable" students moving on to more systematic inquiry in their second activity.

2   Learning Context

We conducted a user study in which students used two different inquiry environments to uncover the model of two natural phenomena.

Learning Environments. The Beer's Law Lab simulation[1] (*beer's law lab*) simulates the phenomenon of light absorbance, while the Capacitor Lab Simulation[2] (*capacitor lab*) emulates a capacitor attached to a battery. We modified the original simulations to make them isomorphic in terms of the number of independent variables and their relationship with the dependent variable. We embedded both simulations into an environment with a recording tool and a graphing tool. Students could explore the simulations by changing the values of the different independent variables and observing their effect on the dependent variable. Furthermore, they were encouraged to keep track of their trials using the record tool. Once recorded, the students were free to change the order in which the trials were displayed, remove the trials to shorten the table, restore

---

[1] https://phet.colorado.edu/sims/html/beers-law-lab/latest/beers-law-lab_en.html.

[2] https://phet.colorado.edu/sims/html/capacitor-lab-basics/latest/capacitor-lab-basics_en.html.

the simulation to the state of the trial of their choices or plot the data onto a graph.

User Study. 148 undergraduate students started with a pre-survey gathering background information. They then completed two inquiry activities. Each began with a 5 min pre-test followed by a 15 min activity in which they were asked to describe the relationships between the independent variables with the dependent variable to the best of their knowledge. Students took a short break between both inquiry activities. The two activities differed in their topic and simulation (Beer Law vs. Capacitor), and order was randomized. We obtained a total of 144 log files from the Beer's law lab and 137 from the Capacitor lab. Complete log files from both activities were available for only 135 students. Out of those 135 students, 45 identified as males, 88 as females, and 2 did not answer or identified as non binary. 103 were first year students, 20 second years, 8 third years and 4 were fourth years. In our experiments, we cluster the interaction sequences of all available logs for the respective simulations, but compute cluster stabilities over the 135 students with both complete logs only.

3   Clustering Framework

In order to analyze and compare students' inquiry strategies across the two labs, we employ the clustering framework described in this section. We first transform the extracted interaction data into action sequences where an action is a combination of an inquiry phase and a variable of the simulation. We then employ three different paradigms ranging from detailed, simulation-specific to coarse, general encoding to create students' transition matrices. Finally, we cluster students separately for each lab, based on the aggregation of the three transition matrices.

Action Extraction. We extract for each student and activity a sequence of actions from the recorded log data. Each action is categorized according to its phase as either 1) changing a parameter in the simulation (*explore*), 2) recording the state of the simulation (*record*), 3) rearranging the data points in the table (*analyze-table*), or 4) interacting with the plot tool (*analyze-plot*). Items are qualified by their context: *explore* actions are qualified by either the independent variables of the corresponding lab or "other". *Record* items are qualified by the independent variables whose values differ from the previous record. Finally, *analyze-plot* actions are qualified by the plotted variables. Otherwise, the item is qualified by "other". Because we have indication about students' reasoning when they are manipulating the table, we only qualify *analyze-table* by *general*. Many pauses between actions are the result of mouse movements. Therefore, we merge *explore* actions separated by less than 3 seconds of inaction (elbow of pause length distribution). Furthermore, we decide to implement a standard duration of 20 seconds (beginning of tail pause length distribution) for a *break*.

Feature Creation. We use three different perspectives to qualify our actions ranging from a detailed and lab-specific qualification to a more coarse and simulation-agnostic encoding: 1) variables, 2) complexity and 3) heterogeneity. The variable dimension is used to encode the component of the inquiry labs the students interacted with; That dimension is specific to the employed simulation. The complexity dimension is used to characterize the relationship of that component to the dependent variable; The interaction sequences resulting from that second perspective are generalizable across isomorph simulations. Finally, we use the heterogeneity dimension to characterize how focused students are on one component; The latter is simulation agnostic. Using these three perspectives, we created three new sequences, one per dimension. Similarly to [7], we then transform each of these sequences into a Markov Chain which we then use to create transition matrices where each cell corresponds to the probability of a student transferring from one action to another.

3.1 Behavior Identification

For each paradigm, we start by computing the pairwise similarity matrices using the Euclidean norm and a radial basis function (rbf) kernel. We then sum the per-paradigm similarity matrices to form a global adjacency matrix. Next, we use spectral clustering to find groups of students behaving similarly.

We optimize the width γ of the rbf kernel per dimension via a grid search maximizing the clusters' *Silhouette* score. We also use the Silhouette score to determine the optimal number of clusters n_c.

4 Results

In each simulation, two different patterns of inquiry emerge from the clustering. We call them `systematic` inquiry and `free` inquiry. The larger clusters contained students who systematically explore and record the same variables, systematically across all variables. The smaller clusters group students who freely explore the parameters of the inquiry labs without restricting themselves to the well-known *Control the Variables Strategy* used for inquiry. We name the two found inquiry strategies `systematic` and `free`. Figure 1 illustrates a subset of the averaged transition matrices for both the Capacitor and the Beer's Law labs. More specifically, we observe that `systematic` students show strong transitions between *record* and *explore*. When a student explores a specific variable, they tend to also record a sample in which only the explored variable varied from the last recorded sample (a transition from *explore-x* to *record-x*). Students also graph the width and the concentration samples separately, or at least do not change axis frequently. Finally, `systematic` students take long breaks. Indeed, we observed little transitions from "actions" to *pauses*, but many going from *pauses* to *pauses*. The *explore-x/record-x* transitions are still present, but much less pronounced in the `free` clusters. Students in these clusters tend to frequently take breaks, while exploring a variable. As if they reflect on what to do next or

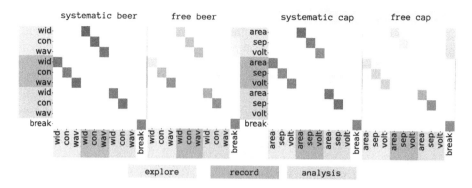

Fig. 1. Subset of the averaged transition matrices for the clusters `systematic beer` (n=81), `systematic cap` (n=82), `free beer` (n=63), `free cap` (n=55).

whether to record the sample before doing it. It also seems like students in the `free` cluster do more context switching, particularly in the capacitor lab. In other words, they are less focused on one variable and like to context switch, often changing from one parameter to another, analyzing different variables in one phase. In general, we observe that exploring freely helps to identify the independent variables of the system. On the other hand, being systematic helps recovering the mathematical relations of those independent variables.

In a second analysis, we are interested in measuring the overlap among the `systematic` and `free` clusters (in terms of students) across simulations. We therefore compute the cluster stability across the corresponding clusters by measuring the percentage of students who are classified in the same clusters across activities (either both assigned to `free-beer` and `free-cap` or to `systematic-beer` and `systematic-cap`). We find that it is as high as 0.74. This shows that most students who use a `systematic` (or `free`) strategy on the beer's law lab, also use a `systematic` (or `free`) strategy on the capacitor lab. In other words, this indicates that most students use a similar strategies for both activities.

5   Summary and Discussion

In this paper, we cluster students based on their log data on two subsequent activities based on two different inquiry labs. We use three dimensions to encode the log data, each of them capturing different elements of the interactions made on the simulations. We find that students either explore the simulations through `systematic` cycles of *explore* and *record* actions `systematically` over all variables, or `freely` explore the simulation. By using different dimensions to represent students' interactions, we are able to derive generalizable findings across subsequent simulations. Specifically, the encoding of the complexity of variables can be standardized across diverse environments, while the heterogeneity of the interactions remains invariant. This opens the door for topic-general representations of student actions. The characterization of student interaction is restricted to specific contexts in studies where only one simulation is used. However, our observation that the same individuals exhibit consistent patterns of interaction across different inquiry labs suggests that the captured behaviors reflect deeper knowledge and habitual engagement, rather than situational or contextual factors. This finding has important implications for the assessment and tutoring of inquiry competencies using simulations.

References

1. Alfieri, L., Brooks, P.J., Aldrich, N.J., Tenenbaum, H.R.: Does discovery-based instruction enhance learning? J. Educ. Psychol. **103**(1), 1 (2011)
2. Cock, J., Marras, M., Giang, C., Käser, T.: Early prediction of conceptual understanding in interactive simulations. In: International Educational Data Mining Society (2021)
3. Conati, C., Fratamico, L., Kardan, S., Roll, I.: Comparing representations for learner models in interactive simulations. In: Conati, C., Heffernan, N., Mitrovic, A., Verdejo, M.F. (eds.) AIED 2015. LNCS (LNAI), vol. 9112, pp. 74–83. Springer, Cham (2015). https://doi.org/10.1007/978-3-319-19773-9_8
4. Fratamico, L., Conati, C., Kardan, S., Roll, I.: Applying a framework for student modeling in exploratory learning environments: Comparing data representation granularity to handle environment complexity. Int. J. Artif. Intell. Educ. **27**, 320–352 (2017)
5. Funke, J.: Analysis of minimal complex systems and complex problem solving require different forms of causal cognition (2014)
6. Kirschner, P., Sweller, J., Clark, R.E.: Why unguided learning does not work: an analysis of the failure of discovery learning, problem-based learning, experiential learning and inquiry-based learning. Educ. Psychol. **41**(2), 75–86 (2006)
7. Klingler, S., Käser, T., Solenthaler, B., Gross, M.: Temporally coherent clustering of student data. In: International Educational Data Mining Society (2016)
8. Mayer, R.E.: Should there be a three-strikes rule against pure discovery learning? Am. Psychol. **59**(1), 14 (2004)
9. Perez, S., et al.: Control of variables strategy across phases of inquiry in virtual labs. In: Penstein Rosé, C. (ed.) AIED 2018. LNCS (LNAI), vol. 10948, pp. 271–275. Springer, Cham (2018). https://doi.org/10.1007/978-3-319-93846-2_50
10. Perkins, K., Moore, E., Podolefsky, N., Lancaster, K., Denison, C.: Towards research-based strategies for using PhET simulations in middle school physical science classes. In: AIP Conference Proceedings. vol. 1413, pp. 295–298. American Institute of Physics (2012)
11. Sao Pedro, M.A., De Baker, R.S., Gobert, J.D., Montalvo, O., Nakama, A.: Leveraging machine-learned detectors of systematic inquiry behavior to estimate and predict transfer of inquiry skill. User Model. User-Adap. Inter. **23**, 1–39 (2013)
12. Tschirgi, J.E.: Sensible reasoning: a hypothesis about hypotheses. Child Dev. **51**(1), 1–10 (1980)
13. Wang, K.D., Cock, J.M., Käser, T., Bumbacher, E.: A systematic review of empirical studies using log data from open-ended learning environments to measure science and engineering practices. Br. J. Educ. Technol. **54**(1), 192–221 (2023)
14. Wieman, C.E., Adams, W.K., Loeblein, P., Perkins, K.K.: Teaching physics using PhET simulations. Phys. Teach. **48**(4), 225–227 (2010)

Improving the Item Selection Process with Reinforcement Learning in Computerized Adaptive Testing

Yang Pian[1], Penghe Chen[1,2]([⊠]), Yu Lu[1,2], Guangchen Song[2], and Pengtao Chen[3]

[1] School of Educational Technology, Faculty of Education, Beijing Normal University, Beijing 100875, China
chenpenghe@bnu.edu.cn
[2] Advanced Innovation Center for Future Education, Beijing Normal University, Beijing 100875, China
[3] Geophysical Exploration Academy of China Metallurgical Bureau, Beijing 071051, China

Abstract. Item selection is the key process for computerized adaptive testing (CAT) to effectively assess examinees' knowledge states. Existing item selection algorithms mainly rely on information metrics, suffering two issues: one is that the implicit cognitive information like relations between testing items as well as knowledge components cannot be captured by the information-based methods, and the other one is that the information-based algorithms computes item's suitableness depending on examinees' knowledge states which are estimated and imprecise inherently. To address these two issues, this work proposes to employ reinforcement learning technology to learn the item selection algorithm automatically in a data-driven manner. It is also able to properly capture the implicit cognitive relations between different testing items and avoid unnecessary item testing, and does not depend on examinees' estimated knowledge states at all.

Keywords: Reinforcement learning · Computerized adaptive testing · Item selection · User simulator · Knowledge space

1 Introduction

Computerized Adaptive Testing (CAT) is a form of computer-based test that aims to estimate examinees' knowledge states based on the selected testing item sequence and corresponding responses [9]. The item selection algorithm is the key component of CAT, which controls how the best suitable testing items are identified and selected based on examinees' historical responses. Empirical evidence shows that well-designed item selection algorithms can provide examinees with more appropriate testing items, improve testing and learning efficiency, and create a more flexible test-taking experience [6].

N. Wang et al. (Eds.): AIED 2023, CCIS 1831, pp. 230–235, 2023.
https://doi.org/10.1007/978-3-031-36336-8_35

In order to find the most suitable items for testing, different technologies have been employed to build the item selection algorithm. Information-based algorithms, which prioritize items with higher uncertainty of response correctness, are the most typical one. These algorithms compute uncertainty based on examinees' estimated knowledge states and employ various information metrics [3,11,16]. However, there are two main issues with information-based algorithms. One is that these algorithms mainly utilize explicit information to find the "best" next item without considering implicit cognitive information like cognitive structure between testing items and knowledge components, which can help derive examinees' knowledge state of unseen knowledge components based on tested ones and avoid unnecessary item testing. The other one is that computing the information of testing items heavily relies on the examinees' knowledge state which is estimated and imprecise inherently. As a result, the computed information of the testing items may not be precise, which may result inappropriate item selection.

To address these two issues, we employ reinforcement learning (RL) technology to automatically learn the item selection algorithm in a data-driven manner. RL is a computational framework that models sequential decision-making tasks [15] and has been explored to solve various kinds of sequence-related educational tasks, such as educational activity scheduling and learning materials recommendation [17,18]. All these sequential decision-making tasks have demonstrated the effectiveness of RL technology, which motivates us to solve the item selection task with RL technology.

The suitableness of RL technology can be three-fold. First, item selection is a sequential decision-making task in nature, which matches RL's capability. Second, the data-driven manner of RL makes the learned item selection algorithm considers both explicit and implicit information of testing items. Third, the RL-based item selection algorithm does not depend on exminees' knowledge states, which avoids the influence of imprecise estimation.

2 Related Work

2.1 Item Selection in Computerized Adaptive Testing

As the key component of CAT, item selection algorithm design has received a lot of attention, and a variety of selection techniques have been proposed. Classical selection methods mainly focus on local or global information provided by the testing items. Related statistics include the Fisher's information (MFI) [11], the Shannon Entropy [16], the Kullback-Leibler (KL) information [3], etc. Furthermore, non-statistical metrics such as item exposure, content balance and time control have also been integrated into item selection to ensure test equality [7]. In addition, more recent studies have combined multiple measures or criteria into a single selection algorithm. [8], resulting in improved performance.

Overall, current item-selecting algorithms mostly rely on explicit item traits or examinee's knowledge states, neglecting valuable pedagogical information. Since the implicit relations between testing items and knowledge components would provide valuable information for more efficient item selection algorithms, we propose to incorporate this characteristic in the new algorithm design.

2.2   Reinforcement Learning for Education

Reinforcement learning (RL) is a learning paradigm that models sequential decision-making processes [15]. It has been adopted to solve different types of sequencing tasks in education [14].

First, RL algorithms have been employed to generate personalized curriculum learning sequences, such as conducting the pedagogical policy induction task [18] and organizing the curriculum learning sequence arrangement [13]. Previous work shows that RL-induced instructional policies significantly outperform baselines in optimizing the sequences of instructional activities [4]. Second, RL algorithms have been used to provide automated learning feedback in the form of hints and scaffolding. For example, Efremov, Ghosh and Singla have utilized RL to generate different hints based on programming codes [5]. Third, RL algorithms have also been employed for learning material recommendations, including sequencing studying content in online educational systems [2], and generating course recommendations [17].

It can be seen that RL has been adopted to solve various educational tasks that have sequential characteristics inherently. However, less work has been conducted to address the item selection task in CAT, which motivates us to improve the efficiency and accuracy of item selection with RL methods.

3   Framework Illustration

In this work, we try to build a new item selection algorithm utilizing RL technology in CAT. As depicted in Fig. 1, the framework consists of three modules: 1) the RL-based item selection module that takes charge of learning the item selection policy and selects the best next testing item; 2) the state estimation module that is responsible for estimating examinees' knowledge states based on responses; and 3) the user simulator module that generates data required to train the RL-based item selection model and state estimation model.

At each round of interaction, based on the latest system state, the item selection module would make decision on either selecting a new testing item or issuing the end of testing (EOT). In the case of a new testing item, a corresponding response is computed based on the examinee's knowledge state, and is returned back to the item selection module for another round of interaction. Meanwhile, the examinee also compare this testing item with her answering history, and generate corresponding rewards to the DQN model in item selection module. In the case of EOT, the history of selected items and responses are fed into the state estimation module to derive the examinee's knowledge state, which is subsequently compared to the ground truth and generates rewards to the DQN model in the item selection module.

RL-Based Item Selection Module. Reinforcement learning is a computational framework targeting on sequential decision-making problems [15], in which

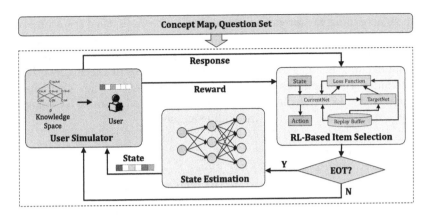

Fig. 1. Structure of Proposed Framework.

Q-learning is a typical model-free algorithm that tries to find the optimal policy maximizing expected reward. This work utilizes the DQN model [12] of RL to learn item selection algorithm. Integrating with deep learning networks, the DQN model is able to handle large state space. Generally speaking, a DQN model consists of four main elements: states, actions, rewards, and policy. *State*, denoted by s_t, represents examinee's answering results on differnt testing items; *Action*, denoted by a_t, represents the next item to be tested issued by the system; *Reward*, denoted by r_t, represents immediate reward obtained by the system at state s_t after taking the action a_t, and are utilized to estimate the model parameters by minimizing the loss function; *Policy*, denoted by $\pi(s_t)$, computes the probability distribution over all testing items given current agent state s_t.

State Estimation Module. The purpose of this module is to infer examinees' knowledge states based on their testing items and responses. In this work, we adopt the Multi-Layer Perceptron (MLP) model to conduct this estimation. Specifically, a three-layer MLP model is employed, including input layer, hidden layer, and output layer. In this work, we train the state estimation model by minimizing the maximum difference between true and estimated knowledge states, which is more sensitive to accurate estimation of all knowledge components compared to the commonly used mean squared error loss.

User Simulator Module. As real-world public datasets lack data category or quantity requirements, we built a user simulator [10] to generate data for training and evaluating the DQN and MLP models. The proposed user simulator consists of three parts: the knowledge structure generator (KSG), the examinee state generator (ESG), and the response sequence generator (RSG). Specifically, KSG generates all the possible knowledge structures based on knowledge components and their relations, using a depth-first search algorithm guided by Knowledge Space Theory [1]. ESG assigns the knowledge structures generated by KSG to examinees according to a predefined cognitive distribution, then transforms each into a knowledge state vector using a threshold of mastery. RSG constructs a item pool and generates testing items with difficulty values for all the knowledge components, then generates multiple test sequences of varying lengths that are distributed to examinees. Examinees answer correctly when their mastery level exceeds the item difficulty based on ESG's knowledge state. By doing this, response sequences are obtained from all examinees.

4   Exemplary Case

We use an exemplary case to illustrate the item selection process. In this example, five knowledge components from junior high school mathematics and 100 different testing items are composed. As shown in Fig. 2, given the examinee's true knowledge state as $[0.84, 0.75, 0.68, 0.8, 0.71]$, the system can estimate the knowledge state as $[0.82, 0.75, 0.65, 0.82, 0.73]$ with responses of 40 testing items. Taking the first knowledge component as an example, items $[16\ 13\ 10\ 11\ 4\ 15\ 19\ 17\ 12]$ belong to this knowledge component. Examinee's response is $[✓, ✓, ✓, ✓, ✓, ✓, ✗, ✗, ✓]$. It can be seen that the examinee answers item 16 with difficulty of 0.8 correctly and item 17 with a difficulty of 0.85 incorrectly. Hence, the system reasonably estimates the examinee's mastery level of the first knowledge component as 0.82.

Item	Value																			
TrueValue	[0.84, 0.75, 0.68, 0.8, 0.71]																			
Model Prediction	[0.82, 0.75, 0.65, 0.82, 0.73]																			
Item Selected (1~20)	16	66	50	81	23	27	26	53	30	35	13	43	25	82	10	84	72	29	70	54
Response (1~20)	✗	✓	✓	✓	✓	✓	✓	✗	✓	✗	✓	✓	✓	✓	✓	✓	✓	✓	✓	✗
Item Selected (21~40)	11	78	73	4	57	64	28	48	95	92	15	24	19	83	33	99	17	12	46	51
Response (21~40)	✓	✗	✓	✓	✗	✓	✓	✓	✗	✓	✓	✓	✗	✓	✓	✗	✗	✓	✓	✓

Fig. 2. Exemplary Case.

5   Conclusion

In this work, we adopted the RL technology to learn the item selection algorithm of CAT automatically in a data-driven manner. Specifically, we utilized the DQN model to decide how to choose the most appropriate testing item in each round. Based on all the selected testing items and responses, we adopted an MLP network to estimate the examinees' knowledge states. In addition, in order to train the proposed RL model and MLP model, we designed a user simulator to generate training data. The proposed method could make better use of implicit cognitive process information and avoid the influence of imprecise knowledge state estimation, which provides more efficient item selection process.

Acknowledgements. This research is supported by the National Natural Science Foundation of China (No. 62177009, 62077006, 62007025).

References

1. Albert, D., Lukas, J.: Knowledge Spaces: Theories, Empirical Research, and Applications. Psychology Press, London (1999)
2. Azhar, A.Z., Segal, A., Gal, K.: Optimizing representations and policies for question sequencing using reinforcement learning. In: International Educational Data Mining Society (2022)
3. Chang, H.H., Ying, Z.: A global information approach to computerized adaptive testing. Appl. Psychol. Measur. **20**(3), 213–229 (1996)
4. Doroudi, S., Aleven, V., Brunskill, E.: Where's the reward? a review of reinforcement learning for instructional sequencing. Int. J. Artif. Intell. Educ. **29**(4), 568–620 (2019)
5. Efremov, A., Ghosh, A., Singla, A.: Zero-shot learning of hint policy via reinforcement learning and program synthesis. In: Proceedings of Educational Data Mining (EDM) (2020)
6. ETS: Graduate record examinations 1996–97 information and registration bulletin (1996)
7. Fan, Z., Wang, C., Chang, H.H., Douglas, J.: Utilizing response time distributions for item selection in cat. J. Educ. Behav. Stat. **37**(5), 655–670 (2012)
8. Han, K.T.: An efficiency balanced information criterion for item selection in computerized adaptive testing. J. Educ. Measur. **49**(3), 225–246 (2012)
9. Kingsbury, G.G., Zara, A.R.: Procedures for selecting items for computerized adaptive tests. Appl. Measur. Educ. **2**(4), 359–375 (1989)
10. Li, X., Lipton, Z.C., Dhingra, B., Li, L., Gao, J., Chen, Y.N.: A user simulator for task-completion dialogues. arXiv preprint arXiv:1612.05688 (2016)
11. Lord, F.M.: Applications of item response theory to practical testing problems. Routledge, Milton Park (2012)
12. Mnih, V., et al.: Human-level control through deep reinforcement learning. Nature **518**(7540), 529–533 (2015)
13. Narvekar, S., Peng, B., Leonetti, M., Sinapov, J., Taylor, M.E., Stone, P.: Curriculum learning for reinforcement learning domains: A framework and survey. arXiv preprint arXiv:2003.04960 (2020)
14. Singla, A., Rafferty, A.N., Radanovic, G., Heffernan, N.T.: Reinforcement learning for education: Opportunities and challenges. arXiv preprint arXiv:2107.08828 (2021)
15. Sutton, R.S., Barto, A.G.: Reinforcement learning: An introduction. MIT press, Cambridge (2018)
16. Tatsuoka, C.: Data analytic methods for latent partially ordered classification models. J. R. Stat. Soc. Ser. C (Appl. Stat.) **51**(3), 337–350 (2002)
17. Vedavathi, N., Bharadwaj, R.S.: Deep flamingo search and reinforcement learning based recommendation system for E-learning platform using social media. Procedia Comput. Sci. **215**, 192–201 (2022)
18. Zhou, G., Azizsoltani, H., Ausin, M.S., Barnes, T., Chi, M.: Hierarchical reinforcement learning for pedagogical policy induction. In: Isotani, S., Millán, E., Ogan, A., Hastings, P., McLaren, B., Luckin, R. (eds.) AIED 2019. LNCS (LNAI), vol. 11625, pp. 544–556. Springer, Cham (2019). https://doi.org/10.1007/978-3-030-23204-7_45

The Role of Social Presence in MOOC Students' Behavioral Intentions and Sentiments Toward the Usage of a Learning Assistant Chatbot: A Diversity, Equity, and Inclusion Perspective Examination

Songhee Han[1]([⊠]) [iD], Jiyoon Jung[2] [iD], Hyangeun Ji[3] [iD], Unggi Lee[4] [iD], and Min Liu[1] [iD]

[1] The University of Texas at Austin, Austin, TX 78705, USA
song9@utexas.edu
[2] Valdosta State University, Valdosta, GA 31698, USA
[3] Temple University, Philadelphia, PA 19122, USA
[4] Korea University, Seongbuk-gu Seoul, South Korea

Abstract. Improving social presence has been popularly researched for massive open online courses (MOOCs). Despite this, little consideration has been given to how students intend to utilize new technology from a diversity, equity, and inclusion (DEI) standpoint. In this study, we examined the role of social presence in diverse MOOC students' behavioral intentions and sentiments toward using a learning assistant chatbot. Considering potentially disparate perceptions of the technology depending on the students' demographic factors (age, gender, region, and native language), we investigated the relationships among their social presence, age, behavioral intentions (before and after use), sentiment levels, and numbers of turn-takings with the chatbot. The investigation showed positive and moderate correlations between social presence and post-behavioral intention and between social presence and pre-behavioral intention. Also, the sentiment and social presence toward the chatbot turned out to be different according to the students' region factor based on 13 different geographical categorizations. The native language factor was also influential in creating different sentiment levels between native and non-native English user groups. The findings implicate the importance of promoting a better understanding of the role of social presence in students' learning experiences with AI-based applications from a DEI perspective.

Keywords: Behavioral intentions · Chatbot · Demographic factors · Social presence

1 Introduction

Research on massive open online courses (MOOCs) has consistently highlighted the importance of improving social presence to address the issues of isolation and inadequate support. In this regard, prior research has focused on examining the application

N. Wang et al. (Eds.): AIED 2023, CCIS 1831, pp. 236–241, 2023.
https://doi.org/10.1007/978-3-031-36336-8_36

of various technologies in facilitating social presence in MOOCs. These studies have demonstrated that the integration of technology in course activities leads to increased social presence among students (e.g., [2]). However, the importance of students' acceptance of the technology used was often overlooked or minimized in these studies, as they treated the technology as a requirement for completing the course.

The effectiveness of technology implementation in MOOCs is highly dependent on the voluntary participation and acceptance of students. While MOOCs typically offer support resources, such as FAQ pages, students frequently express dissatisfaction with the lack of available help [5]. To bridge the gap between the potential benefits of technology and students' actual behavioral intentions to use it, it is essential to assess user acceptance when introducing new technology. However, the current literature on online learning lacks such a multifaceted approach.

Furthermore, it is crucial to adopt a diversity, equity, and inclusion (DEI) perspective while researching the implementation of new technology in MOOCs. With MOOCs moving towards job-embedded professional skills, DEI-focused professional learning experiences have become increasingly important [1]. Therefore, we propose that any technology implementation in MOOCs, including learning assistant chatbots, should prioritize promoting DEI by taking into account the diverse students' potentially disparate perceptions of the technology.

2 Research Questions

The main purpose of this study is to examine the role of social presence in diverse MOOC students' behavioral intentions and sentiments toward using a learning assistant chatbot. The research questions are as follows.

1. What are the relationships among social presence, age, behavioral intentions (before and after use), sentiment levels, and numbers of turn-takings when MOOC students use a learning assistant chatbot?
2. Which demographic factors influence students' social presence, behavioral intentions, and sentiment toward the chatbot?

3 Method

3.1 Participants

The research site is an online course provider for journalists' professional development in the southwestern U.S. We recruited 237 students from two MOOCs who participated in the survey. Two courses were available during the data collection period, one focused on promoting mental health, while the other provided an introduction to photogrammetry. All study participants were given a course completion fee waiver ($30 equivalent). The participants' ages were normally distributed among the six ranges we offered in the survey (18–24, 25–34, 35–44, 45–54, 55–64, and Over 64), and genders were equally balanced between the binary categories. Nationalities were spread among 62 different countries and territories, including the U.S. (20.7%), Brazil (7.6%), and India (5.5%) as the top three nationalities. Native languages were also found to be diverse, totaling

56 different languages, with English (35%), Spanish (12.2%), and Portuguese (10.5%) as the top three. During the data analysis, we exclusively utilized the data from those participants whose chatbot usage had been validated based on the log data ($n = 170$) among the total survey participants.

3.2 Learning Assistant Chatbot

A knowledge-based chatbot was employed as a learning assistant in this study. These chatbots rely on natural language processing techniques to search their knowledge bases for relevant information when users input queries or questions. The knowledge bases can be open or closed domains focused on specific areas of knowledge. In this particular study, the chatbot was designed, by one of the authors, with a closed-domain knowledge base to provide targeted support for learners. Based on Dialogflow, a framework for natural language understanding chatbots developed by Google, the chatbot tries to determine a user's intent by comparing their inquiry with a set of predetermined intents, then generates a response accordingly [6].

3.3 Experiment Procedure

Volunteers for this IRB-approved experiment were recruited via email. In the experiment, we asked the participants to 1) ask the chatbot the three questions that MOOC students at the research site asked most frequently (main course activities, method to achieve course completion certificates, and synchronous meeting times) and fill in the forms with the information retrieved and 2) test the chatbot freely (prompt: "Talk to the chatbot as much as you like about our courses. Fill in the blank with the subject you discussed.").

3.4 Data Sources and Collection Instruments

In this study, two data sources produced two and one different types of data, respectively: a survey (quantitative scale measurements and participant demographic data) and the Google Cloud Console (chatbot text log data).

In the survey, the participants were asked before the experiment how much they would like to use a learning assistant in their course (pre-behavioral intention, one-item scale). Next, they were asked about their experiences with the chatbot after the experiment, including their social presence level (five-item scale, Cronbach alpha was $\alpha = .87$) and post-behavioral intention level (three-item scale, Cronbach alpha was $\alpha = .80$). The final section of the survey requested demographic information about the participant (age, gender, region, and native language).

The scales measuring students' social presence and post-behavioral intention to use a knowledge-based chatbot in MOOCs [4] were utilized during the survey. A single-item measure was chosen for measuring a participant's pre-behavioral intention since the construct was unidimensional and specific in scope [3]. The item was: "Before interacting with the chatbot, please rate your preference for using a chatbot to know more about our courses." For each survey item, participants rated their agreement score on a Likert scale from one (completely disagree) to five (completely agree).

Following that, participants' demographic data were collected based on the instruments used in a previous study [5]: age (six categories), gender (male, female, non-binary, prefer not to say, or self-describe), region (short text), and native language (short text). The chatbot log text data were downloaded from Google Cloud Console. After removing all computer codes and timestamps, we only used the text data.

3.5 Data Analysis

Each student's social presence and pre- and post-behavioral intentions were calculated by averaging the items' scores on each scale (six, one, and three items, respectively). As a result, social presence ranged from 1.4 to 5, and pre- and post-behavioral intentions ranged from 1 to 5. A student's sentiment score about the chatbot interaction was calculated based on Valence Aware Dictionary and sEntiment Reasoner, a Natural Language Toolkit's sentiment analysis library [7]. It ranged from −0.7219 to 0.9372 (where −1 indicates complete negative sentiment and 1 indicates complete positive sentiment). The number of turn-takings was measured by counting how many student-chatbot interactions happened in one session; it ranged from 1 to 14 times. For all the quantitative analyses, we utilized the students' survey measurement data which could be matched with chatbot log data based on their Qualtrics response's IP address and the NGINX access log of the webpage embedded with the learning assistant chatbot. As secondary references, students' free interactions in the second phase of the experiment were compared with the chatbot log for user verification. A total of 170 students' data were matched and used in this study.

Students' inputs about their current residence were categorized into 13 different regions based on the United Nations Statistics Division's 18 sub-regional codes [8]. Their native languages were classified into two groups considering the courses were provided in English: native and non-native English user groups.

4 Results

4.1 Relationships Among Social Presence, Age, Behavioral Intentions, Sentiment, and Numbers of Turn-Takings

Spearman's rank-order correlations indicated that there were positive and moderate correlations between social presence and post-behavioral intention ($r_s = .54$, $n = 170$, $p < .01$) and between social presence and pre-behavioral intention ($r_s = .47$, $n = 170$, $p < .01$). There were also positive and weak correlations between the number of turn-takings and sentiment ($r_s = .26$, $n = 170$, $p < .01$) and between pre- and post-behavioral intentions ($r_s = .24$, $n = 170$, $p < .01$). Additionally, we found negative and weak correlations between age and post-intention ($r_s = -.20$, $n = 170$, $p < .01$) and age and social presence ($r_s = -.17$, $n = 170$, $p < .05$). Table 1 shows the correlations among those variables. Next, we compared student experiences across multiple demographic groups, focusing on the meaningful relationships among social presence, pre- and post-behavioral intentions, and sentiment toward the chatbot.

Table 1. Correlation matrix.

	1	2	3	4	5	6
1. Pre-behavioral intention						
2. Post-behavioral intention	.24**					
3. Turn-takings	− .11	.06				
4. Sentiment	.04	.08	.26**			
5. Social presence	.47**	.54**	− .05	.06		
6. Age	− .06	− .20**	.10	.03	− .17*	

** p < .01, * p < .05 (2-tailed).

4.2 Demographic Factors Influencing Students' Social Presence, Behavioral Intentions, and Sentiments Toward the Chatbot

Age and gender groups did not show statistically significant differences in the levels of social presence, behavioral intentions, sentiment, and turn-taking, according to Kruskal-Wallis test results. On the other hand, a Kruskal-Wallis test on the regional groups indicated that sentiment ($\chi2$ (12, $N = 170$) $= 25.50$, $p = .013$) and social presence ($\chi2$ (12, $N = 170$) $= 22.10$, $p = .036$) of the 13 regions were significantly different in statistics standpoint. In detail, across the 13 regional groups (1: Australia and New Zealand, 2: Eastern Asia, 3: Eastern Europe, 4: Latin America and the Caribbean, 5: Northern Africa, 6: Northern America, 7: Northern Europe, 8: South-eastern Asia, 9: Southern Asia, 10: Southern Europe, 11: Sub-Saharan Africa, 12: Western Europe, 13: Western Asia), the students' sentiment was the highest in Northern Europe ($Mdn = .63$, $n = 2$) and the lowest in Southern Europe ($Mdn = 0$, $n = 17$). The social presence was the highest in Sub-Saharan Africa ($Mdn = 4.40$, $n = 20$) and the lowest in Australia and New Zealand ($Mdn = 3.10$, $n = 2$).

Additionally, a Mann-Whitney U test of the native language factor revealed a statistically significant difference in sentiment between native and non-native English user groups. The sentiment mean rank was higher in the non-native English user group than in the native English user group, $U = 3940.00$, $z = -2.01$, $p = .044$, with a small effect size under 0.10.

5 Conclusion

In this study, we examined how social presence affected the behavioral intentions and sentiments of diverse MOOC students toward using a learning assistant chatbot from a DEI perspective. The results revealed positive and moderate correlations between social presence and both pre- and post-behavioral intentions, with post-intention showing a slightly stronger correlation with social presence. However, social presence did not show any significant relationship with sentiment toward using the chatbot and a negative correlation to students' age, as derived from the chatbot log. Instead, the sentiment was weakly but positively related to turn-takings. Moreover, the sentiment and social presence toward the chatbot varied according to regional factors based on 13 different geographical categorizations, and the native language appeared to have an influence in creating different sentiment levels between native and non-native English user groups. These findings underscore the crucial role of social presence in influencing students' behavioral intentions toward adopting new technology, as supported by existing literature (e.g., [2]).

Overall, our findings highlight the strong association between social presence and MOOC students' intentions and sentiments toward using new technology, such as a learning assistant chatbot. However, it is noteworthy that new technology experiences, including social presence and sentiment, differ based on students' geographical location, language use, and age, underscoring the importance of considering DEI principles in the research and development of learning assistant technologies. Further investigations are warranted to gain a deeper understanding of how social presence is constructed during chatbot interaction sessions for diverse students.

References

1. Borneman, E., Littenberg-Tobias, J., Reich, J.: Developing digital clinical simulations for large-scale settings on diversity, equity, and inclusion: design considerations for effective implementation at scale. In: Proceedings of the Seventh ACM Conference on Learning @ Scale, pp. 373–376. ACM, Virtual Event USA (2020)
2. Ebadi, S., Amini, A.: Examining the roles of social presence and human-likeness on Iranian EFL learners' motivation using artificial intelligence technology: a case of CSIEC chatbot. Int. Learn. Environ. 1–19 (2022). https://doi.org/10.1080/10494820.2022.2096638
3. Fuchs, C., Diamantopoulos, A.: Using single-item measures for construct measurement in management research: conceptual issues and application guidelines. Die Betriebswirtschaft. **69**, 195 (2009)
4. Han, S., Cai, Y., Shao, P., Liu, M.: Scale development to measure students' learning experiences with knowledge-based chatbots in massive open online courses. Presented at the Annual conference of American Educational Research Association 2023 (AERA 2023), Chicago, Illinois (2023)
5. Han, S., Lee, M.K.: FAQ chatbot and inclusive learning in massive open online courses. Comp. Educ. **179**, 104395 (2022). https://doi.org/10.1016/j.compedu.2021.104395
6. Han, S., Liu, M., Pan, Z., Cai, Y., Shao, P.: Making FAQ chatbots more inclusive: an examination of non-native English users' interactions with new technology in massive open online courses. Int. J. Artif. Intell. Educ. (2022). https://doi.org/10.1007/s40593-022-00311-4
7. NLTK: Natural Language Toolkit. https://www.nltk.org/
8. UNSD — Methodology. https://unstats.un.org/unsd/methodology/m49/

Audio Classifier for Endangered Language Analysis and Education

Meghna Reddy and Min Chen(✉) ⓘD

University of Washington Bothell, Bothell, WA 98011, USA
{meghnard,minchen2}@uw.edu

Abstract. Around 42% of the world languages are considered endangered due to the decline in the number of speakers. MeTILDA (Melodic Transcription in Language Documentation and Application) is a collaborative platform created for researchers, teachers, and students to interact, teach, and learn endangered languages. It is currently being developed and tested on the Blackfoot language, an endangered language primarily spoken in Northwest Montana, USA and Southern Alberta, Canada. This study extends MeTILDA functionality by incorporating machine learning framework in documenting, analyzing, and educating endangered languages. Specifically, this application focuses on two main components, namely audio classifier and language learning. Here, the audio classifier component allows users to automatically obtain instances of vowels and consonants in Blackfoot audio files. The language learning component enables users to visually study the pitch patterns of these instances and improve their pronunciation by comparing with that of native speakers using a perceptual scale. This application reduces manual efforts and time-intensive tasks in locating important segments of Blackfoot language for research and educational purpose.

Keywords: Audio Classification · Endangered Language Education · MeTILDA

1 Introduction

42% of the world languages [1] are currently endangered. Language endangerment is considered one of the most urgent problems facing humanities [2] because endangered languages represent a vast repository of human knowledge about the natural world and cultural traditions which is irreplaceable. Linguists have responded to this issue through a renewed commitment towards language documentation and education [3].

This paper presents a collaborative research study that aims to document and educate Blackfoot, an endangered language with around 3,250 active speakers [4]. Blackfoot is a pitch accent language where words with the same sound sequence can convey different meanings when changing in pitches, e.g., ápssiw means 'it is an arrow' while apssíw means 'it is a fig; snowberry.' Studies have shown the important role of rhythm and melody in language acquisition, especially with pitch-accent endangered languages like Blackfoot [5]. In our work, we developed an audio classifier that automatically identifies vowels and consonants instances, and a cloud-based platform called MeTILDA (Melodic Transcription in Language Documentation and Application) to help Blackfoot language education.

N. Wang et al. (Eds.): AIED 2023, CCIS 1831, pp. 242–247, 2023.
https://doi.org/10.1007/978-3-031-36336-8_37

2   Related Work

Many studies have been done to classify audio for various purposes, such as speech classification [6], event detection [7], and music recommendation [8]. Some work focuses on extracting useful audio features including low-level spectral and temporal features [6], mid-level featured such as Mel-Frequency Cepstral Cooefficients (MFCCs), spectrograms and mel-spectrograms [8]. Other work aims to develop effective machine learning or deep learning models. For example, Artificial Neural Networks (ANN) [7], Convolution Neural Networks (CNN) and Deep Belief Networks [9] have demonstrated promising results in automating the audio classification process. However, there has been limited research in audio classification for language education. In addition, as a pitch accent language, Blackfoot audio classification and language education faced several unique challenges including limited audio dataset for training, varied lengths of vowels and consonants, and impact of pitch on word meanings. Therefore, models developed for other purposes fail to support effective Blackfoot vowel/consonant classification.

In addition, existing software systems offered limited support to address the urgent need of endangered language education. Most existing linguistics tools such as FLEx[1], ELAN[2], and Praat[3] provide essential support for linguistic research but are less effective for language education, while language learning tools such as Babbel[4], Rosetta Stone[5] largely focus on commonly spoken languages. The support is even more inadequate for pitch accent languages. Research has shown that pronunciation learning technique is significantly understudied in Indigenous languages [10]. Pitch movements are often not explicitly represented in instructions and remain unclear to learners.

3   Audio Classifier and MeTILDA Framework

In our study, we developed an audio classifier to automatically extract vowels and consonants from Blackfoot speech recordings, which can be input into the MeTILDA platform for users to visually study pitch patterns and improve pronunciations. MeTILDA uses a perceptual scale developed in our earlier study [11] to help researchers and learners "see what they hear." It consolidates and automates the workflow of generating perceived pitch changes of words in visual aids, called Pitch Art. This visual diagram would allow Blackfoot teachers and learners to understand how their pronunciation compares to that of native speakers.

---

[1] fieldworks.sil.org
[2] https://archive.mpi.nl/tla/elan
[3] https://www.fon.hum.uva.nl/praat/
[4] https://www.babbel.com/
[5] https://www.rosettastone.com/

3.1 Audio Classifier

The audio classifier is trained and tested on Blackfoot speech recordings and their associated TextGrid annotation files provided by our collaborators[6].

Data Processing and Feature Extraction. The first step is to process the data and extract audio features for machine learning model development. To automate this process, we incorporate "praatIO"[7] in our system to split training data into segments of vowels and consonants based on timestamps specified in the TextGrid annotation files. For testing data, a challenge is to split speech recordings into small segments with appropriate lengths whereas the lengths of vowels and consonants can be varied in Blackfoot speech. In the literature, attempts were made to identify the lengths of vowels and consonants in languages spoken in Australia and Philippines [12]. Most fall within a band of 50 ms (ms) to 250 ms. This information is useful and matches with our observations on Blackfoot vowels and consonants based on speech recordings. Specifically, most vowels vary between 100–150 ms in length and most of the consonants between the 100–180 ms. Therefore, currently 100 ms is used as the window size for the testing data.

Various features are then extracted from audio segments including MFCCs, spectrograms and low-level features. Here, MFCCs are a set of coefficients that make up the mel-frequency spectrum to represent an audio signal's overall shape and frequency-based features [13]. A spectrogram is a visual representation of an audio unit with its most important frequency-based features taken into consideration. It works best as an input for deep learning models as image data. In addition, we also explored a set of low-level features including chroma STFT, root-mean-square (RMS) value, spectral centroid, spectral bandwidth, spectral rolloff and zero crossing rate [14]. We developed tools that allow users to select different combinations of audio features from the above-mentioned set for further processing and re-training models.

Model Training. After feature extraction, we explore different deep learning models including ANN, CNN, deep belief network for vowel/consonant classification. ANNs have proven to perform well for regression and classification applications ranging from banking to time-series forecasting and medicine. Sigmoid activation function is used in our study as it works best with binary classifiers [7] such as ours with 2 possible output classes (vowel or consonants). CNNs use images as inputs for training. In our application, spectrograms of audio units are a good visual representation to be used to train a CNN. In terms of deep belief network, in our previous study, it has been used successfully in identifying important Blackfoot sounds of 'h' and 'okii' as part of the DeepAudioFind project [15]. It is therefore used in this study to serve as a baseline. The classification results can be passed into MeTILDA for language learning.

3.2 MeTILDA

In MeTILDA, with the goal to help researchers and learners "see what they hear," we incorporated a perceptual scale developed by our previous study [11]. This scale provides

---

[6] Dr. Mizuki Miyashita, a professor of linguistics at University of Montana (UM) and Mr. Naatosi Fish, a Blackfoot community linguist.

[7] https://github.com/timmahrt/praatIO

a common reference for comparing pitch across recordings, regardless of the speakers' natural pitch. MeTILDA is hosted on the cloud[8] and the current features are organized into three components: Create, Learn, and Login/Team Project.

Create: The *Create* page allows users to upload recordings to the cloud database and select sound file(s) from the database for further processing. As shown in Fig. 1 (a), when a sound file is uploaded or selected, its waveform and spectrogram are shown on the screen. Users can then identify vowel or consonant instances based on the input from audio classifier or adjust the duration of the instances using the tools provided. F0 measurements are then computed using the perceptual scale. User can select the Average button, shown on a pie menu or enter orthographic symbols for the syllables, based on which, Pitch Art is automatically generated in a drawing window. MeTILDA allows the measurement of multiple recordings on the same page (as in Fig. 1 (b)) and prints word melodies of all selected recordings. This feature is useful for comparing the pitch movements produced by native speakers and learners. Other features offered by the *Create* page include saving Pitch Art images, listening to only the tones of the word melody, and toggling a variety of appearance options (e.g., displaying syllable text, showing pitch in an F0 contour instead of averaged, showing pitch in hertz instead of the psychoacoustic scale). Figure 1 (c-d) show the My Files and History pages: Once the Pitch Art is drawn, users can save the uploaded sound files as well as measurement data (c) and Pitch Art images (d) for future access, all in the cloud database.

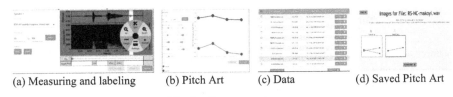

(a) Measuring and labeling   (b) Pitch Art   (c) Data   (d) Saved Pitch Art

Fig. 1. *Create* page images

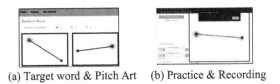

(a) Target word & Pitch Art   (b) Practice & Recording

Fig. 2. *Learn* page images

Learn: The *Learn* page supports practice on word pronunciations. As shown in Fig. 2 (a), users may choose a syllable pattern, each containing words for users to select and practice. Users can listen to a native speaker's pronunciation or pitch tones for reference, and they can record their own pronunciations. As in Fig. 2 (b), the user's pitch track appears in a dotted line, enabling visual comparison with the model Pitch Art. Each previously recorded sound is presented, with an option to play or pause. The *Learn* page

---

[8] https://metilda.herokuapp.com/

also has a View Students option so the tool can be used as a course supplement, allowing teachers to view their students' pitch tracks.

Login/Team Project: Users can obtain access to MeTILDA by logging in based on their user category: researcher, teacher, student, or other. Different users have varied access privileges to data in the system. For example, researchers and teachers can access all features: measurement tools, Pitch Art creation, saving, and stored data. Students have access to play, record, and submit in the *Learn* page only. Teachers can view their students' submitted work, while students cannot see each other's work.

4 Experimental Results

Performance comparison is done on different audio classifiers using various machine learning models with a combination of features. The dataset provided by the collaborators includes 4,852 audio segments with 2,211 vowels and 2,641 consonants. Using 10-fold cross validation scheme, two top performers are listed in Table 1. As a comparison, using the baseline model from our previous study [15] only achieved 77% in Precision and Recall. We want to acknowledge the dataset is relatively small to thoroughly reflect the model performance, but the preliminary results show great potential. The collaborators are working to provide more data for us to further improve and verify the models.

Table 1. Summary of trained models with top performance

Features + Model Type	Precision	Recall	Features + Model Type	Precision	Recall
MFCCs + ANN	89%	89%	Spectrograms + CNN	88%	90%

MeTILDA is currently used by linguistics researchers in finding patterns among Blackfoot speakers and in documenting the language. It has also been used by teachers and students in evaluating a learner's pronunciation as compared to that of native speakers. A recent survey was conducted to gather feedback on usefulness, ease of learning, and satisfaction from representative user groups. Totally 14 users participated with 3 self-identified as linguists, 10 students, and 1 teacher. By average, the rating for each question is above 4.0 out of 5 and over half of the questions received more than 4.5 ratings.

5 Conclusions

With the goal to promote research and education in endangered languages, this project develops an audio classification application to automate the process of obtaining instances of vowels and consonants in Blackfoot. We also present a working prototype called MeTILDA that has shown promising results towards analyzing and learning Blackfoot speech. As a result, linguistics researchers are provided with multiple tools to document and analyze language recordings. Teachers have access to collections of words and recordings from native speakers to teach students. Students are given the tool to compare their own pronunciation of words to that of native speakers. In the future work, we plan to further extend our work to classify and educate other sounds in Blackfoot and other endangered languages.

Acknowledgments. This work is supported by National Science Foundation (NSF BCS-2109654). We also appreciate the late Mr. Earl Old Person for his audio recording as a native speaker and the learners of the Blackfoot language.

References

1. Eberhard, D.M., Simons, G.F., Fennig, C.D. (eds.): Ethnologue: Languages of the World. SIL International (2022)
2. Rogers, C., Campbell, L.: Endangered Languages. *obo* in Linguistics (2011)
3. State of the Art of Indigenous Languages in Research: A Collection of Selected Research Papers. ISBN: 978-92-3-100521-3 (2022)
4. Kaneko, I.: A metrical analysis of Blackfoot nominal accent in optimality theory. Doctoral dissertation, University of British Columbia (2000)
5. Bird, S., Miyashita, M.: Teaching phonetics in the context of Indigenous language revitalization. In: International Symposium on Applied Phonetics, pp. 39–44 (2018)
6. Bhattacharjee, M., Prasanna, S.M., Guha, P.: Time-frequency audio features for speech-music classification. arXiv:1811.01222 (2018)
7. Eutizi, C., Benedetto, F.: On the performance improvements of deep learning methods for audio event detection and classification. In: International Conference on Telecommunications and Signal Processing, pp. 141–145 (2021)
8. Chen, K., Liang, B., Ma, X., Gu, M.: Learning audio embeddings with user listening data for content-based music recommendation. In: IEEE International Conference on Acoustics, Speech and Signal Processing, pp. 3015–3019 (2021)
9. Scarpiniti, M., et al.: Deep Belief Network based audio classification for construction sites monitoring. In: Expert Systems with Applications, vol. 177 (2021)
10. McIvor, O.: Adult Indigenous language learning in western Canada: what is holding us back? In: Michel, K., Walton, P., Bourassa, E., Miller, J. (Eds.) Living Our Languages: Papers from the 19th Stabilizing Indigenous Languages Symposium, pp. 37–49 (2015)
11. Chen, M., Borad, J., Miyashita, M., Randall, J.: Integrated cloud-based system for endangered language documentation and application. In: IEEE 4th International Conference on Multimedia Information Processing and Retrieval, pp. 235–238 (2021)
12. Aoyama, K., Reid, L.A.: Cross-linguistic tendencies and durational contrasts in geminate consonants: an examination of Guinaang Bontok geminates. J. Int. Phon. Assoc. **36**(2), 145–157 (2006)
13. Prabakaran, D., Sriuppili, S.: Speech processing: MFCC based feature extraction techniques-an investigation. J. Phys. Conf. Ser. **1717**(1), 012009 (2021)
14. McFee, B., et al.: librosa: Audio and music signal analysis in python. In: 14th Python in Science Conference, vol. 8, pp. 18–25 (2015)
15. Sandeep, R.: Unsupervised Feature Extraction for Data Mining of Endangered Language Audio Data. University of Washington Bothell (2016)

Quantifying Re-engagement in Minecraft

Jonathan D. L. Casano(✉) [ID], Mikael Fuentes [ID], and Maria Mercedes T. Rodrigo [ID]

Ateneo de Manila University, Quezon City, Philippines
jcasano@ateneo.edu

Abstract. This paper explored how re-engagement (a learner's unprompted and voluntary re-interaction with a learning intervention) may be measured in an open-world, game-based learning environment such as WHIMC, a custom Minecraft world designed to show conditions of altered Earths. Ways to describe/quantify to what extent certain WHIMC elements engender re-engagement among its players were also investigated. Through survey answers that underwent systematic coding, we found that social play, free exploration and interactive learning elements are reported to be the most-liked features of WHIMC and hence, potential triggers for re-engagement. Using the logs of player position, observation making behavior, and Science tools usage, we operationalized these re-engagement triggers and found that the majority of the interactions with them happened outside of the formal testing hours. An average of 8–16 concurrent players were observed during the first 6 nights outside testing hours and 75% of all nights registered concurrent users (*social play*). About a quarter of the respondents (28.22%) had higher exploration times outside testing hours and had visited more worlds outside testing hours (*free exploration*). Around a quarter to a half (27.17–54.21%) of the respondents were interacting with the NPCs outside testing hours (*NPC interactions*). These findings and proposed quantification methods provide inceptive but non-generalizable ways at understanding and describing re-engagement in open-world game-based learning contexts.

Keywords: Engagement · Minecraft · WHIMC

1 Introduction

Re-engagement is defined as a learner's unprompted and voluntary re-interaction with a learning intervention [3]. It is the last stage in the Process Model of Engagement composed of four stages namely, *point of engagement, period of sustained engagement, disengagement* and then (possibly) *re-engagement* [10]. Re-engagement is viewed to be important because this type of *continuation desire* as [11] calls it, is an indicator of intrinsic motivation and intrinsically motivated students have been shown to have better learning effectiveness [15].

The What-If Hypothetical Implementations using Minecraft (WHIMC; https://whimcproject.web.illinois.edu/) project is a set of Minecraft worlds that learners can explore as supplementary activities to learn more about science, mathematics, and astronomy among others [16]. Learners are immersed in alternate versions of Earth, logging both

N. Wang et al. (Eds.): AIED 2023, CCIS 1831, pp. 248–253, 2023.
https://doi.org/10.1007/978-3-031-36336-8_38

the ways in which learners traverse these words (*position logs*) and the observations that learners make during their explorations (*observation logs* and *use of Science tools logs*). Measuring re-engagement in open-ended environments such as Minecraft can be quite challenging. The work of [12] for instance sought the assistance of parents to see if the students were accessing Minecraft from their homes and if the students looked concentrated while interacting with the game. Other researchers like [13] and [14] tried to see if re-engagement could be measured via the use of follow-up surveys that measure residual interest in STEM.

In this work we demonstrated an approach to measuring and describing re-engagement beyond the use of personnel and pen-and-paper surveys. Our formal research questions are the following: **RQ1**: How can re-engagement be measured in an open-world game-based learning environment such as Minecraft? **RQ2**: Are there ways of quantifying to what extent WHIMC elements engender re-engagement among its players?

2   Review of Related Literature

In the context of the Process Model of Engagement, recent experiments that used Minecraft to deliver custom learning experiences focused on measuring engagement at the *point of engagement* and *period of sustained engagement*, that is, from the moment the learning intervention was introduced (*usually at the start of the class period*) up to the point where interaction with the intervention is halted and players are asked to answer questionnaires (*usually at the end of the class period*) [5, 8]. There is currently a lack of similar studies that include the *re-engagement* stage as part of the scope of analysis [2].

Recent attempts at measuring re-engagement were carried out by either asking parents if their children are interacting with the Minecraft learning intervention when they get home or by asking the students to answer a delayed post-test that measures residual interest in STEM. At the time of writing and to the best of our knowledge, this paper shows the first attempt at using Minecraft log-data to investigate and quantify re-engagement. Other recent experiments that also used Minecraft log-data focused on analyzing STEM interest [7], student outcomes [1] and affect [6].

3   Methodology

Three WHIMC features namely *Social Play* (*the affordance of being able to play with a classmate/s in real-time*), *Free Exploration* (*the ability to explore and visit different worlds*), and *Interactive Learning Elements* (interacting with *NPCs, observation making, and use of Science tools*) were identified to be the top most liked features of WHIMC as per prior work [4, 9]. Hence, we focused on these features as potential re-engagement triggers [3]. The subsequent paragraphs described how we have quantified them in the logs available in WHIMC.

Social Play Outside Testing Hours. We used the player positions data to see the number of overlapping play sessions during the night-time intervals within each of the school's module implementation duration. The average number of overlapping player sessions per night is then shown as a simple time-series graph in the results section.

Free Exploration Outside Testing Hours. The number of minutes each student spent playing WHIMC as well as the number of worlds each student visited both during the live testing hours and outside testing hours were counted to operationalize free exploration. As an attempt to represent all possible archetypes, the data was clustered via K-means at k = 6 as it is the optimal value for k returned by the Elbow method (*consecutive sum of squares*). Finally, principal component analysis (PCA) was done to reduce the 6 K-means dimensions into 2 dimensions to conveniently plot the clusters in a 2D graph.

Interactive Learning Elements Outside Testing Hours. To get a sense of the NPC interactions outside testing hours, top-view maps of the WHIMC worlds were created using a program called *uNmINeD*. A *fixation* was operationalized as staying in the same area for 12 s or more When this fixation happens within one *Minecraft hop* near an NPC (which is the distance required to interact with NPC's in Minecraft), the fixation is considered as an interaction with the NPC. Observation logs were processed to find the observations that were created during and outside the testing hours per school. These observations were aggregated and normalized to remove outliers. The number of Science Tools usage were also counted during live testing and outside testing hours.

4   Results and Discussion

The following subsections present graphs generated from the processing and analysis of the player positions, player observations and Science tools logs to show how re-engagement in WHIMC may be quantified (**RQ1**). We then use insights drawn from these graphs to describe the extent to which the WHIMC elements explored may have contributed to re-engagement (**RQ2**).

Social Play. The average number of concurrent players in the first 3 to 4 nights of the module implementations across the schools fell in the range of around 8 to 16 concurrent players per night. 25 out of the 33 total night intervals for all Schools (~75%) recorded concurrent players. In the context of our RQ's, this shows two important things. First, it gives preliminary confirmation that there was indeed activity inside the WHIMC worlds during the night intervals of the module implementation. These interactions are understood to be out of the students' own volition. Second, it reveals that there were indeed traces of Social Play during the students' unprompted use of WHIMC outside class hours.

Free Exploration. The clusters generated by the K-means algorithm plotted via PCA for all schools showed that the clusters representing *high exploration times outside testing hours and high total worlds explored outside testing hours* constitute 28.22% of the data set indicating that about ¼ of respondents roamed around and visited the WHIMC worlds outside the prescribed testing hours. Given this, we speculate that being able to access WHIMC outside the class hours allowed the students to explore each world a little bit more in depth (*high exploration times outside testing hours*) compared to the testing sessions where the modules are guiding them towards advancing the game and therefore visiting more worlds in shorter amounts of time (*low exploration times outside testing hours*).

Interactive Learning Elements. The *fixation percentage* (**fa**) is operationalized as the percent of the *fixations near NPCs* (**fnn**) over the *total number of fixations* (**tf**). Similarly, the NPCs visited per map ratio (**nvr**) is the ratio of the *total maps visited where a fixation near an NPC was recorded* (**mvf**) over the *total maps visited* (**tmv**). The computed **fa** and **nvr** values were averaged per school to get an idea of the average NPC interactions per school. The average **fa** per school provides an idea of how much NPC interactions happened outside testing hours and the average **nvr** per school illustrates the distribution of these NPC interactions across the maps visited. An average **nvr** closer to 1.0 indicates that there was an NPC interaction for all maps visited. Table 1 shows each of the school's **fa** and **nvr** values.

Table 1. The average fixation and average **nvr** per school.

School	Average fa (%)	Average nvr
School 1	34.92	0.88
School 2	54.21	0.76
School 3	27.17	0.72
School 4	40.74	0.74

The average **fa** and **nvr** values communicate that around a quarter (27.17%) to a half (54.21%) of the respondents were interacting with the NPCs outside testing hours. This behavior is also distributed among the NPCs located in the different worlds of WHIMC (0.72–0.74 **nvr**) as opposed to being concentrated in just a few of the worlds. These values validate the analysis done in 3.2 and shows that NPCs in WHIMC can be considerable triggers for re-engagement.

Observation Making & Science Tools Use. With the exception of School 1, most of the observation making behavior was actually done outside testing hours. This particular behavior seems to align with the Free Exploration patterns found where majority of the high exploration times and high number of worlds explored were also noted to have happened outside testing hours. Most Science tools usage was also observed outside testing with an average of around 73 recorded Science tools commands across the nights for the duration of the module implementation.

5 Conclusion and Future Work

This paper explored how re-engagement (*a learner's unprompted and voluntary interaction with a learning intervention*) may be measured in an open-world game-based learning environment such as WHIMC (RQ1). Ways to describe/quantify to what extent the WHIMC elements engender re-engagement among its players were also investigated (RQ2).

(**RQ1**) We demonstrated that re-engagement can be quantified by first gathering data through a survey to determine the most liked features of WHIMC (*which according to Cairn are possible triggers for re-engagement*). These features were then systematically coded to reveal what elements could be the focus of quantification and analysis. In this work, three distinct elements came up as possible triggers for re-engagement namely, social play, free exploration and interactive learning elements inside WHIMC. Using the record of the player's positions, observations and Science tools usage, simple heuristics were developed to check if these possible triggers are present outside testing hours. Social play for instance, was operationalized as overlapping playtimes which simply translates to position readings that co-occurred with each other outside testing hours. The common thing with these quantifications is they leveraged the design of the module implementations. The affordance of students being able to access WHIMC outside testing hours allowed the analysis to consider these outside-testing-hours readings as unprompted out of the students' own volition.

(**RQ2**) A few inceptive ways of reporting the extent to which a certain element may have contributed to re-engagement were demonstrated. For example, the extent of Social play as a possible re-engagement trigger may be described using the average number of concurrent users outside testing hours. These values may further be plotted into a time-series graph to get an idea of how much social play is observed over time. Free exploration as a re-engagement element may be described using the ratio between the worlds visited and the amount of time spent exploring outside testing hours. We do recognize that the findings of this work are at this point non-generalizable and highlight our contribution of designing inceptive re-engagement quantification ideas which other researcher may use as a starting point to design more complex and comprehensive re-engagement quantification techniques.

At this point we feel it is important to seek out other Schools' datasets and see how they compare and contrast with what was found in this initial effort to quantify re-engagement with Minecraft. The current plan is to coordinate with researchers at the University of Illinois Urbana-Champaign to see if their testing data may be used for such purposes. As suggested by a reviewer, we are also considering finding other contexts where the idea of quantifying re-engagement may be valuable such as MOOCS where the attrition rates are high.

Acknowledgments. The authors thank H Chad Lane and Jeff Ginger for their enthusiastic collaboration, Dominique Marie Antoinette Manahan, Maricel Esclamado, and Ma. Rosario Madjos for their support, the Ateneo Laboratory for the Learning Sciences, the Ateneo de Manila University, our funding agency Department of Science and Technology (DOST) for the grant entitled, *"Nurturing Interest in STEM among Filipino learners using Minecraft"*, and our Monitoring Agency the Philippine Council for Industry, Energy, and Emerging Technology Research and Development (DOST-PCIEERD).

References

1. Alawajee, O., Delafield-Butt, J.: Minecraft in education benefits learning and social engagement. Int. J. Game-Based Learn. **11**(4), 19–56 (2021)

2. Baek, Y., Min, E., Yun, S.: Mining educational implications of Minecraft. Comput. Sch. **37**(1), 1–16 (2020)

3. Cairns, P.: Engagement in digital games. In: Why Engagement Matters, pp. 81–104. Springer, Cham (2016)

4. Casano, J.D., Rodrigo, M.M.T.: A comparative assessment of US and PH learner traversals and in-game observations within minecraft. In: International Conference on Artificial Intelligence in Education, pp. 267–270. Springer, Cham (2022)

5. de Andrade, B., Poplin, A., Sousa de Sena, Í.: Minecraft as a tool for engaging children in urban planning: a case study in Tirol Town Brazil. ISPRS Int. J. Geo-Inf. **9**(3), 170 (2020)

6. Esclamado, M.A., Rodrigo, M.M.T.: Are all who wander lost? An exploratory analysis of learner traversals of minecraft worlds. In: International Conference on Artificial Intelligence in Education, pp. 263–266. Springer, Cham (2022)

7. Gadbury, M., Lane, H.C.: Mining for STEM interest behaviors in minecraft. In: International Conference on Artificial Intelligence in Education, pp. 236–239. Springer, Cham (2022)

8. Hobbs, L., et al.: Using Minecraft to engage children with science at public events. Research for All **3**(2), 142–160 (2019)

9. IJsselsteijn, W.A., De Kort, Y.A., Poels, K.: The game experience questionnaire (2013)

10. O'Brien, H.L., Toms, E.G.: What is user engagement? A conceptual framework for defining user engagement with technology. J. Am. Soc. Inform. Sci. Technol. **59**(6), 938–955 (2008)

11. Schoenau-Fog, H.: The player engagement process-an exploration of continuation desire in digital games. In: Digra Conference, Sep. (2011)

12. Tromba, P.: Build engagement and knowledge one block at a time with minecraft. Learn. Lead. Technol. **40**(8), 20–23 (2013)

13. Yi, S., Lane, H.C.: Videogame play and STEM: perceived influences of a sandbox videogame on college major choice. In: The 20th International Conference on Artificial Intelligence in Education (2019)

14. Yi, S.: The impacts of a science-based videogame intervention on interest in stem for adolescent learners. Doctoral dissertation, University of Illinois at Urbana-Champaign (2021)

15. Zaccone, M.C., Pedrini, M.: The effects of intrinsic and extrinsic motivation on students learning effectiveness. Exploring the moderating role of gender. Int. J. Educ. Manage. (2019)

16. WHIMC: What-if hypothetical implementations in minecraft. Accessed from the WHIMC website: https://whimcproject.web.illinois.edu/(n.d.)

Teamwork Dimensions Classification Using BERT

Junyoung Lee[1]([✉])[iD] and Elizabeth Koh[2][iD]

[1] Nanyang Technological University, Singapore, Singapore
junyoung002@e.ntu.edu.sg
[2] National Institute of Education, Singapore, Singapore
elizabeth.koh@nie.edu.sg

Abstract. Teamwork is a necessary competency for students that is often inadequately assessed. Towards providing a formative assessment of student teamwork, an automated natural language processing approach was developed to identify teamwork dimensions of students' online team chat. Developments in the field of natural language processing and artificial intelligence have resulted in advanced deep transfer learning approaches namely the Bidirectional Encoder Representations from Transformers (BERT) model that allow for more in-depth understanding of the context of the text. While traditional machine learning algorithms were used in the previous work for the automatic classification of chat messages into the different teamwork dimensions, our findings have shown that classifiers based on the pre-trained language model BERT provides improved classification performance, as well as much potential for generalizability in the language use of varying team chat contexts and team member demographics. This model will contribute towards an enhanced learning analytics tool for teamwork assessment and feedback.

Keywords: Teamwork dimensions · Natural language processing · Learning analytics tool

1   Introduction

Teamwork is a necessary competency for students in the current education system in preparation for today's global and complex environment, as well as the working world. Through collaboration with others, students can contribute their own knowledge and also learn from others. Analysis of the language used in discussions could shed light on aspects of behaviour that the students might be unaware of during this collaborative process, and help them understand their roles and contributions in the setting. Moreover, teamwork is often inadequately assessed in schools as it can be complex and difficult when there are many teams. Analysis of the teamwork competencies of students can also complement teachers with an insight to the discussion process, and it can also provide formative assessment to students to help them improve their teamwork.

The framework used to identify teamwork competencies from students' dialogue was derived from [3]. This work focuses on the following four dimensions:

N. Wang et al. (Eds.): AIED 2023, CCIS 1831, pp. 254–259, 2023.
https://doi.org/10.1007/978-3-031-36336-8_39

- **Coordination (COD)**: organizing timely completion of task;
- **Mutual performance monitoring (MPM)**: keeping performance of other members in check;
- **Constructive conflict (CCF)**: resolving and diffusing adversarial situations between members via discussion and clarification; and
- **Team emotional support (TES)**: supporting members emotionally and psychologically.

As part of a larger research study on teamwork, the extension of the work in [3] was to identify teamwork dimensions from online chats of students, to provide a formative assessment of teamwork via a web-based application. The study's previous works on automatic analysis of teamwork dimensions on chat text have focused on rule-based classification [7], which requires labor-intensive rule writing, and machine learning-based classification [7,8], which raised concerns of generalizability to new data. [8] had also proposed feature engineering as a way to improve classification performance.

Developments in the field of natural language processing and artificial intelligence resulted in novel approaches namely pre-trained language models (PLMs), such as the Bidirectional Encoder Representations from Transformers (BERT) model [2], that allow for more in-depth understanding of the context of the text through attention models. Contextual classification and sentiment analysis tasks in the field of artificial intelligence in education, such as automated short answer grading or assessing learner sentiments from online discussion forums, have also been successfully supplemented with BERT. [1,5]

This work aims to build upon previous iterations in the study in improving the classification performance of chat messages into teamwork dimensions, and to explore the applicability of a BERT-based classifier over traditional machine learning models. Furthermore, generalizability of the proposed model would be evaluated on unseen data, as it would be ideal in developing a teamwork analytics system for learning groups of different age groups and demographics, with regard to the resource-consuming nature of manually annotating a large corpus.

2   Experimental Setup

2.1   Corpus

The corpus consists of chatlog gathered from 76 teams of a total of 272 14-year-old students in a Singapore secondary school. The students were given approximately 45 min for an ice-breaking activity, in which they were asked to describe their ideal teacher, followed by a dilemma task regarding environmental conservation. The members of the team were physically separated to prevent verbal communication, and the team discussions regarding the activity were carried out via an online chat system.

A total of 19762 raw chat messages were gathered from the participants, inclusive of spam lines. A portion of the chat messages from 7 teams was labelled independently according to the aforementioned teamwork dimensions by both of the two annotators, with Cohen's Kappa ≈ 0.65, indicating substantial inter-rater agreement. The rest of the dataset was divided and labelled individually by each of the annotators. Each message could be annotated for any number or none of the four teamwork dimensions. The number of non-disjoint positive labels for each category is as follows: COD - 2653; MPM - 1357; CCF - 2980; TES - 3506. A sample of the annotated dataset is shown in Table 1.

Table 1. Sample of the annotated dataset

User	Message	COD	MPM	CCF	TES
Student A	Bob are you okay with it	1	1	0	0
Student B	ideal teacher as like um can work well with students and listen to students ideas	0	0	1	0
Student C	yes	0	0	0	1
Student C	So, caring, attentive to our needs and humourous?	0	0	1	0
Student A	humour yes	0	0	0	1
Student B	ok	0	0	0	1

Text Pre-processing. The chat messages were then pre-processed according to the steps outlined in Table 2, as proposed in [7] to simplify the text. A challenge in the analysis of the chatlogs for teamwork competencies comes from prevalence of *Singlish*, a colloquial English spoken in Singapore, as well as textese in chat message environment among youths [6]. Hence, pre-processing also served to improve compatibility with conventional English vectorizers.

Table 2. Example of text pre-processing rules

Procedure	Raw text	$\longrightarrow$	Preprocessed text
Emotion & punctuation tagging	`?`	$\longrightarrow$	`{{question_mark}}`
	`:)`	$\longrightarrow$	`{{pos_emo}}`
Abbreviation expansion	`ikr`	$\longrightarrow$	`I know right`
	`omg`	$\longrightarrow$	`oh my goodness`
Local terms replacement	`macam`	$\longrightarrow$	`similar`
	`chim`	$\longrightarrow$	`difficult`
Named Entity Recognition	`Bob`	$\longrightarrow$	`{{NAME}}`

Table 3 summarizes the feature engineering carried out on the corpus. As proposed by [8] to improve the chat message classification performance, features

Table 3. Feature engineering carried out on the corpus [8]

Feature	Description	Example
F_TIME	Messages that mention task-related time constraint	"faster lah" "we have like 15 mins left"
F_INSTRUCTION	Messages that instruct other team members for the task	"see the url" "guys can we discuss now"
F_PROGRESS	Messages that indicate their progress on the activity	"we r done" "we completed"
F_ELABORATION	Messages that elaborate on discussed ideas	"plants dont reduce smoke" "the teacher should be kind"
F_GREETING	Messages that are greetings	"sup" "good morning guys"
F_POSEMO	Messages that express positive emotions or humor	"just kidding" "LOL"
F_AGREEMENT	Messages that indicate agreement	"yes ok" "yes thats possible"

based on context-sensitive rules were created using indicative terms, part-of-speech tagging, and regular expressions.

For the purpose of this work in assessing the generalizability of a BERT-based classifier, the classification performance on both versions of the corpus pre- and post-feature engineering have been evaluated.

2.2 Model Training

The pre-processed, feature-engineered corpus described above has been split in the 6:2:2 ratio for training, validation, and testing with a preset random seed. The model was set up using the PyTorch Lightning library with the bert-base-cased pre-trained model and tokenizer from the HuggingFace Transformers library [9]. Maximum sequence length for the transformer was set to 200. A linear layer was added as a final layer for multilabel classification. AdamW optimiser [4] was used with warm-up steps of one-third epoch to a learning rate of 2×10^{-5} and linear decay. The criterion was BCEWithLogitsLoss, which combines a sigmoid layer with binary cross-entropy loss for multilabel classification. The maximum epoch was set at 100. The model was trained in the Google Colab environment with 16 GB of RAM on a single Tesla T4 GPU.

3 Results

3.1 Classification Performance

Due to the imbalance in number of positive labels for each class, a comparison of absolute accuracies between models would not be an appropriate gauge for model performance. Instead, macro-averaged precision, recall, F1 score, as well as Hamming distance (also known as Hamming loss) have been calculated for comparison with the best-performing machine learning model proposed by [8], a random forest (RF) classifier with term frequency-inverse document frequency (TF-IDF) vectorizer. Table 4 reports the results.

Table 4. Classification performance comparison

	Without feature engineering		With feature engineering	
	RF	BERT	RF	BERT
Precision	**0.824**	0.810	**0.833**	0.801
Recall	0.406	**0.702**	0.594	**0.747**
F1 Score	0.527	**0.734**	0.684	**0.757**
Hamming distance	0.124	**0.076**	0.090	**0.070**

On test sets with or without additional features, the BERT-based classifier shows comparable precision score to the RF classifier, while outperforming in the other three metrics. The BERT-based classifier also shows nearly consistent performance on both versions of the test data, while the RF classifier displays greater degradation when tested on data without engineered features.

3.2   Inference on Unseen Data

To assess the generalizability of the trained model, it is beneficial to evaluate on a set of unseen data which differs in participant demographic, language use, and task context. In a separate setting, 2 groups of 15 students aged 19 to 42 have been invited to utilize the online chat system to brainstorm for a community service project as an in-class activity. An anonymized chatlog of 129 messages in total was collected, and manually annotated with the four teamwork dimensions. It was pre-processed using the steps in Table 2 with no additional features. The trained BERT-based classifier achieved a Cohen's Kappa score of **0.640** compared to 0.467 achieved by the RF classifier, showing a higher inter-rater reliability.

4   Conclusion

In this work, multilabel classification performance of a BERT-based classifier was evaluated on both seen and unseen corpus of chat messages generated by students discussing a team task, to analyse their teamwork competencies. The results show much potential for scalability of such models without need for extensive feature engineering, and their generalizability with different demographics, in building a reliable teamwork analytics system from online chats. This improved classifier model ultimately contributes towards an enhanced formative learning analytics tool for teamwork assessment, allowing students and teachers to become more aware of the teamwork competencies in the discussion process and receive more holistic feedback.

Acknowledgements. This paper refers to data and analysis from OER62/12EK and OER09/15EK, funded by the Education Research Funding Programme, National Institute of Education (NIE), Nanyang Technological University, Singapore and the Incentivising ICT use Innovation Grant (I3G 9/19EK) from NIE. IRB: IRB-2020-08-006.

References

1. Chanaa, A., El Faddouli, N.E.: BERT and Prerequisite Based Ontology for Predicting Learner's Confusion in MOOCs Discussion Forums. In: Bittencourt, I.I., Cukurova, M., Muldner, K., Luckin, R., Millán, E. (eds.) Artificial Intelligence in Education. pp. 54–58. Springer International Publishing, Cham (2020)
2. Devlin, J., Chang, M.W., Lee, K., Toutanova, K.: BERT: Pre-training of Deep Bidirectional Transformers for Language Understanding. In: Proceedings of the 2019 Conference of the North American Chapter of the Association for Computational Linguistics: Human Language Technologies, Volume 1 (Long and Short Papers), pp. 4171–4186. Association for Computational Linguistics, Minneapolis, Minnesota (2019). https://doi.org/10.18653/v1/N19-1423, https://aclanthology.org/N19-1423
3. Koh, E., Hong, H., Tan, J.P.L.: Formatively assessing teamwork in technology-enabled twenty-first century classrooms: exploratory findings of a teamwork awareness programme in Singapore. Asia Pacific J. Educ. **38**(1), 129–144 (2018)
4. Loshchilov, I., Hutter, F.: Decoupled Weight Decay Regularization. In: International Conference on Learning Representations (2017)
5. Ndukwe, I.G., Amadi, C.E., Nkomo, L.M., Daniel, B.K.: Automatic grading system using sentence-BERT network. In: Bittencourt, I.I., Cukurova, M., Muldner, K., Luckin, R., Millán, E. (eds.) AIED 2020. LNCS (LNAI), vol. 12164, pp. 224–227. Springer, Cham (2020). https://doi.org/10.1007/978-3-030-52240-7_41
6. Ong, K.K.W.: Textese and Singlish in multiparty chats. World Englishes **36**(4), 611–630 (2017)
7. Shibani, A., Koh, E., Lai, V., Shim, K.J.: Assessing the language of chat for teamwork dialogue. J. Educ. Technol. Society **20**(2), 224–237 (2017)
8. Suresh, D., Lek, H.H., Koh, E.: Identifying teamwork indicators in an online collaborative problem-solving task: A text-mining approach. In: Yang, J.C., Chang, M., Wong, L.H., Rodrigo, M.M.T. (eds.) Proceedings of the 26th International Conference on Computers in Education (ICCE) 2018, pp. 39–48. Asia-Pacific Society for Computers in Education (APSCE), Manila, Philippines (2018)
9. Wolf, T., et al.: Transformers: State-of-the-Art Natural Language Processing. In: Proceedings of the 2020 Conference on Empirical Methods in Natural Language Processing: System Demonstrations, pp. 38–45. Association for Computational Linguistics, Online (Oct 2020)

Data Augmentation with GAN to Improve the Prediction of At-Risk Students in a Virtual Learning Environment

Tomislav Volarić[1]([✉])[iD], Hrvoje Ljubić[1][iD], Marija Dominković[1], Goran Martinović[2][iD], and Robert Rozić[1][iD]

[1] University of Mostar, Trg Hrvatskih Velikana 1, Mostar, Bosnia and Herzegovina
{tomislav.volaric,hrvoje.ljubic,marija.dominkovic, robert.rozic}@fpmoz.sum.ba
[2] Faculty of Electrical Engineering, Computer Science and Information Technology, Josip Juraj Strossmayer University of Osijek, Osijek, Croatia
goran.martinovic@ferit.hr

Abstract. In this paper, we explore the use of data augmentation through generative adversarial networks (GANs) for improving the performance of machine learning models in detecting at-risk students in the context of e-learning institutions. It is well known that balancing datasets can have a positive effect on improving the performance of machine learning models, especially for deep neural networks. However, undersampling can potentially result in the loss of valuable data, so data augmentation seems to be more meaningful solution when the dataset is relatively small. One of the most popular data augmentation approaches is the use of GAN networks due to their ability to generate high-quality synthetic samples that belong to the distribution of the original dataset. On the other hand, detecting at-risk students is a hot topic in learning analytics, and ability to detect these students early with high accuracy enables e-learning institutions to take necessary steps to motivate and retain students during the course. We apply this approach to the OULA dataset, a commonly used dataset in learning analytics that includes labeled at-risk students. The OULA dataset is not highly-imbalanced, making it more challenging to improve model performance through these techniques.

Keywords: Data augmentation · Generative adversarial networks · Learning analytics · Virtual learning environment · At-risk students

1 Introduction

Generative adversarial networks [1] - a relatively new and exciting form of deep neural networks - have been widely used for data augmenting purposes in the recent years. Their ability to produce photorealistic images has enabled them to be used in many real-world applications such as visual recognition [2], image

© The Author(s), under exclusive license to Springer Nature Switzerland AG 2023
N. Wang et al. (Eds.): AIED 2023, CCIS 1831, pp. 260–265, 2023.
https://doi.org/10.1007/978-3-031-36336-8_40

manipulation [3], video prediction [4] etc., but they can also be used for generating tabular and other structured data. In this paper, we describe the use of generative adversarial networks for augmenting a tabular dataset, which is challenging in many ways. For starters, tabular datasets are often made from more than one type of data so it is necessary to preprocess it and make it more "meaningful" for the GAN. To be able to generate a mix of discrete and continuous data columns at the same time, GANs need to apply softmax and tanh on their output [5]. Unlike the images, whose pixels have a Gaussian distribution and can be normalized using a min-max transformation, continuous data, most-often, do not have a Gaussian distribution so using a min-max function could cause problems, such as vanishing gradient. Imbalance inside of categorical columns could also lead to problems. GAN trained on such datasets could neglect a minor category and result in mode collapse. Popularization of the platforms for e-learning has led to the abundance of educational data. This has made it possible to monitor student activities, analyze the learning patterns of each student individually and use this knowledge to improve the learning process. This data can also be used to predict those students who are at higher risk of getting bad grades or dropping out of studies. Term used to describe this type of students is "at-risk students". We base our experiment on OULA dataset, which is an educational dataset containing information of students and their interaction with Virtual Learning Environment. We predict the "at-risk" students, firstly on the original and then on the augmented dataset. Term "at-risk students" refers to the students who are more likely to drop out of course or not to fulfill their academic requirements. The on time recognition of these students can be beneficial both for the students and for the university.

2 Experiment Setup

For the experiment, we have used OULA dataset to try to better prediction of the at-risk students. At-risk students are students considered to be at greater risk of not graduating or failing to meet other educational goals. There are several reasons why it is important to predict these types of students. First, students who fail exams and are, therefore, forced to repeat the year are at increased risk of losing confidence, problems with depression and dropping out. Second, universities tend to maintain a certain level of reputation. If too many students drop out, it could cause damage to university's reputation. Therefore, it is in everyone's best interest to develop the most accurate means of predicting high-risk students. In our experiment, we do the predictions on two datasets: one on the original dataset and one on the augmented dataset. The whole process can be roughly divided into three phases. In the first phase, we prepare the data. This includes merging the tables, checking for NaN values and filling them or deleting the columns. Second phase is augmenting the dataset using GAN architecture, and the third and final phase is classification of the students based on the original dataset, classification of the students based on the augmented dataset and comparison of the results.

As already stated, presented analysis is based on the OULA dataset. OULA or Open University Learning Analytics [6] is a set of educational data. It consists of seven CSV files that can be divided in three groups. First group are CSV files with the data about module presentation. Those are assessments, courses and vle. Second group includes CSV files with the data about student activities on the Virtual Learning Environment, and those are studentAssessment, studentRegistration and studentVle. Third group includes only one CSV file and it contains the data of student demographics - studentInfo.

2.1   Data Preparation

For our analysis, we used Google Colaboratory as a working environment. After importing all of the needed libraries, we also imported all of the CSV files from the OULA dataset. In the vle file's columns week_from and week_to there are 5243 empty values, that is, only 1121 non-empty values, so we dropped those columns right away. Next step was to do a right merge of the files student_assessment and assessments based on the column id_assessment so that we get assessments' names inside of a student_assessment table. We then pivoted the resulting table so that the columns are id_student, code_module, code_presentation, and the assessment types CMA, TMA and Exam. Inside of the assessment type columns are mean scores of a student on that type of the assessment, empty values are replaced with the zeros, and the index is reset. The resulting table is called performance. After that, we aggregated another table. This is a student_vle table grouped by the id_student and id_site columns with the total number of all clicks of each student separately for each site separately. This table was then extended using the left merge with the table vle based on the id_site. We pivoted the resulting table and got a table with columns id_student, code_module, code_presentation and all of the sites. We aggregated the values of the site columns using the sum function for the values from the sum_click columns. We also filled empty values with zeros, and reset the index. We then filtered the resulting site columns, that is, we dropped all of the columns whose median is equal to 0. This resulting table is called interaction. Finally, we got our model table by left merging the student_info table with the performance table based on the columns code_module, code_presentation and id_student. The resulting table was then left merged with the interaction table based on the same columns, and as a result, we got our model table. There were still some empty values inside of the imd_band column so we filled them using a mode function from statistics library and removed the percentage symbol. The rest of the empty values were inside of the site columns and were replaced with a 0. We also deleted the identifier columns from which the GAN could hardly learn.

2.2   Data Augmentation

Out of 32593 students, 10156 students dropped out of the course, 7052 failed the course, 12361 passed and 3024 are distinction students, that is, 17208 students are considered at-risk students and 15385 are non-risk. This means that this dataset is already fairly uniform. For simplicity reasons, we encoded the final_result column right away. This was done using a lambda function, which marked the value in the label column with a 0 if it was equal to "Pass" or "Distinction", and marked the rest of the values, that is "Fail" or "Withdrawn", with a 1. Since we wanted to unify the dataset additionally, we trained the GAN using only the non-risk students' data. Therefore, for the training we only used data that has 0 in the label column, but cut the label column out of that training set. Numerical columns were scaled using the Sci-kit learn function $MinMaxScaler$, and categorical columns were encoded using Pandas function get_dummies. We also separated dummy columns into categories: one category for all dummy columns derived from the same category column (gender, region, highest_education, imd_band, age_band and disability). Generator network function as input takes each one of these groups of dummy columns separately, and numerical data as a whole. The generator input is data from latent space, which it uses for generating new samples. A good practice is to use leaky rectified linear activation unit for handling some negative values, both in the generator and the discriminator model. We used it with the recommended value of 0.2 and with the weight initializer "he uniform". We also used batch normalization with the momentum of 0.8 for every layer. For every category of dummy columns, we made a separate branch of hidden layer, and for the numerical columns, only one branch was used. Each of the branches had its own output, which were then combined in the unique output. The discriminator model takes a sample from the generated data as an input, and as an output, it gives a classification prediction whether the given sample is real or fake. In the output layer, we used sigmoid activation function and binary cross-entropy loss function, considering that this is a binary classification problem. We also used the Adam optimization algorithm with the learning rate of 0.0002 and beta1 momentum value of 0.5. Generally speaking, for GAN training it is better to have a small batch size, and a larger number of epochs, but it is also necessary to not overdue the training because of the problems with the training that were already discussed earlier in this paper. We experimented with lots of different batch sizes ranging from 8 to 256 and also with different number of epochs, but for our case batch size of 32 paired with over 20000 epochs proved to be the best. Generator loss at the beginning was very high - over 2, and discriminator loss was very low - under 0.5, but they immediately started to converge. By the end of the training, the generator had learned data distribution very well, and it could produce the data with the distribution similar to the original data distribution.

3   Results

After the training, we generated a set of 1823 samples to get an equal number of non-risk and at-risk students' samples. We rounded and inversed the scaling on the continuous columns of this set, and on the categorical dummy columns we used idxmax function to get them into original shape. We then encoded the categorical columns using the label encoding function from Sci-kit learn preprocessing package, and set all the values in the final result column to be 0, since we used only non-risk students' data to train our GAN. To evaluate performance of our GAN model, we made a comparison of the performance of classification algorithms on the original and augmented dataset. Categorical columns in the original dataset were also encoded using the label encoding function from Sci-kit learn preprocessing package, and for the target column we used lambda function that labelled the values "Pass" and "Distinction" as 0, and "Fail" and "Withdrawn" as 1. We split the datasets on training and test sets and then scaled the original and generated data separately using the standard scaler from Sci-kit. For classification, we used the following algorithms: Xgboost classifier, Random Forest classifier, K Nearest Neighbors, Support Vector Machine and Deep Neural Networks, and for evaluation, we used accuracy, precision, recall, ROC-AUC and F-1 metrics. The results are shown in Table 1.

Table 1. Performance of the various model evaluated on the listed metrics on the original (O) and augmented (A) dataset

Classifier	Xg-boost		Random forest		k-NN		DNN		SVM	
Dataset	O	A	O	A	O	A	O	A	O	A
Accuracy	87.63%	**88.35%**	86.24%	**88.78%**	82.26%	**84.19%**	85.11%	**86.76%**	84.37%	**86.95%**
Precision	89.59%	**90.58%**	88.99%	**90.23%**	**84.76%**	82.73%	86.94%	**88.61%**	**85.99%**	85.48%
Recall	**86.63%**	85.61%	**84.34%**	82.94%	**80.89%**	79.88%	**84.45%**	79.49%	**84.04%**	83.99%
ROC-AUC	94.71%	**95.01%**	86.35%	**88.07%**	82.33%	**83.68%**	85.15%	**85.88%**	84.36%	**86.59%**
F1	**88.08%**	88.02%	**86.61%**	86.43%	**82.78%**	81.34%	**85.68%**	83.80%	**85.00%**	84.73%
Average	89.33%	**89.51%**	86.51%	**87.29%**	**82.60%**	82.27%	**85.47%**	84.91%	84.76%	**85.55%**

4 Conclusions and Future Work

In this paper, we described our attempt to better the performance of the classification algorithms by using generative adversarial networks on the OULA dataset. This dataset is already fairly balanced as it is, but we tried to unify it additionally by generating another 1823 samples of the non-risk students' data. The results of the classification on the augmented dataset were very similar to the results of the classification on the original dataset. This idea seems to be a promising for future work, but at the moment experiment results were both a success and a failure. On the one hand, we did manage to generate data that is realistic similar to the original data and make dataset completely balanced. However, we failed to notably improve the performance of the classifiers. The results may have been more noticeable if we used the dataset that is more imbalanced. Interestingly, all machine learning models using the ROC-AUC metric performed better on the augmented dataset. One of the steps that could be taken to better the results additionally is to further tune the parameters of the both generator and discriminator network and also using the another type of GAN network. It may be useful to train through a greater number of epochs, to try different variation of the generative adversarial networks or some other technology suitable for data augmentation.

References

1. Goodfellow, I., et al.: Generative adversarial nets. In: Advances in Neural Information Processing Systems vol. 27 (2014)
2. Wang, X., Shrivastava, A., Gupta, A.: A-fast-rcnn: Hard positive generation via adversary for object detection. In: Proceedings of the IEEE Conference On Computer Vision And Pattern Recognition, pp. 2606–2615 (2017)
3. Lee, C.H., Liu, Z., Wu, L., Luo, P.: Maskgan: Towards diverse and interactive facial image manipulation. In: Proceedings of the IEEE/CVF Conference on Computer Vision and Pattern Recognition, pp. 5549–5558 (2020)
4. Mathieu, M., Couprie, C., LeCun, Y.: Deep multi-scale video prediction beyond mean square error (2015). arXiv preprint arXiv:1511.05440
5. Xu, L., Skoularidou, M., Cuesta-Infante, A.,Veeramachaneni, K.: Modeling tabular data using conditional gan. In: Advances in Neural Information Processing Systems vol. 32 (2019)
6. Kuzilek, J., Hlosta, M., Zdrahal, Z.: Open university learning analytics dataset. Sci. Data **4**(1), 1–8 (2017)

Prediction of Students' Self-confidence Using Multimodal Features in an Experiential Nurse Training Environment

Caleb Vatral[1]([⊠]) [iD], Madison Lee[2] [iD], Clayton Cohn[1] [iD], Eduardo Davalos[1] [iD], Daniel Levin[2] [iD], and Gautam Biswas[1] [iD]

[1] Institute For Software Integrated Systems, Vanderbilt University, Nashville, TN, USA
caleb.m.vatral@vanderbilt.edu
[2] Peabody College, Vanderbilt University, Nashville, TN, USA

Abstract. Simulation-based experiential learning environments used in nurse training programs offer numerous advantages, including the opportunity for students to increase their self-confidence through deliberate repeated practice in a safe and controlled environment. However, measuring and monitoring students' self-confidence is challenging due to its subjective nature. In this work, we show that students' self-confidence can be predicted using multimodal data collected from the training environment. By extracting features from student eye gaze and speech patterns and combining them as inputs into a single regression model, we show that students' self-rated confidence can be predicted with high accuracy. Such predictive models may be utilized as part of a larger assessment framework designed to give instructors additional tools to support and improve student learning and patient outcomes.

Keywords: Experiential Learning · Simulation-based Training · Multimodal Learning Analytics (MMLA) · Self Confidence · Machine Learning

1 Introduction

In recent years, experiential learning has gained popularity as an effective approach to training for specialized skills, especially in nursing and healthcare. Experiential learning emphasizes hands-on experiences and reflection [3]. In nursing education, experiential learning has seen application through simulation-based training programs. These nursing simulations use high-fidelity manikins to expose students to realistic patient scenarios in a safe and repeatable environment.

Simulation-based experiential learning environments have many advantages. For example, they provide students opportunities to increase their confidence through deliberate repeated practice in a safe environment [4], which is a critical component of an effective nursing curriculum. It influences students' engagement, motivation, and overall performance, directly impacting patient outcomes [5]. However, measuring and monitoring self-confidence is challenging because it has multiple interpretations; it can be measured as a *personality trait* or as a *metacognitive process* [1].

In this paper, we propose a novel approach to predicting students' metacognitive self-confidence in an experiential nurse training environment by combining information from student eye gaze and speech patterns. We develop predictive models of students' self-rated confidence in their simulations, which can contribute to the development of new methods for assessing and enhancing metacognitive self-confidence. This has implications for developing data-driven performance monitoring systems that could be used by students and instructors to improve learning outcomes and better characterize student readiness.

2   Background

Previous work has shown that careful consideration must be made when measuring students' self-confidence to ensure that the correct construct is being measured. Burns et al. [1] showed that self-confidence can be broken down into a spectrum between an online metacognitive judgement and a personality trait based on how it is measured. The metacognitive self-confidence is linked to cognitive and metacognitive processes and is typically measured online as a post-task question; i.e. "How confident are you that your answers/actions are correct?" or "How confident are you that you were successful in completing your assigned task?" Personality trait self-confidence, on the other hand, is linked to personal experience and emotional tendencies and tends to be less related to specific task performance [1]. In our study, we measure the metacognitive aspects of self-confidence by having students rate their confidence as part of an individual performance rating after they review and reflect on a video of their training exercise (see Sect. 3.2). Because of the task-specific nature of this question, the measurement can be interpreted as students' metacognitive self-confidence. Therefore, when building our predictive models, we used students' self-reported confidence as the ground truth for their metacognitive self-confidence (see Sects. 3.3 and 4).

3   Methods

3.1   Experiential Nursing Simulation

Student nurses trained in a simulated hospital room containing standard medical equipment and a manikin patient simulator. Students entered the room and performed routine evaluations of the manikin patient, and then performed relevant prescribed treatments based on their evaluation. For more details on the simulation environment, see [7]. All students provided their informed consent to collect video and audio data as they performed their training activities, and some students volunteered to wear Tobii 3 eye-tracking glasses. In this paper, we analyze the data from 14 students who used eye-tracking glasses.

3.2   Individual Guided Reflection Debriefing

After participating in their instructional simulations, students were given the opportunity to engage in guided reflection designed to promote metacognitive reflection on their performance. Initially, we showed the students their own egocentric eye-tracking footage from the simulation in which they participated. After this, the students re-watched this footage while identifying meaningful event units by pressing a key when they detected a transition from one event to another [10]. Students then reviewed the marked events repeatedly and answered six reflection questions based on that event. One of these questions evaluated teamwork, asking students to rate "To what degree were you working individually versus as a team during this event segment?" on a Likert scale from 1 to 5, and this rating is used later in this paper for feature selection. After answering the questions for each event segment, to conclude the reflection, the students were asked to reflect on the entire simulation experience. They were given a 10-point scale asked, "Please rate YOURSELF on the following measures:" engagement, confidence, patient safety, positive patient outcomes, and scenario objective completion. This paper's main focus is predicting the "Confidence" item in this overall assessment.

3.3   Machine Learning Modeling

We analyzed students' captured eye gaze and speech behavior as an indicator of their overall confidence in the simulation. Using the multimodal eye gaze and speech data collected from the students as features and students' responses to the guided self-reflection as a ground truth for their confidence, we trained a regression model to predict students' self-rated confidence.

We initially developed 27 features derived from the eye gaze and speech data. For each of the students' event segments, we computed these 27 features from the observed data. These initial features were selected in a somewhat post-hoc fashion, partially based on previous work with similar nursing student data [7], and partially based on the features which were easily available from the sensor systems. Because of this post-hoc strategy, not all of these features may be relevant to the prediction of students' self-confidence, so further refinement of the feature set through feature selection processes was necessary.

We performed feature selection by building a mixed effects linear model to measure the fixed effects of the features on self-confidence when controlling for participants. However, in the guided reflection, students only rated their metagcognitive confidence for the overall simulation, not for each event segment. So, we utilized a proxy target variable instead. Utilizing the relationship between teamwork and self-confidence [7], we built the mixed-effects model with students' self-rated teamwork in each segment as the target variable and measured the fixed effects between each of the features and students' self-rated teamwork.

Twelve features shown in Table 1 showed statistically significant effects on teamwork in our feature selection model ($p \leq 0.05$). Seven features were produced automatically by the Tobii glasses 3. One additional eye gaze feature, *PersonGaze*, was computed by the researchers by measuring the overlap between

Table 1. The 12 sequence features extracted from eye gaze and speech data used in the final regression model

Feature	Description
PersonGaze	Percentage of time spent looking at another person
AvgSacHz	Average number of saccades per second
MinSacAmp	Minimum amplitude over all saccades
AvgSacAmp	Average amplitude over all saccades
AvgSacPeakVel	Average peak velocity of over all saccades
StdSacPeakVel	Standard deviation of peak velocity over all saccades
AvgFixHz	Average number of fixations per second
AvgFixPupilDiameter	Average pupil diameter during fixations
MinValence	Minimum emotional speech valence
MaxArousal	Maximum emotional speech arousal
AvgArousal	Average emotional speech arousal
MaxDominance	Maximum emotional speech dominance

the Tobii gaze coordinates and any person-class bounding box produced by the YoloV5L object detection model. The other four features, computed using a trained deep-learning model on sections of the students' speech audio, measured emotional valence, arousal, and dominance of student speech [7,8].

Having selected these 12 features, we then return to the task of predicting metacognitive self-confidence. However, these 12 features are computed for each event, and different students segmented events in different ways. Since our goal was to predict self-confidence over the entire simulation, we formulated the regression as a sequence-to-one regression problem. While several techniques can be used to perform sequence-to-one regression, due to the small sample size of this study we chose to extract basic statistics of the feature sequences to use as the final input features of the regression. For each student's sequence of the 12 features previously identified, we extracted the minimum, maximum, mean, and standard deviation as features to describe the sequence. These four statistical features were calculated for each of the 12 sequence features, leading to an overall 48-dimensional input feature vector for the final regression.

4   Results

For the regression of students' self-confidence scores, because of the small sample size and class imbalance, we used Gradient Boosted Regression Trees with leave-one-out cross-validation. For evaluation, we examined the average root mean squared error (RMSE) and R^2 correlation coefficient compared to the students' self-reflections. The model achieved 0.53 ± 0.17 RMSE and $R^2 = 0.81$. Considering the range of prediction and other limitations, this performance represents a fairly high level of accuracy, which could be informativein a variety of ways.

To explore the model further, we performed a local explainable AI feature contribution analysis using the *Decision Contribution* method [2]. We found 5 unique feature ranking patterns that covered all 14 students. It is most notable that all 5 rankings had the same top-ranked feature: Minimum of AvgSacAmp. which accounted for significantly more of the decision than any of the other features, scoring an absolute sum of decision contributions of 11.99. This was much greater than even the second highest ranked feature, which scored 0.65. However, re-running the regression with only the Minimum of AvgSacAmp feature yielded 1.07 ± 0.16 RMSE and $R^2 = 0.58$, suggesting that while they contributed less, other features still contributed significantly to the overall model performance.

5   Discussion

The analysis presented here was fairly exploratory in nature, given the small sample size and initial post-hoc feature selection methodology. However, the preliminary results suggest several important implications and should be used to drive future research on multimodal prediction of metacognitive self-confidence.

5.1   Saccade Behavior

Saccade behavior seems to be very important in the predictive model's ability to determine students' self-confidence, suggesting that saccade behavior, and its associated cognitive processes, are related to metacognitive self-confidence in some way. 4 out of the 5 top-ranked features were derived from saccade behavior. Extending this, we find a moderate positive Spearman rank correlation between minimum average saccade amplitude and self-confidence ($0.40 \leq \rho \leq 0.92, n = 14$ with Fisher z-score transformation). In other words, larger average saccade amplitudes are linked to higher self-confidence. Prior work has shown relationships between higher-amplitude saccades and goal-directed ideation behavior [9]. Since these simulations tasked students with identifying an unknown problem and coming up with a solution, it is very likely that more confident students spent more time in goal-directed ideation to come up with problem solutions as compared to their peers. However, further work should focus on identifying this relationship more concretely.

5.2   Implications for Instructors

The model presented here also represents a data-driven objective method for instructors to examine and evaluate students' metacognitive self-confidence. With further development, this kind of evaluation could allow instructors to provide more in-depth debriefing and targeted interventions to improve self-confidence, especially for students who have low confidence. Extending this idea, the work is a small step toward a more holistic objective assessment of performance. By aiding instructors' evaluations using data-driven assessments, bias and errors in subjective judgment can be reduced, and the burden of assessment on instructors can be lessened. While self-confidence is only one measure that such data-driven assessments would generate, this work helps to illustrate the longer-term goal and demonstrate that such assessments can be made with multimodal data.

6   Conclusions

In this paper, we showed how multimodal data can be leveraged to model students' self-rated metacognitive confidence scores that are connected to their ability to make metacognitive judgments of their performance. Some limitations of the current study include the small sample size for training the model, as well as the lack of demographic data. In order to show the generality of the methods, future work should repeat this modeling with more students, including students from different populations. Since this model combines self-report with objective measurement, such larger populations would present an excellent opportunity to study diversity and inclusion issues in nursing education. Additionally, future work should apply predictive modeling to other performance concepts, which would allow for a more holistic automated assessment of nurse performance.

References

1. Burns, K.M., Burns, N.R., Ward, L.: Confidence-more a personality or ability trait? it depends on how it is measured: A comparison of young and older adults. Front. Psychol. **7**, 518 (2016)
2. Delgado-Panadero, A., Hernández-Lorca, B., García-Ordás, M.T., Benítez-Andrades, J.A.: "Implementing local-explainability in gradient boosting trees: Feature contribution. Inform. Sci. **589**, 199–212 (2022)
3. Durlach, P.J., Lesgold, A.M.: Adaptive technologies for training and education. Cambridge University Press (2012)
4. Labrague, L.J., McEnroe-Petitte, D.M., Bowling, A.M., Nwafor, C.E., Tsaras, K.: High-fidelity simulation and nursing students' anxiety and self-confidence: A systematic review. In: Nursing Forum, vol. 54, pp. 358–368. Wiley (2019)
5. Lundberg, K.M.: Promoting self-confidence in clinical nursing students. Nurse Educ. **33**(2), 86–89 (2008)
6. Vatral, C., Biswas, G., Cohn, C., Davalos, E., Mohammed, N.: Using the dicot framework for integrated multimodal analysis in mixed-reality training environments. Front. Artifi. Intell. **5**, 941825 (2022)
7. Vatral, C., et al.: A tale of two nurses: Studying groupwork in nurse training by analyzing taskwork roles, social interactions, and self-efficacy. In: 2023 International Conference on Computer Supported Collaborative Learning (2023), (In Press)
8. Wagner, J., et al.: Dawn of the transformer era in speech emotion recognition: closing the valence gap (2022). https://doi.org/10.48550/ARXIV.2203.07378
9. Walcher, S., Körner, C., Benedek, M.: Looking for ideas: Eye behavior during goal-directed internally focused cognition. Conscious. Cogn. **53**, 165–175 (2017)
10. Zacks, J.M., Swallow, K.M.: Event segmentation. Curr. Dir. Psychol. Sci. **16**(2), 80–84 (2007)

Learning from Auxiliary Sources in Argumentative Revision Classification

Tazin Afrin[(✉)] and Diane Litman

University of Pittsburgh, Pittsburgh, PA 15260, USA
{taa74,dlitman}@pitt.edu

Abstract. We develop models to classify desirable reasoning revisions in argumentative writing. We explore two approaches – multi-task learning and transfer learning – to take advantage of auxiliary sources of revision data for similar tasks. Results of intrinsic and extrinsic evaluations show that both approaches can indeed improve classifier performance over baselines. While multi-task learning shows that training on different sources of data at the same time may improve performance, transfer-learning better represents the relationship between the data.

Keywords: Writing · Revision · Natural Language Processing

1 Introduction

Our research focuses on the *automatic classification of desirable revisions of reasoning*[1] in argumentative writing. By reasoning, we refer to how evidence is explained and linked to an overall argument. Desirable revisions (e.g., reasoning supporting the evidence) are those that have hypothesized utility in improving an essay, while undesirable revisions do not have such hypothesized utility [1]. Identifying desirable revisions should be helpful for improving intelligent feedback generation in automated writing evaluation (AWE) systems [5]. In this study, we focus on improving our model learning by taking advantage of auxiliary data sources of revisions. For example, we would like to see how college-level essay data might be beneficial for elementary-school essays. We train two types of models – a *multi-task learning* (MTL) model, and a *transfer learning* (TL) model. In our MTL experiment, we allow information sharing during training using different source data. In our TL experiment, we fine-tune a model pre-trained on source data to see which type of source data might improve performance on the target data. Our results show that both MTL and TL are beneficial for the datasets written by more novice writers. However, for more expert writers (e.g., college students), it is difficult to further improve classifier performance.

Prior Natural Language Processing (NLP) research in academic writing has focused on classifying argumentative revision purposes [8] and understanding

---

Supported by the National Science Foundation under Grant #173572.
[1] Such revisions of text *content* are considered more useful in revising [8].

N. Wang et al. (Eds.): AIED 2023, CCIS 1831, pp. 272–277, 2023.
https://doi.org/10.1007/978-3-031-36336-8_42

Table 1. Comparison of essay corpora used in this study.

Corpus	#Students	Grade Level	Feedback Source	Essay Drafts Used	Essay Score Range	Improvement Score Range
E	143	5th & 6th	AWE	1 and 2	[1, 4]	[0, 3]
H1	47	12th	peer	1 and 2	[0, 5]	[−2, +3]
H2	63	12th	peer	1 and 2	[17, 44]	[−14, +12]
C	60	college	AWE	2 and 3	[15, 33]	-1, +1

revision patterns [5]. While some have classified revisions in terms of *quality* [7], [2] were the first to consider a revision's utility in improving an essay in alignment with previously received feedback. They also investigated state-of-the-art models to identify desirable revisions of evidence and reasoning on three different corpora [1]. We extend their revision annotation framework on one additional corpus, as well as *leverage the corpora as auxiliary sources for MTL and TL*. MTL in NLP is widely used to learn from a limited data source (e.g., [6] used the same primary task of argument classification for multiple low-resource datasets framed as MTL). Following [6], we explore MTL to classify desirable reasoning revisions for multiple datasets. To the best of our knowledge this is the *first exploration of MTL for revision classification*. Transfer learning in NLP is used to reduce the need for labeled data by learning representations from other models [3,4]. Unlike previous works, we first train our model using *source revision data*, then fine-tune the model for a *target revision data*.

2 Data and Annotations

Our data consists of reasoning revisions from four corpora of paired drafts of argumentative essays used in previous revision classification tasks [1,8]. All essays were written by students in response to a prompt, revised in response to feedback, and graded with respect to a rubric. A corpus comparison is shown in Table 1. Their diversity along multiple dimensions makes it challenging to train one model for all. However, since our target is to classify revisions following one annotation framework, it is compelling to investigate how these datasets might be related. In the *elementary* (E) school corpus, students wrote Draft1 about a project in Kenya, then received feedback from an automated writing evaluation (AWE) system. All essay pairs were later graded on a scale from 0 to 3 to indicate improvement from Draft1 to Draft2 in line with the feedback [2]. Two corpora contain essays written by *high-school* students and revised in response to peer feedback – H1 and H2. Drafts 1 and 2 of each high-school corpus were graded using separate rubrics by expert graders. We create an improvement score for each essay pair, calculated as the difference of the holistic score between drafts. The *college* (C) essays unlike the other corpora involving proprietary data, were downloaded from the web [8]. Students received general feedback after Draft1, revised to create Draft2, then revised again to create Draft3 after receiving essay-specific feedback from an AWE system. We create a binary improvement score, calculated as 1 if Draft3 improved compared to Draft2, -1 otherwise.

Table 2. Average number of revisions over 10-fold cross-validation is shown before and after data augmentation (D = Desirable, U = Undesirable).

		Before Augmentation			Augmented for MTL			Augmented for TL		
Corpus	N	Total	D	U	Total	D	U	Total	D	U
E	143	389	186	203	5120	2376	2744	7725	3881	3844
H1	47	387	202	185	5120	2750	2370	5780	2963	2817
H2	63	329	169	160	5120	2770	2350	10986	5997	4989
C	60	207	114	93	5120	2894	2226	5515	3186	2329

For all essays in each corpus, sentences from the two drafts were aligned manually based on semantic similarity. Aligned sentences represent one of four operations between drafts – no change, modification, sentence deleted from Draft1, sentence added to Draft2. Each pair of changed aligned sentences was then extracted as a *revision* and annotated for its *purpose* (e.g., revise reasoning), using the scheme introduced for the college corpus [8]. From among the full set of annotations, we only use reasoning revisions for the current study because they are the most frequent across the four corpora. The reasoning revisions were then annotated for its desirability [1]. We leverage the annotated E, H1, and C data from the previous study [1] and extend the annotation of a new data, H2.

Deep learning requires more than our limited amount of data for training. We use the synonym replacement data augmentation strategy from our prior work [1] to generate more training examples. Since MTL is trained batch by batch for each data in a round robin fashion, we selected a fixed number of instances from each dataset to stay consistent. TL used all available data. Table 2 shows the number of desirable and undesirable revisions for each corpus.

3 Models

Single-Task Learning Model (STL). Our STL model is a neural network model used in previous desirable revision classification task [1]. The input to the model is the revision sentence pair. The model uses the pre-trained BERT ('bert-base-uncased') embedding with a BiLSTM and a Dense layer. The output is a sigmoid activation function for the binary classification task. Classifying desirable reasoning in each corpus is considered an individual task due to the difference in corpora summarized in Table 1. Following previous work, we also select the learning rate $1e^{-3}$ and batch size 16, and apply the same to all data.

Multi-task Learning Model (MTL). The individual tasks in STL are combined in MTL with a *shared BiLSTM layer*. After encoding the revision using the off-the-shelf BERT encoder, we send this to the BiLSTM layer. The BiLSTM layer learns shared information between different tasks. Each task has an individual Dense layer and a Sigmoid output layer to learn task-specific information. During training, we use the same settings as in STL. In MTL, we train the model in the sequence of C, H1, H2, and E data in a round robin fashion for each batch.

This sequence is repeated for all the batches for 10 epochs. Since our batch size is very small, we believe the training will not be affected substantially by the order of selecting the data. We apply 10-fold cross-validation. During testing, we use the respective task for the respective data.

Union Model. Unlike STL (where we train a separate model for each of the four corpora/tasks), for the Union Model we train only one STL model. We use the union of all task data as input. The training is performed following the MTL (e.g., batch by batch) training process. Compared to STL, the Union model will help us understand if using extra data as a source of information is beneficial. Comparison of the Union and MTL models will help us validate that any MTL improvement is not just due to more training data.

Transfer Learning Models (TL). In transfer learning, we learn from a source dataset and apply it to a target dataset to understand the relationship between our data, which may not be obvious from MTL. The source and target data are taken from all possible combinations of our datasets (Table 1). TL also adopts the STL model to first train with the source data, then fine-tune with the target data using all the augmented data available (shown in Table 2).

4 Results

In our *intrinsic* evaluation, classification performance was compared to baseline models in terms of average unweighted F1-score over 10-folds of cross-validation. We compare MTL with two baselines (STL and Union), while TL was compared against one baseline (STL trained on target data). *Extrinsic* evaluation checked how often desirable and undesirable revisions (gold annotations) are related to improvement score using Pearson correlation. We then replicate the process for the predicted revisions to see if they are also correlated in the same way.

MTL Evaluation. Intrinsic evaluation in Table 3a shows that in-general MTL has higher average f1-scores. However, MTL and baseline results are close with no significant difference. Further investigation showed MTL outperformed baselines in identifying undesirable revisions. MTL also showed improvement over Union baseline, indicating that MTL's success over STL is not just due to more training data. Union performed worse than STL, emphasizing the importance of data usage. In extrinsic evaluation, MTL showed significant positive correlations for predicted desirable reasoning for the E data (Table 3b), which is consistent with the Gold correlation. In other cases, either MTL is not consistent with Gold, or the correlation is not to be significant. In contrast, both STL and Union often showed significant correlation to essay improvement. Our results suggest treating our datasets as individual tasks to better relate to student writing improvement (extrinsic evaluation). However, we found sharing features (via MTL) useful for identifying desirable reasoning (intrinsic evaluation).

TL Evaluation. Elementary-school students can be considered as the least experienced writers in our datasets considering the age group. Hence the result

Table 3. MTL and TL evaluation. Best results are bolded. ↑ indicates TL improved over STL. Extrinsic evaluation shows Desirable results only. Significant predicted correlations consistent with using gold labels are bolded. * p< .05.

(a) MTL Intrinsic Evaluation

Corpus	STL	Union	MTL
E	0.597	0.583	**0.607**
H1	**0.649**	0.631	0.627
H2	0.633	0.622	**0.658**
C	0.613	0.539	**0.619**

(b) MTL Extrinsic Evaluation

	Gold	STL	Union	MTL
E	**0.450***	**0.339***	**0.347***	0.317*
H1	**0.351***	0.249	0.266	0.222
H2	**0.301***	**0.274***	**0.300***	0.232
C	0.029	0.039	-0.057	0.003

(c) TL Intrinsic Evaluation.

		Target		
	E	H1	H2	C
STL	0.597	0.649	0.633	0.613
Source E		0.661↑	**0.652↑**	0.606
Source H1	0.607↑		0.636↑	**0.644↑**
Source H2	0.606↑	**0.678↑**		**0.644↑**
Source C	**0.641↑**	0.638	0.598	

(d) TL Extrinsic Evaluation.

		Target		
	E	H1	H2	C
Gold	**0.450***	**0.351***	**0.301***	0.029
STL	**0.339***	0.249	**0.274***	0.039
Source E		0.262	**0.262***	0.033
Source H1	0.337*		**0.308***	0.008
Source H2	0.360*	0.376*		-0.060
Source C	0.350*	0.292*	0.250*	

in Table 3c may indicate that model for elementary-school students needed to learn the structure from better-written essays. Unlike how the H2 data as a source helped the H1 data as the target, the reverse is not entirely true. Moreover, transfer from C decreased performance for H2. Finally, for C data, transfer learning the weights from both high-school datasets helped improve performance over the baseline (trained only on the target data). Although E as the target domain improved most when learning from the C data, the reverse is not true. College-level students were comparatively experienced writers in our corpora, so inexperienced student writing may have not helped.

Table 3d shows that when our target is the E data, TL results are consistent with Gold annotation results for desirable reasoning. Undesirable revisions are not significant. When H1 is the target, H2 yields the highest correlation, which might be because it is also high-school data. C data also showed significant correlation in this case. While all models are consistent with Gold annotation for H2 as the target, H1 showed the highest correlation of desirable revision to essay improvement score. This is surprising because H1 did not improve in intrinsic evaluation. Finally, C as target did not see any significant correlations.

Overall, transfer learning shows that the availability of more data or more information is not enough. Rather, which data is used to pre-train or how it is being used to train the model (e.g., MTL) is also important. Extrinsic evaluation also supports the fact that more data does not mean improvement. For example, transfer from other high-school data yield stronger results for H1 or H2 data compared to transfer from E or C data. Overall, our results from the transfer learning experiments show that for each target data there were one or more source corpora that improved the classifier performance.

5   Discussion of Limitations and Conclusion

Our corpora were originally annotated using a detailed revision scheme [2], then used a simplified scheme [1] to create binary revision classification task. In a real-world scenario, an end-to-end AWE system deploying our model would have errors propagated from alignment and revision purpose classification and perform lower than the presented model. Moreover, we need to examine our additional but less frequent revision purposes too. We also plan to explore other data augmentation techniques to experiment with more complex models. Although we do not have demographic information, the students in the college corpus include both native English and non-native speakers [8]. Another limitation is that the MTL model training process is slow. The current methods also require GPU resources. Moreover, we only investigated one sequence of training the MTL.

We explored the utility of predicting the desirability of reasoning revisions using auxiliary sources of student essay data using multi-task learning and transfer learing. Both experiments indicate that there is common information between datasets that may help improve classifier performance. Specifically, the results of our intrinsic and extrinsic evaluations show that while desirable revision classification using auxiliary sources of training data can improve performance, the data from different argumentative writing tasks needs to be utilized wisely.

References

1. Afrin, T., Litman, D.: Predicting desirable revisions of evidence and reasoning in argumentative writing. In: The 17th Conference of EACL (May 2023)
2. Afrin, T., Wang, E.L., Litman, D., Matsumura, L.C., Correnti, R.: Annotation and classification of evidence and reasoning revisions in argumentative writing. In: Workshop on Innovative Use of NLP for Building Educational Applications (2020)
3. Chakrabarty, T., Hidey, C., Muresan, S., McKeown, K., Hwang, A.: AMPERSAND: Argument mining for PERSuAsive oNline discussions. In: Proceedings of the 2019 EMNLP-IJCNLP. ACL (November 2019)
4. Ghosh, D., Beigman Klebanov, B., Song, Y.: An exploratory study of argumentative writing by young students: A transformer-based approach. In: Proceedings of the Fifteenth Workshop on Innovative Use of NLP for Building Educational Applications. ACL (July 2020)
5. Roscoe, R.D., Snow, E.L., Allen, L.K., McNamara, D.S.: Automated detection of essay revising patterns: Applications for intelligent feedback in a writing tutor. Technol. Instr. Cogn. Learn. **10**(1), 59–79 (2015)
6. Schulz, C., Eger, S., Daxenberger, J., Kahse, T., Gurevych, I.: Multi-task learning for argumentation mining in low-resource settings. In: Proceedings of NAACL - HLT, vol. 2 (Short Papers). ACL (June 2018)
7. Tan, C., Lee, L.: A corpus of sentence-level revisions in academic writing: A step towards understanding statement strength in communication. In: Proceedings of the 52nd ACL, vol. 2: Short Papers (June 2014)
8. Zhang, F., Hashemi, H., Hwa, R., Litman, D.: A corpus of annotated revisions for studying argumentative writing. In: Proceedings of the 55th Annual Meeting of the Association for Computational Linguistics, vol. 1: Long Papers (2017)

Exploring the Effect of Autoencoder Based Feature Learning for a Deep Reinforcement Learning Policy for Providing Proactive Help

Nazia Alam[✉], Behrooz Mostafavi, Min Chi, and Tiffany Barnes

North Carolina State University, Raleigh, NC 27695, USA
{nalam2,bzmostaf,mchi,tmbarnes}@ncsu.edu

Abstract. Providing timely assistance to students in intelligent tutoring systems is a challenging research problem. In this study, we aim to address this problem by determining when to provide proactive help with autoencoder based feature learning and a deep reinforcement learning (DRL) model. To increase generalizability, we only use domain-independent features for the policy. The proposed pedagogical policy provides next-step proactive hints based on the prediction of the DRL model. We conduct a study to examine the effectiveness of the new policy in an intelligent logic tutor. Our findings provide insight into the use of DRL policies utilizing autoencoder based feature learning to determine when to provide proactive help to students.

Keywords: Intelligent Tutoring Systems · Deep Reinforcement Learning · Autoencoder

1 Introduction

Determining when to provide assistance or help is a research problem also known as the assistance dilemma. Providing timely assistance is important as students are not always able to effectively seek help when working with intelligent tutors [3]. If student's do not get timely help as they need, it can obstruct their overall learning process. Effective prediction of when students need help and providing assistance accordingly can play a crucial role in helping students learn better. There have been a few studies on when to provide proactive hints to students. In one of the latest studies Maniktala et al. [5] proposed the HelpNeed (HN) model which predicted when to provide proactive hints to students and provided next-step hints accordingly. The results showed improvement in student performance when students were provided with next-step hints based on prediction from the HN model. However, the HN model used domain-independent features as well as domain-dependent features which makes it not easily generalizable to other

This material is based upon work supported by the National Science Foundation under Grant No. 726550 and 2013502.

domains. We are interested in developing a model which will be generalizable to different domains of intelligent tutors.

In this study, we aim to address the assistance dilemma for the open-ended domain of logic. Deep reinforcement learning (DRL) has been used to achieve phenomenal results in many domains, including intelligent tutoring systems [1,2]. For this study, we investigate the use of a DRL policy to determine when to provide proactive help to students. Using features that are good representations of the student state is vital for any pedagogical policy. Autoencoder models have been successfully used in many domains for feature learning and feature transformation leading to reduced input state space and capturing only the most relevant information [4,7,9]. Here we explore the use of autoencoder models for effective feature learning. Overall, the main contribution of our work is a pedagogical policy to determine when to provide proactive hints to students and provide next-step hints accordingly.

2 Methods

We use an intelligent logic tutor to conduct the experiments. The tutor is given to undergraduate students of a discrete mathematics course in a public university in the US. The tutor is for practicing formal propositional logic proofs. There are four sections in the tutor- introduction, pretest, training, and posttest. Students receive proactive hints only in the training section problems.

Proposed Prediction Model (AE-DRL). Reinforcement learning (RL) deals with how an intelligent agent takes action to maximize cumulative reward in an environment. Q-learning is a model-free RL algorithm that tries to find an optimal policy that maximizes the expected cumulative reward, and DQN [6] is an off-policy variant of the Q-learning model. A variation of DQN is double DQN [8] which uses two different neural networks for action selection and action evaluation. We use a off-policy and offline version of the double DQN model for our experiments. To develop a policy to determine when to provide proactive help to students, we define a Markov decision process. We define the components of the process as below:

State: The state is made of domain-independent student log features. There are 42 features in the log data for each of the students.

Action: At each step of the tutor, there are two possible actions: 1) provide proactive hint 2) do not provide a proactive hint.

Reward: The reward is a combination of the time and steps taken, accuracy, and the number of hints already received. It is designed such that the reward is of higher value when the accuracy is high and the number of hints received, time, and steps taken are low.

$$Reward = \frac{accuracy}{number\ Of\ Hints\ Received\ *\ time\ *\ number\ Of\ Steps}$$

Autoencoder (AE) is a neural network model used to learn a compressed representation of input data. It consists of an encoder which compresses the input data into a latent space representation and a decoder whose purpose is to regenerate the input data from the latent space representation. The encoder part of the autoencoder creates a compressed representation of the input data which contains the most salient features of the data and ignores the insignificant data. The compressed representation with the salient features when used as state for DRL model should help the model focus on the most relevant aspects of the student state and desensitize the irrelevant and insignificant parts. With the use of autoencoders, we are hoping to develop feature sets that are better representative of student states so that we can have more effective policies. The autoencoder model takes the student features as input and creates a compressed representation which is then taken as input by the DQN model and a prediction is generated. In each step of a problem, the AE-DRL model makes a prediction on whether a student needs help in that particular step. Based on the prediction, a next-step hint is given proactively to students.

Research Questions. We want to understand whether the DRL model trained on autoencoder generated latent features is effective in determining when to provide proactive help to students, and thus help students learn better and improve students' performance. Therefore we have the following research questions:

- **RQ1:** Do the adaptive condition students have better posttest performance compared to the control condition?
- **RQ2:** Do the adaptive condition students have comparable posttest performance compared to the HN condition?

3 Experimental Design

We conducted two experiments in our study to answer the two research questions. For the first experiment, we conducted a controlled study where the adaptive condition students received next-step hints upon prediction from AE-DRL policy in the training section of the tutor. The control condition students did not receive any proactive hints. However, students from both conditions can request on-demand hints. For the adaptive condition, the next-step hints are shown as a message in the bottom part of the tutor. For the second experiment, the adaptive condition students received next-step hints upon prediction from AE-DRL policy in the problem-solving part of the training section of the tutor, and the HelpNeed (HN) condition students received next-step hints upon prediction from HN policy. For the adaptive condition, the hints were shown as a message in the bottom part of the tutor. For the HelpNeed condition, the hints were shown inside the tutor interface as a goal that the students need to achieve.

For experiment 1, the number of participants was 52 and 54 in the control and adaptive condition, respectively. For experiment 2, the number of participants was 77 and 72 in the HelpNeed (HN) and adaptive condition, respectively.

For measuring student performance, we look at the time and number of steps taken. Time is the total time taken to complete a particular section of the tutor. As students may leave the tutor system open, we cap each step time to 5 min.

Table 1. Student performance comparison between the adaptive and the control condition

	Control	Adaptive	High Justification Adaptive
	Mean time (Std Dev) in minute		
Final posttest	58.7 (53.9)	50.5 (32.7)	40.4 (22.1)
Total tutor	262.2 (136.4)	253.2 (88.8)	224.3 (95.4)
	Mean number of steps (Std Dev)		
Final posttest	112.4 (82.8)	98.2 (49.8)	97.9 (46.2)
Total tutor	405.6 (159.9)	385.7 (155.8)	407.7 (193.2)

The number of steps is the sum of the number of statements derived to reach the conclusion for a problem.

4 Results and Discussion

Here we evaluate the performance of the students in different conditions. We perform significance analysis using the Mann-Whitney test.

Experiment 1. Table 1 shows the student performance in the posttest section and total tutor for the control and the adaptive condition. For the posttest section, we see that the average time and number of steps in the adaptive condition appear to be less than the control condition. However, it is found that the differences are not significantly different ($p\text{-}value > 0.05$).

Next, we want to analyze how using the given proactive hints affects student performance. Therefore, we look at the students who justified (used the given hint in the solution) the given proactive hint within 2 steps of receiving them. Based on the number of times students justified the given proactive hints within the next 2 steps, we divided the students into two groups: high and low justification groups. We are particularly interested in the high justification group as this group highlights how noticing and using the given proactive hints affect student performance. We compare this group's performance to the control group. It can be seen from Table 1 that the average tutor time in the high justification adaptive group is 38 min lower than the control group and 29 min lower than the overall adaptive group. However, we do not find any significant difference ($p\text{-}value > 0.05$) between the different groups in terms of time and number of steps taken.

Experiment 2. We evaluate the performance of the adaptive condition compared to the HelpNeed condition, shown in Table 2. We find a significant difference ($p\text{-}value < 0.05$) between the performance of the two conditions in the final posttest section for both time and number of steps. One thing to note is that the way the hints were given in the adaptive and the HN conditions were different. The hints were shown as a message outside the screen in the adaptive

Table 2. Student performance comparison between the adaptive and the HelpNeed condition

	HelpNeed	Adaptive	High Agreement Adaptive
	Mean time (Std Dev) in minute		
Final posttest	49.3 (34.5)	64.7 (38.5)	71.1 (33.8)
Total tutor	239.1 (89.6)	290.5 (105.4)	282.6 (86.4)
	Mean number of steps (Std Dev)		
Final posttest	112.7 (87.9)	115.0 (62.2)	124.1 (59.6)
Total tutor	381.0 (140.2)	430.3 (208.4)	395.8 (130.8)

condition, and the hints were shown inside the screen as a popup in the Help-Need condition. As a result of that, there might have been an effect of how the hints were shown on the student performance of both conditions.

Next, We try to determine the agreement between the two hint-giving policies. We run the HelpNeed policy in an offline setup on the data from the adaptive condition to get the HelpNeed predictions. Then we match the predictions from AE-DRL and HelpNeed policy. For each student, we find the percentage of agreement between the predictions from AE-DRL and the HN policy. We find the median percentage of agreement, and all the students who have a higher agreement than the median are put in the high agreement group. There should be no significant difference in performance between these high agreement group students and HN condition students. If there is any difference then it is possible that it is because of the difference in how the hints were shown. When comparing the performance between the high agreement group and the HN condition group, we still find that there is a significant difference ($p\text{-}value < 0.05$) between the adaptive and HN conditions in the final posttest section.

Discussion. We compared the performance of students in the adaptive and the control conditions in terms of time and steps taken. However, we did not find any significant difference. Next, we looked at the high justification group (students who used the given proactive hints within the next 1–2 steps a high number of times) in the adaptive condition. These are the students who actually noticed and utilized the proactive hints given in the training section. We find that the high justification students used even less time and steps in the posttest section compared to the overall adaptive condition. However, there is no significant difference compared to the control group. These results do not provide conclusive evidence that adaptive condition student performance is better than the control group.

From the comparison of the adaptive and the HelpNeed condition, we find that the HelpNeed condition performed significantly better in posttest section of the tutor. However, there was a difference in how the hints were shown in the adaptive and HelpNeed conditions. The way proactive hints were shown in the HelpNeed condition was more noticeable and more likely to be noticed and used by the students. From a comparison of performance between the high agreement group in the adaptive condition and the HelpNeed condition, we find that the

HelpNeed condition students were performing better compared to the adaptive condition high agreement students. Therefore, it is likely that the way the hints were shown in the adaptive condition is a reason for the inferior performance compared to the HN condition. In the future, we plan to repeat the experiment while giving proactive hints from both policies in the same way.

5   Conclusion

In this paper, we proposed a pedagogical policy to determine when to provide proactive hints in an intelligent logic tutor in order to address the assistance dilemma. The pedagogical policy made use of autoencoder based feature learning and DRL model to make the prediction. We conducted a study comparing the proposed policy against a control policy which did not provide any proactive hints, and against a state-of-the-art pedagogical policy to determine when to provide proactive hints. The findings demonstrate the effect of using DRL policy with autoencoder based feature learning for determining when to provide proactive help and shows that there is need to do further research in order to improve.

References

1. Abdelshiheed, M., Hostetter, J.W., Barnes, T., Chi, M.: Bridging declarative, procedural, and conditional metacognitive knowledge gap using deep reinforcement learning. In: CogSci (2023)
2. Abdelshiheed, M., Hostetter, J.W., Barnes, T., Chi, M.: Leveraging deep reinforcement learning for metacognitive interventions across intelligent tutoring systems. In: AIED (2023). https://www.aied2023.org/accepted_papers.html, https://arxiv.org/pdf/2304.09821.pdf
3. Aleven, V., McLaren, B., Roll, I., Koedinger, K.: Toward meta-cognitive tutoring: a model of help seeking with a cognitive tutor. Int. J. Artif. Intell. Educ. **16**(2), 101–128 (2006)
4. Ding, M., Yang, K., Yeung, D.Y., Pong, T.C.: Effective feature learning with unsupervised learning for improving the predictive models in massive open online courses. In: Proceedings of the 9th International Conference on Learning Analytics & Knowledge, pp. 135–144 (2019)
5. Maniktala, M., Cody, C., Isvik, A., Lytle, N., Chi, M., Barnes, T.: Extending the hint factory for the assistance dilemma: a novel, data-driven helpneed predictor for proactive problem-solving help. arXiv preprint arXiv:2010.04124 (2020)
6. Mnih, V., et al.: Human-level control through deep reinforcement learning. nature **518**(7540), 529–533 (2015)
7. Nishizaki, H.: Data augmentation and feature extraction using variational autoencoder for acoustic modeling. In: 2017 Asia-Pacific Signal and Information Processing Association Annual Summit and Conference, pp. 1222–1227. IEEE (2017)
8. Van Hasselt, H., Guez, A., Silver, D.: Deep reinforcement learning with double q-learning. In: Proceedings of the AAAI Conference on Artificial Intelligence (2016)
9. Yousefi-Azar, M., Varadharajan, V., Hamey, L., Tupakula, U.: Autoencoder-based feature learning for cyber security applications. In: 2017 International Joint Conference on Neural Networks (IJCNN), pp. 3854–3861. IEEE (2017)

Who and How: Using Sentence-Level NLP to Evaluate Idea Completeness

Martin Ruskov[✉][ID]

Department of Languages, Literatures, Cultures and Mediations, Università degli
Studi di Milano, 20123 Milan, Italy
martin.ruskov@unimi.it

Abstract. Real-time feedback is very important, yet challenging to provide for free-text learner contributions in Technology-Enhanced Learning. We study whether a generic NLP pipeline can identify completeness
features of learner ideas during security training. We apply PoS Tagging and Dependency Parsing on contextualised short texts, collected
within a dedicated learning environment and we compare the results to
an expert-annotated ground truth. We scan these contributions for the
absence of responsible stakeholder (*who*) or featured action (*how*). A
total of 1174 contributions in two security domains were analysed. We
report precision on *who* ($PPV = 0.929$) and on *how* ($PPV = 0.691$).
We consider the first result to be sufficient to provide real-time formative feedback for the case of absent *who*. Our results suggest that for
the purposes of providing feedback in free input problem-solving exercises, generic transformer pipelines without fine-tuning can achieve good
performance on stakeholder identification.

Keywords: Dependency parser · PoS tagger · Real-time feedback ·
Technology-enhanced learning · Security training

1 Introduction

Real-time feedback is important in order to allow Technology-Enhanced Learning to scale and become accessible to wider audiences. However, providing such
feedback is not straightforward in learning experiences where interactions are
complex and are subject to interpretation. One particular challenge in such a
context is assessing the quality in terms of actionability of learner contributions
proposed during training. The CCO Toolkit is an online environment engaging learners to analyse recurring information security problems and write short
texts to propose potential interventions to contribute to a complex solution [9].
To illustrate the difficulty of assessing such interventions, consider a learner contribution formulated as "CCTV" (meaning video surveillance). It is generally a
meaningful intervention, yet it is too vague and thus hard to put in practice.
Rather, a more complete and actionable intervention would specify who is supposed to do it and how, e.g. "security department needs to identify risk hotspots
for CCTV". Grammatically, *how* is provided by the verb, and *who* – by the
subject of a sentence.

N. Wang et al. (Eds.): AIED 2023, CCIS 1831, pp. 284–289, 2023.
https://doi.org/10.1007/978-3-031-36336-8_44

In this paper, we investigate whether grammatical classification performed by a generic NLP transformer pipeline[1] can be used for real-time assessment of completeness of contributions made within the CCO Toolkit. We define two specific binary classification tasks: absence of a relevant action (*how*) and absence of a corresponding responsible stakeholder (*who*). To address these tasks, we propose two corresponding simple grammatical heuristics, meant to be integrated into the CCO Toolkit to provide real-time feedback to learners and we evaluate the approach on contributions, generated by students using the toolkit in a class setting. Our approach does not rely on any fine-tuning of the transformer-based pipeline. Thus, we expect it to be generalisable to other contexts that employ problem-solving via short text contributions.

2 Background

In crime science, Ekblom defines 8 prevention competences to include *know who to involve* and *know how to put in practice* [4]. More generally, in research on brainstorming, one proposed set of measures for completeness of generated ideas are the 5Ws+H questions – *who, what, why, when, where* and *how* [3]. In either case, identifying relevant actions (*how*) and responsible stakeholders (*who*) for practical ideas is important to make them complete and actionable.

Recent research in Automated Short Answer Grading typically uses lexical semantic similarity to known solutions, but this approach is known to perform poorly on free-text input [8]. This limitation becomes even more prominent in contexts where idea innovation is sought to be encouraged, as is in our case. In one notable exception to the lexical trend, Bagaria et al. [2] work towards the identification and extraction of triplets of *subject-verb-object* from learner-generated sentences, but do not measure the accuracy of their approach.

Broader advancements in NLP offer a grammatical toolset that could potentially address the challenge of providing feedback in unconstrained discussion. In particular Dependency Parsers and PoS Taggers could allow for partial real-time interpretation of learner contributions. Using transformers, Gupta and Nishu [6] have developed custom heuristics regarding location (*where*) extraction from news articles. Kalia et al. perform verb extraction with dependency parsers in another domain and report somewhat positive results [7]. Grammatical analysis has been used to interpret information security reports [5] and software requirements [1].

3 Approach

We propose heuristics detecting if learner contributions in the CCO Toolkit lack action (*how*) and stakeholder (*who*). The toolkit guides learners through a staged brainstorming process with the goal to solve given security problems. The process starts by introducing learners to a problem scenario. Then it asks learners to

---

[1] Pipeline documentation at: https://spacy.io/models#design-trf.

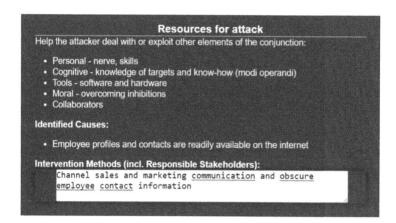

Fig. 1. The prompt in the CCO toolkit asking for intervention ideas.

first identify causes, and then – interventions that could be part of a solution (see Fig. 1). Some other steps follow and at the end learners are provided with feedback and score of their own intervention ideas, based on automated analysis and feedback by peers. For a better explanation of the toolkit, please refer to Ruskov et al. [9]. The combination of the underlying theoretical model, the scenarios and the causes previously identified by learners [10] define a specific context for each contribution, which is only implicitly referenced by the contribution itself.

We analyse a dataset[2] consisting of 1174 intervention contributions written in English by 91 graduate students from two universities. The contributions were generated for 3 different problem scenarios – two from information security ($n = 1041$) and one from community safety ($n = 133$). Their length had a median of 58 characters, and a range from 4 (see S6 in Table 1) to 253. To determine the ground truth, two security experts annotated interventions indicating the presence of *how* and *who*. The *featured action* (74.9% of the entire dataset) annotation specifies whether the intervention contains *how* it is to be applied? Typically – but not always – this is indicated be the verb in one of the contained sentences (e.g. S3 from Table 1, S1 is an example of a verbless indication of *how*). Counterexamples are contributions that are not actionable (S6, S7), possibly describing what is to be achieved, but not how. The *responsible stakeholder* (17.3%) annotation concerns whether it is clear (e.g. S1) or not (S2) *who* is enacting the intervention. Experts were also asked to indicate *ungrammatical* contributions (12.3% from the dataset), commonly caused e.g. by implicit missing parts (S8) or by mistakes (S3). Some contributions were marked as *ambiguous* by experts (30.5%) when they were considered difficult to interpret, e.g. due to punctuation (S3) or being too generic (S2). A cross-annotator comparison was made and differences were discussed between experts to find an agreement. In certain cases, the agreement led to a decision that the contribution is ambiguous.

---

[2] Full dataset and heuristics code at: https://cco.works/opendata.

Table 1. Examples of actual proposed contributions with ground truth annotations. Key: Absence of (H)ow and (W)ho; (U)ngrammatical; (A)mbiguous.

#	Sample	Annotation			
	Intervention	W	H	U	A
S0	discourage staff from committing crime by regularly giving incentives or bonus. hr will be responsible for that				
S1	Up to date checks on potential offenders so they will be stopped from being in proximity of a crime. Security staff, HR managers, IT staff responsible				
S2	bringing awareness of control		✓		✓
S3	Educating employees on the law/privacy this should be implemented by managers			✓	✓
S4	tighter security methods implemented by network administrator and engineers				
S5	Checks to ensure only the most enthusiastic and genuine people are recruited	✓	✓		
S6	CCTV	✓	✓		
S7	Better time-Management	✓	✓		✓
S8	extensive updates, so few are enough	✓	✓	✓	✓
S9	career management team	✓			

We experiment with a solution for two corresponding identification tasks: of the featured action (*how*) and of the corresponding responsible stakeholder (*who*). For each task, using a transformer-based pipeline, we perform PoS Tagging and Dependency Parsing on each intervention and are interested in the absence of the relevant information. Our heuristic algorithms consider candidates from all possible sentences in the contribution. As seen in Table 2, candidates for a relevant action (*how*) are verbs in the sentence `root` (active voice) or parents of an `agent` (passive voice). Candidates for a responsible stakeholder (*who*) are tokens that are 1) `subjects`, but not `pronouns` (active), or 2) `nouns`, child of an `agent` (passive).

4   Experiments

We compare the results of our *how* and *who* heuristics to the expert-annotated ground truth. The CCO Toolkit is intended to provide feedback only when the features of interest are missing from a contribution. As a consequence, our true positives are interventions where neither the ground truth, nor the algorithm identify presence of our answers of interest. Thus, precision (PPV) is the count of correctly classified missing information in contributions over all where the heuristic does not detect it and recall (TPR) is the same value over all contributions where the heuristic classification matches the ground truth. Also, because the objective is to signal only when the features are missing, precision is more important to us than recall. Correspondingly, we employ a generalised F_β-score with $\beta = 0.5$. Due to an imbalanced dataset, we also report the calculated balanced accuracy (bACC).

Results are shown in Table 3. Our heuristics achieve a precision of ($PPV = 0.929$) on the *who* task and ($PPV = 0.691$) on the *how* task. Even the

Table 2. Classification heuristics.

Task	Voice	PoS	Dependency
How	active	VERB	ROOT
	passive	VERB	parent of agent
Who	active	not PRON	nsubj
	passive	NOUN	child of agent

Table 3. Overview of result metrics.

Task	Dataset	PPV	TPR	$F_{0.5}$	bACC
How	**full (1174)**	0.691	0.712	0.695	0.802
	clean (747)	0.738	0.771	0.744	0.844
Who	**full (1174)**	0.929	0.841	0.910	0.768
	clean (747)	0.963	0.856	0.940	0.830

lower score of the latter task outperforms the scores of all heuristics reported by Kalia et al. [7] on their respective different domain and with their bespoke ensemble approach. These results are encouraging, considering that our heuristics use untuned general-purpose models on a real-world dataset. To assess the impact of contributions that experts assessed as ungrammatical or ambiguous, we also report the performance of the heuristics on a part of the dataset ($n = 747$) that excludes these noisy contributions. As expected, the performances were better on the clean data subset. Results were consistent across the two application domains of information security and community safety.

The heuristics identify correctly contributions containing grammatically correct phrases with subject and verb (e.g. S0 from Table 1). When the verb – but not the subject – is present, it is also correctly identified. When a verb is omitted in a contribution (S1), interpretation becomes more difficult. In particular, when a contribution contains only a noun phrase, the heuristics fail to distinguish whether it is a subject (S9) or object (S6). In cases when passive voice was used (S4), the subject was also correctly identified via the agent dependency. Cases of verb nominalisation are a common reason for the heuristic's inability to identify an action that is present, but not put down explicitly as a verb. We consider examples of deverbial nouns (S7), zero-derivation (S8), and gerund (S2, S3). Our heuristic does not identify actions written as deverbial nouns. Due to the exact syntactical correspondence between verb and noun, in some cases of zero-derivation (S8) and gerund (S2) nouns are mistakenly identified as verbs, which leads to the heuristic detecting the presence of *how*. Depending on the situation, this could be correct (S2) or not (S8). For *who* sometimes the object is wrongly identified (e.g. "people" in S5) in a subordinate sentence clause.

5   Future Work

The results clearly indicate that the *who* heuristic is mature enough to be integrated for use in actual learning with the CCO Toolkit to encourage learners as they write. It is probable that in real-world experiences learners will continue to propose incomplete contributions. Yet, we expect that non-intrusive feedback would help learners improve when it is relevant, and would be ignored when perceived as inappropriate. A study with learners is needed to confirm this expectation and to measure the extent to which the new feature would lead them to actually improve the completeness of proposed contributions and to capture their perceptions about the usefulness of the received feedback.

Also, further research could experiment with other pieces of information extraction from learner contributions, measuring other relevant dimensions of idea quality as proposed respectively by Dean et al. [3] and Ekblom [4]. These would not be so directly indicated by grammar, but entity recognition and clustering with previously analysed contributions could enrich the interpretation, as demonstrated by Gupta and Nishu [6]. Last, but not least, given the good performance – particularly on *who* – without any fine-tuning, it is worthwhile measuring how the approach performs in other application contexts and other subject domains.

References

1. Ahmed, S., Ahmed, A., Eisty, N.U.: Automatic transformation of natural to unified modeling language: a systematic review. In: 2022 IEEE/ACIS 20th International Conference on Software Engineering Research, Management and Applications (SERA). IEEE (2022). https://doi.org/10.1109/sera54885.2022.9806783
2. Bagaria, V., Badve, M., Beldar, M., Ghane, S.: An intelligent system for evaluation of descriptive answers. In: 2020 3rd International Conference on Intelligent Sustainable Systems (ICISS), pp. 19–24 (2020). https://doi.org/10.1109/ICISS49785.2020.9316110
3. Dean, D.L., Hender, J.M., Rodgers, T.L., Santanen, E.L.: Identifying quality, novel, and creative ideas: constructs and scales for idea evaluation. J. Assoc. Inf. Syst. **7**(1), 646–699 (2006). https://doi.org/10.17705/1jais.00106
4. Ekblom, P.: Crime Prevention, Security and Community Safety Using the 5Is Framework. Crime Prevention and Security Management. Palgrave Macmillan, UK (2010). https://doi.org/10.1057/9780230298996
5. Gao, P., et al.: Enabling efficient cyber threat hunting with cyber threat intelligence. In: 2021 IEEE 37th International Conference on Data Engineering (ICDE), pp. 193–204 (2021). https://doi.org/10.1109/ICDE51399.2021.00024
6. Gupta, S., Nishu, K.: Mapping local news coverage: precise location extraction in textual news content using fine-tuned BERT based language model. In: Proceedings of the 4th Workshop on Natural Language Processing and Computational Social Science, pp. 155–162. ACL, Online (2020). https://doi.org/10.18653/v1/2020.nlpcss-1.17
7. Kalia, A.K., Batta, R., Xiao, J., Vukovic, M.: Ensemble of unsupervised parametric and non-parametric techniques to discover change actions. In: 2021 IEEE 14th International Conference on Cloud Computing (CLOUD), pp. 572–577 (2021). https://doi.org/10.1109/CLOUD53861.2021.00074
8. Maya, A., Nazura, J., Muralidhara, B.L.: Recent trends in answer script evaluation - a literature survey. In: Proceedings of the 3rd International Conference on Integrated Intelligent Computing Communication & Security, ICIIC 2021, pp. 105–112. Atlantis Press (2021). https://doi.org/10.2991/ahis.k.210913.014
9. Ruskov, M., Celdran, J.M., Ekblom, P., Sasse, M.A.: Unlocking the next level of crime prevention: development of a game prototype to teach the conjunction of criminal opportunity. Inf. Technol. Control **10**(3), 15–21 (2012). http://www.acad.bg/rismim/itc/sub/archiv/Paper3_3_2012.pdf
10. Ruskov, M., Ekblom, P., Sasse, M.A.: Getting users smart quick about security: results from 90 minutes of using a persuasive toolkit for facilitating information security problem solving by non-professionals (2022). https://doi.org/10.48550/ARXIV.2209.02420

Comparing Different Approaches to Generating Mathematics Explanations Using Large Language Models

Ethan Prihar[1]([✉]) [ID], Morgan Lee[1] [ID], Mia Hopman[1] [ID],
Adam Tauman Kalai[2] [ID], Sofia Vempala[1] [ID], Allison Wang[1] [ID],
Gabriel Wickline[1], Aly Murray[3], and Neil Heffernan[1] [ID]

[1] Worcester Polytechnic Institute, Worcester, MA 01609, USA
{ebprihar,mplee,mahopman,svempala,awang9,gwickline,nth}@wpi.edu
[2] Microsoft Research, 1 Memorial Dr, Cambridge, MA 02142, USA
adam.kalai@microsoft.com
[3] UPchieve, Inc., San Francisco, USA
aly.murray@upchieve.org

Abstract. Large language models have recently been able to perform well in a wide variety of circumstances. In this work, we explore the possibility of large language models, specifically GPT-3, to write explanations for middle-school mathematics problems, with the goal of eventually using this process to rapidly generate explanations for the mathematics problems of new curricula as they emerge, shortening the time to integrate new curricula into online learning platforms. To generate explanations, two approaches were taken. The first approach attempted to summarize the salient advice in tutoring chat logs between students and live tutors. The second approach attempted to generate explanations using few-shot learning from explanations written by teachers for similar mathematics problems. After explanations were generated, a survey was used to compare their quality to that of explanations written by teachers. We test our methodology using the GPT-3 language model. Ultimately, the synthetic explanations were unable to outperform teacher written explanations. In the future more powerful large language models may be employed, and GPT-3 may still be effective as a tool to augment teachers' process for writing explanations, rather than as a tool to replace them. The explanations, survey results, analysis code, and a dataset of tutoring chat logs are all available at https://osf.io/wh5n9/.

Keywords: Large Language Models · GPT-3 · Online Learning · Tutoring

We would like to thank NSF (e.g., 2118725, 2118904, 1950683, 1917808, 1931523, 1940236, 1917713, 1903304, 1822830, 1759229, 1724889, 1636782, & 1535428), IES (e.g., R305N210049, R305D210031, R305A170137, R305A170243, R305A180401, & R305A120125), GAANN (e.g., P200A180088 & P200A150306), EIR (U411B190024 & S411B210024), ONR (N00014-18-1-2768), NHI (R44GM146483), and Schmidt Futures. None of the opinions expressed here are that of the funders.

1   Introduction

Online learning platforms offer students tutoring in a variety of forms, such as one-on-one messaging with real human tutors [1] or providing expert-written messages for each question that students are required to answer [5]. These methods, while effective, can be costly and time consuming to scale. However, recent advances in Language Models (LMs) may provide an opportunity to offset the cost of providing effective tutoring to students.

In this work, we explore the effectiveness of using LMs to create explanations of mathematics problems for students within the ASSISTments online learning platform [5]. Recent transformer-based LMs have exhibited breakthrough performance on a number of domains [2,3]. In this work, we perform experiments using one of the most powerful currently available LMs, GPT-3 [2], accessed through OpenAI's API.

Two different approaches to generate this content were explored. The first approach used few-shot learning [2] to generate new explanations from a handful of similar mathematics problems with answers and explanations, and the second approach attempted to generate new explanations by using the LM to summarize message logs between students and real human tutors. After each method was used to generate new explanations, these explanations were compared to existing explanations in the ASSISTments online learning platform through surveys given to mathematics teachers. Comparing teachers' evaluations of the quality of the various explanations enabled an empirical evaluation of each LM-based approach, as well as an evaluation of their applicability in a real-world setting.

2   Background

2.1   Language Models

LMs are a type of deep learning model trained to generate human-like text. They are trained on a massive dataset of millions of web pages, books, and other written documents, and are capable of generating text that is often indistinguishable from human-written text [2,3]. In this work, we focus on GPT-3 since it is a powerful LM that is publicly accessible through a paid API. When using GPT-3, one can specify parameters for the text generation such as Frequency Penalty, which penalizes GPT-3 for repeating phrases in its response, Temperature, which increases the frequency of picking a less-that-most-likely word to include in the response, and Max Tokens, which specifies the maximum length of the response [2].

2.2   Data Sources

The data used to generate explanations using few-shot learning came from the ASSISTments online learning platform [5]. Within ASSISTments, middle-school mathematics students complete mathematics problem sets assigned to them by their teacher. If students are struggling with their assignment, ASSISTments will provide them with an explanation upon their request. When a student requests an explanation, a message that explains how to solve the mathematics problem they are currently struggling with and the solution to the problem is provided to them.

The data used to generate explanations from summaries of tutoring chat logs comes from UPchieve, a provider of online tutoring. UPchieve[1] offers live online tutoring with volunteers through an interface that facilitates sharing of text and images. In ASSISTments, students had the ability to request a chat with a live tutor. When a live tutor was requested, a tutoring session was opened via UPchieve.

3   Data Processing

In order to examine the effects of changes to the prompts on the generate explanations in a way that would not bias the results of the analysis, all of the available data for generating explanations was split in half. Half of the data was used for prompt engineering (development set). This data was used iteratively to examine how small variations in the prompt effected the resulting explanations. Once the generated explanations reached a satisfactory level, the most effective prompts were used on the second half of the data (evaluation set). The analysis of the validity and quality of explanations discussed in the results was performed only on this second half of the data, eliminating any bias from the prompt engineering process.

During the live tutoring partnership period, there were 244 tutoring sessions across 93 students and 110 problems covering various middle-school mathematics skills. Of these tutoring sessions, 2 were excluded because they contained no interaction between student and tutor and 2 were excluded because they were longer than GPT-3's 4,000 token limit. Of the 40,523 problems available, only 914 problems remained after removing the problems that were not of the same skills as the problems in the tutoring sessions and the problems that could not be used in the few-shot learning prompt due to the presence of images within the problems, answers, or explanations. Both the chat logs and the problems and their explanations were evenly partitioned into a development and evaluation set stratified by subject matter.

4   Methodology

4.1   Tutoring Chat Log Summarization

Development data was used to engineer a four step process for generating explanations from tutoring chat logs. The prompts are shown below, with the GPT-3 parameters shown in parentheses as (Frequency Penalty, Temperature, Max Tokens). The text-davinci-003 model was used for all prompts.

---

[1] https://upchieve.org/.

1. Does the tutor successfully help the student in the following chain of messages? [The tutoring chat log.] (0, 0.7, 128)
2. Explain the mathematical concepts the tutor used to help the student, including explanations the tutor gave of these concepts, and ignoring any names. [The tutoring chat log.] (0.25, 0.9, 750)
3. Reword the following explanation to not include references to a tutor or student, and to be in the present tense: [The previously generated explanation.] (0.25, 0.9, 750)
4. Summarize the following explanations, making sure to include the most generalizable math advice. [The previously generated explanations.] (0.25, 0.9, 500)

4.2   Problem-Level Explanation Few-Shot Learning

Before generating explanations for the 53 problems in the summarization development set, problems that were open response or not text-based had to be removed. For each of the 40 remaining problems a prompt was constructed by randomly sampling problems of the same skill from the development set, and appending the phrase below, replacing the content in brackets with the problem content, until there were no more same-skill problems or the max token length was reached. For these prompts, the Frequency Penalty was 0, the Temperature was 0.73, the Max Tokens was 256, and the code-davinci-003 model was used.

Problem: [The text of the problem.]
Answer: [The answer to the problem.]
Explanation: [The explanation for the problem.]

4.3   Empirical Analysis of Generated Explanations

After the summarization and few-shot learning processes were completed for the evaluation data using the processes developed with the development data. The explanations from both processes were manually evaluated by subject-matter experts for both structural and mathematical validity. Structural validity required that the explanation be in the format of an explanation. Mathematical validity required that the explanation be mathematically correct.

The valid generated explanations and teacher-written explanations for the same problems were compiled into a survey. The source of the explanations was blinded, and mathematics teachers were given a picture of each mathematics problem and the text of the explanation and told to rate the explanations on a scale from 1–5. A multi-level model [4] was used to predict the rating of each explanation given random effects for the rater and the mathematics problem, and fixed effects for the source of the explanation. The effects for the sources of explanations were used to determine if there were any statistically significant differences between the sources.

5   Results

Performing the summarization process on the evaluation data resulted in 57 explanations. Expert review found 14 structurally and 17 mathematically invalid explanations. For the 61 problems available in the summarization evaluation set, only 33 explanations could be generated using the few-shot learning approach due to the presence of images in the problems. Expert review found 1 structurally and 26 mathematically invalid explanations.

In total, 26 summarization, 6 few-shot learning, and 10 ASSISTments explanations were included for a total of 42 survey questions. Five current or former middle-school or high-school mathematics teachers completed the survey. Once survey results were collected, two different models were fit; one that only included teachers ratings of valid explanations, and one that included all the generated explanations, with a rating of 1 for explanations that were invalid. The effects and 95% confidence intervals of the different sources of explanations are shown in Fig. 1. ASSISTments explanations are rated the highest, with an average rating of about 4.2. Summarization based explanations were statistically significantly worse than ASSISTments explanations, with an average rating of about 2.6 for the valid explanations and 1.7 for all explanations. Qualitatively, teachers reported that the summarization based explanations used terms that the students did not necessarily know, and tended to give advice that was too general. Few-shot learning based explanations received an average rating of about 3.6 for valid explanations, which was not statistically significantly worse than ASSISTments explanations, but only 6 of the few-shot learning based explanations were valid. Few-shot learning based explanations received an average rating of about 1.6 when invalid explanations were included in the model, which is statistically significantly worse than ASSISTments explanations.

6   Conclusion

Overall it seems that GPT-3 based explanations do not compare in quality to those created by teachers. Fundamentally, GPT-3 was trained to understand language, but not mathematics, and while the structure of what GPT-3 generated made proper use of the English language, it often generated incorrect mathematical content, or simply failed to generate content in the proper format. Summarizing tutors' advice to students created explanations that were significantly worse than teacher-written explanations, and while the valid explanations generated through few-shot learning were not significantly worse than teacher-written explanations, only 10% of the generated explanations were mathematically valid. Ultimately, GPT-3 does not seem to have the grasp of mathematics necessary to generate high-quality explanations.

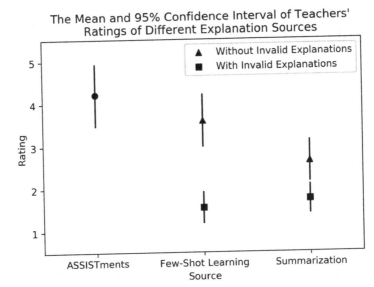

Fig. 1. The mean and 95% confidence interval of teachers' ratings of explanation quality by source, determined using the survey results. Invalid explanations, when included in the model, are assumed to have the lowest rating for quality.

References

1. Upchieve's mission. https://upchieve.org/mission
2. Brown, T., et al.: Language models are few-shot learners. In: Advances in Neural Information Processing Systems, vol. 33, pp. 1877–1901 (2020)
3. Chowdhery, A., et al.: PaLm: acaling language modeling with pathways. arXiv preprint arXiv:2204.02311 (2022)
4. Gelman, A., Hill, J.: Data Analysis Using Regression and Multilevel/Hierarchical Models. Cambridge University Press (2006)
5. Heffernan, N.T., Heffernan, C.L.: The assistments ecosystem: building a platform that brings scientists and teachers together for minimally invasive research on human learning and teaching. Int. J. Artif. Intell. Educ. **24**(4), 470–497 (2014)

Analyzing Response Times and Answer Feedback Tags in an Adaptive Assessment

Jeffrey Matayoshi[(✉)][(iD)], Hasan Uzun, and Eric Cosyn

McGraw Hill ALEKS, Irvine, CA, USA
{jeffrey.matayoshi,hasan.uzun,eric.cosyn}@mheducation.com

Abstract. While many learning and assessment models focus on the binary correctness of student responses, previous studies have shown that having access to extra information—such as the time it takes students to respond to a question—can improve the performance of these models. As much of the previous work in this area has focused on knowledge tracing and next answer correctness, in this study we take a different approach and analyze the relationship between these extra types of information and the overall knowledge of the student, as measured by the end result of an adaptive assessment. In addition to looking at student response times, we investigate the benefit of having detailed information on the responses in the form of answer feedback tags from the adaptive assessment system. After using feature embeddings to encode the information from these feedback tags, we build several models and perform a feature importance analysis to compare the relative significance of these different variables. Although it appears that the response time variable does contain useful information, the answer feedback tags are ultimately much more important to the models.

Keywords: Adaptive assessment · Response times · Answer feedback

1 Introduction and Related Work

Of fundamental importance to learning and assessment models are the responses students give to the problems and questions they are asked. As these responses can vary widely in their content, a natural simplification is to summarize the responses by classifying them as being either correct or incorrect, a procedure currently used in many, if not most, existing student models [11,13]. However, many previous works have examined the use of other sources of information beyond these simple classifications. For example, some studies have shown that the use of student *response times*—i.e., the time a student takes when answering a question—can improve the accuracy of models for detecting engagement [1] or predicting student performance [7,16]. Additionally, other works have analyzed in detail the relationships between response times and student performance on either the next question [12,13] or the entire course [4].

Another goal of previous research has been to improve on the binary correct-or-incorrect classifications that are commonly used. Examples of this include assigning partial credit based on factors such as the number of attempts made or hints accessed [5,17,18], using a more granular classification scheme for the responses [13], or focusing on the specific information contained in wrong answers [12,14,19]. Lastly, other studies have used feature embeddings to capture the complex information contained in code submissions to programming questions [8,15], a relevant technique for our current study.

As much of this previous research focuses on knowledge tracing and next answer correctness, in this work we take a different approach and study the relationship between the overall knowledge of the student—as measured by the end result of an adaptive assessment—and the student responses. In addition to looking at the relationship between the response times of students and the results of the assessment, we investigate the potential for using more specific information about the student responses by leveraging the detailed answer feedback tags returned by the adaptive assessment. After using feature embeddings to encode the information from these feedback tags, we run a feature importance analysis to compare the relative significance of these different types of variables.

2   Background and Experimental Setup

Our study uses a data set obtained from ALEKS, an adaptive learning and assessment system. The ALEKS system contains an untimed, adaptive placement assessment that evaluates a student's mastery of 314 different *topics* from high school mathematics. Each assessment asks at most 30 questions, and the end result is a set of topics that the system believes the student knows. In what follows, we refer to the size of this set of topics as the *final score* of the student, with these scores ranging from a minimum of 0 to a maximum of 314.

During each assessment, an *extra problem* is randomly chosen from the 314 total topics and presented to the student—importantly, the student's response to this extra problem does not affect the final score of the assessment, and the data are instead used to evaluate the system. When presented a question during the assessment, a student can submit a response, which is then graded as correct or incorrect, or they can click on the "I don't know" (IDK) button if they are not familiar with the material. Since the majority of ALEKS topics require open-ended responses, the system uses a library of sophisticated algorithms to process these responses and determine if they are correct. In doing so, the system's algorithms return detailed information about the student responses in the form of *answer feedback tags*. These feedback tags might indicate that a student forgot to simplify their answer, or that they used the wrong unit of measurement.

The ALEKS placement assessment is used at a variety of community colleges and four-year institutions in the U.S., and the majority of the assessments are taken by students in their first year of school. For this study, we extract all available data for placement assessments taken over a time period starting in July 2016 and ending in November 2022, giving us 2,796,640 assessments from a total of 2,198,428 unique students—this amounts to roughly 1.3 assessments per student. We randomly partition the students into training, test, and validation sets of size 85% (1,868,664 students), 10% (219,843 students), and 5% (109,921 students), respectively. This results in a training set of 2,377,294 assessments, a test set of 279,369 assessments, and a validation set of 139,977 assessments.

3   Response Times

We define the response time to be the actual—or, real—time that elapses between the student's initial viewing of the extra problem and their submission of an answer, and we then compare these times to the final scores of the students. In our training data, the mean and median final scores are 148.5 and 140, respectively, with the first and third quartiles having values of 83 and 220, respectively. To normalize the response times, we follow the procedure outlined by Pelánek [12] and convert each response time into a percentile. Specifically, we partition all the extra problems based on the type of student response—correct, incorrect, or IDK—and the particular topic. Then, within the data for each response type and topic pair, we compute the percentiles for the response times to the extra problems. Finally, separately for each response type, we group the data points into bins of width one percentile, compute the average final score in each bin, and then plot the results in Fig. 1.

While our focus is the overall knowledge of the students, in many ways the results are similar to those from previous studies on next answer correctness. For example, Fig. 1 shows that students who spend the least amount of time on incorrect and IDK answers have the lowest average final scores, consistent with the results from other works that focused on wrong answers and their relationship with next answer correctness [12,13]. Additionally, our correct answer plot shows a decreasing trend, something that has been previously observed when studying the relationship between next answer correctness and correct responses to mathematics exercises [12,13].

4   Answer Feedback Tags

Our next analysis looks at the detailed answer feedback tags used by the ALEKS system to describe the student responses. Our data set contains a total of 31,954 unique feedback tags—an average of slightly more than 100 per topic. A small proportion of the responses (about 11%) do not have specific feedback tags, and to these we assign "generic" feedback tags that indicate if the response was classified as correct, incorrect, or IDK. To summarize the information from the feedback tags, we use the Embedding class from the PyTorch library [10]. Specifically, we train a neural network model in which each feedback tag is mapped to an n-dimensional vector containing unique information about the tag [6]. For this initial model, our only features are the feedback tag embeddings, while our target variable is the student's final score. Our neural network is a basic multilayer perceptron with two hidden layers of 10 hidden units each, and we use mean squared error (MSE) as our loss function. To more easily visualize the feedback tags, we use a 2-dimensional embedding—i.e., $n = 2$—for this initial model. The results are shown in Fig. 2 where, to avoid obscuring the details, we restrict the plot to feedback tags with at least 500 data points, and we also exclude the generic feedback tags mentioned previously.

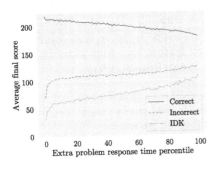

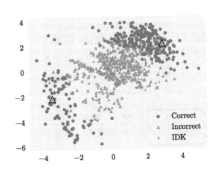

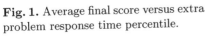

Fig. 1. Average final score versus extra problem response time percentile.

Fig. 2. Plot of 2-dimensional feedback embeddings.

While the neural network model seems to have mostly recovered the response classifications, as the correct feedback tags are fairly well-separated from the rest, there are some incorrect tags that are placed within large clusters of correct tags—two extreme cases are highlighted in Fig. 2 with larger triangles. The highlighted tag in the lower left quadrant of Fig. 2 appears if a student submits a value that properly satisfies a trigonometric equation, but is outside the range of the particular inverse trigonometric function being considered. For these responses, it seems likely the student has some experience with trigonometric functions, which are among the most advanced topics covered by the placement assessment. Notably, the average final score for these students in our training data is fairly high at 213.5 ($N = 590$). The other highlighted feedback in Fig. 2 indicates that a student, for the most part, correctly graphed a strict inequality, but made the small mistake of using a solid line instead of a dashed line. As before, the data indicate that students who receive this feedback tend to have a good understanding of the material in the placement assessment, as their average final score is a relatively high 212.2 ($N = 1119$).

5   Feature Analysis

To evaluate the relative predictive strengths of the different features, we next train several simple neural network models, using the different combinations of features listed in the first column of Table 1. For the Time feature, rather than using the percentile scores we use the actual time taken by students to respond to the extra problem, as this gives better performance on our validation set. The Feedback feature uses the embeddings from the previous section, while the

Response feature is a simplified version of the embedding model that encodes the response only as correct, incorrect, or IDK. For each set of features in the table, we apply a grid search and train 24 different models with various hyperparameters. Using root mean squared error (RMSE) as our chosen measure, from each set of 24 models we find the one that performs the best on our validation set, and we then evaluate this model on our test data. The resulting RMSE values, along with their confidence intervals,[1] are shown in the second column of Table 1. Notably, the models with the Feedback feature are more accurate than the models that use the simple classifications in the Response feature.

Table 1. Root mean squared error (RMSE) values on held-out test data. The feature importance RMSE values are averages computed from 10 iterations.

Model	RMSE (95% CI)	Feature Importance		
		Response	Feedback	Time
Response	69.3 (69.1, 69.4)	—	—	—
Response, Time	68.0 (67.8, 68.1)	95.1 (+27.1)	—	72.6 (+4.6)
Feedback	67.7 (67.6, 67.9)	—	—	—
Feedback, Time	66.6 (66.5, 66.8)	—	96.3 (+29.7)	70.5 (+3.9)

Finally, the right-hand side of Table 1 shows the results from an application of permutation feature importance [2], a method for quantifying the importance of the variables to the models. Since a higher RMSE value indicates a feature is more important, the Time feature is seemingly less important, as the average RMSE values are considerably lower when the Time feature is permuted—either 72.6 or 70.5—in comparison to the RMSE values when either the Response or Feedback values are permuted—95.1 or 96.3, respectively.

6 Discussion

Given that the ALEKS assessment measures a student's overall knowledge, it seems appropriate that the answer feedback tags, which potentially contain detailed information about a student's knowledge of a topic, are more important to the models than the response times. As response times do not directly measure the quality of the submitted answers, the use of response times could potentially lead to predictions that unfairly penalize students. For example, perhaps a student has a long response time because they are diligent and double-check their work; or, alternatively, a student may take longer to parse the instructions— possibly due to accessibility issues or struggles with reading comprehension— while being perfectly capable of performing the mathematical operations. Thus,

---

[1] To account for students with multiple assessments, the confidence intervals are computed using the cluster bootstrap method [3].

for these reasons our future work is focused on potentially using the answer feedback tag feature embeddings, rather than the response times, to improve the ALEKS system's recently introduced neural network assessment engine [9].

References

1. Beck, J.E.: Engagement tracing: Using response times to model student disengagement. In: Artificial Intelligence in Education (2005)
2. Breiman, L.: Random forests. Mach. Learn. **45**(1), 5–32 (2001)
3. Field, C.A., Welsh, A.H.: Bootstrapping clustered data. J. R. Stat. Soc. Ser. B (Stat. Methodol.) **69**(3), 369–390 (2007)
4. González-Espada, W.J., Bullock, D.W.: Innovative applications of classroom response systems: investigating students item response times in relation to final course grade, gender, general point average, and high school ACT scores. Electron. J. Integr. Technol. Educ. **6**, 97–108 (2007)
5. Inwegen, E.V., Adjei, S.A., Wang, Y., Heffernan, N.T.: Using partial credit and response history to model user knowledge. In: Educational Data Mining (2015)
6. Jurafsky, D., Martin, J.H.: Speech and Language Processing (3rd ed. draft) (2021). https://web.stanford.edu/~jurafsky/slp3/
7. Lin, C., Shen, S., Chi, M.: Incorporating student response time and tutor instructional interventions into student modeling. In: User Modeling Adaptation and Personalization (2016)
8. Liu, N., Wang, Z., Baraniuk, R.G., Lan, A.: Open-ended knowledge tracing (2022). https://doi.org/10.48550/ARXIV.2203.03716. https://arxiv.org/abs/2203.03716
9. Matayoshi, J., Uzun, H., Cosyn, E.: Using a randomized experiment to compare the performance of two adaptive assessment engines. In: Educational Data Mining, pp. 821–827 (2022)
10. Paszke, A., et al.: PyTorch: an imperative style, high-performance deep learning library. CoRR abs/1912.01703 (2019). http://arxiv.org/abs/1912.01703
11. Pelánek, R.: Bayesian knowledge tracing, logistic models, and beyond: an overview of learner modeling techniques. User Model. User Adap. Inter. **27**(3), 313–350 (2017)
12. Pelánek, R.: Exploring the utility of response times and wrong answers for adaptive learning. In: Learning @ Scale, pp. 1–4 (2018)
13. Pelánek, R., Effenberger, T.: Beyond binary correctness: classification of students' answers in learning systems. User Model. User Adap. Inter. **30**(5), 867–893 (2020)
14. Pelánek, R., Rihák, J.: Properties and applications of wrong answers in online educational systems. In: Educational Data Mining (2016)
15. Piech, C., Huang, J., Nguyen, A., Phulsuksombati, M., Sahami, M., Guibas, L.: Learning program embeddings to propagate feedback on student code. In: International Conference on Machine Learning, pp. 1093–1102. PMLR (2015)
16. Wang, Y., Heffernan, N.T.: Leveraging first response time into the knowledge tracing model. In: Educational Data Mining (2012)
17. Wang, Y., Heffernan, N.T.: Extending knowledge tracing to allow partial credit: using continuous versus binary nodes. In: Artificial Intelligence in Education (2013)
18. Wang, Y., Heffernan, N.T., Beck, J.E.: Representing student performance with partial credit. In: Educational Data Mining (2010)
19. Wang, Y., Heffernan, N.T., Heffernan, C.: Towards better affect detectors: effect of missing skills, class features and common wrong answers. In: Learning Analytics and Knowledge (2015)

Enhancing the Automatic Identification of Common Math Misconceptions Using Natural Language Processing

Guher Gorgun[1]([⊠])[ID] and Anthony F. Botelho[2][ID]

[1] University of Alberta, Edmonton, AB T6G 2G5, Canada
gorgun@ualberta.ca
[2] University of Florida, Gainesville, FL 32611, USA
abotelho@coe.ufl.edu

Abstract. In order to facilitate student learning, it is important to identify and remediate misconceptions and incomplete knowledge pertaining to the assigned material. In the domain of mathematics, prior research with computer-based learning systems has utilized the commonality of incorrect answers to problems as a way of identifying potential misconceptions among students. Much of this research, however, has been limited to the use of close-ended questions, such as multiple-choice and fill-in-the-blank problems. In this study, we explore the potential usage of natural language processing and clustering methods to examine potential misconceptions across student answers to both close- and open-ended problems. We find that our proposed methods show promise for distinguishing misconception from non-conception, but may need further development to improve the interpretability of specific misunderstandings exhibited through student explanations.

Keywords: misconceptions · sentence-BERT · intelligent tutoring system · natural language processing

1 Introduction

Educators across learning contexts and domains rely on a range of content to assess students' knowledge and understanding of covered concepts. In the domain of mathematics, for example, as is the focus of this paper, it is not uncommon for teachers to assign homework and classwork in the form of problem sets composed of multiple interleaved types of problems [5]. Traditionally, these different types of problems include formats of multiple choice, fill-in-the-blank, and short answer questions, but may also include other types of questions such as drawing charts and graphs, as well as essay questions (though the latter is likely less common in the domain of mathematics). Prior works have described these different types of problems by distinguishing "close-ended" questions from "open-ended" questions (e.g., [1]); while the scoring of student answers to close-ended problems is relatively easy to automate with a matching procedure as there is usually a small number of correct answers and the variation in possible answers to open-ended responses makes this task much more difficult.

In the past, notable research in addressing student misconceptions has been limited to observing student work on close-ended questions through "bugs" and "common wrong answers" (CWA; [5]). While common wrong answers, or particular incorrect responses that are answered by a large proportion of students, can be helpful in understanding student misconceptions, student responses to open-ended problems may provide even greater insights. Teachers often rely on student open-ended work to understand the thought processes and strategies taken by students to find a solution. Therefore, open-ended responses could provide opportunities to identify misconceptions with greater precision.

Recent advancements in natural language processing (NLP) have resulted in the application of deep learning embedding methods such as Sentence-BERT [2], which has been used in educational contexts to identify sets of similar student answers to open response problems (e.g. the method described in [1]). Such methods may be used to identify common incorrect explanations.

This paper represents a proof of concept in using NLP to identify misconceptions through common wrong answers in open-ended explanations. To test the feasibility of our approach, we examine student answers to a single 2-part problem from the ASSISTments learning platform. This 2-part problem consists of a close-ended "Part A" followed by an open-ended "Part B" that prompts students to explain their solution to the preceding part. Through a set of exploratory analyses, we seek to address the following research questions:

1. What are the common answers that emerge when clustering student-written explanations for a single open-ended mathematics problem?
2. Do similar sets of common incorrect answers emerge when comparing across close- and open-ended components of a single mathematics problem?

2 Methods

To conduct our analyses, we select a single 2-part problem from ASSISTments from a large set of student log data collected between 2018 through 2022. From this large set, we identify a candidate set of problems where the problems have at least 2 parts (consisting of a close-ended, followed by an open-ended question) and the second part prompts students to explain their work to the preceding part. Within the system, by default, teachers must manually score open-ended answers on an integer scale from 0–4. We filter problems where there were fewer than 2 teachers who provided scores to students and include only problems where all 5 score values were present in the data. We further filter out any problems where the percentage of unique answers is larger than 75% and 5% for open-ended and close-ended responses, respectively (to identify problems where there is notable variance in student responses to evaluate our methods). From the set of candidate problems, we select the problem that contains the largest sample size. The text of the close-ended portion of this problem is depicted in Fig. 1.

Fig. 1. The close-ended problem prompts for the selected problem set. The problem was followed by an open-ended question prompting students to explain their work.

2.1 Identifying Common Wrong Answers

In the first step of our analysis, we use the student answers to the close-ended portion of the problem to extract the most common incorrect answers. As a "select all that apply" question type, as seen in Fig. 1, students are asked to select all of the values equivalent to the ratio of 12:3. As this type of problem is graded by the computer by matching each student's answer against the known correct answer, we identify all incorrect student answers. From this, we identify the set of unique student answers and calculate a simple frequency to measure each incorrect answer's commonality (see Table 1).

To identify common explanations for open-ended component of the problem set selected, we conduct a second analysis on the subsequent part of the problem. With the set of student textual explanations, we first cleaned the set of answers by removing any HTML tags and accented characters (that would not be recognized by most NLP models), and removed any empty student responses. With these, we utilized a pre-trained Sentence-BERT model (SBERT; [2]) which converts each answer into a 768-valued feature vector. The intuition of this and similar embedding methods is that it creates an embedding space where the distance of each textual sample to all others is correlated with the semantic similarity of the language (i.e. similar student responses should cluster closely together within the space). After generating these embeddings, we grouped sets of student answers by the teacher-given score to identify groups of similar answers within each score band.

We used the k-means clustering method [4] to identify clusters within each score band. Yet, due to the high dimensionality of the data, the k-means cluster-

Table 1. The correctness statistics and common wrong answers for the close-ended component of the problem set selected.

N Correct/ N Incorrect	Correct Answer	3 Most Common Wrong Answers	Count of Wrong Answers
1064 (37.7%)/ 1761 (62.3%)	4:1, 24:6, 1200:300	24:6, 1200:300	172 (10.2%)
		2. 4:1, 24:6	140 (8.3%)
		6:1, 4:1, 24:6, 1200:300	89 (5.3%)

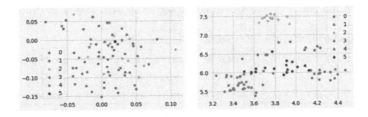

Fig. 2. Clusters identified before and after applying UMAP method.

ing method did not perform well as measured by the resulting clusters' silhouette score [7]. This score suggested poor coherence, indicating that it would be difficult to interpret meaningful differences between cluster groupings. Following a similar procedure to the BERTopic modeling algorithm [3], we applied a dimensionality reduction algorithm, Uniform Manifold Approximation and Projection (UMAP) [6] in an attempt to improve the clustering by removing redundant and irrelevant features from the embedding models (and simplifying the clustering procedure). The result of applying UMAP prior to k-means clustering rendered better clusters with higher values of silhouette coefficients and the sum of squared distances (SSD). This improvement can be seen in Fig. 2 which depicts the resulting cluster cohesion, through a 2D projection of the embedding space based on the most representative axes, before applying UMAP (left) and after (right). Finally, we identified the most frequent bi-grams present in each cluster after removing English stopwords. We also tried using unigrams and trigrams to identify keywords, however, they were not as helpful as using bi-grams in identifying the common theme of each cluster.

3   Results

We analyzed problems and clusters found for each score band in detail to identify common themes and misconceptions, as summarized in Table 2. Within each score band, we typically observed a cluster composed of students indicating that they did not know the answer. The only consistent exception was the score band with the full score where none of the students stated that they did not know the answer or made an accidental slip.

From examining the common wrong answers from the close-ended portion in Table 1, we see a large number of students failing to include the ratio of 4:1 and, to a lesser degree, missing the ratio of 1200:300 or including the incorrect ratio of 6:1. The first missing ratio could indicate that students struggled to represent the ratio in its simplest form; this could point to difficulties representing the ratio as a fraction or other errors when reducing that fraction. Similarly, the second-most-common wrong answer may suggest difficulties for students to identify the larger numbers as multiples of the given ratio. Finally, in the third CWA, the inclusion of the ratio 6:1 suggests a misunderstanding of what a ratio is meant to represent in terms of the relationship between the two numbers.

Table 2. The most frequent bi-grams observed in each cluster in Problem 1.

Score Band	N Samples	Keywords/Bi-grams in Clusters
0	53	Cluster 1: got wrong, didn't know, need help
	20	Cluster 2: idk idk, don't know, know idk
	14	Cluster 3: tp ratio, long ratios, use numbers
1	6	Cluster 1: kinda forgot, sorry got, problem sorry
	6	Cluster 2: scale factors, bc scale, copies multiply
	5	Cluster 3: 12 got, got right, little high
	4	Cluster 4: 1s numbers, added know, timesing numbers
	4	Cluster 5: picked divide, 12 picked, 24 12
2	12	Cluster 1: 12 didn't, 12 know, turn 12
	12	Cluster 2: different multiples, divided just, times different
	12	Cluster 3: don't wrong, got wrong, added don't
	8	Cluster 4: wrong just, knew 24, got wrong
	8	Cluster 5: fractions multiplied, fraction multiplied
	7	Cluster 6: fit 12, lower because, fit numbers
3	29	Cluster 1: number divide, multiply number, divided multiply
	15	Cluster 2: divide multiply, 12 24, factors ratio
	14	Cluster 3: times equals, 12 times, 12 multiplied
4	79	Cluster 1: multiply divide, know multiply, divide number
	64	Cluster 2: 200 300, 1200 300, 12 divided
	46	Cluster 3: 12 know, numbers 12, multiples 12
	36	Cluster 4: equal ratio, original ratio, ratio multiple
	30	Cluster 5: ratio 12, equal 12, equivalent 12
	28	Cluster 6: 12 times, times 100, 24 3 × 2

In comparing these to the clusters and bi-grams in Table 2, it is easy to first realize the large number of clusters containing "non-answers" such as "idk" and "don't know." It is interesting to also recognize that such clusters emerged in several score bands, suggesting that teachers provided some credit to students admitting their lack of understanding. While these types of clusters may indicate very little in terms of misconception, they are quite informative in identifying non-conception. In other words, it is difficult to ascertain from the close-ended problem which answers were deliberate and which were the result of a somewhat random selection. While it is easy to conclude that the inclusion of 6:1 might indicate a misunderstanding of the second number in a ratio or even the relationship between a numerator and denominator, it is also the case that this is the first option provided to students and therefore it may just be the most likely first guess (e.g. the likelihood of students guessing answers is likely not uniform across the selections). Combining the close- and open-ended answers can help distinguish misconception from non-conception. In observing other clusters, we can deduce that some students are exhibiting at least partial knowledge by referencing keywords such as fractions, multiples, and division, but the bi-grams alone are seemingly not the most informative way of identifying specific misconceptions.

4   Conclusion

The analyses presented in this work contribute mixed results in terms of automating the identification of misconceptions in student textual explanations. We found that, although we are able to identify clusters of student answers, the use of bi-grams offers only limited utility in drawing conclusive interpretations as to what each cluster exhibits in terms of student understanding. With that, however, we found that the method we propose here is quite helpful in distinguishing between misconception and non-conception when taking into account the CWAs that emerge from close-ended problems.

Beyond this context, this work also offers contributions in the form of best practices when approaching the clustering of SBERT embeddings. We found that the use of UMAP was instrumental in producing clusters that were interpretable in any capacity. Future works attempting to identify sets of similar student answers should consider such methods to both simplify clustering procedures and improve the cohesion of resulting groups. Of course, as this work represents a simple proof-of-concept, future work is planned to scale these analyses to understand whether our findings generalize to larger sets of problems.

References

1. Baral, S., Botelho, A.F., Erickson, J.A., Benachamardi, P., Heffernan, N.T.: Improving automated scoring of student open responses in mathematics. International Educational Data Mining Society (2021)
2. Devlin, J., Chang, M.W., Lee, K., Toutanova, K.: BERT: pre-training of deep bidirectional transformers for language understanding. arXiv preprint arXiv:1810.04805 (2018)
3. Grootendorst, M.: BERTopic: neural topic modeling with a class-based TF-IDF procedure. arXiv preprint arXiv:2203.05794 (2022)
4. Hartigan, J.A., Wong, M.A.: Algorithm as 136: a k-means clustering algorithm. J. R. Stat. Soc. Ser. C (Appl. Stat.) 28(1), 100–108 (1979)
5. Heffernan, N.T., Heffernan, C.L.: The ASSISTments ecosystem: building a platform that brings scientists and teachers together for minimally invasive research on human learning and teaching. Int. J. Artif. Intell. Educ. 24(4), 470–497 (2014)
6. McInnes, L., Healy, J., Melville, J.: UMAP: uniform manifold approximation and projection for dimension reduction. arXiv preprint arXiv:1802.03426 (2018)
7. Zhou, H.B., Gao, J.T.: Automatic method for determining cluster number based on silhouette coefficient. In: Advanced Materials Research, vol. 951, pp. 227–230. Trans Tech Publications (2014)

User Adaptive Language Learning Chatbots with a Curriculum

Kun Qian[1]([📧])[ID], Ryan Shea[1], Yu Li[1], Luke Kutszik Fryer[2], and Zhou Yu[1,3]

[1] Columbia University, New York, USA
{kq2157,rs4235,yl5016,zy2461}@columbia.edu
[2] The University of Hong Kong, Hong Kong, China
fryer@hku.hk
[3] Articulate.AI Inc, New York, USA

Abstract. Along with the development of systems for natural language understanding and generation, dialog systems have been widely adopted for language learning and practicing. Many current educational dialog systems perform chitchat, where the generated content and vocabulary are not constrained. However, for learners in a school setting, practice through dialog is more effective if it aligns with students' curriculum and focuses on textbook vocabulary. Therefore, we adapt lexically constrained decoding to a dialog system, which urges the dialog system to include curriculum-aligned words and phrases in its generated utterances. We adopt a generative dialog system, BlenderBot3, as our backbone model and evaluate our curriculum-based dialog system with middle school students learning English as their second language. The constrained words and phrases are derived from their textbooks, suggested by their English teachers. The evaluation result demonstrates that the dialog system with curriculum infusion improves students' understanding of target words and increases their interest in practicing English.

Keywords: Lexically constrained decoding · Generative dialog system · User adaptation

1 Introduction

Finding a consistent speaking partner can be challenging for language learners. However, chatbots offer a solution by providing an interactive environment for practice. Traditional chatbots that rely on pre-written scripts often produce utterances that are limited and unresponsive [3,7]. Recent advancements in large pre-trained language models have led to the development of more adaptable conversational AI that can respond more naturally to user input [1,5,6]. However, these systems are primarily focused on casual conversation and lack a structured curriculum. On the other hand, non-native speakers usually learn new languages using textbooks, and it is more helpful if the chatbot generates utterances based on a curriculum.

N. Wang et al. (Eds.): AIED 2023, CCIS 1831, pp. 308–313, 2023.
https://doi.org/10.1007/978-3-031-36336-8_48

Our proposed solution is to create a user-adaptive language learning chatbot that incorporates an English language learning curriculum. The chatbot is designed to assist learners in practicing different language skills, such as grammar and vocabulary, specified by the curriculum. We use the pre-trained language model Blenderbot3 [8] as a foundation and propose a multi-turns grid beam search to include curriculum-specific words and phrases during the decoding stage. To evaluate the effectiveness of our chatbot, we conduct a study with 155 8th-grade students using a curriculum developed based on their textbook and consultation with their teachers. Our experimental results show that:

1. Our curriculum-based chatbot increases the frequency of specified words or phrases used throughout dialogs on both the system side and the user side.
2. The user-adaptive curriculum improves users' engagement and interest in our chatbot.
3. In general, our chatbot helps students correctly understand and use specified words or phrases.

We will release the source code of our chatbot toolkit in the future so that teachers and students can easily design their own curriculum for learning and practicing.

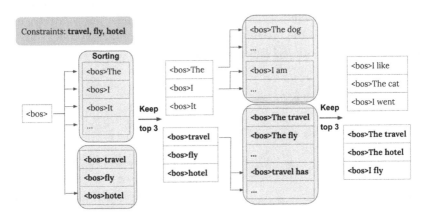

Fig. 1. First two decoding steps of multi-turns grid beam search. In addition to the normal beam (text in normal font), grid beam search adds an extra beam (text bolded) to store candidates that satisfy constraints. The constraint words are *"travel"*, *"fly"*, and *"hotel"* in this case.

2   Methodology

2.1   Multi-turns Grid Beam Search

Previous works, such as machine translation, only apply grid beam search once for each generation. However, for dialog systems, we expect our chatbot to mention constraint words over the course of several dialogue turns, when suitable.

Therefore, we propose a multi-turns grid beam search method, to adapt constrained decoding to a dialog setting.

Simplifed Constraint Beam Box. Figure 1 illustrates an example decoding step of multi-turns grid beam search. The upper beam boxes, where text is in normal font, are unsatisfied candidates. These candidates do not contain any constraint words and are selected following the normal beam search mechanism. The lower beam boxes, containing bolded text, are satisfied candidates including constraint words. They are generated by appending constraint words to the left candidates from the previous step. For example, at the first step, we append constraint words to the start token "< *bos* >" to get satisfied candidates "< *bos* > *travel*", "< *bos* > *fly*" and "< *bos* > *hotel*". Normally, we add extra beams based on the number of satisfied constraints, meaning that candidates in the same beam have the same number of constraint words. However, since one constraint word each turn is enough in our dialog setting, we fix the beam number as two (one for unsatisfied candidates and one for candidates containing one constraint word) and only append constraint words to the upper beam box. This modification helps speed up decoding.

Dynamic Constraint List. Since mentioning constraint words once per dialog is enough, we update our constraint word list dynamically throughout the dialog. As shown in Fig. 2, we check if any constraint word is used at the end of each turn and remove those words from the constraint list. These words are still likely to be generated due to the attention on dialog history, but we do not force the chatbot to generate them in future turns.

Fig. 2. The constraint list is updated through the conversation.

2.2   Evaluation Flow

To collect feedback and evaluate our model with real users, we employ the EduBot [4] platform for deployment. The whole evaluation flow consists of six steps in total:

1. **Instruction.** The instruction page briefly describes the function of our chatbot, the evaluation flow, and the purpose of the test. In addition, constraint words and phrases are also presented.

2. **Pre-Test**. In order to evaluate whether our chatbot can help improve the user's understanding of constraint words, we design five single-choice questions based on the constraint words. These questions are designed to simulate the questions in students' exams.

3. **Conversation**. During the Conversation, an abstracted instruction is shown to users by default to indicate that they should pay attention to the constraint words. A "reset" button is also provided in case user wants to restart the conversation.

4. **Grammar Error Correction**. Grammar error correction is a default function provided by the EduBot platform. It detects grammar errors from the user utterances and presents corrections at the end of conversations automatically. This function is technically supported by [9].

5. **Post-Test**. As a comparison, we ask users to answer the same five questions as presented in the pre-test session. In order to demonstrate that our chatbot helps users learn constraint words/phrases, we expect users to correctly complete the post-test, even if they make mistakes during the pre-test session.

6. **Survey**. Following [2], we design six survey questions for both self-efficacy and user interest. Users are asked to choose a score from one to five for each question and optionally leave comments or suggestions in a text box.

During the evaluation, we translate the instructions (including the abstracted instructions during the conversation session) and survey sessions into Chinese in order to avoid misunderstandings and ensure the quality of the evaluation.

3   Experiment and Results

3.1   Implementation Details

Considering user volume and corresponding server capacity, we use the smallest version of BlenderBot 3, Blenderbot-3B, as our backbone model. From our consultations with students' English teachers, we chose words of seasons (e.g. spring, winter), months (e.g. July, December), and days (e.g. Monday, Sunday) as constraint words. To enforce the model to use these words with their matching preposition, we also include constraint phrases such as "on Monday" or "during winter". We adopt a disjunctive format for constraint words, meaning that only one word from each type (seasons, months, or days) is generated. Once a constraint word is generated, the whole type would be marked as satisfied and removed from the constraint list (Sect. 2.1).

3.2   Results and Analysis

Conversation helps answer test questions. In order to evaluate students' understanding of constraint words in terms of both grammar and vocabulary, we designed the first three pre-test questions for preposition usage of corresponding constraint words. While the last two questions are designed to test whether students understand the semantic meaning of constraint words. We recruit 155

8th-grade Chinese students for evaluation in total. The results in Fig. 3 show that around half of students can correctly answer the first three questions and 80% of students give correct answers for the final two questions. This suggests that most of the students understand the semantic meaning of constraint words but aren't able to use them with correct propositions. In total, 39 out of 267 (14.61%) incorrect answers were corrected after speaking with our chatbot, represented as the red bar. We also count that 21 out of 155 students (13.5%) improve their overall score for all five test questions. This improvement demonstrates that our chatbot help students learn both semantic meaning and usage of constraint words and phrases.

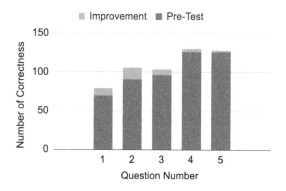

Fig. 3. The correctness of each test question from 155 8th-grade Chinese students. The blue bar represents the number of students that correctly answered questions in the pre-test. The red bar indicates the number of students who failed the corresponding question in the pre-test but correctly answered it after the conversation. (Color figure online)

Constrained Decoding Encourages Users to Practice Target Words More Frequently. We also count the usage frequencies of constraint words in both user utterances and system responses. The result is listed in Table 1. Since we adopt a dynamic constraint threshold (Sect. 2.1) during decoding, which does not force our model to use all constraints at the beginning of a conversation. On the other hand, not all students complete a conversation after seeing all constraint words, which means not all constraint words are necessarily used during the course of a conversation. This leads to the system's usage frequency of constraint words being less than 100%. However, it is obvious that both the system and user sides use words for "season" more frequently than the other two types. This suggests that forcing system to generate constraint words encourages users to use these words in their own utterances. We also correlate the usage of constraint words and the improvement of testing results. Specifically, we compute the frequency of constraint word usage over those students who made wrong answers in the pre-test but correctly answered them after the conversation (same as the students of the red bar in Fig. 3). We find that both the system and users are more likely to use constraint words in the improved test cases. In other words, more frequently seeing and practicing constraint words help users learn those words.

Table 1. Frequencies of constraint word usage for both the system side and user side. Percentages on the left side are counted over all 155 test cases. Percentages on the right half are counted over dialogs after which students correct their test answers. "Seasons", "Months" and "Days" are the three constraint word types. "Any" means any type of constraint words mentioned in utterances counts.

Constraint	All Test Cases				Improved Test Cases			
	Seasons	Months	Days	Any	Seasons	Months	Days	Any
System	81.94%	65.81%	65.16%	90.32%	90.00%	73.33%	66.67%	100.00%
User	38.71%	18.71%	24.52%	50.32%	40.00%	30.00%	26.67%	96.67%

4 Conclusion

In this paper, we propose a user-adaptive generative chatbot for language learning. We use constrained decoding to incorporate a curriculum, which is adaptable based on the user's request. We apply this method to a pre-trained large language model. To evaluate our model, we design an evaluation flow based on constraint words and employ more than 155 students who learn English as a second language. The result demonstrates that our curriculum-incorporated chatbot helps students learn specific words and phrases.

References

1. Brown, T., et al.: Language models are few-shot learners. In: Advances in Neural Information Processing Systems 33, pp. 1877–1901 (2020)
2. Fryer, L.K., Nakao, K., Thompson, A.: Chatbot learning partners: connecting learning experiences, interest and competence. Comput. Hum. Behav. **93**, 279–289 (2019)
3. Kuhail, M.A., Alturki, N., Alramlawi, S., Alhejori, K.: Interacting with educational chatbots: a systematic review. Educ. Inf. Technol. **28**, 973–1018 (2022)
4. Li, Y., et al.: Using chatbots to teach languages. In: Proceedings of the 9th ACM Conference on Learning @ Scale (2022)
5. Radford, A., Wu, J., Child, R., Luan, D., Amodei, D., Sutskever, I., et al.: Language models are unsupervised multitask learners. OpenAI blog **1**(8), 9 (2019)
6. Raffel, C., et al.: Exploring the limits of transfer learning with a unified text-to-text transformer. J. Mach. Learn. Res. **21**(140), 1–67 (2020)
7. Ruan, S.S., et al.: EnglishBot: an AI-powered conversational system for second language learning. In: 26th International Conference on Intelligent User Interfaces (2021)
8. Shuster, K., et al.: BlenderBot 3: a deployed conversational agent that continually learns to responsibly engage. arXiv preprint arXiv:2208.03188 (2022)
9. Yuan, X., Pham, D., Davidson, S., Yu, Z.: ErAConD: error annotated conversational dialog dataset for grammatical error correction. arXiv arXiv:2112.08466 (2021)

Learning About Circular Motion of Celestial Bodies with Interactive Qualitative Representations

Marco Kragten[1]([⊠]) [iD] and Bert Bredeweg[1,2] [iD]

[1] Faculty of Education, Amsterdam University of Applied Sciences, Amsterdam,
The Netherlands
{m.kragten,b.bredeweg}@hva.nl
[2] Faculty of Science, Informatics Institute, University of Amsterdam, Amsterdam,
The Netherlands

Abstract. We investigate how interactive representations can be used to support learners while learning about the circular motion of celestial bodies. We present the developed representation and accompanying lesson, and report on the effect.

Keywords: Interactive Representation · Qualitative Reasoning · System Behavior

1 Introduction

Circular motion is a fundamental concept in physics that describes the motion of an object moving in a circular path. There are several learning difficulties that are related to understanding circular motion in upper secondary physics. For instance, students have difficulties with understanding the causal relationships between force, acceleration and velocity in such systems [1–5]. A common misconception is that student believe that a force needs to be applied to an object to have velocity [1]. Some students do not understand the difference between acceleration and velocity and think that both are always in the same direction [1, 4]. With regard to centripetal force they sometimes mistakenly presume that this force is directed outward or has a curved path [5].

Previous research also demonstrates that students have difficulties in understanding the direction, magnitude, and components of vectors [5]. Circular motion can be described by decomposing the vector quantities in a horizontal and vertical direction. The direction and magnitude of the quantities than follow a simple harmonic motion. Students sometimes incorrectly assume that velocity has a nonzero vector at the turnaround point or that acceleration is zero [4].

We investigate how interactive qualitative representations [6] can be used to support learners while learning about the motion of celestial bodies. We use the vocabulary and algorithms known as qualitative reasoning [7], which make it possible to represent the distinctive features of these systems in a conceptual way.

N. Wang et al. (Eds.): AIED 2023, CCIS 1831, pp. 314–320, 2023.
https://doi.org/10.1007/978-3-031-36336-8_49

2 Qualitative Representation of Circular Motion

In a qualitative representation, *entities* represent the physical objects that constitute the system (e.g., star) [6]. *Quantities* represent the measurable properties of entities. Given that the circular motion of a star orbiting a black hole is projected on a coordinate plane, the horizontal and vertical position, centripetal force, acceleration and velocity are the characteristic quantities (Fig. 1). Each quantity has a *value* and a *direction of change*, represented as a tuple $<v, \partial>$. The possible values are represented in a *quantity space*, also for ∂. For instance, the direction of change can be captured by $\{-, 0, +\}$, referring to decreasing, steady, and increasing, respectively. A similar quantity space can be used for the possible values. In a qualitative representation, each qualitatively distinct behavior of the system is represented as a *state*. Consequently, each state has a unique set of tuples $<v, \partial>$ for the quantities describing the system. The circular motion of a star can be described using eight qualitative distinct states.

Consider the position of the star in state 1 (Fig. 1), in which case it holds that $x = <+, 0>$ and $y = <0, +>$. The star is at its most-right position (somewhere in the positive interval, hence '+') and there is no further change in the horizontal direction, hence $\partial x = 0$. The y-coordinate is '0', but the star is in an upward motion so there is a positive change in the vertical direction, hence $\partial y = +$. Note that, the black hole is located at the origin of the coordinate plane.

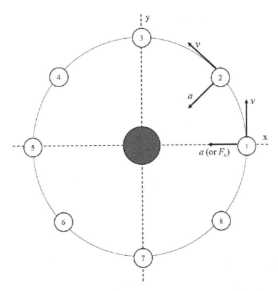

Fig. 1. Circular motion of a star orbiting a black hole. A system manifesting circular motion has eight qualitatively distinct states.

The centripetal force (F_c) and thereby the acceleration (a) is directed to the left. To describe the change of velocity we decompose the vectors of acceleration (and velocity) into a horizontal (a_x) and vertical component (a_y). For the horizontal acceleration in state 1 holds $a_y = <-, 0>$, which represents that a_y is at its most-negative value and momentarily steady. There is no vertical acceleration but there is a negative direction of change, hence $a_y = <0, ->$. There is no horizontal velocity and the change is negative, thus $v_x = <0, ->$. The vertical acceleration is at its maximum, thus $v_y = <+, 0>$.

Two types of causal dependencies are distinguished: *proportionality* and *influence* [6]. When two quantities have a proportional relationship (P−, P+), a change in one quantity (the cause) results in a change in the other quantity. A proportional relationship can be positive, where both quantities change in the same direction, or negative, where the quantities change in the opposite direction. The relationship between the quantities x and F_x is negative proportional (P−) (Fig. 2). The relationship between the quantities F_x and a_x is positive proportional (P+). Note that relationships between the quantities that describe the horizontal motion are similar to those of the vertical direction (upper and lower part in Fig. 2, respectively).

Causal dependencies of type influence (I+, I−) represent the relationship between a process and another quantity. If an influence is positive (I+), a positive value of the process results in a change in the positive direction of the affected quantity, a negative value results in a change in the negative direction. The relationship between a_x and v_x is of the type positive influence (I+) (if $a_x = -$ then $\delta v_x = -$, if $a_x = 0$ then $\delta v_x = 0$ and if $a_x =+$ then $\delta v_x = +$) (Fig. 2). The relationship between v_x and x is also a positive influence (I+) (if $v_x = -$ then $\delta x = -$, if $v_x = 0$ then $\delta x = 0$ and if $v_x = +$ then $\delta x = +$).

Correspondences (C) can be added to describe the relation between co-occurring values to determine the possible states of the system. An important insight is to realize that the pendulum movements in both directions are dependent. In the present system, the values of x and F_x are dependent, they correspond inversely (if $x = -$ then $F_x = +$, if $x = 0$ then $F_x = 0$ and if $x = +$ then $F_x = -$). The values of F_x and a_x are also dependent, they correspond regularly (if $F_x = -$ then $a_x = -$, if $F_x = 0$ then $a_x = 0$ and

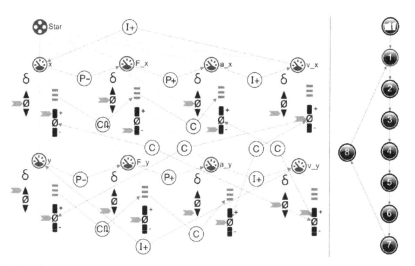

Fig. 2. Qualitative representation of circular motion in the horizontal and vertical direction. The simulation generates 8 states (RHS). Simulation result of state 1 is shown.

if $F_x = +$ then $a_x = +$). The correspondences are bidirectional indicated by an arrow point on both sides.

Simulation of the qualitative representation results in eight distinct states (Fig. 2.)

3 Method

Subjects. Twenty-three students of a K-11 class participated in the study. The evaluation study was conducted during a regular lesson (150 min) at the school.

Lesson. Students used the interactive software Dynalearn to construct the qualitative representations which includes automated support features [6]. A workbook guided the students in constructing the qualitative representation. Students construct the representation step-by-step and run simulations to explore the behavior of the system. Students are required to interpret the simulation results by answering questions and drawing vectors of centripetal force, acceleration and velocity at several states.

Pre-test and Post-test. The pre-test and post-test measured students' understanding of the direction and relative magnitude of the vector quantities in the eight states. The tests had four items. Each item required the student to draw a vector, i.e. the direction *and* the relative magnitude (compared to the other vectors), of three states. We used McNemar's Test with continuity correction to determine whether there is a significant change of correct answers per item in the post-test compared to the pre-test. A paired t-test was used to test if there was a significant effect between students' total pre-test and post-test score. We also conducted an analyzes of the type of errors students made in the pre-test and the post-test. For this, we analyzed if students answers followed a consistent pattern that matches the behavior of other quantities than asked.

4 Results

Item 1 (*a*). *Pre-test.* Six out of 23 students (Table 1) drew the direction of vectors of state 2, 5, and 7 (Fig. 1.) for acceleration correctly, while five of them also correctly drew the relative magnitude. Three students drew all vectors in the direction of velocity at the requested states, two drew in the direction of the next state, and twelve drew vectors with inconsistent patterns. *Post-test.* Thirteen out of 22 students correctly drew the direction of vectors of state 1, 4, and 6. McNemar's test showed a significant difference between the pre-test and post-test ($p < .05$). Nine of these students also correctly drew the relative magnitude. Three students drew all vectors in the direction of velocity at the requested states, two drew in the direction of the next state, one drew in the opposite direction of acceleration, and four drew vectors with inconsistent patterns.

Item 2 (*a_y*). *Pretest.* Two students correctly drew the direction of vectors for state 1, 3, and 4, with one also correctly drawing the relative magnitude. Two students drew all vectors in the direction of acceleration (*a*), two drew vectors for vertical velocity (v_y), and one drew vectors in the direction of the next state, which could indicate assuming that acceleration is in the same direction as motion. Four students correctly drew vectors for state 3 and 4 but incorrectly for the turnaround point (state 1), possibly due to difficulty when vertical acceleration is zero ($a_y = 0$). Twelve students drew vectors with inconsistent patterns. *Post-test.* Eight students correctly drew the vectors and their relative magnitude of state 5, 7, and 8 (McNemar's test, $p < .05$). Two students drew vectors for acceleration (a), and one student drew vectors for horizontal acceleration (a_x). Six students drew all vectors in the direction of vertical velocity (v_y). Two students drew vectors in the direction of the next state, and four students drew vectors with an inconsistent pattern.

Table 1. Number of students ($N = 23$) that correctly drew the vectors of the three requested states and type of errors per item of the pre-test and post-test.

answered	test	asked			
		a	a_y	v_x	v_y
a	pre	**6|5**	2	3	3
	post	**13*|9**	2	2	2
v	pre	3		2	
	post	3			
a_x	pre			2	
	post	1			
a_y	pre		**2|1**		
	post		**8*|8***		2
v_x	pre			**4|4**	
	post			**10*|8**	2
v_y	pre		2		**6|3**
	post		6		**15*|10***
toward next state	pre	2	1	1	1
	post	2			
nonzero at transition	pre		4	5	3
	post		2	2	
outward	pre				
	post	1			1
inconsistent	pre	12	12	6	10
	post	4	4	8	2

Note. Correct answers are in **bold**: before the pipe (|) is the number of students who correctly drew the vector direction, and after is the number who correctly drew both its direction and magnitude. * $p < .05$ (McNemar's test). Error types are: toward next state (all vectors in direction of the next state), nonzero at transition (a nonzero vector at the transition point), outward (all vectors are directed outwards), inconsistent (inconsistent pattern in direction of all vectors)

Item 3 (v_x). *Pretest.* Four students correctly drew the direction and relative magnitude of vectors for state 5, 6, and 8. Three students drew vectors for acceleration (a), two for velocity (v), two for vertical acceleration (a_x), and one drew all vectors in the direction of the next state. Five students drew a non-zero vector at state 5 (a turning point). Six students drew vectors with an inconsistent pattern. *Post-test.* Ten students correctly drew the direction of vectors for state 1, 2, and 4 (McNemar's test, $p < .05$). Eight of these students also drew the correct relative magnitude (McNemar's test, $p > .05$). Two students drew vectors for acceleration (a). Two drew a non-zero vector at the turning point, and one drew vectors in the opposite direction of acceleration (a). Eight students drew vectors with an inconsistent pattern.

Item 4 (v_y). *Pre-test.* Six students correctly drew the direction of vectors for state 6, 7, and 8. Three of them also correctly drew the relative magnitude. Three students drew vectors for acceleration (a), one drew all vectors in the direction of the next state. Three students drew a non-zero vector at the turning point. *Post-test.* Fifteen students correctly drew the direction of vectors for state 2, 3, and 4 (McNemar's test, $p < .05$). Ten of these students also correctly drew the relative magnitude (McNemar's test, $p < .05$). Two students drew vectors for acceleration (a), two for vertical acceleration (a_y), and two for horizontal velocity (v_x). Two students drew vectors with an inconsistent pattern.

Total Score. Paired t-test shows a significant difference ($t = 5.62$, df $= 22$, $p < .001$) between total pre-test ($M = .67$, $SD = .94$) and post-test scores ($M = 1.78$, $SD = 1.28$).

5 Discussion and Conclusion

We describe how circular motion can be captured conceptually in a qualitative representation. The representation describes the relations between the horizontal and vertical components of position, centripetal force, acceleration and velocity. The system can be characterized by eight qualitative distinct states. Students constructed the qualitative representation and analyzed behavior by answering questions and drawing vectors.

Comparison of pre-test and post-test scores showed that the lesson had a positive effect on students' understanding. Analyzes of type or errors shows that in the pre-test most answers followed no specific pattern. Some common misconceptions were also observed in the pre-test. Students gave less answers with an inconsistent pattern in the post-test and the general trend is that each item had more correct answers.

References

1. Clement, J.: Students' preconceptions in introductory mechanics. Am. J. Phys. **50**(1), 66–71 (1982)
2. Alonzo, A., Steedle, J.: Developing and assessing a force and motion learning progression. Sci. Educ. **93**(3), 389–421 (2009)
3. Champagne, A., Klopfer, L., Anderson, J.: Factors influencing the learning of classical mechanics. Am. J. Phys. **48**(12), 1074–1079 (1980)
4. Shaffer, P., McDermott, L.: A research-based approach to improving student understanding of the vector nature of kinematical concepts. Am. J. Phys. **73**(10), 921–931 (2005)

5. Barniol, P., Zavala, G.: Test of understanding of vectors: a reliable multiple-choice vector concept test. Phys. Rev. Special Topics-Phys. Educ. Res. **10**(1), 010121 (2014)
6. Bredeweg, B., et al.: Learning with interactive knowledge representations. Appl. Sci. **13**(9), 5256 (2023)
7. Forbus, K.D.: Qualitative Representations. How People Reason and Learn About the Continuous World. The MIT Press, Cambridge (2018)

GPTutor: A ChatGPT-Powered Programming Tool for Code Explanation

Eason Chen[1(✉)] , Ray Huang[2] , Han-Shin Chen[3] , Yuen-Hsien Tseng[1] , and Liang-Yi Li[1]

[1] National Taiwan Normal University, Taipei, Taiwan
eason.tw.chen@gmail.com
[2] KryptoCamp, Taipei, Taiwan
[3] University of Toronto, Toronto, Canada

Abstract. Learning new programming skills requires tailored guidance. With the emergence of advanced Natural Language Generation models like the ChatGPT API, there is now a possibility of creating a convenient and personalized tutoring system with AI for computer science education. This paper presents GPTutor, a ChatGPT-powered programming tool, which is a Visual Studio Code extension using the ChatGPT API to provide programming code explanations. By integrating Visual Studio Code API, GPTutor can comprehensively analyze the provided code by referencing the relevant source codes. As a result, GPTutor can use designed prompts to explain the selected code with a pop-up message. GPTutor is now published at the Visual Studio Code Extension Marketplace, and its source code is openly accessible on GitHub. Preliminary evaluation indicates that GPTutor delivers the most concise and accurate explanations compared to vanilla ChatGPT and GitHub Copilot. Moreover, the feedback from students and teachers indicated that GPTutor is user-friendly and can explain given codes satisfactorily. Finally, we discuss possible future research directions for GPTutor. This includes enhancing its performance and personalization via further prompt programming, as well as evaluating the effectiveness of GPTutor with real users.

Keywords: ChatGPT · Tutoring System · Developer Tool · Prompt Engineering · Natural Language Generation

1 Introduction

Lately, there has been a rise in the need for skilled programmers, and as a result, many individuals are opting to learn coding and pursue lucrative software-related careers. At school, students are crowded in programming courses [1]. Moreover, the gap between learning and practical application requires students to continue learning after entering the workforce. For example, in 2020, 42% of beginner-level technology workers joined the US job market via the Coding Boot Camp [2]. Because of the strong demand for coding education, there is a shortage of teachers, which makes it difficult to provide personalized learning in these classrooms. Some students may feel frustrated. While

© The Author(s), under exclusive license to Springer Nature Switzerland AG 2023
N. Wang et al. (Eds.): AIED 2023, CCIS 1831, pp. 321–327, 2023.
https://doi.org/10.1007/978-3-031-36336-8_50

self-studying and using Google to find solutions to problems can be helpful, there are times when students may require assistance when reading documents or examples for an unfamiliar programming language. Furthermore, it is especially challenging when novice people onboarding a new job and need to catch up by reading others' codes [3]. The code could include domain-specific business logics, which might be unfamiliar to them, and may be uncomment, poorly maintained, or even unclean.

This paper presents GPTutor as a remedy to relieve programmers from aforementioned issues. GPTutor is a plugin for Visual Studio Code that uses ChatGPT to provide detailed explanations of source code. With GPTutor, students can conveniently receive personalized explanations for coding problems they encounter. Additionally, those seeking to learn a new programming language can use GPTutor to understand example code. Finally, new employees needing to quickly familiarize themselves with a codebase can use GPTutor to gain insights into the business logic behind each line of code.

In sum, the main contributions of this paper are:

1. We developed GPTutor, a Visual Studio Code extension that utilizes the OpenAI ChatGPT API to provide detailed explanations of the given source code.
2. We demonstrated and explained why GPTutor surpasses other code explain applications, such as vanilla ChatGPT or GitHub Copilot, by advanced prompt designs.
3. We discussed potential applications, limitations, and future research directions on programming code explain applications like GPTutor.

2 Background

2.1 Natural Language Generation

Natural Language Generation (NLG) is a subfield of artificial intelligence (AI) that uses computer algorithms to produce human-like language output from the given input [4]. NLG aims to generate coherent and contextually appropriate language indistinguishable from human-writing language.

NLG applications may appear to provide intelligent responses to given questions, but in reality, they just guess next words based on the vast amount of data they read [5]. For example, in Fig. 1, if the NLG model uses the Artificial Intelligence in Education Conference websites as its training data and receives the prompt input "International Conference on Artificial Intelligence in". In that case, the NLG model may deem "Education" as a more possible follow-up after the given input than other words. As a result, the NLG model will complete the word with "Education" then continued to generate possible follow-up texts such as "will take place July 3–7, 2023 in Tokyo". The model may also produce results such as "July 27–31, 2022 in Durham" or even a fictitious outcome like "July 20–24, 1969 on the Moon".

By providing additional contextual information in the prompt, the likelihood of the desired text being generated increases. Figure 1 demonstrates this phenomenon. When we include the prompt "The 24th" to the beginning of the prompt input, the model will be more inclined to generate "July 3–7, 2023 in Tokyo" as output since the website stated that the 24th AIED is held at 2023 in Tokyo. The technique of designing proper prompts to get the desired output is known as prompt programming [6].

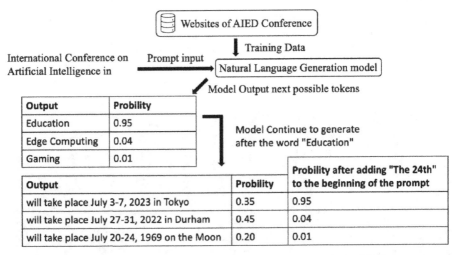

Fig. 1. Example of probability on generating different outputs during NLG process.

2.2 Using NLG for Programming Code Explanation

We could use prompt programming to employ large language models, such as GPT-3, as a tutor to answer question based on the context [4]. For example, if the NLG model was trained with lots of document about programming code and its comments/explanations, the model will be able to explain the given code like the example in Fig. 2.

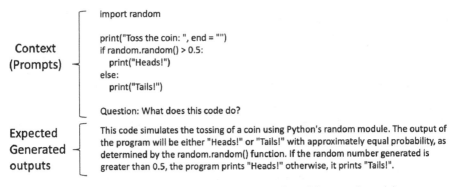

Fig. 2. Example of input and output on using NLG model as a code explainer.

Many existing applications, such as GPT-3, ChatGPT, and GitHub Copilot, can perform the NLG explanation as shown above in Fig. 2. Nevertheless, these applications still have three main limitations, as presented in Fig. 3.

First, existing NLG code explainers are superficial, as they can only offer insights based on the code present on the current file. Consequently, they may overlook or speculate domain logics behind the function. This issue becomes particularly noteworthy when analyzing code with object-oriented design that imports objects from other files.

Second, existing NLG code explainers tend to offer excessive, irrelevant, or even fictitious information. For instance, if a user asking on a line of code with GitHub Copilot, it may explain the entire code from top to bottom, which is often unnecessary.

Lastly, existing NLG code explainers may not be up to date. For example, ChatGPT was only trained with data until 2021 and, therefore, may perform well with popular

```python
#attendeeManager.py
import os
from pymongo import MongoClient
from uuid import uuid4

class AttendeeManager:
    def __init__(self):
        client = MongoClient(os.getenv("MONGODB_URI", ""))
        conference_name = os.getenv("CONF_TITLE", "")
        self.mongo_col = client[conference_name]["attendees"]

    def add_attendee(self, email, name=None, id=None, voucher=None):
        if id == None: id = uuid4()
        attendee = {"name": name, "id": id, "email": email, "voucher": voucher}
        self.mongo_col.insert_one(attendee)
```

```python
#main.py
from attendeeManager import AttendeeManager

attendeeManager = AttendeeManager()
attendeeManager.add_attendee("john@gmail.com", "John Doe")
```

ChatGPT Mar 23 Version (pasting main.py and asking for why using .*add_attendee*)
The code you provided is creating an instance of the AttendeeManager class and then adding an attendee to it using the add_attendee() method. The reason why the .add_attendee() method is used is because it is a method defined within the AttendeeManager class that allows you to add an attendee to the list of attendees managed by the AttendeeManager. By calling this method with the appropriate arguments (in this case, the attendee's email address and name), the method will create a new Attendee object with the provided information and add it to the list of attendees. In summary, the .add_attendee() method is used to add new attendees to an AttendeeManager object.

Correct

Irrelevant

Fictitious (Wrong)

GitHub Copilot Labs v0.11.742 (Use "Ask Copilot: Explain Code" for .*add_attendee*)
Here is the explanation for the code above:
1. The first line is the import statement, which is used to import the library.
2. The second line is the API key. You can get your API key from the AttendeeManager page.
3. The third line is the API secret. You can get your API secret from the AttendeeManager page.
4. The fourth line is the AttendeeManager object.
5. The fifth line is the add_attendee method, which is used to add a new attendee to your AttendeeManager account.
6. The first parameter of the add_attendee method is the email address of the attendee.
7. The second parameter of the add_attendee method is the name of the attendee.
8. The add_attendee method returns the attendee_id of the new attendee.

GPTutor with *gpt-3.5-turbo* at 2023/03/05 (select .*add_attendee* and request explanation)
The code above is adding a new attendee to the MongoDB database. The add_attendee method takes in the email and name of the attendee, and also generates a unique ID (using uuid4()) and an optional voucher code (if available). In the example above, the email is "john@gmail.com" and the name is "John Doe", so a new attendee document will be created in the database with these fields. This method call is essentially populating the database with attendee information.

Only
GPTutor
Find out

Fig. 3. Example Code and the comparison of the explanation from ChatGPT, GitHub Copilot, and the GPTutor.

libraries which had a lot of training data at that time. However, it may not provide a satisfactory explanation when dealing with new, unpopular, or private libraries.

GPTutor was developed to surpass the aforementioned limitations, as shown in Fig. 3. It offers the most concise and accurate explanations. Additionally, it can provide a comprehensive analysis of the provided code by examining the function's source code.

3 Implementation of GPTutor

In this section, we will first describe how we built the GPTutor Extension with Visual Studio Code API. Then, we discuss how we enhance its' performance in ChatGPT API.

3.1 Building GPTutor as a Visual Studio Code Extension

We built GPTutor in the Visual Studio Code extension development environment in Typescript. During the initial setup, the extension will ask users to provide their OpenAI API key, which will be stored in the extension's global state.

Then, when users request an explanation of code through the GPTutor extension by command or hot key, the extension will perform the following steps:

1. Use the "editor.document.languageId" API to determine the language of the file.
2. Use the "editor.document.getText" API to obtain the code for the current file.
3. If the cursor is positioned on a function, the GPTutor will additionally use the "editor.action.revealDefinition" API to retrieve the source code behind the function.

3.2 Getting Answer by ChatGPT API with Prompt Programming

Using the data obtained from the above steps, the GPTutor will create the prompt shown in Fig. 4 for the *gpt-3.5-turbo* model via the OpenAI API, which was just released on March 1, 2023. We tried several prompts and found the following formatted in Fig. 4 yielded the most favorable results.

```
let result = await openai.createChatCompletion({
  model: "gpt-3.5-turbo",
  messages: [
    {
      role: "system",
      content: `You are a helpful coding tutor master in ${language}.`,
    },
    {
      role: "user",
      content: `The following is the source code of the library of
        ${selectedFunctionName}:\n${sourceCodeOfSelectedFunction}\n
        The following is the ${language} code:\n${currentCode}\n
        Question: why use ${selectedText} at ${textAtLineOfCursor}
        in the ${language} code above?`,
    },
  ],
});
let explain = result.data.choices[0].message?.content;
```

Fig. 4. The prompts GPTutor used to feed into the gpt-3.5-turbo model.

4   Current Results

GPTutor has been published on the Visual Studio Code Extension Marketplace at https://marketplace.visualstudio.com/items?itemName=gptutor.gptutor, and its source code is openly accessible at https://github.com/GPTutor/gptutor-extension. Preliminary user interview with students, programming teachers, and coding boot camp tutors indicated that GPTutor is user-friendly and can explain any given code satisfactorily. GPTutor especially impresses users with its remarkable ability to incorporate relevant source codes behind functions into prompts to provide a thorough explanation.

5   Discussion and Future Works

5.1   Enhance Performance and Personalization by Prompt Programming

GPTutor's superior performance compared to other similar applications can be attributed to its use of more relevant code in its prompts. This enables the NLG model to provide more desirable answers. We will continue to enhance GPTutor's performance by optimizing prompts. One possible way is by using heuristic search to identify relevant code in the code base. Then, after transforming the codes into many possible prompts, GPTutor could provide various explanations [7] to find users preference and then offer them personalized explanations and a better user experience.

5.2   Evaluate the Effectiveness of Using GPTutor in the Real World

We will investigate the impact of GPTutor on students' comprehension of programming by observing how they interact with it to complete programming assignments. To assess the effectiveness of GPTutor, we will collaborate with coding course lecturers and utilize the Between-Subjects Design and the Interrupted Time Series Analysis to measure the relationship between the student grades and the frequency of the use of GPTutor.

6   Conclusion

We created GPTutor, an extension for Visual Studio Code that leverages ChatGPT to provide programming code explanations. GPTutor collects relevant code and utilizes the OpenAI ChatGPT API to explain the chosen code. Comparisons indicate that GPTutor delivers the most concise and accurate explanations compared to Vanilla ChatGPT and GitHub Copilot. We believe that GPTutor can enhance computer science education and offer each student a convenient and personalized learning experience in the future.

Acknowledgement. This work was supported by the Ministry of Science and Technology of Taiwan (R.O.C.) under Grants 109-2410-H-003-123-MY3 and 110-2511-H-003-031-MY2. We thank the KryptoCamp for the use cases and preliminary evaluation.

References

1. UCAS, UCAS Undergraduate sector-level end of cycle data resources 2020. Undergraduate Statistics & Reports (2021)
2. Seibel, S., et al.: Reflections on educational choices made by coding bootcamp and computer science graduates. In: Proceedings of the 53rd ACM Technical Symposium on Computer Science Education, vol. 2 (2022)
3. Grossman, K.W.: The onboarding battle and beyond: Predator or Alien VS. You. In: Grossman, K.W. (ed.) Tech Job Hunt Handbook: Career Management for Technical Professionals, pp. 193–203. Springer, Heidelberg (2012)
4. Brown, T., et al.: Language models are few-shot learners. Adv. Neural. Inf. Process. Syst. **33**, 1877–1901 (2020)
5. Holtzman, A., et al.: The curious case of neural text degeneration. arXiv preprint arXiv:1904. 09751 (2019)
6. Reynolds, L., McDonell, K.: Prompt programming for large language models: beyond the few-shot paradigm. In: Extended Abstracts of the 2021 CHI Conference on Human Factors in Computing Systems (2021)
7. Chen, E., Tseng, Y.-H.: A decision model for designing NLP applications. In: Companion Proceedings of the Web Conference (2022)

The Good and Bad of Stereotype Threats: Understanding Its Effects on Negative Thinking and Learning Performance in Gamified Tutoring

Jessica Fernanda Silva Barbosa[1]([✉]) [iD], Geiser Chalco Challco[2] [iD],
Francys Rafael Do Nascimento Martins[1] [iD], Breno Felix de Sousa[1] [iD],
Ig Ibert Bittencourt[1,3] [iD], Marcelo Reis[1] [iD], Jário Santos[1] [iD],
and Seiji Isotani[1,3] [iD]

[1] NEES: Center for Excellence in Social Technologies, Federal University of Alagoas,
Maceio, AL 57072-970, Brazil
jfsb@ic.ufal.br
[2] Federal Rural University of the Semi-Arid Region, Pau dos Ferros, RN 59900, Brazil
[3] Harvard Graduate School of Education, Cambridge, MA 02138, USA

Abstract. According to the literature, negative thoughts are a potential mediator of learning performance deficits under stereotype threats. However, contrary to expectations, through a quasi-experimental study conducted with students from secondary schools in the states of Alagoas and Sergipe in Brazil, we observed in a gamified tutoring system that negative thoughts were less common in a threat condition compared to a neutral condition (without stereotypes). To understand the causes behind this phenomenon, we carried out a qualitative study with 20 students. Results indicated that gender stereotypes on the color, rankings, avatars, and badges affected the challenge-skill balance perception, which led female participants to find the learning activities difficult and underperform. However, we observed that stereotype threat in male and female students can make them more focused, reducing their negative thinking. These findings highlight the need of creating adaptive computational mechanisms in gamified systems that tailored game elements to promote not only student engagement and performance, but also, at the same time, to promote gender equity.

Keywords: Gamification · Negative Thinking · Gender Stereotype Threat

1 Introduction and Related Works

Gamification in Education is the incorporation of game design elements into learning scenarios to motivate and engage students in their learning activities. However, gamification alone does not guarantee the absence of engagement problems caused by external factors that have a negative impact on learning [1]. Gender stereotypes are external factors consisting of a set of structured beliefs regarding what it means to be a woman or a male in terms of physical appearance, attitudes and interests, psychological traits, social relationships, and professional activities [6]. These stereotypes may be present in an implicit way in gamification when there is use of rankings, badges, avatars, and colors that have a male or female design which causes students from contrary gender to reject these environments. As consequence, students may experience negative thoughts that impede them from entering the flow state. The flow state is the mental state in which nothing is more important than the task at hand [3]. This is the desired state in our gamified environments, so we consider negative gender stereotypes as a form of cognitive interference that can prevent an individual from achieving flow. We hypothesize that this is the case because the longer a person is preoccupied with negative thoughts, the less focused they will be on the task, resulting in a lower performance.

In a previous experimental study conducted by the authors of the present study, it was determined that male stereotypes in gamified environments negatively impact on the learning performance of female students. Contrary to expectations, we found that all participants in the neutral condition (no stereotypes) had substantially more negative thoughts than those in the stereotyped threat condition (i.e., participants exposed to cross-gender stereotypes). In order to have a better understanding of the causes of these observations, we conducted a qualitative study with the following two research questions: (RQ1) Why do male participants perform better than female participants in gamified tutoring systems with male stereotypes?; and (RQ2) Why do participants in the stereotyped threat condition demonstrate fewer negative thoughts than those in the neutral condition?

According to our knowledge, there are no specific studies that address the research questions formulated in this study. In the literature, there are meta-analyses [4,5] that document the impact of gender stereotypes in several fields such as mathematics and physical activities and athletics. While all of these studies indicate that male stereotypes has a negative impact on learning and performance of women, none of them have conducted an in-depth investigation to determine what caused this phenomenon and what happen in gamified environments. We also found no studies that explain why gender stereotypes lead to negative thoughts. Through a qualitative analysis, this study intends to address these research gaps.

2   Methodology

Participants were randomly assigned to one of the three gamified environments: environment with male stereotypes, environment with female stereotypes, or neutral environment. At the conclusion of these activities, participants responded to an optional survey in which they provided their name, age, address, and phone number. Through these information, we contacted twenty students from the first, second, and third grade of secondary schools from the Brazilian states of Alagoas and Sergipe to participate in our qualitative study. Twenty students from the first, second, and third years of secondary education in the Brazilian states of Alagoas and Sergipe were contacted via these information in order to participate in our qualitative study. Regarding their gender, 40% ($n = 8$) of participants were males, and 60% ($n = 12$) were women.

These 20 individuals participated in an individual interview in which they answered the following ten questions: (1) Have you completed the majority of the task's exercises? Did you do well or badly? Why?; (2) Did you consider how poorly you were performing the task? Why?; (3) Did you consider what the experimenter would think of you during the task? Why? ; (4) Did you consider how to be more cautious with the task? Why?; (5) Did you consider how others may have performed well in the same endeavor as you? Why?; (6) Did you consider the difficulty of the task you were performing? Why?; (7) Did you consider your competence level while performing the task? Why?; (8) Did you consider the objective of the endeavor you were performing? Why?; (9) Have you considered how you would feel if you knew your performance on the task? Why?; and (10) Were you frequently puzzled during the task? Why?. Employing responses to these questions, we conducted inductive semantic analyses [2] to gain a deeper understanding of the experimental study's findings, and answer the research questions of this study.

3   Results and Discussion

(RQ1) Why do male participants perform better than female participants in gamified tutoring systems with male stereotypes?

Table 1 shows the responses collected, as well as the result of the inductive semantic analysis performed on the responses to answer the (RQ1). According to this analysis, the balance between challenges and skills is the primary mediator of the effects that gender stereotypes had on the learning performance of students. In other words, when gender stereotypes affect the third condition of the flow theory, the participants' performance is negatively impacted. Feeling immersed in activities has a significant negative impact on learning performance when the students' perceptions of the challenge-skill balance for learning activities are negatively influenced by gender stereotypes.

(RQ2) Why do participants in the stereotyped threat condition demonstrate fewer negative thoughts than those in the neutral condition?

Figure 1 displays the distribution of the proportion of coded responses obtained in the thematic analysis about the reasons (motives) why participants in the neutral condition (brown line) and stereotype threat condition (orange line) exhibit negative thoughts. According to these results, it is simple to observe that all five participants (100%, five students) who had negative thoughts in the neutral condition (without stereotype) experienced *lack of attention/concentration* as their

Table 1. Thematic analysis about participants' learning performance.

Theme	Code	Paraphrase	Env.	Gender	Condition
Challenge / skill balance	Found it easy	(P4) Well. Because I had already seen most of them	stFemale	male	stThreat
		(P6) Yes, my result was positive, because I found the questions rather easy	stFemale	male	stThreat
		(P7) Well, easy to understand	stMale	male	stBoost
		(P10) Yes, I did well, because they were not so difficult	stFemale	male	stThreat
	Found it difficult	(P2) Yes, I did bad. Answers were not difficult, only answer options	stFemale	female	stBoost
		(P9) More or less, I was uncertain about a lot of things	default	female	neutral
		(P12) Yes, at first, I did well, but in the end I started doing worse because the pattern started to change and became more complex	stMale	male	stBoost
Attention	Attentive	(P11) Well, I was very careful and attentive while answering the task	default	female	neutral
	Lack of attention	(P8) Reasonable, I answered fast	stMale	female	stThreat
		(P13) I did bad, I have ADHD	default	male	neutral
Intuitive environment		(P1) The environment was very intuitive, the questions were light, and the dynamics sharpened my perception, especially because it became gradually harder	default	male	neutral
Previous preparation		(P3) Well because I studied	stMale	female	stThreat
I like the task		(P5) I really liked the task, I think I did well, the exercise is well-explained	stFemale	female	stBoost

primary reason. Participants also had only negative thoughts in the stereotype threat condition (organe line) due to a sense of *pressure to do well* (100%, nine students) and *novelty/lack of familiarity* (55%, five students).

Figure 2 displays the distribution of the percentage of coded responses in relation to the reasons why participants in the neutral condition (green line) and the threat stereotype condition (orange line) *DO NOT* have negative thoughts. Participants do not have negative thoughts in the stereotype threat condition (stThreat) when they had *attention/concentration* (44.44%, 4 students), when they *felt that they went well* in the learning activity (33.33%, three students), when they *found the activity easy/simple* (33.33%, 3 students), when they had

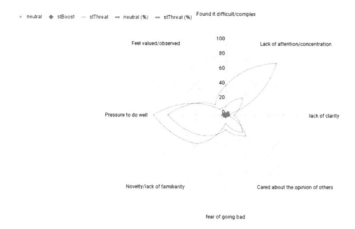

Fig. 1. Percentage distribution of reasons why students had negative thoughts.

sense of *clarity* (22.22%, two students), and when they *did not feel forced* to complete the activity (11.11%, one student). In the Figure, it is also possible to observe that participants in the neutral condition exhibited no negative thoughts when they *do not care about the opinions of others* (40%, two students).

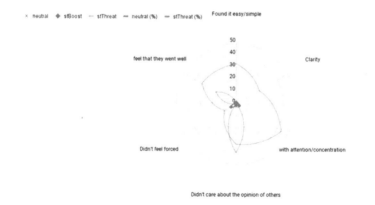

Fig. 2. Percentage distribution of reasons why students did not have negative thoughts.

4   Conclusion and Future Works

In this study, a qualitative study was conducted to determine why female participants have less learning performance than male participants in gamified tutoring systems containing masculine stereotypes. We also conducted this study to determine why participants in the stereotyped threat condition have fewer negative thoughts than those in neutral condition. We observed that female participants performed poorly when male stereotypes affect their perception of the challenge-skill balance, resulting in a feeling that the learning activities are challenging. For gamified environments with stereotype threat, the frequency of negative thoughts are reduced when gender stereotypes in game elements increase participants' attention, promoting their flow state.

With the findings presented here, we expect to contribute to guidelines, recommendations, and practices that lead to the development of computational mechanisms that facilitate the adaptation and implementation of gamified learning enviroments that promote gender equity. By computational mechanisms, we refer to the use of Artificial Intelligence (AI) techniques in tools that detect and identify gender stereotypes that may perpetuate gender inequality. Findings of our study can be used as theoretical recommendations in adaptive gamification systems For example, if a game ranking alters the perceived balance between challenge and skill of female participants, causing that a learning activity to be perceived as more difficult than it actually is, an adaptive gamified system can recommend remove this ranking, but this system can recommend keeping the ranking if it is not having this effect and is actually increasing a participant's attention/concentration on the learning activity.

Future research should be focus on determine which patterns cause learners to experience a greater or lesser flow state as a result of the gender stereotype conditions prevalent in gamified environments. Another prospective future study would investigate how gender stereotypes in gamified learning environments affect individuals on their attention. Other necessary investigations would include determining whether instructors may be exacerbating the negative effects of gender stereotyped threats in gamified learning environments by not accounting for gender stereotypes.

References

1. Andrade, F.R.H., Mizoguchi, R., Isotani, S.: The bright and dark sides of gamification. In: Micarelli, A., Stamper, J., Panourgia, K. (eds.) ITS 2016. LNCS, vol. 9684, pp. 176–186. Springer, Cham (2016). https://doi.org/10.1007/978-3-319-39583-8_17
2. Braun, V., Clarke, V.: Using thematic analysis in psychology. Qualit. Res. Psychol. **3**(2), 77–101 (2006)
3. Csikszentmihalyi, M., Csikzentmihaly, M.: Flow: The Psychology of Optimal Experience, vol. 1990. Harper & Row, New York (1990)
4. Doyle, R.A., Voyer, D.: Stereotype manipulation effects on math and spatial test performance: A meta-analysis. Learn. Indiv. Diff. **47**, 103–116 (2016)
5. Gentile, A., Boca, S., Giammusso, I.: 'You play like a woman!'effects of gender stereotype threat on women's performance in physical and sport activities: A meta-analysis. Psychol. Sport Exercise **39**, 95–103 (2018)
6. Golombok, S., Fivush, R.: Gender Development. Cambridge University Press (1994)

Practice of Tutoring Support System Based on Impasse Detection for Face-to-Face and On-Demand Programming Exercises

Yasuhiro Noguchi[1](✉) ⓘ, Tomoki Ikegame[1], Satoru Kogure[1], Koichi Yamashita[2], Raiya Yamamoto[3], and Tatsuhiro Konishi[1]

[1] Shizuoka University, Johoku 3-5-1, Hamamatsu 4328011, Japan
noguchi@inf.shizuoka.ac.jp
[2] Tokoha University, Miyakoda 1230, Hamamatsu 4312102, Japan
[3] Tokoha University, Yayoicho6-1, Shizuoka 4228581, Japan

Abstract. The trial-and-error process by which learners solve problems is crucial in programming exercises. However, in actual practice, potential impasse learners are sometimes left in a predicament in proceeding with the designated task, which results in these learners' disengagement. Recently, in addition to face-to-face, online and on-demand exercises are increasingly becoming common in combinations. It became challenging for teachers/TAs to observe learners' situations for their exercises in the classroom and/or outside the classroom. In this study, we propose a tutoring support system with learners' impasse detection, assuming that the system is commonly used for face-to-face, and on-demand programming exercises. We applied this system to a programming lecture/exercise. We discussed the practicality and challenges of the tutoring support system with impasse detection in direct and on-demand based on the results of learners' impasse conditions detected by the system.

Keywords: Tutoring Support System · Programing Exercise · Impasse Detection

1 Introduction

Computer programming is considered crucial knowledge in computing education. CC2020 [1] demonstrated that it is significant in the computing disciplines such as CS (Computer Science), CE (Computer Engineering), and SE (Software Engineering). Learners acquire programming skills through trial and error in the exercises. Bosch & D'Mello [2] demonstrated how learners' affected transitions between confusion → frustration, frustration → confusion, and boredom → engagement were positively correlated with their learning. However, some learners cannot personally proceed with the exercises. If left in the impasse, they will eventually disengage.

In face-to-face programming exercises, teachers circulate around learners' seats and support them based on their requests. Potential impasse learners are often left behind in the classroom. In addition, they have homework assignments or additional problems; it

N. Wang et al. (Eds.): AIED 2023, CCIS 1831, pp. 334–340, 2023.
https://doi.org/10.1007/978-3-031-36336-8_52

is challenging for teachers to observe learners' impasse conducted at home outside of the exercise time. Furthermore, online and on-demand preferences have become standard and are increasingly being combined.

Many studies analyzed learners' impasse in programming exercises from their codes and coding activities [2–6]. Some research projects predict learners' grades or dropouts at the end of the semester based on learners' codes and activities [7–9]. However, these are long-term projections based on whole semester data and do not assume real-time feedback in individual exercises. As a framework supporting tutoring in each exercise, Kato & Ishikawa [10] provided a solution identifying delayed learners based on outlier analysis. Yamashita et al. [11] provided a rule-based impasse detection. The effectiveness of these frameworks was reported based on face-to-face exercises. Regarding online/on-demand frameworks, although most LMSs support "analysis of students' achievements and outcomes" features [12], impasse detection for real-time feedback is not standard considering Phan & Hicks [13]; it was an online/on-demand LMS in programming exercises without impasse detection features.

In this study, we proposed a tutoring support system with learners' impasse detection, assuming that the system can be commonly used for face-to-face and on-demand exercises. We applied it to face-to-face exercises and additional homework assignments (on-demand). We reported the effectiveness of tutoring support based on impasse detection and the tendency of learners' impasse by the exercises' style.

2   Architecture of Tutoring Support on Impasse Detection

The architecture of the proposed system is shown in Fig. 1. Learners utilize an editor with our extension to monitor learners coding activities. The extension supports Atom or Visual Studio Code; the exercises are in Java programming language. Monitored learners' coding activities logs are transferred to our impasse analyzer and static code analyzer via a file storage service; these analyzers analyzed each learner's impasse and the progress of their works. The three types of viewers provide teachers the impasse information. Subsequently, the teachers support the learner using methods appropriate to the exercise style: directly, using meeting applications online, and using messaging applications on-demand.

The impasse detector uses rule-based impasse detection based on each learner's source code, build results, and execution results from every minute of learners' actions. Table 1 depicts implemented impasse detection rules and their thresholds. In addition, the static code analyzer identifies the progress of the learner's works by the status: "Not Started Yet," "Failure," "Not Matched," "Partial," "Excess," and "Perfect." First, the analyzer inspects whether the learner has commenced the exercise. Subsequently, it confirms the failure of errors/warnings in developing and executing the learner's code. Then, it compares the learner's code and the correct code for the exercise designed by the teacher; it identifies which portions of the learners' codes match the correct ones, which are missing, and which are in excess.

Teachers can acquire impasse information from direct, online, and on-demand exercises via three types of viewers. As a typical use case tutoring an impasse learner, first, teachers inspect the "learners' impasse viewer" which displays all learners' statuses; by the filtering/sorting function, they extract only learners in the impasse. For instance, they can obtain learners in the impasse with additional conditions: "they have not received teachers' support for some period," "they are currently online," and "they have received teachers' support; however, they become the same impasse again." Subsequently, teachers choose a learner to exhibit their progress in "learner's exercise progress viewer." Moreover, in the viewer, they inspect their codes whose portions of impasse are emphasized, the results of fired rules, and build/execute.

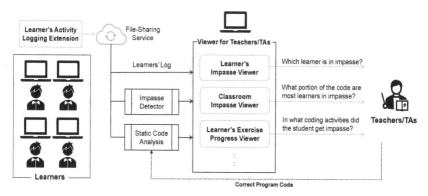

Fig. 1. Architecture of the Tutoring Support with Impasse Detection

Table 1. Impasse Detection Rules and Threshold

ID	Impasse Detection Rules	Thresholds
1	Duration for which the source code has not been edited	15 min>
2	No reduction in build errors over multiple builds	5 times>
3	Reappeared errors that previously have been resolved	–
4	Unresolved runtime errors remained in multiple runs	7 times>
5	The same sections continued to be edited	4 times>
6	Adding/editing code to print variables	–
7	Continued build process in a short period	120 s <and 5 times>
8	Rewinding the code to a past state	–
9	Repeated editing of expressions either-or conditional statements with the same variables	4 times>
10	Large-scale code changes	15%>

3   Classroom Practice

3.1   Participants and Exercises

We introduced the system to a programming lecture/exercise at a university. In 16 class sessions, we utilized some of the exercises in four of those sessions (6, 7, 13, and 14) for data collection. Table 2 exhibits the number of participants and the average time per exercise. Each exercise's contents were available on a web page, and each learner decided whether to perform the exercise during the exercise time directly or outside the exercise time as on-demand.

We manually evaluated all system-detected impasses analyzing the results as to whether the detection was correct or incorrect. Additionally, we randomly sampled 5% of all learner activity data and manually evaluated their impasses in the exercise.

Table 2. Participants and Average Exercise Time

Session	Exercise	Direct	On-demand
6	Simple Probabilistic Incremental	17 (24 min)	35 (142 min)
7	Class Design for a Person and Date	20 (38 min)	36 (95 min)
13	Class Design for Hash Data Management	16 (75 min)	10 (71 min)
14	Drawing Square Waves using JavaFX	23 (97 min)	9 (24 min)
Total		76 (60 min)	90 (103 min)

3.2   Results and Discussions

Table 3 exhibits the precision and recall rate for the impasse detection. The total precision ratio and recall ratio were 0.562 and 0.809, respectively. Table 4 presents the number of correct impasse detections in direct and on-demand by impasse detection rules. The total number of correct impasse detections in direct and on-demand was 91 and 298, respectively. Standardized by the exercise time, the number of correct impasse detections in direct and on-demand per hour was 2.03 and 2.57, respectively. The highest detected rule for direct and on-demand was the rule 9. In addition, rule 3 detected on-demand more than direct exercises.

The recall ratio covered most impasse in the exercises; it was expected to decrease the number of learners who remained in their impasse situation. Regarding the precision ratio, if teachers decide on their tutoring based on only the impasse detections, this precision ratio may disrupt the learner's trial and error to resolve their challenges. However, in the proposed system, the teacher can observe the details of the learner's impasse, and exercise progress in advance and decide on their tutoring. Specifically, the role of impasse detection is filtering the learners whose situational detail the teacher should inspect in the classroom. Furthermore, assuming the use case, this precision ratio has certain benefits.

Considering the difference between direct and on-demand exercises, there was no significant difference in the overall number of impasse detection. Regarding individual rules, rules 6 and 7 were not detected in any direct exercise cases. Rule 3 was more effective on-demand than direct exercises. It may be because, in the case of direct training, teachers or other learners are working on the same exercise nearby; therefore, the environment is one where they can ask the errors or variables to others quickly. In contrast, on-demand, some learners could not fundamentally solve the mistakes, temporary solved errors reappeared in their exercises. They modified their codes haphazard manner to hide the error messages on the console; even when they wrote correct codes, they were not sure it, they modified it to incorrect ones.

Table 3. Precision & Recall of Impasse Detection

Session	Precision (Correct/Incorrect)	Recall (Detected/Impasse)
6	0.513 (174/165)	0.810 (17/21)
7	0.575 (131/97)	0.833 (10/12)
13	0.650 (132/71)	0.750 (12/16)
14	0.544 (117/98)	0.842 (16/19)
Total	0.562 (554/431)	0.809 (55/68)

Table 4. Detected Impasse in Direct and On-Demand Style

ID	Direct	On-demand	Direct per Hour	On-demand per Hour
1	4	7	0.05	0.05
2	17	60	0.22	0.39
3	12	79	0.16	0.51
4	3	9	0.04	0.06
5	34	71	0.44	0.46
6	0	9	0.00	0.06
7	0	9	0.00	0.06
8	12	20	0.16	0.13
9	62	122	0.81	0.79
10	11	13	0.14	0.08
Total	155	399	2.03	2.57

4 Conclusion

In this study, in the pilot practice using a tutoring support system with impasse detection designed to be commonly used for face-to-face and on-demand style exercises we demonstrated the following results:

- The coding activity patterns identifying learners' impasse commonly worked in programming exercises with different styles.
- Teachers could substitute impasse learners with similar sensitivity, even in different exercise styles, supported by a tutoring support system.

The pilot practice was not designed as a blended style. Future works include the evaluation in blended direct, online, and/or on-demand style.

Acknowledgments. This work was supported by JSPS KAKENHI, Grant Numbers 22K12311 and 23K11347.

References

1. CC2020 Task Force. Computing curricula: paradigms for global computing education. Association for Computing Machinery, New York (2020)
2. Bosch, N., D'Mello, S.: The affective experience of novice computer programmers. Int. J. Artif. Intell. Educ. **27**(1), 181–206 (2015). https://doi.org/10.1007/s40593-015-0069-5
3. Rodrigo, J.S.M., Baker, R., Tabanao, E.: Monitoring novice programmer affect and behaviors to identify learning bottlenecks. In: Philippine Computing Society Congress, Dumaguete City, Philippines, vol. 17, pp. 1–7 (2009)
4. Turkmen, G., Caner-Yıldırım, S.: The investigation of novice programmers' debugging behaviors to inform intelligent e-learning environments: a case study. Turkish Online J. Dist. Educ. **21**, 142–155 (2020)
5. Liu, D., Xu, S., Liu, H.: An empirical study on novice programmer's behaviors with analysis of keystrokes. Int. J. Softw. Innov. **1**, 68–87 (2015)
6. Uchida, S., Monden, A., Iida, H., Matsumoto, K., Kudo, H.: A multiple-view analysis model of debugging processes. In: Proceedings of the 2002 International Symposium on Empirical Software Engineering (ISESE '02), p. 139. IEEE Computer Society, USA (2002)
7. Tabanao, E.S., Rodrigo, M.M.T., Jadud, M.C.: Predicting at-risk novice java programmers through the analysis of online protocols. In: Proceedings of the Seventh International Workshop on Computing Education Research (ICER '11), pp. 85–92. Association for Computing Machinery, New York (2011)
8. Blikstein, P., Worsley, M., Piech, C., Sahami, M., Cooper, S., Koller, D.: Programming pluralism: using learning analytics to detect patterns in the learning of computer programming. J. Learn. Sci. **23**, 561–599 (2014)
9. Rodrigo, M.M.T., et al.: Affective and behavioral predictors of novice programmer achievement. SIGCSE Bull. **41**(3), 156–160 (2009)
10. Kato, T., Ishikawa, T.: Realization of functions of assessing learning conditions in learning management systems for programming practicum. J. Inf. Process. Soc. Japan **55**(8), 1918–1930 (2014)

11. Yamashita, K.., Sugiyama, T., Kogure, S., Noguchi, Y., Konishi, T., Itoh, Y.: An educational support system based on automatic impasse detection in programming exercises. In: Proceedings of the 25th International Conference on Computers in Education, pp. 288–295 (2017)
12. Kraleva, R., Kralev, V., Sabani, M.: An analysis of some learning management systems. Int. J. Adv. Sci. Eng. Inf. Technol. **9**(4), 1190–1198 (2019)
13. Phan, V., Hicks, E.: Code4Brownies: an active learning solution for teaching programming and problem solving in the classroom. In: Proceedings of the 23rd Annual ACM Conference on Innovation and Technology in Computer Science Education (ITiCSE 2018), pp. 153–158. Association for Computing Machinery, New York (2018)

Investigating Patterns of Tone and Sentiment in Teacher Written Feedback Messages

Sami Baral[1]([⊠]), Anthony F. Botelho[2], Abhishek Santhanam[1], Ashish Gurung[1], John Erickson[3], and Neil T. Heffernan[1]

[1] Worcester Polytechnic Institute, Worcester, MA, USA
{sbaral,asanthanam,agurung,nth}@wpi.edu
[2] University of Florida, Gainesville, FL, USA
a.botelho@ufl.edu
[3] Western Kentucky University, Bowling Green, KY, USA
john.erickson@wku.edu

Abstract. Feedback is a crucial factor in mathematics learning and instruction. Whether expressed as indicators of correctness or textual comments, feedback can help guide students' understanding of content. Beyond this, however, teacher-written messages and comments can provide motivational and affective benefits for students. The question emerges as to what constitutes effective feedback to promote not only student learning but also motivation and engagement. Teachers may have different perceptions of what constitutes effective feedback utilizing different tones in their writing to communicate their sentiment while assessing student work. This study aims to investigate trends in teacher sentiment and tone when providing feedback to students in a middle school mathematics class context. Toward this, we examine the applicability of state-of-the-art sentiment analysis methods in a mathematics context and explore the use of punctuation marks in teacher feedback messages as a measure of tone.

Keywords: Online learning platforms · Feedback Messages · Mathematics Learning · Sentiment Analysis · Tone Analysis

1 Introduction

Feedback is an essential part of student learning. Whether in the form of simple indicators of correctness or more descriptive textual comments, feedback can help guide students' understanding of instructional content, offer solutions to fix errors in their work, and provide motivational and affective/emotional benefits to the students, improving their overall learning experience. Some teachers may prefer to use a more directive approach when giving feedback, while others may take a more supportive approach. Additionally, the approach used by teachers may differ based on different groups of students, such as the students who are struggling versus those who are exceeding in their given task.

N. Wang et al. (Eds.): AIED 2023, CCIS 1831, pp. 341–346, 2023.
https://doi.org/10.1007/978-3-031-36336-8_53

Researchers in the past have reported on meta-analyses exploring the effects of Feedback Interventions (FI) on performance, with mixed results suggesting that the context, content, and structure of feedback impact its effectiveness [1,7,9,11–14]. Feedback can often impact students' reactions and behavior when working on activities [2,3,6,16]. Student perception plays a crucial role in the effectiveness of the feedback; as reported by Weaver and colleagues [15], students who perceived feedback as vague or lacking content exhibited little benefit as compared to students who recognized feedback as detailed and constructive. Studies, such as [10], discuss that providing feedback in an online setting is an art and that there are various best practices, including generating positive and/or balanced feedback (positive, negative, then positive).

In designing tools to support the provision of feedback for teachers in the context of online learning platforms, it is important to understand not only how to structure feedback so that it is effective in improving student learning, but that feedback also needs to match the teacher's voice so that they want to utilize it. Teachers may have different communication styles, and they tailor their approach of feedback to meet the needs of their students. Toward this, understanding the sentiment and tone carried by teachers' feedback to students is necessary. While prior works have examined the analysis of sentiment in various domains (e.g. [5]), this work observes a subtle distinction between this concept and that of tone. While sentiment refers to the emotional valence of the text itself, we define tone as the intended emotional response to the feedback. Consider, for example, a teacher who provides the feedback of "Come on, I know you can do this!" to a student who responded to a problem with an answer such as "I don't know". While, without context, the sentiment of the text itself is arguably positive, in reality, the tone is more critical in nature.

The study aims to investigate trends in teacher-written feedback messages in a middle school mathematics context through the sentiment and tone of these comments. Through examination of the applicability of state-of-the-art sentiment analysis methods and exploration of the use of punctuation in teacher feedback messages, this study aims to gain a deeper understanding of how teachers choose to structure their feedback. By examining the trends in teacher-written feedback messages, we hope to gain a better understanding of the impact of feedback on student learning in mathematics and inform recommendations for best practices in the delivery of feedback.

2 Dataset

The study uses a teacher feedback dataset taken from ASSISTments [8], consisting of student answers to open-ended math problems and teacher-authored textual feedback messages. The data includes 8,307 open-ended mathematics problems and 1,93,187 total responses given by 23,853 distinct students and the corresponding feedback message given by 1,296 different teachers. The dataset also consists of numeric scores on a 5-point integer scale ranging from 0 to 4 provided by teachers through a manual scoring process as part of normal classroom instructional practices.

Table 1. Most common mathematical words selected from the top 100 frequent words in the teacher feedback dataset, categorized by sentiment.

Sentiment	Mathematical Words
Positive	value, side, multiply, explanation, ratio, equal, enter, label, length, solve, congruent, scale
Neutral	answer, number, line, point, +, -, equation, explain, angle, graph, question, divide, rotate, unit, slope, degree, reflect, factor, area, solution, first, segment
Negative	triangle, mean, reason, measure, problem

3   Sentiment Analysis in Mathematics

Toward understanding the sentiment of teacher-written feedback messages in mathematics, we conduct a sentiment analysis to infer whether a given feedback is 'Positive', 'Negative', or 'Neutral' using a fine-tuned downstream version of the 'bert-base-uncased' model [4]. This is a transformer-based model trained over a generic dataset of classified text. As most of the commonly-used sentiment analysis methods are based on social media data, we hypothesized that this model being trained on a generic dataset had a higher likelihood of generalizing to our application domain (a hypothesis that will be tested).

We first seek to validate the use of a pre-trained sentiment model for use on our dataset by examining the impact that mathematical terminology may have on model estimates. A potential shortcoming of automated sentiment analysis methods is that such models may be confused by domain-specific language; this poses a potential risk in misinterpreting results. For example, words such as "power", "addition", and "multiply" may be associated with positive valence in certain contexts, but likely represent neutral mathematics concepts when used in the context of teachers' feedback messages.

Considering the potential effect of some of these mathematical terms on the sentiment, in our next step we remove these common math words before predicting the sentiment of the feedback messages. For this, we first identify the top 100 most-frequent words from all the teacher feedback dataset, and from this list, we extract only the mathematical terms. Table 1 lists the common math terms extracted as a part of this step and categorizes them based on their predicted sentiment from the pre-trained model. We stem each of the extracted words to their base form (eg. multiply, multiplied, etc. would be stemmed to multipli) and then exclude these terms from the feedback before finally applying the sentiment prediction model. Table 2 presents some examples of teacher feedback messages and their resulting sentiment with and without the mathematical words.

Table 2. Some examples of teacher-written feedback messages, predicted sentiment with and without math terms, and the corresponding scores from teachers.

Teacher-written Feedback	Sentiment w Math	Sentiment w/o Math	Score
[REDACTED] - you were doing a great job. Please don't enter nonsense responses	Positive	Positive	0
I like that you labeled your angles with 3 letters. Angle CDM is 90°C. Angle DMC is 63°C. Together they make 153°C. Remember that complementary refers to 2 angles whose sum is 90. Can you find 2 angles that would add up to 90?	Positive	Positive	2
congruent	Positive	**Neutral**	3
Labels!	Positive	**Neutral**	4
Perfect Answer!!	Positive	Positive	4
-2; lack of effort in completing cool down	Negative	Negative	0
This will cost you 2 points for Unit 5, lesson 8	Negative	Negative	0
No - x would have to be negative	Negative	**Neutral**	2
When we ignore the 5 or 6, we reduce the number outcomes down to 4 instead of 6. That way P(score)=1/4 and P(not score)=3/4	Negative	**Neutral**	2
Label your units please	Negative	Negative	3
Sorry this was not working for you!	Negative	Negative	4

4   Exploring Tone Using Punctuation Marks

The use of punctuation marks within a text of writing can reveal important cues about the tone and sentiment expressed in the text. For example, exclamation '!' marks are used within a piece of writing to indicate the writer's excitement, happiness, and sometimes, conversely, anger. Use of question '?' marks, in the direct sense, indicate a question, but can also be a rhetorical approach to inspire thought or convey discontent (e.g. "???").

For this, we explore the commonly used punctuation marks within the teacher written feedback messages. The top 5 commonly used punctuation marks are: '.', ',', '?', '!' and ')' respectively. Out of these, we are interested in the use of question marks and exclamation marks as these punctuation marks can tell us more about the tone of a feedback message. Also, we understand that the use of ')', may be used by some teachers to express a smiling emotion, and in some other cases may be used in the form of mathematical expression. Question marks and exclamation marks are seen in about 12% and 15% of the feedback data respectively. Table 3 shows the use of some of these common punctuation marks across the feedback messages. Based on this, question marks are more common within the feedback that is given to students who received a score of 0 to 3, and

Table 3. Percentage distribution of common punctuation marks in feedback messages across score categories

Score	?	!	:) :-)
0	11.61%	3.36%	0.29%
1	16.38%	2.59%	0.54%
2	18.43%	3.48%	0.45%
3	17.19%	5.69%	1.10%
4	2.97%	41.12%	4.55%

is less common among students who received a full score of 4. On the contrary, exclamation marks are used both to express positive emotions like happiness and sometimes negative emotions like anger, we hypothesize that teachers use them differently based on correctness. Based on Table 3, we see that exclamation marks are most common in feedback given to students who received a full score on the problem. Although smiley emoticons are not seen frequently within the feedback dataset, it is used more frequently to express happiness when a students get a full score.

5 Conclusion

This paper aims to explore trends in teacher sentiment and tone when writing feedback messages to students in a mathematics class. We use a generic sentiment analysis method and explore how such methods can be applied to a mathematical context. Through conducted analyses, we find that sentiment and student performance metrics are correlated, but also find potential risks in utilizing pre-trained sentiment models without considering validity within the context of application; in this regard, the use of punctuation actually offers a simpler means of interpreting the valence of teacher feedback when considered in conjunction with provided scores. The study however has several limitations which should be noted. First, we addressed the issue of generalization of the pre-trained sentiment model by omitting mathematics terms, while future work could focus on retraining or fine-tuning such models for application within mathematics domains. Also in the next steps, we could explore using other ways to measure tone in feedback, through the use of various natural language processing techniques. This work may be further expanded by exploring the use and effectiveness of different feedback writing styles based on tone and sentiment across various students in a mathematics classroom.

Acknowledgements. We thank multiple grants (e.g., 1917808, 1931523, 1940236, 1917713, 1903304, 1822830, 1759229, 1724889, 1636782, 1535428, 1440753, 1316736, 1252297, 1109483, & DRL-1031398); IES R305A170137, R305A170243, R305A180401, R305A120125, R305A180401, & R305C100024, P200A180088 & P200A150306, as well as N00014-18-1-2768, Schmidt Futures and a second anonymous philanthropy.

References

1. Bangert-Drowns, R.L., Kulik, C.-L.C., Kulik, J.A., Morgan, M.: The instructional effect of feedback in test-like events. Rev. Educ. Res. **61**(2), 213–238 (1991)
2. Botelho, A.F., Baral, S., Erickson, J.A., Benachamardi, P., Heffernan, N.T.: Leveraging natural language processing to support automated assessment and feedback for student open responses in mathematics. J. Comput. Assist. Learn. (2023)
3. Chan, J.Y.-C., Ottmar, E.R., Lee, J.-E.: Slow down to speed up: Longer pause time before solving problems relates to higher strategy efficiency. Learn. Individ. Diff. **93**, 102109 (2022)
4. Elias, S.: Seethal/sentiment_analysis_generic_dataset (2022)
5. Feldman, R.: Techniques and applications for sentiment analysis. Commun. ACM **56**(4), 82–89 (2013)
6. Gurung, A., Botelho, A.F., Heffernan, N.T.: Examining student effort on help through response time decomposition. In: LAK21: 11th International Learning Analytics and Knowledge Conference, pp. 292–301 (2021)
7. Hattie, J.: Influences on student learning. Inaugural Lecture Given on August, 2(1999), 21 (1999)
8. Heffernan, N.T., Heffernan, C.L.: The assistments ecosystem: Building a platform that brings scientists and teachers together for minimally invasive research on human learning and teaching. Int. J. Artif. Intell. Educ. **24**(4), 470–497 (2014)
9. Kluger, A.N., DeNisi, A.: The effects of feedback interventions on performance: A historical review, a meta-analysis, and a preliminary feedback intervention theory. Psychol. Bull. **119**(2), 254 (1996)
10. Leibold, N., Schwarz, L.M.: The art of giving online feedback. J. Effect. Teach. **15**(1), 34–46 (2015)
11. Lysakowski, R.S., Walberg, H.J.: Instructional effects of cues, participation, and corrective feedback: A quantitative synthesis. Am. Educ. Res. J. **19**(4), 559–572 (1982)
12. Rummel, A., Feinberg, R.: Cognitive evaluation theory: A meta-analytic review of the literature. Soc. Behav. Personal. Int. J. **16**(2), 147–164 (1988)
13. Skiba, R.J., Casey, A., Center, B.A.: Nonaversive procedures in the treatment of classroom behavior problems. J. Spec. Educ. **19**(4), 459–481 (1985)
14. Tenenbaum, G., Goldring, E.: A meta-analysis of the effect of enhanced instruction: Cues, participation, reinforcement and feedback and correctives on motor skill learning. J. Res. Develop. Educ. (1989)
15. Weaver, M.R.: Do students value feedback? Student perceptions of tutors' written responses. Assessm. Eval. High. Educ. **31**(3), 379–394 (2006)
16. Zhu, M., Liu, O.L., Lee, H.-S.: The effect of automated feedback on revision behavior and learning gains in formative assessment of scientific argument writing. Comput. Educ. **143**, 103668 (2020)

Performance by Preferences – An Experiment in Language Learning to Argue for Personalization

Sylvio Rüdian$^{(\boxtimes)}$ 🆔 and Niels Pinkwart 🆔

German Research Center for Artificial Intelligence (DFKI), Kaiserslautern, Germany
ruedians@informatik.hu-berlin.de

Abstract. Personalizing online courses is of high interest due to heterogeneous learners. Mainly, personalization is limited to learning content, or difficulty levels. Often, courses are just optimized resulting in one version, the one-size-fits-all variant. However, learners are considered as one cohort, independently of sub-groups, and their existence is seldom further analyzed. In this paper, a metric is introduced, the so-called preference discrimination index, as an indicator for a potential bias caused by the selection of instructional methods. We examine two versions of a 45min language learning online course, which cover the same learning content, but one version differs in instructional methods, enriched by simulations. The result shows that the "traditional" course without simulations discriminates more for considered preferences, indicating the need for improvements.

Keywords: Online courses · Adaption · Preferences · Bias

1 Introduction

Exams in online courses can be evaluated after learners have completed them. Item-based parameters cover difficulty, guessing, and discrimination regarding high and low performers, which are fundamental metrics. The item discrimination relates to the indent that high performers (learners with high ability levels) have fewer difficulties in solving a specific task than low performers (learners with low ability levels). Hence, item discrimination is an internal metric for instructors to analyze course tasks. Therefore, two groups are formed: the upper, and lower groups. Both are the extrema. More concretely, the upper group represents high-performing learners of the test. Learners are sorted by total score, the upper 27% define the upper group, and the lower 27% define the lower group [1]. The mean difference in performance for one specific task between both groups is measured, defined as the item discrimination index (IDI). This metric can be interpreted to understand whether single tasks discriminate well between good and low-performing learners. Hence, based on resulting scores, exams can be optimized, and incorrect, inconsistent, or confusing tasks can be identified.

In this paper, a new item parameter is introduced which is called preference discrimination. Preference discrimination follows the concept of item discrimination but

N. Wang et al. (Eds.): AIED 2023, CCIS 1831, pp. 347–352, 2023.
https://doi.org/10.1007/978-3-031-36336-8_54

changes the internal metric (sum of total score) to an external one (preferences) to form groups. It is based on the assumption that learners have certain preferences and if they get tasks those instructional methods fit, they perform better than those who do not get tasks that suit. This is called the "meshing hypothesis". To give an example, it is assumed that learners that prefer collaboration and get group tasks have more correct answers to the examined item than others. The metric shows whether there is a relation between performance and preference to argue for the need to adapt course items preference-based. Although such a consideration makes sense from the first view, it is controversially discussed by researchers. Kirschner & Merriënboer [2] argue, that the teaching/learning method must not match the learner's preferred style. Some learners may even learn better if they get tasks that do not match well and are more challenging. Despite existing discussions, even since learners with the same preferences for certain instructional methods perform better, or worse than others, an additional consideration can be beneficial to avoid structural biases. Hence, a preference discrimination (PDI) parameter can be an indicator for differentiation, and further adaption. It is intended that effects are seldom detected, but if preference discrimination can be seen, tutors should pay attention to avoid the exclusion of subgroups. Learners are grouped by dichotomous preference levels. The task-wise mean performance of both groups is measured. Their difference represents the PDI in a range of $[-1, 1]$, similar to item discrimination, but with a focus on learner preferences. In an optimal world, if all learners perform equally, PDIs will always be zero.

2 Preference Discrimination Index

More formally, let N_u be the number of learners that prefer trait j, and N_v be the number of learners that reject trait j. Let U be the number of learners in group N_u who solved item i correctly and, in addition, prefer trait j, called "upper group". Let V be the sum of learners' correct responses in group N_v who reject trait j, called "lower group". Then PDI_{ij} is defined as the preference discrimination index for item i considering trait j with $PDI_{ij} := \frac{U}{N_u} - \frac{V}{N_v}$. The decision of whether an IDI value is good – or not, is taken by theoretical insights [3] or based on experiences in classroom tests [1]. However, the optimal IDI value depends on the selection of the upper and lower group. But sometimes it is not possible to select both extrema if the metric to form groups is binary or class-based (like a high, or low preference, or no preference at all), or the number of learners is rather low that no statistically significant difference can be detected if the extrema are considered only. Then, the resulting discrimination values must be adjusted for interpretation that is similar to the selection of extrema values only. To map IDI values from the extrema-only approach ($x = .27$) to dichotomous values ($x = .50$), a mapping from $IDI_{x=.27}$ to $IDI_{x=.50}$ is given.

$$IDI_{x=.27} := \frac{U_{x=.27} - V_{x=.27}}{0.27N}; \quad IDI_{x=.50} := \frac{U_{x=.50} - V_{x=.50}}{0.5N}$$

Then, a scaling parameter G is searched for:

$$IDI_{x=.27} = G * IDI_{x=.50}$$

$$\Leftrightarrow \frac{1.852(U_{x=.27} - V_{x=.27})}{0.5N} = G * \frac{(U_{x=.50} - V_{x=.50})}{0.5N}$$

Assuming that there is no difference between $(U_{x=.27} - V_{x=.27})$ and $(U_{x=.50} - V_{x=.50})$, with $(U_{x=.50} - V_{x=.50}) \; ! = 0$, and ignoring the difference caused by the mean values, follows:

$$G = 1.852 \Rightarrow IDI_{x=.50} \approx \frac{IDI_{x=.27}}{1.852}$$

Hence, the interpretations from the indicator scale [1] can be mapped: for $IDI_{x=.27} = .2 \Rightarrow IDI_{x=.50} \gtrsim .11$, for $IDI_{x=.27} = .3 \Rightarrow IDI_{x=.50} \gtrsim .16$ and for $IDI_{x=.27} = .4 \Rightarrow IDI_{x=.50} \gtrsim .22$. If the concept of the IDI is transferred to the PDI, its interpretation is slightly different. Assuming that learners with a preference for collaboration perform better on a concrete task than learners who do not have this preference. Then, the PDI will be positive as the upper group is represented by learners with high preference levels for the considered trait. Nevertheless, the opposite side can also be the case. Learners that have a low up to no preference for the trait perform better than others. In that constellation, a negative PDI will be the result. In contrast to the IDI, this is not a negative signal. It shows that a certain group performs better than the other one. Finally, the interpretation can easily be derived (Table 1). However, the term "improvement" is not related to the quality to discriminate between high and low performers, but rather to adjusting related (or previous) tasks to lead to better performances.

Table 1. Interpretation of PDI values with x = .50.

Interpretation	Range	Group Performance
very good item	$\gtrsim .22$	Group U performs better than group V
space for improvement	.16 to .22	
possibly need to improvement	.11 to .16	
No need to consider	$-.11$ to .11	
possibly need to improvement	$-.16$ to -.11	Group V performs better than group U
space for improvement	$-.22$ to -.16	
very good item	$\lesssim -.22$	

Using all samples to derive the PDI with $x = .50$ has the advantage that the resulting value can be interpreted as the relative performance difference between the two groups. Under the assumption that preferences are related to the performance, whose reasons can be manifold, effects above .22 are of high interest as it means: For half of the learners, the task can be optimized with an expected optimization potential of .22 (if $PDI = .22$). Reaching those values is of high interest, and then, the one-size-fits-all environment cannot be the best choice as learners perform differently if they are considered group-wise based on preference levels.

3 Course Setup

For the study, a self-regulated online course for beginners to learn Spanish as a foreign language is created. It follows the task-based oriented approach [4] with some introductions and hints. The technical base is Moodle with H5P [5]. The course uses a fixed learning progression, where learners are guided through. Three preferences are taken by the preference-based questionnaire [6]. For them, simulations are created aiming to relate to preferences for group work (GW), competitions (CP), and a teacher-student relation (PD).

- The *Teacher Simulation* (PD) contains translation tasks, enriched by a video container. Hints were pre-recorded, which are run under the condition that the learner makes at least one mistake. As people may behave differently if they know that they are observed (Hawthorne Effect [7]), this may affect learner performance, in relation to the preference for a teacher-student relationship.
- In the *Group work simulation* (GW), the task has to be solved in collaboration with a virtual chat partner. Both "learners" get hints and they can only find the correct solution if they collaborate and process hints from the other peer.
- Tasks of the *Competition simulation* (CP) are enriched by a stopwatch. After submitting the response, a ranking of the competition with points is shown. With that setting, it is aimed that a competition-like atmosphere is created.

The Spanish beginner course consists of 19 tasks for beginners. The estimated course duration is 45 min. Version 1 of the course is enriched by the simulations. It is assumed that some learners prefer those settings, indicating that matching with the instructional methods tends to better performances ("meshing hypothesis" [8]). Version 2 of the course is the "more classical" variant containing the same tasks as in version 1, without simulations.

4 Methodology and Results

Students complete one of the two versions of the online course, which cover the same learning content but differ in instructional methods. 103 students completed version 1, and 54 completed version 2. All participants must have a sufficient skill level for German, as the course is in German. To filter all participants that do not have the required level that they claim to have, a first attention check is done. The 10-item preference questionnaire [6] is used to derive preference levels. Levene's test and t-tests are done to identify whether preference level means are similar in both variants. Levene's tests are not significant in all cases, hence, equal variances can be assumed. For all preferences, no statistically significant difference can be found in t-tests. For further analysis, this is a great sign as it can be assumed that both cohorts do not differ in preferences.

For the analysis, groups are formed by preferences. Then, the PDI is derived for all tasks of both course variants, and each trait individually, including correlations between learner performances, and preferences. Interpretable results are searched for. Absolute values of the PDIs are summarized preference-wise, intending that there is no difference between both variants, neglecting the meshing hypothesis.

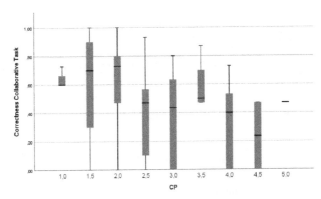

Fig. 1. Correctness in GW simulation, and preference for CP.

Correlations with absolute values higher than .2 are scarce. However, one task could be identified, that has a weak correlation between the correctness within the collaborative task and the preference for CP: $\mu = -.259, p = .02$. A visualization can be found in Fig. 1 for a continuous preference scale (non-dichotomous). Learners with a higher preference level for competition (1 = high, 5 = low) perform better in the group simulation task than those who reject it. Under $p \leq .05$, the correlation is statistically significant, arguing for the meshing hypothesis. For other tasks, no statistically significant difference between groups can be found. However, if there are no further effects, the sum of PDIs must be similar in both course versions. Results are summarized in Table 2.

Table 2. $\sum_{i=1}^{n} |PDI(i)|$ for two course versions and three preferences.

	PD	GW	CP
Version 1	5.15	5.42	5.65
Version 2	6.63	7.21	7.33

There, it can be seen that sum values are rather similar in version 1 for all preferences. Version 2, which is the more "traditional" course without simulations, has higher values, with more deviation. It indicates that the second variant discriminates more related to preferences than the first version. If instructional methods vary, less preference discrimination can be seen. For the second version, the bias based on preference levels is higher, so it can be worth considering those groups more in-depth.

5   Discussion and Conclusion

Similar to the IDI, the PDI does not consider whether the results are statistically significant. Hence, if remarkable effects are identified due to high PDI values ($|PDI| \gtrsim$ .22), further statistical tests (like t-tests) should take place to avoid misinterpretations reasoned by a small group of learners.

The interpretation is done task-wise, but the effects can be manifold. The indicator shows, that a possible effect can be measured at a certain point of the learning process, but previous tasks may have influenced learning performance. Hence, intending to personalize tasks related to instructional methods, the indicator must not necessarily be related to the concrete item. Thus, adapting course sequences based on PDI only is not recommended, as other influencing factors may not be considered.

The study is limited to short-term effects within a task. Further research should focus on other effects using final exams. Moreover, other psychometric scales should be used to form groups (e. g. personality by Big-5, cultural traits), or learning history. Besides, analyzing tasks with different instructional methods in other domains would be beneficial to detect other group-based optima.

To sum up, in research, the term learning styles and the meshing hypothesis are controversially discussed [2]. Even if there are different opinions, we can conclude that learners have different preferences to learn. The PDI can indicate potential bias. Hence, tutors can use the metric to identify tasks, that may be personalized in the future.

Acknowledgments. This work was supported by the German Federal Ministry of Education and Research (BMBF), grant number 16DHBQP058 (KI Campus 2.0).

References

1. Ebel, R.L., Frisbie, D.A.: Essentials of educational measurement (1972)
2. Kirschner, P.A., Merriënboer, J.J.V.: Do learners really know best? Urban legends in education. Educ. Psychol. **48**(3), 169–183 (2013)
3. Kelley, T.L.: The selection of upper and lower groups for the validation of test items. J. Educ. Psychol. **30**(1), 17–24 (1939)
4. Müller-Hartmann, A., Schocker-von Ditfurth, M.: Teaching English: task-supported language learning, vol. 3336, UTB (2011)
5. Homanová, Z., Havlásková, T.: H5P interactive didactic tools in education. IATED (2019)
6. Rüdian, S., Pinkwart, N.: Do learners really have different preferences? In: International Conference on Advanced Learning Technologies. IEEE (2022)
7. Jones, S.R.: Was there a Hawthorne effect? Am. J. Sociol. **98**(3), 451–468 (1992)
8. Pashler, H., McDaniel, M., Rohrer, D., Bjork, R.: Learning styles: concepts and evidence. Psychol. Sci. Public Interest **9**(3), 105–119 (2008)

Emotionally Adaptive Intelligent Tutoring System to Reduce Foreign Language Anxiety

Daneih Ismail[(✉)] and Peter Hastings[(✉)]

DePaul University, Chicago, IL 60614, USA
dismail1@depaul.edu, peterh@cdm.depaul.edu

Abstract. Quality education is associated with two necessary factors: learners' cognitive and emotional states. An adaptive system that takes the emotional state into account can enhance learning. We implemented an emotionally adaptive intelligent tutoring system that detects foreign language anxiety and provides appropriate intervention as needed. Note that the current experiment is still ongoing, and the results reported here are based on the data we currently have. We compared the adaptive approach with fixed feedback strategies. Our preliminary results revealed a statistically significant effect of using adaptive feedback to reduce foreign language anxiety. Also, we found a statistically significant improvement in the learning gains with a moderate effect size when using adaptive feedback.

Keywords: Affect detection · Foreign language anxiety · Machine learning · Animated agent · Emotional support · Adaptive feedback

1 Introduction

Modeling learners' affect plays a critical role in shaping Artificial Intelligence in Education. In recent years, researchers built emotionally intelligent tutoring systems to support STEM [14] and linguistic [15] fields. Within these systems, researchers detect emotions such as motivation, engagement, confusion, frustration, and anxiety. They use both sensor-full and sensor-free metrics to detect these emotions.

Previous research demonstrated that an adaptive affective system was effective within a science domain [7]. Here, we evaluate the effectiveness of adaptive feedback to enhance learning and reduce foreign language anxiety (FLA).

Previous research mentioned that an emotionally intelligent tutoring system can simulate a human tutor which helps the learner to perform better and reach a positive emotional state [19]. A successful learning environment is built on cognitive and affective support because all individuals are different and adapt to various emotional support, needs, and personalities [21]. To study the effectiveness of adaptive feedback for English as a foreign language, we analyzed the following research questions:

N. Wang et al. (Eds.): AIED 2023, CCIS 1831, pp. 353–358, 2023.
https://doi.org/10.1007/978-3-031-36336-8_55

Research Question 1: "How effective is an adaptive feedback approach relative to a fixed feedback approach for reducing FLA?"

Research Question 2: "How effective is an adaptive feedback approach relative to a fixed feedback approach for increasing learning?"

2   Related Work

Foreign language anxiety (FLA) is one of the main impediments to learning a new language. It has a long-term effect on willingness to communicate in the foreign language [3,16,18]. Also, it inhibits language acquisition, especially by increasing the learner's reluctance to practice [10,11,16,17]. Moreover, it hinders performance [16,17] and achievements [8].

The goal of assessing when to provide emotional support when teaching English as a second language is to improve positive emotions, reduce negative emotions, and enhance achievements. Providing an adaptive system that takes into account the current emotional state can be a key factor for students to succeed academically [9,15]. A system that understands when to provides support can accelerate learning [6]. Adaptive support can be beneficial because each intervention could have positive or negative side effects, like providing emotional support when it is not needed [6].

Previous research mentioned that getting benefits from an animated agent that provides emotional support depends on different factors, such as the learner's current emotion, gender, or achievement [2]. The effectiveness of an adaptive, supportive animated agent for reducing FLA and increasing learning acquisition is an active research question. Christudas et al. previously found that personalized e-learning improves learning achievement and satisfaction [4]. They took into account learner behavior and feedback; though they didn't measure the learner's emotional state. Other research mentioned that, when comparing adaptive and non-adaptive e-learning systems, the adaptive system which took into account the students' knowledge outperformed the non-adaptive system [1]. There was an increase in motivation, engagement, and learning for learners who used the adaptive system [1]. This study did not, however, take into account the learner's anxiety state. Therefore, we hypothesize that using emotionally adaptive feedback would enhance both the emotional state and learning.

3   Method

To build an emotionally intelligent tutoring system, we developed a machine learning model to detect FLA. We implemented a Random Forest chain regressor model using the scikit-learn Python machine learning library [20]. We predicted FLA [13], change in FLA, and which intervention would be most beneficial. In a previous experiment, we tested 6 different approaches for delivering feedback: Motivational supportive feedback and explanatory feedback presented by text, voice, and agent. The explanatory feedback explains the correct answer without alluding to the incorrect answer. The motivational supportive feedback provides explanatory feedback between two positive comments. To build the machine learning model, we divided the dataset, which consisted of 3940 records, into 6 groups based on the intervention used. For each group, we implement a Random Forest Chain regressor algorithm to calculate FLA [13], and the change in FLA after receiving the feedback. In each model, we used 10-fold cross-validation and 100 random generations for the chain order. There were 9 independent features [13] and two dependent variables. Then we predicted the intervention based on Random Forest algorithm, which chose the intervention that caused the maximum reduction in FLA. This means finding the intervention which causes the maximum difference between anxiety self-report and anxiety predicted by the model.

Based on [12], motivational supportive feedback presented by the agent could reduce FLA the most followed by explanatory text feedback. Thus we used motivational supportive agent and explanatory text as a fixed strategy conditions. Previous research mentioned that using voice with text helps foreign language learners to focus and increase learning gain [5], thus we used explanatory feedback presented by voice and text as a fixed strategy. In our preliminary experiment, participants were randomly assigned to four different groups. The adaptive feedback provided explanatory or motivational supportive feedback presented by the text, voice with text, or agent with voice and text. The machine learning algorithm decided which intervention would better support the learner's emotional state. This algorithm was built based on previous research [12]. The fixed feedback were either motivational supportive feedback presented by agent, voice, and text; explanatory feedback presented by voice and text; or explanatory feedback presented by the text. We had 80 participants (Adaptive N = 20, Voice Explanatory N = 15, Agent Supportive N = 22, Text Explanatory N = 23).

First, the participants read and agreed to the informed consent. Then they provided some demographic information (age, gender, English level, educational degree, native language, employment status, and number of years studying English). After that, they filled FLCAS questionnaire to measure their anxiety during English class. Then they started the pre-test, which consists of five sections (vocabulary, listening, grammar, reading, and writing). After that, they did 26 exercises with the intelligent tutoring system. After each exercise, the machine learning algorithm calculated the anxiety level and then provided feedback based on the assigned group. If the participant was in the adaptive group they received feedback based on their anxiety level. Otherwise, if the participant was in one of the fixed groups, they received the same feedback but delivered according to their assigned group (motivational supportive agent, explanatory voice, or explanatory text). Then the participants filled out a self-report about their current anxiety level. After finishing the 26 exercises the participants answered 5 post-test questions.

4   Results

4.1   Reducing FLA

RQ1 asked about the effectiveness of adaptive feedback for reducing FLA. To address this, we did an ANOVA to compare the reduction of learners' FLA when using adaptive feedback vs. fixed feedback. We found a significant reduction in anxiety in the adaptive condition $F(3, 2071) = 9.454, p < .001$. Table 1 presents the mean and standard deviation for the change in FLA after receiving the feedback. To investigate this in more detail, we did a separate ANOVA with two factors: adaptive vs. agent supportive. Although there was a difference, it did not reach the $\alpha < 0.05$ threshold: $F(1, 1086) = 3.631, p = .057$. We also did an ANOVA with adaptive vs. voice explanatory feedback as factors. We found a significant result, $F(1, 906) = 6.054, p = .014$. Finally, we did an ANOVA with adaptive vs. text explanatory as factors. We found a significant difference, $F(1, 1112) = 28.931, p < 0.001$.

Table 1. Difference in FLA between groups

	Mean	SD
Adaptive	6.88	32.47
Voice explanatory	1.56	31.91
Agent supportive	3.01	34.33
Text explanatory	-3.32	30.7

4.2   Learning Achievement

To ensure that there is no difference in the prior knowledge between the four groups, we did an ANOVA and found no significant difference $F(3, 399) = .254, p = .859$. To answer the second research question about the effectiveness of adaptive feedback in increasing learning gain, we evaluated the results of the pre- and post-test using paired t-test analysis. We found that post-test scores were significantly higher (M = 58.31, SD = 37.24), than pre-test scores (M = 44.66, SD= 30.24), $t(99) = -3.487, p < .001$. The effect size for this learning gain ($d = .4$) is considered moderate. We found that post-test scores were significantly higher than the pre-test for the fixed groups: emotional supportive feedback present by agent $t(109) = -2.655, p = .009$, explanatory text feedback $t(114) = -3.544, p < .001$, but the effect size for was small (d= .27, d= .36). However, for explanatory voice feedback $t(74) = -3.755, p < .001$ the effect size was moderate (d = .47).

5   Discussion and Conclusion

Our first research question was whether adaptive feedback could reduce FLA or not. To answer this question, we compared adaptive feedback with fixed feedback. We found that an adaptive emotionally intelligent tutoring strategy reduced anxiety more than fixed strategies. This is aligned with [15] who found that an affective intelligent tutoring system reduces anxiety when learning Japanese as a foreign language. However, we did not reach the $\alpha < 0.05$ threshold between the adaptive feedback and the fixed emotionally supportive agent. The presence of the emotionally supportive agent helped in reducing FLA regardless if the system is adaptive or not. This echoes previous research, which found that emotionally supportive agents reduce FLA [3,12]. The adaptive feedback significantly reduced FLA compared to text explanatory feedback. This implies that it is important to provide appropriate feedback when needed [6].

When looking at the pre- to post-test results, the adaptive feedback increased the learning gain more effectively than the fixed strategies. It was not a big change, but it may not be surprising considering that the tutoring system's content is relatively difficult, on par with TOEFL and IELTS English language standardized tests. This may limit the amount by which the learners' anxiety levels would be reduced over the course of the experiment and, in turn, limit the extent of their learning achievement.

One limitation of this work is the small sample size N=80. Another is that there was a high dropout rate, which may also be due to the difficulty of the content. It should be noted that this experiment is still ongoing, and we hope that having more data for the conditions will allow us to draw more concrete conclusions.

References

1. Alshammari, M., Anane, R., Hendley, R.J.: Design and usability evaluation of adaptive e-learning systems based on learner knowledge and learning style. In: Abascal, J., Barbosa, S., Fetter, M., Gross, T., Palanque, P., Winckler, M. (eds.) INTERACT 2015. LNCS, vol. 9297, pp. 584–591. Springer, Cham (2015). https://doi.org/10.1007/978-3-319-22668-2_45
2. Arroyo, I., Woolf, B.P., Cooper, D.G., Burleson, W., Muldner, K.: The impact of animated pedagogical agents on girls' and boys' emotions, attitudes, behaviors and learning. In: 2011 IEEE 11th International Conference on Advanced Learning Technologies, pp. 506–510. IEEE (2011)
3. Ayedoun, E., Hayashi, Y., Seta, K.: Adding communicative and affective strategies to an embodied conversational agent to enhance second language learners' willingness to communicate. Int. J. Artif. Intell. Educ. **29**(1), 29–57 (2018). https://doi.org/10.1007/s40593-018-0171-6
4. Christudas, B.C.L., Kirubakaran, E., Thangaiah, P.R.J.: An evolutionary approach for personalization of content delivery in e-learning systems based on learner behavior forcing compatibility of learning materials. Telemat. Informat. **35**(3), 520–533 (2018)

5. Clark, R.C., Mayer, R.E.: E-learning and the Science of Instruction: Proven Guidelines for Consumers and Designers of Multimedia Learning. Wiley (2016)

6. D'Mello, S., Graesser, A.: AutoTutor and affective AutoTutor: Learning by talking with cognitively and emotionally intelligent computers that talk back. ACM Trans. Interact. Intell. Syst. (TiiS) **2**(4), 1–39 (2013)

7. Faivre, J., Nkambou, R., Frasson, C.: Integrating adaptive emotional agents in ITS. In: Intelligent Tutoring Systems, pp. 996–997 (2002)

8. Farid, R.M.B.B.: Influence of language anxiety and prior knowledge on EFL students' performance in essay writing. Arab J. Educ. Psychol. Sci. **5**(23), 599–617 (2021)

9. Harley, J.M., et al.: Examining the predictive relationship between personality and emotion traits and students' agent-directed emotions: towards emotionally-adaptive agent-based learning environments. User Model. User-Adapt. Interact. **1**, 177–219 (2016). https://doi.org/10.1007/s11257-016-9169-7

10. Ismail, D., Hastings, P.: Identifying anxiety when learning a second language using e-learning system. In: Proceedings of the 2019 Conference on Interfaces and Human Computer Interaction, pp. 131–140 (2019)

11. Ismail, D., Hastings, P.: A sensor-lite anxiety detector for foreign language learning. In: Proceedings of the 2020 Conference on Interfaces and Human Computer Interaction, pp. 19–26 (2020)

12. Ismail, D., Hastings, P.: Way to Go! Effects of motivational support and agents on reducing foreign language anxiety. In: Roll, I., McNamara, D., Sosnovsky, S., Luckin, R., Dimitrova, V. (eds.) AIED 2021. LNCS (LNAI), vol. 12749, pp. 202–207. Springer, Cham (2021). https://doi.org/10.1007/978-3-030-78270-2_36

13. Ismail, D., Hastings, P.: Toward ubiquitous foreign language learning anxiety detection. In: Artificial Intelligence in Education. Posters and Late Breaking Results, Workshops and Tutorials, Industry and Innovation Tracks, Practitioners' and Doctoral Consortium: 23rd International Conference, AIED 2022, Durham, UK, July 27–31, 2022, Proceedings, Part II. pp. 298–301. Springer, Cham (2022). https://doi.org/10.1007/978-3-031-11647-6_56

14. Jiang, Y., et al.: Expert feature-engineering vs. deep neural networks: Which is better for sensor-free affect detection? In: Penstein Rosé, C., et al. (eds.) Artificial Intelligence in Education, pp. 198–211. Springer International Publishing, Cham (2018)

15. Lin, H.-C.K., Chao, C.-J., Huang, T.-C.: From a perspective on foreign language learning anxiety to develop an affective tutoring system. Educ. Technol. Res. Develop. **63**(5), 727–747 (2015). https://doi.org/10.1007/s11423-015-9385-6

16. Liu, M.: Anxiety in EFL classrooms: Causes and consequences. TESL Reporter **39**(1), 13–32 (2006)

17. Liu, M., Huang, W.: An exploration of foreign language anxiety and English learning motivation. Educ. Res. Int. 2011 (2011)

18. Liu, M., Jackson, J.: An exploration of Chinese EFL learners' unwillingness to communicate and foreign language anxiety. Modern Lang. J. **92**(1), 71–86 (2008)

19. Mohanan, R., Stringfellow, C., Gupta, D.: An emotionally intelligent tutoring system. In: 2017 Computing Conference, pp. 1099–1107. IEEE (2017)

20. Pedregosa, F., et al.: Scikit-learn: Machine learning in python. J. Mach. Learn. Res. **12**, 2825–2830 (2011)

21. Petrovica, S., Ekene, H.K.: Emotion recognition for intelligent tutoring. In: CEUR Workshop Proceedings 1684 (2016)

Amortised Design Optimization for Item Response Theory

Antti Keurulainen[1,2(✉)], Isak Westerlund[2], Oskar Keurulainen[2], and Andrew Howes[1,3]

[1] Aalto University, Espoo, Finland
[2] Bitville Oy, Espoo, Finland
antti.keurulainen@bitville.com
[3] University of Birmingham, Birmingham, UK

Abstract. Item Response Theory (IRT) is a well known method for assessing responses from humans in education and psychology. In education, IRT is used to infer student abilities and characteristics of test items from student responses. Interactions with students are expensive, calling for methods that efficiently gather information for inferring student abilities. Methods based on Optimal Experimental Design (OED) are computationally costly, making them inapplicable for interactive applications. In response, we propose incorporating amortised experimental design into IRT. Here, the computational cost is shifted to a precomputing phase by training a Deep Reinforcement Learning (DRL) agent with synthetic data. The agent is trained to select optimally informative test items for the distribution of students, and to conduct amortised inference conditioned on the experiment outcomes. During deployment the agent estimates parameters from data, and suggests the next test item for the student, in close to real-time, by taking into account the history of experiments and outcomes.

Keywords: Item Response Theory (IRT) · Experimental Design · Deep Reinforcement Learning (DRL)

1 Introduction

Item Response Theory (IRT) is a method for inferring student abilities and test item characteristics by observing test outcomes conducted with students [5]. In IRT, the relationship between an individual's ability and their response to a test item is typically modeled with logistic regression, describing the probability that an individual with a certain ability will give a correct response to a test item with a given difficulty. Inferring student abilities and test item characteristics offers benefits such as user skill assessment, test item calibration, and optimization of learning experiences for students by Intelligent Tutoring Systems (ITS) or human tutors.

To fully exploit the benefits of IRT in real-time interactions with students, each interaction should provide as much information as possible. Optimal Experimental Design (OED) aims to select an experiment that maximizes an information

© The Author(s), under exclusive license to Springer Nature Switzerland AG 2023
N. Wang et al. (Eds.): AIED 2023, CCIS 1831, pp. 359–364, 2023.
https://doi.org/10.1007/978-3-031-36336-8_56

criterion [11,13]. Combining OED with IRT enables designing experiments that are maximally informative about student abilities.

We incorporate amortised experimental design into IRT and construct a system in which the next test for the student, conditioned on the previous test outcomes, can be given in near-real time. Our approach involves training a Deep Reinforcement Learning (DRL) agent, which allows for amortising both the OED and parameter estimation tasks simultaneously.

2 Background and Related Work

The most simple IRT variant is the 1PL model, also known as the Rasch model [10]. The probability that a student i gives a correct answer for the item j is defined as:

$$P(y_{i,j} = 1|\theta_i, b_j) = \frac{1}{1 + e^{-(\theta_i - b_j)}} \tag{1}$$

where $y_{i,j}$ is the Bernoulli-distributed outcome, θ_i is the ability of student i, and b_j is the difficulty of the item j. The task is to infer the parameters θ and b from the observations, i.e., from the test outcomes. We consider a setting where only the student ability θ is estimated, while the item parameter b is treated as a design parameter. This formulation models an experimental design setting where items can be selected for students to obtain information about their latent abilities.

Recent work has focused on inferring IRT parameters using deep neural network variants. Wu et al. [14] suggest IRT inference using Variational Inference (VI). Paassen et. al [7] combine sparse factor analysis with Variational Autoencoders (VA). Deep-IRT [15] combines key-value networks with IRT to improve the explainability of deep learning-based knowledge tracing. Closely related to IRT is Knowledge Tracing (KT), which tracks and models students' knowledge over time based on their responses during a set of interactions [2]. Li et al. [6] suggest a value-based RL method for an adaptive learning system for optimal teaching. Ghosh et al. [4] suggest BOBCAT, casting experimental design for IRT as a metalearning problem. However, their question selection algorithm assumes questions to be discrete variables.

Advances in Deep Learning have opened up new possibilities for OED. It allows for amortising the OED process by pretraining a deep neural network before deployment using existing or synthetic data. During execution, a pretrained neural network can potentially generalize to unseen models within the prior distribution used in the amortisation stage. A deep neural network can also be conditioned on the outcomes of previous interactions for learning an adaptive policy where all previous interactions affect the selection of the next design. However, optimizing the exact mutual information directly might lead to an intractable solution because of the need to calculate nested expectations [9]. Recently, various suggestions have been made to solve these challenges. Foster et al. suggest Deep Adaptive Design (DAD) [3], which provides non-myopic, amortised experimental design by maximising an approximate lower bound of the mutual information instead of an exact mutual information objective. Blau et al. [1] introduce an RL-based method, highlighting the strong exploration and exploitation tradeoff capability.

3   Amortised Design Optimisation for IRT (ADOIRT)

In this section, we propose an amortised design optimisation approach to IRT, which we name Amortised Design Optimisation for IRT (ADOIRT).

3.1   POMDP Formulation for ADOIRT

We formulate design selection and parameter estimation as a Partially Observable Markov Decision Process (POMDP) given by a tuple $\langle S, A, T, R, O, f, \rho_0, \gamma \rangle$, and find an approximately optimal policy using Reinforcement Learning. The agent selects actions $a_t = (d_t, \hat{\theta}_t) \in A$, where A is the action space consisting of all possible actions, including design requests and estimated student abilities.

At each time step t, the environment is in state $s_t \in S$, where S is the state space consisting of all possible states. The state $s_t = \theta_t$ is the true student ability. The transition function $T : S \times A \times S \to [0, 1]$ specifies deterministic dynamics: $p(\theta_{t+1}|a_t, \theta_t) = 1$ if $\theta_{t+1} = \theta_t$ and 0 otherwise. The initial state s_0 is sampled from the prior distribution $\rho_0 = p(\theta)$, and γ is the discount factor.

Observations are sampled from the observation function $f : S \times A \to O$ such that $o_t = (\hat{d}_t, y_t) \sim f(s_t, a_t)$. The observation function consists of two parts: mapping the requested design d_t to a corrupted item $\hat{d}_t$, and obtaining the experiment outcome y_t. The corrupted item results from selecting the best item during deployment or adding Gaussian noise during training. The outcome is given by $y_t \sim p(y_t|\theta_t, \hat{d}_t)$.

The reward function $R : S \times A \to \mathcal{R}$ specifies the reward received by the agent for taking a particular action in a particular state. The reward is the squared error between the true and estimated student ability: $R(s_t, a_t) = (\theta_t - \hat{\theta}_t)^2$. Consequently, the agent is rewarded for selecting experiments that lead to accurate parameter estimations.

3.2   ADOIRT Architecture and Training

The training architecture is illustrated in the left panel of Fig. 1. At the start of each episode, the simulator is initialised by sampling new model parameters from the prior. ADOIRT makes requests for the item difficulties, and the simulator produces an outcome using the obtained design and sampled student ability. The outcome and obtained design are concatenated to the history of observations.

In the deployment phase, the trained ADOIRT can be used with existing item response data. First, a MLE estimate for the item difficulties is produced using stochastic gradient descent (SGD). Once the estimates of the item difficulties are available, the item requests by the ADOIRT are mapped to the closest items from the existing dataset. Based on the outcomes of the student, the goal of the agent is to select informative items from the existing collection of items.

The policy network implementation is a multilayer perceptron (MLP) network with four layers followed by mean pooling across experiment trials, an output layer, and separate heads for action distribution and value estimation,

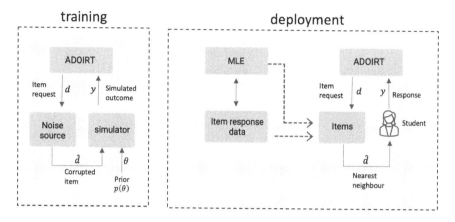

Fig. 1. ADOIRT architecture. Left panel (training): During training, synthetic data is generated by sampling the student abilities and item difficulties from a prior distribution. The simulator uses the obtained item parameters and the IRT model to produce outcomes. Right panel (deployment): In the deployment phase, the item parameters can be estimated beforehand with a standard method, such as MLE, using any existing data. The item requests by ADOIRT are mapped to the closest estimated item difficulty in the collection of available items.

which are two-layer MLP networks. ReLU activation functions are used after each MLP layer. We train the policy function with the Proximal Policy Optimisation (PPO) algorithm [12] using the Stable Baselines 3 (SB3) library [8].

4   Results

We assess the performance of ADOIRT in estimating student abilities in a scenario where the unknown item difficulties are inferred using maximum likelihood estimation. We simulate synthetic datasets with 200 students and 50 items for training and evaluation. The item request by the agent is then mapped to the closest estimated difficulty in this collection.

We compare the performance with a situation where design values are randomly chosen. To gain further insight into the performance, we also trained the agent by concealing experiment outcomes from the observation until the final time step of the episode. In this case, the agent learned a well performing, non-adaptive design strategy.

Figure 2 illustrates the key results. It shows that experiments chosen by ADOIRT result in lower error in student ability estimation compared to the baselines (panels a-c). In the case of non-adaptive designs (panel c), the inferred student abilities are clustered in ten clusters. This is a reasonable non-adaptive design strategy, as it effectively covers the design space with a limited number of points. Furthermore, the clusters are denser closer to zero, as incorrect predictions for student abilities far from the mean lead to higher penalties. The quantitative results are presented in Table 1.

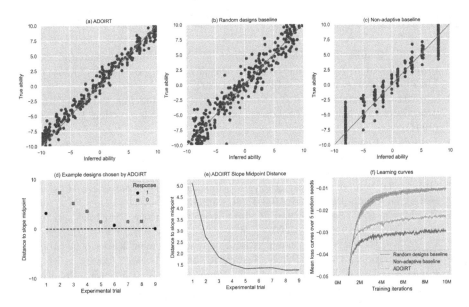

Fig. 2. Panel (a): True vs inferred ability for ADOIRT. Panel (b): True vs inferred ability for random designs. Panel (c): True vs inferred ability for non-adaptive optimized designs. Panel (d): An example case where 10 items are presented to the student and the agent successfully converges on designs close to the midpoint of the sigmoid function. Panel (e): ADOIRT's design values converging to sigmoid midpoint (1000 models). Panel (f): Agent training learning curves with 1 std shaded area. The shaded area represents 1 std over 5 training runs. Panels (a)-(e) use the lowest training error seed.

Table 1. ADOIRT-based inference surpasses non-adaptive and random designs in inferring student abilities. This is evaluated using 50 datasets, MLE estimates, and averaging performance over 1000 episodes per dataset. Mean and standard error are computed by repeating the training with five random seeds.

Method	MSE mean	MSE standard error
Inference with ADOIRT	**1.91**	**0.03**
Inference with non-adaptive designs	3.05	0.07
Inference with random designs	3.39	0.06

5   Discussion and Further Work

In this article, we introduced a novel method for amortised experimental design and parameter estimation for the IRT setting, named ADOIRT. Our studies showed that ADOIRT outperforms non-adaptive and random design strategies, thus being capable of inferring student abilities from a small number of interactions in near real-time. Interesting future work would involve testing the system in a real-world setting with human participants.

References

1. Blau, T., Bonilla, E.V., Chades, I., Dezfouli, A.: Optimizing sequential experimental design with deep reinforcement learning. In: International Conference on Machine Learning, pp. 2107–2128. PMLR (2022)
2. Corbett, A.T., Anderson, J.R.: Knowledge tracing: Modeling the acquisition of procedural knowledge. User modeling and user-adapted interaction **4**(4), 253–278 (1994)
3. Foster, A., Ivanova, D.R., Malik, I., Rainforth, T.: Deep adaptive design: Amortizing sequential bayesian experimental design. In: International Conference on Machine Learning, pp. 3384–3395. PMLR (2021)
4. Ghosh, A., Lan, A.: Bobcat: Bilevel optimization-based computerized adaptive testing. arXiv preprint arXiv:2108.07386 (2021)
5. Hambleton, R.K., Swaminathan, H.: Item Response Theory: Principles and Applications. Springer, Dordrecht (2013). https://doi.org/10.1007/978-94-017-1988-9
6. Li, X., Xu, H., Zhang, J., Chang, H.H.: Deep reinforcement learning for adaptive learning systems. J. Educ. Behav. Statist. 10769986221129847 (2020)
7. Paaßen, B., Dywel, M., Fleckenstein, M., Pinkwart, N.: Sparse factor autoencoders for item response theory. In: Proceedings of the 15th International Conference on Educational Data Mining, p. 17 (2022)
8. Raffin, A., Hill, A., Gleave, A., Kanervisto, A., Ernestus, M., Dormann, N.: Stable-baselines3: Reliable reinforcement learning implementations. J. Mach. Learn. Res. **22**(268), 1–8 (2021). http://jmlr.org/papers/v22/20-1364.html
9. Rainforth, T., Cornish, R., Yang, H., Warrington, A., Wood, F.: On nesting monte carlo estimators. In: International Conference on Machine Learning, pp. 4267–4276. PMLR (2018)
10. Rasch, G.: Probabilistic models for some intelligence and attainment tests. ERIC (1993)
11. Ryan, E.G., Drovandi, C.C., McGree, J.M., Pettitt, A.N.: A review of modern computational algorithms for Bayesian optimal design. Int. Statist. Rev. **84**(1), 128–154 (2016)
12. Schulman, J., Wolski, F., Dhariwal, P., Radford, A., Klimov, O.: Proximal policy optimization algorithms. arXiv preprint arXiv:1707.06347 (2017)
13. Smucker, B., Krzywinski, M., Altman, N.: Optimal experimental design. Nat. Methods **15**(8), 559–560 (2018)
14. Wu, M., Davis, R.L., Domingue, B.W., Piech, C., Goodman, N.: Variational item response theory: Fast, accurate, and expressive. arXiv preprint arXiv:2002.00276 (2020)
15. Yeung, C.K.: Deep-irt: Make deep learning based knowledge tracing explainable using item response theory. arXiv preprint arXiv:1904.11738 (2019)

Early Prediction of Student Performance in Online Programming Courses

Enqi Liu[✉], Irena Koprinska, and Kalina Yacef

School of Computer Science, The University of Sydney, Sydney, NSW 2006, Australia
eliu5850@uni.sydney.edu.au, {irena.koprinska,
kalina.yacef}@sydney.edu.au

Abstract. Early prediction of student grades is important for both teachers and students. It can help teachers take timely remedial actions to avoid drop-out and poor learning outcomes, and also help students improve their engagement, motivation and achievement of desired results. In this study, we present a machine learning approach to predict the final student grade from information available in the middle of the course, in the context of introductory programming courses for primary and high school students. We define and extract suitable features from the raw data and use a decision tree classifier to produce a compact set of rules, which are both accurate and interpretable by teachers and students. The decision tree rules provide insights about the important factors for success and highlight key programming tasks that predict the final student performance.

Keywords: Student performance prediction · Online programming courses · Decision trees

1 Introduction

Early prediction of students' final grades can be very helpful to both teachers and students. It allows teachers to detect potentially underperforming or dropping-out students and take the necessary measures to provide targeted support. Similarly, it allows students to adjust their learning behaviour to achieve the desired results, manage their time and resources more efficiently, and may also improve their motivation and engagement.

We consider introductory computer programming courses for primary and high school students. Computer programming is an essential skill in today's technological society and there has been a growing interest in teaching programming in schools. Previous research exists on early prediction of student grades for university students [1–9] but not for primary and high school students. Our goal is to investigate the behaviour of primary and high school students enrolled in introductory programming courses, and to provide actionable insights to teachers and students by applying interpretable machine learning methods. In particular, we aim to: (i) investigate whether the final student grade can be predicted in the middle of the course and how accurate the prediction is, and (ii) determine the important features for student's success.

© The Author(s), under exclusive license to Springer Nature Switzerland AG 2023
N. Wang et al. (Eds.): AIED 2023, CCIS 1831, pp. 365–371, 2023.
https://doi.org/10.1007/978-3-031-36336-8_57

2 Related Work

A number of studies have sought to predict student's final grades in their university degree, based on attributes such as previous grades and personal characteristics. Asif et al. [1] used various classifiers to predict the graduation performance in a four-year university degree from pre-university marks and marks from the first and second year at the university. At a course level, Elbadrawy et al. [2] investigated different recommender system techniques to predict the students' next term course grades as well as the grades within the current term. Thai-Nghe et al. [9] proposed methods for predicting student performance in algebra courses based on personalized multi-linear regression and matrix factorization. Meier et al. [5] proposed an algorithm to predict the final student's grade based on the student's marks in early assessments such as homeworks and mid-term exams, in the context of an introductory signal processing course. Kotsiantis et al. [6] predicted the final exam performance based on assignment marks throughout the semester in a distance education environment, achieving accuracy of 79% with an ensemble classifier, but the prediction was performed just prior the exam, not earlier.

Romero et al. [7] predicted final student grades based on Moodle usage data (number of quizzes passed and failed, assignments done, activity on the discussion board, as well as the time spent on each), obtaining 67% accuracy with decisions trees. In [8] they investigated the prediction of student grades based on the student's participation in the discussion forum in the middle and end of the semester, achieving accuracy of 75% and 90% respectively. Analysing student data during the semester, Koprinska et al. [3] built a decision tree classifier using data from three sources: submission steps and outcomes from an automated marking system, assessment marks during the semester and student engagement with the discussion forum Piazza. Their classifier could predict the exam grade (high, average and low) with an accuracy of 66.5% and 72.7% in the middle and at the end of the semester respectively. They also found that it is possible to predict the final exam outcome (pass or fail) mid-semester, with an accuracy of 87% [4].

Relevant prior work related to programming courses for high school students includes predicting progress [10], investigating student engagement and its impact on the final grade [11] and using progress networks for analysing student difficulties [12].

3 Context of the Study

3.1 Data

In this paper we investigate the behaviour of primary and high-school students aged 10–18 years participating in online programming courses. We used data from three different courses offered via the Grok Academy platform (https://grokacademy.org/cha llenge): Newbies, Beginners and Intermediate. The Newbies is conducted in the Blockly visual environment, while the Beginners and Intermediate use Python. All required ethics approvals to use the data have been obtained.

Each course runs over 5 weeks and includes several modules. A module contains *slides* - explanatory material, introducing the programming concepts, and *problems* – programming tasks that students need to submit. The programming tasks are auto marked and students can submit multiple times. After each submission students are provided with feedback about the passed and failed cases; they can correct their solution and submit again. Marks are allocated for each problem, with a maximum mark of 400 (40 tasks × 10 points) if students pass all test cases. Marks are deduced after every 5 incorrect submissions to a maximum deduction of 5 points per problem.

3.2 Predicted Variable

We predict the final score, based on information available in the middle of the course. As a middle-of-course point in a 5-weeek course we selected week 3, just before the final task for this week. At this stage, students have already completed about half of the tasks in the course. This is an optimal point, providing sufficient data for analysis, as well as sufficient time for remedial action before the end of the course. The final score is discretized into four equal ranges: Poor [0–100), Fair [100–200), Good [200–300) and Excellent [300, 400].

As Table 1 shows, there were 144 students in the Newbies, 842 in the Beginners, and 418 in Intermediate courses.

Table 1. Score distributions for the three courses

Score	Newbies	Beginners	Intermediate
0–100 (Poor)	14	115	14
100–200 (Fair)	14	118	167
200–300 (Good)	47	225	237
300–400 (Excellent)	69	384	0
Total	144	842	418

3.3 Features

Table 2 presents the features we defined and extracted to characterize student behavior.

Table 2. Features extracted from the raw data

Features	Description
Demographic	Demographic information about the student – year at school and gender (male, female, other)
Attempts	A feature vector representing the number of attempts the student needed to pass each of the previous tasks, before the middle-of-course point. There were 23 tasks for Newbies, 24 for Beginners and 14 for Intermediate, which corresponds to the dimensionality of the feature vector. Students who did not successfully complete a task were assigned a value equal to the largest number of attempts for that task + 1
Inactive time	The time from when the task is released to when the student runs code for this task for the first time
Early start	A Boolean representing whether the student started the task earlier relative to the average of the other students who attempted this task. A value of 1 means an early start, 0 – not an early start
Edit distance	The average edit distance between the consecutive student submissions of the previous tasks, which have passed at least one more test case. The edit distance is calculated between the abstract syntax trees of the submissions before the middle-of- course point
Problems passed	Number of problems passed before the middle-of-course point
Problems failed	Number of problems failed before the middle-of-course point
Problems viewed	Number of problems viewed before the middle-of-course point

4   Results and Discussion

4.1   How Accurately Can We Predict the Final Grade in the Middle of the Course?

To investigate how accurately we can predict the final grade from information available in the middle of the course, we employed a Decision Tree (DT) classifier. DTs have shown excellent results in many machine learning applications. They are suitable for our task as they are explainable classifiers – the produced rules can be easily understood by educators and students, and directly applied in practice.

In total, we conducted 20 experiments for each course, experimenting with different combinations of features (different combinations of feature groups from Table 2). We split the data for each course into training and test set using a stratified 70/30% split. Then we applied 10-fold stratified cross-validation on the training set to select the best DT hyperparameters for each feature combination. An exception was the Newbies dataset where we used 8-fold cross validation for hyperparameter tuning due to the smaller sample size. The searched DT hyperparameters included "criterion" (information gain and Gini index), "splitter", "max_depth" and max_leaf_nodes" using the sklearn machine learning library in Python. The best hyperparameter combination for each feature set was evaluated on the test set.

Table 3 shows the results for the best feature set and hyperparameters for each course. The DTs achieved accuracy of 75% for Newbies, 61.66% for Beginners and 76.19%

for Intermediate. These results are considerably higher than the baseline accuracies of predicting the majority class in the training set which are 47.93% for Newbies, 45.61% for Beginners and 56.70% for Intermediate.

4.2 Which Are the Most Important Features for Student Success?

Figure 1 shows the produced DTs for the three courses. As we can see the DTs are very compact, containing a small number of rules and features. The selected features for all three DTs are similar, with each DT utilizing the feature 'Problems passed' (number of previous problems passed) and a time feature – either "Inactive time" or "Early start".

Table 3. Accuracy [%] and selected features

	Accuracy	Num. Rules	Features
Newbies	75.00	8	Attempts + Early start + Problems passed
Beginners	61.66	4	Inactive time + Problems passed
Intermediate	76.19	5	Attempts + Inactive time + Problems passed

Newbies	Beginners
Problems_passed <= 7.5 | Early_start = 0: Poor | Early_start = 1 | | Attempts_9 <= 1.5: Excellent | | Attempts_9 > 1.5: Good Problems_passed > 7.5 | Early_start = 0: | | Attempts_17 <= 5.5: Good | | Attempts_17 > 5.5: Fair | Early_start = 1 | | Attempts_17 <= 25: | | | Attempts_11 <= 2: Excellent | | | Attempts_11 > 2: Good | | Attempts_17 > 25: Good	Inactive time <= 13.52 days | Problems_passed <= 8.5:Good | Problems_passed > 8.5:Excellent Inactive time > 13.52 days | Problems_passed <= 7.5:Poor | Problems_passed <= 8.5:Fair
	Intermediate
	Problems_passed <= 2.5:Poor Problems_passed > 2.5 | Inactive time <= 9.48 days | | Attempts_2 <= 8: Good | | Attempts_2 > 8: Fair | Inactive time > 9.48 days | | Inactive time <= 15.33 days: Fair | | Inactive time > 15.33: Poor

Fig. 1. DTs produced for each course

The most important feature overall is "Problems passed" – it appears in all DTs, and in two of them (Newbies and Intermediate) is also selected as the root of the trees. The rules including this feature show that the more problems a student passes before the middle of the course, the higher the final grade is. The number of attempts on specific tasks is another important feature. The rules suggest that students who require fewer attempts to pass key tasks (e.g. Task 2 for Intermediate and Tasks 9, 11 and 17 for Newbies) tend to have higher grades. The information about the specific tasks will be useful to educators to flag the key places where students may need help and timely intervention. Another important factor is when students start working on the task (when they first run code). The features 'Early start' and 'Inactive time' illustrate the importance of starting work early. Students who achieved better grades usually started earlier than the other students.

In summary, the results show that it is possible to predict accurately the final grade based on student behaviour information available in the middle of the course. The features used are easy to extract: previous problems passed, number of attempts and time features. The results also highlight the importance of starting early after the problems are released, and working steadily, attempting and passing the problems every week.

5 Conclusion

In this paper, we investigated whether final student grades can be accurately predicted in the middle of the course, for timely intervention. We conducted our analysis in the context of online introductory programming courses for primary and high school students at three different levels: Newbies, Beginners and Intermediate. We defined and extracted suitable features from the raw data and used a DT classifier to produce a compact set of rules, which are easily interpretable by teachers and students. We achieved accuracy results of up to 76%, considerably higher than the baseline used for comparison. The learned DT rules highlighted the importance of starting early and attempting and passing the tasks every week, and also flagged specific key tasks which predict the final performance. Importantly, our DT based approach is easy to implement and can be applied to other courses to provide insights about the final student grade based on information available in the middle of the course, allowing students to adjust their learning behaviour and teachers to take timely remedial actions.

References

1. Asif, R., Merceron, A., Ali, S.A., Haider, N.G.: Analyzing undergraduate students' performance using educational data mining. Comput. Educ. **113**, 177–194 (2017)
2. Elbadrawy, A., Polyzou, A., Ren, Z., Sweeney, M., Karypis, G., Rangwala, H.: Predicting student performance using personalized analytics. Computer **49**, 61–69 (2016)
3. Koprinska, I., Stretton, J., Yacef, K.: Predicting student performance from multiple data sources. In: Conati, C., Heffernan, N., Antonija Mitrovic, M., Verdejo, F. (eds.) Artificial Intelligence in Education: 17th International Conference, AIED 2015, Madrid, Spain, June 22-26, 2015. Proceedings, pp. 678–681. Springer, Cham (2015). https://doi.org/10.1007/978-3-319-19773-9_90
4. Koprinska, I., Stretton, J., Yacef, K.: Students at risk: detection and remediation. In: Proceedings of International Conference on Educational Data Mining, pp. 512–515 (2015)
5. Meier, Y., Xu, J., Atan, O., Van der Schaar, M.: Predicting grades. IEEE Trans. Signal Process. **64**, 959–972 (2015)
6. Kotsiantis, S., Patriarcheas, K., Xenos, M.: A combinational incremental ensemble of classifiers as a technique for predicting students' performance in distance education. Knowl.-Based Syst. **23**(6), 529–535 (2010)
7. Romero, C., Ventura, S., Espejo, P.G., Hervás, C.: Data mining algorithms to classify students. In: International Conference on Educational Data Mining, pp. 8–17 (2008)
8. Romero, C., López, M.-I., Luna, J.-M., Ventura, S.: Predicting students' final performance from participation in on-line discussion forums. Comput. Educ. **68**, 458–472 (2013)

9. Thai-Nghe, N., Drumond, L., Horváth, T., Nanopoulos, A., Schmidt-Thieme, L.: Matrix and tensor factorization for predicting student performance. In: International Conference on Computer Supported Education, pp. 69–78 (2011)

10. Zhang, V., Jeffries, B, Koprinska, I.: Predicting progress in a large-scale online programming course. In: International Conference on Artificial Intelligence in Education (2023)

11. Polito, S., Koprinska, I., Jeffries, B.: Exploring student engagement in an online programming course using machine learning methods. In: Rodrigo, M.M., Matsuda, N., Cristea, A.I., Dimitrova, V. (eds.) Artificial Intelligence in Education. Posters and Late Breaking Results, Workshops and Tutorials, Industry and Innovation Tracks, Practitioners' and Doctoral Consortium. Lecture Notes in Computer Science, pp. 546–550. Springer, Cham (2022). https://doi.org/10.1007/978-3-031-11647-6_112

12. McBroom, J., Paassen, B., Jeffries, B., Koprinska, I., Yacef, K.: Progress networks as a tool for analysing student programming difficulties. In: Australasian Computing Education Conference, pp. 158–167 (2021)

Classifying Mathematics Teacher Questions to Support Mathematical Discourse

Debajyoti Datta[1]([✉])(ID), James P. Bywater[2](ID), Maria Phillips[1](ID), Sarah Lilly[1](ID), Jennifer L. Chiu[1](ID), Ginger S. Watson[3](ID), and Donald E. Brown[1](ID)

[1] University of Virginia, Charlottesville, VA, USA
dd3ar@virginia.edu
[2] James Madison University, Harrisonburg, VA, USA
[3] Old Dominion University, Norfolk, VA, USA

Abstract. This paper examines whether natural language processing technologies can provide teachers with high-quality formative feedback about questioning practices that promote rich inclusive mathematical discourse within classrooms. This paper describes how a training dataset was collected and labeled using teacher questioning classifications that are grounded in the mathematics education literature, and it compares the performance of four classifier models fine-tuned using that dataset. Of the models tested, we find that RoBERTa, an open-source LLM, had a 76% accuracy in classifying questions. These modern transfer-learning based approaches require significantly fewer data points than traditional machine-learning methods and are ideal in low-resource scenarios like question classification. The paper concludes by discussing potential use cases within the field of mathematics teacher education and describes how the classifier models created can be publicly accessed.

Keywords: Mathematics · Teachers · Questioning · Text classification

1 Rationale

The National Council of Teachers of Mathematics (NCTM) define mathematical discourse [15] as the "purposeful exchange of ideas through classroom discussion, as well as through other forms of verbal, visual, and written communication" (p. 29). Through discussion students can develop their identities as knowers and doers of mathematics [2] and have more equitable and inclusive opportunities to develop deep mathematical understandings [16]. One core teaching practice that promotes mathematical discourse in classrooms is asking purposeful questions [15]. Purposeful questions elicit student thinking and then explore or probe that thinking. In doing so, the teacher guides the classroom discourse by asking students to, for example, justify their thinking, consider different approaches, assess how alternative approaches are different and similar, or clarify when their thinking might or might not be applicable. These questions stand in contrast

N. Wang et al. (Eds.): AIED 2023, CCIS 1831, pp. 372–377, 2023.
https://doi.org/10.1007/978-3-031-36336-8_58

to those that only seek to assess procedural or factual knowledge and that do not explicitly discuss, clarify, or develop student thinking [3]. To support the development of questioning skills, teacher preparation courses typically provide opportunities for pre-service teachers to rehearse asking questions that promote mathematical discourse, and the pre-service teachers receive feedback and advice from a coach. A key barrier to providing these opportunities is that observation feedback is time-consuming and difficult to scale. Automated classification of teacher questions might help mitigate this barrier. To this end, this paper will ground our teacher questionings classifications in currently existing mathematics education literature, describe how our dataset was collected, labeled, and used to train and test a variety of classifier models, and then report and discuss answers to the research question: How accurately do different classifier models classify teacher questions?

2   Background

A variety of frameworks have been developed by mathematics education researchers to describe mathematics teacher questioning [3,10,11,21]. To support the development of teacher questioning practices, validated observational measures of teaching practice are used to provide specific and actionable empirically grounded formative feedback. The Instructional Quality Assessment (IQA) [4] has been shown to measure teacher questioning with validity, and focuses on the degree to which teachers use probing and exploration questions in addition to procedural questions and questions that are not related to student mathematical reasoning. Since our work is intended to support scalable formative feedback about teacher questioning, the categories we adopt are based on those used by the IQA and by coaches of pre-service mathematics teachers. Specifically, the categories we use are: probing and exploring, procedural, and expository, with an additional category of other for questions that did not fall within the prior three.

3   Method

3.1   Dataset

The dataset used in this study consists of 4,413 labeled teacher questions, with 2,061 unique questions. The questions were drawn from anonymized transcripts of classroom discourse between high school mathematics teachers and individual or small groups of students engaged in inquiry tasks [5]. The teacher questions were labeled by four annotators. Three of the annotators were enrolled in an education Ph.D. program and had prior classroom teaching experience, and the other was an undergraduate education major. All four had prior annotating experience. Two techniques were used to label the teacher questions, with two annotators using classical labeling and two using model-assisted labeling. The Labelbox/Labelstudio [12] platform was used for this purpose. During classical labeling, the annotators were presented with a question to classify and asked to pick one of the question categories. The inter-annotator agreement (kappa) was 0.52. During model-assisted labeling, annotators were again presented with a question to classify, but one of the question categories was pre-selected and the annotator either accepted or changed the pre-selected choice. We use a pre-trained weak-supervision model for this purpose [17] and the inter-annotator agreement (kappa) was 0.61. Prior work has shown that these two techniques produce similar annotator accuracy and agreement, but that model-assisted labeling can significantly ($p < 0.001$) decrease the average time taken to label questions from 15.2 s for classical labeling to 10.4 s for model-assisted labeling, and that this was not due to annotators uncritically accepting all of the model-assisted recommendations. [6]

3.2   Classifier Models

With the advent of transfer learning, the standard approach to solving natural language processing tasks like question classification is to fine-tune a pre-trained LLM [7]. This approach often requires less data than fine-tuning a model from scratch. In this study we used four models: BERT-base [7], RoBERTa-base [14], Microsoft DeBERTa [9], and DistilBERT [18]. As their names indicate, all these models are variants of the original BERT model but they vary in their architectures and unsupervised pre-training paradigm. We chose to use these BERT-based models so that we could compare these different versions and understand how domain transfer (transferring from an unsupervised task to education specific question classification tasks) results in performance variations. In some cases, models that performed better on traditional benchmarks do not perform better in specialized domains.

3.3   Analysis of Classifier Accuracy

In order to address our first research question, we used a train-test split of 70–30 and each question was truncated to the max configuration length specific to each of the four models used. We used accuracy as the evaluation metric as it is common in classification tasks and benchmarks. Accuracy is defined as the number of correct predictions divided by the total number of predictions. The evaluation runtime refers to the running time for each model which is also important since we plan to use it downstream in a conversational agent and faster inference time will enable a better user experience. We used models from the HuggingFace Transformers library and performed training using Google Colab Pro. Each model was trained until there was no improvement in validation loss by using EarlyStopping which was between 3 and 4 epochs. We used the default learning rate of 2e-5 with a weight decay of 0.01. The carbon footprint for the entire experiment was 0.28 kgCO2eq calculated using [13].

4   Results

As show in Table 1, the model accuracies varied between 74% for the DistilBERT to 76% for the RoBERTa model. The evaluation accuracy for the pre-trained model is similar indicating the difficulty of the task.

Table 1. Classifier model performance

Model Name	Accuracy	Runtime	Hidden Size	Vocab Size
bert-base	0.75	6.3532	1024	50272
roberta-base	0.76	5.6464	768	50265
microsoft/deberta-v3-base	0.75	7.8258	768	128100
distilbert-base-uncased	0.74	1.5125	768	30522

Examining the confusion matrices for each model (Fig. 1), we see that the different classifiers have different strengths at classifying different categories, however in general, *procedural* questions had higher classification accuracy than the *expository* and *probing and exploring* question types. The most common confusion was between *probing and exploring* and *procedural* questions.

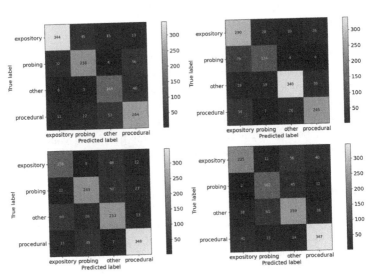

Fig. 1. Confusion matrices for BERT (top left), RoBERTa (top right), DeBERTa (bottom left), and DistilBERT (bottom right).

5   Limitations

While direct comparison of these results with others is not possible because of the unavailability of shared datasets, studies that have used similar data found similar results when using larger training datasets. For example, [20] reported an F1 score of 0.65 from a training dataset of 27,434 teacher sentences that were labeled with one of six teacher talk moves, and [8] reported a weighted F1 score of 0.69 from a training dataset of 10,080 teacher utterances that were labeled as either containing or not containing a question. Another limitation of this study is the small size of the dataset.

6   Access and Potential Use Cases

Results indicate that the RoBERTa teacher question classifier will be valuable for teacher educators and researchers interested in developing tools to support teachers develop high-quality questioning skills. For example, tools similar to [19] could allow teachers to upload transcripts of class discussions and receive feedback about the types of questions they asked. Given the growing interest in using teaching simulations, the classifier could also be used to provide immediate post-simulation feedback. We have open-sourced the models described in the study at this URL [1].

7   Conclusion

The study describes the accuracy of teacher question classifiers that are grounded in categories described within mathematics education literature and used by practitioners who work to support teachers adopt equitable and inclusive mathematical teaching practices. By describing the accuracy of our classifiers and illustrating how our most accurate classifier decides on its classifications we hope our classifier can be used in a variety of tools for providing valid, high-quality automated feedback about mathematical questioning.

References

1. https://huggingface.co/debajyotidatta/roberta-base_edu
2. Aguirre, J., Mayfield-Ingram, K., Martin, D.B.: The Impact of Identity in K-8 Mathematics Learning and Teaching: Rethinking Equity-Based Practices. National Council of Teachers of Mathematics, Incorporated (2013)
3. Boaler, J., Brodie, K.: The importance, nature and impact of teacher questions. In: Proceedings of the Twenty-Sixth Annual Meeting of the North American Chapter of the International Group for the Psychology of Mathematics Education, vol. 2, pp. 774–782 (2004)
4. Boston, M.: IGA mathematics lesson observation rubrics and checklists (2012). https://peabody.vanderbilt.edu/docs/pdf/tl/IQA_RaterPacket_LessonObservations_Fall_12.pdf

5. Bywater, J.P., Chiu, J.L., Hong, J., Sankaranarayanan, V.: The teacher responding tool: Scaffolding the teacher practice of responding to student ideas in mathematics classrooms. Comput. Educ. **139**, 16–30 (2019)
6. Datta, D., et al.: Evaluation of mathematical questioning strategies using data collected through weak supervision. arXiv preprint arXiv:2112.00985 (2021)
7. Devlin, J., Chang, M.W., Lee, K., Toutanova, K.: Bert: Pre-training of deep bidirectional transformers for language understanding. arXiv preprint arXiv:1810.04805 (2018)
8. Donnelly, P.J., Blanchard, N., Olney, A.M., Kelly, S., Nystrand, M., D'Mello, S.K.: Words matter: Automatic detection of teacher questions in live classroom discourse using linguistics, acoustics, and context. In: Proceedings of the Seventh International Learning Analytics & Knowledge Conference, pp. 218–227 (2017)
9. He, P., Gao, J., Chen, W.: Debertav 3: Improving deberta using electra-style pre-training with gradient-disentangled embedding sharing. arXiv preprint arXiv:2111.09543 (2021)
10. Herbel-Eisenmann, B.A., Breyfogle, M.L.: Questioning our patterns of questioning. Math. Teach. Middle School **10**(9), 484–489 (2005)
11. Kazemi, E., Stipek, D.: Promoting conceptual thinking in four upper-elementary mathematics classrooms. J. Educ. **189**(1–2), 123–137 (2009)
12. Labelbox: Labelbox: Training data platform (2020). https://labelbox.com/
13. Lacoste, A., Luccioni, A., Schmidt, V., Dandres, T.: Quantifying the carbon emissions of machine learning. arXiv preprint arXiv:1910.09700 (2019)
14. Liu, Y., et al.: Roberta: A robustly optimized bert pretraining approach (2019)
15. NCTM: Principles to actions: Ensuring mathematical success for all. NCTM, Reston (2014)
16. Oakes, J., Joseph, R., Muir, K.: Access and achievement in mathematics and science: Inequalities that endure and change. In: Handbook of Research on Multicultural Eucation, vol. 2, pp. 69–90 (2004)
17. Ratner, A.J., De Sa, C.M., Wu, S., Selsam, D., Ré, C.: Data programming: Creating large training sets, quickly. In: Advances in Neural Information Processing Systems, pp. 3567–3575 (2016)
18. Sanh, V., Debut, L., Chaumond, J., Wolf, T.: Distilbert, a distilled version of bert: smaller, faster, cheaper and lighter. arXiv preprint arXiv:1910.01108 (2019)
19. Suresh, A., et al.: Using transformers to provide teachers with personalized feedback on their classroom discourse: The talkmoves application. arXiv preprint arXiv:2105.07949 (2021)
20. Suresh, A., Sumner, T., Jacobs, J., Foland, B., Ward, W.: Automating analysis and feedback to improve mathematics teachers' classroom discourse. In: Proceedings of the AAAI Conference on Artificial Intelligence, vol. 33, pp. 9721–9728 (2019)
21. Wood, D.: Aspects of teaching and learning. In: Woodhead, M., Faulkner, D., Littleton, K. (eds.) Cultural Worlds of Early Childhood, pp. 157–177 (1998)

Task-Based Workplace English: An Adaptive Multimodal Tutoring System

Pravin Chopade[1] ⓘ, Shi Pu[2(✉)] ⓘ, Michelle LaMar[2], and Christopher Kurzum[2]

[1] Independent Researcher, Princeton, USA
[2] Educational Testing Service, 660 Rosedale Rd, Princeton, NJ 08540, USA
{spu,mlamar,ckurzum}@ets.org

Abstract. Personalized language learning (PLL) is central to educational and industrial communities worldwide. In our ongoing work, we developed a new multimodal intelligent tutoring system (ITS) with dynamic adaptation for a task-based workplace English language learning. Our work focuses on helping users improve their speaking and behavioral communication skills in the context of a fictional workplace environment while interacting with character-avatars(e.g., project supervisors and fellow team members). The system captures the users' multimodal interaction footprint, including clickstream, audio, and video data. Our learner model measures user skill mastery based on users' response data and provides dynamic adaption on item difficulty.

Keywords: Artificial Intelligence · Personalization · Language Learning · Intelligent Tutoring System · Speaking Domain Model · Spoken Dialogue System

1 Introduction

In recent years, several companies have created foreign language learning platforms, bringing together learners' interactions with system adaptivity and a personalized learning experience to meet learners' needs [7]. Language learning is a complex domain where many different cognitive processes and skills interact in the learner's production and reception of language. Language learning follows an expected path or progression, where more complex or advanced skills are built upon foundational, enabling skills. For example, language learners acquire simple, basic vocabulary first to describe familiar, concrete objects or people before they can use more sophisticated words or expressions to talk about abstract concepts.

It is critical to note that not all language learning develops in a predictable linear pathway. The current thinking in the field of second language acquisition suggests that language educators need to consider the non-linear, complex, dynamic nature of language development which requires different cognitive and linguistic skills and competencies [1]. This implies that when developing language

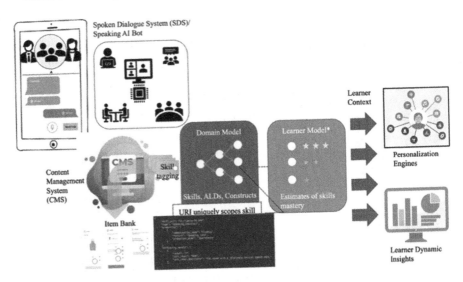

Fig. 1. New multimodal ITS framework operationalizing content creation and personalization for task-based language learning

learning solutions, designers and developers should consider teaching iteratively with mindful repetition for engagement in learning; learners need to learn how to adapt their language resources to different contexts of language use, communicative purposes, and changing situations.

On the other hand, in the 21st century, there is a growing need for effective workplace communication skills to improve employee engagement and productivity and to build a positive work environment. Thus, the speaking domain model (SDM) designed to support the development of digital language learning solutions or to provide personalized instruction should include the key linguistic skills and subskills as well as the different workplace communicative activities, functional skills, and contexts of language use. These activities and contexts provide instructional designers with the foundations for creating content that reflects the dynamism and non-linearity of language competence. These activities and contexts also help create opportunities for learners to practice in different learning tasks or scenarios, optimizing conditions for language learning. The next section outlines our multimodal ITS framework with the SDM and learner model.

2   Framework Design

Figure 1 shows our multimodal ITS framework for content creation and personalization in task-based language learning. A domain model defines a given learning domain's skills, relationships, and other properties of interest. A learner model (aka diagnostic model or user proficiency model) tracks an individual learner's mastery of the skills in the domain model. Learner models are closely related to

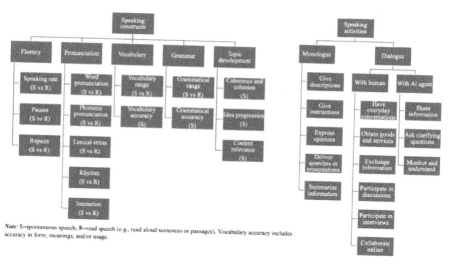

Fig. 2. Speaking domain model speaking constructs and speaking activities

the domain model but are not necessarily the same. This distinction provides flexibility between what is theoretically defined and what a system may attempt to diagnose.

2.1 Speaking Domain Model

Figure 2 illustrates the high-level language skills and communicative activities conceptualized in our speaking domain model. The speaking constructs consist of fluency, pronunciation, vocabulary, grammar, and topic development. Each top-level construct is further qualified into more specific sub-dimensions. For example, the sub-constructs for vocabulary are vocabulary range and vocabulary accuracy. Similarly, speaking activities consist of top-level contexts of monologue and dialogue, with dialogue including a user's spoken interactions with a human or with an AI agent. The categories listed below provide a starting point for defining lower-level "skills" worth tracking for learners' exposure and mastery.

2.2 Learner (User Proficiency) Model

A learner model is the heart of the "Workplace" application's backend; it takes in a user's prior knowledge state or study history and outputs a user's skill level. Recent research in learner modeling has explored various deep learning approaches, including Recurrent Neural Networks [4], Transformer-based models [6], and Graph Neural Networks [3], to improve the accuracy of these models. In our current development work, we use Deep Performance Factors Analysis for Knowledge Tracing (DPFA) [5] as the user-proficiency model, which estimates the users' proficiency in the underlying skills required to complete each item. For example, the model will estimate the user's proficiency in reading accuracy,

Table 1. Verbal and non-verbal communication features

Communication Skill	Features
Verbal	Speaking fluency, Speaking accuracy, Long pauses, Filler words,
	Grammar errors, Staying on topic, Speaking confidence
Non-verbal	Eye gaze, Head posture

fluency, long pauses, and filler words for a read-aloud item. Our system then utilizes the estimated user proficiency to determine the appropriate difficulty level of presented items.

3 Workplace App Product Development

Our "Workplace" app provides English language learners opportunities to practice workplace related English communication skills. The app is organized into projects with multiple stages, each containing multiple (usually three) tasks and multiple questions. Tasks consist multiple-choice questions, audio questions where students are required to produce an oral response, video questions where students must record short videos while producing oral responses, and text questions where students are required to produce written responses.

Once a user produces an audio or video recording in response, our systems estimate the user's verbal and non-verbal communication skills using a variation of the Speech Rator scoring engine [2]. Table 1 illustrate the features estimated by the engine.

4 Feedbacks

Figure 3 illustrates the feedbacks we provide to the users:

1. Skill performance feedback ranges from low to medium to high based on verbal communication features (e.g., speaking rate, grammar, staying on topic) and non-verbal communication features (eye gaze and head posture) listed in Table 1.
2. Stage insights feedback with a skill proficiency meter to denote how well the user performed on specific learning objectives, areas for improvement, and valuable tips.
3. Overall project-level feedback with skill proficiency, frequent strengths, and frequent areas for improvement, as well as personalized insights and recommendations.

5 Personalization

We plans to dynamically change item difficulty based on users estimated skill proficiency. This feature is part of our ongoing work and will be implemented

Fig. 3. Learners personalized dynamic insights for task-based performance

in future versions. In the following section we presents the training data we collected and briefly describe the learner model we use to estimate users' skill proficiency.

5.1   Functional Usability Study

A functional usability study was conducted in December 2022. A total of 119 Indian users (Male = 60, Female = 58, Prefer to self-describe = 1) in the age group 19–29 were recruited and were asked to finish the same project in the "Workplace" app. We use DPFA [5] to estimate users' skill proficiency. The heat map in Fig. 4 visualizes the estimated mastery of speaking and multimodal skills for a selected user over time. Each column in the heat map represents a question

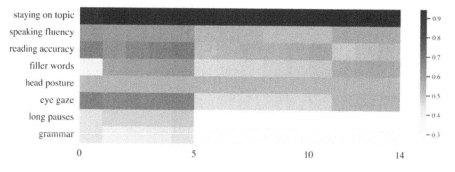

Fig. 4. The heat map visualizes the estimated mastery of speaking and multimodal skills.

the user answers within the app, while each row represents a particular skill. The color of each cell represents the probability that the user has mastered the skill, with darker shades indicating higher probabilities. These estimations are updated in real-time after each question is answered, allowing for a dynamic representation of the user's progress and improvement. Based on the users' skill mastery progress, the personalization engine will offer skill-specific feedback and recommendations for improvement.

6 Conclusion

This paper presented a new multimodal ITS with AI-based personalization capability. Our key results from an initial Alpha study show that we can track and identify learners' task-based verbal and non-verbal workplace communication skills over three stages in each project. In future work, we plan to use these initial results and Indian leaner's insights from the functional usability study to improve our personalization model performance. We also plan to develop this "Workplace" app for global users with all accessibility aspects to improve their workplace English language communication skills.

References

1. Cameron, L., Larsen-Freeman, D.: Complex systems and applied linguistics. Int. J. Appl. Linguist. **17**(2), 226–240 (2007)
2. Chen, L., et al.: Automated scoring of nonnative speech using the speechrater SMV. 5.0 engine. ETS Res. Rep. Ser. **2018**(1), 1–31 (2018)
3. Nakagawa, H., Iwasawa, Y., Matsuo, Y.: Graph-based knowledge tracing: Modeling student proficiency using graph neural network. In: IEEE/WIC/ACM International Conference on Web Intelligence, pp. 156–163 (2019)
4. Piech, C., et al.: Deep knowledge tracing. Adv. Neural Inf. Process. Syst. **28** (2015)
5. Pu, S., Converse, G., Huang, Y.: Deep performance factors analysis for knowledge tracing. In: Roll, I., McNamara, D., Sosnovsky, S., Luckin, R., Dimitrova, V. (eds.) AIED 2021. LNCS (LNAI), vol. 12748, pp. 331–341. Springer, Cham (2021). https://doi.org/10.1007/978-3-030-78292-4_27
6. Pu, S., Yudelson, M., Ou, L., Huang, Y.: Deep knowledge tracing with transformers. In: Bittencourt, I.I., Cukurova, M., Muldner, K., Luckin, R., Millán, E. (eds.) AIED 2020. LNCS (LNAI), vol. 12164, pp. 252–256. Springer, Cham (2020). https://doi.org/10.1007/978-3-030-52240-7_46
7. Shawky, D., Badawi, A.: Towards a personalized learning experience using reinforcement learning. In: Machine Learning Paradigms: Theory and Application, pp. 169–187 (2019)

A Bayesian Analysis of Adolescent STEM Interest Using Minecraft

Matthew Gadbury[✉] [iD] and H. Chad Lane [iD]

University of Illinois Urbana-Champaign, Champaign, USA
{gadbury2,hclane}@illinois.edu

Abstract. Minecraft continues to be a popular digital game throughout the world, and the ways in which adolescents play can provide insight into their existing interests. Through informal summer camps using Minecraft to expose middle school students to concepts in astronomy and earth science, we collected self-reports of STEM and Minecraft interest, as well as behavioral log data through player in-game interactions. Finding relationships between in-game behaviors and individual interest can provide insight into how educational experiences in digital games might be designed to support learner interests and competencies in STEM. Bayesian model averaging of data across camps was implemented to address the relatively small sample size of the data. Results revealed the important role of existing interest and knowledge for developing and sustaining interest.

Keywords: Digital games · informal learning · interest · Bayesian analysis

1 Introduction and Background

One conception of interest is that of a psychological state marked by heightened attention and focus, as well as a motivational force pushing an individual to explore and find additional information about an object or domain [3]. Interests can be activated or "triggered" by events in the environment, such as novelty, autonomy, personal relevance, or community [8], which in turn can lead to individuals developing greater competence and interest in the domain [5]. Adolescence tends to be a period when students are beginning to form identities, and this is also a time when interest in Science, Technology, Engineering, & Math (STEM) falters for many [7]. Utilizing technology to trigger and maintain interest in STEM for adolescents shows some evidence of being effective for developing interest.

Technology has been identified as both a trigger of interest learner interest [8] and a conduit by which other facets of interest development can be applied through formal or informal educational spaces. Minecraft remains one of the most played digital games in the world. Given its popularity, Minecraft stands as a low barrier educational tool for many students, in that students should be able to engage content without spending much time learning controls and how to navigate the environment. Research conducted using Minecraft has shown promise for increasing motivation to learn STEM content [1]. For these reasons, we use Minecraft to explore the following research question: To what extent does a STEM-focused Minecraft summer camp influence adolescent interest in STEM and what is the role of Minecraft interest?

2 Methods

2.1 Participants

Participants ($n = 96$) included middle school and early high school students ages 11–15 taking part in week-long summer camp programs implemented by our research team. A total of 7 separate camps across 3 separate sites (3 in the Western United States, 2 in the Midwestern U.S., and 2 in the Eastern U.S.) were held during the summer of 2022. Due to illness and absences amounting to more than half the camp days in a week, data of 22 participants were removed from the final count, resulting in ($n = 74$, 32% female). Breakdown of race/ethnicity showed: 29% White, 24% Black, 13% Hispanic, 3% Asian, 2% Native American, 8% Other and 21% Prefer Not to Answer (PNA). Written or online consent was obtained from at least one parent or guardian of each participant prior to participation.

2.2 Materials and Procedure

Each participant was provided with a laptop, mouse, and access to Minecraft: Java Edition. Most of each camp was spent playing custom-built worlds introducing themes in Astronomy and Earth Science in Minecraft. Worlds were inspired by "What if" questions posed by astronomers, such as "What if Earth had no moon?". In total, our server hosts 2 orientation worlds, 5 hypothetical "What if" worlds, and 5 known exoplanets.

Each camp consisted of 5, 3-h meetings spanning one week. A pre-survey on STEM interest and Minecraft interest, validated by our lab but unpublished, was administered before the start of each camp. On the first day of each camp, participants were led through the 2 orientation worlds. The following 2 days had participants explore the rest of the curated worlds and make in-game observations inferring what is happening and why. Additionally, after exploring each world for 25–30 min, a set of 3–4 open-ended self-explanations were pushed out to each participant. The final 2 days of each camp had participants build habitats on Mars, considering known challenges to human survival on Mars. Participants also completed post-surveys on STEM and Minecraft interest, which were identical to the pre-survey.

2.3 Analysis

Assessing interest in STEM and Minecraft, models used either STEM interest or Minecraft interest as the dependent variable. The STEM survey consisted of 20 Likert-type questions asking students how interested they are in STEM activities, with 1 = "Not at all interested" to 5 = "extremely interested". An example of a STEM survey question is, "How interested are you in using numbers to confirm ideas and solve problems?". Scores were aggregated across questions. The same approach was taken with the Minecraft survey, which consisted of 20 Likert-type questions asking students how interested in STEM-related Minecraft activities they are. An example of a Minecraft interest question is, "How interested are you in learning what colors each biome has and why?". Demographic data, such as age and race/ethnicity, are also included in the final models.

Self-explanations (SEs)

Participants completed a total of 22 open-ended self-explanations. An example of a self-explanation prompt is "How would you define habitability?" or "What might be a good way to generate energy on Earth without a moon and why?". Answers were scored based on correctness and from 0 to 3. All self-explanations from the summer were scored by two graduate students. The aggregate percent agreement was 73%. All disagreements were resolved through discussion. After finalizing scores, an average score across answers was calculated for each participant.

Observations

In-game observations were collected for each participant. Participants were encouraged repeatedly throughout the camp to make scientific observations about what they saw in the game and why they think it is the case. An example observation participant might make is, "Because there are two moons, more light is reflected to Earth". Only the frequency of in-game observations made is used in this paper, which is considered a measure of behavioral engagement with the game and content.

Exploration

Location data was captured for each participant every 3 s. Exploration was calculated as the number of squares a participant crossed into on a 10x10 grid overlay of each map. Exploration was averaged across all maps visited to provide a mean exploration measurement for each participant.

Bayesian Analysis

Given the small sample size of the data, a Bayesian analysis is a more robust approach at analyzing the data than using frequentist statistics [6]. A non-informative prior is used in this approach, however the posterior probability distribution estimated from this research can used for future interest research as the starting prior. We can establish a stronger connection between the predictors and dependent variable by estimating the probability of contribution, and this is accomplished using Bayesian Model Averaging (BMA). BMA is a way to estimate parameters that averages the predictions of different models being considered, and each model is given a weight based on its probability [4].

Correlations were run first to examine the relationship between the variables. All variables are then centered to avoid multicollinearity. All analyses were conducted in R using the "BMA" package.

3   Results

A paired samples t-test was first conducted to see if there were any changes in STEM interest and Minecraft interest before and after the summer camp. The results from the pre-survey ($M = 73.82$, $SD = 15.9$) and post-survey ($M = 74.39$, $SD = 14.77$) of STEM interest showed there was no change in STEM interest as a result of the camp $t(61) =$ 0.11, $p = 0.914$. Looking at Minecraft interest, the difference between the pre-survey ($M = 70.93$, $SD = 14.68$) and the post-survey ($M = 68.13$, $SD = 14.71$) was also non-significant $t(60) = 1.45$, $p = 0.152$.

Table 1 shows correlations between analyzed variables. All continuous variables were centered before using Bayesian analysis.

Table 1. Correlation table of measured variables

	SInt	MInt	SE	Obs	Exp
SInt	--				
MInt	0.61	--			
SE	0.29	0.31	--		
Obs	0.12	0.24	0.48	--	
Exp	0.16	0.19	0.23	0.28	--

Note: SInt is STEM Interest, MInt is Minecraft Interest, SE is self-explanation score, Obs is observation frequency, and Exp is exploration metric.

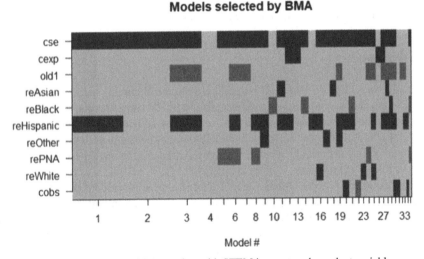

Fig. 1. Bayesian Model Averaging with STEM interest as dependent variable

BMA was implemented to understand how much each predictor contributes to the overall model, and which predictors are candidates for removal. See Fig. 1 for BMA results. Blue bars represent a positive predictor included in a model and red represents a negative predictor included in a model. What is clear from BMA is that self-explanations appear most frequently in models. The only other predictor that appears frequently is the positive effect of Hispanic on STEM interest. Interestingly, age negatively predicted STEM interest, where older learners showed less interest in STEM.

Models selected by BMA

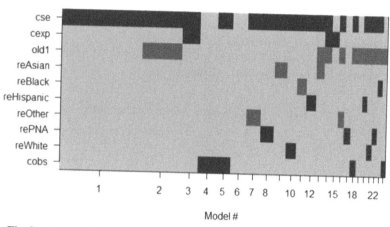

Fig. 2. Bayesian Model Averaging with Minecraft interest as dependent variable

As with STEM interest, Age is a negative predictor of Minecraft interest. Taken together, this might suggest that as learners age their interest in STEM decreases along with interest in Minecraft. Figure 2 shows the results of BMA. Again, self-explanations prove to be the most common predictor across models.

4 Discussion

No change in STEM or Minecraft interest between the beginning and end of the camp experience was observed. Interest in Minecraft and STEM showed moderate correlations with each other, which may point to inherent STEM activities and content built into Minecraft. Finding out what activities and camp aspects contribute most to STEM interest is needed to allay uncertainty as to how interest is triggered in the experience and what learners find most engaging.

Data collected reflects unique parts of the informal environment that can be measured using a digital game: observation frequency, exploration patterns, and in-game self-explanation responses. Given the novelty of the "What if" scenarios, we thought exploration would be quite extensive and would be reflective of participant information-seeking behavior for STEM content. However, exploratory behavior proved to be a very weak predictor of STEM interest, as did frequency of observations. The possibility exists that participants did not explore extensively but did take time to interact with non-playable characters (NPCs) who provide detailed information about each hypothetical. The only predictor that bore any substantial relationship to STEM interest was average score on the self-explanations, which was essentially a knowledge assessment for each world aimed at measuring cognitive interest. Higher knowledge in a domain is usually an indicator of higher interest [2].

Bayesian analysis is a powerful tool for exploring the triggering and development of interest. The data analyzed here can be used to inform a prior distribution for future analysis of camp data. Also, given the frequent small sample sizes we encounter when running summer camps and after school programs, Bayesian analysis can be used with relatively small sample sizes to estimate the posterior probability distribution. The power of BMA arises from the idea that no predictor ever dominates the entire model but also each predictor does contribute to some degree [4].

Future analyses will divide participants by high or low interest to see if learners with low interest experience any changes in interest, and if the magnitude of change is greater than those entering with high interest. In addition, analysis can benefit from precise measurements that capture situational interest from specific aspects of the environment, and that measure interest in domains dominating camp content.

References

1. Baek, Y., Min, E., Yun, S.: Mining educational implications of minecraft. Comput. Schools **37**(1), 1–16 (2020). https://doi.org/10.1080/07380569.2020.1719802
2. Fastrich, G.M., Murayama, K.: Development of interest and role of choice during sequential knowledge acquisition. 16 (2020)
3. Hidi, S.E., Ann Renninger, K.: On educating, curiosity, and interest development. Current Opinion Behav. Sci. **35**, 99–103 (2020). https://doi.org/10.1016/j.cobeha.2020.08.002
4. Hinne, M., Gronau, Q.F., van den Bergh, D., Wagenmakers, E.-J.: A conceptual introduction to Bayesian model averaging. Adv. Methods Pract. Psychol. Sci. **3**(2), 200–215 (2020)
5. Hulleman, C.S., Thoman, D.B., Dicke, A.-L., Harackiewicz, J.M.: The promotion and development of interest: the importance of perceived values. In: O'Keefe, P.A., Harackiewicz, J.M. (eds.) The Science of Interest, pp. 189–208. Springer, Cham (2017). https://doi.org/10.1007/978-3-319-55509-6_10
6. König, C., van de Schoot, R.: Bayesian statistics in educational research: a look at the current state of affairs. Educ. Rev. **70**(4), 486–509 (2018). https://doi.org/10.1080/00131911.2017.1350636
7. Maltese, A.V., Tai, R.H.: Pipeline persistence: examining the association of educational experiences with earned degrees in STEM among U.S. students. Sci. Educ. **95**(5), 877–907 (2011). https://doi.org/10.1002/sce.20441
8. Ann Renninger, K., Bachrach, J.E., Hidi, S.E.: Triggering and maintaining interest in early phases of interest development. Learn. Cult. Soc. Interact. **23**(2019), 100260 (2019). https://doi.org/10.1016/j.lcsi.2018.11.007

Automatic Slide Generation Using Discourse Relations

Teppei Kawanishi[(✉)] and Hiroaki Kawashima[iD]

Graduate School of Information Science, University of Hyogo, Kobe, Japan
tep.kawanishi@gmail.com, kawashima@gsis.u-hyogo.ac.jp

Abstract. Slides are frequently used in educational situations such as lectures and presentations as a means of quickly and concisely conveying information. Automatic slide generation from documents has recently been studied to reduce the workload of creating slides. However, the generated slides in existing studies have their text as bullets on the same level, without any indentation or text layout as in slides created by humans. We propose automatic slide generation with a layout that reflects the structure of input documents. Specifically, we utilize a fine-tuned BERT classifier to predict the discourse relation of each sentence pair in a document. The evaluation results show that the trained discourse relation classifier achieves higher accuracy than the model used in the existing method. Moreover, combining keyword extraction and automatic text summarization, the proposed method is able to automatically generate slides with layouts considering discourse relations in a document.

Keywords: automatic slide generation · discourse relation · slide layout · BERT model

1 Introduction

Slides are commonly used in various educational situations, such as lectures and presentations, to convey information quickly and succinctly. However, creating slides requires significant time and effort. For instance, when creating slides from a document (e.g., textbook, article, and paper), one must have a complete understanding of the content, consider the text to include, and lay out the slides in a clear and understandable manner.

Recent studies have proposed methods that automatically generate slides from research papers (e.g., D2S [6] and DOC2PPT [3]). While slides contain several media, such as text and images, we here focus on the text in generated slides in the existing studies and identify two challenges. The first challenge is that all sentences are on the same level of bullet points, and the second challenge is that all the text is the same color and size. These issues make it difficult for readers or listeners to know where to focus and what is important when viewing the slides.

© The Author(s), under exclusive license to Springer Nature Switzerland AG 2023
N. Wang et al. (Eds.): AIED 2023, CCIS 1831, pp. 390–395, 2023.
https://doi.org/10.1007/978-3-031-36336-8_61

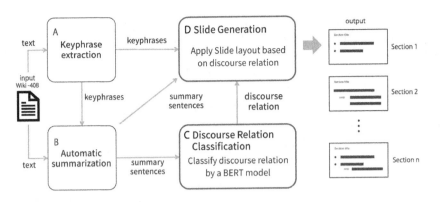

Fig. 1. The overview of the proposed slide generation method, which consists of four main modules (A to D).

In this study, we address these challenges by generating slides that reflect the relation and structure of input text in the slide layout. By incorporating keyphrase extraction and discourse relation classification into our automatic slide generation system, in addition to automatic text summarization, we aim to overcome the challenges in existing research and generate slides that are easier to understand the content.

2   Proposed Slide Generation Method

Our proposed method consists of four main modules, as shown in Fig. 1. The input is a single English Wikipedia article, using the text from the Wiki-40B dataset [4]. The output is an automatically generated slide. We use Wikipedia articles as input in order to set a certain context for the evaluation, although there are no strong methodological restrictions.

Overall steps of the proposed method is as follows. First, a text document (i.e., an article) from the Wiki-40B is input to the Keyphrase Extraction module (Fig. 1(A)). Next, the same document text, its section titles, and keyphrases predicted by the Keyphrase Extraction module (A) are input to the Automatic Text Summarization module (B). The summary sentences are then input to the Discourse Relation Classification module (C) to predict the relations of summary sentence pairs obtained by (B). Finally, using the keyphrases, summary sentences, and sentence relations obtained from these three modules, a slides with an appropriate layout is generated (D).

2.1   Keyphrase Extraction

The first module of our proposed method extracts phrases that best describe the subject of a document from an input document using keyphrase extraction. Keyphrase extraction is the task of extracting phrases that best describe the subject of an input document.

Our proposed method utilizes MultipartiteRank [1] for keyphrase extraction. MultipartiteRank is a graph-based unsupervised ranking algorithm influenced by Google's PageRank algorithm. It selects and ranks candidate phrases from the input document based on a weighted, fully directed graph. We use the top five ranked phrases in two ways: 1) as part of the input query for the automatic summarization module (Sect. 2.2), and 2) to determine which words should be emphasized in the generated slides.

2.2 Automatic Text Summarization

In this second module, we use an automatic summarization model to condense the input document into several sentences. Specifically, we utilize the QA module in the D2S system [6] for document summarization.

D2S was designed to perform the abstractive document-to-slide generation task as a query-based single-document text summarization (QSS) task. The QA module is responsible for generating summarized sentences based on a given Query and Context. The Query consists of a Question and Keywords, which together determine the focus of the summarization. The Keywords support the Question and are composed of multiple words. In addition, the QA module also takes sentences from the input document as Context and generates answers based on the Query and Context. The number and length of sentences of generated answers could be controlled by the maximum output tokens of the QA module. In the experiment, we set the maximum number of tokens as 200 per slide.

For this work, we input the section title as the Question and the keyphrases extracted in Sect. 2.1 as the Keywords. By incorporating not only the question but the keywords, we aim to generate more accurate and informative summaries.

2.3 Discourse Relation Classification

This module is the key component of our method as it predicts discourse relations between summarized sentences. Discourse Relation Classification is a task that involves classifying the relation between two input sentences.

To create the Discourse Relation Classification model, we fine-tuned BERT [2] using a dataset extracted from English Wikipedia [5]. This training dataset comprises 400,000 adjacent sentence pairs (with and without a connective) extracted from the English Wikipedia and collected for 20-class connective classification. However, our research goal is to predict the relation between two sentences rather than the connective between them. To adapt the dataset to our research goal, we created a new dataset based on the connective classification dataset. We referred to the manual of Penn Discourse Treebank 3.0 (PDTB3) [7], which is commonly used in the tasks related to discourse relation, and reconstructed the dataset. For example, the original dataset contains two conjunction labels "then" and "finally". Referring to the PDTB3 dataset, it was found that these labels mostly represent the discourse relation of Temporal.Asynchronous.Precedence. Based on this information, the two conjunction labels were integrated into a single discourse relation class. A similar approach

Table 1. Comparison of discourse relation classification performance

Model	Macro F1	Accuracy	Precision	Recall
Ours (14 classes)	**0.586**	**0.589**	**0.630**	**0.591**
Ours (20 classes)	0.478	0.488	0.536	0.488
Existing Model [5] (20 classes)	0.318	0.327	–	–

was applied to other conjunction labels, resulting in the reconfigured 14 relation class labels and used as a training dataset to fine-tune BERT.

2.4 Slide Generation

Slide generation (Fig. 1(D)) is carried out in the following steps. First, the summary text (Sect. 2.2) is placed within the slide using a predetermined layout based on the results of the discourse relation classification (Sect. 2.3). Then, the keyword extraction results (Sect. 2.1) are used to change the color of matching text within the article. A single slide is generated for each section of the input article from Wiki-40B [4], with the section title displayed at the top of the slide.

3   Evaluation and Results

3.1 Discourse Relation Classification

We conducted a comparative evaluation between the proposed method described in Sect. 2.3 and the existing model [5] (Decomposable Attention Model) using the test data from the dataset [5]. In Table 1, we show the results for both 14-class and 20-class classifications, where 20-class was used to maintain the same conditions as [5]. The model we created in this study achieved higher accuracy than the model [5] in all metrics and performed even better in the 14-class classification.

3.2 Slide Generation

Figure 2 shows some examples of generated slides using the proposed method described in Sect. 2. As can be seen in the generated slide in Fig. 2b, keyphrases are highlighted in red, and the layout is determined by the temporal relation "Precedence" of each sentence pair inferred by the discourse relation classification model in Sect. 2.3. More specifically, the inferred relation means that the first sentence precedes the second sentence and the second sentence precedes the third sentence. Figure 2d is an example of "Arg2-as-instance," which considers the second sentence describes an instance or example of the first sentence. Figure 2f shows more complicated relationships; in these examples, some predicted relations are not appropriate. In particular, the last (i.e., fourth) sentence in Fig. 2f describes the fact that can be inferred by the first to third sentences.

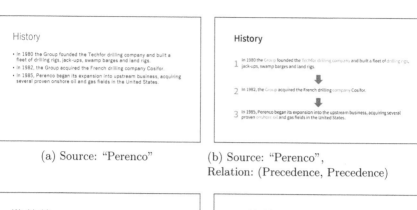

(a) Source: "Perenco"

(b) Source: "Perenco",
Relation: (Precedence, Precedence)

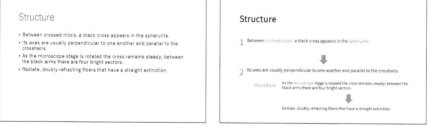

(c) Source: "Shared lane marking"

(d) Source: "Shared lane marking",
Relation: (Arg2-as-instance, None)

(e) Source: "Spherulite"

(f) Source: "Spherulite",
Relation: (Precedence, Result, Precedence)

Fig. 2. Examples of generated slides with only automatic summarization (left column) and the proposed method (right column). Captions denote each source article and the estimated relations between adjacent sentences. For example, (Relation1, Relation2) means the first and second sentences have Relation1, and the second and third sentences have Relation2.

The adjacent-pairwise prediction in the proposed approach cannot handle such relationship involving more than two sentences, and this limitation should be addressed in future work.

4   Conclusion

In this study, we proposed an automatic slide generation method incorporating keyphrase extraction and discourse relation classification to address the issues related to the text content in the output slides of existing studies. By applying discourse relation classification to slide generation, we confirmed the ability to create slides with layouts such as indentation. Moreover, we showed that the discourse relation classifier based on the BERT model achieved higher scores than the existing method. However, if any of the modules, such as automatic text summarization or discourse relation classification, output incorrect results, it may lead to the generation of slides with an inappropriate content layout. Therefore, our future work needs to improve each module's accuracy and also incorporate the generation of several possible candidates from which users can choose an appropriate slide.

Acknowledgements. This work was supported by JSPS KAKENHI Grant Number JP19H04226 and JP21H05302.

References

1. Boudin, F.: Unsupervised keyphrase extraction with multipartite graphs. In: Proceedings of the Conference of the North American Chapter of the Association for Computational Linguistics: Human Language Technologies, pp. 667–672 (2018). https://doi.org/10.18653/v1/N18-2105
2. Devlin, J., Chang, M.W., Lee, K., Toutanova, K.: BERT: pre-training of deep bidirectional transformers for language understanding. In: Proceedings of the Conference of the North American Chapter of the Association for Computational Linguistics: Human Language Technologies, vol. 1, pp. 4171–4186 (2019). https://doi.org/10.18653/v1/N19-1423
3. Fu, T.J., Wang, W.Y., Mcduff, D., Song, Y.: DOC2PPT: automatic presentation slides generation from scientific documents. In: Proceedings of the AAAI Conference on Artificial Intelligence, pp. 634–642 (2022)
4. Guo, M., Dai, Z., Vrandečić, D., Al-Rfou, R.: Wiki-40B: multilingual language model dataset. In: Proceedings of the 12th Language Resources and Evaluation Conference, pp. 2440–2452 (May 2020)
5. Malmi, E., Pighin, D., Krause, S., Kozhevnikov, M.: Automatic prediction of discourse connectives. In: Proceedings of the Eleventh International Conference on Language Resources and Evaluation (LREC 2018), pp. 1643–1648 (2018)
6. Sun, E., Hou, Y., Wang, D., Zhang, Y., Wang, N.X.R.: D2S: document-to-slide generation via query-based text summarization. In: Proceedings of the Conference of the North American Chapter of the Association for Computational Linguistics: Human Language Technologies, pp. 1405–1418 (2021). https://doi.org/10.18653/v1/2021.naacl-main.111
7. Webber, B., Prasad, R., Lee, A., Joshi, A.: The Penn Discourse Treebank 3.0 Annotation Manual (2019). https://catalog.ldc.upenn.edu/docs/LDC2019T05/PDTB3 Annotation-Manual.pdf

RoboboITS: A Simulation-Based Tutoring System to Support AI Education Through Robotics

S. Guerreiro-Santalla[1], H. Crompton[2], and F. Bellas[1(✉)]

[1] CITIC Research Center, Universidade da Coruña, A Coruña, Spain
{sara.guerreiro,francisco.bellas}@udc.es
[2] Old Dominion University, Norfolk, VA, USA
crompton@odu.edu

Abstract. This paper presents a novel tutoring system to educate pre-university students about AI, a key issue to develop AI in Education for Sustainable Society. With the aim of following a learning-by-doing approach to AI, we decided to focus on robotics as the main application domain for the students' activities. Specifically, the tutoring system is based on the Robobo educational robot, and its simulation environment. A prototype version of the tutoring system, called RoboboITS, has been released and tested in two in-person sessions with 17 students in a secondary school at Virginia (USA), leading to and promising outcomes for future development.

Keywords: AI education · AI for K12 · Personalized Learning · Tutoring Systems · Learning by Doing · Autonomous Robotics · Robotic Simulation

1 Introduction

Continued education about Artificial Intelligence (AI) is a key necessity for all the nations in their digital education plans [1, 2]. This aspect is highlighted in the report published by UNICEF [3] which emphasises that children's AI literacy and participation are being considered among their fundamental rights. If we aim to develop an "AI in Education for Sustainable Society" to provide inclusive and equitable quality education and to promote lifelong learning opportunities, developing formal, reliable, and responsible AI literacy for general education is critical [4].

Within this challenging scope, the authors of the current work have been developing an educational project called AI+ [5] between 2019 and 2022 with the aim of defining an AI curriculum for high school students in Europe. From the experience gained, and according to recent references in this scope like [6] or [7], we can conclude that there are two main limitations to start teaching AI at this level in the short-term. The first one is related to the heterogeneous students' background in programming required to follow a practical perspective to teach AI. The second one is the teachers' confidence towards teaching about AI. To face these two issues, we propose here to develop an Intelligent

Tutoring System (ITS) [8], which supports teachers in students' training on AI topics, and which also provides an adaptive and personalized teaching to students according to their programming skills.

Fig. 1. Left: Robobo real robot. Right: RoboboSim interface

The development of ITS in digital subjects has a long history [9], and it has also been included in the 2015–2030 Sustainable Development Goals (SDGs), namely, 4c, to increase the supply of qualified tutors. However, specific ITS for teaching about AI like the one presented here are still scarce and under development.

2 The RoboboITS

RoboboITS is a simulation based ITS designed to provide a personalized training in AI fundamentals to pre-university students, following a practical approach by solving autonomous robotics challenges. It is based on the Robobo educational robot [10] (see Fig. 1 left) and its 3D simulator called RoboboSim [11] (see Fig. 1 right), which has been improved in the realm of this research. The architecture of the RoboboITS follows the basic pattern of this type of systems, including three basic models [8]: tutor (pedagogy), student (learner), and domain (AI). In addition, an interface model has been considered, because the simulation-based interface is a key aspect here. RoboboITS has been implemented in Python, as independent modules in libraries that are integrated with the Unity3D interface of the simulator.

The flowchart of the system architecture is shown in Fig. 2, where the previous three models have been implemented as modules. The *Domain* Module contains the specific knowledge about AI that students should learn. In this case, it is based on the teaching units developed within the AI+ [5] project, which follow a project-based learning methodology. Accordingly, this module includes the goals of each project (called activity), the learning objectives for the students, and the code with the solutions.

The *Tutor* Module represents the knowledge on teaching and learning approaches obtained from teaching experts (real teachers). In this case, the module includes the division of each activity into small tasks, so different guidance levels can be achieved. For each of the tasks, the correction of the solutions is performed in this module too, by comparing them with the correct one obtained from the *Domain* Module.

The third module is the *Student* Module. It includes everything related to the student: initial and acquired knowledge at each moment, interactions with the system, developed

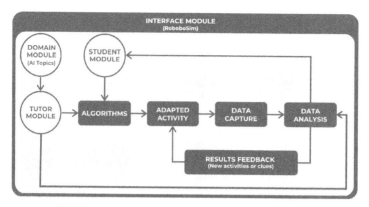

Fig. 2. Flowchart representing RoboboITS architecture.

programs, number of trials and time needed to solve the tasks, and the "level" of each student, which makes up an estimation of his/her competence in the project completion.

Following the diagram of Fig. 2, the *Algorithms* block rely on the three previous modules to adapt the activity and explanation to each student. Next, the learner performs the activity (*Adapted Activity*), and the system automatically captures all the necessary *Data* both to evaluate the degree of completion, and to detect all the problems he/she may have in achieving it. In the current version of the RoboboITS, these data are: (m1) time required to solve the activity, (m2) number of attempts needed, (m3) student's level at the beginning and end of each of the activities, (m4) time required to read the statements, (m5) number of times the student opens the documentation, and (m6) number of times the student reads the statements.

Once captured, the *Data Analysis* is performed, including the correction of the code, and the corresponding *Feedback* is provided through the simulator interface to support the student' progress in the activity. Two main types of feedback have been included:

1. *Activity division*: if the same failure occurs a predefined number of consecutive times (m2), the student is considered not capable of solving the proposed activity in its current level and is downgraded. In lower levels, the challenge is divided into easier tasks, so the student' solution should improve.
2. *Visual clues*: with independence of the level, the time required to solve the activity (m1) and the number of attempts (m2) trigger the presentation of visual clues to the user through the simulator interface.

Finally, the *Interface* Module is devoted with the connection between students and the system. This connection is a bidirectional communication and, as shown in the flowchart (Fig. 2), it includes all the previous modules and blocks. In this case, this module is made up of the 3D simulator interface. A key property of the RoboboITS is to make this interface as attractive to students as possible, and some design features have been included to "gamify" it. The goal is to take advantage of the robotics field and include support to students by means of graphical clues, direct interaction from the Robobo robot, and others. Students program their solutions using Python or Scratch, as supported by RoboboSim [11], but such programming interfaces are independent.

3   Perception and Actuation for Autonomous Robotics

To test the RoboboITS prototype with students, a specific AI topic was implemented for the first educational intervention: *perception and actuation* in AI. What is relevant in this case from the AI perspective is that students learn how to program the robot to be autonomous, that is, relying on sensor measures and avoiding predefined times or thresholds. This version of the RoboboITS is accessible to download and test[1].

After an initial questionnaire is completed, the student is assigned to a level, and the final challenge for this lesson is presented: *programming Robobo so it can autonomously avoid an obstacle, regardless of its position and size.* The expected robot response is shown with a video[2], so students have a clear idea of what their program should achieve. For all of them, the final challenge has been organized in two main activities, to highlight the relevance of using sensors in autonomous robotics.

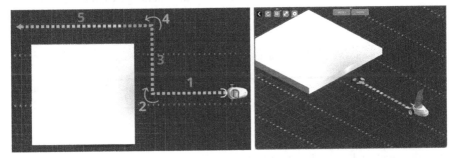

Fig. 3. Left: Movements to solve the challenge. Right: RoboboITS trial stage with visual clues

The objective of this first activity is to create a program that allows to avoid the obstacle *without* the use of sensors, so they realize that solving the task this way requires a lot of effort in trial-and-error manual adjustments. The simulated world where the challenge must be faced is shown in Fig. 3, where visual clues are shown.

For those students of level 3 (the highest), the ITS does not divide the activity into tasks. In level 2, the system proposes two tasks: a first one in which they must program movements 1 and 2 (Fig. 3 left), and a second one in which they must program the remaining three (3, 4 and 5 in the figure). In addition, the specific Scratch blocks they must use are also shown in the interface. Finally, if the student is in level 1, he/she faces the activity as five independent tasks, one for each movement. In addition, visual clues are included to indicate both the path to follow and the direction of rotation, depending on the movement to be made, as it can be seen in Fig. 3 right.

After completing Activity 1, the statement of Activity 2 automatically appears. In it, the robot must reproduce the previous behaviour (5 movements), but in an autonomous way. For students to realize what this means, now the position and size of the obstacle changes every time the world is restarted (every time they run their code). Therefore, to

---

[1] https://cutt.ly/I20RYDQ.

[2] https://cutt.ly/m2LfYEe.

solve this problem, two sensors must be continuously used: the infrared sensor to detect distance to the obstacle, and the orientation sensor to perform the turns.

4 Test with Students

The first testing stage within the educational research we have followed was carried out at Virginia (USA), with a group of 17 students. The group was heterogeneous, ranging in age from 14 to 17 years, and it was composed of 10 boys, 6 girls and 1 person who preferred not to say his or her gender. Of all of them, only 8 had previous programming training, and with different levels. The experimental test at the centre, guided by the researchers, was divided into two sessions: (1) Introduction (2 h): a short lecture with a general introduction to AI and the topics of perception and actuation was carried out as well as an introduction to the Robobo programming with Scratch and to the RoboboSim simulator. (2) ITS Test (2 h): each student faced the challenge explained in Sect. 4 following the tasks proposed by RoboboITS.

Table 1. Four answers to the initial and final questionnaire, more can be consulted at[3]

Question	Initial questionnaire % of successes	Final questionnaire % of successes
What is an autonomous robot?	88,2	100
To program an autonomous robot, it is necessary to avoid... (options)	41,2	70,6
The IR sensor of RoboboSim measures.. (options)	70,6	82,4
When you program a robot, is it possible to use exact values in the comparisons of sensor values?	41,2	82,4

With this intervention, we addressed the following research questions: is the RoboboITS useful for learning AI fundamentals? Does it support students with different programming skills? This way, we are facing the two main limitations in the short-term education about AI explained in Sect. 1. To answer them, an online questionnaire was filled by students at the beginning and ending of the test. Moreover, a log file with the task execution trace followed by each of them in the ITS was recorded.

From the results of the questionnaires, it was possible to check that most of the students improved their knowledge about perception and actuation in AI. This is specially relevant in those more technical issues (like questions 2 and 4 in Table 1), which cannot be easily solved by simple study, as they require practical experimentation.

After analysing the files containing the execution trace of the task, it was detected that, out of the 17 students, only 3 failed to overcome the final goal. Figure 4 shows an example of the progress of three students, one from each level, and the data extracted for each of them in terms of task/activity completion, and clues received. Level 3 student

---

[3] https://cutt.ly/N2L3gh5.

took a total of 70 min to solve the challenge, and only needed two clues to do it, one in each activity. On the other hand, the level 2 student managed to pass activity 1 at the initial level, but in activity 2 the clues were not enough, so he/she was downgraded. Finally, the level 1 student, after two hours and the support of many clues, managed to successfully pass the whole challenge.

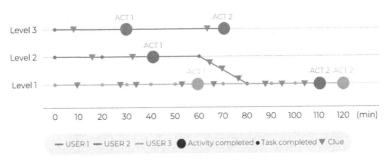

Fig. 4. Progress of three students on the RoboboITS to solve the challenge

5 Conclusions

Introducing education about AI in pre-university levels in the short-term implies supporting teachers and students in this new scope within their digital training. To this end, the current work presents an educational research focused on developing and testing an Intelligent Tutoring System for AI fundamentals called RoboboITS. The system relies on autonomous robotics projects to introduce the different concepts in a practical fashion through a 3D simulator. The results have shown an improvement in the AI knowledge during one session using the RoboboITS, as well as an adequate adaptation to the students' skills with the aim of reaching a homogeneous level in all the group. Future work will be focused on including more teaching units and topics about AI and including more levels and clues by means of machine learning models.

Acknowledgments. Grant TED2021-131172B-I00 funded by MCIN/AEI/10.13039/5011000 11033 and by the "European Union NextGenerationEU/PRTR".

References

1. Miao, F., Holmes, W.: AI and education: guidance for policy-makers, UNESCO (2021). https://unesdoc.unesco.org/ark:/48223/pf0000376709. Accessed 13 Jan 2023
2. European Commission: Digital Education Plan 2021–2017 (2020). https://ec.europa.eu/edu cation/education-in-the-eu/digital-education-action-plan_en. Accessed 13 Jan 2023
3. Dignum, V., Penagos, M., Pigmans, K., Vosloo, S.: Policy Guidance on AI for Children, UNICEF (2021). https://www.unicef.org/globalinsight/reports/policy-guidance-ai-children. Accessed 13 Jan 2023

4. Schiff, D.: Education for AI, not AI for education: the role of education and ethics in national AI policy strategies. Int. J. Artif. Intell. Educ. (2021)

5. Bellas, F., Guerreiro-Santalla, S., Naya, M., Duro, R.J.: AI curriculum for european high schools: an embedded intelligence approach. Int. J. Artif. Intell. Educ. (2022). https://doi.org/10.1007/s40593-022-00315-0

6. Yang, W.: Artificial Intelligence education for young children: why, what, and how in curriculum design and implementation, Comput. Educ. Artif. Intell. **3** (2022). https://doi.org/10.1016/j.caeai.2022.100061

7. UNESCO: K-12 AI curricula: a mapping of government-endorsed AI curricula (2022). https://unesdoc.unesco.org/ark:/48223/pf0000380602.locale=en. Accessed 13 Jan 2023

8. Holmes, W.: Artificial intelligence in education. In: Tatnall, A. (eds.) Encyclopedia of Education and Information Technologies. Springer, Cham (2020). https://doi.org/10.1007/978-3-030-10576-1_107

9. Mousavinasab, E., Zarifsanaiey, N., Niakan Kalhori, S.R., Rakhshan, M., Keikha, L., Ghazi Saeedi, M.: Intelligent tutoring systems: a systematic review of characteristics, applications, and evaluation methods. Interact. Learn. Environ. **29**(1), 142–163 (2021)

10. Bellas, F., et al.: The robobo project: bringing educational robotics closer to real-world applications. In: Lepuschitz, W., Merdan, M., Koppensteiner, G., Balogh, R., Obdržálek, D. (eds.) RiE 2017. AISC, vol. 630, pp. 226–237. Springer, Cham (2018). https://doi.org/10.1007/978-3-319-62875-2_20

11. RoboboSim simulator. https://github.com/mintforpeople/robobo-programming/wiki/Unity. Accessed 13 Jan 2023

Towards Analyzing Psychomotor Group Activity for Collaborative Teaching Using Neural Networks

Jon Echeverría(✉) and Olga C. Santos

Artificial Intelligence Department, UNED, Madrid, Spain
{jecheverria,ocsantos}@dia.uned.es

Abstract. Collaborative teaching in psychomotor skills requires the ability to detect and analyze the movements and postures of multiple participants simultaneously. In this paper we explore the feasibility of analyzing group activity in psychomotor contexts using neural networks. We apply our approach to *kihon kumite*, a training exercise in karate where two participants perform predetermined techniques. The overall activity is measured as the sum of individual performances, with the support of computer vision techniques to label individuals and classify postures. Furthermore, interactions between participants are also analyzed, which allows for a more comprehensive understanding of the group activity. Our approach shows promising results for improving collaborative teaching in psychomotor skills and can be extended to other physical activities.

Keywords: Psychomotor Learning · Martial Arts · Computer Vision · Personalized Learning · Collaborative Learning

1 Introduction

Artificial intelligence systems to develop psychomotor skills and collaborative learning are aligned with UN Sustainable Development Goal 4, as they can improve the quality and accessibility of education as well as promote teamwork and collaboration among students with personalization support. The pandemic has also highlighted the need for online systems in physical education and highlighted how scarce they are [1].

There are several studies [2–4] that suggest that collaborative learning can improve motivation, performance, and retention of knowledge compared to individual education in psychomotor contexts. The immediate feedback and social interaction that occur in collaborative learning can significantly improve the quality of learning and student performance. There are examples in different areas that corroborate this idea: karate [5], dance [6], gymnastic [7] and sports in general [8]. Despite the examples, there are not many research works yet that take place in this area.

Limited collaborative work in physical education has been a concern [9–11] for many years, and research has suggested a variety of reasons for this. Teachers may prioritize individual skill development over collaborative learning, leading to a lack of emphasis on social interaction and teamwork in physical education. To enable effective collaborative learning in psychomotor contexts, it is crucial to detect and analyze the movements and postures of multiple participants simultaneously. This presents several challenges, such as (i) monitoring the movements of more than one participant at the same time, and (ii) analyzing these movements in an interrelated manner. To achieve this goal, we began our research in the context of karate martial art, where two karate practitioners (i.e. karatekas) perform what is called the one-step attack (*kihon kumite*), which is a training group activity of predetermined paired attack and defense techniques. Thus, we have explored how to analyze the group activity of several individuals by identifying their postures with the aim of assessing if the overall group activity can be measured as the sum of individual performances.

2 Related Works

Computer-Supported Collaborative Learning (CSCL) investigates the use of technology to support and enhance collaborative learning in a wide range of educational contexts. While there has been considerable research in areas such as mathematics [12, 13], science [14, 15], and language learning [16, 17] there has been comparatively less research on the use of technology to support collaborative learning in psychomotor contexts, such as physical education and sports training.

Previous studies have investigated different types of relationships in group activities. For example [18, 19] explored the effects of different types of competition on group collaboration in an online game context. They found that cooperative competition (i.e., competing towards a common goal) was more effective in promoting collaboration than individual competition or no competition. Another study [20] investigated the effects of complementary collaboration (i.e., when group members have different roles or tasks that contribute to a common goal) on the creativity and innovation of a team. Finally, there are studies [21] that analyzed the effects of oppositional collaboration (i.e., competing against each other towards different goals) on learning outcomes in a collaborative game. They found that oppositional collaboration had a positive effect on learning outcomes, as it stimulated learners to think critically and adapt to different situations.

In turn, computer vision and neural networks serve for human movement computing [22, 23] but usually focus on detecting individual movements [24] as result of the complexity of the person re-identification problem [25]. However, research related to physical activity, sports and collaborative education is emerging [26].

3 Material and Methods

Computer vision techniques and neural networks have been used to identify and classify the postures of participants in *kihon kumite* training, and to analyze the group activity. We have used the OpenPose [27] algorithm for human pose estimation. A dataset is being created for this research. Currently, our dataset contains 6 different postures in *kihon kumite* recorded during the training sessions of a pair of karatekas using cameras in different positions (two opposite profiles). Specifically, the dataset contains three attacks (*kamae*, *gedan barai* and *oi tsuki*) and the three defensive postures (*kamae*, *soto uke* and *gyako tsuki). Kamae* is featured in both because it is the starting posture both for the *tori* (attacker) and *uke* (defender) when situated in opposing sides on the tatami to start the *kumite*. This also allows that *tori* and *uke* are recognized, It is important to highlight that there are challenges and limitations when using computer vision techniques and neural networks to analyze group activities in psychomotor contexts. In particular, the person re-identification problem can be very complex and can have a significant impact on the accuracy of the results. Therefore, in this study this problem has been addressed by including the kamae stance in both, which allows to recognize both tori and uke and, therefore, to solve the problem of re-identification in this situation. For motion classification, the neural network in Table 1 was designed to allow flexibility for scaling and varying its structure.

Table 1. NN1 architecture and accuracy results

Model	Layer (type)	Trainable params	Total params	Train acc	Valid acc	Test acc
NN1	Flatten + Dense + BatchNor-malization + Drop + 2Dense	2.881	2.981	98,61	98,59	98,68

As a starting point for our investigation, we have focused on classifying the movements of each karateka in terms of: i) the identification of persons (labelling), ii) the categorization of the movements (i.e., which attack or defensive movement was performed), iii) the order of the movements (not all sequence of movements are possible), and iv) the final performance result (the sequence of movements performed analyzed globally).

4 Results and Discussion

Classification results show that it was possible to classify the six different postures with the following evaluation measures: precision ≥ 0.98, recall ≥ 0.96, F1-score ≥ 0.95. The accuracy results for the classification of each stance are shown in Table 2. Furthermore, the algorithm was able to identify the stances in a multi-person manner (i.e., the kumite partner), as shown in Table 2. Moreover, the algorithm was able to identify the postures in a multi-person way (i.e., the kumite couple).

Regarding the movements order, they are predefined and stored in memory in such a way that the sequenced classes are verified to match the established kihon kumite pattern. Considering independent collaboration, the percentage of success is measured as the total of well-executed motions based on the class prediction provided by the computer vision algorithm, where the minimal confidence criterion is set at 95%. Following Eq. (1) the degree of success of the activity (A) is calculated by dividing the weighted average of pose classification (P) by total number of movements.

$$A\% = \frac{\Sigma_1^N(P)}{N} \tag{1}$$

Results in Table 2 show that the six individual postures (Pi) are identified by saving their prediction value and thus make a final weighting of the activity with the average of all. If the postural identification is below the defined threshold (95%), the identified posture would be "other". This fact would mean that the series of movements would not be completed due to the lack of identification of any of them, and the average would not be obtained, nor would the activity have been considered satisfied.

Table 2. Values the precision of individual postural classification (Pi) and averaging the 6 postures (A) for five training exercises between the two karatekas

Activity Performance

P1-attack (*kamae*)	P2-def (*kamae*)	P3-attack (*g. barai*)	P4-attack (*oi tsuki*)	P5-def (*soto uk.*)	P6-def. (*gyaku t.*)	A (% of success)
0,9857	0,9589	0,9687	0,9635	0,9552	0,9666	*96,64%*
0,9974	0,9869	0,9564	0,9582	0,9634	0,9513	*96,89%*
0,9888	0,9909	0,9984	0,9624	0,9574	0,9513	*97,49%*
0,9588	0,9761	0,9583	0,9789	0,9883	0,9513	*96,86%*
0,9659	0,9562	0,9577	0,9787	0,9768	0,9513	*96,44%*

The proposed approach for analyzing group activity in psychomotor contexts focuses on detecting and analyzing the movements and postures of multiple participants simultaneously to enable effective collaborative teaching of psychomotor skills. The use of computer vision techniques to label individuals and classify postures allows for automated data analysis, which can provide more objective and comprehensive feedback to both learners and instructors. This has the potential to improve the quality of teaching and learning experiences in a range of physical activities beyond karate, including team sports and dance. For example, this approach could be applied to different group activities such as: i) activities where movements are independent (e.g., a group of dancers each performing their own poses or movements without direct interaction), ii) activities where movements are collaboratively dependent (e.g. dance movements where one dancer helps another develop their pose), and iii) activities where movements are competitively dependent (e.g., a karateka launching an attack at another, such that if she executes it poorly, the opponent can make a better defense, and if she executes it well, the defender cannot defend herself). Thus, in our research we aim to improve group personalized learning focusing on group interactions. In order to improve group dynamics, we are exploring individual and group metrics of teams with various sorts of goals, roles and types of interactions (competition, collaboration, opposition). Overall, our work aims to contribute to the development of intelligent systems that can support personalized collaborative learning in physical education and sports.

In this approach, it is essential to control the postural identification, since it is the basis for the digitization of the activity creating a digital environment (smart learning environment) where students can find motivation in innovation, new forms of communication in community, sensorized content and group activity monitoring. In addition, identification control opens the door for its application to other areas, being able to generalize the tool. In the opposite direction, it will also allow the process of specialization, for instance applying computer vision to identify the postures of the hands for example, in a medical context with multiple hands performing a surgery.

Therefore, despite the encouraging results obtained in this study, it is important to take into account that there are limitations in the speed of graphical processing that could affect the efficiency of the system in a real-time environment. Therefore, further research is needed to assess the generalizability and scalability of this approach to other contexts and physical activities.

5   Conclusions

The use of computer vision and neural networks in karate training has shown a promising path to provide precise tracking and feedback to practitioners in a collaborative learning context. Our collaborative psychomotor approach allows for monitoring and analysis of joint movements in group activity, which can motivate users to continue training and improve their performance. In particular, we are working in offering new possibilities for evaluating and monitoring student performance in group contexts. The positive outcomes of this research open up new avenues for advancing psychomotor learning with CSLC theory in the AIED research community.

Acknowledgments. This work is part of the project HUMANAID (TED2021-129485BC1) funded by MCIN/AEI/10.13039/501100011033 and the European Union "NextGenerationEU"/PRTR.

References

1. Saini, A.: Changing paradigms of teaching amid Covid-19: a SWOT analysis of online classes in physical education and sports sciences. In: 2021 International Conference on Computational Performance Evaluation (ComPE), pp. 115–119 (2021). https://doi.org/10.1109/ComPE53109.2021.9751886
2. Chiviacowsky, S.: Self-controlled feedback: does it enhance learning because performers get feedback when they need it? Res. Q. Exerc. Sport **73**, 408–415 (2002). https://doi.org/10.1080/02701367.2002.10609040
3. Powersm, H.S., Conway, T.L., McKenzie, T.L., Sallis, J.F., Marshall, S.J.: Participation in extracurricular physical activity programs at middle schools. Res. Q. Exerc. Sport **73**(2), 187–192 (2002). https://doi.org/10.1080/02701367.2002.10609007
4. Fernández-Espínola, C., et al.: Effects of cooperative-learning interventions on physical education students' intrinsic motivation: a systematic review and meta-analysis. Int. J. Environ. Res. Public Health **17**(12) (2020). https://doi.org/10.3390/ijerph17124451
5. Qasim, S., Ravenscroft, J., Sproule, J.: The effect of Karate practice on self-esteem in young adults with visual impairment: a case study. Aust. J. Educ. Dev. Psychol. **14**, 167–185 (2014)
6. Barr, S.: Collaborative practices in dance research: unpacking the process. Res. Danc. Educ. **16**(1), 51–66 (2015). https://doi.org/10.1080/14647893.2014.971233
7. Bayraktar, G.: The effect of cooperative learning on students' approach to general gymnastics course and academic achievements. Educ. Res. Rev. **6**, 62–71 (2011)

8. Fernandez-Rio, J., Casey, A.: Sport education as a cooperative learning endeavour. Phys. Educ. Sport Pedag. **26**(4), 375–387 (2021)

9. Alton-Lee, A.: Cooperative learning in physical education: a research-based approach edited by ben dyson & ashley casey. Qual. Res. Educ. **1**, 228 (2012). https://doi.org/10.17583/qre.2012.453

10. Ryan, R.M., Deci, E.L.: Intrinsic and extrinsic motivations: classic definitions and new directions. Contemp. Educ. Psychol. **25**(1), 54–67 (2000)

11. Drewe, S.B.: Competing conceptions of competition: implications for physical education. Eur. Phys. Educ. Rev. **4**(1), 5–20 (1998). https://doi.org/10.1177/1356336X9800400102

12. Khanlari, A., Scardamalia, M.: Knowledge building, robotics, and math education (2019). https://doi.org/10.22318/CSCL2019.963

13. Kollar, I., Fischer, F., Hesse, F.W.: Collaboration scripts - a conceptual analysis. Educ. Psychol. Rev. (2), 159–185 (2006). http://nbn-resolving.de/urn/resolver.pl?urn=nbn:de:bvb:19-epub-12924-1

14. Suthers, D.: Technology affordances for intersubjective meaning making: a research agenda for CSCL. Int. J. Comput. Collab. Learn. **1** (2006). https://doi.org/10.1007/s11412-006-9660-y

15. Edelson, D.: Design research: what we learn when we engage in design. J. Learn. Sci. **11**, 105–121 (2002). https://doi.org/10.1207/S15327809JLS1101_4

16. Gebre, E., Bailie, A.: Learning with multiple representations and student engagement in secondary education: a preliminary review of literature (2019). https://doi.org/10.22318/CSCL2019.941

17. Lala, R., et al.: Enhancing free-text interactions in a communication skills learning environment (2019). https://doi.org/10.22318/CSCL2019.877

18. Peng, W., Hsieh, G.: The influence of competition, cooperation, and player relationship in a motor performance centered computer game. Comput. Hum. Behav. **28**(6), 2100–2106 (2012). https://doi.org/10.1016/j.chb.2012.06.014

19. Morschheuser, B., Riar, M., Hamari, J., Maedche, A.: How games induce cooperation? A study on the relationship between game features and we-intentions in an augmented reality game. Comput. Hum. Behav. **77**, 169–183 (2017)

20. Green, B.N., Johnson, C.D.: Interprofessional collaboration in research, education, and clinical practice: working together for a better future. J. Chiropr. Educ. **29**(1), 1 (2015). https://doi.org/10.7899/JCE-14-36

21. Gonçalves, B., Marcelino, R., Torres-Ronda, L., Torrents, C., Sampaio, J.: Effects of emphasising opposition and cooperation on collective movement behaviour during football small-sided games. J. Sports Sci. **34**(14), 1346–1354 (2016). https://doi.org/10.1080/02640414.2016.1143111

22. Nandakumar, N., et al.: Automated eloquent cortex localization in brain tumor patients using multi-task graph neural networks. Med. Image Anal. **74**, 102203 (2021). https://doi.org/10.1016/J.MEDIA.2021.102203

23. Aggarwal, J.K., Cai, Q.: Human motion analysis: a review. Comput. Vis. Image Underst. **73**(3), 428–440 (1999). https://doi.org/10.1006/CVIU.1998.0744

24. Dong, H., et al.: Towards multi-pose guided virtual try-on network (2019)

25. Rani, J.S.J., Augasta, M.G.: PoolNet deep feature based person re-identification. Multimed. Tools Appl. (2023). https://doi.org/10.1007/s11042-023-14364-7

26. García-González, L., Santed, M., Escolano-Pérez, E., Fernández-Río, J.: High-versus low-structured cooperative learning in secondary physical education: Impact on prosocial behaviours at different ages. Eur. Phys. Educ. Rev., 1356336X221132767 (2022)

27. Cao, Z., Hidalgo, G., Simon, T., Wei, S.-E., Sheikh, Y.: OpenPose: realtime multi-person 2D pose estimation using part affinity fields. IEEE J. Mag. **43** (2021). Accessed 06 June 2021

Warming up the Cold Start: Adaptive Step Size Method for the Urnings Algorithm

Bence Gergely[1,2,3(✉)] [iD], Han L.J. van der Maas[4], Gunter K. J. Maris[5], and Maria Bolsinova[6] [iD]

[1] Eötvös Lóránt University, Doctorate School of Psychology, Budapest, Hungary
[2] Eötvös Lóránt University, Institute of Psychology, Budapest, Hungary
[3] Károli University of the Reformed Church, Budapest, Hungary
gergely.bence@kre.hu
[4] University of Amsterdam, Amsterdam, The Netherlands
[5] TATA Consultancy Services, Zaventem, Belgium
[6] University of Tilburg, Tilburg, The Netherlands

Abstract. Adaptive learning systems (ALS) tailor educational material to the level of the users. In ALS ability should be continuously estimated based on the users' responses to adaptively selected practice items. However, the large-scale, adaptive, and dynamic nature of ALS poses challenges for traditional estimation methods. The Urnings algorithm [1] has been recently proposed to address these challenges. However, the original algorithm does not address the cold-start problem which ALS suffer from: Initially, it is difficult to adapt item selection to the users' abilities based on limited available information. We develop a modification of the Urnings algorithm aiming to alleviate the cold-start problem by increasing the step size of the algorithm when a systematic change in the ratings is detected, and decreasing it when the ratings are relatively stable. The results of our simulation studies showed that the modified algorithm moves away from the initial values faster, responds to sudden changes in ability better, and results in overall higher accuracy than the original algorithm.

Keywords: adaptive learning systems · trackers · rating systems · Urnings algorithm · cold-start problem

1 Introduction

The goal of an adaptive learning system (ALS) is to tailor the learning material to the behaviour and needs of its users [7,12]. To achieve this, the performance of the users needs to be monitored to continuously estimate the users' ability and items should be selected with difficulty matching the current ability [3, ?].

One line of methods used in ALS are connected to the Elo rating system (ERS), which despite its practicality, lacks desirible statistical properties [2,4].

N. Wang et al. (Eds.): AIED 2023, CCIS 1831, pp. 409–414, 2023.
https://doi.org/10.1007/978-3-031-36336-8_64

Recently a new rating system called "Urnings" has been introduced to address the limitations of the ERS [1]. Urnings estimates the ability and difficulty (under the Rasch model [11]) in an online fashion by tracking these parameters on the probability scale using ratings defined on a grid of discrete proportions. After every response the ratings are updated based on the difference of the observed and simulated outcome (using the current ratings) of the user-item pair. The ratings can adapt to the changes in the true ability, have a tractable limiting distribution and provide unbiased estimates with known standard errors when there is no change in true values [1].

Initially, there is very little or no information about the users' ability in a rating system, thus it takes a long time to achieve an accurate estimate. In this phase, the predictions and the item recommendations can be sub-optimal, failing to adapt to the user's needs. This issue is referred to as the cold-start problem [9,10,14]. It affects new users and makes them more likely to abandon the system due to the inappropriately selected items, which might be experienced as frustrating, demotivating or tedious [7,8,14].Only a few studies provided solutions for the cold-start problem in ALS applications, all of them focusing on the ERS, and its alternatives like Glicko system [5]. These solutions either tried to decrease the prior uncertainty by predicting better starting values based on the background characteristics of the users [10], or implemented continuous control methods to change the size of the updates [5,7,15].

The Urnings algorithm currently does not address the cold-start problem. Therefore, in this paper, we present a modification of the Urnings capable of changing the size of the updates adaptively by analysing the direction and rate of change of the ratings in order to reach the true ability level faster but maintain low standard errors when the estimate is close to the true value.

2   Adaptive Step Size

Opposing to the ERS which tracks the parameters on the logit scale and uses continuous ratings, the ratings in Urnings algorithm are defined in a discrete grid of proportions $\{\frac{0}{n}; \frac{1}{n}; \frac{2}{n}; ...; \frac{n}{n}\}$, where n is the granularity of the grid and in the original algorithm it is kept constant. After each response, ratings can either change by $\frac{1}{n}$ or they constant. The updates are symmetric, meaning that if the student rating increases, the item rating necessarily decreases. The choice of n is very important: A smaller n allows for moving quickly through the parameter space and is better suited for following sudden changes in ability, whereas a large n allows for more precise measurement when the ability is relatively stable.

We build the modified algorithm on the idea that if the ratings are systematically changing in one direction they are likely far from the true ability, whereas if they fluctuate around a constant value (i.e. the chain of rating is stationary), they are likely close to the true ability. Consequently, the objective of an adaptive step size algorithm is to monitor systematic changes in the ratings, and when it occurs increase the step size.

To monitor the rate of change in the ratings, we will consider the chain of differences between the consecutive ratings, which we refer to as the "differential process."[1] To analyse the change we define a moving window of length $l \in \mathbb{N}$. If the chain of user's ratings is a weak-sense stationary Markov chain (i.e., if the chain has reached its limiting distribution), then the expectation of the differential process in the given time window is 0. If the mean of the differential process is indeed 0, then the step size can decrease to decrease the standard error. If it is not 0, then the chain is not yet in the neighbourhood of the true value and, so the step size should be increased.

One possible method to test whether the expected value of the differential process is different from 0 is using a one-sample permutation test [13]. The permutation test can provide an exact p-value for small window sizes and has no distributional assumptions. If the permutation test is significant, there is evidence that the ratings are systematically changing in one direction.

The permutation test can be applied as follows: First, for small window size (i.e. 10) we create all possible combinations of $\{-1, 1\}$ of length l, for larger l we create these arrays by sampling from this set with replacement. Second, we multiply each of these arrays with the values of the differential process within the given window (i.e., random signs are assigned to these values), and calculate the mean in each permutation. This forms act the null distribution whereas the mean of the differential process is the observed statistic. The p-value is computed by computing the proportions of permutations in which the mean is at least as extreme in the absolute value[2] as the observed mean.

We define a minimum and maximum granularity n_{min} and n_{max}. The algorithm starts from n_{min} (i.e., the largest step size). Then, the step size is modified based on the result of the permutation test. If the permutation test is not significant, n is increased (doubled) after each time window until it reaches n_{max}.[3] Since we change n (the denominator of the rating), we are also doubling the numerator to keep the rating constant. If the permutation test is significant, i.e., when systematic change is detected, we set n back to n_{min}, while the rating's numerator is adjusted accordingly.

3   Methods

Using simulated data we illustrate the cold-start problem in the case of the original Urnings algorithm and show how its modification offers a solution for this problem. Two simulation studies were conducted: 1) with constant user ability; 2) with abilities of all users making a sudden change in the middle of the simulation.

---

[1] Note, that in this paper we focus only on adapting step size for the users, as there are typically much more responses available per item and item difficulty is less prone to sudden changes which makes the cold-start problem and the problem of following change less important on the item side.

[2] i.e., the two-sided p-value is computed

[3] Note that n_{min} and n_{max} should be chosen in such a way that the $nmin$ can be reached by dividing the n_{max} by a power of 2

In the first scenario, we simulated an ALS with 300 items and 1500 users, with the true ability and difficulty parameters drawn from $\mathcal{U}(0,1)$. Every user responded to 300 adaptively selected items with the same adaptive mechanism as in [6]. For the original Urnings, we considered $n = 8, 64$ to demonstrate that the bias-variance trade-off of the ratings depends on the $n =$ and how the cold-start problem is affected by it. We expect that while $n = 8$ allows us to move through the parameter space faster (i.e., bias is reduced in the beginning), the total error in the estimation of ability would be higher in the long run because the variance of the ratings is larger than for $n = 64$. The adaptive step size algorithm aims at reducing the bias in the beginning by having a large step size but reducing the variance when the ratings are stabilised and the step size is small. For the modified algorithm we used $n_{min} = 8$ and $n_{max} = 64$ to match the range of the non-adaptive conditions. The step size for the items was equal to $\frac{1}{64}$ in all cases. All user ratings were initialised at 0.5. Item starting values (the numerators of the ratings) were drawn from their limiting distribution [1] which is a binomial distribution with the probability parameter equal to the true value. This mimics a calibrated ALS, as it removes all additional uncertainty of the users' ability coming from the parallel calibration of the items. The window size l was set to 10. The α-level for the permutation test was set to 0.1.

In the second scenario, we examine what happens when true abilities change. The setup of the simulation was the same as before, but after each user responded to 150 items we changed their true value by again drawing from $\mathcal{U}(0,1)$. Here we considered only $n = 64$ for the original algorithm.

We assess the performance of the algorithms by calculating the absolute distance between the estimate and the true value at the given time point for each user and then calculating the mean. This measure (mean absolute difference, MAD) captures the total error in the system. We also look at how fast the MAD reaches the reference value of 0.1, which we consider small enough for the rating to be a good measure of ability (hitting time, HT). We compute the hitting time, defined as the first iteration at which the MAD is at or below 0.1.

4   Results

First, we compare the results of the two cases with fixed n in Fig. 1a. As expected, $n = 8$ allowed for a faster decrease in error, but the MAD stabilises around a higher value based on the last 150 iterations (MAD = 0.11) than the $n = 64$ case (HT = 98, MAD = 0.07). The algorithm with adaptive step size allows for a rapid decrease in the beginning (HT = 27) and at the same time, it was more accurate in the second half of the analysis than the original algorithm (MAD = 0.06). The overall MAD was 0.10, 0.11, and 0.07 for $n = 64, 8$ and the modified algorithm respectively. The modified Urnings algorithm reached the reference level of 0.1 71 iterations faster than the original Urnings. Both the overall error level and the error level of the two halves of the iterations were lower for the modified algorithm than the fixed step size cases.

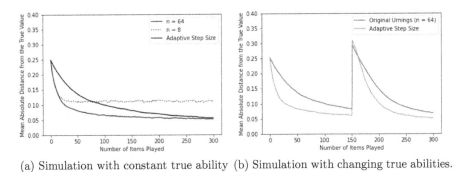

(a) Simulation with constant true ability (b) Simulation with changing true abilities.

Fig. 1. Mean absolute difference between user ratings and true values over 300 items.

In the second simulation, we presented what happens when the true ability of the students changes with a discrete jump. The modified algorithm (MAD = 0.09, $\text{MAD}_{firsthalf} = 0.10$, $\text{MAD}_{secondhalf} = 0.10$) outperformed the original Urnings (MAD = 0.13, $\text{MAD}_{firsthalf} = 0.13$, $\text{MAD}_{secondhalf} = 0.13$) in terms of total error, both before and after the true values change. Similarly to the first simulation, the modified algorithm showed a steeper decrease in error, than the original Urnings algorithm. The steep decrease in error in the modified case in the first half is due to the large initial step size, the modified algorithm requires some time to detect the change, after which it reduces error compared to the original algorithm.

5   Discussion

In this study, we presented a possible way to implement adaptive step size into the Urnings algorithm, by changing the granularity of the discrete grid of the ratings based on the detection of their systematic change. Our simulation demonstrated that this modified Urnings algorithm can reduce the length of the cold-start problem, meaning that users need to solve fewer items to get close to their true ability. Based on the simulation study the length of the cold-start period with Urnings is about one and a half times longer than with the modified algorithm. The average of the mean absolute difference was approximately one and a half times larger in the original Urnings version.

Further study of the proposed method is needed to investigate how they perform under a wide range of conditions. For example, when different trajectories of ability change are present in the system. Furthermore, the performance of the algorithm in real data needs to be evaluated. However, analysing learning data from a system where item selection was based on ability estimates obtained through a different algorithm than the one that is used can produce biased results since the way items are administered (i.e., which data would be observed) is dependent on the ability estimates (i.e., on the tracking algorithm).

Fine-tuning the parameters of the developed algorithms like the size of the window, the α-level for the permutation test, and minimum and maximum step size is crucial for successful applications. A way to empirically decide which parameter is suitable for the given problem is yet to be developed [4].

References

1. Bolsinova, M., Maris, G., Hofman, A.D., van der Maas, H.L., Brinkhuis, M.J.: Urnings: a new method for tracking dynamically changing parameters in paired comparison systems. J. Roy. Stat. Soc. Ser. C (Appl. Stat.) (2022)
2. Brinkhuis, M.J., Maris, G.: Dynamic parameter estimation in student monitoring systems. Measurement and Research Department Reports (Rep. No. 2009–1). Arnhem: Cito 146 (2009)
3. Brinkhuis, M.J., Savi, A.O., Hofman, A.D., Coomans, F., van Der Maas, H.L., Maris, G.: Learning as it happens: a decade of analyzing and shaping a large-scale online learning system. J. Learn. Anal. 5(2), 29–46 (2018)
4. Elo, A.E.: The Rating of Chessplayers, Past and Present. Arco Publications (1978)
5. Glickman, M.E.: Dynamic paired comparison models with stochastic variances. J. Appl. Stat. 28(6), 673–689 (2001)
6. Hofman, A.D., Brinkhuis, M.J., Bolsinova, M., Klaiber, J., Maris, G., van der Maas, H.L.: Tracking with (un) certainty. J. Intell. 8(1), 10 (2020)
7. Klinkenberg, S., Straatemeier, M., van der Maas, H.L.: Computer adaptive practice of maths ability using a new item response model for on the fly ability and difficulty estimation. Comput. Educ. 57(2), 1813–1824 (2011)
8. Ostrow, K.: Motivating learning in the age of the adaptive tutor. In: Conati, C., Heffernan, N., Mitrovic, A., Verdejo, M.F. (eds.) AIED 2015. LNCS (LNAI), vol. 9112, pp. 852–855. Springer, Cham (2015). https://doi.org/10.1007/978-3-319-19773-9_131
9. Pankiewicz, M.: Assessing the cold start problem in adaptive systems. In: Proceedings of the 26th ACM Conference on Innovation and Technology in Computer Science Education, vol. 2, pp. 650–650 (2021)
10. Pliakos, K., Joo, S.H., Park, J.Y., Cornillie, F., Vens, C., Van den Noortgate, W.: Integrating machine learning into item response theory for addressing the cold start problem in adaptive learning systems. Comput. Educ. 137, 91–103 (2019)
11. Rasch, G.: Studies in mathematical psychology: I. Probabilistic models for some intelligence and attainment tests (1960)
12. Shemshack, A., Spector, J.M.: A systematic literature review of personalized learning terms. Smart Learn. Environ. 7(1), 1–20 (2020). https://doi.org/10.1186/s40561-020-00140-9
13. Tritchler, D.: On Inverting Permutation Tests. J. Am. Stat. Assoc. 79(385), 200–207 (1984)
14. Wauters, K., Desmet, P., Van Den Noortgate, W.: Adaptive item-based learning environments based on the item response theory: possibilities and challenges. J. Comput. Assist. Learn. 26(6), 549–562 (2010)
15. Wauters, K., Desmet, P., Van Noortgate, W.: Monitoring learners' proficiency: weight adaptation in the elo rating system. In: Educational Data Mining 2011 (2010)

---

[4] The analysis script is hosted at https://github.com/mrpogge/Urnings_AIED.git

Gamiflow: Towards a Flow Theory-Based Gamification Framework for Learning Scenarios

Geiser Chalco Challco[1]([✉]) [iD], Ig Ibert Bittencourt[2,3] [iD], Marcelo Reis[2] [iD], Jario Santos[2] [iD], and Seiji Isotani[3] [iD]

[1] Federal Rural University of the Semi-Arid Region, Pau dos Ferros, RN 59900, Brazil
geiseres@ufersa.edu.br

[2] NEES: Center for Excellence in Social Technologies, Federal University of Alagoas, Maceio, AL 57072-970, Brazil

[3] Harvard Graduate School of Education, Cambridge, MA 02138, USA

Abstract. Participating in gamified learning scenarios increases students' motivation and engagement, leading them to improve their learning outcomes. Nevertheless, these benefits are only accomplished when gamification is grounded in theories of human motivation and behavior. In this paper, we present a gamification design framework based on Csikszentmihalyi's flow theory, as well as its empirical validation in eight different real-world contexts. These results indicate that our framework is usable, pertinent, and yields high-quality outputs, enabling gamification of learning scenarios that effectively promote the flow state and learning. his framework also facilitates the systematization of the application of flow theory, which is the first step in the development of intelligent gamification recommendation systems.

Keywords: Gamification · Framework · Flow State · Learning Scenario

1 Introduction and Related Works

One of the challenges in Gamification of Education is using game elements to make students feel totally involved and focused on the completion of learning activities. This mental state, referred by Cziksentmihalyi as the flow state [4], is a state in which nothing matters more than completing the activities, and when students are in this state, the activities lead to improved learning outcomes. However, to ensure that Csikszentmihalyi's flow theory is effectively applied in the gamification process, few gamification design frameworks have been developed [3,9]. Gamification frameworks frequently adjust only the difficulty of learning activities, ensuring that the condition of balance between ability and challenge is met [2,5,7,8], but none of them indicate how a game design can be utilized to achieve this balance without altering the content of learning activities. Only the SGI framework [1] is a design framework in which SDT theory and Pink's

N. Wang et al. (Eds.): AIED 2023, CCIS 1831, pp. 415–421, 2023.
https://doi.org/10.1007/978-3-031-36336-8_65

theory of motivation are aligned with the conditions of flow theory, presenting a conceptual model to design game elements that promote the flow state, but this framework lacks evaluation and specific guidelines that can allow for the systematization of flow theory's application. In this article, we present a systematic gamification design framework based on the three antecedent conditions that facilitate the attainment of flow state. In addition, we presented the results of empirical validations we conducted to determine the effectiveness of our framework in real-world contexts.

2 Gamiflow: A Gamification Design Framework

Figure 1 summarizes the phases and steps of our proposed framework.

Fig. 1. Phases and steps of the Gamiflow design framework.

Analysis. At the end of this phase, the gamification design's roles and overall direction are clearly defined. In this sense, we must first *describe the learning scenario* by identifying the target-audience, the entities involved (people, objects, and agents that interact with the target-audience), and the non-game context objectives (pedagogical objectives), such as knowledge acquisition and skill development. The second step is to *identify the engagement problem*, which corresponds to identifying what lack of involvement entails in terms of behavioral interactions, cognitive and/or affective outcomes. The third step is to *delineate the target-behavior*, which is a description of how to achieve the non-game context objectives, including events, actions, roles, and their resources. *Identify student player profiles* utilizing a player type model (e.g., Bartle or Yee's model) is the optional fourth step.

Design: Content Gamification. According to Csikszentmihaly's flow theory, the environment must be designed according to the following three principles in order to accomplish the flow state: (1) a balance between skill and challenge, (2) clear objectives and short-term goals, and (3) clear and immediate feedback. In this regard, the first step in adhering to the first principle is to *define the game dynamics of DMB, DAF, and DAB*. The scenario must have a game-dynamic of progression, narrative, or emotion to maintain equilibrium (DMB), a game-dynamic of constraints or relationships to prevent boredom (DAB), and a game-dynamic of constraints or relationships to prevent frustration (DAF).

The second principle is met by *applying the SMART criteria* - Specific, Measurable, Achievable, Realistic, and Timely objectives - to the challenges in each of the game-dynamics (DMB, DAF, and DAB) implemented by game components (such as missions, levels, and quests). Lastly, in order to adhere to the third principle, all game dynamics must *provide immediate feedbacks*.

Figure 2 depicts a content gamification example for the activity of writing figure names. The challenges (as game mechanics) of identifying and writing each letter of the figure's name were aligned with a game-dynamic of progression (illustrated in the bottom-left portion of the figure) as the DMB. These challenges were implemented as achievements and points, and they were incorporated into an interface that displays the number of points that the user can earn for each letter, and the maximum number of points that can be obtained. This interface satisfies the SMART criteria, and it is also used to provide immediate feedback.

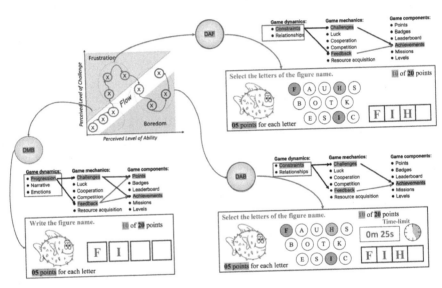

Fig. 2. Example of content gamification for writing figure names.

To avoid frustration in our example, the DAF was defined as a game-dynamic of constraints that limit the letters students can choose. This constraint is depicted in Fig. 2 (top-right) as an interface containing letters that the student must select to designate the next letter of the figure's name. Clearly, this interface reduces the level of difficulty, and it satisfies the SMART criteria, being the immediate and unambiguous feedback: red letters for incorrect selection, and green letters for correct selection. As depicted in Fig. 2 (bottom-right), to prevent boredom, a game-dynamic of constraints was implemented as the DAB in which there is a time limit for completing the activity. This constraint obviously increases the perception of difficulty because the time limit interface satisfies the SMART criteria, and it is also used to provide clear and immediate feedback.

Design: Structural Gamification. In gamified learning scenarios, extrinsic motivators must be added when students lack intrinsic motivation. Based on SDT theory, these extrinsic motivators should be aligned to transform unmotivated students into intrinsically motivated students; therefore, these motivators are added via game mechanics such as resource acquisition, transactions, and rewards to implement the dynamics. Goods and spectacles (such as video, animation, and music) are game components that should be introduced to students who are unmotivated and externally regulated. In the game dynamics, we can include trophies, virtual or physical gifts, and leaderboards for introjected-regulated students, which will impact their ego or sense of self-worth. We can use certificates and badges as game components to motivate students who are identified-regulated. For learners who are self-regulated and intrinsic motivated, we can use unlockable content, praise, and progression as game elements. Points (XP - experience points), coins, or other game-elements can be used in conjunction with extrinsic motivators as intermediate elements that can be traded for physical/virtual goods. The alignment of all game-dynamics (DMB, DAF, and DAB) and extrinsic motivators according to the player types identified in the analysis phase is an optional step that can be performed during this phase.

3   Methodology

Our study utilized Design Science Research (DSR) as its research methodology [10]. In this regard, we initially validated our framework through a case study involving fourteen (14) postgraduate students in Computation Applied to Education at the University of São Paulo who used our framework and provided us with feedback based on their responses to the TAM questionnaire [11]. To determine the efficacy of our framework, eight undergraduates gamified their own learning scenarios and conducted empirical studies as part of their undergraduate theses. Table 1 provides a summary of the findings of these investigations.

Table 1. Empirical studies conducted to evaluate the effectiveness.

	Type of Study	Domain	Ed. Level	Duration
s1	Quasi-experiment	Physics	Upper Secondary	04 weeks
s2	Quasi-experiment	Math	Lower Secondary	01 week
s3	Quasi-experiment	Digital Literacy	Upper Secondary	01 week
s4	Quasi-experiment	Writing	Upper Secondary	01 week
s5	Quasi-experiment	Programming	Undergraduate	02 weeks
s6	Single subject	Syllabic Literacy	Pre-primary education	02 days
s7	Correlational	Cybersecurity	Undergraduate	02 days
s8	Qualitative	Ergonomics	Continue Teacher Formation	04 weeks

4   Results and Discussion

After eliminating casual responses, the TAM questionnaire responses were as follows: Nine (09) of eleven (11) respondents indicate that our framework has good and very-good Perceived Usefulness ($\chi^2 = 11.63$, $p = .003$); and eight (08) respondents indicate that our framework has good and very-good Relevance ($\chi^2 = 7.81$, $p = .027$) and Output Quality ($\chi^2 = 7.81$, $p = .024$).

Five (05) of the eight empirical studies (detailed in Table 1) were quasi-experiments that contrasted the effects of gamified versions of learning scenarios (obtained through our framework) to the original versions (non-gamified scenarios). Figure 3 depicts the forest plots in which the findings of empirical studies are summarized. Regarding (a) the flow state of the students, the overall effect size was $g = 0.8512$ ($z = 2.21$ and $p = 0.273$) and the heterogeneity was substantial, with $I^2 = 80.2\%$ and $Q(5) = 25$. Regarding (b) the learning achievement of students, the aggregate effect size of was $g = 0.9365$ ($z = 3.21$ and $p = 0.0013$), with moderate heterogeneity indicated by $I^2 = 55.6\%$ and $Q(5) = 11$.

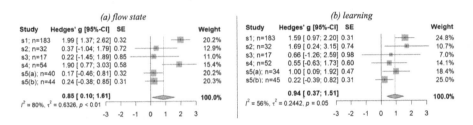

Fig. 3. Forest plots of the effect size to students' flow state (left) and learning (right).

In the domain of syllabic literacy, one empirical study (s6) was conducted as a single-subject experiment with an A-B-A-B design and positive effects on flow state and learning outcomes [6]. The participant was a 5-year-old child, and using a think-aloud method, we determined that the gamified scenario obtained by employing our framework reduces undesirable behaviors and increases students' interest and focus. The correlational study (s7) conducted on the domain of cybersecurity to support CTF (Capture The Flag) events showed that Achievement and Immersion profiles positively affected flow state [12]. A qualitative study (s8) involving 10 participants in the field of teaching ergonomics discovered no significant differences between the flow state and learning outcomes.

5   Conclusion and Future Works

This paper presents a gamification framework based on flow theory, SDT theory, and SMART criteria, along with the results of a case study and eight experimental studies conducted to validate and evaluate its effectiveness. It has been demonstrated that our framework is useful and relevant for practitioners who wish to produce high-quality output. Students' flow state and learning outcomes have been improved by gamified learning scenarios generated by our framework.

Our framework is a set of systematic guidelines instructional designers can use to meet the design conditions postulated by flow theory, and thus to accomplish well-thought-out gamification of learning scenarios based on patterns of game-dynamics, known as DMB, DAF, and DAB. In the future, we will incorporate data derived from the execution of gamified scenarios into our framework. With these results, we expect to create intelligent recommendation systems to support evidence-driven gamification by combining theoretical knowledge provided by our framework with data derived from its application to provide adaptive gamification of learning scenarios with these data.

References

1. AlMarshedi, A., wills, G., Wanick Vieira, V., Ranchhod, A.: SGI: a framework for increasing the sustainability of gamification impact. Int. J. Infonom. **8**(12), 1044–1052 (2015)
2. Böckle, M., Micheel, I., Bick, M., Novak, J.: A Design Framework for Adaptive Gamification Applications (2018)
3. Bouzidi, R., De Nicola, A., Nader, F., Chalal, R.: A systematic literature review of gamification design. In: Proceedings of the 20th Annual Simulation and AI in Games Conference (GAME-ON 2019). EUROSIS, Belgium (2019)
4. Csikszentmihalyi, M.: Flow and the Foundations of Positive Psychology: The Collected Works of Mihaly Csikszentmihalyi. Springer, Heidelberg (2014). https://doi.org/10.1007/978-94-017-9088-8
5. García, F., Pedreira, O., Piattini, M., Cerdeira-Pena, A., Penabad, M.: A framework for gamification in software engineering. J. Syst. Softw. **132**, 21–40 (2017). https://doi.org/10.1016/j.jss.2017.06.021
6. Jogo, D.A., Challco, G.C., Bittencourt, I.I., Reis, M., Silva, L.R., Isotani, S.: Investigating how gamified syllabic literacy impacts learning, flow and inappropriate behaviors: a single-subject study design. Int. J. Child-Comput. Interact., 100458 (2022). https://doi.org/10.1016/j.ijcci.2022.100458
7. Khoshkangini, R., Valetto, G., Marconi, A., Pistore, M.: Automatic generation and recommendation of personalized challenges for gamification. User Model. User-Adapt. Interact. **31**(1), 1–34 (2021)
8. Pedreira, O., García, F., Piattini, M., Cortiñas, A., Cerdeira-Pena, A.: An architecture for software engineering gamification. Tsinghua Sci. Technol. **25**(6), 776–797 (2020). https://doi.org/10.26599/TST.2020.9010004
9. Priyadi, O., Ramadhan, I., Sensuse, D.I., Suryono, R.R.: Kautsarina: gamification in software development: systematic literature review. In: Ben Ahmed, M., Abdelhakim, B.A., Ane, B.K., Rosiyadi, D. (eds.) Emerging Trends in Intelligent Systems & Network Security. Lecture Notes on Data Engineering and Communications Technologies, pp. 386–398. Springer, Heidelberg (2023). https://doi.org/10.1007/978-3-031-15191-0_37

10. Vaishnavi, V.K., Kuechler, W.: Design Science Research Methods and Patterns: Innovating Information and Communication Technology. CRC Press, Boca Raton (2015)
11. Venkatesh, V., Bala, H.: Technology acceptance model 3 and a research agenda on interventions. Decis. Sci. **39**(2), 273–315 (2008)
12. Vitorino, D., Bittencourt, I.I., Chalco, G.: StarsCTF: a capture the flag experiment to hack player types and flow experience. In: Rocha, Á., Fajardo-Toro, C.H., Rodríguez, J.M.R. (eds.) Developments and Advances in Defense and Security. SIST, vol. 255, pp. 467–477. Springer, Singapore (2022). https://doi.org/10.1007/978-981-16-4884-7_39

Using Large Language Models to Develop Readability Formulas for Educational Settings

Scott Crossley[1]([✉]) [iD], Joon Suh Choi[2] [iD], Yanisa Scherber[2] [iD], and Mathis Lucka[3]

[1] Vanderbilt University, Nashville, USA
scott.crossley@vanderbilt.edu
[2] Georgia State University, Atlanta, Georgia
{jchoi92,yscherber1}@gsu.edu
[3] deepset, Berlin, Germany
mathis.lucka@deepset.ai

Abstract. Readability formulas can be used to better match readers and texts. Current state-of-the-art readability formulas rely on large language models like transformer models (e.g., BERT) that model language semantics. However, the size and runtimes make them impractical in educational settings. This study examines the effectiveness of new readability formulas developed on the CommonLit Ease of Readability (CLEAR) corpus using more efficient sentence-embedding models including doc2vec, Universal Sentence Encoder, and Sentence BERT. This study compares sentence-embedding models to traditional readability formulas, newer NLP-informed linguistic feature formulas, and newer BERT-based models. The results indicate that sentence-embedding readability formulas perform well and are practical for use in various educational settings. The study also introduces an open-source NLP website to readily assess the readability of texts along with an application programming interface (API) that can be integrated into online educational learning systems to better match texts to readers.

Keywords: Text Readability · Large Language Models · Natural Language Processing

1 Introduction

Instructional texts aligned with students' reading levels are beneficial for gradually improving reading skills [1]. Providing students with texts that are matched to their reading abilities can ensure a stronger understanding of material and increased knowledge. Readability formulas are often used to match texts to readers. However, traditional readability formulas are based on weak proxies of language features related to text comprehension and do not measure many text features that are essential to determining text difficulties such as text cohesion and semantics. Additionally, many traditional readability formulas are biased because they were normed based on readers from specific demographics or on small text samples from specific domains.

N. Wang et al. (Eds.): AIED 2023, CCIS 1831, pp. 422–427, 2023.
https://doi.org/10.1007/978-3-031-36336-8_66

To address these concerns, the CommonLit Readability Prize (CRP) was hosted on Kaggle (https://www.kaggle.com/c/commonlitreadabilityprize) in the summer of 2021. The CRP tasked data scientists on the Kaggle platform to build open-source readability formulas to predict the reading ease of ~5,000 reading excerpts normed for grade 3–12 classrooms. The reading ease scores scalar values unique to each individual text and were the result of teacher judgments of text difficulty using pairwise comparisons. Over 70,000 readability formulas were developed. The winning formulas were strongly predictive of text readability for the excerpts in the CommonLit Ease of Readability (CLEAR) Corpus [2, 3], explaining around 90% of the variance. All winning formulas were ensemble models based on large language models (LLMs) which use contextualized token embeddings approaches to model language phenomena (e.g., question answering, sentence classification, and sentence-pair regression). However, while the winning models were highly predictive, they were not efficient. For example, the second-place model was 105 gigabytes in size and reported runtimes of 39 min to process seven texts. The inefficiency of the winning models makes it unlikely that they will be readily adapted by teachers, administrators, materials developers, and learning engineers regardless of their predictive power.

The goal of this study is to develop and assess the efficiency of new readability formulas derived from the CLEAR corpus that are based on contextual token embedding approaches similar to those found in the LLMs that won the CRP but are more efficient, allowing for quicker run times and less data storage. To do this, we develop three efficient readability formulas based on contextual token embedding approaches. These approaches include Doc2Vec [4], Universal Sentence Encodings (USE) [5], and SentenceBert (SBERT) [6]. We compare the performance of these readability formulas to traditional readability formulas (e.g., Flesch-Kincaid Grade Level [7], Flesch Reading Ease [8]), more advanced linguistic features formulas (The Crowdsourced Algorithm of Reading Comprehension [9], The Coh-Metrix Second Language Readability Index [10], and the model reported in Crossley et al. 2022 [3]), and the winning formulas from the CRP competition.

2   Method

2.1   Traditional Readability Formulas

We derived traditional readability scores using the Automatic Readability Tool for English (ARTE) [11]. ARTE is a freely available tool that automatically calculates a number of existing readability formulas on batches of text files. The formulas selected were not trained or normed on the CLEAR corpus. The formulas included Flesch - Kincaid Grade Level [7], Flesch Reading Ease [8], Automated Readability Index ARI), The Simple Measure of Gobbledygook (SMOG) [12], and the New Dale-Chall [13].

2.2   NLP-Informed Readability Formulas

We calculated two NLP-informed readability formulas using ARTE that were not trained or normed on the CLEAR corpus: The Crowdsourced Algorithm of Reading Comprehension (CAREC) [9] and The Coh-Metrix Second Language Readability Index

(CML2RI) [10]. We also report on an NLP-informed readability formula that was developed specifically for the CLEAR corpus and was reported in Crossley, 2022 [3].

2.3 CRP Readability Formulas

We replicated the top four readability formulas reported in the CRP. All formulas were based on transformer models, which depend on contextualized token embeddings that capture semantic information about a language. Transformer models use neural networks with multiple hidden layers that include millions of parameters that interact in complex ways, making it difficult to fully explain or interpret what the model is doing at each pass Table 2. Transformer models take into consideration the order in which words appear (e.g., *peace* and *war* would be represented differently than *war* and *peace*) and have attention mechanisms which allow input weights to be based on importance in a task [15]. These self-attention mechanisms allow the models to dynamically select which words are important for its calculations from a wider context window (the full length of the input). This approach allows transformer models to distinguish differences between the uses of the word *bank* in the sentences *The man robbed the bank* and *The man sat on the river bank*.

Because of the cost associated with developing transformer models, it is common to use pre-trained language models like BERT [15] and fine-tune them for different tasks. Fine-tuning leverages knowledge about diverse language-related tasks to influence the pre-trained weights of the model by providing labeled training data that are specific to the downstream task (i.e., the final target task). In the case of the CLEAR corpus, this would be the prediction of the ease of readability scores. All model winners from the CRP competition were fine-tuned on the CLEAR corpus.

2.4 Efficient Word Embedding Readability Formulas

We derived three sentence-embedding readability formulas based on Doc2Vec, Universal Sentence Encodings, and SentenceBert (SBERT) models. The three models map sentences and paragraphs (rather than individual words) to a multi-dimensional vector space, which can be used to train and predict readability scores. These models are more efficient based on size and run-time. It was not possible to fine-tune the Doc2vec and Universal Sentence Encoding models on the CLEAR corpus. However, the SBERT model was fine-tuned on the CLEAR corpus.

To develop readability formulas for the sentence embeddings models above, we did the following:

- For the SBERT model only, fine-tune the core BERT model using the CLEAR corpus.
- Calculate the sentence embeddings for each excerpt in the training and test set.
- For each excerpt in the test set, calculate the cosine similarity of each except in the train set to determine which excerpt from the train set is the most similar (i.e., has a cosine similarity score closest to one).
- Assign the known reading ease score for the most similar training excerpt to the test excerpt.
- Repeat for each excerpt in the test set until all items have been assigned a predicted reading ease score.

2.5 Statistical Analysis

Correlation analyses were conducted to compare how the scores from the traditional readability formulas, the NLP-informed readability formulas, the winning CRP models, and the light-weight sentence-embeddings models correlated with the reading ease scores from the CLEAR corpus. Only the test set for the CLEAR corpus as defined in the CRP was used.

3 Results

3.1 Traditional and NLP-Informed Readability Formulas

Correlations for traditional readability formulas reported medium to strong effects with reading ease scores ($r \sim .5$) for the test set. The strongest correlation was reported for the New Dale Chall Readability Formula while the weakest correlation was reported for ARI. The variance explained by the formulas varied from 23% to 31%. Correlations for NLP-informed models reported strong effects with reading ease scores (r between .546 to .711) for the test set. The strongest correlation was found for the Crossley et al. model (2022) that was trained specifically on the CLEAR corpus. The weakest correlation was reported for CML2RI formula. The variance explained by the formulas varied from 29% to 51% (see Table 1).

Table 1. Correlations between ease of readability score and traditional/NLP-informed readability formulas

Variable	2	3	4	5	6	7	8	9
1. BT Ease	0.540	-0.517	-0.484	-0.551	-0.556	-0.588	0.546	0.711
2. FRE	1	-0.912	-0.860	-0.936	-0.837	-0.725	0.703	0.742
3. FKGL		1	0.987	0.835	0.682	0.587	-0.691	-0.677
4. ARI			1	0.778	0.631	0.537	-0.688	-0.635
5. SMOG				1	0.808	0.708	-0.659	-0.742
6. DC					1	0.738	-0.667	-0.738
7. CAREC						1	-0.583	-0.723
8. CML2RI							1	0.791
9. Crossley 2022								1

*FRE = Flesch Reading Ease, FKGL = Flesch Kincaid Grade Level, ARI = Automated Readability Index, SMOG = Simple Measure of Gobbledygook, DC = New Dale Chall, *CAREC = The Crowdsourced Algorithm of Reading Comprehension, CML2RI = The Coh-Metrix Second Language Readability Index, Crossley 2022 = Crossley et al. (2022) Model

3.2 Transformer-Based Readability Formulas

Correlations for the CRP models reported large effects with reading ease scores ($r = .90$). The strongest correlation was reported for the third-place model ($r = .903$) while the weakest correlation was reported for the fourth-place model (.901). The variance explained by all the models was $\sim .81$. Correlations for the sentence embedding models reported large effects with reading ease scores. The strongest correlation was reported for SBERT ($r = .84$) while both Doc2Vec and the Universal Sentence Encoder reported correlations of .51. The variance explained by the SBERT model was .71.

Table 2. Correlations between ease of readability score and transformer-based readability formulas

Variable	2	3	4	5	6	7	8
1. BT Ease	0.902	0.902	0.903	0.901	0.512	0.513	0.843
2. First place	1	0.994	0.991	0.991	0.578	0.571	0.931
3. Second place		1	0.993	0.994	0.582	0.568	0.932
4. Third place			1	0.992	0.572	0.563	0.927
5. Fourth place				1	0.578	0.566	0.931
6. Doc2Vec					1	0.456	0.553
7. USE						1	0.542
8. Sbert							1

*USE = Universal Sentence Encoder Model, Sbert = SentenceBert Model

4 Discussion and Conclusion

Readability formulas are an important component for assigning appropriate texts to readers to help enhance the development of reading skills. Previous research has indicated that NLP-informed formulas, which measure more advanced features of texts than traditional readability formulas, are strong predictors of readability. However, as NLP techniques have advanced, concurrent advances in readability formulas had not, at least until the release of the CommonLit Readability Prize (CRP) competition. The BERT models from the CRP that measured semantic similarity far outperformed traditional readability formulas that are based on weak proxies of word decoding and syntactic parsing. As well, BERT-based readability formulas outperform readability formulas derived from more advanced NLP features. However, the BERT models required large amounts of storage and long run-times, making them generally impractical in most educational settings.

The more computationally efficient readability formula derived from SBert performed on par with the less efficient BERT models from the CRP competition. Lower performance was reported for the Doc2Vec and Sentence Encoder models. The SBERT model also outperformed traditional and NLP-informed readability formulas. Importantly, the SBERT model has a storage space that is 200 time smaller than the first-place CRP model and its runtime is a 100 time faster. Thus, the SBERT model's size and runtime make it feasible to scale in learning technologies that include assessments of text readability as well as for providing feedback to teachers, matching texts to students, or assessing student products. To assist with the integration of the SBERT model, we have developed an application programming interface (API) for the Automatic Readability Tool for English (ARTE) that is freely available (please see https://nlp.gsu.edu/APIdoc) so that learning engineers can integrate the SBERT readability formula into learning technologies to better match texts to readers. In addition, we have developed

ARTE into an intuitive and easy to use web tool that will automatically read in texts and provide readability results based on the SBERT formula so that researchers and teachers interested in reliably assessing text readability can do so on their devices (https://nlp. gsu.edu).

References

1. Mesmer, H.A.E.: Tools for Matching Readers to Texts: Research-Based Practices. Guilford Press (2008)
2. Crossley, S.A., Heintz, A., Choi, J.S., Batchelor, J., Karimi, M., Malatinszky, A.: The CommonLit ease of readability (CLEAR) corpus. In: Proceedings of the 14th International Conference on Educational Data Mining, pp. 755–760 (2021)
3. Crossley, S., Heintz, A., Choi, J.S., Batchelor, J., Karimi, M., Malatinszky, A.: A large-scaled corpus for assessing text readability. Behav. Res. Methods, 1–17 (2022)
4. Le, Q., Mikolov, T.: Distributed representations of sentences and documents. In: Proceedings of the 31st International conference on machine learning, pp. 1188–1196 (2014)
5. Cer, D., et al.: Universal Sentence Encoder. arXiv preprint arXiv:1803.11175 (2018)
6. Reimers, N., Gurevych, I.: Sentence-BERT: sentence embeddings using siamese BERT-networks. arXiv preprint arXiv:1908.10084 (2019)
7. Kincaid, J.P., Fishburne Jr., R.P., Rogers, R.L., Chissom, B.S.: Derivation of new readability formulas (automated readability index, fog count and Flesch reading ease formula) for navy enlisted personnel. Naval Technical Training Command Millington TN Research Branch (1975)
8. Flesch, R.: A new readability yardstick. J. Appl. Psychol. 32(3), 221–233 (1948)
9. Crossley, S.A., Skalicky, S., Dascalu, M.: Moving beyond classic readability formulas: new methods and new models. J. Res. Read. 42(3–4), 541–561 (2019)
10. Crossley, S.A., Greenfield, J., McNamara, D.S.: Assessing text readability using cognitively based indices. TESOL Q. 42(3), 475–493 (2008)
11. Choi, J.S., Crossley, S.A.: Advances in readability research: a new readability web app for English. In: Proceedings of the 22nd International Conference on Advanced Learning Technologies, pp. 1–5 (2022)
12. McLaughlin, G.H.: SMOG grading-a new readability formula. J. Read. 12(8), 639–646 (1969)
13. Chall, J.S., Dale, E.: Readability Revisited: The New Dale-Chall Readability Formula. Brookline Books (1995)
14. Rudin, C.: Stop explaining black box machine learning models for high stakes decisions and use interpretable models instead. Nat. Mach. Intell. 1(5), 206–215 (2019)
15. Devlin, J., Chang, M.W., Lee, K., Toutanova, K.: BERT: pre-training of deep bidirectional transformers for language understanding. arXiv preprint arXiv:1810.04805 (2019)

A Quantitative Study of NLP Approaches to Question Difficulty Estimation

Luca Benedetto[✉] [iD]

Department of Computer Science and Technology, University of Cambridge,
Cambridge, UK
luca.benedetto@cl.cam.ac.uk

Abstract. Question Difficulty Estimation from Text (QDET) received an increased research interest in recent years, but most of previous work focused on single *silos*, without performing quantitative comparisons between different models or across datastes from different educational domains. To fill this gap, we quantitatively analyze several approaches proposed in previous research, and compare their performance on two publicly available datasets. Specifically, we consider reading comprehension Multiple Choice Questions (MCQs) and maths questions. We find that Transformer-based models are the best performing in both educational domains; models based on linguistic features perform well on reading comprehension questions, while frequency based features and word embeddings perform better in domain knowledge assessment.

Keywords: Difficulty estimation · Natural language processing · Survey

1 Introduction

Estimating the difficulty of exam questions is an important part of any educational setup. In recent years, Question Difficulty Estimation from Text (QDET) by means of NLP techniques has become a fairly popular research topic, as a way to target the limitations of traditional approaches to difficulty estimation (manual calibration and pretesting). Most previous work focused on specific *silos*, without performing comparisons between different approaches, across datasets from different educational domains, or with questions of different types. In this work, we quantitatively evaluate approaches proposed in previous research for QDET, and compare their performance on two publicly available datasets from different educational domains, to understand if and how the accuracy of different models varies across different types of questions. Specifically, we experiment on reading comprehension MCQs (*RACE++* [13,14]), and math questions (*ASSISTments* [10]); all the questions are in English. We find that Transformer-based models [18] are the best performing across different educational domains (DistilBERT [16] performing almost as well as BERT [8]), even on smaller

---

This paper reports on research supported by Cambridge University Press & Assessment. We thank Dr Andrew Caines for the feedback on the manuscript.

. Wang et al. (Eds.): AIED 2023, CCIS 1831, pp. 428–434, 2023.
https://doi.org/10.1007/978-3-031-36336-8_67

datasets. As for the other models, the hybrid ones often outperform the ones based on a single type of features, the ones based on linguistic features perform well on reading comprehension questions, while frequency based features (TF-IDF) and word embeddings (word2vec) perform better in domain knowledge assessment. Code and supplementary material is available at https://github.com/lucabenedetto/qdet-comparison.

2 Related Work

A recent survey [1] analyzed high level trends of QDET, and another [6] presented an overview of previous approaches and proposed a taxonomy to categorize them. However, neither performed a quantitative experimental evaluation. Previous work experimented with a variety of model architectures and features. The most commonly used machine learning algorithms are Random Forests [4,5,19], SVM [9], and linear regression [11,17]. As for the features, previous works experimented with linguistic features [11,17], word2vec embeddings [9], frequency-based features [5], and readability indexes [12]. Also, multiple papers proposed hybrid approaches created by combining some of the aforementioned features (e.g., [4,19]). Previous research also used end-to-end neural networks: BERT [8] was used in [3,20], and DistilBERT [16] in [3].

3 Evaluated Models

1) Linguistic features [7,11,17] are measures related to the number and length of words and sentences in the question. We use seventeen linguistic features, taking them from previous research, and use them as input to a Random Forest regression model. **2) Readability indexes** [2,4,12] are measures designed to evaluate how easy a reading passage is to understand; we evaluate the same features used in [4] and input them to a Random Forest regression model. **3)** Considering frequency-based features [5], we use **TF-IDF**, which measures how important a word is to a document in a corpus, and input the TF-IDF weights to a Random Forest regression model. As in [5], we consider three approaches to encode the questions: i) Q_O considers only the question, ii) Q_C appends the text of the correct option to the question, and iii) Q_A concatenates all the options (right and wrong) to the question. **4) Word2vec** has been the most common technique for building non-contextualized word embeddings in previous research [9]. With word2vec, we experiment with the same three encodings as with TF-IDF (Q_O, Q_C, and Q_A), and use the embeddings as input to a Random Forest regression model. **5)** We also experiment on several **Hybrid Approaches** proposed in previous research, which are obtained by concatenating features from two (or more) of the approaches presented above, and using them as input to a single Random Forest regression model. **6)** Lastly, we consider **Transformers**: following the examples of [3,20], we experiment with **BERT** and **DistilBERT**, fine-tuning the publicly available pre-trained models on the task of QDET. Again, we evaluate the same three approaches for encoding: Q_O, Q_C and Q_A.

4   Experimental Evaluation

4.1   Datasets

RACE++ is a dataset of reading comprehension MCQs, built by merging RACE [13], with RACE-c [14]. Each question is made of a reading passage, a prompt, and four answer options, one of them being correct. Each question is manually assigned one out of three difficulty levels (0, 1, 2), considered as *gold reference* for QDET. The *train* dataset contains 100,568 questions from 27,572 reading passages, and *test* contains 5,642 questions from other 1,542 passages. Although the difficulty levels in *RACE++* are discrete, all models are trained as regressors and output a continuous difficulty, which is then converted to one of the three levels with a threshold (i.e., by mapping to the closest level).

ASSISTments [10] is a dataset of questions mostly about maths; no answer options are available. Questions are assigned a difficulty obtained with pretesting, which is continuously distributed in the range $[-5; 5]$ in a Gaussian-like shape, with $\mu = -0.32$ and $\sigma = 1.45$. *Train* contains 6,736 questions, and *test* 2,245.

4.2   Results

Comparison with Ground Truth Difficulty. QDET is a text regression task, traditionally evaluated by comparing the estimated values with gold references. We use the most common metrics from the literature [6]: Root Mean Squared Error (RMSE), R2 score, and Spearman's ρ (for RMSE lower is better, for R2 and ρ higher is better). We perform five independent training runs and show the mean and standard deviation of the metrics on the *test* set.

Table 1 presents the results for *RACE++*. All models outperform the two baselines, Transformers are the best performing (most likely, the attention can capture the relations between the passage and the question), and there is not a clear advantage of BERT over DistilBERT. For Transformers, using the text of the answer choices (Q_A) seems to improve the accuracy of the estimation, but the difference with Q_O and Q_C is often minor. As for the other features, the Linguistic perform better than Readability, word2vec, and TF-IDF. Whilst it makes sense that word2vec and TF-IDF are not very effective here, as *RACE++* is a language assessment dataset, the fact that the Readability features perform poorly partially comes as a surprise. Considering word2vec and TF-IDF, in most cases we can see no significant difference between the results obtained with the three different encodings (Q_O, Q_C, and Q_A). Using models with hybrid features brings some advantages, which are greater when using more types of features.

The results for *ASSISTments* are shown in Table 2. The Transformers are again the better performing models, neither being clearly better, and they are followed closely by TF-IDF and word2vec. As this is a domain knowledge assessment dataset, question difficulty mostly depends on terms and topics, thus Linguistic and Readability features do not perform well. Also, since the hybrid approaches that we experiment with are composed of a model involving one between Linguistic and Readability features, they are outperformed by TF-IDF and word2vec.

Table 1. Experimental results ($\mu \pm \sigma$) on *RACE++*. We group the models in four sections: i) random and majority baselines, ii) linguistic, readability, word2vec and TF-IDF features, iii) hybrid approaches, and iv) Transformers. Q_O, Q_C, and Q_A are the three encodings described in Sect. 3.

Model	RMSE	R2	Spearman's ρ
Random	1.026 ±0.004	−1.899 ±0.022	−0.005 ±0.011
Majority	0.616 ±0.000	−0.046 ±0.000	−
Ling.	0.471 ±0.004	0.388 ±0.011	0.653 ±0.005
Read.	0.552 ±0.003	0.160 ±0.008	0.507 ±0.005
W2V Q_{Only} (Q_O)	0.513 ±0.002	0.276 ±0.005	0.570 ±0.003
W2V $Q_{Correct}$ (Q_C)	0.518 ±0.002	0.263 ±0.006	0.559 ±0.003
W2V Q_{All} (Q_A)	0.507 ±0.005	0.291 ±0.013	0.580 ±0.008
TF-IDF Q_O	0.516 ±0.013	0.265 ±0.038	0.568 ±0.025
TF-IDF Q_C	0.508 ±0.005	0.290 ±0.015	0.585 ±0.010
TF-IDF Q_A	0.511 ±0.011	0.280 ±0.031	0.577 ±0.022
Ling., Read.	0.478 ±0.002	0.371 ±0.006	0.644 ±0.003
W2V Q_A, Ling.	0.467 ±0.003	0.399 ±0.008	0.658 ±0.005
Ling., Read., TF-IDF Q_C	0.463 ±0.002	0.409 ±0.005	0.668 ±0.004
W2V Q_A, Ling., TF-IDF Q_C	0.464 ±0.003	0.408 ±0.004	0.666 ±0.005
DistilBERT Q_O	0.391 ±0.009	0.578 ±0.019	0.778 ±0.009
DistilBERT Q_C	0.391 ±0.007	0.579 ±0.015	0.777 ±0.007
DistilBERT Q_A	0.381 ±0.009	0.600 ±0.019	**0.790** ±0.007
BERT Q_O	0.383 ±0.008	0.597 ±0.021	0.778 ±0.011
BERT Q_C	0.410 ±0.007	0.537 ±0.018	0.752 ±0.008
BERT Q_A	**0.372** ±0.012	**0.619** ±0.028	0.789 ±0.014

Table 2. Evaluation of the models on *ASSISTments*. As baselines, we use a random predictor, and the average difficulty of the *train* set. The text of the answer options is not available here, therefore we can use only Q_O.

Model	RMSE	R2	Spearman's ρ
Random	3.222 ± 0.037	−4.248 ± 0.119	0.010 ± 0.020
Average	1.408 ±0.000	−0.002 ±0.000	−
Ling.	1.380 ± 0.002	0.038 ± 0.002	0.209 ± 0.006
Read.	1.372 ± 0.002	0.049 ± 0.002	0.228 ± 0.013
W2V Q_O	1.278 ± 0.002	0.175 ± 0.002	0.375 ± 0.009
TF-IDF Q_O	1.274 ± 0.004	0.179 ± 0.005	0.346 ± 0.004
Ling., Read.	1.363 ± 0.004	0.061 ± 0.005	0.259 ± 0.010
W2V Q_O, Ling.	1.296 ± 0.006	0.151 ± 0.008	0.343 ± 0.013
W2V Q_O, Ling., TF-IDF Q_O	1.289 ± 0.005	0.160 ± 0.006	0.359 ± 0.010
DistilBERT Q_O	**1.267** ± 0.009	**0.189** ± 0.012	0.402 ± 0.012
BERT Q_O	1.272 ± 0.034	0.182 ± 0.045	**0.441** ± 0.013

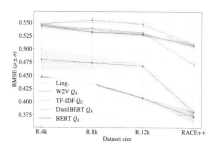 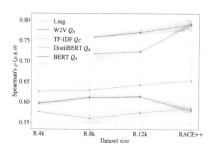

Fig. 1. RMSE and Spearman's ρ ($\mu \pm \sigma$) for different sizes of the training set.

Variation in Performance Depending on Training Dataset Size. To study how the model accuracy varies depending on the training set size, we built three reduced versions of the *train* set of *RACE++* (*R.4k*, *R.8k*, and *R.12k*), which are obtained by randomly sampling 4000, 8000, and 12000 questions respectively. Figure 1 plots, for different models, how the evaluation metrics on the test set vary depending on the training size. As expected, models tend to get better for larger training sizes, and in particular there is a significant step between *R.12k* and the whole *RACE++* (100,000 questions). Also, even when trained on *R.4k*, Transformers are better than the other models trained on the whole dataset.

5   Conclusions

In this paper, we quantitatively compared previous approaches to QDET on two publicly available datasets from different educational domains. We found that Transformers consistently outperform the other approaches, and that there is not always a large advantage of BERT over DistilBERT, suggesting that in many practical applications it might make sense to use the latter. We observed that Transformers are better even on smaller datasets, but never used datasets with less than 4000 questions. As for the other approaches, linguistic features perform well on reading comprehension MCQs, but are not effective on domain knowledge questions, for which TF-IDF and word2vec embeddings perform better. Future work might also employ different techniques for mapping from the continuous estimation to the discrete difficulty levels, experiment with other hybrid models, and explore integrations with research on unsupervised QDET [15].

References

1. AlKhuzaey, S., Grasso, F., Payne, T.R., Tamma, V.: A systematic review of data-driven approaches to item difficulty prediction. In: Roll, I., McNamara, D., Sosnovsky, S., Luckin, R., Dimitrova, V. (eds.) AIED 2021. LNCS (LNAI), vol. 12748, pp. 29–41. Springer, Cham (2021). https://doi.org/10.1007/978-3-030-78292-4_3

2. Beinborn, L., Zesch, T., Gurevych, I.: Candidate evaluation strategies for improved difficulty prediction of language tests. In: Proceedings of the Tenth Workshop on Innovative Use of NLP for Building Educational Applications, pp. 1–11 (2015)

3. Benedetto, L., Aradelli, G., Cremonesi, P., Cappelli, A., Giussani, A., Turrin, R.: On the application of transformers for estimating the difficulty of multiple-choice questions from text. In: Proceedings of the 16th Workshop on Innovative Use of NLP for Building Educational Applications, pp. 147–157 (2021)

4. Benedetto, L., Cappelli, A., Turrin, R., Cremonesi, P.: Introducing a framework to assess newly created questions with natural language processing. In: Bittencourt, I.I., Cukurova, M., Muldner, K., Luckin, R., Millán, E. (eds.) AIED 2020. LNCS (LNAI), vol. 12163, pp. 43–54. Springer, Cham (2020). https://doi.org/10.1007/978-3-030-52237-7_4

5. Benedetto, L., Cappelli, A., Turrin, R., Cremonesi, P.: R2DE: a NLP approach to estimating IRT parameters of newly generated questions. In: Proceedings of the Tenth International Conference on Learning Analytics & Knowledge, pp. 412–421 (2020)

6. Benedetto, L., et al.: A survey on recent approaches to question difficulty estimation from text. ACM Comput. Surv. (CSUR) 55, 1–37 (2022)

7. Culligan, B.: A comparison of three test formats to assess word difficulty. Lang. Test. 32(4), 503–520 (2015)

8. Devlin, J., Chang, M.W., Lee, K., Toutanova, K.: BERT: pre-training of deep bidirectional transformers for language understanding. In: Proceedings of the 2019 Conference of the North American Chapter of the Association for Computational Linguistics: Human Language Technologies (2019)

9. Ehara, Y.: Building an English vocabulary knowledge dataset of Japanese English-as-a-second-language learners using crowdsourcing. In: Proceedings of the Eleventh International Conference on Language Resources and Evaluation (2018)

10. Feng, M., Heffernan, N., Koedinger, K.: Addressing the assessment challenge with an online system that tutors as it assesses. User Model. User-Adapt. Interact. 19(3), 243–266 (2009)

11. Hou, J., Maximilian, K., Quecedo, J.M.H., Stoyanova, N., Yangarber, R.: Modeling language learning using specialized Elo rating. In: Proceedings of the Fourteenth Workshop on Innovative Use of NLP for Building Educational Applications (2019)

12. Huang, Y.T., Chen, M.C., Sun, Y.S.: Development and evaluation of a personalized computer-aided question generation for English learners to improve proficiency and correct mistakes. arXiv preprint arXiv:1808.09732 (2018)

13. Lai, G., Xie, Q., Liu, H., Yang, Y., Hovy, E.: RACE: large-scale reading comprehension dataset from examinations. In: Proceedings of the 2017 Conference on Empirical Methods in Natural Language Processing, pp. 785–794 (2017)

14. Liang, Y., Li, J., Yin, J.: A new multi-choice reading comprehension dataset for curriculum learning. In: Asian Conference on Machine Learning. PMLR (2019)

15. Loginova, E., Benedetto, L., Benoit, D., Cremonesi, P.: Towards the application of calibrated Transformers to the unsupervised estimation of question difficulty from text. In: RANLP 2021, pp. 846–855. INCOMA (2021)

16. Sanh, V., Debut, L., Chaumond, J., Wolf, T.: DistilBERT, a distilled version of BERT: smaller, faster, cheaper and lighter. arXiv preprint arXiv:1910.01108 (2019)
17. Trace, J., Brown, J.D., Janssen, G., Kozhevnikova, L.: Determining cloze item difficulty from item and passage characteristics across different learner backgrounds. Lang. Test. **34**(2), 151–174 (2017)
18. Vaswani, A., et al.: Attention is all you need. In: NIPS (2017)
19. Yaneva, V., Baldwin, P., Mee, J., et al.: Predicting the difficulty of multiple choice questions in a high-stakes medical exam. In: Proceedings of the Fourteenth Workshop on Innovative Use of NLP for Building Educational Applications (2019)
20. Zhou, Y., Tao, C.: Multi-task BERT for problem difficulty prediction. In: 2020 International Conference on Communications, Information System and Computer Engineering (CISCE), pp. 213–216. IEEE (2020)

Learning from AI: An Interactive Learning Method Using a DNN Model Incorporating Expert Knowledge as a Teacher

Hattori Kohei[✉], Tsubasa Hirakawa, Takayoshi Yamashita, and Hironobu Fujiyoshi

Chubu University, Kasugai, Aichi, Japan
tr20011-6745@sti.chubu.ac.jp, hirakawa@mprg.cs.chubu.ac.jp, takayoshi@isc.chubu.ac.jp, fujiyoshi@isc.chubu.ac.jp

Abstract. Visual explanation is an approach for visualizing the grounds of judgment by deep learning, and it is possible to visually interpret the grounds of a judgment for a certain input by visualizing an attention map. As for deep-learning models that output erroneous decision-making grounds, a method that incorporates expert human knowledge in the model via an attention map in a manner that improves explanatory power and recognition accuracy is proposed. In this study, based on a deep-learning model that incorporates the knowledge of experts, a method by which a learner "learns from AI" the grounds for its decisions is proposed. An "attention branch network" (ABN), which has been fine-tuned with attention maps modified by experts, is prepared as a teacher. By using an interactive editing tool for the fine-tuned ABN and attention maps, the learner learns by editing the attention maps and changing the inference results. By repeatedly editing the attention maps and making inferences so that the correct recognition results are output, the learner can acquire the grounds for the expert's judgments embedded in the ABN. The results of an evaluation experiment with subjects show that learning using the proposed method is more efficient than the conventional method.

Keywords: attention branch network · visual explanation · human-in-the-loop · learning from AI · educational applications

1 Introduction

Visual explanation is an approach that visually interprets the grounds of a judgment by visualizing the region of attention [2,9,12] during inference by a convolutional neural network (CNN) [3–5,10,11]. As one visual-explanation method, an attention branch network (ABN) [2] improves classification accuracy by introducing an attention mechanism that visualizes an attention map as a grounds

© The Author(s), under exclusive license to Springer Nature Switzerland AG 2023
N. Wang et al. (Eds.): AIED 2023, CCIS 1831, pp. 435–446, 2023.
https://doi.org/10.1007/978-3-031-36336-8_68

for making decisions about classification results in an image-classification task and weights feature maps in regard to gaze regions.

Although an ABN makes it possible to visually understand the grounds for decisions, if the number of training data for the target task is insufficient or if label noise is caused by annotation errors, two problems arise: (i) classification accuracy is reduced and (ii) and obtaining an attention map that provides the correct grounds for decisions becomes difficult. In such cases, the reliability of the attention map decreases, and it becomes difficult to understand the correct grounds for decisions. In response to these problems, a method for introducing expert knowledge into the network was proposed by Mitsuhara et al. [7]. As for this method, the ABN is fine-tuned by using an attention map modified by an expert in image-classification tasks, and on the grounds of the modified attention map and attention mechanism of the ABN, visual explanation and recognition performance are improved.

Various studied have attempted to help people understand AI behavior and intervene in AI learning by introducing knowledge through such visual explanations. In contrast, this study aims to create a "human-in-the-loop" mechanism, by which people learn from AI by not only humans intervening in AI learning but also AI intervening in human learning. To achieve that aim, in this paper, we propose a method by which humans learn the grounds for their decisions from AI (i.e., they "learn from AI") by using a deep-learning model (incorporating the knowledge of experts) as a teacher. As for the proposed method, the learner manually edits an attention map and uses it to perform inference of an ABN (incorporating an expert's knowledge) and check the results. The attention map is repeatedly and interactively edited so that the inference result of the ABN is correct and attains a high score. Humans can learn the grounds for the judgments of the expert embedded in the ABN through attention map. To implement this learning method an educational application was created. In an experiment, we used an educational application that we created to evaluate the proposed method. In particular, we evaluated subjects tasked with judging disease in fundus images. As a result of the experiment, we evaluated the learning effect of the proposed method and the educational effect of the created application.

The contributions of this study is summarized as follows:

- A method by which AI intervenes in human learning is proposed. In detail, the learner edits an attention map and checks changes in the inference results given by an ABN incorporating an expert's knowledge; as a result, the learner learns the grounds for the expert's judgment.
- To enable effective learning from AI, an educational application by which a learner learns while interacting with AI through editing an attention map was created, and the effectiveness of using the app was verified.

2   Related Studies

Visual explanation is a method for visualizing the grounds for decisions of deep-learning models, including CNNs, by using an attention map. Typical visual-explanation methods include class activation mapping (CAM) [12],

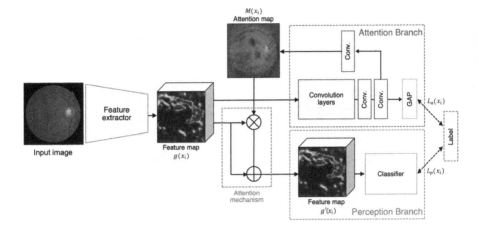

Fig. 1. Configuration of an attention branch network

gradient-weighted CAM [9], and attention branch network (ABN) [2]. CAM first applies global average pooling (GAP) [6] to the feature maps acquired by convolution. It then uses the average value of the feature maps on each channel output by GAP as weights. It finally generates an attention map from the weighted sum of each feature map. As shown in Fig. 1, the ABN applies a 1×1 convolution to the feature map immediately before applying GAP to generate an attention map. Moreover, an attention mechanism that weights the feature maps on the grounds of the acquired attention maps is introduced in a manner that simultaneously improves recognition performance.

Visual explanation can visually interpret the grounds of a decision of deep learning models by using an attention map. However, if training is insufficient owing to insufficient training data, bias, or label noise caused by annotation errors, the appropriate gaze regions are not generated. To address this problem, a method for manually correcting the attention maps of samples misrecognized by trained ABN and relearning them as teaching data was proposed by Mitsuhara et al. [7]. As for this method, an expert in image-classification tasks modifies the attention map by applying their own knowledge and retrains the network using the modified attention map. In this way, the network can learn the appropriate gaze regions and import expert knowledge. In the case of a detailed-image identification task, the expert's knowledge was used to modify the attention map so that it focuses on more-detailed regions of the object. As a result, recognition performance and visual explanatory power were both improved.

Visual explanation is the presentation of the grounds for the AI's judgment from the AI to the human, and the incorporation of expert knowledge in the ABN used by the AI is an effort by the human to intervene in the AI's learning. On the contrary, this study aims to have AI intervene in a human's learning. As for the proposed method, the attention map is used as a cue to connect the human and AI, and the learner edits the attention map while checking changes in the recognition results of the ABN (which incorporates the expert's knowledge). Repeating this interactive operation makes it possible to learn the grounds of the judgments of the expert embedded in the AI in a manner that is more effective than normal learning.

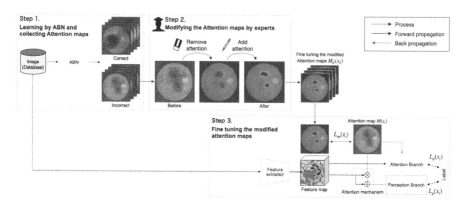

Fig. 2. Flow of incorporation of expert knowledge

3   Proposed Method

As for the proposed learning method, the learner interactively learns the grounds of the decision by using an ABN, which incorporates the expert's knowledge, as a teacher. First, expert knowledge is imported into the ABN to create a teaching AI model. The learner then edits the attention map and applies the edited map to the attention mechanism of the ABN to output classification results, which the learner then learns from the AI. For this learning, we created an educational application for editing attention maps and making inferences. Using the created application, the learner can learn the exact gaze region by editing the attention map interactively in a manner that allows novice learners with no expertise to learn the expert's knowledge on their own. In this study, diabetic retinopathy, an ocular fundus disease, was the subject of judging the presence of disease in fundus images.

3.1   Creating an AI Model that Incorporates Knowledge of Experts

First, expert knowledge is incorporated into the ABN to create the teaching AI. The flow of incorporating expert knowledge into ABN is shown in Fig. 2. Incorporating the expert knowledge consists of the following three steps:

Step 1.   Use the learned ABN to collect misidentified training samples and corresponding attention maps.

Step 2. The attention map of a misidentified sample is modified on the grounds of the expert's knowledge so that the gaze is on the region of the sample that is the grounds for the judgment of disease or no disease. $\mathcal{D}$ is taken as the set of training samples, and $x_i \in \mathcal{D}$. Also $M_e(x_i)$ is taken as the attention map of x_i as modified by the expert.

Step 3. The modified attention map $M_e(x_i)$ is used for fine tuning the ABN, where learning error $L(x_i)$ is defined as

$$L(x_i) = L_a(x_i) + L_p(x_i) + \lambda L_m(x_i) \tag{1}$$

where λ is a scale parameter for $L_m(x_i)$, and $L_a(x_i)$ and $L_p(x_i)$ are the training errors (obtained from cross-entropy errors) for the classification results from the attention branch and perception branch, respectively, which are used in the usual training of the ABN. In contrast, L_m is the error when outputting the same attention map $M_e(x_i)$ as the corrected attention map $M(x_i)$, and it is defined as

$$L_m(x_i) = \|M_e(x_i) - M(x_i)\|_2 \tag{2}$$

Note that only the parameters of the attention and perception branches are updated when fine tuning the modified attention maps, not the entire ABN network.

As described above, incorporating expert knowledge into the ABN makes it possible to focus on the same regions as experts and output correct classification results.

3.2   Inference Processing Using Edited Attention Maps

As for the proposed learning method, the learner learns by editing the attention maps and checking the classification results of the ABN by referring to the edited attention maps. Hereafter, the inference process of the ABN using the attention maps edited by the learner is described.

The attention map edited by the learner for training sample x_i is taken as $M'(x_i)$. Note that the elements of a normal attention map output by an ABN are continuous values $[0, 1]$, whereas each element of the edited attention map is a binary value $\{0, 1\}$. The edited attention map $M'(x_i)$ is utilized in the attention mechanism of the ABN in a manner that outputs classification results that take into account the regions focused on by the learner. If $g(x_i)$ is taken as the feature map output from the feature extractor of the ABN, feature map $g'(x_i)$ weighted by using $M'(x_i)$ is defined as

$$g'(x_i) = g(x_i) \cdot (1 + M'(x_i)) \tag{3}$$

The weighted feature map $g'(x_i)$ is input to the perception branch, which outputs the classification result when the regions of the edited attention map are focused on.

In this way, the learner can edit the attention map and see the corresponding classification results. By repeatedly editing the attention maps and checking the scores of the classification results, the learner learns the relationship between the fluctuation of the classification results and the corresponding gaze regions, and they can learn the appropriate gaze regions (namely, the knowledge of the experts) by themselves.

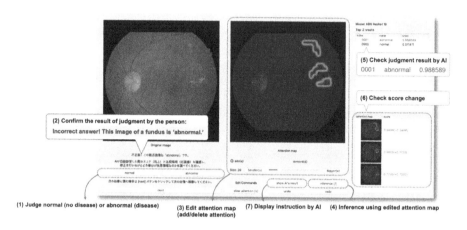

Fig. 3. Overview of educational application

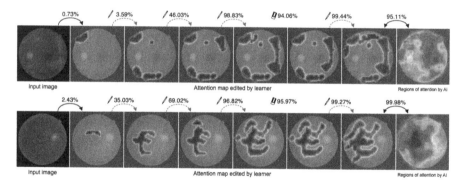

Fig. 4. Editing of attention maps by using educational app and changes in classification results

3.3 "Learning from AI" Educational Application

To effectively implement the proposed learning method, we created an educational application ("app" hereafter). The app aims to enable interactive learning between the learner and the AI by allowing the learner to easily operate the aforementioned functions. The educational app based on the proposed learning method is overviewed in Fig. 3, and the operation procedure of the app is explained below.

Step 1. A randomly selected sample from the training dataset is displayed, and the learner judges the classification ("normal" or "diseased").

Step 2. Check whether the selected judgment result is correct.

Step 3. (If the judgment in Step 2 is incorrect.) The learner edits the attention map so that it matches the correct-judgment label and emphasizes the regions that could be the grounds for the judgment.

Step 4. Using the edited attention map, the learner makes inferences with ABN.

Step 5. Confirm the classification result by the ABN.

Step 6. Repeat Steps 3 to 5. During this repetition, considering the changes in the scores of the classification results, the learner learns the gaze regions in which the scores increase by comparing their own edited attention map with that of the ABN over time.

Step 7. After Steps 3 to 6 are completed, the attention maps of the ABN are displayed as study samples. The learner then compares their edited attention map with the one shown by the ABN; accordingly, they learn which regions to pay attention to.

Step 8. Return to Step 1 and repeat the training with different samples.

By repeating the process from Step 3 to Step 6, the learner learns interactively with the AI. If the learner responds correctly to the label, Steps 3 to 6 are unnecessary, and training continues with a different sample. Moreover, by following Step 7, the learner is expected to learn the grounds of the judgment based on the expert's knowledge by checking the attention map output by the ABN. Two examples of editing an attention map by using the educational app and the corresponding changes in classification results are shown in Fig. 4. It can be seen that adding appropriate regions increases the score, and adding inappropriate regions decreases the score; thus, active learning is expected to be efficient and highly effective.

4   Experimental Results

The effectiveness of the proposed teaching method was verified by an evaluation experiment using human subjects. In the experiment, to mitigate differences in judgment ability due to age, 36 subjects in their 20s or so were used. Moreover, people with neither health problems nor prior knowledge were recruited as subjects by means of a questionnaire on their health status and prior knowledge at the time of participation in the experiment. Before the experiment, the 36 subjects were divided into the following three groups. The first group was trained without using the learning app, the second group was trained with an AI model without expert knowledge, and the third group was trained with an AI model incorporating expert knowledge. The results obtained by these three different learning methods used by the three groups confirmed the effectiveness of the proposed learning method.

4.1   Experimental Procedure

First, as a preliminary test before the start of the learning process, 60 images were presented to the subjects, who judged the presence or absence of disease. A time limit was not imposed on the subjects when responding to the images. According to the percentage of correct answers in this preliminary test, the subjects were divided into three groups with similar averages. The subjects then learn by using the learning method assigned to their group. The learning time was set to 30 min for all groups. After the learning was completed, a test was conducted in the same setting as the preliminary test. Using the same setting makes it is possible to determine if samples that were misclassified before learning were correctly classified by using each learning method. Each learning method is described as follows.

Table 1. Correct answer rate achieved by each subject (%)

	No education app		Proposed method			
			AI model without knowledge		AI model with knowledge	
	Before learning	After learning	Before learning	After learning	Before learning	After learning
Subject	56.67	70.00	55.00	51.67	50.00	75.00
	48.33	76.67	63.33	56.67	50.00	75.00
	46.67	78.33	43.33	60.00	45.00	78.33
	61.67	76.67	41.67	66.67	63.33	78.33
	43.33	75.00	60.00	70.00	38.33	78.33
	43.33	76.67	71.67	73.33	55.00	80.00
	38.33	78.33	33.33	73.33	56.67	80.00
	68.33	78.33	53.33	75.00	51.67	81.67
	73.33	76.67	66.67	76.67	56.67	81.67
	70.00	80.00	68.33	76.67	58.33	81.67
	53.33	81.67	58.33	78.33	41.67	81.67
	65.00	85.00	75.00	80.00	75.00	81.67
Average	55.69 ± 11.86	77.78 ± 3.65	57.50 ± 12.82	69.86 ± 9.20	53.47 ± 9.88	$\mathbf{79.44 \pm 2.50}$

Learning Without the App. The educational app does not provide interactive learning; instead, it presents information about the fundus images and their correct labels, and the subject learns from that information alone.

Learning Using the App with AI Model Incorporating Expert Knowledge. The app is used for learning, and an AI model with no expert knowledge is used as the teacher. Here, an AI model with no expert knowledge means an ABN obtained only by ordinary supervised learning using training data and correct labels.

Learning Using the App with AI Model Incorporating Expert Knowledge. The AI model and educational app described in Sect. 3 are used for learning.

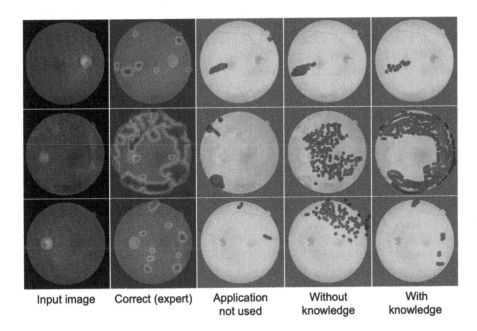

| Input image | Correct (expert) | Application not used | Without knowledge | With knowledge |

Fig. 5. Comparison of attention maps

4.2 Datasets

The effectiveness of learning by using the proposed method was experimentally evaluated on the basis of the percentage of correct responses in binary classification for judgment of disease in fundus images. The effectiveness of the proposed method could be appropriately evaluated because the images are difficult to classify accurately without expert knowledge and cannot be observed on a daily basis.

As datasets of fundus-disease images, the Messidor Dataset [1] for diabetic retinopathy grading and the Indian Diabetic Retinopathy Image Dataset (IDRiD) [8] were used. The Messidor Dataset has four levels of grading, from "0" to "3," with the larger values indicating greater severity of disease. For the pretest and post-test conducted by the subjects, 30 grade-0 images and 10 images each of grades 1 to 3 were taken from the Messidor Dataset. The IDRiD dataset was used for learning; in particular, 124 normal-fundus images and 81 diseased-fundus images were used as the learning materials. Classification accuracy of the model for the test with the IDRiD dataset was 94.44% for the model without expert knowledge and 97.22% for the model incorporating expert knowledge.

4.3 Evaluation by Classification Accuracy

Classification accuracies of each group are listed in Table 1. According to the results in the table, compared with the other learning methods, the proposed method using the AI model with expert knowledge results in higher post-learning scores. Moreover, there is large variation in scores of the group that did not use the app and the group that used the app with the AI model without expert knowledge. This variation might be due to the fact that learning and subsequent solving of the problem (judgment of disease or no disease) is dependent on individual ability. On the contrary, the groups that learned with the proposed method using the AI model with expert knowledge showed less variation, and that trend indicates that the proposed method is effective for teaching multiple learners.

Table 2. Class IoU scores for attention maps of subjects

| | No education app | | Proposed method | | | |
| | | | AI model without knowledge | | AI model with knowledge | |
	Before learning	After learning	Before learning	After learning	Before learning	After learning
Subject	0.0974	0.1194	0.0728	0.1346	0.1083	0.1931
	0.1479	0.1378	0.0740	0.0670	0.0983	0.0949
	0.0314	0.0983	0.1090	0.0659	0.1782	0.1262
	0.0983	0.0762	0.0938	0.0811	0.0555	0.0761
Average	0.0958	0.1079	0.0895	0.0835	0.1107	**0.1314**

4.4 Evaluation by Attention Maps

Next, the attention maps edited by the subjects were compared so as to verify the learning effect of the developed app based on the proposed method, namely, whether the subjects are paying attention to the appropriate regions. Examples of attention maps compiled by a subject in each group are shown in Fig. 5. It is clear from the figure that the subject who learnt with the proposed method compiled attention maps similar to those of the expert. Although the subject using the proposed method using expert knowledge added some parts to match the expert's attention, the subject did not add other regions of attention.

As for a quantitative evaluation using the attention maps, the class intersection-over-union (IoU) scores for the disease regions of the attention maps for the 30 disease images of the subjects are listed in Table 2. The proposed method with the expert-knowledge AI model achieves the highest class IoU after learning, and that result indicates that it enables the subject to compile an attention map that is closest to that of the expert. The above results demonstrate that the subject interactively learned which regions to focus on by learning with the AI model incorporating the knowledge of an expert. They also demonstrate that a highly effective learning can be achieved by showing the grounds of judgment of disease to the learner by means of the attention map shown by the AI.

4.5   Feedback from Subjects

As a qualitative evaluation of the developed educational application, feedback from the subjects after they used the application to learn using the AI model incorporating expert knowledge is listed below:

- "I think I could learn actively."
- "I think it is an appropriate educational application."
- "Just learning how to use the educational application helped me understand the essence of the disease."

In contrast, feedback from the subjects trained using the AI model incorporating no expert knowledge is listed below:

- "The results of editing suspected disease regions were not linked to evaluation by AI."
- "I failed to focus on detailed regions."
- "It was difficult to correct my assumptions about the characteristics of the disease because the focus was on regions differing from the ones that I had judged as disease."

These results demonstrate that the educational application with the AI model that incorporates expert knowledge clarifies the regions of disease as a grounds for judgment and facilitates the learning effect of the proposed method.

5   Concluding Remarks

A method that enables a learner to "learns from AI" by means of a deep-learning model incorporating expert knowledge as a teacher was proposed. In detail, the attention maps and attention mechanism of an attention branch network (ABN) are applied in a manner that imports expert knowledge into the network. Using the ABN incorporating expert knowledge, the learner can edit the attention maps and check the classification results by applying the map to the attention mechanism; as a result, the leaner can learn on the basis of the judgment grounds of the AI model. To implement the proposed learning method, an educational application was created and used to evaluate the method in experiments with subjects. The results of the experiment using images of ocular fundus disease showed that the proposed learning method was more effective than the standard method and enabled the subjects (learners) to focus on the same disease regions as the expert when judging the images.

As for future work, we hope to perform larger-scale subject experiments, improve the application, and apply the proposed "learn from AI" method to tasks that require a higher level of expertise (such as visual inspection) in addition to judgment of disease in medical images as described in this report.

Acknowledgments. This paper is based on results obtained from a project JPNP18002, commissioned by the New Energy and Industrial Technology Development Organization (NEDO).

References

1. Decencière, E., et al.: Feedback on a publicly distributed image database: the Messidor database. Image Anal. Stereol. **33**(3), 231–234 (2014)
2. Fukui, H., Hirakawa, T., Yamashita, T., Fujiyoshi, H.: Attention branch network: learning of attention mechanism for visual explanation. In: CVPR, pp. 10705–10714 (2019)
3. He, K., Zhang, X., Ren, S., Sun, J.: Deep residual learning for image recognition. In: CVPR, pp. 770–778 (2016)
4. Hu, J., Shen, L., Sun, G.: Squeeze-and-excitation networks. In: CVPR, pp. 7132–7141 (2018)
5. Krizhevsky, A., Sutskever, I., Hinton, G.E.: ImageNet classification with deep convolutional neural networks. In: NIPS, vol. 25, pp. 1–9 (2012)
6. Lin, M., Chen, Q., Yan, S.: Network in network. In: ICLR, pp. 1–10 (2014)
7. Mitsuhara, M., et al.: Embedding human knowledge into deep neural network via attention map. In: VISAPP, pp. 626–636 (2021)
8. Porwal, P., et al.: Indian Diabetic Retinopathy Image Dataset (IDRiD): a database for diabetic retinopathy screening research. Data **3**(3), 1–8 (2018)
9. Selvaraju, R.R., Cogswell, M., Das, A., Vedantam, R., Parikh, D., Batra, D.: Grad-CAM: visual explanations from deep networks via gradient-based localization. In: ICCV, pp. 618–626 (2017)
10. Simonyan, K., Zisserman, A.: Very deep convolutional networks for large-scale image recognition. In: ICLR, pp. 1–14 (2015)
11. Szegedy, C., et al.: Going deeper with convolutions. In: CVPR, pp. 1–9 (2015)
12. Zhou, B., Khosla, A., Lapedriza, A., Oliva, A., Torralba, A.: Learning deep features for discriminative localization. In: CVPR, pp. 2921–2929 (2016)

AI Cognitive - Based Systems Supporting Learning Processes

Urszula Ogiela and Marek R. Ogiela[✉]

AGH University of Krakow, 30 Mickiewicza Ave., 30-059 Kraków, Poland
{ogiela,mogiela}@agh.edu.pl

Abstract. One of the most important issues in modern educational activities is equal opportunities in learning for different students and persons, having various social, technical, and economic conditions or limitations. To achieve such goal, we propose to use different classes of cognitive information systems, which perform cognitive resonance processes, and allow to support visual data understanding and semantic content evaluation. Such systems can facilitate learning activities at different levels, and have a positive influence on education processes, by improving perception abilities for students. In particular cognitive information systems will be proposed as a tool for creation of user-oriented learning protocols. Such protocols consider users' perception capabilities, and can be dependent on external features, in which person try to acquire some new knowledge, abilities and competence.

Keywords: Cognitive information systems · artificial intelligence · learning processes

1 Introduction

In modern socio-technical activities and educational processes we can use artificial intelligence methods to support teaching and learning activities. Many games, quizzes, surveys and test are available thank to the application of multimedia systems for these purposes [1, 2]. Also, AR/VR technologies allow to create immersive educational experiences in which student or learners can be a part of virtual world, in which he or she can gain some knowledge or test cognitive skills [3]. Beside mentioned multimedia technologies we can also consider other tools based on AI procedures i.e. cognitive information systems [4, 5]. Such systems were developed especially for visual pattern semantic evaluation, and to support cognition abilities. This means that it can also be implemented to support learning processes [6]. The main idea of cognitive information systems lays on application of cognitive resonance processes, which, allow to imitate the natural way of human thinking. Such processes, and its application for learning procedure will be presented in following sections. Also, the way of its application in creation of thematic-based visual games will be presented. Application of such systems allow to define a user-oriented visual quizzes or surveys, considering expertise knowledge or thematic preferences, age, gender, educational level, and even perception abilities in evaluation of external world.

© The Author(s), under exclusive license to Springer Nature Switzerland AG 2023
N. Wang et al. (Eds.): AIED 2023, CCIS 1831, pp. 447–452, 2023.
https://doi.org/10.1007/978-3-031-36336-8_69

2 Cognitive Resonance Processes

Cognitive information systems implement computer resonance processes and try to mimic, to some extent, the natural way humans think [7, 8]. It can play important role in socio-technical activities and social structures. In such processes system should compare some real feature or parameters registered during analysis with some knowledge store in databases. During comparison is perform a kind of hypothesis verification, and decision if it is true with expectation generated during resonance processes. Cognitive information systems by performing resonance processes imitate the brain functions and are based on one of the models of human perception i.e. knowledge-based perception. According to this model, the human mind cannot correctly recognize any object, persons or situation if we haven't any knowledge or previous experiences in recognizing such pattern. It also remains true if we know object or pattern, but we have to recognize it in unusual or not typical situation. So, having such knowledge-based perception model we can implement new classes of intelligent information systems, which perform cognitive resonance processes, and can be called as cognitive information systems. In our previous research we defined several different classes of such systems especially oriented for visual pattern analysis and semantic content understanding [9]. Such systems connected with appropriately selected user interfaces like eye-tracking, leap motion or Kinect devices allow to construct efficient tools for supporting learning activities, which can be thematically oriented on group of learners and consider their perception skills.

3 Visual Thematic Patterns Supporting Learning Activities

In this section will be described solutions based on visual patterns, which for learning activities require specific information from a particular area of interest or teaching subject. In such thematic-based cognitive games it is possible to test or increase cognitive skills by proper selection by student visual parts in proper manner or in requested order. In such games, it is necessary to correctly select the semantic combination of elements, which fulfil particular requirements or have a specific meaning. For example, we can select all or only a small number of visual parts presenting specific information. Figure 1 presents an example in which young learners has to select different combinations of correct parts, dependent on the asked questions. In such approach it is possible to specify very simple or more complex answers, which required basic or expert knowledge about presented patterns.

Figure 2 presents another example with moving attention procedures and multi-level answers. Depending on the whole image division on parts, it is possible to find different possible answers. When we consider only biggest parts of this image it is possible to find only one answer (red rectangle). For more detailed division it is also possible to find other possible solutions based on yellow net or green parts.

Considering thematic-based or multi-level cognitive sequences, it is also possible to create a procedure which required basic or very specific knowledge from different areas or subjects like history, medicine, engineering, art etc.

Fig. 1. Examples of visual games presenting different rare animals (A – kakapo, B – panda, C – platypus, D – echidna, E- kea). Proper selection will depend on the questions e.g.: Where do you see egg-laying mammals? (C, D); Where do you see parrots? (A, E) etc.

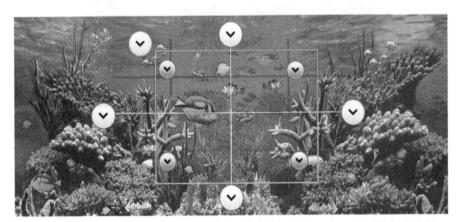

Fig. 2. Examples of visual pattern with multi-level answers e.g.: On which part(s) do you see fish? Depending on the image division it is possible to find only one answer (the red rectangle), four (yellow parts) or sixteen (green parts). The question may also be connected with the creatures having particular size, color etc. (Color figure online)

4   Eye Tracking Sensors in Visual Thematic Sequences

In presented visual thematic games we can use eye tracking technology as a contactless interface, which allow users to quickly select the correct patterns from a given set of elements. Generated patterns are selected considering the subject areas and level of students. Verification of cognitive skills or knowledge can be done in two different ways. The first way lays on checking user's information or knowledge, which require selection of patterns subset depending on the verification questions. The second way lays on presentation of single pattern, with several elements, which should be determined according semantic path. On each stage a question with increasing details level is presented. Such questions should concern elements visible in different areas of the pattern, which can be understood only by users having knowledge related to the content of patterns.

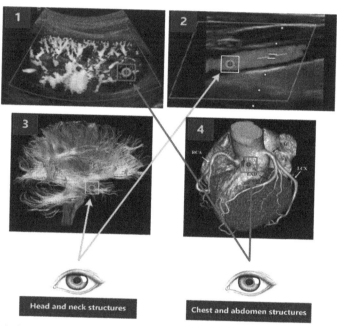

Fig. 3. Visual thematic sequence based on medical images. Image presents a set of medical visualizations: (1) kidney vessels, (2) carotid arteries, (3) brain tractography, (4) heart muscle. The first verification stage can be connected with finding chest and abdomen structures (images 1&4). The second stage may be connected with finding head and neck visualization (images 2&3). After finding appropriate visualization a question about lesion can be formulated etc.

In Fig. 3 are presented example with several medical structures presenting different internal organs. When medical students try recognize them they can quite easily recognize particular visualization. After the recognition he can still be asked about the places of illness or level for visible lesions. In subsequent iterations he/she can be asked for selection of sets of images, on which are presented particular structures located in particular body area like chest or abdomen structures or head and neck vessels.

In such procedures it is possible to determine whether the user has specific knowledge from particular areas of interest, but also verify if he/she can understand the questions, and is able to create right pattern sequences reflecting semantic meaning of the asked question.

5   Conclusions

In this paper we've described a new possible application of cognitive information systems and eye tracking devices in creation of thematic visual cognitive sequences.

Presented procedures can be used in education and cognition processes to facilitate learning procedures at different expertise levels. It also allows to test perception skills, by checking how learners can understand visual pattern with the content from different areas of expertise [10, 11]. Application of eye tracking devices allow to introduce portable systems, and mobile application working in different conditions, so it facilitate the educational processes and reduce the differences and barriers, offering the same educational conditions for different students or learners.

Acknowledgment. The research project was supported by the program "Excellence initiative – research university" for the AGH University of Krakow.

This work has been partially supported by the funds of the Polish Ministry of Education and Science assigned to AGH University of Krakow.

References

1. Chen, X., Xie, H., Hwang, G.-J.: A multi-perspective study on artificial intelligence in education: grants, conferences, journals, software tools, institutions, and researchers. Comput. Educ. Artif. Intell. **1**, 100005 (2020)
2. Guan, C., Mou, J., Jiang, Z.: Artificial intelligence innovation in education: a twenty-year data-driven historical analysis. Int. J. Innov. Stud. **4**(4), 134–147 (2020)
3. Yang, S.J.H., Ogata, H., Matsui, T., Chen, N.-S.: Human-centered artificial intelligence in education: seeing the invisible through the visible. Comput. Educ. Artif. Intell. **2**, 100008 (2021)
4. Ogiela, M.R., Ogiela, U.: Secure information splitting using grammar schemes. In: Nguyen, N.T., Katarzyniak, R.P., Janiak, A. (eds.) New Challenges in Computational Collective Intelligence. SCI, vol. 244, pp. 327–336. Springer, Heidelberg (2009). https://doi.org/10.1007/978-3-642-03958-4_28
5. Ogiela, M.R., Ogiela, L., Ogiela, U.: Biometric methods for advanced strategic data sharing protocols. In: 2015 9th International Conference on Innovative Mobile and Internet Services in Ubiquitous Computing IMIS 2015, pp. 179–183 (2015). https://doi.org/10.1109/IMIS.2015.29
6. Chen, X., Xie, H., Zou, D., Hwang, G.-J.: Application and theory gaps during the rise of artificial intelligence in education. Comput. Educ. Artif. Intell. **1**, 100002 (2020)
7. Ogiela, M.R., Ogiela, U.: Shadow generation protocol in linguistic threshold schemes. In: Ślęzak, D., Kim, T.-H., Fang, W.-C., Arnett, K.P. (eds.) SecTech 2009. CCIS, vol. 58, pp. 35–42. Springer, Heidelberg (2009). https://doi.org/10.1007/978-3-642-10847-1_5
8. Ogiela, M.R., Ogiela, U.: Linguistic extension for secret sharing (m, n)-threshold schemes. In: SecTech 2008 - 2008 International Conference on Security Technology, Hainan, Island, Sanya, China, 13–15 December 2008, pp. 125–128 (2008). https://doi.org/10.1109/SecTech.2008.15
9. Ogiela, L., Ogiela, M.R.: Advances in Cognitive Information Systems. Cognitive Systems Monographs, vol. 17. Springer, Heidelberg (2012). https://doi.org/10.1007/978-3-642-25246-4

10. Ogiela, L.: Computational intelligence in cognitive healthcare information systems. In: Bichindaritz, I., Vaidya, S., Jain, A., Jain, L.C. (eds.) Computational Intelligence in Healthcare 4. SCI, vol. 309, pp. 347–369. Springer, Heidelberg (2010). https://doi.org/10.1007/978-3-642-14464-6_16

11. Ogiela, L.: Syntactic approach to cognitive interpretation of medical patterns. In: Xiong, C., Huang, Y., Xiong, Y., Liu, H. (eds.) ICIRA 2008. LNCS (LNAI), vol. 5314, pp. 456–462. Springer, Heidelberg (2008). https://doi.org/10.1007/978-3-540-88513-9_49

Modeling Problem-Solving Strategy Invention (PSSI) Behavior in an Online Math Environment

Nidhi Nasiar[⊠], Ryan S. Baker, Yishan Zou, Jiayi Zhang, and Stephen Hutt

Graduate School of Education, University of Pennsylvania, Philadelphia, PA, USA
nasiar@upenn.edu

Abstract. This study uses Knowledge Engineering (KE) to develop an automated model of problem-solving strategy invention (PSSI) behavior (defined as inventing a new strategy for solving a math problem, outside of system-offered default strategies). The PSSI model identified the students inventing new strategies, and examined the relationship between PSSI behavior and existing fine-grained detectors of self-regulation. The findings suggest that students inventing new strategies to use for problems, are more likely to transform the information provided in the question, and to reason around the problem's contextual information.

Keywords: Strategy Invention · Self-Regulated learning · SMART model

1 Introduction

It is essential to learn to choose problem-solving appropriate strategies, and online learning platforms often scaffold students by recommending strategies. However, these supports are just a starting point for students to learn the use of strategies before creating their own, a valuable 21st century skill involving creativity and critical thinking. The invention of new strategies relies heavily on self-regulation, a key component of effective problem-solving [7]. Previous research in modeling self-regulated learning (SRL) behavior has typically considered high-level strategies ([4] is an exception), but fine-grained behaviors can provide insights into the underlying processes [11] and help us understand the role cognitive operations play in the emergence of SRL strategies.

The increasing availability of fine-grained interaction data made it feasible to model student behavior and strategies. Behavior modeling can be done using Knowledge Engineering (KE), also known as rational modeling; Machine Learning (ML) based on prediction modeling; or as an output from a bottom-up approach such as cluster analysis or sequential pattern mining. In this study, we use Knowledge Engineering (KE) to develop an automated model of problem-solving strategy invention (PSSI) behavior, defined as the learner inventing new strategies to solve the math problem, in addition to default strategies provided by the learning system. We adopt this term from [8], which defines invention activities as problem-solving tasks where learners invent novel procedures as solutions to unfamiliar problems, focusing on the development of novel problem-solving strategies. This study is conducted with CueThink, a learning platform designed to support middle school students in developing mathematical problem-solving skills. A KE

N. Wang et al. (Eds.): AIED 2023, CCIS 1831, pp. 453–459, 2023.
https://doi.org/10.1007/978-3-031-36336-8_70

model was used to identify the students exhibiting PSSI behavior in CueThink, and study the relationship of PSSI with other indicators of SRL behavior.

1.1 Approaches to Behavior Modeling in AIED Systems

Supervised Machine Learning has emerged as the predominant approach to behavior modeling. Despite their efficacy and ease of validation, the resultant detectors often lack interpretability, although explainable AI has attempted to remedy that. Another approach is Knowledge Engineering (KE), or rational modeling, which has a rich history of modeling complex behavior in AIED systems. For example, [1] developed a KE model of help-seeking within Cognitive Tutor with production rules. The transparency of KE models has enabled insights into behaviors such as gaming the system [6]. One popular method to build KE models is Evidence-centered design (ECD), which gathers evidence of competencies within assessment of complex behavior [5]. ECD involves expert-designed tasks and statistical models to evaluate learner competencies. ECD can handle complex input data while maintaining transparency and interpretability [9].

A KE model relies on domain experts' knowledge to establish a set of rules based on existing literature, and/or formalizations of experts' decision-making processes to capture the behavior [1]. KE models have higher levels of transparency and interpretability in decision-making, and capture deeper underlying features of behavior [6], making them sometimes able to generalize better to new contexts. KE models are usually validated by verifying that the rules fully capture the behavior, by comparison to theory, and iteratively by researchers, domain experts, designers, and teachers, which is different from calculating performance metrics for machine-learned models– but construct validity is important to both paradigms. PSSI is apt for a KE model as it is a straightforward construct with an explicit decision-making process. The system logs the choice of students to create a new strategy. The decisions taken to identify PSSI behavior are direct, independent, and each one is conceptually explainable. No arbitrary cut-offs or calculations, which are difficult to identify rationally, are needed to identify PSSI.

1.2 SRL Behavior, and Existing SMART Models on CueThink

Self-regulated learning (SRL) is a series of learner-generated thoughts and behaviors for goal attainment through information seeking, strategy planning, and effort alignment with objectives [10]. Log data from AIED systems can be used to model SRL behaviors such as help-seeking [1]. This paper focuses on PSSI, an SRL behavior involving an individual-level decision-making process to create and utilize new strategies beyond those provided by the system. PSSI is situated in the SMART operations of Winne & Hadwin's COPES model [10], as the process of inventing new strategies, which requires problem identification, monitoring and evaluation of available default strategies against the requirements to solve the problem, and creating new strategies if necessary. Learners engage in translating their existing knowledge and manipulating known information into a new representation to find a solution. A recent paper by Zhang et al. [11] built ML detectors of four SRL constructs also based on the SMART model [10] for the CueThink system: numerical representation (NR), contextual representation (CR), outcome orientation (OO), and data transformation (DT). NR and CR are Assembling behaviors within the SMART framework, assembling information into new representations, and occurring early in problem-solving. OO involves estimating the final answer before starting to solve, and DT is manipulating information into different representations to support solving problem. The automated detectors were successful at capturing the 4 constructs; achieving AUC ROC from 0.76–0.89 under 10-fold student-level cross-validation [11]. We analyzed the relationships between the PSSI behavior and the predictions of these 4 detectors on the same students, to explore the fine-grained cognitive operations that students employ when inventing strategies.

2 Methods

2.1 Math Learning Environment: CueThink

The study uses a dataset from CueThink, an online learning platform that presents problems as Thinklets, to be solved as a four-phase process—Understand, Plan, Solve, and Review. The platform design is based on Winne & Hadwin's SMART model of cognitive operations [10], facilitating analyses using that model. The Understand phase involves reading the problem, extracting information, and creating a representation. The Plan phase involves selecting predefined strategies (for instance, model with an equation; work backwards), or create new strategies, and outlines their use. In the Solve phase, the student gives a solution and explanation, while Review phase is for reflection on the answer's clarity and logic. This study focuses on the Plan phase where PSSI behavior occurs. The dataset, also used by [11], consists of 79 grade 6 and 7 students at a diverse suburban school in the southwestern U.S. during 2020–2021. The log data captured action-level student usage of the application and their text entries. Students spent an average of 5.2 h using CueThink, spending 1.8 h per Thinklet. After removing duplicates, 181 attempted Thinklets on 24 unique problems remained.

Building automated models of PSSI using KE included a formal operationalization of PSSI, establishing conditions and rules for detecting, and identifying edge cases affecting detection. The resultant operational definition was straightforward and did not involve subjective interpretation. Consequently, as with other studies employing similar operationalizations in KE models [1], inter-rater coding was unnecessary.

2.2 Developing the KE Model

Problem-solving strategy invention (PSSI) is defined as the behavior where a learner invents a new strategy to solve a given problem. In CueThink, where the learner is provided an initial set of 8 strategies to select from during the Plan phase, PSSI is operationally defined as the student inventing a new strategy not in the provided list and not suggested by the teacher. If a new strategy is created, the log data records this choice along with the exact text of that new strategy.

KE model development was an iterative process involving SRL and KE experts and CueThink developers. The rules and conditions that were developed to describe PSSI behavior went through multiple rounds of revision to ensure robustness and validity. The rules emerged through discussion and were conceptually coherent both in theory and practice, with the potential to generalize to future problems (via an automated model), and then the final set of rules was formalized and programmed.

Data Extraction & Preparation. Data extraction and preparation began by extracting log data from the Plan phase. Initially, 3380 logged actions were recorded; 2326 strategy-specific actions were extracted for further analysis.

Making the Rules & Conditions. The following rules identified PSSI and edge cases:

Compare with Existing Strategies: In the 1st step, every student strategy was compared against the 8 default strategies: "Draw a picture," "Make a table," "Solve with an easier problem," "Work backwards," "Guess, check and revise," "Model it with manipulatives," "Look for a pattern," "Model with an equation." The log data had instances of differently capitalized versions of default system strategies; these were treated as system strategies. Non-matching strategies were tagged as new.

Spelling Errors: Some added strategies were the same default strategies with spelling errors. Misspellings were checked automatically by computing the Levenshtein distance, which measures character differences between words, and removing strategies with a distance of 4 or less from the default list.

Filter out Class-Wide Strategies: In many cases, the same new strategies were added by multiple students from one class, likely due to the teacher discussing and sharing strategies in class. As our focus is on student invention, strategies added by more than 5 students in one class were tagged as 'class-wide', and tagged as non-PSSI behavior. However, one class's class-wide strategy could still be considered new in another class.

3   Results

3.1   PSSI Behavior and Using New Strategies

The KE model classified total of 37 students (out of 79 students) as exhibiting PSSI behavior; 70.27% of these students added more than 1 unique strategy. Students collectively added 85 new strategies on 55 math problems. Examples of invented strategies included "Use direction arrows when adding and subtracting", "Use multiplication rules & relationships", "and "Chunk the problem into smaller problems." Creating a new strategy alone is insufficient for improving problem-solving; students must also use the new strategy. The 1st and 3rd authors qualitatively coded the recorded videos from the Solve phase, where students make their process visible using the virtual whiteboard to write and draw their solutions, to see if students used their new strategies. Inter-rater reliability (IRR) was acceptable (kappa = 0.79). After removing two instances with a poor-resolution video and a missing file, students used their newly added strategies in 97.7% of cases (85 out of 87 instances). Thus, we can say that students showing PSSI behavior are quite likely to use their invented strategy in solving problems.

3.2 Association with SMART Models

The predictions of four SRL detector from Zhang et al. [11] were initially confidence probabilities, which were converted into a binary variable for comparison to the binary assessments of PSSI. The outputs were analyzed at the level of an entire math problem. Risk ratio (also called relative risk) was calculated to examine the associations between variables. The risk ratio value indicates how many times more likely construct 2 is if construct 1 is present, than if construct 1 is absent (e.g. a risk ratio of 1.0 indicates that construct 2 is neither more nor less likely if construct 1 is present). The results of the co-occurrence of PSSI with SMART models from [11] are shown in Table 1. PSSI behavior showed very strong association with DT; with learners engaging in DT 3.62 times as likely to show PSSI behavior. Learners engaging in CR had just over twice the probability of exhibiting PSSI behavior. On the other hand, the association between PSSI and NR, and PSSI and OO were not significantly different from chance.

Table 1. Risk Ratio values of association between PSSI and SMART models

Construct 1	Construct 2	Risk Ratio	Confidence Interval (95%)
PSSI	Numerical Representation (NR)	0.96	0.61–1.51
PSSI	Contextual Representation (CR)	2.12	1.12–4.01
PSSI	Outcome Orientation (OO)	1.05	0.61–1.80
PSSI	Data Transformation (DT)	3.62	1.39–9.40

4 General Discussion, Applications, and Limitations

We present a KE model of inventing new strategies (PSSI), a key skill for effective problem-solving in math. Since less than half of students show this behavior, it is crucial to offer support, such as scaffolds to explain the role of strategies and invention practices. Multiple reasons may drive PSSI behavior, including seeking efficiency, recognizing limitations in default strategies, or using familiar techniques. Correlating PSSI with other SRL constructs, we found a high co-occurrence between PSSI and DT, where learners manipulate information that is presented to them to find solutions. The link for PSSI - DT is plausible, as both involve the cognitive operation of translation (from the SMART model). As the learners invent a new strategy, we can expect learners to manipulate how information is presented to them to use a new strategy to reach solution. Additionally, a strong association was found between PSSI and CR, when students create their internal problem representation using contextual details (e.g., settings, characters, situations, etc.), a process that typically occurs early in problem-solving [10]. Though the exact causality of this relationship remains unclear, representing a problem contextually might imply a deeper understanding of the task, thereby equipping learners to assess if the strategies available are insufficient and invent new ones.

The KE model for PSSI identifies students who are creating strategies. It informs formative feedback for teachers to promote asset-based approaches like sharing students' creative strategies with the class. It can also be used to scaffold students who struggle with strategy invention (as in [2]), and provide metacognitive prompts for reflection and planning, like pop-up messages encouraging students to deliberately choose the strategies before solving, reminders that they can create their own ones, or reminding students of the new strategies they decided to use. Further scaffolds based on SRL processes of orientation and reflection, can improve the quality of students' invented strategies [3]. Investigating cases where students don't use their invented strategies could be informative. Identifying problems with less frequent use of default strategies indicates a lack of suitable default strategies, which could be added in the future.

Limitations of this study include operationalizing PSSI only in the Plan phase, and not considering other spontaneous strategy inventions that might have occurred in the Solve phase. Future work should explore broader PSSI definitions and assess PSSI's impacts on learning. Additionally, utilizing NLP for semantic analysis could provide insights into new strategy types and their effectiveness in problem-solving.

In conclusion, inventing new strategies is essential for effective math problem-solving, as PSSI develops transferrable skills for tackling new contexts. It is crucial to foster PSSI skills since less than half of learners engage in it. This study used KE to develop an automated PSSI model to identify students engaging in this behavior, and findings indicate that the self-reported strategy invention led to the actual use of those strategies. The model enabled us to analyze associations of PSSI with other SRL behaviors, to inform teacher reports and scaffolds for strategy invention.

References

1. Aleven, V., Mclaren, B., Roll, I., Koedinger, K.: Toward meta-cognitive tutoring: a model of help seeking with a cognitive tutor. Int. J. AIED **16**(2), 101–128 (2006)
2. Gidalevich, S., Kramarski, B.: The value of fixed versus faded self-regulatory scaffolds on fourth graders' mathematical problem solving. Instr. Sci. **47**(1), 39–68 (2018). https://doi.org/10.1007/s11251-018-9475-z
3. Holmes, N.G., Day, J., Park, A.H.K., Bonn, D.A., Roll, I.: Making the failure more productive: scaffolding the invention process to improve inquiry behaviors and outcomes in invention activities. Instr. Sci. **42**(4), 523–538 (2013). https://doi.org/10.1007/s11251-013-9300-7
4. Hutt, S., et al.: Investigating SMART models of self-regulation and their impact on learning. EDM (2021)
5. Mislevy, R.J., Haertel, G.D.: Implications of evidence-centered design for educational testing. Educ. Meas. Issues Pract. **25**(4), 6–20 (2006)
6. Paquette, L., de Carvalho, A., Baker, R.S.: Towards understanding expert coding of student disengagement in online learning. Learn. Instr. **15**(2), 123–139 (2014)
7. Perels, F., Gürtler, T., Schmitz, B.: Training of self-regulatory and problem-solving competence. Learn. Instr. **15**(2), 123–139 (2005)
8. Roll, I., Aleven, V., Koedinger, K.R.: Analysis of students' actions during online invention activities-seeing the thinking through the numbers. In: Proceedings of the 9th International Conference of the Learning Sciences-Volume 2, pp. 49–52, June 2010
9. Shute, V., Kim, Y.J., Razzouk, R.: ECD for dummies. Working Examples (2010)

10. Winne, P.H., Hadwin, A.F.: Metacognition in Educational Theory and Practice. Chapter in Studying as Self-Regulated Learning, pp. 277–304 (1998)
11. Zhang, J., et al.: Using machine learning to detect SMART model cognitive operations in mathematical problem-solving process. J. Educ. Data Min. **14**(3), 76–108 (2022)

A SHAP-Inspired Method for Computing Interaction Contribution in Deep Knowledge Tracing

Enrique Valero-Leal[1,2] , May Kristine Jonson Carlon[1] ,
and Jeffrey S. Cross[1(✉)]

[1] Tokyo Institute of Technology, Tokyo 152-8550, Japan
cross.j.aa@m.titech.ac.jp
[2] Universidad Politécnica de Madrid, 28040 Madrid, Spain

Abstract. Deep knowledge tracing (DKT) consists of predicting the probability of correctly answering a test or quiz question using the history of a particular learner's previous question-answer interactions. The probability of a correct answer is computed using a complex recurrent neural network. In this work, an approach similar to Shapley Additive exPlanations (SHAP) to better understand DKT was used. The number of skills a learner must master to lead to improved learning outcomes in an explainable manner was first reduced. Then, the impact of subsequences rather than every single interaction is studied, as simpler results are expected to be easier to understand. Results help to highlight both subsequences in which the student acquired knowledge and in which its progress stagnated.

Keywords: Deep knowledge tracing · Adaptive learning · Explainable AI · SHAP

1 Introduction

Explainable artificial intelligence (XAI) has grown in popularity in recent years, as it is seen as a solution to the problems posed by the difficulty of understanding how black box models work. One of the most prominent methods is Shapley Additive exPlanations (SHAP) [6], which measures feature attribution in a prediction taking inspiration from game theory.

XAI grew in popularity partly because of the emerging dominance of deep learning techniques that appear to be advantageous for tasks with larger amounts of data. However, most of the advances have been developed in the field of predictive modeling and the existing methods for other tasks are limited. One of them is knowledge tracing, which can be defined as the task of modeling the learner's behavior and interactions with the system (performance, forgetting concepts, and happenstance such as slips and guesses, among others).

Earlier models relied on Bayesian learning and inference and were referred to as Bayesian knowledge tracing [3], which focus on modeling the learner mastery of the skills related to the study plan, most commonly known as knowledge

J. Wang et al. (Eds.): AIED 2023, CCIS 1831, pp. 460–465, 2023.
https://doi.org/10.1007/978-3-031-36336-8_71

concepts (KC). A more modern approach to tackle the problem is to use deep knowledge tracing (DKT) [7] where the goal is to predict the outcome of the next interaction of the learner with the system given the previous interactions, processing them using a (recurrent) deep neural network.

Knowledge tracers are typically used in adaptive learning technologies, where each learner's experience is personalized based on various data sources to improve learning outcomes. An illustration of this is when a knowledge tracer can give suggestions to the learner on what to do next based on the activities of previously successful learners and the specific learner's current disposition. A concern with black box models such as DKT and its variants is more attention is given to improving its prediction ability than its explainability. This is a significant letdown for adaptive learning technologies since a huge contributor to learning is timely feedback, which cannot be provided if it is hard to justify the model prediction. Due to the number of parameters, the question "Why this outcome was predicted?" does not have an explicit answer for DKT. In this work, we aim to measure the impact of past interactions of the learner on the prediction using a SHAP-inspired methodology for the field of knowledge of tracing.

2   Preliminaries

Given a sequence of pairs of question-answer (q_t, a_t), $t \in 1, .., T$ that represents T interactions of a user with the system, DKT can be understood as a function that takes as input the aforementioned sequence and outputs a vector $\boldsymbol{p} = (p_1, .., p_N)$ that represents the probability of correctly answering the i-th question for all N questions in the system in the next interaction $(T + 1)$. Since we usually work with skills identifiers instead of individual question identifiers, the vector $\boldsymbol{p}$ can be understood as the skill mastery after T interactions.

The presence of a significant amount of parameters and the processing of the data by a black box calls for explainability in the prediction. A very mathematically sound method to achieve this is to compute Shapley values, which, in machine learning models, represent the contribution of each feature value to improving the prediction of their team (referred to as coalitions) over the mean prediction if feature values are to form such teams that compete with each other in terms of making the best predictions. By having some notion of feature importance, how black box models arrive at predictions can be less difficult to understand. However, given that the number of coalitions is exponential, other methods are needed. An interesting innovation formalized in SHAP is to understand Shapley values as an additive model

$$f(\boldsymbol{c}) = \beta_0 + \sum_i^N \beta_i c_i,$$

where $\boldsymbol{c} \in \{0, 1\}^N$ is a coalition vector, in which a 0 represents the absence of the i-th feature and a 1 its presence for the N features of the model (players in the game). The coefficients β of the model are Shapley values and this new vision makes its computation faster.

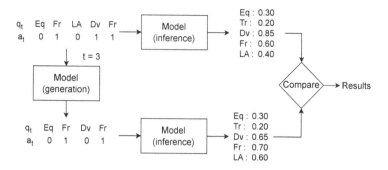

Fig. 1. Computing interaction importance by generating a new sequence by removing the third exercise and simulating the answers for the rest of the exercises

Previous works focused on predicting the test score using the skill mastery computed by the DKT model and then evaluating with SHAP the impact of each skill on the predicted score [5]. Another proposal by [2] explained using SHAP to identify the variables that have an impact on a learner guessing correctly a question, thus including in the model additional information such as the hints used and the number of attempts. The former base the explanation of the $T+1$ interaction on skill mastery, which is computed by a black box that is never interpreted, whereas the latter relies on metadata that not all datasets include.

3   Methodology

We propose to explain the skill mastery of a learner attending exclusively to the previous sequence of interactions with the system using a SHAP-like methodology to measure the importance of past interactions. However, defining the set of possible coalitions and the concept of similarity between them in the field of knowledge tracing is unfeasible, since a target dataset may contain sequences of different lengths and the order in which the questions are answered differs greatly. Since the neural network outputs a set of probabilities of correctly answering the next question, these probabilities will be used to generate synthetic sequences through a sliding window.

Given DKT as presented in Sect. 2, when evaluating the importance of the i-th interaction (q_i, a_i), said interaction would be redacted from the dataset and sample new answers, $a_t \forall t \in i+1, .., N$ using a sliding window from the instant i, assuming the sequence of questions $a_t \forall t \in i+1, .., N$ remains the same. The obtained *counterfactual* sequence, when compared with both the real and hypothetical skill mastery, can be used to study the *conditional contribution* of interaction i to the skill mastery estimation (its impact is conditioned on the rest of the sequence). This idea can be visualized in Fig. 1.

Since a single sample is not reliable enough for this study, it is recommended to generate multiple copies of sequences and weight them using the likelihood of the sequence. The likelihood is computed by multiplying the probability of (in)correctly answering each question from the interaction i until the end of the learning process.

The problem can be further simplified using a three-step procedure:

1. Given a sequence of question-answers (q_t, a_t), $t \in 1..T$, the skill mastery is computed for each instance after every interaction, obtaining a dataset of size $T \times N$. This dataset can also be understood as a multivariate time series with N features and T time instants.
2. Next, feature agglomeration is applied over the dataset, reducing the number of skills from N to M, with $M < N$. Features that change similarly over time are combined, reducing their number in an explainable manner.
3. Finally, the sequence is divided into subsequences using the Pelt algorithm [4], studying the impact of a subsequence rather than a single interaction and showing the clustered features.

Applying the above-described simplification, the conditional contribution of a subsequence can be computed. The next step is to compute the *marginal contribution* of every subsequence, using a SHAP-like methodology. The skill mastery increment for every possible (ordered) combination of subsequences is computed using the procedure described in Fig. 1. Then, a linear model is trained using as features the presence or absence of a subsequence and as targets for the different skill clusters (a linear model is needed per cluster). To obtain Shapley additive values for the features, the SHAP kernel will be used as the weight for the instances.

Although we consider our proposal to be rather experimental, it is possible to see value in different scenarios. A straightforward idea is to use the system to gain greater insight into how knowledge acquisition was produced. First, dividing the process into subsequences and clustering features greatly simplifies evaluation for tutors. Second, the SHAP-inspired procedure allows for a more mathematically complete analysis of how the knowledge was acquired, rather than limiting tutors to observe the evolution of the skill mastery. If our proposal is coupled with a learner profiling system, it can be used for curriculum design. The learning of certain skills could be optimized for a similar learner by recommending him (sequences of) exercises with a high Shapley value.

4 Experiments

In the following proof of concept, the Assistment 2009 dataset[1] and the DKT model from the "Knowledge Tracing Collection with PyTorch" implementation[2] were used for the experiments. These experiments are described with better detail in the public repository of the project[3].

---

[1] sites.google.com/site/assistmentsdata.

[2] https://github.com/hcnoh/knowledge-tracing-collection-pytorch.

[3] https://github.com/MetaCL/KT-XAI.

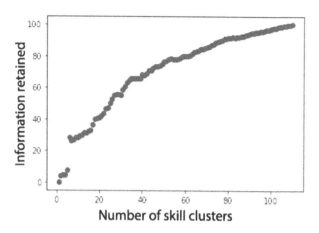

Fig. 2. Feature clustering. The percentage of area under the curve is roughly 70%

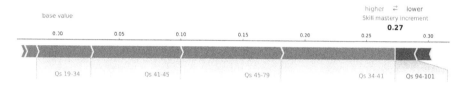

Fig. 3. Force plot for a cluster of features and the sixth learner

The focus is on the sixth learner of the dataset (ordered by the time in which they first interact with the system), who answered a total of 101 questions. First, 20 skill clusters were created, reducing the number of skills to less than one-fifth of its original size while retaining 40% of the information (see Fig. 2). Information in this case is defined as the complement of the distance between each point and its cluster centroid divided by the distance of each data point and the centroid of all data.

The Pelt algorithm was applied with a penalty of 2.5 yields. The cluster of features {Estimation, Ordering Integers, Ordering Real Numbers} was the subject of investigation.

After applying the proposed SHAP procedure, the results can be visualized in Fig. 3. The plot shows the positive and negative impacts on the learning of every subsequence. Interestingly, the last questions were actually detrimental to the learning process, and thus such subsequence should not be recommended to a learner with a similar profile. Given the local nature of our method, applying this methodology to another learner might result in a different clustering of the features and different subsequence impacts. Furthermore, since it is based on sampling, the random state of the machine will also affect the output.

5   Conclusion and Future Research

An XAI method was used to assess the interaction attribution in DKT. This research was undertaken with the intention to help educators to better understand the reasoning behind tutoring systems with potential use of metacognitive tutoring systems [1]. While the methodological portion of the research has been concluded, there are some aspects that require future work, as noted below:

- The DKT model learned probabilities was around 50% (pure uncertainty). This limitation could be addressed by modifying the function of the last layer of the network (for instance, adding a Boltzmann function).
- Finding more generic sub-sequences that also appear in other learners' learning could be useful to describe the overall functioning of the system. Pattern mining or focusing on embedding techniques are plausible approaches. The results shown in our proof of concept were rather preliminary.
- Demonstrating the proposed research's effectiveness on other datasets from different domains and populations can improve its inclusivity and generalizability.

Acknowledgements. This work was supported by the Japan Society for the Promotion of Science (JSPS) via the Grants-in-Aid for Scientific Research (Kakenhi) Grant Number JP20H01719. It is also partially funded by the Technical University of Madrid and the Tokyo Institute of Technology's Summer Exchange Research Program (SERP) (2020/21 call).

References

1. Carlon, M.K.J., Cross, J.S.: Knowledge tracing for adaptive learning in a metacognitive tutor. Open Educ. Stud. **4**(1), 206–224 (2022)
2. Clavié, B., Gal, K.: Deep embeddings of contextual assessment data for improving performance prediction. In: Proceedings of The 13th International Conference on Educational Data Mining (EDM 2020), pp. 374–380 (2020)
3. Corbett, A.T., Anderson, J.R.: Knowledge tracing: modeling the acquisition of procedural knowledge. User Model. User-Adap. Inter. **4**(4), 253–278 (1994)
4. Killick, R., Fearnhead, P., Eckley, I.A.: Optimal detection of changepoints with a linear computational cost. J. Am. Stat. Assoc. **107**(500), 1590–1598 (2012)
5. Kim, S., Kim, W., Jang, Y., Choi, S., Jung, H., Kim, H.: Student knowledge prediction for teacher-student interaction. In: Proceedings of the AAAI Conference on Artificial Intelligence, pp. 15560–15568 (2021)
6. Lundberg, S.M., Lee, S.I.: A unified approach to interpreting model predictions. In: Advances in Neural Information Processing Systems, vol. 30 (2017)
7. Piech, C., et al.: Deep knowledge tracing. In: Advances in Neural Information Processing Systems, vol. 28 (2015)

Analyzing Users' Interaction with Writing Feedback and Their Effects on Writing Performance

Yang Jiang(✉) ⓘ, Beata Beigman Klebanov, Oren E. Livne, and Jiangang Hao

Educational Testing Service, Princeton, NJ 08541, USA
{yjiang002,bbeigmanklebanov,olivne,jhao}@ets.org

Abstract. With the advance of technology, computer-based tools are increasingly used for writing instruction. However, there is a gap between their practical application and research on how they are used and their effects on writing skills. In the present study, we examined the use of Writing Mentor® (WM), a free Google Docs add-on designed to support academic writing through automated feedback. We used event logs to explore the activities that users-in-the-wild engaged in while revising their submissions. We found that the quality of users' written products significantly improved from the first submission to the last. Viewing feedback related to the writing being well-edited more frequently and more time spent in WM were significantly associated with a bigger improvement in writing quality. Our findings have implications for the development of writing feedback and the design of AI-assisted tools to support writing.

Keywords: Writing · feedback · automated scoring · writing analytics · natural language processing · automated writing evaluation · Writing Mentor

1 Introduction

Writing is considered integral to educational success at all levels and to success in workplace. However, results from large-scale assessments indicated that low literacy is a global challenge and many students lack sufficient skills to be good writers.[1] With the advances in technology, writing increasingly happens on computers, and computer-based tools are developed for writing instruction and support [1, 2].

Many computer-based tools support writing by providing formative feedback as people iteratively revise their essays [3, 4]. However, findings on the effects of feedback on performance are mixed. A meta-analysis by Kluger and DeNisi [5] suggested that while feedback improved performance on average, over 1/3 of the feedback interventions had negative effects on performance. In her review of literature on formative feedback, Shute [3] pointed out that factors such as task characteristics, instructional contexts, and student characteristics could all potentially moderate the effects of feedback.

---

[1] OECD: https://www.oecd.org/pisa/pisa-2015-results-in-focus.pdf. NAEP: https://www.nationsreportcard.gov/highlights/reading/2022/

N. Wang et al. (Eds.): AIED 2023, CCIS 1831, pp. 466–471, 2023.
https://doi.org/10.1007/978-3-031-36336-8_72

Considering the equivocal findings [3–6], it is important to examine how writing feedback is used and its effects on writing skills, as well as how students engage with different types of feedback (e.g., on English conventions vs. on coherence of the writing) when interacting with writing tools and the impact of each type of feedback. The present study aims to fill the gap by analyzing the process data and essays submitted by users of Writing Mentor (**WM**). We explored users' activities in WM and the quality of their submitted written products. The research questions we aim to answer are: 1) Did the quality of writing improve after using WM? and 2) What were the relationships between users' interaction with WM feedback and their writing quality?

2   Writing Mentor

WM (Fig. 1) is a free Google Docs add-on designed to provide feedback along multiple dimensions, guiding writers to think whether their writing is convincing (e.g., use of claims and citation of sources), well-developed (e.g., topic development), coherent (e.g., flow of ideas), and well-edited (e.g., knowledge of English conventions) (see [1] for a detailed description of WM). Users can receive feedback related to the four dimensions of their writing by clicking on the appropriate feedback category and the content under the category. Feedback in WM is generated by leveraging ETS's NLP capabilities and is presented by a friendly, non-binary persona named "Sam" [1]. Users are expected to improve their writing through a dynamic and recursive loop of writing, submitting, receiving feedback, and making revisions.

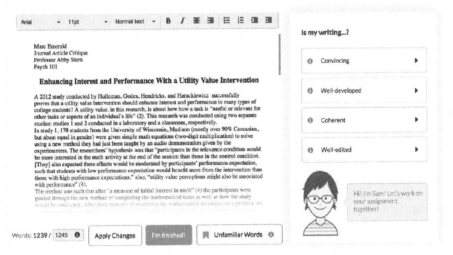

Fig. 1. Screenshot of Writing Mentor.

3   Methods

3.1   Data

Data used for this study was collected from *users-in-the-wild* who used WM and submitted their documents from its launch in November 2017 until October 2018. WM users worked on various documents for various purposes; in order to focus the sample for this study somewhat, we extracted submissions with "Ms/Miss/Mr/Mrs" in the header, reasoning that these could be teacher names and therefore could indicate submissions in a secondary school context. We also excluded submissions with "Book Review", "Literary Analysis", and "Summary" in document names to focus the set on argumentative and narrative essays. Further, documents 1) with only one submission, 2) with fewer than two logged WM events, 3) whose first and/or last submissions were excessively short or long and therefore difficult for automated evaluation, and 4) that were revised by more than one user, were removed. This left 596 documents with at least two submissions (a first draft and at least one revision) for data analysis.

3.2   Measures

Writing Quality. To evaluate the quality of the writing, we first manually scored some submissions and then built automated models to predict human scores on more essays. The scoring models were developed using a bigger data set, including documents with only one submission and those submitted within a broader time range.

After adapting the scoring rubrics from the Smarter Balanced Assessment Consortium [7], four raters scored a total of 1,927 essays along three traits: Purpose/Organization (range 0–4), Development/Elaboration (0–4), and Conventions (0–3). After training, the raters scored 155 calibration essays independently and then met to resolve any discrepancies. The quadratic weight kappa (QWK) ranged from 0.58 to 0.84 for Purpose/Organization, 0.53 to 0.80 for Development/Elaboration, and 0.64 to 0.80 for Conventions, for all possible rater pairs. Once inter-rater reliability was established, the raters scored the remaining essays independently. We then summed up the three trait scores to obtain a total composite score ranging from 0–11. We further rescaled the total score into integer scales of 0–4 by coding scores ranging from 0–2 as 0, scores 3–4 as 1, scores 5–6 as 2, scores 7–8 as 3, and scores 9–11 as 4.

Automated scoring models were constructed by using the Rater Scoring Modeling Tool (RSMTool) (https://rsmtool.readthedocs.io/en/stable/), a Python package for building and evaluating automated scoring models. Different sets of linguistic features that have shown previous success in predicting writing quality were evaluated. These features included a suite of 10 features generated by e-rater [8], an automated scoring engine deployed for automated essay scoring. E-rater features cover grammar, language use, mechanics, fluency, style, organization, and development. We also included features capturing personal writing [e.g., 9].

For building automated scoring models, data was partitioned by user id to make sure that no data from the test users appears in the training data; 86% of the essays were used for training, and 14% were set aside for testing. Linear regression models were developed using 5-fold cross-validation on the training set. The performance of the final model on the test set was QWK $= 0.60$ and Pearson's $r = 0.61$. While the model performance was relatively low compared to some operational high-stakes assessments, we considered it acceptable given the lack of standardization of the writing task (e.g., lack of knowledge about the specific writing assignment) and writing context. This final model was used to generate predictions of the scores of users' first and last submissions.

Interaction with WM Feedback. We used event logs to generate a list of features representing user interaction with WM feedback between their first and last submissions. These features included the total number of events that are associated with each of the four categories of feedback (ones related to the writing being convincing, well-developed, coherent, or well-edited) between the first and last submissions of writing. Higher values in the features indicated more frequently engaging with the feedback and more feedback was provided to users. In addition, we calculated the total amount of time (in minutes) users spent in WM between the first and last submissions. A 99% winsorization was applied to these measures with the top 1% extreme values of each measure being replaced by the value at the 99th percentile to exclude extreme cases.

4   Results

4.1   Changes in Writing Quality

Results from paired Wilcoxon signed-rank test indicated that users' last submissions ($N = 596$) scored significantly higher than their first submissions ($Ms = 2.28$ and 2.06, $U = 147$, $p < .001$). On average, the last submissions were also significantly longer than the first submissions ($Ms = 511.56$ and 452.71 words, $U = 309$, $p < .001$).

Table 1. Descriptive statistics of variables on writing quality of the first and last submissions, changes between the first and the last submissions, and writing and interaction processes.

Variable	Mean	SD	Median	Min	Max
Score first	2.06	0.70	2	0	3
Score last	2.28	0.57	2	0	3
Score change	0.22	0.50	0	−1	3
Word count first	452.71	266.92	430	26	1007
Word count last	511.56	242.06	523	32	1007
Word count change	58.86	144.55	0	−556	742
n_Convincing	0.97	1.29	1	0	17
n_Well-developed	0.89	1.10	1	0	16
n_Coherent	2.11	3.27	0	0	30
n_Well-edited	2.05	2.57	1	0	25
Total time (in minutes)	20.18	24.86	10.38	0.53	145.26

4.2 Relationships Between Interaction Behaviors and Writing Quality

On average, use of the four categories of writing feedback was low (see Table 1). Users showed a relatively higher average frequency of viewing feedback targeted to make the writing more coherent and more well-edited compared to the other feedback categories.

Ordinal logistic regression was conducted to explore the relationships between users' interaction with WM and the quality of their writing. In this model, the dependent variable was the score of the user's last submission, and the predictors were the frequency of events for the four WM feedback categories. Time spent in WM (with square root transformation) and the score of the first submission were included as covariates. Results indicated that receiving more "well-edited" feedback was significantly positively associated with the quality of the last submissions after controlling for the quality of the first submissions (*odds ratio* (OR) = 1.34, t = 4.61, p = .003). The frequencies of events for the other feedback categories were not significantly associated with writing quality after controlling for other variables. Total time spent in WM (OR = 1.06, t = 6.33, p < .001) and the score of the first submission (OR = 70.64, t = 14.79, p < .001) were significantly positively associated with the quality of the last submission.

5 Discussion and Conclusion

With the surge of AI-based tools for writing support, it is important to evaluate their effectiveness in helping users improve their writing. In the present study, we combined process data (event logs) and response data (submitted user writing) to assess the effect of Writing Mentor, an application that provides feedback on the writing being convincing, well-developed, coherent, and well-edited, on writing quality. Results indicated that the quality of writing improved significantly from the first to the final submissions. Users viewed the "coherent" and "well-edited" feedback more frequently than other categories. Receiving more "well-edited" feedback was significantly positively associated with a bigger improvement in writing quality. Feedback in the "well-edited" category is generated based on NLP features such as errors in grammar, mechanics, word choice, which probably helped users identify grammar or spelling errors and revise accordingly, thus improving the overall quality of the submissions. In addition, more time spent in WM was also associated with a bigger improvement in writing quality. Our findings thus show evidence that NLP-based feedback can support users in improving the quality of their written products.

On the other hand, the number of events for the other categories (convincing, well-developed, coherent) was not related to writing quality. One explanation was the lower extent of use of some of these categories. It is also possible that such feedback is less immediately actionable than "well-edited" (e.g., feedback on spelling errors can be easily used for corrections). It will be interesting to examine the relationship with writing quality among users who engaged more extensively with these types of feedback.

5.1 Limitations and Future Work

We used ordinal logistic regressions to explore the relationships between engagement with feedback and writing quality. We note that no causal inferences could be established from the present study. A larger number of events for a specific feedback category indicated that the user received more feedback in the category. However, it does not necessarily indicate that user responded to the feedback. It is therefore important for future work to examine whether users made revisions in response to the feedback and whether the revisions improved the quality of the writing.

Acknowledgments. We would like to thank Jill Burstein, Mengxiao Zhu, Sophia Chan, James V. Bruno, Eowyn Winchester, Hillary Molloy, Josh Crandall, Lisa Bergman, Nitin Madnani, and Martin Chodorow for their contributions to the project.

References

1. Burstein, J., et al.: Writing mentor: writing progress using self-regulated writing support. J. Writ. Anal. **2**, 285–313 (2018). https://doi.org/10.37514/jwa-j.2018.2.1.12
2. Foltz, P.W., Rosenstein, M.: Data mining large-scale formative writing. In: Lang, C., Siemens, G., Wise, A., Gašević, D. (eds.) Handbook of Learning Analytics, pp. 199–210. Society for Learning Analytics Research (2017)
3. Shute, V.J.: Focus on formative feedback. Rev. Educ. Res. **78**, 153–189 (2008). https://doi.org/10.3102/0034654307313795
4. Wang, E.L., et al.: eRevis(ing): students' revision of text evidence use in an automated writing evaluation system. Assess. Writ. **44**, 100449 (2020). https://doi.org/10.1016/j.asw.2020.100449
5. Kluger, A.N., DeNisi, A.: The effects of feedback interventions on performance: a historical review, a meta-analysis, and a preliminary feedback intervention theory. Psychol. Bull. **119**, 254–284 (1996). https://doi.org/10.1037/0033-2909.119.2.254
6. Nunes, A., Cordeiro, C., Limpo, T., Castro, S.L.: Effectiveness of automated writing evaluation systems in school settings: a systematic review of studies from 2000 to 2020 (2022)
7. Smarter Balanced Assessment Consortium: Smarter Balanced Performance Task Scoring Rubrics Grades 3–11 (2014)
8. Attali, Y., Burstein, J.: Automated essay scoring with e-rater v.2. J. Technol. Learn. Assess. **4**, 3–30 (2006)
9. Beigman Klebanov, B., Burstein, J., Harackiewicz, J.M., Priniski, S.J., Mulholland, M.: Reflective writing about the utility value of science as a tool for increasing STEM motivation and retention – can AI help scale up? Int. J. Artif. Intell. Educ. **27**(4), 791–818 (2017). https://doi.org/10.1007/s40593-017-0141-4

Annotating Educational Dialog Act with Data Augmentation in Online One-on-One Tutoring

Dapeng Shan[1], Deliang Wang[2(✉)], Chenwei Zhang[2], Ben Kao[1], and Carol K. K. Chan[2]

[1] Faculty of Engineering, The University of Hong Kong, Hong Kong, China
u3006522@connect.hku.hk
[2] Faculty of Education, The University of Hong Kong, Hong Kong, China
wdeliang@connect.hku.hk

Abstract. During the COVID-19 pandemic, educational activities have shifted online, providing opportunities for researchers to analyze interaction data between teachers and students. In this study, we focus on automatically annotating dialog acts in one-on-one tutoring on online platforms. We address the challenge of limited training data, particularly for "rare codes", by proposing a data augmentation pipeline that leverages GPT-3.5's generative ability to create synthetic, multi-labeled dialog data. Experiments with real online tutoring platform data demonstrate the effectiveness of our approach in enhancing the machine annotator's accuracy.

Keywords: Online tutoring · Educational dialog act · Multi-label annotation · Data augmentation · GPT-3.5

1 Introduction

Educational dialog acts (EDAs) are specific functions or purposes of spoken or written utterances in educational contexts. They help researchers understand communication and learning processes, informing educational interventions. EDAs are used to study language strategies [4], feedback [1], and the development of intelligent tutoring systems [5]. A dialog act coding scheme is usually applied to categorize dialog elements based on their functional roles in teacher-student interactions. However, annotating EDAs faces challenges like cost, scalability, and multi-labeling, making it labor-intensive and time-consuming. This paper aims to design a machine annotator to label EDA data using a small-scale multi-labeled dataset as training data. Our method employs a transformer-based pre-trained language model as the base classifier. With a coding scheme and a small manually-labeled training set, we identify rare codes and propose a data augmentation pipeline to improve classification accuracy. This pipeline utilizes GPT-3.5's generative ability to create training conversations with coherent contexts including rare codes. Experiments with real online tutoring platform data show significant improvements in the classifier model's prediction accuracy, especially for rare codes.

© The Author(s), under exclusive license to Springer Nature Switzerland AG 2023
N. Wang et al. (Eds.): AIED 2023, CCIS 1831, pp. 472–477, 2023.
https://doi.org/10.1007/978-3-031-36336-8_73

2 Related Work

Our study focuses on automatic EDA annotation. Similar to our objective, there have been a few attempts to design such annotators. Some studies employ traditional machine learning models, such as hidden Markov models (HMM), random forests (RF), naive Bayes (NB), and support vector machines (SVM), with manually crafted syntax and lexical features (e.g., [10,12]). These models are limited in three ways. First, only syntactic features are used so they are less capable in handling semantic information in the EDAs. Second, the scarcity of data for "rare codes" poses challenges in training a dependable annotator. Third, these models do not handle "multi-labeling"; Each utterance is given only a single code. Other studies handle semantic information by applying pre-trained language models (PLMs) (e.g., [8,11]). However, they are also limited to inadequate "rare codes" training data and "single-labeling" classification.

As we have mentioned, we propose a data augmentation pipeline to enhance the classification accuracy on rare codes. Other common techniques for textual data augmentation include *noise injection* (adding random noise to original data), *back translation* (translating an original sentence to another language and then back), and *synonym replacement* (replace a word or phrase with a synonym) [7]. Our work focuses on how to use a large language model (LLM) to generate dialog conversations for rare codes. In particular, we propose a pipeline that can effectively generate high-quality conversations using GPT-3.5.

3 Data and Coding Scheme

We use data collected from an online task-based tutoring platform. The dataset consists of 29,278 tutorial sessions between 350 amateur tutors and 745 students in Singapore. The tutor-student conversation in each session is recorded in the dataset. The annotation task is to label each utterance in tutorial conversations with EDAs given a coding scheme. Table 1 presents the coding scheme used in this study. The coding scheme was specifically adapted for online, one-on-one tutoring, and is based on previous research on educational dialog acts [2,6]. There are 15 codes that represent different learning and teaching behaviors, which are designed to capture the cognitive and meta-cognitive progress of learning and the pedagogical approaches used by tutors. These codes are divided into six groups, each capturing a certain kind of educational interaction between tutors and students.

We manually annotated 593 math tutoring sessions by labeling each utterance in those sessions according to our coding scheme. Two undergraduate students studying education were trained to conduct the labeling task separately. The inter-rater reliability of this annotation process, as measured by Krippendorff's Alpha, was 0.83. The distribution of codes in the dataset is shown in the Ratio column in Table 1. It can be seen that tutor knowledge sharing and explanation (KS) and off-topic (OT) are the most common codes in the dataset, while student behaviors such as proposing procedures or steps for a solution (P) and

Table 1. This table presents the coding scheme utilized in this study. The column **Group** comprises six high-level interactive behavior groups. The column **Code** denotes the code name for each Educational Dialogue Act (EDA). The **Role** column specifies whether an EDA is exclusive to tutors (T), students (S), or applicable to both. The Column **Description** specifies the definition of each EDA. The column **Ratio** presents the percentage of utterances that contain the corresponding EDA.

Group	Code	Role	Description	Ratio
Question asking	CQ	S&T	Ask for ideas and opinions that advances the task resolution	2.6%
	PQ	S&T	Ask specifically for the procedure, process or a series of steps to solve a task or achieve a certain outcome	1.4%
	REQ	S&T	Ask for logical reasoning to prove, justify or infer the cause-and-effect relationship with evidence or examples	1.6%
Constructive Participation	C	S	Provide ideas or opinions that advances the task resolution	3.5%
	P	S	Provide the procedure or process of to the whole or partial solution, achieving certain outcomes	0.8%
	RE	S	Explain logical reasoning, justifications or explanations with evidence or examples	0.9%
	KS	T	Provide information, share knowledge or explain solutions	26.9%
Evaluation and Feedback	EVQ	S&T	Ask for the listener to evaluate the speaker's expressions	2.6%
	A	S&T	Affirm agreement or confirm correctness	3.0%
	FB	S&T	Provide informative or challenging feedback on the listener's expression	2.3%
Metacognition	RFQ	T	Ask for the student's reflection of self-understanding, performance or progress	3.9%
	RF	S	Reflection of self-understanding, performance or progress	3.2%
	RFN	S	Reflection of a partial or lack of understanding	1.9%
Active participation	AP	S	Simple acknowledgement without additional constructive action	3.8%
Off topic	OT	S&T	Off-topic expressions or questions	47.0%

making logical reasoning to explain, justify, or infer with evidence or examples (RE) are much rarer. Tutor elicitation behaviors such as CQ, REQ, and PQ are also relatively uncommon. Informative comments from tutors and students are also relatively rare in the dataset. Finally, it is worth noting that 4.9% of total utterances, or 22.1% of utterances excluding KS and OT, were labeled with more than one code.

4   Methods

We employ RoBERTa [9] as our base pre-trained language model (PLM) classifier; however, it could be substituted with alternative PLM models such as BERT [3]. To improve performance on under-represented codes, we propose a data augmentation pipeline using GPT-3.5 to create synthetic dialog utterances covering rare codes. The pipeline involves four steps: **[Prioritizing]** Identify codes needing the most augmentation based on the PLM base classifier's performance, such as low F1 scores. Also, consider code correlations, giving lower priority to codes that frequently co-occur with others to avoid introducing noise. **[Prompting]** Design prompts to guide GPT-3.5 in generating desirable conversations. The best prompts we found include a context description to provide instruction context, a code description to elaborate the definition of the desired code, referencing examples obtained from real data, and a diversity instruction to increase generated text diversity. **[Filtering]** Remove redundant text using a 4-gram Self-BLEU score threshold of 0.5 to prevent overfitting. Calculate the

Self-BLEU measure by computing the BLEU score of each generated text using all others as references, with a lower score indicating higher diversity. [**Multi-labeling**] Generate multi-labels for the augmenting data. Use the PLM base classifier to classify and add missing codes to the generated utterances, ensuring that the augmenting data is multi-labeled and doesn't affect the model's accuracy on non-target codes.

For each selected rare code, we collect 2,000 GPT-3.5-generated utterances, generating a total of 8,000 utterances. About 76% of them are removed by the filtering step, leaving 1,918 synthetic utterances (an average of 479.5 per prioritized code) to augment the training set and boost the base classifier's performance on rare codes.

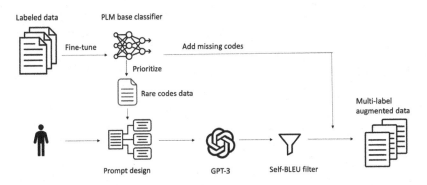

Fig. 1. Pipeline for generating multi-label synthetic conversations with rare codes

5   Results and Discussion

We evaluate classifier performance under different data augmentation conditions using micro-average and macro-average F1 scores. Micro-average F1 reflects the classifier's performance on majority codes, while macro-average F1 gives equal weight to each code. Code-level F1 scores are incorporated into the analysis to provide a more comprehensive understanding of the performance disparities. We compare four data augmentation conditions: 1) No augmentation (NoAug(Roberta-base)), 2) Oversampling[1], 3) Our pipeline without the final multi-labeling step (GPT(Single)), and 4) Our full pipeline (GPT(Multi)). To address limited testing data for some codes, we use 5-fold cross-validation for more robust results. All F1 scores are presented in Table 2, with standard deviations in brackets.

---

[1] We chose oversampling, which attains the highest Macro-F1 and Micro-F1 scores among Back Translation, Random Insert, and Random Replace, as a representative of traditional perturbation-based data augmentation techniques.

Table 2. Micro-F1, Macro-F1 and Code-level F1 scores over various augmentation strategies (s.d. shown in brackets)

	NoAug (Roberta-base)	Over-sampling	GPT (Single)	GPT (Multi)
CQ	**0.787**$_{(0.089)}$	0.754$_{(0.086)}$	0.765$_{(0.083)}$	0.762$_{(0.022)}$
PQ	0.679$_{(0.215)}$	0.769$_{(0.256)}$	0.737$_{(0.095)}$	**0.813**$_{(0.071)}$
REQ	0.841$_{(0.101)}$	0.872$_{(0.062)}$	0.795$_{(0.071)}$	**0.870**$_{(0.049)}$
EVQ	0.710$_{(0.023)}$	**0.806**$_{(0.091)}$	0.791$_{(0.075)}$	0.724$_{(0.078)}$
RFQ	0.848$_{(0.037)}$	**0.853**$_{(0.041)}$	0.840$_{(0.066)}$	0.851$_{(0.053)}$
C	0.759$_{(0.110)}$	0.764$_{(0.066)}$	0.732$_{(0.056)}$	**0.767**$_{(0.033)}$
P	0.562$_{(0.147)}$	0.597$_{(0.226)}$	0.613$_{(0.166)}$	**0.679**$_{(0.095)}$
RE	0.706$_{(0.161)}$	0.740$_{(0.104)}$	0.753$_{(0.061)}$	**0.825**$_{(0.091)}$
A	**0.802**$_{(0.052)}$	0.778$_{(0.049)}$	0.752$_{(0.032)}$	0.778$_{(0.051)}$
FB	0.643$_{(0.154)}$	0.648$_{(0.116)}$	0.701$_{(0.117)}$	**0.741**$_{(0.081)}$
RF	0.661$_{(0.037)}$	0.670$_{(0.072)}$	0.652$_{(0.047)}$	**0.668**$_{(0.039)}$
RFN	0.786$_{(0.031)}$	**0.808**$_{(0.110)}$	0.799$_{(0.034)}$	0.790$_{(0.044)}$
AP	**0.833**$_{(0.066)}$	0.813$_{(0.106)}$	**0.833**$_{(0.058)}$	0.821$_{(0.046)}$
KS	0.953$_{(0.013)}$	0.951$_{(0.008)}$	**0.955**$_{(0.008)}$	0.954$_{(0.012)}$
OT	0.940$_{(0.007)}$	0.940$_{(0.004)}$	0.936$_{(0.003)}$	**0.943**$_{(0.014)}$
Micro-F1	0.890$_{(0.012)}$	0.893$_{(0.006)}$	0.891$_{(0.008)}$	**0.899**$_{(0.003)}$
Macro-F1	0.767$_{(0.043)}$	0.784$_{(0.047)}$	0.778$_{(0.029)}$	**0.799**$_{(0.013)}$

In Table 2, it is evident that the PLM base classifier without data augmentation attains a high overall accuracy, with a Micro-F1 of 0.890 and Macro-F1 of 0.767. A closer examination of individual codes reveals that the model performs exceptionally well for prevalent codes such as KS (F1 = 0.953) and OT (F1 = 0.940). This indicates that PLMs without data augmentation can achieve high prediction accuracy when provided with ample data. However, the model faces challenges in accurately predicting codes with limited data, as demonstrated by the lower F1 scores for P, FB, and RF. Employing a simple oversampling data augmentation strategy improves both Micro and Macro F1 scores, suggesting that data imbalance is a significant contributor to the low prediction performance. The GPT(Single) condition, which only employs GPT-3.5 for single-label augmentation, enhances the overall Micro and Macro-F1 compared to the NoAug condition. Nevertheless, it underperforms relative to the oversampling approach. A more in-depth analysis of code-level F1 scores reveals that, compared to oversampling, GPT(Single) performs better on rare codes P, RE, and FB but worse on most non-rare codes. This implies that the utterances generated by GPT-3.5 contain non-targeted codes that should not be disregarded. Lastly, the GPT(Multi) condition, which utilizes the entire data augmentation pipeline, achieves the highest Micro-F1 and Macro-F1 among all four conditions. It also exhibits the highest accuracy for all four rare codes with a substantial margin compared to the second-best condition. GPT(Multi) improves the accuracy on 10 out of 15 codes compared to GPT(Single) and 11 out of 15 codes compared to Oversampling. Consequently, it can be reasonably concluded that the data augmentation pipeline effectively addresses the data imbalance issue in the limited training data situation.

6 Conclusion

We demonstrated the effectiveness of using pre-trained language models for automatic annotation of multi-label educational dialog acts. By leveraging the large-scale knowledge stored in GPT-3.5, we proposed a pipeline for generating synthetic data that outperformed traditional perturbation-based data augmentation methods. Our analysis highlighted that annotating non-targeted codes within the generated text substantially influences prediction accuracy.

References

1. Carless, D.: Feedback as dialogue. In: Peters, M. (eds.) Encyclopedia of Educational Philosophy and Theory, pp. 1–6. Springer, Singapore (2016). https://doi.org/10. 1007/978-981-287-532-7_389-1
2. Chi, M.T., Wylie, R.: The ICAP framework: linking cognitive engagement to active learning outcomes. Educ. Psychol. **49**(4), 219–243 (2014)
3. Devlin, J., Chang, M.W., Lee, K., Toutanova, K.: BERT: pre-training of deep bidirectional transformers for language understanding. arXiv preprint arXiv:1810.04805 (2018)
4. Graesser, A.C., Person, N.K.: Question asking during tutoring. Am. Educ. Res. J. **31**(1), 104–137 (1994)
5. Graesser, A.C., VanLehn, K., Rosé, C.P., Jordan, P.W., Harter, D.: Intelligent tutoring systems with conversational dialogue. AI Mag. **22**(4), 39–39 (2001)
6. Hennessy, S., et al.: Developing a coding scheme for analysing classroom dialogue across educational contexts. Learn. Cult. Soc. Interact. **9**, 16–44 (2016)
7. Li, B., Hou, Y., Che, W.: Data augmentation approaches in natural language processing: a survey. AI Open (2022)
8. Lin, J., et al.: Is it a good move? Mining effective tutoring strategies from human-human tutorial dialogues. Futur. Gener. Comput. Syst. **127**, 194–207 (2022)
9. Liu, Y., et al.: Roberta: a robustly optimized BERT pretraining approach. arXiv preprint arXiv:1907.11692 (2019)
10. Rus, V., Maharjan, N., Banjade, R.: Dialogue act classification in human-to-human tutorial dialogues. In: Innovations in Smart Learning. LNET, pp. 185–188. Springer, Singapore (2017). https://doi.org/10.1007/978-981-10-2419-1_25
11. Suresh, A., Jacobs, J., Harty, C., Perkoff, M., Martin, J.H., Sumner, T.: The Talk-Moves dataset: K-12 mathematics lesson transcripts annotated for teacher and student discursive moves. arXiv preprint arXiv:2204.09652 (2022)
12. Vail, A.K., Boyer, K.E.: Identifying effective moves in tutoring: on the refinement of dialogue act annotation schemes. In: Trausan-Matu, S., Boyer, K.E., Crosby, M., Panourgia, K. (eds.) ITS 2014. LNCS, vol. 8474, pp. 199–209. Springer, Cham (2014). https://doi.org/10.1007/978-3-319-07221-0_24

Improving Code Comprehension Through Scaffolded Self-explanations

Priti Oli[1(✉)], Rabin Banjade[1], Arun Balajiee Lekshmi Narayanan[2], Jeevan Chapagain[1], Lasang Jimba Tamang[1], Peter Brusilovsky[2], and Vasile Rus[1]

[1] University of Memphis, Memphis, TN 38152, USA
{poli,rbnjade1,jchpgain,ljtamang,vrus}@memphis.edu
[2] University of Pittsburgh, Pittsburgh, PA 15260, USA
{arl122,peterb}@pitt.edu

Abstract. Self-explanations could increase student's comprehension in complex domains; however, it works most efficiently with a human tutor who could provide corrections and scaffolding. In this paper, we present our attempt to scale up the use of self-explanations in learning programming by delegating assessment and scaffolding of explanations to an intelligent tutor. To assess our approach, we performed a randomized control trial experiment that measured the impact of automatic assessment and scaffolding of self-explanations on code comprehension and learning. The study results indicate that low-prior knowledge students in the experimental condition learn more compared to high-prior knowledge in the same condition but such difference is not observed in a similar grouping of students based on prior knowledge in the control condition.

Keywords: Program Comprehension · Scaffolding · Computer Science Education · Java Programming · Intelligent Tutoring System

1 Introduction

Computer programming is a critical skill in today's world. However, learning to program is challenging, as shown by high attrition rates (30–40%, or even higher) in introductory CS courses [1]. The premature focus on *writing* code rather than *reading* code may be part of the reason why learning to program has historically been challenging [3]. Code comprehension activities such as code tracing and code reading have been argued to be critical in the early stage of learning because they allow beginners to develop programming skills with a lower cognitive load than writing code itself [2].

In this paper, we present an Intelligent Tutoring System (ITS) *DeepCodeTutor* that helps students master code comprehension skills by providing automatic assessment and scaffolding for self-explanation of code examples. Self-explanation has been proven to be an effective strategy for learning computer programming concepts [6]; however, the presence of a human tutor is usually required to make

© The Author(s), under exclusive license to Springer Nature Switzerland AG 2023
N. Wang et al. (Eds.): AIED 2023, CCIS 1831, pp. 478–483, 2023.
https://doi.org/10.1007/978-3-031-36336-8_74

this strategy work. In *DeepCodeTutor*, students' self-explanations are automatically assessed, appropriate feedback is provided (positive, neutral, negative), and, if required, a sequence of scaffolding hints follows. The hints are in the form of questions that prompt the student to think. We follow constructivist theories of learning with early hints in the sequence being vague and then more and more informative if students are still floundering.

In the following sections, we introduce the *DeepCodeTutor*, explain its approach to assessment and scaffolding of self-explanations, and present a randomized controlled study of its effectiveness in helping learners better understand code and master programming concepts.

2   DeepCodeTutor: Automatic Scaffolding of Code Explanations

DeepCodeTutor aims to help students comprehend and explain the logical step and logical step details of a given code example. The system engages students in a dialog-based approach, prompting students to explain the code of the worked-out program example and reacting to student explanations. If a student provides a correct and complete explanation, she will receive positive feedback and a summary explanation of the code. If the student's explanation is incomplete or incorrect, the system uses scaffolding questions to guide their comprehension and learning and correct misunderstandings. The number of hints provided varies depending on the student's individual needs, understanding, and articulation. The system will show the model explanation if the student fails to explain the concept correctly, even after scaffolding.

The user interface of DeepCodeTutor consists of the following components. The goal description for the Java code example is displayed in the top left corner of the app. It is highlighted in red for immediate attention and easy visibility for students (Fig. 1, A). The interactive code editor (Fig. 1, B) displays the target code example the student should read, comprehend, and articulate. The code example is divided into logical blocks/chunks separated by empty lines. When a question is asked about a specific block/line of code, as shown in the figure, the target block is highlighted in yellow. On the right side of the interface (Fig. 1, C) is a display box that shows the entire dialogue history displaying the student's response in blue on the right, while the tutor's response is in green on the left. The student input box is at the bottom right corner of the interface (Fig. 1, D).

The tutor and the student discuss the code block by block. At the start of each task, the students are asked to explain the program in their own words. The student's initial explanation is then automatically assessed using automated semantic similarity methods, which compare the student's explanation to a benchmark, expert-provided explanation (e.g., the expert explanations in the DeepCode). The semantic similarity is computed by extending word-to-word semantic similarity measures to sentence and paragraph level [5]. The semantic similarity is calculated at the sentence and paragraph level by comparing a variety of features, including an alignment score based on the optimal alignment of the sentences using chunks and a branch-and-bound solution to the quadratic assignment problem, word embeddings, unigram overlap with synonym check, bigram overlap and BLEU scores [5]. To be considered complete and correct, students must effectively convey all the key ideas in experts' self-explanations. If a student articulates a major misconception which DeepCodeTutor will detect, using a bank of major misconceptions, and correct it immediately.

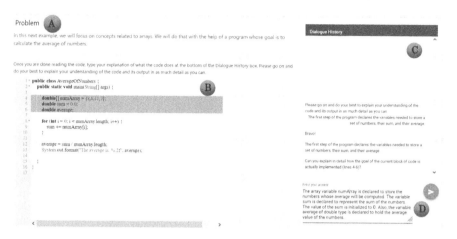

Fig. 1. A screenshot of the DeepCodeTutor Interface: It includes (A) The goal description for each task, (B) an interactive Java code editor that shows the current Java code example, and (C) a dialogue history of the interaction between the tutor and learners, and (D) an input box for the learner to type their responses. (Color figure online)

3 Experimental Design

To assess the value of DeepCodeTutor, we conducted a randomized controlled trial experiment in which participants were randomly assigned to two approximately equal groups: an experimental or treatment group that interacted with DeepCodeTutor (scaffolded self-explanations) and a control group in which participants were asked to read the expert-annotated code examples worked-out examples (no self-explanations elicited, no scaffolding offered). Both groups were presented with identical Java code examples, ensuring equal content exposure. The main outcome variable we focused on was the effectiveness of the two conditions in inducing learning gain, measured as an improvement in pre-to-post-test scores. The pre-and post-tests were identical and consisted of five short Java programs for which the student had to predict the output.

We recruited 90 students from an introductory Java Programming class in an undergraduate Computer Science program at a large public university in the United States. The participants were compensated with gift cards and extra credit for their participation. The study was conducted online by providing clear instructions about accessing and navigating the system. Students were asked to share their screens to ensure they followed the instructions correctly. A research team member was available to assist with any questions during the experiment. Out of all the participants, 14 identified themselves as non-native English speakers. Out of the 90 participants, 89 completed the task, 47 in the control group and 42 in the experimental group.

The overall experiment protocol was as follows. First, students were informed about the experiment, offered a chance to ask questions, and asked to sign a consent form if they agreed to proceed. Then, they completed a background questionnaire about their primary language of communication, programming experience, and current major. This was followed by a self-efficacy questionnaire and a pretest, which assessed students' prior knowledge of the targeted programming concepts. After that, the subjects proceeded to the main task, where they interacted with five worked code examples doing either code-reading or scaffolded self-explanation, depending on the group. After completing the main task, the students took a posttest targeting the same concepts that were covered in the pretest and main task. Finally, the students completed an evaluation survey to provide their perceptions of TutorApp. The system logged all student inputs and tracked the time associated with each action without recording any identifiable information.

We used the Normalized Learning Gain (NLG) as the main performance metric to evaluate Learning Gain. It allows for consistent analysis of diverse student populations with varying prior knowledge [4]. For the calculation of NLG, if posttest score >pretest score, NLG = (posttest-pretest)/(5-pretest). If posttest <pretest, NLG = (posttest-pretest)/pretest. We discard all the cases where the student's score is perfect, i.e., 5 in both pretest and posttest. While the pretest and post-test scores are not perfect but equal, then the NLG = 0.

4 Results

For the normalized learning gain analysis, data from 21 participants in the control group and 11 participants in the experimental group were excluded due to perfect pretest and posttest scores, and also excluded one participant for scoring 0 on both tests.

Table 1 shows the average pretest, posttest, normalized learning gain, and learning gain (posttest-pretest) for the experiment and control group. The distribution of pretest scores indicated a bias in our sample towards high-prior knowledge students, with more students having a high pretest score. Also, the self-efficacy scores of the students in the study were found to be generally high, with a mean of 3.96 (S.D = 0.48) for all students, 3.92 (S.D = 0.48) for the experimental group, and 4.01 (S.D = 0.48) for the control group. This suggests that our data is skewed toward high-prior knowledge and high self-efficacy participants.

4.1 How Effective is Automated Scaffolded Self-explanation for Code Comprehension and Learning?

We first examined the differences in learning between the experimental and control group using a t-test, which showed no significant difference in the normal-

Table 1. Mean and standard deviation of pretest, posttest, and normalized learning gain (NLG) and learning gain (posttest-pretest).

Group	n	Pretest		Posttest		NLG		Learning Gain	
		mean	SD	mean	SD	mean	SD	mean	SD
Experimental Group	30	3.0	1.41	3.56	1.19	0.26	0.40	0.56	1.08
Control Group	26	3.19	1.23	3.53	1.20	0.22	0.54	0.34	1.14

ized learning gain (t = 0.33, p = 0.36) or the post-test score (t = 0.08, p = 0.465) between the two groups. To better understand the effect size of DeepCodeTutor, we calculated Cohen's d for the learning gain, which was found to be a small effect size of 0.19 in favor of scaffolded self-explanation. While we hypothesized that using DeepCodeTutor would lead to better learning, the study has not confirmed this hypothesis. The lack of significant differences could result from several factors, for e.g., our sample was biased toward high-prior knowledge and high self-efficacy students. Furthermore, we conducted the experiment at the end of the semester, which means students had many chances to master the concepts.

4.2 Does the Use DeepCodeTutor Result in Different Learning Outcomes for Students with High and Low Prior Knowledge?

To examine the impact of DeepCodeTutor on students with different levels of prior knowledge, we split students in each experimental condition into two subgroups based on their prior knowledge (Med = 3.5 for the experimental group and Med = 3 for the control group). We conducted a t-test on the pretest score between low-prior knowledge (N = 15, M = 1.93, S.D = 1.27) and high-prior knowledge students (N = 15, M = 4.06, S.D = 0.25) in the experimental condition. The results show a significant difference in prior knowledge between the two groups (t = 6.32, p < 0.05), which validates our split using the median.

Table 2. Independent sample t-test for learning gain of low-prior and high-prior knowledge student in experimental and control group

Group	Prior Knowledge	N	mean	SD	t-val	Sig
Experimental	Low	15	0.46	0.33	2.91	**0.003**
	High	15	0.07	0.39		
Control	Low	14	0.22	0.52	0.02	0.49
	High	12	0.22	0.58		

As we can see in Table 2, there is a significant difference in the normalized learning gain between low and high prior knowledge students in the experimental group, whereas the average learning gain is the same for the control group. This suggests that the scaffolded self-explanation may be particularly helpful for students with lower levels of prior knowledge.

5 Conclusion and Future Work

In this paper, we presented a novel approach to engaging students in studying program examples and reported the results of its experimental evaluation. The key idea of our approach is to support student self-explanation of code fragments with automatic assessment and scaffolding. The results of the experiment show a statistically significant learning gain in low-prior-knowledge students in the experimental condition compared to high-prior-knowledge. In future work, we plan to investigate further the effectiveness of our technology.

Acknowledgements. This work has been supported by the following grants awarded to Dr. Vasile Rus: the Learner Data Institute (NSF award 1934745); CSEdPad: Investigating and Scaffolding Students' Mental Models during Computer Programming Tasks to Improve Learning, Engagement, and Retention (NSF award 1822816), and Department of Education, Institute for Education Sciences (IES award R305A220385). The opinions, findings, and results are solely those of the authors and do not reflect those of NSF or IES.

References

1. Beaubouef, T., Mason, J.: Why the high attrition rate for computer science students: some thoughts and observations. ACM SIGCSE Bull. **37**(2), 103–106 (2005)
2. Lopez, M., Whalley, J., Robbins, P., Lister, R.: Relationships between reading, tracing and writing skills in introductory programming. In: Proceedings of the Fourth International Workshop on Computing Education Research. ACM (2008)
3. McCracken, M., et al.: A multi-national, multi-institutional study of assessment of programming skills of first-year CS students. In: Working Group Reports from ITiCSE on Innovation and Technology in Computer Science Education, pp. 125–180 (2001)
4. Nissen, J.M., Talbot, R.M., Thompson, A.N., Van Dusen, B.: Comparison of normalized gain and Cohen's d for analyzing gains on concept inventories. Phys. Rev. Phys. Educ. Res. **14**(1), 010115 (2018)
5. Rus, V., D'Mello, S., Hu, X., Graesser, A.: Recent advances in conversational intelligent tutoring systems. AI Mag. **34**(3), 42–54 (2013)
6. Tamang, L.J., Alshaikh, Z., Khayi, N.A., Oli, P., Rus, V.: A comparative study of free self-explanations and Socratic tutoring explanations for source code comprehension. In: Proceedings of the 52nd ACM Technical Symposium on Computer Science Education, pp. 219–225 (2021)

Using Large Language Models to Provide Formative Feedback in Intelligent Textbooks

Wesley Morris[1]([⊠]) [iD], Scott Crossley[1] [iD], Langdon Holmes[1] [iD], Chaohua Ou[2] [iD], Danielle McNamara[3] [iD], and Mihai Dascalu[4] [iD]

[1] Vanderbilt University, Nashville, USA
wesley.g.morris@vanderbilt.edu
[2] Georgia Tech, Atlanta, USA
[3] Arizona State University, Tempe, USA
[4] University Politehnica of Bucharest, Bucharest, Romania

Abstract. As intelligent textbooks become more ubiquitous in classrooms and educational settings, the need arises to automatically provide formative feedback to written responses provided by students in response to readings. This study develops models to automatically provide feedback to student summaries written at the end of intelligent textbook sections. The study builds on Botarleanu et al. (2022), who used the Longformer Large Language Model, a transformer Neural Network, to build a summary grading model. Their model explains around 55% of holistic summary score variance when compared to scores assigned by human raters on an analytic rubric. This study uses a principal component analysis to distill scores from the analytic rubric into two principal components – content and wording. When training the models on the summaries and the sources using these principal components, we explained 79% and 66% of the score variance for content and wording, respectively. The developed models are freely available on HuggingFace and will allow formative feedback to users of intelligent textbooks to assess reading comprehension through summarization in real-time. The models can also be used for other summarization applications in learning systems.

Keywords: intelligent textbooks · large language models · automated summary scoring · transformers

1 Introduction

An essential component of intelligent textbooks is the capacity to provide formative feedback to students in real time regarding text comprehension. Recent developments in Natural Language Processing (NLP) allow more sophisticated feedback approaches based on open-ended assessments like text summarization. For instance, Crossley et al. [1] developed a summarization model to predict ratings of main idea integration in student summaries using lexical diversity features, a word frequency metric, and word2vec semantic similarity scores between summaries and source material. The model explained 53% of the variance in ratings. Botarleanu et al. [2] used large language models (LLMs)

© The Author(s), under exclusive license to Springer Nature Switzerland AG 2023
N. Wang et al. (Eds.): AIED 2023, CCIS 1831, pp. 484–489, 2023.
https://doi.org/10.1007/978-3-031-36336-8_75

to predict overall student summarization scores and explained ~55% of score variance. These NLP models show the potential for open-ended assessments of text comprehension through summarization in intelligent textbooks.

The goal of this paper is to introduce more robust LLMs to provide formative assessment for summaries written at the end of chapter sections within an intelligent textbook framework. The models presented in this study are more robust in two ways. First, they are trained on data developed specifically for the model. Second, instead of training them to predict each scale of an analytic rubric for summarization, the analytic scores in a rubric are aggregated into two criteria using a principal component analysis. This study aims to develop summarization models that can be integrated into intelligent textbooks to make them more interactive by providing actionable feedback to students about their summaries to increase comprehension of course material. In doing so, we develop two LLMs to provide formative assessment on summaries: 1) a model based on RoBERTa to predict scores based on the summary itself, and 2) a model using Longformer to predict summary scores while including text from the textbook as context. The research questions that guide this study are

- Are more robust LLMs able to provide more accurate models that can be used to guide student understanding as they read an intelligent textbook?
- Does the inclusion of text from the textbook improve the accuracy of the LLMs?

2   Methods

2.1   Data

Our summary corpus comprises 4,233 summaries of 101 source texts written by high school, university, and adult writers recruited through Amazon's Mechanical Turk service between 2018 and 2021. Source texts were on a variety of different topics, including the dangers of smoking, computer viruses, and the effect of UV radiation. The sources had a mean word count of 308.5 (SD = 130.49), while the summaries had a mean word count of 75.18 (SD = 50.51). Each summary was scored by expert raters using a 1–4 scaled analytic rubric to score 7 criteria important in understanding the quality of summarizations. The criteria included the main point (i.e. to what extent the summary captured the main idea of the source), details of language beyond the source (i.e. how well all relevant details were included in the summary), paraphrasing (i.e. avoiding plagiarism by paraphrasing the original material), objective language use (i.e. reflecting the point of view of the source), and cohesion (i.e. how well the summary was rationally and logically organized). Inter-rater reliability showed acceptable agreement among raters ($r > .8$ and $\kappa > .7$). A subset of this data set was used in Crossley et al. [1] and Botarleanu et al. [2].

A principal component analysis (PCA) was conducted to assess the potential for dimension reduction for the analytic scores in the rubric. The PCA revealed strong covariance allowing six of the scores to be combined into two principal components. The analytic scales of details, main point, and cohesion were combined into a weighted score designated as Content. The analytic scales for paraphrasing, objective language use, and language beyond the source were combined into a weighted score designated as Wording. The component scores were normalized to a scale from 0 to 1 using min-max normalization and used as outcome variables in our large language models.

2.2 Large Language Models

LLMs are neural network architectures for natural language processing which use the principle of self-attention to generate large, pre-trained models which can then be further finetuned for downstream tasks. These pre-trained models are trained on large corpora using masked language modeling, in which the text is tokenized, but some tokens are masked. The task of masked language modeling is to predict the masked tokens based on all the tokens that come before and after the masked token. After many epochs of training on very large corpora, the parameters of the model come to represent a general knowledge of the language domain on which they were trained.

The pretrained model can be further refined in two ways. The primary method of model refinement is through finetuning. The model is trained on the target task using the training data and labels in finetuning. Encoder-only transformers, such as those used in this study, include a special classification token at the beginning of the sequence. As the model processes the language data, the embedding of the classification token comes to represent semantic information about the text as a whole and can be used with the labels supplied in the training data to train a traditional machine learning algorithm. In finetuning, the model's parameters and the machine learning classification head are trained.

The other method, domain adaptation, is used when there is a large amount of unlabeled data but a relatively small amount of labeled data. In this case, the model is trained using a masked language model on language data from the target language domain in order to allow the model greater familiarity with the target domain. After domain adaptation, the resultant model is then finetuned on the labeled data for classification, regression, or other specific tasks.

We used RoBERTa [3] as our initial LLM, which is an encoder-only transformer model pretrained on the English Wikipedia corpus and Bookcorpus. The transformer neural network architecture relies on attention mechanisms in which, at every layer, each token embedding is modified by each other token embedding. As a result, the computational requirements grow quadratically as a function of the input sequence length. In RoBERTa, the length of the input sequence is limited to 512 tokens to ensure computing efficiency. While this length is sufficient for many summaries, it is not long enough to include text from the textbook in the model input.

The Longformer LLM [4] is capable of handling longer input sequences by utilizing sparse attention, in which not all tokens are compared with every other token. Instead, Longformer uses a sliding attention window so that each token only attends to the tokens a certain number of positions to its left and right. Sparse attention mitigates the problem of limited sequence length by reducing the computational complexity of the attention mechanism. Additionally, Longformer utilizes global attention in which certain tokens are attended to by every other token. Combining these two types of sparse attention allows Longformer to increase the max sequence length from 512 tokens to 4,096 tokens while remaining efficient. The Longformer max sequence length allows us to include both the summary and text from the textbook into the input sequence.

We divided the summary corpus into training, validation, and test sets. To help ensure generalizability across source texts and prompts, we selected 15 out of the 101 sources text to comprise the test set only (i.e., these source texts were not used in training or validation) After splitting the data in this way, the training, validation, and test sets comprised 3,285, 703, and 702 summaries respectively. During finetuning, each summary in RoBERTa was tokenized and fed into the model. For Longformer, the summary and the source text for the summary were concatenated using a special separator token and then tokenized together to generate the input sequences. These token sequences were used as input data for their respective models, and the final classification token was used to train a linear regression head. After training, we tested the performance of each model by predicting the Content and Wording scores for the summaries. We evaluate model performance in terms of correlation with the human rater judgments and explained variance (R^2).

In addition to the finetuning procedures described above, we also domain-adapted the Longformer pre-trained model on a set of 93,484 summaries written by middle and high-school students. The summaries were collected from six online sources through the Commonlit platform (commonlit.org) [5]. This is a large, unlabeled dataset in the target language domain, and we considered it a reasonable candidate for domain adaptation. After constructing the domain-adapted model, we finetuned it using the same methods described above and evaluated its performance by calculating the correlation between predicted scores and human rater judgments.

3 Results

For Content scores, the Longformer model, in which both the summary and the source were included in the input, achieved higher accuracy ($r = .89$, $R^2 = .79$) than the RoBERTa model, which only included the summary ($r = .82$, $R^2 = .67$). For Wording

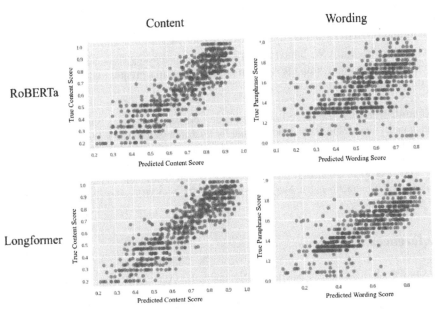

Fig. 1. Predicted Scores Plotted Against Actual Scores for the Four Models

scores, the Longformer model outperformed ($r = .81$, $R^2 = .66$) the RoBERTa model ($r = .60$, $R^2 = .36$) (see Fig. 1). The domain-adapted Longformer model performed worse than the non-domain adapted model for both content ($r = .85$, $R^2 = .72$) and wording ($r = .78$, $R^2 = .60$).

4 Discussion

The Longformer models developed in this study outperform RoBERTa models and previous LLM-based automatic summary evaluation models. For instance, previous NLP models have achieved $R^2 \sim 55\%$ (citations). In addition, finetuning directly on the pretrained model produced better results than finetuning on the domain-adapted model. This may be because the Commonlit dataset, which included many summaries, only included six sources, which did not provide the variance needed for the problem space. The small number of sources may have resulted in catastrophic forgetting [6], where the model overfitted to those sources and forgot part of the parameters set during pretraining.

5 Conclusions and Future Work

Although the data used in the training and test sets for this model are not exactly the same as the task of grading summaries of textbook sections, the accuracy rates for Content scores (and likely Wording scores) are strong enough for inclusion into intelligent textbooks to provide students with opportunities for open-ended comprehension assessment and interactive feedback. We plan to integrate this model into a prototype intelligent textbook currently in development. Students will be required to write a short summary at the end of each section before moving on to the next. After passing through a filter ensuring that the summaries are more than fifty words, in English, not plagiarized, and on topic, the summaries will be automatically graded on the two criteria, and students will receive their grades instantly. We hope to demonstrate that providing students with feedback on their summaries along with the revision process leads to a greater understanding of texts within intelligent textbooks.

Acknowledgements. This material is based upon work supported by the National Science Foundation under Grant 2112532 and the Institute for Education Science under Grant R305A180261. Any opinions, findings, and conclusions or recommendations expressed in this material are those of the author(s) and do not necessarily reflect the views of the National Science Foundation.

This work was supported by the Ministry of Research, Innovation, and Digitalization, project CloudPrecis, Contract Number 344/390020/ 06.09.2021, MySMIS code: 124812, within POC.

References

. Crossley, S.A., Kim, M., Allen, L., McNamara, D.: Automated summarization evaluation (ASE) using natural language processing tools. In: Isotani, S., Millán, E., Ogan, A., Hastings, P., McLaren, B., Luckin, R. (eds.) AIED 2019. LNCS (LNAI), vol. 11625, pp. 84–95. Springer, Cham (2019). https://doi.org/10.1007/978-3-030-23204-7_8

2. Botarleanu, R.M., Dascalu, M., Allen, L.K., Crossley, S.A., McNamara, D.S.: Multitask summary scoring with longformers. In: Rodrigo, M.M., Matsuda, N., Cristea, A.I., Dimitrova, V. (eds.) AIED 2022. LNCS, vol. 13355, pp. 756–761. Springer, Cham (2022). https://doi.org/10.1007/978-3-031-11644-5_79

3. Liu, Y., et al.: RoBERTa: a robustly optimized BERT pretraining approach. arXiv:1907.11692 [Cs]. http://arxiv.org/abs/1907.1169 (2019)

4. Beltagy, I., Peters, M.E., Cohan, A.: Longformer: The Long-Document Transformer arXiv: 2004.05150. arXiv. https://doi.org/10.48550/arXiv.2004.05150 (2020)

5. Crossley, S.A., Heintz, A., Choi, J., Batchelor, J., Karimi, M., Malatinszky, A.: The CommonLit ease of readability (CLEAR) corpus. In: Educational Data Mining (2021)

6. Ramasesh, V.V., Lewkowycz, A., Dyer, E.: Effect of scale on catastrophic forgetting in neural. In: International Conference on Learning Representations, September 2021

Utilizing Natural Language Processing for Automated Assessment of Classroom Discussion

Nhat Tran[✉], Benjamin Pierce, Diane Litman, Richard Correnti, and Lindsay Clare Matsumura

University of Pittsburgh, Pittsburgh, USA
{nlt26,bep51,dlitman,rcorrent,lclare}@pitt.edu

Abstract. Rigorous and interactive class discussions that support students to engage in high-level thinking and reasoning are essential to learning and are a central component of most teaching interventions. However, formally assessing discussion quality 'at scale' is expensive and infeasible for most researchers. In this work, we experimented with various modern natural language processing (NLP) techniques to automatically generate rubric scores for individual dimensions of classroom text discussion quality. Specifically, we worked on a dataset of 90 classroom discussion transcripts consisting of over 18000 turns annotated with fine-grained Analyzing Teaching Moves (ATM) codes and focused on four Instructional Quality Assessment (IQA) rubrics. Despite the limited amount of data, our work shows encouraging results in some of the rubrics while suggesting that there is room for improvement in the others. We also found that certain NLP approaches work better for certain rubrics.

Keywords: Classroom discussion · Quality assessment · NLP

1 Introduction and Background

Instructional quality has been of great interest to educational researchers for several decades. Due to their cost, measures of instructional quality that can be obtained at-scale remain elusive. Previous work [5] has shown that providing automated feedback on teachers' talk moves can lead to positive instructional changes. We report here on initial attempts to apply Natural Language Processing (NLP) methods such as pre-trained language models or sequence labeling with BiLSTM [4] to automatically produce rubric scores for individual dimensions of classroom discussion quality from transcripts, building upon two established measures that have shown high levels of reliability and validity in prior learning research - the *Instructional Quality Assessment (IQA)* and the *Analyzing Teaching Moves (ATM)* rubrics [2,6].

Our corpus consists of 170 videos from 31 English Language Arts classrooms in a Texas district. 18 teachers taught fourth grade, 13 taught fifth grade, and on

Supported by a grant from the Learning Engineering Tools Competition.

Table 1. Data distribution and mean (**Avg**) of 4 focused *IQA* rubrics for Teacher (*T*) and Student (*S*) with their relevant *ATM* codes. An *IQA* rubric's distribution is represented as the counts of each score (1 to 4 from left to right) (n=90 discussions).

Rubric			Relevant ATM code	
Short Description	Distribution	Avg	Code Label	Count
S1: *T* connects *S*s	[51, 19, 9, 11]	1.8	Recap or Synthesize S Ideas	75
S2: *T* presses *S*	[7, 7, 9, 67]	3.6	Press	927
S3: *S* builds on other's idea	[65, 6, 8, 11]	1.6	Strong Link	101
S4: *S* support their claims	[28, 12, 8, 42]	2.7	Strong Text-based Evidence	403
			Strong Explanation	286
-			Others	52687

average had 13 years of teaching experience. The student population was considered low income (61%), with students identifying as: Latinx (73%), Caucasian (15%), African American (7%), multiracial (4%), and Asian or Pacific Islander (1%). The videos were manually scored holistically, on a scale from 1 to 4, using the *IQA* on 11 dimensions for both teacher and student contributions. They were also scored using more fine-grained talk moves at the sentence level using the *ATM* discourse measure. The current work focuses on only the **90** discussion transcripts that have already been converted to text-based codes.

As summarized in column 1 of Table 1, our classifiers are trained to predict 4 of the 11 *IQA* rubrics containing aspirational teacher (T) and student (S) 'talk moves' - *T Links Student Contributions* (score S1), *T Presses for Information* (S2), *S Link Contributions* (S3), *S Support Claims with Evidence* (S4). Besides the 5 *ATM* codes (column 4 in Table 1) related to these 4 *IQA* rubrics the rest are labeled as *Others*. The distributions of *IQA* scores for each rubric and of relevant *ATM* codes are summarized in columns 2 and 5 of Table 1, respectively. We notice that the frequencies of ATM codes related to S1 and S3 are very low (less than or approximately 1 per transcript). Below is an example excerpt with annotated *ATM* codes from our data:

Teacher: [The girls get the water and the boys do the herds, right?]Others [Where did you get that from the text?]Press

Student: [Other people, mostly women and girls who had to come fill their own containers, many kinds of birds, all flap, twittering and cawing. Herds of cattle had been brought to good grazing by the young boys who looked after them.]StrongText−BasedEvidence

In this paper we present several *IQA* classifiers, and show that using predicted *ATM* codes as features outperforms an end-to-end model. The long-term goal of our work is to use such classifiers in a tool for automated *IQA* assessment so that teachers and coaches can evaluate classroom discussion quality in real-time.

2   Methods

We train different models for *IQA* assessment to explore tradeoffs between scoring performance, explainability, and training set. Our baseline is a neural **end-to-end model**, as neural models often have high performance and do not require feature engineering. However, since the *IQA* score of a specific rubric can be inferred from the number of times the relevant *ATM* codes are used (Sect. 1), we also develop **IQA prediction models using ATM codes** as predictive features. This in turn requires *ATM* models to predict the relevant 6 *ATM* codes. Specifically, for each sentence, these models will predict 1 of 6 *ATM* code labels in column 4 of Table 1. This is a 6-way classification task.

Hierarchical *ATM* Classification. We hypothesize that it would be easier to separate *Others* from the 5 focal *ATM* codes as they have specific usages. We perform a 2-step hierarchical classification at sentence level as follows. Step 1, binary classification, is to classify *Other* versus *5 focal ATM Codes*. If the code is not *Others*, step 2 is to perform another 5-way classification to identify the final label. We train separate BERT-based classifiers for each step. The input for the classifiers is the current sentence concatenated with previous sentences in the same turn and sentences from one previous turn. Because each *ATM* code except *Others* can be only from one speaker, either Teacher or Student, we train two classifiers for the 5-way classification of Step 2, one classifier used to predict teacher codes (*Recap or Synthesize S Ideas* and *Press*) and one classifier specialized in student codes (*Strong Link, Strong Text-Based Evidence* and *Strong Explanation*). Depending on the speaker, only one of them is called for Step 2.

ATM Sequence Labeling. A classroom discussion can be considered as a sequence of sentences. This approach assigns a label (1 out of the 6 *ATM* codes) to each sentence in a conversation. Unlike the Hierarchical Classification approach that predicts the label of each sentence independently, in this approach, the label of a given sentence is more dependent on the labels of nearby sentences. We use BERT-BiLSTM-CRF as our sequence labeling model. BiLSTM-CRF has been widely used for sequence labeling tasks [4] and BERT [3] provides a powerful tool for sentence representation that can work well with that architecture.

Additional Techniques. During *ATM* classification, since *Others* constitutes more than 90% of the total labels, we **downsample** the training data to reduce the imbalance. For *IQA* classification, annotators tend to group consecutive sentences sharing the same functionality in one turn as one *ATM* code (e.g., one *Strong Texted-Based Evidence* code is used for two sentences in the excerpt in Sect. 1), but our prediction is done on sentence level. **Merging adjacent *ATM* predictions that are the same into one code** in the inference phase helps preserve this nature. Also, since the range of the *IQA* scores is very small (1 to 4), translating the absolute counting of *ATM* codes to *IQA* scores (see Sect. 1) can drastically shift the *IQA* scores due to misclassification of the *ATM* codes. To alleviate this sensitivity, we build separate **linear regression models to estimate each *IQA* score from the counting of relevant *ATM* codes**, then use the nearest integers as the *IQA* scores.

3   Results

Table 2. *ATM* Codes 6-way Classification Results (F_1 scores over 5-fold cross-validation with standard deviations in parentheses).

Method	Step 1	6-way
Non-hierarchical Classification (All Data)	–	0.29 (0.04)
/w 60% downsampling	–	0.49 (0.03)
Hierarchical Classification (All Data)	0.56	0.41 (0.04)
/w 60% downsampling	0.72	0.65 (0.02)
Sequence Labeling (All Data)	–	0.45 (0.03)
/w 60% downsampling	–	0.62 (0.01)
Hierarchical with Oracle for Step 1 (All Data)	1	0.57 (0.02)
/w 60% downsampling	1	0.68 (0.04)

***ATM* Code Prediction** results (macro average F_1 scores) are shown in Table 2. The Non-hierarchical baseline is a BERT-based 6-way classifier given the same input as our hierarchical approach. Both Hierarchical Classification and Sequence Labeling outperform the Non-hierarchical baseline, whether using all data or downsampled data for training. The numbers also show that downsampling the proportion of the most popular class (*Others*) to 60% increases the performance of all models[1]. For the 2-step Hierarchical Classification approach, it improves the performances of both step 1 and the final 6-way classification. Additionally, the Sequence Labeling and Hierarchical approaches perform similarly. Although Sequence Labeling has a slightly higher score when all training data is used (0.45 vs. 0.41, $\rho = 0.039$), it is inferior to Hierarchical Classification with 60% downsampling (0.62 vs. 0.65, $\rho = 0.046$). Using a perfect Oracle model with 100% accuracy for Step 1 (*Others* versus 5 *ATM codes*) does not lead to a large gain in the 6-way classification results compared to our fully automated Hierarchical approach in the downsampled version (0.68 vs 0.65). We thus use the non-oracle models built with 60% downsampling rate for the inference of the classroom discussion quality (*IQA* scores) below.

***IQA* Score Prediction** is performed based on the models for *ATM* codes prediction. The Quadratic Kappa (QK) scores for the estimations of the four rubrics of classroom discussion quality are reported in Table 3. The baseline for each rubric is an end-to-end Longformer model [1] which directly predicts the *IQA* scores given the raw text transcripts using a linear layer on top of the hidden representation of [CLS], ignoring the *ATM* codes. The results show that all variations of Hierarchical Classification and Sequence Labeling outperform the baselines, which emphasizes the importance of utilizing *ATM* codes to infer *IQA* scores. Besides increasing performance, the ATM-based models also increase model interpretability, useful for generating formative feedback in the future.

---

[1] We tried different ratios and 60% provides the best results.

Table 3. *IQA* Scores Estimation Results in Quadratic Kappa (QK) averaged over 5-fold cross-validation, inferred from Absolute Counting (A.Count) and Linear Regression (Regression) after *ATM* prediction. **Bold** numbers are the best results for each rubric.

Rubric	Baseline	Hierarchical		Sequence	
		A.Count	Regression	A.Count	Regression
S1: Teacher connects Students	0.34	0.43	0.54	0.50	0.55
/w merged codes		0.48	0.55	0.52	**0.57**
S2: Teacher presses Student	0.35	0.60	0.65	0.55	0.62
/w merged codes		0.64	**0.68**	0.57	0.63
S3: Student builds on each other	0.30	0.42	0.51	0.47	0.51
/w merged codes		0.49	**0.54**	0.50	0.53
S4: Student support their claims	0.36	0.60	0.65	0.57	0.61
/w merged codes		0.65	**0.70**	0.61	0.63

One notable observation is that using regression to estimate the *IQA* scores is always better than using absolute counting. This supports our assumption that regression will alleviate the sensitivity of miscounting and provides a smoother transition from the number of times *ATM* codes appear to the actual *IQA* scores. Using regression, the highest gain in QK scores for Hierarchical Classification (0.09) and Sequence Labeling (0.07) are from S3 and S2, respectively.

Merging consecutive same *ATM* codes into one also improves the performance of *IQA* score estimation as expected. For the same approach, the increases from this technique are mostly larger when absolute counting is used. The *ATM* classification was on sentence level and absolute counting is more sensitive to over-counting, so the merging technique is more effective for this method.

While Sequence Labeling yields the best S1 results, the best results for the other rubrics come from Hierarchical Classification. Our reasoning is that for S1, the relation to adjacent sentences plays a more important role to identify the relevant *ATM* code as there should be multiple students speaking out their ideas before the teacher can connect/synthesize them. Thus, Sequence Labeling, which focuses more on dependencies between sentences, performs better. This suggests that certain approaches are more suitable for certain rubrics.

Finally, Fig. 1 demonstrates the performance of our best models (with *regression* and *merged codes*) over training size. While the baselines do not improve much after a certain size of training data, the lines of Hierarchical Classification and Sequence Labeling maintain upward trends, suggesting that these models will continue to benefit from more data as we complete our video transcription. Even using only 90 discussions, the QK results show that *ATM*-based model reliability is already substantial for S2 and S4, and moderate for S1 and S3, even though there were infrequent instances of relevant codes in the corpus.

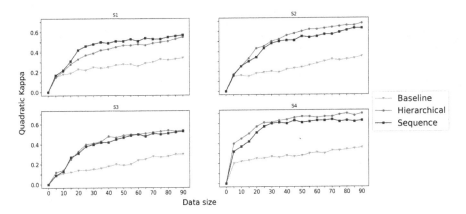

Fig. 1. *IQA* Score Estimation Results (QK) in relation to the amount of training data.

4   Conclusion and Future Directions

We experimented with NLP approaches to automatically assess discussion quality using the *IQA*, and to deal with imbalanced *ATM* data and imperfect *ATM* code prediction. Our results show that *IQA* models using either Hierarchical Classification or Sequence Labeling to first predict *ATM* codes outperform baseline end-to-end *IQA* models, while each of the ATM-based *IQA* models performs better than the other in certain *IQA* rubrics. Once the full corpus is available, we will generate a validity argument for whether automated scoring replicates known associations in the corpus, and incorporate the demographic information into our analyses. We will also utilize *ATM* codes beyond the 5 relevant to the focused rubrics to add more context for *ATM* prediction. To mitigate the limited size of even the full corpus, we will explore whether techniques such as transfer learning that can take advantage of classroom discussion data from math [8] or high school [7] that are now being made available to the community.

References

1. Beltagy, I., Peters, M.E., Cohan, A.: Longformer: The long-document transformer. arXiv:2004.05150 (2020)
2. Correnti, R., Matsumura, L.C., Walsh, M., Zook-Howell, D., Bickel, D.D., Yu, B.: Effects of online content-focused coaching on discussion quality and reading achievement: Building theory for how coaching develops teachers' adaptive expertise. Read. Res. Q. **56**(3), 519–558 (2021)
3. Devlin, J., Chang, M.W., Lee, K., Toutanova, K.: BERT: Pre-training of deep bidirectional transformers for language understanding. In: Conference of the North American Chapter of ACL: Human Language Technologies, pp. 4171–4186, Minneapolis (2019)
4. Huang, Z., Xu, W., Yu, K.: Bidirectional LSTM-CRF models for sequence tagging. CoRR arXiv: abs/1508.01991 (2015)

5. Jacobs, J., Scornavacco, K., Harty, C., Suresh, A., Lai, V., Sumner, T.: Promoting rich discussions in mathematics classrooms: Using personalized, automated feedback to support reflection and instructional change. Teach. Teach. Educ. **112**, 103631 (2022)
6. Matsumura, L.C., Garnier, H.E., Slater, S.C., Boston, M.D.: Toward measuring instructional interactions "at-scale". Educ. Assess. **13**(4), 267–300 (2008)
7. Olshefski, C., Lugini, L., Singh, R., Litman, D., Godley, A.: The Discussion Tracker corpus of collaborative argumentation. In: Language Resources & Evaluation (2020)
8. Suresh, A., et al.: Using AI to promote equitable classroom discussions: The Talk-Moves application. In: International Conference on Artificial Intelligence in Education, pp. 344–348 (2021)

It's Good to Explore: Investigating Silver Pathways and the Role of Frustration During Game-Based Learning

Nidhi Nasiar[1]([✉]), Andres F. Zambrano[1], Jaclyn Ocumpaugh[1], Stephen Hutt[2], Alexis Goslen[3], Jonathan Rowe[3], James Lester[3], Nathan Henderson[3], Eric Wiebe[3], Kristy Boyer[4], and Bradford Mott[3]

[1] University of Pennsylvania, Philadelphia, USA
nasiar@upenn.edu
[2] University of Denver, Denver, USA
[3] North Carolina State University, Raleigh, USA
[4] University of Florida, Gainesville, USA

Abstract. Game-based learning offers rich learning opportunities, but open-ended games make it difficult to identify struggling students. Prior work compares student paths to a single expert's "golden path." This effort focuses on efficiency, but additional pathways may be required for learning. We examine data from middle schoolers who played Crystal Island, a learning game for microbiology. Results show higher learning gains for students with exploratory behaviors, with interactions between prior knowledge and frustration. Results have implications for designing adaptive scaffolding for learning and affective regulation.

Keywords: Game-based learning · pathways · frustration

1 Introduction

Game-based learning enhances engagement and learning, but its flexibility makes identifying struggling students challenging [16]. Researchers seeking to develop adaptive scaffolding for these environments have studied these issues using sequence mining, random walk analysis [15], and comparisons to an expert's most efficient solution—called a "golden path" [16]. However, efficient gameplay does not necessarily improve learning [14]. This study examines "silver pathways" likely to enhance learning by comparing [16]'s "golden path" to the paths taken by students with high and low learning gains in a game called Crystal Island (CI). We also examine how prior knowledge and frustration affect the relationship between these pathways and learning.

2 Related Work

2.1 Inquiry Learning in Game-Based Environments

Inquiry-based learning, a foundational pedagogical approach, can be effective, but concerns about interest and cognitive load make it difficult to develop adaptive scaffolds. Prior work in [2] has modeled scientific inquiry skills in immersive virtual lab environments. This study is situated in [13]'s framework, which emphasizes the role of exploration and experimentation. [16] compares students' trajectories in CI to a golden path, hypothesizing that expert solutions are more efficient than novices, and acting as a proxy to determine whether a student's path reflects such expertise (Fig. 1).

Fig. 1. Overview of Crystal Island with expert "golden path" as operationalized by [16].

2.2 Frustration and Learning

Frustration is known to be crucial to learning, and has been investigated using various tools [8]. The NASA TLX survey [6] (used in this work), was developed to explore astronauts' cognitive load, but has recently been used to measure negative emotions in multiple domains [7]. It assesses students' retrospective perception of their cognitive load, a measure affected by prior knowledge, time pressure, and mental effort (which can also impact frustration). Empirical studies using affect detectors show that the affective transitions in the most popular theoretical model [5] are not common, both the epistemic emotions themselves (e.g., frustration) and their hypothesized transitions (when present) appear to have strong relationships with learning [3, 10]. In general, frustration appears to have a 'Goldilocks Effect,' where either too much or too little leads to lower learning [5, 11]. Successful interventions have been designed [4], suggesting that understanding the relationship between frustration and learning in complex problem-solving can be used to scaffold learning further.

3 Methods

The study uses previously analyzed data from Crystal Island (CI) [12], a learning game for middle school microbiology that promotes inquiry-based learning by having students assume the role of a medical detective investigating an outbreak on a remote island. In this single-player game, students are tasked with identifying the disease and its source of transmission. To do so, they must travel to different locations on the island, collecting data and examining other information sources. The original data included 92 middle-school students enrolled in an urban public school in the southeastern US. Due to missing post-tests, 26 were excluded, leaving 66 students for analysis.

3.1 Measures of Learning and Frustration

Each student completed identical pre and post-tests (scaled from 0 to 13), which were used to calculate normalized learning gain using Eq. 1 [18].

$$\text{Norm\_gain} = \begin{cases} \frac{post-pre}{1-pre}post > pre \\ \frac{post-pre}{pre}post \leq pre \end{cases} \quad (1)$$

Students also completed surveys of interest and engagement, which included an adapted version of the NASA-TLX. This study examines the NASA-TLX frustration scale, which asks students to self-report a range of negative emotions on a scale of 0–100. We dichotomize these measures (i.e., high vs. low) by splitting on the median: learning gains (*0.15*), pretest (*8*), and frustration (*31.5*).

3.2 Operationalizing and Comparing Student Learning Pathways

In order to exclude brief erratic movements between locations, we only consider pathways to locations where students stayed for at least 20 s. We also consider self-loops, defined as locations where the student stays for at least 10 min (a threshold chosen by dividing the mean of total gameplay time (65 min) by the 6 locations). Finally, we identify paths followed by at least half of the students. To compare the paths of different student groups, we use two metrics: *similarity* and *density*. Similarity measures the number of common transitions between two graphs over the total number of possible transitions [17]. It ranges from 0–1, where 1 indicates identical graphs. Density also ranges from 0–1, and measures student exploration by dividing the number of transitions by the total possible transitions [19].

4   Analysis and Results

Our goal is to identify silver pathways—those that are less efficient but improve learning. We first compare students with high and low learning gains, but learning gains are impractical for triggering real-time interventions as students must complete the game for it to be calculated. Therefore, we compare pathways between students with high and low prior knowledge (i.e., pre-test scores), as moderated by frustration.

4.1 Pathways Differences in Students with High and Low Learning Gains

Figure 2 shows the common paths for students with high and low learning gains, respectively. We note that the full golden path appears for those with high learning gains, but 5/6 golden path transitions are also found among students with low learning gains, suggesting this path is not sufficient for improving learning. Both graphs in Fig. 2 have high similarity to the golden path (*0.71 vs. 0.80* for high vs. low learners). Instead, the main difference between these groups is in density scores (*0.47 vs. 0.42* for high vs. low learners). Low learners also have more self-transitions.

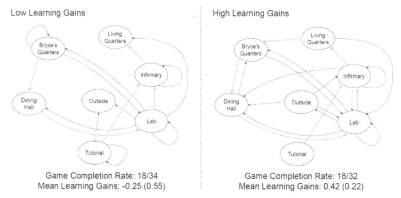

Fig. 2. Differences in pathways based on learning gains. Arrows denote paths taken by at least 50% of the students in that group; those that mirror [16]'s golden path are in yellow.

4.2 Pathway Differences Based on Prior Knowledge and Frustration

We next considered interactions with prior knowledge and frustration. Table 1 shows the similarity and density scores for different groups, while Fig. 3 shows their common learning pathways. Students are divided by both prior knowledge, (i.e. pre-test) and frustration. Specifically, Table 1 shows the similarity of different groups' learning pathways to the golden path and high learning gains path. These results show that students with low frustration and high prior knowledge have greater similarity to the golden path. Students with high prior knowledge also show greater similarity to the high learning graph. Likewise, Fig. 3 shows that high prior knowledge learners complete most of the golden path, but also other paths. Notably, the group with the highest learning gains (low pretest and low frustration) completed only 4/6 golden path transitions—but not those from the infirmary to the living quarters or from Bryce's quarters to the dining hall. This group also shows a self-loop (i.e., >10 min) in the tutorial.

When controlling for frustration, high prior knowledge learners tend to have slightly higher density scores than the low prior knowledge learners (Table 1). Higher frustration seems to increase density (*0.39 vs. 0.42* for low and high frustration with low pretests, respectively; *0.44 vs. 0.56* for low and high frustration with high pretests). Apparent exploratory behaviors (i.e., more paths) do not necessarily optimize learning, as learning gains trend slightly higher among those with low frustration than high frustration (*0.10 vs. −0.15* learning gains for low and high frustration learners with high pre-tests; *0.24 vs. 0.13* learning gains for low and high frustration learners with low pre-tests.

Table 1. Density and similarity scores of pathway analyses by frustration and prior knowledge.

	Low Frustration		High Frustration	
	Low PK	High PK	Low PK	High PK
Density	0.39	0.44	0.42	0.56
Similarity (Golden Path)	0.69	0.78	0.67	0.69
Similarity (High Learning Path)	0.78	0.86	0.80	0.94

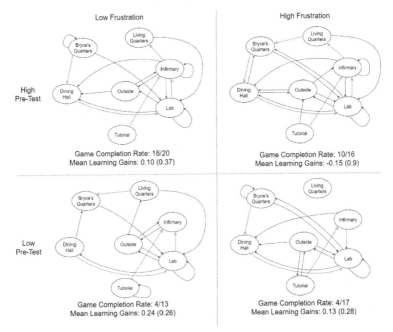

Fig. 3. Differences in pathways based on pre-test and frustration. Arrows denote paths taken by at least 50% of students in that group; those that mirror [16]'s golden path are in yellow.

5   Discussion and Conclusion

Deciding when to support student learning in a game-based environment is challenging. Prior work comparing student behaviors to experts [16] is important, but we should not expect novices to mirror expert behavior fully. This study builds on research showing that those with high prior knowledge replicate parts of [16]'s golden path through CI, but we show that exploration may be better for learning than efficiency. We also show that frustration interacts with learning gains by increasing what otherwise looks like exploratory behaviors. Future work should evaluate other frustration measures, including *in situ* measures used in a large body of work [9]. The single, retrospective measure administered in this study makes it difficult to evaluate the degree to which students' ultimate success (or failure) in the game has influenced their memory of this construct.

In situ frustration measures would also provide more nuance, allowing us to differentiate between intentional and haphazard exploratory behaviors. Such distinctions are important when developing adaptive scaffolds. Finally, *in situ,* measures would allow us to explore whether the Goldilocks effect (i.e., "just right" levels of frustration) varies across student groups. Given the range of variables that frustration interacts with, some learners may tolerate lower levels of frustration than others (e.g., [1]). This study has implications for improving interventions within gameplay based on frustration.

Acknowledgement. This study was supported by NSF under IIS grant Award #2016943, Award #2016993, and #1409639. Any opinions, findings, and conclusions or recommendations expressed in this material are those of the author(s) and do not necessarily reflect the views of the NSF.

References

1. Andres, J.M., Hutt, S., Ocumpaugh, J., Baker, R.S., Nasiar, N., Porter, C.: How anxiety affects affect: a quantitative ethnographic investigation using affect detectors and data-targeted interviews. In: International Conference on Quantitative Ethnography, pp. 268–283 (2021)
2. Baker, R., Clark-Midura, J., Ocumpaugh, J.: Towards general models of effective science inquiry in virtual performance assessments. J. Comput. Assist. Learn. **32**(3), 267–280 (2016)
3. Baker, R., et al.: Affect-targeted interviews for understanding student frustration. In: Proceedings of the International Conference on Artificial Intelligence & Education (2021)
4. DeFalco, J., et al.: Detecting and addressing frustration in a serious game for military training. Int. J. Artif. Intell. Educ. **28**(2), 152–193 (2018)
5. D'Mello, S., Graesser, A.: Dynamics of affective states during complex learning. Learn. Instr. **22**(2), 145–157 (2012)
6. Hart, S.G., Staveland, L.E.: Development of NASA-TLX (task load index): results of empirical and theoretical research. Adv. Psychol. **52**, 139–183 (1988)
7. Hart, S.G.: NASA-task load index (NASA-TLX); 20 years later. In: Proceedings of the Human Factors & Ergonomics Society Annual Meeting, vol. 50, no. 9, pp. 904–908 (2006)
8. Hutt, S., Grafsgaard, J., D'Mello, S.: Time to scale: Generalizable affect detection for tens of thousands of students across an entire school year. In: Proceedings of the 2019, CHI, pp. 1–14 (2019)
9. Jensen, E., Hutt, S., D'Mello, S.: Generalizability of sensor-free affect detection models in a longitudinal dataset of tens of thousands of students. In: International EDM (2019)
10. Karumbaiah, S., Baker, R.S., Ocumpaugh, J.: The case of self-transitions in affective dynamics. In: Proceedings of the 20th International Conference on Artificial Intelligence in Education, pp. 172–181 (2019)
11. Liu, Z., Pataranutaporn, V., Ocumpaugh, J., Baker, R.S.J.d.: Sequences of frustration and confusion, and learning. In: Proceedings of the 6th International Conference on Educational Data Mining, pp. 114–120 (2013)
12. Min, W., et al.: Multimodal goal recognition in open-world digital games. In: Thirteenth Artificial Intelligence and Interactive Digital Entertainment Conference (2017)
13. Pedaste, M., et al.: Phases of inquiry-based learning: definitions and the inquiry cycle. Educ. Res. Rev. **14**, 47–61 (2015)
14. Sabourin, J., Mott, B., Lester, J.: Discovering behavior patterns of self-regulated learners in an inquiry-based learning environment. In: International Conference on Artificial Intelligence in Education, pp. 209–218 (2013)

15. Snow, E., Likens, A., Jackson, T., McNamara, D.: Students' walk through tutoring: Using a random walk analysis to profile students. In: Proceedings of the International Conference on Educational Data Mining (2013)
16. Sawyer, R., Rowe, J., Azevedo, R., Lester, J.: Filtered time series analyses of student problem-solving behaviors in game-based learning. In: International EDM (2018)
17. Tversky, A.: Features of similarity. Psychol. Rev. **84**(4), 327 (1977)
18. Vail, A.K., Grafsgaard, J.F., Boyer, K.E., Wiebe, E.N., Lester, J.C.: Predicting learning from student affective response to tutor questions. In: Micarelli, A., Stamper, J., Panourgia, K. (eds.) ITS 2016. LNCS, vol. 9684, pp. 154–164. Springer, Cham (2016). https://doi.org/10. 1007/978-3-319-39583-8_15
19. Voloshin: Introduction to graph theory. Nova Science Publishers (2009)

Ghost in the Machine: AVATAR, a Prototype for Supporting Student Authorial Voice

Jasbir Karneil Singh[(✉)] [iD], Ben K. Daniel [iD], and Joyce Hwee Ling Koh [iD]

University of Otago, Dunedin, New Zealand
jaz.singh@postgrad.otago.ac.nz

Abstract. This case study reports the results of a newly developed academic writing software for developing an authorial voice for first-year undergraduates in Fiji. The AVATAR prototype is a student-facing learning dashboard designed to help first-year university students become aware of and reflect on the concept of authorial voice in English argumentative essays. This study asked 23 non-native speakers of first-year English university students to use AVATAR. Pre- and post-use authorial confidence data was captured via questionnaires. Key findings reveal that students' use of AVATAR significantly increased their awareness of the value of authorial voice. The results have implications for academic writing and the development of student authorial voice.

Keywords: authorial voice · academic writing · educational technology

1 Authorial Voice in Non-native English Argumentative Essays

1.1 Introduction

There are always ghosts in the machine. Who is it we hear when we are reading something written? How do we 'hear' the person behind the words? These questions revolve around the concept of authorial voice. Voice in writing is critical to successful academic writing and has been the subject of constant pedagogical research and development. Given its significance in higher education, developments in educational technology must provide affordances for the holistic development of authorial voice in students. As a basic idea, an authorial voice can be described as the metaphorical 'voice' readers can hear when they read a text (Elbow 2007). Authorial voice has become highly influential in current conceptions of literacy practice and research as writing is seen as a socially situated act, making authorial voice an expression of authorial identity.

In terms of pedagogical research, there appears to be two broad inter-related aspects of authorial voice: 1. The performance of authorial voice on paper, which has been mostly explored using corpus analyses of student texts, and 2. Student self-perception of their authorial voice and authorial identity, which has been usually investigated through interviews, surveys, and questionnaires. The performance aspect of authorial voice predominantly features linguistic functions that fall under what Hyland's (2005) model of voice describes as writer-oriented features (i.e., stance) or reader-oriented (i.e., engagement) features. Meanwhile, the self-perception aspect of authorial voice is often associated with internalized interpretations that a writer may have about themself. Both these aspects intertwine (Ivanić 2005) and form underpinnings for this study.

1.2 Authorial Voice Importance and Challenges for NNSE Argumentative Essay Writing

Authorial voice development is critical for non-native speakers of English (NNSE) students because expressing one's stance and engaging with readers is vital in university academic writing. This is particularly true for the argumentative essay genre. The performance of a solid authorial voice (Lee and Deakin 2016) and authorial confidence (Prat-Sala and Redford 2012) leads to higher quality argumentative essays. However, NNSE student writers at university face challenges developing and expressing their authorial voice. These include performance-related challenges such as a lack of knowledge and practice using linguistics features to express a robust authorial voice (Ellery 2008) and related self-perception issues such as low authorial confidence and self-efficacy beliefs (Ellery 2008). Studies exploring NNSE authorial voice challenges in argumentative essays suggest addressing NNSE students' confidence and beliefs about themselves as writers (Prat-Sala and Redford 2012). However, small Pacific Island nations are underrepresented in pedagogical research in this area. Thus, the context of this study is the Pacific Island nation of Fiji. Its higher education institutions comprise almost entirely NNSE students and the argumentative essay genre is an essential assessment item for first-year university students. Significantly, the existing research about Fijian university students' challenges (Goundar and Bogitini 2019) strongly implies the need to scaffold the development of authorial voice confidence.

1.3 Existing Authorial Voice-Specific Learning Software

Another important authorial voice issue is keeping abreast of technological developments in writing pedagogy. The elusive nature of the authorial voice first complicates the development of authorial voice-specific technology. Most writing support software offer authorial voice support through a range of affordances. For example, automated writing support software may offer suggestions with word choices. While intelligent and affective tutoring systems may address student sentiment, these technologies still lean heavily towards the performance of authorial voice on paper and do not usually put student self-perception as the central prompt. Indeed, recent developments such as AI-generated texts have significant implications for how students perform their authorial voice and perceive themselves as writers. As such, we have developed an authorial voice-specific learning technology that integrates student reflexivity and self-perception with linguistic features for authorial voice.

2   Design and Development of AVATAR Prototype

2.1   Theoretical Frameworks and Functionality

AVATAR (Authorial Voice Toolkit for Authorship Reflection) is an intelligent tutoring system that uses a student-facing self-reflexive approach to encourage the awareness and use of authorial voice in NNSE argumentative essay writing at university. The first underpinning of AVATAR is its operationalization of authorial voice using Hyland's (2005) model of stance and engagement that specifies the language used to express authorial voice. Following Yoon (2017), AVATAR performs automated corpus analyses using regular expressions of authorial voice language, i.e., words/phrases used to express authorial voice. This feature identifies the authorial voice language already present in an uploaded text. Moreover, AVATAR encourages self-reflection about authorial voice by adapting Zhao's (2013) voice analytic rubric.

Zhao's (2013) voice rubric was designed for teachers to assess voice in student writing. AVATAR uses an adapted version that transforms Zhao's (2013) voice rubric into a student-facing self-assessment rubric. Thus, when students upload their essays into AVATAR, the system encourages them to reflect on their authorial voice by asking them to self-assess their uploaded essays for voice strength. AVATAR then generates visualizations to show the student what authorial voice words/phrases they already have in their essay, alongside how they scored themselves for authorial voice. AVATAR then offers the student suggestions of words/phrases that can help strengthen the voice dimension for which they do not feel confident, i.e., the voice dimension for which they scored themselves low. The student can then re-insert these authorial voice words/phrases and self-score to see if they feel more confident. This cyclical affordance for authorial voice awareness and development is presented in Fig. 1.

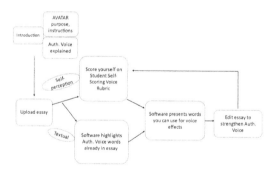

Fig. 1. Simplified AVATAR functionality.

3   User Experience Test

3.1   Research Aim

This study is a one-group pre-post-test experiment to determine if Fiji first-year NNSE undergraduate students changed their authorial confidence after a short usage session with the AVATAR prototype software.

RQ: Does using AVATAR affect the authorial confidence of Fiji first-year NNSE undergraduate students writing English argumentative essays?

3.2 Participants

The participants for this study were NNSE Fijian first-year undergraduate students enrolled in the English for Academic Purpose (EAP) course at a large public university in Fiji in Semester 1, 2022. These students had all passed Fiji Year 13 English and were proficient English writers at pre-university levels. Their age range was 18–20 (mean age: 18.96, SD: 0.71). While gender and ethnic balance were sought, this was impossible given the snowball and convenience sampling.

3.3 Instrument

The principal instrument for this study is a subscale from the Student Attitudes and Beliefs About Authorship Scale (SABAS) (Cheung, Stupple and Elander 2017). SABAS consists of 17 Likert scale items ranging from 1 (strongly disagree) to 6 (strongly agree). The authorial confidence sub-scale consists of eight of the 17 statements. All sub-scales of SABAS are internally validated, and the score for the authorial confidence sub-scale is given by the mean of the eight items comprising it (Cheung et al. 2017).

3.4 Data Collection

Data collection was done in one-on-one sessions. The pre-use session consisted of participants answering the SABAS questionnaire. Participants had been instructed to bring the argumentative essay they had recently submitted to the EAP course. Participants were then given a laptop to use the AVATAR web app. The user sessions were participant-led, with participants free to explore the software and ask questions. Upon indicating they were finished using the system, the participants answered a post-use SABAS questionnaire.

3.5 Data Analyses

All pre- and post-use scores the participants gave for the authorial confidence sub-scale were recorded (n = 23). Each participant contributed a score for pre- and post-use authorial confidence. The mean for all participants' pre- and post-use authorial confidence was calculated for comparison. Using SPSS (Version 29.0.0), the data set was found to be not normally distributed. Therefore, a Wilcoxon Signed Ranks test was performed, and the effect size was calculated to evaluate the change in participant beliefs. The study had limitations, as the small sample size influences the interpretation of effect size and limits generalizability. The study also could not explore demographic data such as gender and ethnicity at this stage and focused on the overall writing experiences.

4 Findings

The Wilcoxon Signed Ranks test results indicated a statistically significant difference between the mean pre-use (4.71) and post-use (5.32) scores of participants for authorial confidence (p = < 0.01). The difference in mean scores was 0.61 (SD: 0.49), with an effect size of 1.25. Results indicate a positive effect size but with high variability.

5 Discussion

The finding of this study has implications for how educational technology can foster student authorial identity development and AI-assisted production of self. The effect size from this study compares favorably with Hattie's (2018) ranking of effect sizes of influences on learning outcomes. For example, self-assessment has a reported mean effect size of 1.33, self-efficacy 0.92, and intelligent tutoring systems 0.48 (Hattie, 2018). Compared to these, the effect size of this study seems significant. However, this needs to be interpreted contextually as the small sample size and high variability in data prompt the need for further research. Nevertheless, the proven significant difference in pre- and post-use authorial confidence scores does suggest that changes in authorial confidence were brought about by the use of AVATAR for most participants. The items comprising the authorial confidence subscale specifically deal with authorial voice and agency. For example, SABAS Item #1, "I have my own style of academic writing", or Item #11, "I have my own voice in my writing", deal with voice awareness. The authorial agency is tested in Item #4, "What I write communicates my confidence about the area to the reader", and Item #12 "I feel in control when writing assignments". The overall increase in these authorial confidence scores for the participants of this study points to them becoming aware of their voice in their writing, and it is significant that they also feel more in control of their writing. These self-efficacy issues are particularly challenging for novice NNSE student writers at university (Ellery 2008).

Moreover, post-use, students also felt better about indicating their ideas in their writing. This was tested, for example, by SABAS Item #3, "I am able to document my ideas clearly in my writing". These items are related to authorial reflection and awareness. AVATAR provides affordances for these aspects in a student-facing approach that may help resolve concerns about students' production of written assessments using technology. Confidence in academic writing is also an issue for NNSE student-writers such as Pacific Island students. There is a dearth of pedagogical research and development in terms of authorial voice and educational technology in Pacific contexts, and the positive findings in our study call for continued research in this area. This would be significant for equity in educational technology. This inclusivity in automated educational technology can also help fulfil broader educational sustainability goals.

6 Conclusions

The authorial voice is a critical aspect of NNSE student success at university as it helps them express their stance, engage with their readers, and express the student's authorial identity. However, a lack of awareness about authorial voice and low authorial confidence are critical challenges that educational technology can help address. This study has shown that using the AVATAR prototype software's reflexive affordances for authorial voice positively changed student feelings about authorial confidence. It is a promising start that can be used for further research on how current concerns about automated writing technology is not a threat but an opportunity. The paradigm shift that AI has brought about in the production of selves needs the inclusion of interactive systems that engage, in new ways, the human factor, the ghost in the machine.

References

Cheung, K.Y.F., Stupple, E.J.N., Elander, J.: Development and validation of the student attitudes and beliefs about authorship scale: a psychometrically robust measure of authorial identity. Stud. High. Educ. 42(1), 97–114 (2017). https://doi.org/10.1080/03075079.2015.1034673

Elbow, P.: Voice in writing again: embracing contraries. Coll. Engl. 7 (2007). http://scholarworks.umass.edu/eng_faculty_pubs/7

Ellery, K.: Undergraduate plagiarism: a pedagogical perspective. Assess. Eval. High. Educ. 33(5), 507–516 (2008). https://doi.org/10.1080/02602930701698918

Goundar, P.R., Bogitini, L.J.K.D.: First year fijian undergraduate university students academic essay writing error analysis. Int. J. Appl. Linguist. Engl. Lit. 8(1), 172–177 (2019). https://doi.org/10.7575/aiac.ijalel.v.8n.1p.172

Hattie, J.: Hattie ranking: 252 influences and effect sizes related to student achievement. Vis. Learn. (2018). https://visible-learning.org/hattie-ranking-influences-effect-sizes-learning-achievement/. Accessed 15 Apr

Hyland, K.: Stance and engagement: a model of interaction in academic discourse. Discourse Stud. 7(2), 173–192 (2005). https://doi.org/10.1177/1461445605050365

Ivanić, R.: The discoursal construction of writer identity. In: Beach, R., Green, J., Kamil, M., Shanahan, T. (eds.) Multidisciplinary perspectives on literacy research, pp. 391–416. Hampton Press (2005)

Lee, J.J., Deakin, L.: Interactions in L1 and L2 undergraduate student writing: interactional metadiscourse in successful and less-successful argumentative essays. J. Second. Lang. Writ. 33, 21–34 (2016)

Prat-Sala, M., Redford, P.: Writing essays: does self-efficacy matter? The relationship between self-efficacy in reading and in writing and undergraduate students' performance in essay writing. Educ. Psychol. 32(1), 9–20 (2012). https://doi.org/10.1080/01443410.2011.621411

Yoon, H.-J.: Textual voice elements and voice strength in EFL argumentative writing. Assess. Writ. 32, 72–84 (2017). https://doi.org/10.1016/j.asw.2017.02.002

Zhao, C.G.: Measuring authorial voice strength in L2 argumentative writing: the development and validation of an analytic rubric. Lang. Test. 30(2), 201–230 (2013). https://doi.org/10.1177/0265532212456965

Evaluating Language Learning Apps for Behaviour Change Using the Behaviour Change Scale

Ifeoma Adaji[(✉)] (iD)

The University of British Columbia, Okanagan, BC, Canada
ifeoma.adaji@ubc.ca

Abstract. The proliferation of affordable smartphones has led to a significant increase in the number of apps that teach languages. These apps are designed to positively influence their users to learn a new language and are easily accessible in the app store with many of them having a free version. Research suggests that learning is a process that can lead to permanent behaviour change or potential behaviour change. Thus, understanding the strength or potential of these apps to lead to the behaviour change of learning a new language is important in determining the success or otherwise of the app. To contribute to research in the area of app-based learning, in this paper, we present the preliminary results of reviewing learning apps in the Google Play Store. We use the App Behaviour Change Scale (ABACUS) that measures the behaviour change potential of smartphone apps to determine how likely an app will lead to the behaviour change of learning a new language. The results of this study can guide application developers in designing apps that will lead to a positive change in behaviour when learning new languages by highlighting the aspects of the app that can be improved based on the ABACUS scale.

Keywords: Behaviour Change · Language Learning · Mobile Apps · Google Play Store

1 Introduction

Technology has become a pertinent part of the learning process within and outside the classroom [3]. With the increase in the use of mobile phones over the years, there has been a steady increase in the number of people who use apps for learning new languages. Several apps exist in the app store for learning new languages with some similarities and differences. The main goal of these apps is to influence their users to adopt a new behaviour of learning a new language. Reviews have been carried out to study these apps. For example, Gangaiamaran et al. [5] carried out a review of language learning apps to identify the design, method, theory and pedagogical features of these apps. Similarly, Khurram [6] reviewed apps for learning and identified the state of the art in learning apps as well as design flaws such as weak theoretical support and lack of a validation process. While the results from these evaluations are important, it is also important to

© The Author(s), under exclusive license to Springer Nature Switzerland AG 2023
F. Wang et al. (Eds.): AIED 2023, CCIS 1831, pp. 510–514, 2023.
https://doi.org/10.1007/978-3-031-36336-8_79

examine the behaviour change motive of these apps. Research suggests that learning is a process that can lead to permanent behaviour change or potential behaviour change [7, 11]. In the case of learning languages, the behaviour change is communicating in a new language. How these apps perform with respect to changing the behaviour of their users is important in determining if the user will learn the new language or not. Thus, reviewing the apps for their ability to change behaviour is pertinent and is the main aim of this paper. To achieve this, we reviewed mobile apps on the Google Play Store using the App Behaviour Change Scale (ABACUS) of McKay et al. [8]. This scale is commonly used to review apps for their potential to encourage behaviour change. Thus, our research question is:

RQ1: How likely will the apps encourage the behaviour change of learning a new language?

This research question is important because it will shed light on if an app will likely lead to a behaviour change or not. If an app will not lead to behaviour change, this information can be useful to the stakeholders of the app in the re-design or upgrade of the app. In this paper, we present the results of the analysis of only five apps which we have reviewed so far. We are still in the process of analyzing the over forty apps that we have identified as being relevant to our study and will include those results in future work.

2   Methodology

2.1   App Search and Selection

To select the apps that were included in this study, we searched the Google Play Store using the term "Learn Language". We only selected apps that have reviews, have been downloaded at least ten thousand times and have an overall rating of at least 4 stars. This is important to filter out apps that are not popular in the app store. Furthermore, we selected apps that focused on speaking a language and not writing the language. We are still in the process of data collection and will review apps from the Apple Store in future work.

Although our search resulted in 46 apps, we randomly selected five for this paper while we are in the process of reviewing the other 41.

2.2   ABACUS Review

The ABACUS scale [8] is a 21-item instrument that measures the potential of an app to encourage behaviour change. As suggested by the developers of the scale, each of the five apps reviewed was downloaded and used for about 10 min. This is to give the rater time to familiarize themselves with how the app functions. Two reviewers reviewed each app and attempted to use all parts of each app. Where there were differences in functionality scores, the reviewers considered the app together and agreed on a final score for each app.

The summary of the five selected apps is shown in Table 1.

Table 1. Summary of apps

Name of app	Average Rating	Number of Reviews	Number of Downloads
Memrise	4.6	1.48 million	Over 10million
Lingo Legend Language Learning	4.8	962	Over 10,000
Duolingo	4.3	14.5 million	Over 100 million
Drops Language Learning	4.4	253,000	Over 5 million
Babbel	4.6	837,000	Over 50 million

3   Results

3.1   ABACUS Review Results

As shown in Table 2, most of the apps check most of the boxes on the ABACUS scale. For all five apps, users can sign in with their existing Gmail account. This reduces the signup process and has been shown to be persuasive in making users register for a service [9]. Customization and personalization of features is another finding that was common across all apps. Research suggests that systems that offer personalized content are likely to be more persuasive in influencing a user to carry out a target behaviour [9], in this case learning a new language. In the apps we reviewed, users were asked to provide some information about themselves such as the language they wanted to learn, the level they were currently at in the language (for example, beginners, experts, etc.), the time of day they prefer to learn, where they would use the new language (for example, at work, on vacation etc.). This information was then used to create a personalized learning experience for the learner. Reminders were another feature that all the apps had. Users can specify when they want to be reminded to use the app for the day.

Not all the apps were similar in terms of providing general encouragement. For example, Duolingo used occasional prompts to encourage users to keep learning. For example, while using the app, one of the reviewers received the prompt below:

We're in this together! You're practicing French now with 183,963 other learners right now.

This is a form of social comparison where one's behaviour is compared to that of others in order to encourage the user to carry out a target behaviour [4, 9]. Research suggests that social comparison can lead to a change in behaviour [4, 9].

Also, not all apps encouraged users to practice the use of the app before signing up or logging in with their Gmail account. Lingo was the only app that allowed users to "rehearse" using the app without committing to signing up. This strategy known as rehearsal is a persuasive strategy that can influence people to use an app or technology if they know they do not have to commit to it [9].

Our results have shown that all the apps we reviewed in this study used some form of persuasive strategy (for example, reduction, rehearsal, social comparison, personalization etc.) to make the apps persuade users to carry out the target behaviour of learning

Table 2. Summary of ABACUS review of apps. A tick indicates a score of at least 3 by both reviewers in the stated category.

Name of app ABACUS items	Memrise	Lingo Legend Language Learning	Duolingo	Drops language learning	Babbel
Allow or encourage practice or rehearsal in addition to daily activities		✓			
Allow the user to easily self-monitor behaviour	✓	✓	✓		
Provide instruction on how to perform the behaviour	✓	✓	✓	✓	✓
Customize and personalize some features	✓	✓	✓	✓	✓
Reminders and/or prompts or cues for activity	✓	✓	✓	✓	✓
Give user feedback (person or automatic)		✓			
Encourage positive habit formation	✓	✓	✓	✓	✓
Provide general encouragement			✓		
Goal setting	✓	✓	✓	✓	✓
Material or social reward or incentive		✓			
Information provided about the consequences of continuing and/or discontinuing behaviour,			✓		

a new language. Persuasive strategies are nudges which when used at the right time can influence a user to carry out a target behaviour [4]. They are commonly applied in the design and development of technology which aims to guide users to carry out a target behaviour. Persuasive strategies are used in several domains including learning, e-commerce [2], health [1] and fitness [10]. Thus, its use in the area of learning a new app is not surprising. The results from this study show the importance of using persuasive strategies in application development. The high ratings of the users in Table 1 are an indication of the positive perception of users towards the apps.

4   Conclusion

The increase in the availability and affordability of smartphones has led to a significant increase in the use of mobile apps to learn languages. These apps are designed to positively influence their users to learn a new language and are easily accessible in the app store with many of them having a free version. Because learning is a process that can lead to behaviour change, it is pertinent to review existing apps for their potential to lead to a change in the behaviours of their users. To contribute to research in this area, using the App Behaviour Change (ABACUS) Scale, we review five apps from the Google Play Store. Our findings suggest that all apps are developed using persuasive strategies which are just in-time nudges that often result in a desired behaviour change. We are still in the process of reviewing other apps and will present our results in future work. The results of this study can guide application developers in designing apps that will lead to a positive change in behaviour when learning new languages. In the future, we plan to investigate if non popular apps also used persuasive strategies and which ones were used if any.

References

1. Adaji, I., Vassileva, J.: A gamified system for influencing healthy e-commerce shopping habits. In: UMAP 2017 - Adjunct Publication of the 25th Conference on User Modeling, Adaptation and Personalization (Bratislava, 2017), pp. 398–401 (2017)
2. Adaji, I., Vassileva, J.: Tailoring persuasive strategies in E-commerce. In: International Workshop on Personalized Persuasive Technology, Amsterdam, 2017, pp. 57–63 (2017)
3. Ahmadi, D., Reza, M.: The use of technology in English language learning: a literature review. J. Res. Engl. Educ. **3**(2), 115–125 (2018)
4. Cialdini, R.B.: Influence: Science and Practice. Pearson Education Boston (2009)
5. Gangaiamaran, R., Pasupathi, M.: Review on use of mobile apps for language learning. Int. J. Appl. Eng. Res. **12**(21), 11242–11251 (2017)
6. Khurram, M.: A systematic literature review on learning apps evaluation. J. Inf. Technol. Educ. Res. **21**(2023), 663–700 (2023)
7. Kostrubiec, V., Zanone, P.G., Fuchs, A., Scott Kelso, J.A.: Beyond the blank slate: routes to learning new coordination patterns depend on the intrinsic dynamics of the learner-experimental evidence and theoretical model. Front. Hum. Neurosci. (2012). https://doi.org/10.3389/FNHUM.2012.00222/FULL
8. McKay, F.H., Wright, A., Shill, J., Stephens, H., Uccellini, M.: Using health and well-being apps for behavior change: a systematic search and rating of apps. MIR mHealth uHealth **7**(7), e11926 (2019)
9. Oinas-Kukkonen, H., Harjumaa, M.: Persuasive systems design: key issues, process model, and system features. Commun. Assoc. Inf. Syst. **24**(1), 28 (2009)
10. Oyibo, K., Adaji, I., Olagunju, A.H., Deters, R., Olabenjo, B., Vassileva, J.: Ben'fit: design, implementation and evaluation of a culture-tailored fitness app. In: ACM UMAP 2019 Adjunct - Adjunct Publication of the 27th Conference on User Modeling, Adaptation and Personalization, June 2019, pp. 161–166 (2019). https://doi.org/10.1145/3314183.3323854
11. PSYCHOLOGY 101 A Ten Chapter Overview of Psychology

Evaluating the Rater Bias in Response Scoring in Digital Learning Platform: Analysis of Student Writing Styles

Jinnie Shin[1]([✉]), Zeyuan Jing[1], Lodi Lipien[2], April Fleetwood[2], and Walter Leite[1]

[1] University of Florida, Gainesville, FL, USA
{Jinnie.Shin,Jingzeyuan,Walter.Leite}@ufl.edu
[2] Florida Virtual Schools, Orlando, FL, USA
{llipien,afleetwood}@flvs.net

Abstract. Automated essay scoring and short-answer scoring have shown tremendous potential for enhancing and promoting large-scale assessments. Challenges still remain in the equity and the implicit bias in scoring that is ingrained in the scoring system. One promising solution to mitigate the problem is the introduction of a measurement model that quantifies and evaluates the degree to which the raters are showing biases in students' written responses. This paper presents an adoption of a generalized many-facet Rasch model (GMFRM, [8]) to evaluate the rater biases regarding students' writing styles. We modeled students' writing styles using a rich set of computational linguistic indices from LingFeat [7] that are empirically and theoretically associated with the text difficulty. The findings showed that the rater bias exists in scoring responses that are explicitly covering a variety of topics with less elaborative descriptions compared to the other writing styles captured in students' responses. The rater severity was the only type of bias characteristic that was statistically significant. We discussed the implications and future applications of our findings to advance automated essay scoring and teacher professional development.

Keywords: Constructed-response item scoring · Natural Language Processing (NLP) · Discourse analysis · Generalized many-facet Rasch model · Rater bias

1 Introduction

Constructed-response (CR) items are increasingly understood as a valuable tool for evaluating students' science competencies in online STEM education. The use of open-ended CR items allows students to share their authentic responses in applying science concepts in real-life situations. However, one of the important, but not well studied, innate problems of open-ended CR items is the potential human scorer biases introduced with their subjective scoring. In recent years, a growing amount of awareness is in understanding whether the biases teachers bring in the scoring processes may threaten the validity and reliability of formative and summative science assessments [1]. Even more worrisome, teacher biases can negatively impact the adoption of advanced educational technology, such as automated scoring systems [2] and unintentionally affecting students' performance outcomes [3]. Human raters hold important key roles to resolve this issue and to promote assessment reliability and validity.

Consistent effort in the previous literature focused on evaluating the internal sources of bias that are associated with the assessment and the rater characteristics such as the types of prompts in the exam, language background and the experiences of raters [4, 5]. Recently, an increasing number of studies are putting more attention on the effect of such biased ratings in automated essay scoring [6]. One of the primary sources of bias, *yet* given limited attention in the research community, concerns the varying writing styles and competencies that students demonstrate in their responses. Rater bias refers to the types and patterns of scoring where the additional factors other than just the responses affect the scoring divisions. Hence, we focus on evaluating and addressing these potential biases in the scoring processes of CR items collected from one of the largest K-12 online science learning platform. We focus on addressing how student's unique writing styles may lead to biases in human scoring. The following two research questions are introduced to guide our study: (1) *Do raters show scoring biases based on students' response writing styles?* (2) *If yes, what are the characteristics of the rater bias in terms of the rater severity, consistency, and score restrictions?*

2 Methods

2.1 Data and Sample

A total of 2,899 responses on five science assessment questions were collected from a total of 798 students from Florida Virtual School (FLVS). Each constructed-response item required students to provide explanations and rationales to support their learning in the corresponding lesson. The course M/J Comprehensive Science 1 has a total of 15 assessments that contain open-ended questions, referred to as essay items. A total of 35 teachers graded the responses with a single teacher rating the response of each student. A teacher reads the students' response to each essay item, grades their response, then leaves the student a comment. The data from each of the items on all the assessments were pulled and exported into a spreadsheet. In this study, we selected a total of five items from the same assessment that had the highest response rates for further analysis. Specifically, the five items were answered by a total of 560 to 659 students (Item 1 = 560, min = 0, max = 7, M = 5.66; Item 2 = 659, min = 0, max = 4, M = 3.06; Item 3 = 560, min = 0, max = 7, M = 5.80; Item 4 = 560, min = 0, max = 7, M = 5.55; Item 5 = 560, min = 0, max = 7, M = 5.67). For a scoring range consistency, we rescaled the response scores to five categories each representing 1 = very low, 2 = low, 3 = medium, and 4 = high, and 5 = very high level of understanding (The original score range is identified as min-max scores).

2.2 Clustering of Students' Writing Styles

We first clustered students' varying writing styles to address our first research question and explore whether students' writing styles may bias the grading patterns of the teachers. We used the advanced semantic and syntactic feature extraction introduced by [7]. The linguistic and psycholinguistic indices are primarily categorized into five groups including: Syntactic (i.e., measuring the complexity of grammar and structure), Advanced Semantic (i.e., measuring the complexity of meaning structure), Lexicon Semantic (i.e., measuring word/phrasal-specific difficulty), Discourse (i.e., measuring coherence/cohesion), and Descriptive features (i.e., traditional features forming text difficulty). The indices provide a theoretical and empirical representations of the readability or the difficulty of the text. The Python tool extracted a total of 160 features which are reduced to the dimension of the final 23 features. The dimensionality reduction of the final features was conducted based on removing variables with no standard deviation and removing variables with negligible correlation with the final score category (<0.1). The final features are introduced in Table 1.

2.3 Generalized Many Faceted Rasch Model

We adopted the generalized many-facet Rasch model (GMFRM, [9]) to measure and evaluate the rater bias. The model estimates P_{ijrkg}, which is the probability that a rater r gives a particular score to an examinee i, in group g, in each task j, which has k score category. The model will estimate the degree of rater biases in terms of the three characteristics of the rater, including scoring consistency (α_r; ;; rater's tendency to provide a similar score to the response of similar quality), scoring severity (β_r; ;; rater's tendency to provide lower ratings than are justified by the other responses), and the scoring restrictiveness (d_{rm}; rater's tendency to overuse certain score category, m). The scoring characteristics will be estimated while accounting for the cultural representation of the argument clusters (i.e., which writing style group the argument cluster is mostly adopted by). In our study, we modified the parameter explanations in order to incorporate student's writing style groups, such that α_{rg} represented the rater r's consistency when interacting with the writing style group g. Lower α_{rg} suggest that the score given by rater r is strongly biased as they will tend to assign different scores to the students in writing style group g with similar latent ability, θ. Subsequently, β_{rg} represented the rater r's severity when interacting with the writing style group g, where higher β_{rg} suggest that the rater tends to consistently assign low scores to the student in writing style group g, hence more "severe" in grading. Lastly, d_{rmg} are measures of score restrictions, representing the rater r's threshold to assign the score category m versus the adjacent categories of score m to writing group g. The differences between the multiple d_{rmg} identifies which score category the rater r tends to overuse to writing group g (Eq. 1). The data cleaning was conducted in R-4.2.2 [9][1]. The parameters were estimated using a Markov Chain Monte Carlo (MCMC)

$$P_{ijrkg} = \frac{exp\sum_{m=1}^{k}[\alpha_{rg}\alpha_i(\theta_j - \beta_i - \beta_{rg} - d_{rmg})]}{\sum_{l=1}^{k}exp\sum_{m=1}^{l}[\alpha_{rg}\alpha_i(\theta_j - \beta_i - \beta_{rg} - d_{rmg})]} \qquad (1)$$

Table 1. The GMFRM Results.

Category	Feature	Description
Syntactic	*Phrasal*	Ratio of Noun phrases count to Adj phrases count (ra_NoAjP_C)
		Ratio of Noun phrases count to Adv phrases count (ra_NoAvP_C)
		Ratio of Noun phrases count to Prep phrases count (ra_NoPrP_C)
		Ratio of Noun phrases count to Verb phrases count (ra_NoVeP_C)
		Ratio of Noun phrases count to Subordinate Clauses (ra_NoSuP_C)
		Total count of Adjective phrases (to_AjPhr_C)
		Ratio of Adj phrases count to Adv phrases count (ra_AjAvP_C)
		Total count of Adverb phrases (to_AvPhr_C)
		Ratio of Adverb phrases count to Subordinate Clauses (ra_AvSuP_C)
		Total count of Subordinate Clauses (to_SuPhr_C)
		Ratio of Subordinate Clauses to Prep phrases count (ra_SuPrP_C)
		Average AoA of words per token (at_AAKuW_C)
	Tree structure	Average Tree height per sentence
Advanced Semantic	*Semantic Knowledge*	Semantic Richness, 200 topics extracted from WeeBit (BRich20_S)
		Semantic Noise, 100 topics extracted from Wikipedia (WNois10_S)
		Semantic Richness, 50 topics extracted from Wikipedia (WRich05_S)
	Word Familiarity	Average SubtlexUS count value per token (at_SbCDC_C)
		Average SubtlexUS FREQcount value per token (at_SbFrQ_C)
Lexicon Semantic	*Psycholinguistic*	Average lemmas AoA of lemmas, Bird norm per token(at_AABiL_C)
		Average lemmas AoA of lemmas, Bristol norm per token (at_AABrL_C)
Discourse	*Text difficulty*	Log (Total count of tokens)/Log(Total count of sentences) (TokSenL_S)
		Flesch Kincaid Grade level (FleschG_S)
Descriptive	*Entity*	Total number of Entities Mentions counts (to_EntiM_C)

3   Results and Discussion

3.1   Writing Style Analysis Results

We used the K-means clustering approach to identify the underlying writing styles from the total of 23 extracted features. K-means clustering algorithm using the elbow method using the distortion score and the silhouette measures (coefficient $= 0.69$) to identify the optimal number of clusters, which resulted in $k = 4$ writing style clusters. The cluster centroids are provided in Table 2. The common patterns of writing styles in each cluster could be represented as C1: *elaborative in plain language*, C2: *elaborative in difficult language*, C3: *focused and elaborative*, and C4: *brief and dense* information responses.

Table 2. The Writing Style Clustering Centroids

Feature		C1	C2	C3	C4	Feature		C1	C2	C3	C4
Noun Phrasal Features	ra_NoAjP_C	5.83	6.95	**7.31**	2.32	AoA verbs	at_AAKuW_C	5.27	**5.81**	5.66	5.08
	ra_NoAvP_C	6.89	7.15	**8.15**	3.55	Tree	as_TreeH_C	7.89	7.59	**7.97**	5.62
	ra_NoSuP_C	4.69	**6.68**	5.78	2.81	Semantic Knowledge	BRich20_S	14.3	14.4	**14.7**	14.5
	ra_NoPrP_C	**5.69**	4.92	5.18	2.78		WNoise10_S	3.97	4.01	**4.06**	3.78
	ra_NoVeP_C	1.34	1.52	**1.53**	1.14		WRich05_S	1.96	1.93	1.92	**5.60**
Adverbs & Adjectives	ra_AVSuP_C	0.61	**0.72**	0.62	0.37	Word Familiarity	at_SbCDC_C	**2.61**	2.16	2.35	1.61
	ra_AjAvP_C	0.27	0.25	**0.35**	0.09		at_SbFrQ_C	**1.25**	0.65	0.95	0.25
	to_AvPhr_C	1.94	1.79	**1.96**	0.89	Psycholing-uistic (AoA lemmas)	at_AABiL_C	**3.35**	3.08	3.16	2.58
Subordinate Clauses	ra_SuPrP_C	**1.19**	0.89	0.98	0.40		at_AABrL_C	1.89	**2.26**	2.09	1.78
	to_SuPhr_C	**3.15**	2.47	3.15	1.11						

Note. **Bold** = highest value, underlined = lowest value

Specifically, C4: *brief and dense* showed responses that are limited in use of noun phrases in ratio of auxiliary components, such as adverbial, adjectives, and subordinate clauses compared to the other clusters. Given that noun phrases in the responses often correspond to the science concepts (e.g., gravity, earth, space, weight), the high ratio represents a more delegate effort to explain and elaborate on the description of the science concept. Simpler sentence structures with verbs with slightly lower age-of-acquisition (AoA) entailed these responses. They often had lower coherence scores (less elaborative) but present large distinctive science topic coverage, indicating that the responses were often "matter-of-fact" and outlining a few science concepts rather than describing them in detail. C3: *focused and elaborative*, by contrast, encompasses responses that are limited in distinctive science topic coverage with an elaborative response that describes the science concept (noun phrases) with a comparably large number of adjectives, adverbial phrases, thus entailing complex sentence tree structures. Additionally, C1 and C2 shared similarities in providing elaborative descriptions of the concepts supported by their high use of phrasal features but differ in terms of the familiarity and the acquisition level of the words (i.e., verbs and lemma).

3.2 Rater Bias Evaluation Results

Table 3 provides the final estimation results from the GMFRM shown in Eq. 1 about the relationship between rater bias characteristics and the probabilities of the rater assigning specific scores. The results showed that consistency in rating (α_{rg}) between the four writing style groups had no significant differences across student groups. The parameters showed relatively moderate values (0.36–0.46), which suggest that the consistent rating scores were provided for the item responses with similar qualities in terms of its underlying latent ability. The severity of the rating, β_{rg}, by contrast, indicated a bias in interacting with writing style group 4. The severity of the rater was significantly higher in writing style group 4 (-0.90 (.10)) compared to the rest of the writing style groups. The higher value indicates that the rater was most "severe" in scoring this group and assigning consistently low scores to the students. The item parameters indicate that the five items showed moderate difficulty and discrimination with no statistical differences among each other (Item 1: $\beta = 0.10$ (.08), $\alpha = 1.10$ (.07); Item 2: $\beta = -0.03$ (.09), $\alpha = 0.75$ (.05); Item 3: $\beta = 0.04$ (.07), $\alpha = 1.26$ (.09); Item 4: $\beta = 0.06$ (.08), $\alpha = 0.98$ (.06); Item 5: $\beta = -0.16$ (.09), $\alpha = 0.99$ (.07)).

Table 3. The GMFRM Results: Rater Bias Characteristics

Student group	Severity β_{rg}	Consistency α_{rg}	Score Range Restriction d_{r2g}	d_{r3g}	d_{r4g}	d_{r5g}
Writing style group 1	−1.46 (.16)	0.46 (.07)	−0.82 (.34)	0.36 (.23)	2.45 (.26)	−1.99 (.22)
Writing style group 2	−1.49 (.11)	0.36 (.03)	0.12 (.31)	0.86 (.27)	2.32 (.25)	−3.31 (.34)
Writing style group 3	−1.42 (.11)	0.38 (.04)	0.43 (.41)	-0.09(.33)	2.51 (.33)	−2.85 (.37)
Writing style group 4	**−0.90** (.10)	0.44 (.07)	−0.19 (.45)	1.28 (.40)	1.41 (.39)	−2.50 (.46)

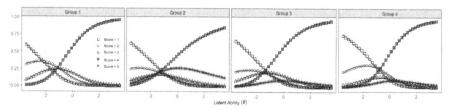

Fig. 1. Item Characteristics Curves of the Four Writing Groups. Model fit evaluation (BIC = 11364.77, AIC = 6431.794, WAIC = 4781.549).

The range restriction results in Table 3 indicated that all writing style groups showed extreme response score tendency, indicating that the rater tends to overuse the extreme rating categories (5) while avoiding the middle score categories. Figure 1 provides the item characteristics curves for one of the items for each writing style group. Across all groups, there was a strong tendency and preference for extreme scores (1 or 5) while avoiding the scores in the middle. Specifically, students in writing style groups 2 and 3 tended to end up with the lowest score (= 1) than the score category 2, for the students who are in the same ability range. Students in writing group 4 tended to have a very limited chance to score category 3 compared to the other scoring groups when comparing students with the same ability range.

4   Conclusion

Evaluating rating biases in CR scoring is a challenging task that affects the creation of unintended results in student evaluation. This study explores one hypothesized source of bias, how students are presenting their response, to understand rater biases in terms of the severity, consistency, and score use restrictions. To address RQ1, a significant severity difference was identified where student responses that are comparably brief and dense in explanations may be given lower scoring when compared to the students with similar ability levels but different response presentation styles. Answering RQ2, the analysis uncovered a consistent pattern of extreme score use. Our findings add to the increasing literature focusing on the presence and the effect of rater biases in automated essay scoring. The novelty of our study lies in the use of writing style analysis to evaluate the source of bias in scoring open-ended responses. This framework may be adopted to extend the capacity and accuracy of the AES, where the establishment of scoring invariance regarding students' writing styles is necessary.

4.1   Limitations and Future Research

One limitation of this study is the lack of data in evaluating inter-rater reliability as the essays were only graded by one teacher Further research is needed to understand the generalizability of the findings for the scenarios where more than two teacher scorings are obtained, for other assessment types and domains, and whether the rater characteristics (e.g., backgrounds, experience) may interact with the bias patterns.

References

1. Ercikan, K., Guo, H., He, Q.: Use of response process data to inform group comparisons and fairness research. Educ. Assess. 25(3), 179–197 (2020)
2. Amorim, E., Cançado, M., Veloso, A.: Automated essay scoring in the presence of biased ratings. In: Proceedings of the 2018 Conference of the North American Chapter of the Association for Computational Linguistics: Human Language Technologies, vol. 1, Long Papers, pp. 229–237 (2018)
3. Stecher, B.M., et al.: The effects of content, format, and inquiry level on science performance assessment scores. Appl. Measur. Educ. 13(2), 139–160 (2000)

4. Ahmadi Shirazi, M.: For a greater good: bias analysis in writing assessment. SAGE Open **9**(1), 2158244018822377 (2019)
5. Mohd Noh, M.F., Mohd Matore, M.E.E.: Rater severity differences in English language as a second language speaking assessment based on rating experience, training experience, and teaching experience through many-faceted Rasch measurement analysis. Front. Psychol. **13**, 941084 (2022)
6. Uto, M., Okano, M.: Learning automated essay scoring models using item-response-theory-based scores to decrease effects of rater biases. IEEE Trans. Learn. Technol. **14**(6), 763–776 (2021)
7. Lee, B.W., Jang, Y.S., Lee, J.H.J.: Pushing on text readability assessment: a transformer meets handcrafted linguistic features (2021). arXiv preprint arXiv:2109.12258
8. Uto, M., Ueno, M.: A generalized many-facet Rasch model and its Bayesian estimation using Hamiltonian Monte Carlo. Behaviormetrika **47**(2), 469–496 (2020). https://doi.org/10.1007/s41237-020-00115-7
9. Stan Development Team. "RStan: the R interface to Stan." R package version 2.21.8 (2023). https://mc-stan.org/
10. R Core Team. R: A language and environment for statistical computing. R Foundation for Statistical Computing, Vienna, Austria (2022). https://www.R-project.org/

Generative AI for Learning: Investigating the Potential of Learning Videos with Synthetic Virtual Instructors

Daniel Leiker[1,3](✉) , Ashley Ricker Gyllen[2,3] , Ismail Eldesouky[3] ,
and Mutlu Cukurova[1]

[1] UCL Knowledge Lab, London WC1N 3QS, UK
daniel.leiker.16@ucl.ac.ui
[2] Metropolitan State University, Denver, CO 80204, USA
[3] EIT InnoEnergy, Eindhoven, Netherlands

Abstract. Recent advances in generative artificial intelligence (AI) have captured worldwide attention. Tools such as Dalle-2 and ChatGPT demonstrate that
tasks previously thought to be beyond the capabilities of AI may now augment
the productivity of creative media in various new ways. To date, there is limited
research investigating the real-world educational value of AI-generated media. To
address this gap, we examined the impact of using generative AI to create learning videos with synthetic virtual instructors. We took a mixed-method approach,
randomly assigning adult learners (n = 83) into one of two micro-learning conditions, collecting pre- and post-learning assessments, and surveying participants on
their learning experience. The control condition included a traditionally produced
instructor video, while the experimental condition included an AI-generated learning video with a synthetic virtual instructor. Learners in both conditions demonstrated significant improvement from pre- to post-learning (p <.001), with no
significant differences in gains between the two conditions (p = .80), and no
qualitative differences in the perceived learning experience. These findings suggest that AI-generated learning videos have the potential to be a viable substitute
for videos produced via traditional methods in online educational settings, making
high quality educational content more accessible across the globe.

Keywords: AI-Generated Learning Content · Generative AI · AI in Education

1 Introduction

Despite the growing body of evidence demonstrating the positive impacts of using AI
to support learning, engagement, and metacognitive development [1–3], the use of generative AI in learning contexts remains largely unexamined. Recent advancements in
generative AI tools have enabled the synthesis of realistic digital content, including photorealistic images, voice cloning, and face animation [4–6]. These tools are being leveraged across several industries including entertainment, customer services and marketing

N. Wang et al. (Eds.): AIED 2023, CCIS 1831, pp. 523–529, 2023.
https://doi.org/10.1007/978-3-031-36336-8_81

[7]. If used responsibly, generative AI has the potential to help address global educational challenges, improve online learning, and reshape our creative and knowledge-based workforces. Taking an evidence-based approach to creating learning videos using generative AI is an appealing way to meet the growing global demand for online learning content and address the challenges associated with producing high quality learning videos (e.g., lectures' lack of on-screen experience, time and resources needed, and the under-resourced nature of educational institutions). Furthermore, AI-generated virtual instructors may be particularly useful, as previous research shows that adding pedagogical agents or avatars to virtual learning positively impacts learners' behaviors, attitudes, and motivation [8–11].

This study focuses on using generative AI to create learning videos resembling traditional lecture videos. The study was conducted in collaboration with EIT InnoEnergy, a European company promoting innovation and entrepreneurship in the fields of sustainable energy. The subject matter was sampled from an InnoEnergy's Skills Institutes course aimed at accelerating workforce transition toward a clean energy economy. A significant challenge of generating learning products to achieve this aim is the rapid pace of research and technology in sustainable energy, requiring fast and frequent iterations of the curriculum. A potential solution to address this challenge is creating asynchronous online learning content using generative AI, guided by multimedia learning research [9, 12, 13], as an alternative to traditional production methods. This study examines the viability of this approach by addressing the following research questions: 1) To what extent does the use of AI-generated learning videos with synthetic virtual instructors differentially impact learning when compared to traditionally produced instructor videos when embedded in an online micro-learning course? 2) What are the perceived differences between videos with synthetic virtual instructors and those with traditional instructors in an online educational setting?

2 Methodology

2.1 Participants

The sample included 83 adult learners (M_{age} = 41.5 years; range = [18, 64]), of which 73% identified as male, 19% identified as female, 0% identified as non-binary, and 8% preferred not to disclose their gender identity. All the participants in the sample reported some level of higher education (4% associate degree; 15% bachelor's degree; 58% master's degree; 23% doctorate degree). Additionally, 68% were unfamiliar with the course subject matter, while 32% reported having prior subject matter knowledge.

2.2 AI-Generated Video Creation and Learning Design

The focus of this experiment is the introduction of AI-generated learning videos with synthetic virtual instructors. A traditionally produced instructor video served as a control for our experiment. The AI video creation platform Synthesia was used to generate text-to-videos with photo-realistic synthetic actors using the verbatim transcript of the control video. To establish the photo realistic quality for these generated videos, Synthesia recorded footage of a live actor and then established a synthetic clone of the actor.

Neural video synthesis was then applied to add realistic gestures and movements to the synthetically generated clone to produce the final production asset. For this study, the AI-generated virtual instructor was matched to the traditional instructor as closely as possible in age, gender, and race. This resulted in two videos with identical content but different visual representations of the instructor (see Online Appendix). Using these videos, two micro-learning courses were designed to create an experimental and control condition. The main goal of the course was to provide an effective learning experience centered around the instructional video content. The course consisted of an introduction page, a pre-learning assessment, the instructional video content (4.5 min in length), a click/reveal application activity, and a post-learning assessment identical to the pre-learning assessment. The courses were delivered to learners via EdApp a learning platform designed with a focus on supporting micro-learning formats.

2.3 Procedure

Participants were invited through an email campaign by EIT InnoEnergy. A mixed-method approach was utilized, examining both quantitative and qualitative data. The analyses for all quantitative data were completed using relevant packages in R, and all qualitative data were evaluated using NVivo. After completing informed consent, participants were randomly assigned into either the experimental or control condition and completed the micro-learning course described above. Finally, they completed a learning experience survey, responding to the open-ended question *"Do you have any overall suggestions for improving this course?"* and indicating their agreement on a 5-point Likert Scale to the following statements:

1. My overall experience with this micro-learning course was positive.
2. The use of video in the course met my expectations.
3. The use of video in the course improved my understanding of the material.
4. I would be interested in taking other courses like this.

3   Results

3.1   Impacts of AI-Generated Learning Videos on Learning Performance

Difference scores (post-minus pre-learning) were calculated to represent knowledge gains. Skew and kurtosis were estimated, and the data were visually inspected for normality. A paired sample t-test revealed significant improvement from pre- ($M = 0.53$ $SD = 0.65$) to post-learning ($M = 1.51, SD = 1.03$) across the entire sample, t (82) = 8.31, $p <.001, d = 0.91$. Indicating that the micro-learning course was effective at facilitating learning. Regression analyses indicated that there was no effect of condition on knowledge gains ($\beta = .03, p = .80, r = .03$), such that gains were not significantly different in the experimental condition ($M_{gains} = 1.00, SD = 1.04, n = 52$) vs. the control condition ($M_{gains} = 0.94, SD = 1.13, n = 31$). This finding was unchanged when controlling for participants' pre-learning performance or self-reported prior knowledge.

3.2 Learner Perceptions of AI-Generated Learning Videos

Quantitative Findings. For the closed-ended questions, 80% of the sample provided agreement ratings. Pearson Chi-Squares indicated no significant differences existed between the experimental and control conditions for overall experience ($\chi 2$ (3) = 0.19, $p = .98$, $V = .05$); video met my expectations ($\chi 2$ (4) = 4.65, $p = .34$, $V = .27$); video improved my understanding ($\chi 2$ (4) = 2.59, $p = .63$, $V = .20$); or interest in additional courses ($\chi 2$ (3) = 0.08, $p = .99$, $V = .03$). These findings, illustrated in Fig. 1, suggest that participants who viewed the AI-generated virtual instructor were just as likely as participants who viewed the traditional instructor video to agree that the overall learning experience was positive, the video met their expectations, the video improved their understanding of the content, and that they would be interested in taking other micro-learning courses like the one they took.

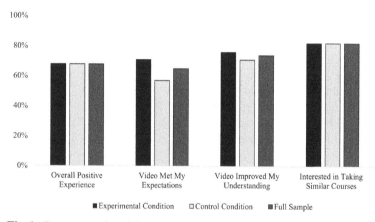

Fig. 1. Percentage of participants that agreed with each closed-ended statement.

Qualitative Findings. For the open-ended survey question, 20% of the sample provided a response. A qualitative thematic analysis identified five common themes in the open-ended responses, each of which were present in both the experimental and control conditions (see Fig. 2). Some responses to this question (e.g., *"I do not see where the AI generated content comes in"*) indicated that some participants in the experimental condition were not aware that the virtual instructor was synthetic. These qualitative findings suggest that the general perceptions of the instructional video content were similar for participants who viewed AI-generated virtual instructor and participants who viewed the traditional instructor video.

4 Discussion

The current study explored the efficacy of AI-generated learning videos with synthetic virtual instructors, comparing them to traditionally produced instructor videos. Knowledge gains from pre-to post-assessments were observed across the sample, with no differences between conditions. Further, no differences were observed in learner perceptions

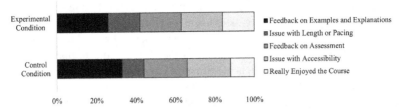

Fig. 2. Proportion of open-ended responses across the five major themes by condition.

of the AI-generated virtual instructor compared to the traditional instructor video (i.e., responses to closed-ended questions were nearly identical, and open-ended responses demonstrated no thematic differences between conditions). While previous research shows improved learning outcomes with pedagogical agents and avatars [8, 11], this study found equal learning between the traditional instructor video and AI-generated virtual instructor. This is likely an artifact of the decision to mimic talking head formats used in traditional instructor videos, a notably different approach from research focused on including meaningful gestures and emotional cues to improve learning [e.g., 9]. Nonetheless, this study highlights the viability of AI-generated virtual instructors as an alternative approach to human-made pedagogical agents and avatars and adds to the research on their utility in virtual learning.

Unlike traditional learning videos that require hours-to-days of human labor and are costly to produce, the time and cost of production for AI-generated learning videos is greatly reduced. For example, production for our traditional instructor video was completed over several days, and required hiring a studio location, camera crew, and video editors to process the footage at an approximate cost of $300 per finished minute. In contrast, all production elements for the AI-generated learning video (studio, filming, and editing) were flattened into one step that required a single editor and was completed in less than an hour at a cost of $15 per finished minute. Similarly, it is time-consuming to correct errors and update information in traditional learning videos, whereas generative AI tools can accomplish this in minutes. However, the fear of false or inaccurate information being presented to learners is one of the greatest challenges to leveraging these tools in support of education [7]. To address this concern and safeguard accuracy, we applied a human-in-the-loop approach using transcripts created with subject matter experts to populate the AI-generated learning videos. It should also be noted that AI-generated learning videos may not be sufficient for effective learning on their own. Rather, they should be integrated within lessons grounded in sound learning science and instructional design principles [13]. Taking a holistic approach like that in this study (i.e., presenting AI-generated learning videos alongside other pedagogical techniques such as assessments and opportunities for application), can help ensure that the AI-generated learning videos support overall learning.

This preliminary study was not without limitations. Due to the desire for brevity in the micro-learning courses the pre- and post-learning assessments were brief, limiting investigations into how AI-generated virtual instructors might differentially support various aspects of learning (e.g., difficulty of the material or complexity of concepts). Additionally, the study was likely underpowered to detect more nuanced differences in how learners perceive the two videos given our sample sizes. Future studies are warranted with a more robust learning assessment and larger sample sizes, to allow for a deeper dive into the efficacy and perceptions of AI-generated virtual instructors.

Despite its limitations, the current study is the first to indicate that learners have equal knowledge gains and experiences with an AI-generated virtual instructor as they do with traditional instructor videos. It demonstrates that using generative AI to create learning videos can be a cost-effective and time-efficient substitute for traditional video production in online education. These advantages can help deliver high quality educational content to learners across the globe even for fields with rapidly evolving research and technology such as sustainable and clean energy.

References

1. VanLehn, K., Banerjee, C., Milner, F., Wetzel, J.: Teaching algebraic model construction: a tutoring system, lessons learned and an evaluation. Int. J. Artif. Intell. Educ. **30**(3), 459–480 (2020). https://doi.org/10.1007/s40593-020-00205-3
2. Azevedo, R., Cromley, J.G., Seibert, D.: Does adaptive scaffolding facilitate students' ability to regulate their learning with hypermedia? Contemp. Educ. Psychol. **29**(3), 344–370 (2004). https://doi.org/10.1016/j.cedpsych.2003.09.002
3. Luckin, R., Holmes, W., Griffiths, M., Forcier, L.B.: Intelligence Unleashed: An argument for AI in Education. Pearson Education, London (2016)
4. Karras, T., Laine, S., Aittala, M., Hellsten, J., Lehtinen, J., Aila, T.: Analyzing and improving the image quality of StyleGAN. In: Proceedings of the IEEE/CVF Conference on Computer Vision and Pattern Recognition, pp. 8110–8119 (2020). https://doi.org/10.1109/cvpr42600.2020.00813
5. Zhang, Y., et al.: Learning to speak fluently in a foreign language: multilingual speech synthesis and cross-language voice cloning. In: Proceedings of the Interspeech, pp. 2080–2084 (2019). https://doi.org/10.21437/Interspeech.2019-2668
6. Mirsky, Y., Lee, W.: The creation and detection of deepfakes: a survey. ACM Comput. Surv. **54**, 1–41 (2012). https://doi.org/10.1145/3425780
7. Whittaker, L., Letheren, K., Mulcahy, R.: The rise of deepfakes: a conceptual framework and research agenda for marketing. Australas. Mark. J. **29**(3), 204–214 (2021). https://doi.org/10.1177/1839334921999479
8. Schroeder, N.L., Adesope, O.O., Gilbert, R.B.: How effective are pedagogical agents for learning? A meta-analytic review. J. Educ. Comput. Res. **49**(1), 1–39 (2013). https://doi.org/10.2190/EC.49.1.a
9. Wang, F.X., Li, W.J., Mayer, R.E., Liu, H.S.: Animated pedagogical agents as aids in multimedia learning: effects on eye fixations during learning and learning outcomes. J. Educ. Psychol. **110**, 250–268 (2018). https://doi.org/10.1037/edu0000221
10. Hudson, I., Hurter, J.: Avatar types matter: review of avatar literature for performance purposes. In: Lackey, S., Shumaker, R. (eds.) VAMR 2016. LNCS, vol. 9740, pp. 14–21. Springer, Cham (2016). https://doi.org/10.1007/978-3-319-39907-2_2
11. Mabanza, N., de Wet, L.: Determining the usability effect of pedagogical interface agents on adult computer literacy training. In: Ivanović, M., Jain, L.C. (eds.) E-Learning Paradigms and Applications. SCI, vol. 528, pp. 145–183. Springer, Heidelberg (2014). https://doi.org/10.1007/978-3-642-41965-2_6

12. Mayer, R.E.: Cognitive theory of multimedia learning. Cambridge Handb. Multimedia Learn. **41**, 31–48 (2005). https://doi.org/10.1017/cbo9780511816819.004
13. Clark, R.C., Mayer, R.E.: E-learning and the Science of Instruction: Proven Guidelines for Consumers and Designers of Multimedia Learning. John Wiley & Sons, New York (2016).https://doi.org/10.1002/9781119239086

Virtual Agent Approach for Teaching the Collaborative Problem Solving Skill of Negotiation

Emmanuel Johnson[(✉)], Jonathan Gratch, and Yolanda Gil

University of Southern California, Los Angeles, CA 90007, USA
ejohnson@isi.edu

Abstract. Collaborative Problem Solving(CPS) skills are increasingly needed for the future of work. However, most programs do not offer opportunities for students to develop such skills. A recent assessment of students' ability to solve problems collaboratively showed a significant global deficit. Previous research suggests that virtual agents are a promising tool for providing training on interpersonal skills, which is a crucial aspect of CPS. We believe that virtual agents can be adapted to teach specific CPS skills. However, there are still unanswered questions about the effectiveness of using virtual agents to train individuals on CPS, particularly in relation to teaching specific interpersonal skills. In this paper, we demonstrate how virtual agents can be used to teach the CPS skill of negotiation effectively. To do this, an online course was created using virtual agent based role-playing partners to teach the CPS skill of negotiation. Results show that participants who received instruction and feedback were able to think more globally about the negotiation and shift from focusing on improving outcomes of one issue towards making trade-offs across other issues of interest. These findings suggest that virtual agents have promise for teaching CPS skills. Our approach could be extended to teach other specific CPS skills using virtual agents.

Keywords: negotiation training · collaborative problem solving skills · virtual agents

1 Introduction

The modern workforce demands that individuals possess the ability to collaboratively solve problems within diverse groups. The 2015 Programme for International Student Assessment (PISA) conducted by the Organization for Economic Cooperation and Development (OECD) revealed only 8% of students globally performed at the highest level of proficiency and 29% scored at the lowest level of proficiency in Collaborative Problem Solving (CPS) [13]. These results underscore an urgent need to teach students CPS if we are to ensure their preparation for the future of work. Graesser et al. [6] recognize the need for teaching these skills and further point to the lack of training currently being provided. It can be difficult to teach CPS because it falls within what Aleven and colleagues define

N. Wang et al. (Eds.): AIED 2023, CCIS 1831, pp. 530–535, 2023.
https://doi.org/10.1007/978-3-031-36336-8_82

as an ill-defined domain [1], and presents a challenge for teaching because these skills lack clear assessment metrics and lack prescribed formulas to guarantee success.

However, CPS involves many cognitive and interpersonal skills that can be taught independently, and the work on assessing these skills provides insight into how to teach CPS. Emerging research suggests that virtual agents hold promise for teaching and assessing a range of interpersonal skills that underlies CPS [5,6,8] such as communication [3], perspective taking [6] and negotiation [7,10,12]. Thus, one approach for teaching CPS is to leverage virtual agent base experiential simulations coupled with traditional video lectures to provide personalized online CPS training.

To test the effectiveness of this approach, an online course was created and evaluated for teaching the collaborative problem-solving skill of negotiation. Results show that this approach was effective in improving students' negotiation abilities. Participants who received instruction and feedback were able to think more globally about the negotiation and shift from focusing on increasing their outcome on one issue to making trade-offs across multiple issues. These are important concepts to master in negotiation and are needed for CPS. This paper makes two major contributions. First, we developed an approach for teaching and assessing collaborative problem-solving skills using virtual agents and demonstrated its effectiveness. Second, we demonstrated a practical method for building an online course using virtual agents as role-playing partners.

The paper is structured as follows; In Sect. 2, we provide an overview of our negotiation mini-course. In Sect. 3, we describe our experiment to understand the impact of this course, and in Sect. 4, we present the results of our findings with discussions.

2   Building a Negotiation Training Course

We build a short course that draws on content from a larger semester-long online negotiation course and leverages virtual agent based role-playing simulations. We created videos focusing on the concepts of value claiming and value creating. The value claiming lecture video focuses on teaching tactics for obtaining the largest slice of the pie. It encourages participants to understand their Best Alternative to the Negotiation Outcome (BATNA), to establish a bottom line, to set reasonable aspirations, and to make ambitious initial offers and efficient concessions. The value creation lecture video highlights the importance of learning what one's opponent wants, making trade-offs, and focusing on interest rather than position. In particular, as novice negotiators often wrongly assume their opponent wants the same thing (the "fixed-pie bias"), the video focuses on overcoming this bias by exploring tradeoffs across issues. After watching the lecture videos, students are able to practice their negotiation with a virtual agent and get personalized feedback on their value claiming and value creation abilities.

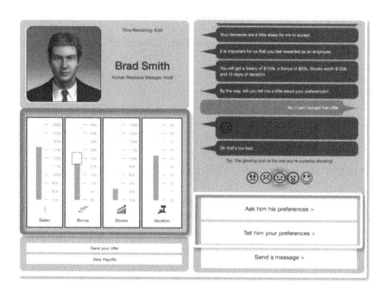

Fig. 1. Hiring Manager Agent built using IAGO. Users can exchange offers (highlighted in red), or exchange information about preferences over issues (highlighted in yellow).

3   Evaluating the Impact of an Online Mini Negotiation Course

To examine the effectiveness of the mini-course, we devised a two condition between-subjects experiment contrasting instruction and personalized feedback with mere practice (i.e., negotiating multiple times without pedagogical feedback).

Participants: A total of 150 (40 female) participants were recruited through Amazon Mechanical Turk to participate in the study. Of that 150, 49 were removed because the system crashed during their session and we could not recover their data or were removed for duplicate entries.

Negotiation Task: Participants engaged in a 4-issue salary negotiation for 10 min with a virtual agent built using the IAGO [11] where they assumed the role of an employee seeking a position at a large technology company and the agent assumed the role of the company's HR manager. The issues to be negotiated were salary, bonuses, stock options, and vacation days. For each issue, participants and the agent had to agree upon one of ten levels. For example, participants could negotiate a salary between $70,000-$160,000 in $10K increments (see Fig. 1). Each party received points based on the level they negotiated for each issue. For example, if participants negotiated a salary of $90,000 (l=3), they would receive 15 points. They were told the total number of points they would receive was the sum across the four issues (i.e., a linear utility function). Neither the agent nor the participant knew the other's preferences but the participant was assigned a preference. In addition to these preferences, participants were told they would receive only six points (their BATNA) if they failed to reach an agreement within the seven minutes allotted.

Experimental Manipulation: Participants were assigned to either the mere practice or personalized learning condition. Both groups completed a total of 4 learning modules but in different orders. All participants first watched an introductory video on negotiation and then completed a baseline negotiation. Those in the mere practice condition completed a second negotiation immediately after. They also had the opportunity to view the lessons and feedback after completing both simulations. Those in the personalized learning condition received instruction followed by practice and personalized feedback. After the baseline simulation, participants watched a value claiming lecture and received feedback on how well their baseline negotiation followed value claiming principles and then a video lesson on value creation as well as feedback on how well they followed those principles in their baseline simulation. Finally, participants in the personalized learning condition completed a second negotiation to compare any learning gains with the mere practice group. The feedback provided was based on metrics presented by Johnson et al. [9] for value claiming and value creating. The goal of this feedback was to improve participants' use of the negotiation principles.

Measures: Prior to the course, we gathered a number of demographic information and self-reported negotiation skill assessments. To assess the quality of the simulated negotiations outcome, we measure the points obtained by a participant and the joint points (i.e., the sum of points obtained by the participant and the agent). User points are a measure of value claiming. Joint points are a measure of value creation. Lastly, we assess how well participants grasped the principles covered in the video lectures by examining their aspirations and trade-offs made in the negotiation. After the second negotiation, we assessed their feelings about the negotiation outcome and their performance using the Subjective Value Inventory [4]. Specifically, we examined their feelings about the negotiation outcome, feelings of self, and feelings toward the relationship.

4   Results and Discussion

Negotiated Outcome: User points and joint points were examined with a 2 (condition: mere practice vs personalized learning) x 2 (time: negotiation 1 versus negotiation 2) repeated measures ANOVA to understand the impact of pedagogy and mere practice on participant outcomes. With regard to the participant's individual outcome (user points), there was a main effect of practice ($F(1,87)$ = 9.586, p = .003), but there was no interaction effect ($F(1,99)$ = .009, p = .924). This means that in both conditions participants are performing relatively the same. Analyzing joint points revealed similar results to user points, there is a main effect of time ($F(1,87)$ = 3.639, p = .059), but there is no interaction with condition ($F(1,87)$ = 1.037, p = .311). This shows that across both the mere practice and personalized learning condition, the participants claimed more joint value in the second negotiation compared to the first

Making Tradeoffs: To test the effects of training, we assessed whether or not participants were making tradeoffs. A 2 (condition: mere practice vs personalized training) x 2 (time: negotiation 1 vs negotiation 2) x 4 (issues: salary vs bonuses vs stocks vs vacation) repeated measure ANOVA was conducted. This analysis revealed a significant main effect of issues ($F(1,261) = 8.717$, $p = .000$) and time ($F(1,87) = 7.179$, $p = .009$). There is an interaction between issue, time, and condition ($F(3,261) = 3.180$, $p = .025$). This result shows a clear shift to thinking more globally in the type of deal participants found in the personalized learning condition. Participants in the personalized learning condition are making trade-offs by claiming more stock at the expense of salary. However, in the mere practice condition, participants are claiming more salary and similar levels of other issues. This suggests that participants in the personalized learning group are learning to make tradeoffs between issues whereas those in the mere practice groups are simply trying to get more salary.

Subjective Value of Negotiation. In assessing participants' feelings toward the negotiation, an independent t-test was conducted. We examined participants' feelings about the negotiation outcome, their performance, and their relationship with the agent. In terms of participants' feelings about the negotiation outcome, there is no significant effect of condition, $t(84) = -1.378$, $p = .305$, although those in the personalized learning condition ($M = 5.162$, $SD = .739$) scoring higher than the mere practice condition ($M = 4.92$, $SD = .866$). Analysis of how they felt about the negotiation revealed no significant effect of condition, $t(84) = -1.725$, $p = .846$. However, participants in the personalized learning condition ($M = 5.02$, $SD = .883$) did score higher than the mere practice condition ($M = 4.67$, $SD = .959$). These two results indicate that participants in both the mere practice and personalized learning groups felt the same regarding their final outcomes and their feelings about their performance. Lastly, we examined their feeling about the relationship with the agent. This revealed a significant effect of condition, $t(83) = -.398$, $p = .003$. Participants in the personalized learning condition ($M = 5.711$, $SD = .984$) rated the feeling of the relationship with the agent much higher than the mere practice condition ($M = 5.596$, $SD = 1.542$).

In this paper, we argue that CPS skills can be taught through an online mini-course that combines automated Participants who watched the training and received personalized feedback improved at making trade-offs between issues as well as aspired for more issues broadly. Although this did not lead to significantly better outcomes, this could be due to how participants in the personalized learning condition are making tradeoffs between salary and stock options. Given that salary is worth 5 points and stock options 2, this would naturally lower the value of the personalized learning group's total points. One explanation is that they may be bringing outside bias into the negotiation. Rather than focusing on the preferences assigned to them, they may be bringing their natural bias into the role-playing exercise. This is one of the problems that can occur in a role-playing exercise highlighted by Alexander and LeBaron [2]. These results suggest that utilizing virtual agents coupled with traditional video lectures could be effective in helping students improve their skills but further research is needed to understand how these results compare to real-world scenarios.

References

1. Aleven, V., Ashley, K., Lynch, C., Pinkwart, N.: Intelligent tutoring systems for ill-defined domains: Assessment and feedback in ill-defined domains. In: The 9th international Conference on Intelligent Tutoring Systems, pp. 23–27 (2008)

2. Alexander, N., LeBaron, M.: Death of the role-play. Hamline J. Pub. L. Pol'y **31**, 459 (2009)

3. Chollet, M., Sratou, G., Shapiro, A., Morency, L.P., Scherer, S.: An interactive virtual audience platform for public speaking training. In: Proceedings of the 2014 International Conference On Autonomous Agents and Multi-agent Systems, pp. 1657–1658. International Foundation for Autonomous Agents and Multiagent Systems (2014)

4. Curhan, J.R., Elfenbein, H.A., Eisenkraft, N.: The objective value of subjective value: A multi-round negotiation study. J. Appl. Soc. Psychol. **40**(3), 690–709 (2010)

5. Fenstermacher, K., Olympia, D., Sheridan, S.M.: Effectiveness of a computer-facilitated interactive social skills training program for boys with attention deficit hyperactivity disorder. Sch. Psychol. Q. **21**(2), 197 (2006)

6. Graesser, A.C., Fiore, S.M., Greiff, S., Andrews-Todd, J., Foltz, P.W., Hesse, F.W.: Advancing the science of collaborative problem solving. Psychol. Sci. Public Interest **19**(2), 59–92 (2018)

7. Gratch, J., DeVault, D., Lucas, G.: The benefits of virtual humans for teaching negotiation. In: Traum, D., Swartout, W., Khooshabeh, P., Kopp, S., Scherer, S., Leuski, A. (eds.) IVA 2016. LNCS (LNAI), vol. 10011, pp. 283–294. Springer, Cham (2016). https://doi.org/10.1007/978-3-319-47665-0_25

8. Häkkinen, P., Järvelä, S., Mäkitalo-Siegl, K., Ahonen, A., Näykki, P., Valtonen, T.: Preparing teacher-students for twenty-first-century learning practices (prep 21): a framework for enhancing collaborative problem-solving and strategic learning skills. Teach. Teach. **23**(1), 25–41 (2017)

9. Johnson, E., DeVault, D., Gratch, J.: Towards an autonomous agent that provides automated feedback on students' negotiation skills. In: Proceedings of the 16th Conference on Autonomous Agents and MultiAgent Systems, pp. 410–418. International Foundation for Autonomous Agents and Multiagent Systems (2017)

10. Kim, J.M., et al.: Bilat: A game-based environment for practicing negotiation in a cultural context. Int. J. Artif. Intell. Educ. **19**(3), 289–308 (2009)

11. Mell, J., Gratch, J.: Iago: interactive arbitration guide online. In: AAMAS, pp. 1510–1512 (2016)

12. Monahan, S., Johnson, E., Lucas, G., Finch, J., Gratch, J.: Autonomous agent that provides automated feedback improves negotiation skills. In: Penstein Rosé, C., et al. (eds.) AIED 2018. LNCS (LNAI), vol. 10948, pp. 225–229. Springer, Cham (2018). https://doi.org/10.1007/978-3-319-93846-2_41

13. Peña-López, I., et al.: Pisa 2015 results (volume v). collaborative problem solving (2017)

How Useful Are Educational Questions Generated by Large Language Models?

Sabina Elkins[1,2(✉)], Ekaterina Kochmar[2,3], Iulian Serban[2],
and Jackie C. K. Cheung[1,4]

[1] McGill University and MILA (Quebec Artificial Intelligence Institute),
Montreal, Canada
sabina.elkins@mail.mcgill.ca
[2] Korbit Technologies Inc., Montreal, Canada
[3] MBZUAI, Abu Dhabi, United Arab Emirates
[4] Canada CIFAR AI Chair, Montreal, Canada

Abstract. Controllable text generation (CTG) by large language models has a huge potential to transform education for teachers and students alike. Specifically, high quality and diverse question generation can dramatically reduce the load on teachers and improve the quality of their educational content. Recent work in this domain has made progress with generation, but fails to show that real teachers judge the generated questions as sufficiently useful for the classroom setting; or if instead the questions have errors and/or pedagogically unhelpful content. We conduct a human evaluation with teachers to assess the quality and usefulness of outputs from combining CTG and question taxonomies (Bloom's and a difficulty taxonomy). The results demonstrate that the questions generated are high quality and sufficiently useful, showing their promise for widespread use in the classroom setting.

Keywords: Controllable Text Generation · Personalized Learning ·
Prompting · Question Generation

1 Introduction

The rapidly growing popularity of large language models (LLMs) has taken the AI community and general public by storm. This attention can lead people to believe LLMs are the right solution for every problem. In reality, the question of the usefulness of LLMs and how to adapt them to real-life tasks is an open one.

Recent advancements in LLMs have raised questions about their impact on education, including promising use cases [1,8–10]. A robust question generation (QG) system has the potential to empower teachers by decreasing their cognitive load while creating teaching material. It could allow them to easily generate personalized content to fill the needs of different students by adapting questions to Bloom's taxonomy levels (i.e., learning goals) or difficulty levels. Already, interested teachers report huge efficiency increases using LLMs to generate questions [1,8]. These improvements hinge on the assumption that the candidates are high

N. Wang et al. (Eds.): AIED 2023, CCIS 1831, pp. 536–542, 2023.
https://doi.org/10.1007/978-3-031-36336-8_83

quality and are actually judged to be useful by teachers generally. To the best of our knowledge, there has yet to be a study assessing how a larger group of teachers perceive a set of candidates from LLMs.[1] We investigate if LLMs can generate different types of questions from a given context that teachers think are appropriate for use in the classroom. Our experiment shows this is the case, with high quality and usefulness ratings across two domains and 9 question types.

2 Background Research

Auto-regressive LLMs are deep learning models trained on huge corpora of data. Their training goal is to predict the next word in a sequence, given all of the previous words [11]. An example of an auto-regressive LLM is the GPT family of models, such as GPT-3. Recently, GPT-3 has been fine-tuned with reinforcement learning to create a powerful LLM called InstructGPT, which outperforms its predecessors in the GPT family [6]. Using human-annotated data, the creators of InstructGPT use supervised learning to train a reward model, which acts as a reward signal to learn to choose preferred outputs from GPT-3.

An emerging paradigm for text generation is to prompt (or 'ask') LLMs for a desired output [5]. This works by feeding an input prompt or 'query' (with a series of examples for a one- or few-shot setting) to a LLM. This paradigm has inspired a new research direction called *prompt engineering*. One of the most common approaches to prompt engineering involves prepending a string to the context given to a LLM for generation [4]. For controllable text generation (CTG), such a prefix must contain a control element, such as a keyword that will guide the generation [5].

Questions are one of the most basic methods used by teachers to educate. As this learning method is so broad, it uses many organizational taxonomies which take different approaches to divide questions into groups. One popular example is Bloom's taxonomy [3], which divides educational material into categories based on student's learning goals. Another example is a difficulty-level taxonomy, which usually divides questions into 3 categories of easy, medium, and hard [7]. By combining CTG and these question taxonomies, we open doors for question generation by prompting LLMs to meet specifications of the educational domain.

3 Methodology

3.1 Controllable Generation Parameters

Parameter settings used in this paper were guided by preliminary experimentation. Firstly, 'long' context passages (6-9 sentences) empirically appeared to improve generation. Secondly, the few-shot setting outperformed the zero-shot setting, with five-shot (i.e., with 5 context/related question type pairs included

---

[1] [10] show that subject matter experts can't distinguish between machine and human written questions, but state that a future direction is to assess CTG with teachers.

in the prompt) performing best. Few-shot generation is where prompts consist of an instruction (e.g., "Generate easy questions."), examples (e.g., set of n context/easy question pairs), and the desired task (e.g., context to generate from). Thirdly, there was not a large enough sample size to definitively say which question taxonomies are superior to use as control elements for CTG. Two representative taxonomies were chosen for the experiments in Sect. 3.2: Bloom's taxonomy [3] (which includes *remembering, understanding, applying, analyzing, evaluating,* and *creating* question types) and a difficulty-level taxonomy (which includes *beginner, intermediate,* and *advanced* question types) [7]. These taxonomies approach the organization of questions in different ways, by the learning goal and by complexity respectively. This creates an interesting comparison among the taxonomic categories to help explore the limits of the CTG approach.

3.2   Teacher Assessment Experiment

Question Generation. The human assessment experiment was conducted with candidates generated in the machine learning (ML) and biology (BIO) domains. There are 68 'long' context passages (6-9 sentences) pulled from Wikipedia (31 are ML, 37 are BIO). Using hand-crafted examples for 5-shot prompting, InstructGPT was prompted to generate 612 candidate questions.[2] Each passage has 9 candidates, one with each taxonomic category as the control element.

Annotators. There are two cohorts of annotators, BIO and ML. The 11 BIO annotators have biology teaching experience at least at a high school level, and were recruited on the freelance platform Up Work. The 8 ML annotators have CS, ML, AI, math or statistics teaching experience at a university level, and were recruited through word of mouth at McGill and Mila. All of the annotators are proficient English speakers and are from diverse demographics. Their teaching experience ranges from 1-on-1 tutoring to hosting lectures at a university. The experiments are identical for both cohorts. As such, the experiment is explained in a domain-agnostic manner. The results will be presented separately, as the goal of this work is not to show identical trends between the two domains, but that CTG is appropriate for education in general.

Metrics. Each annotator was trained to assess the generated candidates on two of four quality metrics, as well as a *usefulness* metric. This division was done to reduce the cognitive load on an individual annotator. The quality metrics are: *relevance* (binary variable representing if the question is related to the context), *adherence* (binary variable representing if the question is an instance of the desired question taxonomy level); and *grammar* (binary variable representing if the question is grammatically correct), *answerability* (binary variable representing if there is a text span from the context that is an answer/leads to one). The

---

[2] The passages, few-shot examples, prompt format, taxonomic level definitions, annotator demographics and raw results are available: https://tinyurl.com/y2hy8m4p.

relevance, grammar,[3] *answerability,* and *adherence* metrics are binary as they are objective measures, often seen in QG literature to assess typical failures of LLMs such as hallucinations or malformed outputs [5]. The subjective metric assessed, the *usefulness* metric, is rated on a scale because it is more nuanced. This is defined by a teacher's answer to the question: "Assume you wanted to teach about context X. Do you think candidate Y would be useful in a lesson, homework, quiz, etc.?" This ordinal metric has the following four categories: *not useful, useful with major edits* (taking more than a minute), *useful with minor edits* (taking less than a minute), and *useful with no edits.* If a teacher rates a question as *not useful* or *useful with major edits* we also ask them to select from a list of reasons why (or write their own).

Reducing Bias. We first conducted a pilot study to ensure the metrics and annotator training were unambiguous. We randomized the order of candidates presented and asked annotators to rate one metric at a time to avoid conflation. We included unmarked questions in order to ascertain if the annotators were paying attention. These questions were obviously wrong (e.g., a random question from a different context, a candidate with injected grammatical errors). Any annotators who did not agree on a minimum of 80% of these 'distractor' questions were excluded. The annotators' performance on these is discussed in Sect. 4.2.

4 Results and Analysis

4.1 Generation Overlap

We observed overlaps within the generated candidates for this experiment. Specifically, despite having different control elements, sometimes the LLM generates the same question for a given context passage twice. As a result, out of 612 candidates, there are 540 unique ones (88.24% are unique). We believe this overlap is low enough so the generated candidates are still sufficiently diverse for a teacher's needs. It is important to keep in mind that this overlap is not reflected in the following results, as teachers were asked to rank every candidate independently. Future work by the authors will remove this independence assumption.

4.2 Annotator Agreement

All of the participants annotated candidates from 6 context passages. In order to assess their agreement on the task, they annotated a 7th passage that was the same for all annotators in a given domain cohort. The results for each metric are reported in Table 1. In both domains, *relevance, grammar,* and *answerability* have between 85% and 100% observed agreement. The *adherence* metric has lower agreement, between 60% and 80%. Since this metric is more complex than the

---

[3] Despite not being a teacher's opinion, this is evaluated because we want to know the model's success here without relying on automatic assessment.

others and captures the annotators' interpretations of the question taxonomies, we consider this moderate agreement to be acceptable and expected.

Unlike the binary metrics, all candidates were rated on *usefulness* by two annotators. As before, only one context passage, the agreement on which is presented in Table 1, was seen by all annotators. Section 4.4 discusses the aggregation of the *usefulness* scores on the rest of the data. In both cohorts, the observed agreement on *usefulness* is around 63%. This metric is defined according to a teacher's opinion, and as such is subjective. Thus, the lower agreement between annotators is to be expected. Using Cohen's κ to measure the agreement yields a $\kappa = 0.537$ for the ML cohort and a $\kappa = 0.611$ for the BIO cohort, which implies moderate and substantial agreement respectively [2]. Additionally, the agreement of the annotators on the included 'distractor' candidates for this metric (see Sect. 3.2) is $\kappa = 1$ (i.e., perfect agreement), which shows that the annotators agree on the fundamental task but might find different questions useful for their particular approach to teaching.

4.3 Quality Metrics

Three quality metrics, *relevance*, *grammar*, and *answerability*, are consistently high for all generated candidates (see in Table 1). The fourth quality metric, *adherence*, varies across the taxonomic categories as seen in Fig. 1a. This variation is similar within the two domains. As might be expected, the more objective categories are easier for the LLM to generate. For instance, looking only at the 'remembering' category has an *adherence* of 83.3% for the ML cohort and 91.7% for the BIO cohort. This category is intended to ask for a student to recall a fact or definition. This might be simple for the LLM to replicate by identifying a relevant text span, and reflects the traditional QG task. By contrast, asking a LLM to generate a 'creating' question is a more open-ended problem, where a text span from the context may not be the answer. Accordingly, the model struggles on this less constrained task, and has an *adherence* of only 40.0% for the ML cohort and 36.1% for the BIO cohort.

Table 1. The quality metrics' mean (μ), standard deviation (σ), and observed agreement (i.e., % of the time the annotators chose the same label).

Metric	$\mu \pm \sigma$ (ML)	Agreement % (ML)	$\mu \pm \sigma$ (BIO)	Agreement % (BIO)
Relevance	0.967±0.180	100	0.972±0.165	100
Grammar	0.957±0.204	92.6	0.970±0.170	100
Adherence	0.674±0.470	62.2	0.691±0.463	79.9
Answerability	0.914±0.282	94.4	0.930±0.256	86.7
Usefulness	3.509±0.670	62.7	3.593±0.682	62.8

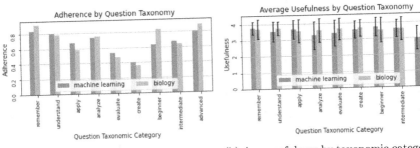

(a) *Adherence* by taxonomic category. (b) Avg. usefulness by taxonomic category.

Fig. 1. Visualizations of the *usefulness* and *adherence* metrics.

4.4 *Usefulness* Metric

The *usefulness* metric's ordinal categories (see Sect. 3.2) are mapped from 1 (*not useful*) to 4 (*useful with no edits*). The average usefulness for all candidates is 3.509 for the ML cohort and 3.593 for the BIO cohort. Note that an individual candidate's usefulness is already the average score between two annotator's ratings, and the whole average usefulness is the average across all candidates. This is a highly promising result showing that on average teachers find that these generated candidates will be useful in a classroom setting.

There is no significant difference between the usefulness scores of any of the question taxonomy categories, though some variation is present (see Fig. 1b). On average, each of the question taxonomies are rated between *useful with minor edits* and *useful with no edits* (i.e., [3,4]). Considering the *adherence* that differs across categories, it is also important to note that a question which does not adhere to its question taxonomy can still be useful in a different way than intended. 56.8% of the time the reason cited for 'not useful' candidates is related to their grammar or phrasing. This can possibly be fixed by a filter that removes malformed questions, but it will lower the available diversity of questions.

Conclusion. This work takes steps to demonstrate the realistic usefulness of applying CTG to generate educational questions. The results show that CTG is a highly promising method that teachers find useful in a classroom setting. We do not include baselines because the goal is not to show these questions are better than others, only to show they are of high enough quality. Limitations include the single LLM considered, the independence assumption seen in Sect. 4.1, and the lack of comparison between human and machine-authored questions. The authors plan to explore these avenues in future work. Applying generated candidates to form real-world lessons and evaluate their impact will demonstrate their ultimate value. CTG could pave the way for a new approach to education and transform the experiences of millions of teachers and students.

Acknowledgements. We'd like to thank Mitacs for their grant for this project, and CIFAR for their continued support. We are grateful to both the annotators for their time and the anonymous reviewers for their valuable feedback.

References

1. Baidoo-Anu, D., Owusu Ansah, L.: Education in the Era of Generative Artificial Intelligence (AI): Understanding the Potential Benefits of ChatGPT in Promoting Teaching and Learning (2023). Available at SSRN 4337484
2. Landis, J.R., Koch, G.G. The measurement of observer agreement for categorical data. Biometrics, 159–174 (1977)
3. Krathwohl, D.R.: A revision of Bloom's taxonomy: An overview. Theory Practice **41**(4), 212–218 (2002)
4. Liu, P., Yuan, W., Fu, J., Jiang, Z., Hayashi, H., Neubig, G.: Pre-train, prompt, and predict: A systematic survey of prompting methods in natural language processing. ACM Comput. Surv. **55**(9), 1–35 (2023)
5. Mulla, N., Gharpure, P.: Automatic question generation: a review of methodologies, datasets, evaluation metrics, and applications. Progress Artif. Intell., 1–32 (2023)
6. Ouyang, L., et al.: Training language models to follow instructions with human feedback. arXiv preprint arXiv:2203.02155 (2022)
7. Pérez, E.V., Santos, L.M.R., Pérez, M.J.V., de Castro Fernández, J.P., Martín, R.G.: Automatic classification of question difficulty level: Teachers' estimation vs. students' perception. In: 2012 Frontiers in Education Conference Proceedings, pp. 1–5. IEEE (2012)
8. Terwiesch, C.: Would Chat GPT Get a Wharton MBA? A Prediction Based on Its Performance in the Operations Management Course. Mack Institute for Innovation Management at the Wharton School, University of Pennsylvania (2023)
9. Wang, X., Fan, S., Houghton, J., Wang, L.: Towards Process-Oriented, Modular, and Versatile Question Generation that Meets Educational Needs. arXiv preprint arXiv:2205.00355 (2022)
10. Wang, Z., Valdez, J., Basu Mallick, D., Baraniuk, R.G.: Towards human-like educational question generation with large language models. In: Artificial Intelligence in Education: 23rd International Conference, AIED 2022, Durham, UK, 2022, Proceedings, Part I, pp. 153–166. Springer International Publishing, Cham (2022)
11. Zhang, H., Song, H., Li, S., Zhou, M., Song, D.: A survey of controllable text generation using transformer-based pre-trained language models. arXiv preprint arXiv:2201.05337 (2022)

Towards Extracting Adaptation Rules from Neural Networks

Ange Tato[1(✉)] and Roger Nkambou[2]

[1] École de Technologie Supérieure, Montreal, Canada
ange-adrienne.nyamen-tato@etsmtl.ca
[2] Université du Québec á Montréal, Montreal, Canada
nkambou.roger@uqam.ca

Abstract. Defining adaptation rules is an important step in the design of adaptive systems. This paper proposes using a constrained multimodal neural network to extract adaptation rules. The proposed approach enhances a serious game's adaptive capability, which aims to help learners improve their socio-moral reasoning skills. The neural network takes learners' multimodal data as input and predicts how it will answer an exercise. The rules extraction is based on reading and interpreting weights learned by the trained network to determine the players' attributes and the system's elements that play an important role in predicting the knowledge involved. The extracted rules are then validated using a decision tree. This validation shows that the proposed technique can support the production of adaptation rules in adaptive systems.

Keywords: adaptation rules · neural networks · decision tree

1 Introduction

The role of adaptation rules is to adapt the content according to the learner's specificity to improve learning. Generally, these rules are defined by domain experts or theories. Several systems provide personalized hints adapted to users using techniques such as Bayesian student models [13,15], or the Item Response Theory [9]. Much of the work on adaptation has focused on knowledge tracing to select the next problem or exercise to present to the learner, including deep knowledge tracing that uses an LSTM to represent the latent knowledge of the learner [12]. The adaptation requires selecting a set of resources for a specific situation (e.g., web content, exercises, a path to follow, etc.). These resources can be selected with rules [10]. These previous works only focus on selecting the next exercise as the key adaptation principle. However, that is not enough as many other factors can be considered, including improving learners' emotional experience or some gameplay items.

Neural networks (NN) can learn any function that maps inputs to outputs and are suitable when a lot of data is available. One reason for the success

Supported by NSERC.

of NN is that several research groups [6,7] have mathematically demonstrated that a NN with just a single hidden layer is a universal function approximator. However, a NN can be complex regarding explainability. First, when NNs are created, the initial values for the weights are set randomly. Thus, it is common for the relative importance of inputs to differ considerably across models [3,5]. Many approaches have been proposed to determine the relative importance of the input variables or for rule extraction in NNs. Rule extraction algorithms belong to three categories: decompositional, pedagogical, and eclectic [2]. Our research falls into decompositional techniques, where rules are extracted at the hidden and output neurons level by analyzing weight values [8,11]. The rules we want to extract from the NN are buried in its structure, and its weights assigned to the links between its nodes. We developed a constrained multi-modal NN that takes into account learner multi-modal information. The main goal of this work is to automatically determine the attributes of the learners and the gameplay elements that impact learning and then specify what actions to take when specific values for these attributes are observed.

2   Lesdilemmes: A Serious Game to Improve Socio-Moral Reasoning Skills

This work aims to make a serious game called LesDilemmes [14,15] adaptive. LesDilemmes is a first-person serious game that aims to assess and improve the development of the social reasoning skills of the player. Players face different socio-moral dilemmas in a 3D environment where they have to make decisions and are asked to provide oral justifications. To introduce feedback and scoring inside the game, simulated social feedback was added, showing a number of "likes" and "friends" depending on the player's responses. The learner model implemented in the learning environment includes three key dimensions: affective state, cognitive profile, and socio-moral reasoning profile.

The original So-Moral task includes five different levels of socio-moral reasoning [1]: (1) Authoritarian-based consequences, (2) Egocentric exchanges, (3) Interpersonal Focus, (4) Societal Regulation and (5) Societal Evaluation.

The domain experts provided adaptation rules focusing mainly on "learning" messages (feedback) and NPCs' reactions after the player's evaluation. Here is an example of 1 rule in LesDilemmes: (1) If the current reasoning level of the learner is r_p and she/he agrees with the justification of an NPC whose reasoning level is $r_n > r_p$, therefore, the NPC responds with a positive emotion (happy face) and do a "thumbs up"; (2) If the current reasoning level of the learner is r_p < 4 then we present an "encouraging" message and a "learning" message that will help him think like a person with a reasoning level $r > r_p$

3   Constrained Multimodal Neural Network

Multilayer perceptron (MLP) consists of an input layer, an output layer, and n hidden layers where n >= 1. The basic computation of a neuron with two inputs

x_1, x_2 can be written as follows: $y = f(w_1 \cdot x_1 + w_2 \cdot x_2 + b)$ where f is an activation function and b is the bias. From this equation, we can see that if the weights w_i are positives, f is a strictly increasing function, and the x_i are positives too, the importance of inputs become easy to determine. x_1 will be more important than x_2 if $w_1 > w_2$. We constrained our NN architecture with weights that can only be positives. It has been proven that constraining the NN's weights to be positive does not affect the Learning process [4]. All the input values have been normalized in $[0,1]$ to keep them positive. f is the Rectified Linear Unit (ReLU) function. The designed NN predicts the social moral reasoning level from learner multimodal data. The purpose of such a NN is to determine the learner's attributes and the gameplay elements that positively or negatively influence the knowledge element in play.

The model has three branches, each for a specific modality. A branch for the emotions (the seven basic emotions as well as the valence and the arousal), a branch representing the evaluation made by the player on the justifications of non-player characters (NPCs) (5 binary entries each representing whether the learner agrees (1) or disagrees (0) with the comments of the 5 NPC), and a last branch consisting of the reasoning level of the first NPC visited. The final layer of the NN is the classification layer, where the activation function is the Softmax. The labels to predict are $[1,0]$ (the reasoning level is lower or equal to 3) and $[0,1]$ (the reasoning level is higher than 3). We did not consider predicting each level separately because of the imbalanced data between the five levels. We did several runs to determine precisely the impact of each input on the output, and the results were averaged on these runs.

4   Experiments

4.1   Dataset

We conducted an experiment asking students to play the non-adaptive version of LesDilemmes. Twenty-nine youths participated in this study. During the game, nine dilemmas were presented. Their answers were recorded, and the verbal justification was transcribed using Google speech-to-text API and analyzed using a data mining algorithm [16]. The players were asked to evaluate the 5 NPCs in each dilemma. We also recorded the first NPC visited, as each NPC has a different reasoning level. We measured emotional reactions during the game using the Facereader analysis of the players' facial reactions. We obtained 261 data entries (29 * 9 dilemmas).

4.2   Results and Discussions

Each layer of the learned NN is determined by a matrix of weights and a vector of bias values. The training achieved an accuracy of 65 ~70 %, which is reasonable (given the amount of data and the simplicity of the architecture), but far from perfect. That is why the final results are combined with those from a decision tree. Decision trees represent transparent models because symbolic rules are easily extracted [2]. We used a decision tree trained on our dataset to assess the rules extracted from our NN model and then used its results to support the NN's results. From our decision tree, we can (for instance) extract the following rules: (1) A learner with a low value of Arousal has more chance to have a reasoning level higher than 3; (2) A learner who agrees with an NPC with level 2 has more chance to have a reasoning level higher than three if he feels less disgusted.

Those rules are based on the Gini index of the attributes and the number of samples respecting each of these rules. To extract the rules from the NN, we used the following process. Let W be the weight matrix of the penultimate layer where each row $W_j = [w_1, ..., w_n]$ represents the weights connected to each neuron j of the last layer. $n = 3$ (our three modalities) and $j \in 1,2$ where $j = 1$ means weights are connected to the neuron that fires a value close to 1 when the reasoning level is lower than 3 ([1,0]). To evaluate the importance of the inputs to the output prediction, we did not directly consider the weight matrix but the relative values. Thus, instead of w_i we used a_i, which is the result of applying the softmax function on all the weights connected to the same neuron.

$$a_{i,j} = \frac{e^{w_{i,j}}}{\sum_{i=1}^{n} w_{i,j}} \tag{1}$$

The value $A_{i,j}$ that we used to evaluate the importance of the input i on the prediction of each of the output j, based on all the runs k is computed as follows:

$$A_{i,j} = \frac{\sum_k a_{i,j,k}}{\sum_k \sum_j a_{i,j,k}} \tag{2}$$

In each layer, inputs with a higher value of $A_{i,j}$ are the most important. This process is done for every layer, and the results are shown in Tables 1, 2, 3. Now if we look at the A_i learned from the NN, we can see that: (a) Emotion (Arousal, surprised and happy) is the most important feature compared to other features to predict if the reasoning level is high or not (see the values of $A_{i,1}$ and $A_{i,2}$ in Table 1; (b) The Evaluation of NPC with reasoning levels 2 and 5 is more important than the evaluation of other NPCs (see the values of A_i in Table 2).

Table 1. Excerpt of the NN's weights of the penultimate layer, learned from 3 different training. $A_{i,j}$ is calculated after 20 runs.

Runs	Run 1		Run 2		Run 3		$A_{i,1}$	$A_{i,2}$
Output	[1,0]	[0,1]	[1,0]	[0,1]	[1,0]	[0,1]	[1,0]	[0,1]
Emotions	0.007	0.812	0	0.751	0.925	0.739	0.475	0.689
Eval NPCs	0.066	0	0.3231	0.040	0.043	0.378	0.367	0.276
First NPC visited	0.067	0.607	0.742	0.099	0.116	0.071	0.158	0.035

Table 2. Excerpt of the NN's weights of the neuron representing emotions, learned from 3 different training. A_i is calculated after 20 runs.

Neuron (Emotions)	Run 1	Run 2	Run 3	A_i
Happy	0	0.4748	0	0.120
Angry	0.069	0.702	0	0.090
Valence	0.346	0.006	0.045	0.128
Arousal	0.039	0.404	0.578	0.234

Table 3. Excerpt of the NN's weights of the neuron representing the evaluation on the 5 NPCs, learned from 3 different training. A_i is calculated after 20 runs.

Neuron (Eval NPCs)	Run 1	Run 2	Run 3	A_i
Eval 2	0.0.078	0.755	0	0.215
Eval 4	0.028	0.588	0	0,147
Eval 5	0.920	0.369	0	0.273

Based on these observations, we can see that the most important features to predict the reasoning level are the arousal, valence, and happy emotional states, but also the evaluation done on NPC with reasoning levels 2 and 5. The results are quite similar to that of the decision tree. However, we cannot specify the ideal values for those features directly from the learned weights, which is why we used the decision tree as a support. One advantage of using the NN is that, when the data gets bigger, the model will still be able to point out features that need to be observed and modified, compared to the decision tree. The final rules implemented in LesDilemmes are those from the NN. For example, to keep the player happy, we change the background music according to his emotions. We also force him to explore all the NPC since, in the first version, they could skip the other NPC after having visited 3 of them.

5   Conclusion

We proposed a first solution for extracting adaptation rules from a multimodal neural network. We carried out a preliminary experiment using data collected from LesDilemmes. We considered both the rules produced by the network and those produced by a decision tree built from the same dataset. This allows us not only to validate the adaptation rules extracted from the NN but also to semantically enrich those rules. Another interesting contribution is that the proposed solution also allows the identification of the attributes of the players and the gameplay elements on which the adaptation should focus. The extracted rules have been added to those produced by the experts and integrated into a new version of the game which is being experimented with. Future work will focus on assessing the impact of these rules on real-time learning. We will also carry out more studies in order to improve the accuracy of the proposed solution.

References

1. Beauchamp, M., Dooley, J.J., Anderson, V.: A preliminary investigation of moral reasoning and empathy after traumatic brain injury in adolescents. Brain Inj. **27**(7–8), 896–902 (2013)
2. Bologna, G., Hayashi, Y.: A comparison study on rule extraction from neural network ensembles, boosted shallow trees, and svms. Appli. Comput. Intell. Soft Comput. **2018** (2018)
3. Chakraborty, M., Biswas, S.K., Purkayastha, B.: Rule extraction from neural network trained using deep belief network and back propagation. Knowl. Inf. Syst. **62**(9), 3753–3781 (2020). https://doi.org/10.1007/s10115-020-01473-0
4. Chorowski, J., Zurada, J.M.: Learning understandable neural networks with non-negative weight constraints. IEEE Trans. Neural Netw. Learn. Syst. **26**(1), 62–69 (2015)
5. De Oña, J., Garrido, C.: Extracting the contribution of independent variables in neural network models: a new approach to handle instability. Neural Comput. Applicat. **25**(3-4), 859–869 (2014)
6. Funahashi, K.I.: On the approximate realization of continuous mappings by neural networks. Neural Netw. **2**(3), 183–192 (1989)
7. Hornik, K., Stinchcombe, M., White, H.: Multilayer feedforward networks are universal approximators. Neural Netw. **2**(5), 359–366 (1989)
8. Lu, H., Setiono, R., Liu, H.: Neurorule: A connectionist approach to data mining. arXiv preprint arXiv:1701.01358 (2017)
9. Manouselis, N., Sampson, D.: A multi-criteria model to support automatic recommendation of e-learning quality approaches. In: EdMedia: World Conference on Educational Media and Technology, pp. 518–526. Association for the Advancement of Computing in Education (AACE) (2004)
10. Muñoz-Merino, P.J., Kloos, C.D., Muñoz-Organero, M., Pardo, A.: A software engineering model for the development of adaptation rules and its application in a hinting adaptive e-learning system. Comput. Sci. Inf. Syst. **12**(1), 203–231 (2015)
11. Murdoch, W.J., Szlam, A.: Automatic rule extraction from long short term memory networks. arXiv preprint arXiv:1702.02540 (2017)
12. Piech, C., et al.: Deep knowledge tracing. In: Advances in Neural Information Processing Systems, pp. 505–513 (2015)
13. Suebnukarn, S., Haddawy, P.: A bayesian approach to generating tutorial hints in a collaborative medical problem-based learning system. Artif. Intell. Med. **38**(1), 5–24 (2006)
14. Tato, A., Nkambou, R.: Infusing expert knowledge into a deep neural network using attention mechanism for personalized learning environments. Front. Artif. Intell., 128 (2022)
15. Tato, A., Nkambou, R., Brisson, J., Robert, S.: Predicting learner's deductive reasoning skills using a bayesian network. In: André, E., Baker, R., Hu, X., Rodrigo, M.M.T., du Boulay, B. (eds.) AIED 2017. LNCS (LNAI), vol. 10331, pp. 381–392. Springer, Cham (2017). https://doi.org/10.1007/978-3-319-61425-0_32
16. Tato, A.A.N., Nkambou, R., Dufresne, A.: Convolutional neural network for automatic detection of sociomoral reasoning level. In: Proceedings of the 10th International Conference on Educational Data Mining, EDM (2017)

A Support System to Help Teachers Design Course Plans Conforming to National Curriculum Guidelines

Yo Ehara[✉]

Tokyo Gakugei University, Koganei, Tokyo 1848501, Japan
`ehara@u-gakugei.ac.jp`

Abstract. This study addresses the challenge of ensuring the alignment of educational materials with the objectives of public education guidelines outlined in official documents. In Japan, "Courses of Study (COS)," issued by the government, provides a detailed outline for curricula in public and private education. Teachers ought to design lessons based on this outline. This can be a time-consuming and challenging task, particularly. Accordingly, this paper proposes a new support system for designing curricula that performs an automatic search of parts of COS corresponding to teaching materials. Computational retrieval via advanced deep transfer learning methods such as BERT and DeBERTa is expected to be effective and time-saving in the implementation of COS. Using a novel dataset linking textbook text with the COS, the proposed method annotates relevant sections of the COS. Experimental results showed that the proposed method could present the corresponding sections of the COS with a statistically significantly higher mean average precision (mAP) in both unsupervised and supervised learning compared to methods that do not consider semantic factors, such as edit distance.

Keywords: Support System · Course Plans · Information Retrieval

1 Introduction

The Japanese government has implemented uniform standards throughout the country by thoroughly determining the educational syllabi of compulsory and secondary education in both public and private schools. For example, the concept of IP addresses must be taught in technology classes in the first year of junior high school, which falls under compulsory education.

Uniform standards of education in a densely populated country such as Japan are rare. Conversely, in many countries, educational administration differs regionally. This highly standardized education system is one of the reasons for Japan's high-quality primary and secondary education, ranking second in the world in student skills at the age of fifteen in the OECD's Program for International Student Assessment.

N. Wang et al. (Eds.): AIED 2023, CCIS 1831, pp. 549–554, 2023.
https://doi.org/10.1007/978-3-031-36336-8_85

The center of the standardized education system is Courses of Study (COS), which are guidelines that describe and define the educational content in detail. To meet this high-level standard, Japanese primary and secondary school teachers are required to plan and implement lessons per the COS. However, preparing lesson plans based on the COS guidelines requires significant effort and advanced knowledge. Currently, school teachers manually implement COS in their lesson plans.

Thus, to reduce teachers' workload while maintaining high-quality elementary and secondary education based on the COS, this paper proposes a method for automatically retrieving the sentences present in the COS that correspond to each sentence in lesson plans whilst taking semantics into account.

The volume of the COS is large; therefore, it cannot be easily read and implemented by any single person but can be conveniently processed by a computer. Thus, we expected computational retrieval via advanced deep transfer learning methods such as BERT [1] or DeBERTa [5] to be effective.

Furthermore, to evaluate the retrieval performance, we created a novel dataset linking the textbook text, which is the basis of the text written in the lesson plans, with the COS. Specifically, we annotated, sentence-by-sentence, the relevant sections of the COS that should be presented when the textbook text is input by four qualified Japanese elementary and secondary school students. For each natural sentence query, a dataset was created with multiple corresponding sentences in the COS. Notably, a very short text was used in this search.

Experimental results showed that the proposed method could present the corresponding sections of the COS with a statistically significantly higher mean average precision (mAP) in both unsupervised and supervised learning than methods that do not consider semantic factors, such as edit distance. The highest precision was 0.29, which is practical considering that the search results were sentences.

We used the official "the Explanatory Guide to the Courses of Study for Junior High Schools" in information technology for evaluation [2]. For ease of evaluation, we built our dataset targeting information technology. To the best of our knowledge, only three textbooks in Japan have been officially approved to conform to the COS by the Ministry of Education, Culture, Sports, Science, and Technology (MEXT).

The textbooks used are as follows[1]:

Textbook1. "New Technology and Home Economics: Technology to Create the Future." from Tokyo Shoseki.

Textbook2. "Junior High School Technology Course." from Kyoiku Tosho Shuppan.

Handbook of Textbook2. "Handbook of Technology" from Kyoiku Tosho Shuppan.

Textbook3. "Junior High School Technology" from Kairyudo Shuppan.

To evaluate the performance, we created a dataset in which annotators manually evaluate the kind of search results they would be happy to receive when entering textbook text. All annotators were students holding bachelor's degrees

---

[1] We simply list their title and publisher here.

specializing in technology education in Japanese compulsory education. The sentences in the pages handled by the textbooks and those handled by the COS were separated by punctuation marks and line breaks, respectively, and visually formed into sentence-by-sentence data using a text editor. Next, the annotators manually read each sentence in the sentence-separated textbook and searched for the corresponding parts of sentences in the COS to annotate.

Using these data, we evaluated the performance of the retrieval system by examining the degree of accuracy with which it matched human perception when the system retrieved the COS commentary from textbooks. The retrieval system prepares a function to measure the degree of similarity between the retrieved sentences and the candidate's search results, displaying output as search results in order of increasing similarity. Three methods were compared as a baseline: editing distance, unsupervised deep learning, and supervised deep learning.

The edit distance is the distance between two character strings, which is calculated by repeatedly inserting, deleting, and replacing a single character to determine how much it differs between the two strings and the minimum number of times it is inserted, deleted, and replaced. It does not consider the meaning of the sentence, as it only considers the characters.

For deep learning (unsupervised), we used the pre-trained model "ku-nlp/deberta-v2-base-japanese", which is an improved version of BERT, DeBERTa, created by a team from Kyoto University using 142 GiB Japanese text. The similarity between sentences was calculated using the size of the inner product of the sentence vector, representing the meaning of each sentence output by the model.

For deep learning (supervised), additional learning or fine-tuning was performed on a pre-trained model. Specifically, a binary classifier was trained to identify whether there was a correspondence between pairs of textbooks and commentary sentences. The supervised classifier constructed with BERT could output the probability of correspondence between input pairs, which is used as the similarity. As a pre-trained model, we used "cl-tohoku/bert-base-japanese-whole-word-masking" by a team from Tohoku University.

For evaluating the performance, we retrieved 50 textbook sentences and the corresponding evaluation data from the COS. For deep learning (supervised), other textbook sentences and corresponding instruction manual data were used as training data.

The standard mAP was used to measure the retrieval accuracy. The average precision (AP) for a textbook sentence is the value obtained by averaging the percentage (precision) of comments related to the textbook sentence out of the top M comments of the COS presented by the search system when searching for that textbook sentence while moving M. mAP is the value obtained by averaging the number of comments that were related to the textbook sentence from the top M comments of the COS presented by the search system when searching for the textbook sentence. The mAP is the average of the APs for all textbook texts used in the measurement; that is, the mAP is the average fit rate for each class and the average fit rate for all classes. The mAP has a maximum value of 1.0, where higher values correspond to a higher accuracy.

Table 1. Accuracies of Textbook1

Method	mAP
Baseline	0.062
DeBERTa (Unsupervised)	0.010 (**)
BERT (Supervised)	0.299 (**)

Table 2. Accuracies of Textbook2

Method	mAP
Baseline	0.034
DeBERTa (Unsupervised)	0.083 (**)
BERT (Supervised)	0.151 (**)

Table 3. Accuracies of Handbook of Textbook2

Method	mAP
Baseline	0.063
DeBERTa (Unsupervised)	0.184 (**)
BERT (Supervised)	0.119 (**)

Table 4. Accuracies of Textbook3

Method	mAP
Baseline	0.050
DeBERTa (Unsupervised)	0.104 (**)
BERT (Supervised)	0.202 (**)

Table 1 shows that, of the three methods, edit distance has the lowest accuracy and deep learning (supervised) has the highest accuracy. (**) denotes statistical significance compared to the baseline (Wilcoxon test, $p < 0.01$).

These results indicate that the performance of the retrieval system can be improved using pre-trained models and training data while considering semantics as compared with baselines. Both the supervised and unsupervised methods significantly outperformed the baseline methods. In addition, a value of 0.299 means that, in an actual search system, if a sentence in a textbook is input and the number of similar outputs is 10, approximately three of the ten outputs will show the corresponding part of the sentence. Considering that we are searching for sentences rather than documents and that each sentence can be read by a native Japanese speaker in about 10 s, this result is practical.

Table 2, Table 3, and Table 4 show the accuracies of the other three textbooks and handbooks. Overall, the results consistently show that the proposed methods outperform the baseline methods in all textbooks and handbooks. As shown in Table 3, the accuracy of the supervised method is lower than that of the unsupervised method. This is presumably because this targets the handbook of Textbook 2. The handbook lists only examples rather than technical explanations, and many expressions appear only once in the handbook, which is why the accuracy of "supervised" is lower than that of "unsupervised."

In contrast to general searches, this study uses a sentence-by-sentence search; therefore, users can display the search results. Therefore, considering that the presentation of 10 items corresponds to 10 texts and that the user can read the full text in a short time, this study can be considered to have sufficient practical applicability.

2 Related Work

A detailed English-language reference on technology education as compulsory education in Japan can be found in [6]. In this study, BERT was used to perform the semantic search. Similarly, many systems have been developed to perform semantic searches using BERT. A recent study targeting Japanese users performed a medical semantic search [4].

In Japan, search engines specializing in educational content have also been developed [7]. However, because technologies such as BERT have emerged rather recently, semantic searches using natural sentences have not been conducted in previous studies.

3 Conclusion

In Japan, a high standard of general education is provided throughout the country, in compliance with the COS. However, with a great deal of effort, the teachers must make instructional plans that conform to the COS. In this study, for the first time, an attempt was made to develop a method to automatically search the COS sentence-to-sentence to create an instructional plan that conforms to the COS.

We evaluated the proposed method by creating a dataset of all technology textbooks published in Japan with a focus on information technology. The proposed method exhibited an mAP of 0.29. In this case, because the search target is a sentence, a human can read each sentence in approximately 10 s. Therefore the proposed method, which displays appropriate examples in three out of ten cases, is practical.

Future work will include improving the evaluation data with more thorough annotations and accelerating the semantic search using point-cloud technology. Also, we can make smart and personalized readability assessments [3] considering the related information obtained from COS.

Acknowledgements. This work was supported by JST ACT-X Grant Number JPM-JAX2006, Japan. We thank Ryotaro Hashiguchi, Mako Kihara, Tokiwa Moriya, and Taichi Tomita for their cooperation in preparing the dataset and some parts of the experiments.

References

1. Devlin, J., Chang, M.W., Lee, K., Toutanova, K.: BERT: pre-training of deep bidirectional transformers for language understanding. In: Proceedings of NAACL (2019)
2. Ministry of Education, Culture, Sports, Science and Technology of Japan: The courses of study (2017 version) (2017). https://www.mext.go.jp/a_menu/shotou/new-cs/youryou/eiyaku/1298356.html
3. Ehara, Y.: No meaning left unlearned: predicting learners' knowledge of atypical meanings of words from vocabulary tests for their typical meanings. In: Proceedings of Educational Data Mining (short paper) (2022)
4. Fujishiro, N., Otaki, Y., Kawachi, S.: Accuracy of the sentence-BERT semantic search system for a Japanese database of closed medical malpractice claims. Appl. Sci. **13**(6) (2023). https://doi.org/10.3390/app13064051. https://www.mdpi.com/2076-3417/13/6/4051
5. He, P., Liu, X., Gao, J., Chen, W.: DeBERTa: decoding-enhanced BERT with disentangled attention. arXiv preprint arXiv:2006.03654 (2020)
6. Muramatsu, H.: Trends of technology education in compulsory education in Japan. J. Rob. Mech. **29**(6), 952–956 (2017). https://doi.org/10.20965/jrm.2017.p0952
7. Yoshii, A., Yamada, T., Shimizu, Y.: Development of federated search system for sharing learning objects between NIME-glad and overseas gateways. Educ. Technol. Res. **31**(1–2), 125–132 (2008)

Predicting Student Scores Using Browsing Data and Content Information of Learning Materials

Sayaka Kogishi[1]([✉]), Tsubasa Minematsu[2], Atsushi Shimada[3], and Hiroaki Kawashima[1]

[1] Graduate School of Information Science, University of Hyogo, Kobe, Japan
kogi.kage.vboh@gmail.com, kawashima@gsis.u-hyogo.ac.jp
[2] Data-Driven Innovation Initiative, Kyushu University, Fukuoka, Japan
minematsu@ait.kyushu-u.ac.jp
[3] Faculty of Information Science and Electrical Engineering, Kyushu University, Fukuoka, Japan
atsushi@ait.kyushu-u.ac.jp

Abstract. Recent digital material delivery systems enable teachers not only to upload lecture materials but to analyze students' behavior, such as browsing data with detailed operation logs that record which student performed which operation on which page at which time. While such behavioral data has been elucidated to be useful for predicting students' performance in existing studies, it has yet to be fully verified how content (e.g., learning materials) information can be integrated with behavioral data. This paper proposes methods to utilize content information jointly with behavioral data and compares them with the baseline method using only behavioral data. The results indicate that one of the proposed methods performs better prediction of quiz-score prediction. This suggests that both the browsing behavior of students and the content information have an impact on student performance.

Keywords: browsing data · learning content · score prediction

1 Introduction

Students' behavior in learning situations not only provides information on their commitment to learning but also is an important predictor of their academic performance, such as test scores and final grades. Recent advancements in digital material delivery systems enable teachers not only to upload lecture materials but to analyze students' behavior. The behavioral data are captured as detailed operation logs that record which student performed which operation on which page at which time. Those behavioral data are important clues to understanding students' performances and topics they are finding difficulties.

Some studies reveal that students' behavioral data are important clues to predict their performance [3,5,6], which helps teachers approach each student

N. Wang et al. (Eds.): AIED 2023, CCIS 1831, pp. 555–560, 2023.
https://doi.org/10.1007/978-3-031-36336-8_86

at an early stage and leads to an improvement in students' academic ability. However, it has not been fully verified how content (e.g., learning materials) information can be integrated with behavioral data for such prediction.

In this study, we aim to improve the accuracy of estimating each student's performance by using text information from lecture materials (referred to as "content" in this paper), such as slides used in lectures, in addition to the browsing data that records the learner's behavior. Here, the performance is defined as the score of the weekly quiz after the lecture, and the browsing data is based on the operation logs obtained from an e-Book system. We will verify how much the use of content information contributes to the prediction of the quiz scores.

2   Methods

2.1   Behavioral Data Capturing

In this study, we utilize log data captured by a digital material delivery system (i.e., an e-Book system) called BookRoll. The BookRoll system [7] allows students to browse lecture materials uploaded by faculty members online. At the same time, the system can store various log data regarding students' behavior. For example, operation type is recorded along with information such as the ID of the student who performed the operation, date and time, content ID, and page number. Operation type includes, for example, opening/closing the content, going to the next page, going back to the previous page, adding a marker or memo, and searching within the content.

2.2   Overview of the Proposed Methods

Figure 1 shows the overview of the proposed methods. From the browsing data obtained from BookRoll, the number of operations for each operation type and browsing time on each page in each content are calculated for each student. Browsing time is calculated by extracting only the actions related to page transitions from the browsing data and computing the time interval between actions.

Then, these features are summarized in a vector representation, which we refer to as a *behavioral vector*, for each student. Here, we denote the feature vector of student i's behavior in content c as $\boldsymbol{u}_c^{(i)}$. The dimensionality of $\boldsymbol{u}_c^{(i)}$ is "(the number of pages in content c) × (the number of operating types + 1)," where the addition of one element corresponds to the browsing time, and the number of pages is different for each content.

We consider the score prediction using only the behavioral vector as a baseline (the upper flow in Fig. 1). In addition to the behavioral data, this study uses content information, specifically text data in the slides and quiz questions in the lectures. We embed those text information as vector representations (the lower flow in Fig. 1), as will be explained later. The behavioral and content information are then integrated as a single feature vector for each student in each content and used as a feature vector to predict the learner's score using a machine learning

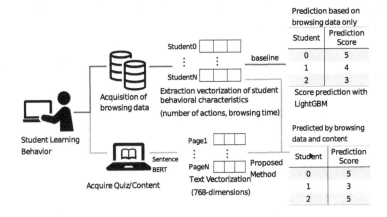

Fig. 1. Overview of the proposed methods

model. Here, we propose two methods to construct feature vectors using both behavioral and content information. Method 1 focuses on using text information in slide pages that students have browsed for a long time. Method 2 focuses on using behavioral features (i.e., operation frequencies) on pages related to quizzes.

2.3   Method1: Using Content Information Weighted by Behavior

In Method 1, the behavioral feature vector is concatenated with the feature vector obtained from the content the student has browsed for a long time. Then, the concatenated feature vector is used for score prediction. Behavioral features are known to be useful in existing studies [3,5,6]. We aim to further improve the prediction accuracy with the information each learner might have watched.

First, the textual information of each page in the content c is vectorized. For the vectorization, we use a pre-trained Sentence-BERT [1][1], which yields a 768-dimensional *page vector* $\boldsymbol{v}_{(c,p)}$ for each page p, where the vector is normalized as a unit vector. Here, to obtain a vector that contains more content information on pages that each student has browsed for a long time, the page vectors are weighted based on $t^{(i)}_{(c,p)}$, the sum of the browsing time of student i for page p, and then added together. Specifically, the browsing time is directly weighted by each page $w^{(i)}_{(c,p)} = t^{(i)}_{(c,p)}$ as the linear combination of the page vectors, i.e., $\boldsymbol{v}^{(i)}_c = \sum_p w^{(i)}_{(c,p)} \boldsymbol{v}_{(c,p)}$, which is then normalized to a unit vector. We refer to $\boldsymbol{v}^{(i)}_c$ as the *browsed-content vector*. For the calculation of browsing time $t^{(i)}_{(c,p)}$ for each page, log data that had the same page open continuously for longer than five minutes were considered to have been left open for an extended period of time and were excluded from the summation.

---

[1] A BERT model, in which the original model [2] is fine-tuned with Multiple Negatives Ranking Loss on an unpublished data set [1].

The obtained behavioral vector $\boldsymbol{u}_c^{(i)}$ and the browsed-content vector $\boldsymbol{v}_c^{(i)}$ are concatenated to obtain the following feature vector.

$$\boldsymbol{f}_{1,c}^{(i)} = (\boldsymbol{u}_c^{(i)}, \boldsymbol{v}_c^{(i)}) \tag{1}$$

Method 1 uses this feature vector $\boldsymbol{f}_{1,c}^{(i)}$ to predict scores. In the evaluation, we also evaluate each behavioral vector $\boldsymbol{u}_c^{(i)}$ (baseline) and the browsed-content vector $\boldsymbol{v}_c^{(i)}$ separately.

2.4 Method 2: Using Behavioral Data in the Pages Related to Quiz

In Method 2, the behavior on the pages related to each quiz is extracted to predict the score. For each question j in the quiz q, we create a *quiz vector* $\boldsymbol{Q}_{(q,j)}$ by vectorizing all question sentences and multiple correct answers, where the Sentence-BERT (see Sect. 2.3) is used for the vectorization. Then, we use the cosine similarity to measure the relationship between quiz vector $\boldsymbol{Q}_{(q,j)}$ ($j = 1, ..., 5$) and page vector $\boldsymbol{v}_{(c,p)}$. We use the computed similarity to weigh the behavior on each page to construct a feature vector. Let the sum of the similarities be the weight $w_{(c,p)}$ of page p. If the cosine similarity is less than 0.4, we do not include it in the summation. Here, $w_{(c,p)}$ indicates how much each page p is related to each quiz for content c. By weighting the per-page behavior feature vector $\boldsymbol{u}_{(c,p)}^{(i)}$ with this $w_{(c,p)}$, the following summarized behavioral feature vector is obtained, which is used to predict scores in Method 2.

$$\boldsymbol{f}_{2,c}^{(i)} = \sum_p w_{(c,p)} \boldsymbol{u}_{(c,p)}^{(i)} \tag{2}$$

3 Evaluation

3.1 Dataset

We use browsing data from the BookRoll system, content data (slides and quizzes used in the lectures), and score data (the results of quizzes given during the lectures), obtained from lectures on information-related subjects at Kyushu University in 2020. The lectures were 90-minute classes for 100 students over seven weeks, and the quiz was given in about 10 min at the end of each class time. The browsing data consisted of a total of 200,818 log lines. The quiz consisted of five multiple-choice questions. When a student submitted the quiz, the text of the question, the selected choice, the correctness of the answer (i.e., correct or not), and the submission time were recorded. The content includes a balanced mix of both text- and image-dominant pages, where images are often used for detailed explanations and concrete examples. Students could browse slides even during quiz time.

3.2   Prediction and Evaluation Methods

Scores are predicted using LightGBM [4] for each quiz and are evaluated by the root mean square error (RMSE), the average of every fold in 5-fold cross-validation. We employ LightGBM because of its computational efficiency for large-dimensional feature vectors and its capability of extracting important elements in feature vectors that contribute to the prediction.

A score out of 5 is used for the evaluation, calculated for each quiz. Since two contents were used in week 2 and two quizzes were given in week 2, the score prediction is divided into week 2(1) and week 2(2).

The vector would become too large if all operation types were used in the behavioral vectors. Therefore, we limited the number of operation types to be included in the feature vectors. Specifically, we excluded operation types that had a zero contribution to the prediction. To compute the contribution, we constructed a model that predicts the total score of all quizzes by LightGBM in advance. As a result, the number of operation types included in the behavioral vector was 12.

The feature vectors are calculated and compared in two conditions: using operations during only the lecture time including one hour before and after the class time (**in-class** condition) and using all operations including outside the class time (**all** condition).

3.3   Results

Scores are predicted for each quiz using the baseline method and each of the proposed methods (Method 1 and Method 2), respectively. The following feature vectors are used to compare the prediction results.

Behavioral vector (baseline) (b): $u_c^{(i)}$, the vector with the number of times each operation is performed on each page by each student as its elements.

Browsed-content vector (c): $v_c^{(i)}$, the weighted sum of page vectors using the browsing time of each page as the weight.

Method 1 (b+c): $f_{1,c}^{(i)}$, the concatenated vector of the behavior vector $u_c^{(i)}$ and the browsed-content vector $v_c^{(i)}$.

Method 2 (b'): $f_{2,c}^{(i)}$, the weighted sum of behavioral vectors on the pages related to the quiz for content c.

The average of the RMSEs obtained for each quiz in each method is shown in Table 1. In both the in-class and all conditions, the RMSE value was smaller when the browsed-content vector was used than when only the behavioral vector was used (baseline), and when both vectors were used (Method 1), the average value of RMSE was the smallest. On the other hand, Method 2 had the largest average value of RMSE, which was worse prediction accuracy than the baseline. This result could be attributed to the fact that content information is not directly used in Method 2 and that the information on per-page behaviors is summed. Meanwhile, Method 1 successfully focuses on which information student looked at more, which would have improved the prediction accuracy.

Table 1. Average of RMSE obtained for each quiz (on a scale of 0 to 5). Bold font indicates the best result in each condition.

Condition (period)	Baseline (b)	Browsed-content vector (c)	Proposed method 1 (b + c)	Proposed method 2 (b')
In-class	1.039	1.025	**1.009**	1.108
All	1.076	0.945	**0.941**	1.138

4   Conclusion

The results show that, compared to the baseline, the RMSE value is lower for the proposed method that utilizes vectorized text information in slide pages each student browses for a long time. This indicates that the presence of content information may have a positive impact on the accuracy of predicting students' quiz scores. Although not examined in this study, higher accuracy can be expected if the page vector includes information of more detailed content information, such as the layout of pages and figures other than text. Furthermore, although the prediction has been conducted with a score out of 5 points for each quiz, predicting correct/incorrect answers for each question may lead to a more in-depth prediction of student understanding.

Acknowledgements. This work was supported by JSPS KAKENHI Grant Number JP19H04226 and JP21H05302.

References

1. https://huggingface.co/sonoisa/sentence-bert-base-ja-mean-tokens-v2. Accessed 11 Jan 2023
2. https://huggingface.co/cl-tohoku/bert-base-japanese-whole-word-masking. Accessed 1 Feb 2023
3. Akçapınar, G., Hasnine, M.N., Majumdar, R., Flanagan, B., Ogata, H.: Developing an early-warning system for spotting at-risk students by using eBook interaction logs. Smart Learn. Environ. **6**(1), 1–15 (2019). https://doi.org/10.1186/s40561-019-0083-4
4. Ke, G., et al.: LightGBM: a highly efficient gradient boosting decision tree. In: Proceedings of Advances in Neural Information Processing Systems (NIPS) (2017)
5. Leelaluk, S., Minematsu, T., Taniguchi, Y., Okubo, F., Shimada, A.: Predicting student performance based on lecture materials data using neural network models. In: Proceedings of the Workshop on Predicting Performance Based on the Analysis of Reading Behavior, pp. 11–20 (2022)
6. Minematsu, T., Shimada, A., Taniguchi, R.: Analytics of the relationship between quiz scores and reading behaviors in face-to-face courses. In: Proceedings of the Workshop on Predicting Performance Based on the Analysis of Reading Behavior, pp. 1–6 (2017)
7. Ogata, H., et al.: Learning analytics for E-book-based educational big data in higher education. In: Yasuura, H., Kyung, C.-M., Liu, Y., Lin, Y.-L. (eds.) Smart Sensors at the IoT Frontier, pp. 327–350. Springer, Cham (2017). https://doi.org/10.1007/978-3-319-55345-0_13

Preserving Privacy of Face and Facial Expression in Computer Vision Data Collected in Learning Environments

T. S. Ashwin[✉][ID] and Ramkumar Rajendran[ID]

Indian Institute of Technology Bombay, Mumbai, India
ashwindixit9@gmail.com, ramkumar.rajendran@iitb.ac.in

Abstract. Learners' affective states are analyzed using their facial expressions obtained from image/video data. Deep learning architectures are the state-of-the-art affective state recognition methods used in the literature. Vast volumes of these data are used in the training, testing, and validation process. Now, sharing these data publicly for research purposes is a challenging task as it has privacy issues. The importance of creating a successful face de-identification algorithm for privacy protection cannot be overstated because the face is the single biometric feature that discloses the most recognizable qualities of a person in an image or a video frame. Existing approaches to face de-identification with facial expressions are not explored in the education domain. In this study, we design a methodology for a face de-identification technique that automatically creates a new face while maintaining the emotion and non-biometric facial characteristics of a target face from an input facial image. We consider a proxy set, which consists of a sizable number of synthetic faces produced by StyleGAN, and choose the proxy set face that most closely resembles the target face in terms of facial expression and position. Since the faces in the proxy settings were formed artificially, the face chosen from this collection is absolutely anonymous. We created a dataset of 10 students to test the methodology. The performance of StyleGAN was measured for standard parameters such as gender, emotion, age, etc., and the results show that the generated face preserved emotional attributes with a de-identified face.

Keywords: Privacy protection · Face de-identification · Facial Expressions · StyleGAN

1 Introduction

The face is one of many biometric aspects that is known to keep important identity information, hence it is crucial to mask facial identifying elements before publishing images or videos online (or through other data transmission medium). De-identification is a method for hiding a subject's identity so that they cannot be recognised by conventional biometric identification systems. Traditional methods of accomplishing privacy protection, including warping, pixelization, blurring, etc., are based on basic image processing techniques that also disguise other significant non-biometric information, like emotion and face expression, in addition to identity. Additionally, these methods do not ensure that the identification attributes have been entirely obscured [1,4].

With the introduction of e-learning tools that make it simple to track learner behaviour and performance, collecting student data has become easier. With the use of such data, educators and researchers can examine student learning and inform instruction by using learning analytics and educational data-mining tools [5]. To preserve student privacy, users of this data must be extremely careful as to how it is being used or shared as more information about students is gathered. The two methods that are frequently taken into account in data privacy are privacy-preserving data mining and privacy-preserving data publication. Sharing private data about people while maintaining their privacy is known as privacy-preserving data publishing. Data mining without using confidential material is referred to as privacy-preserving data mining. This research focuses on publishing data while protecting privacy [4,9].

Learning process can be understood significantly through the recognition of the emotions expressed. These emotions can be best identified through the facial expressions of the learners. This creates a need to collect the data from the learners where their facial expressions need to be captured and analysed [3,5,14]. In the scenario of a learning environment, there arises a situation where the data will be collected from a wide range of individuals like school going kids, Ph.D. scholars or elderly people and may be at times, the data from a renowned person may have to be considered. In such scenarios, there arises the need for de-identification of the face in the collected data sets as there will be ethical issues or misuse of the biometric features of the face. In order to avoid any such complexities, the concept of de-identification can be inculcated where the actual face will be faked as another face which will respect the privacy of the people. These de-identified faces will only have the expressions and the facial emotional features replicated hence facilitating the study [2]. This will be an added advantage to create a dataset without the actual faces or made available publicly at various repositories like PSLC datashop[1] and others. We may add such de-identified faces to these repositories where data from a wide range of regions or places can be collected and stored under a single roof for access for numerous studies. It may also be used as training data where a huge number of such de-identified faces can be generated with varied expressions leading to robustness of a study.

Several automated learning-based face de-identification techniques have been developed as a result of the development of deep neural network architectures, including generative models, differential privacy, and secure multi-party computation. These techniques have been applied in different use cases such as face recognition, object detection, medical imaging. However, these methods are not directly adaptable to the education domain due to variations in image vari-

---

[1] https://pslcdatashop.web.cmu.edu/.

ants such as occlusion in the classroom, race, cultural expressions, and learning environment, which can affect the background. There are libraries such as TensorFlow Privacy that can be used to train machine learning models with strong privacy guarantees [8]. But, these are limited to a particular set of libraries and the programming language used in the development process [2]. Hence, in this study we perform face de-identification by preserving some of the facial expression features.

We concentrate on face de-identification from photos in the current study as well. In order to achieve face de-identification while maintaining non-biometric facial features and expresssion features, we use a deep learning based architecture. The faces produced by this method are fully anonymous, meaning they bear no resemblance to any actual face. The following are the primary contributions:

– Design and develop a methodology that de-identifies faces while preserving the facial expressions related to learning-centered emotions.
– Analysis of methodology using the created data to apply and understand the performance of the face de-identified data.

The rest of the paper is organized as follows. Section 2 discusses the methodology. Section 3 demonstrates the performance of the methodology on created data and its performance metrics. Section 4 concludes the paper with future directions.

2 Methodology

The flow of the methodology is shown in Fig. 1. The methodological approach is adopted from [1]. We made several changes to it as the data belongs to different distributions, and we are considering expressions that are related to learning-centered emotions such as engaged, boredom, confused, and frustrated.

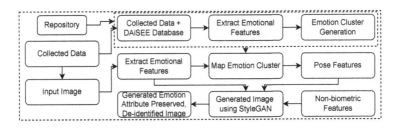

Fig. 1. Flow diagram of the proposed methodology.

We collected and annotated data related to learning-centered emotions in a computer-based learning environment. Here only one person is visible in a single image frame. We also considered the DAiSEE dataset (already annotated with learning-centered emotions and 112 learner videos. We did not use this existing database in the study and collected our data as it contains multiple emotion labels for a single image frame) [6]. We extracted the emotion features and clustered them using existing deep learning architecture. This entire set of clusters is stored in a database. Now two images of a different person from the same emotion cluster and similar pose are taken, and one of the faces is kept intact to extract the non-biometric attributes. A generative model is used to generate a new face with a feature fusion of non-biometric attributes, pose, and emotional features. This generated face is a de-identified face.

We used existing deep learning architecture that is already used for the learning-centered emotion classification and fine-tuned the model for better performance [3]. We trained the classification model using an existing database DAiSEE [6] and a small set of data (30%) from collected input image frames. The affective states considered are learning-centered emotions (engaged, boredom, frustrated, confused). Fine tuning of the emotion classification model is mentioned below. Inception-v3 based FaceNet architecture recognizes the human faces in a single image frame with better accuracy than state-of-the-art methods [11,13]. The FaceNet is trained and tested on the LFW and YouTube face database [7,15]. But, this architecture is not trained for affective state recognition. Transfer learning helps in optimizing and fine-tuning the deep learning architecture, and the FaceNet architecture already contains the trained weights required for face detection. Hence, we considered FaceNet as a base architecture and used it for transfer learning to build the model. To classify the affective states, we used the pretrained weights from the existing model [3]. The optimised model detects the learners' affective state classification using their facial expressions. In this architecture, we performed the convolution, activation, and pooling operations till we obtained a dimension less than 256 * 256 (3 extra pooling layers (2 convolutional layers for every pooling layer, leaky ReLU as activation function) were added to get to the dimension 256 * 256), then we added the inception-v3 based FaceNet architecture which factorizes into smaller convolutions and uses auxiliary classifiers for more stable learning and better convergence. Rest of the parameters and hyperparameters used in the study follows the deep learning architecture used in [3].

We used trained weights of the Inception v3 model as a pre-trained model as it is already trained on learning-centered emotions [3]. From this, emotional attributes are collected. This collected feature has redundant attributes and noise. Hence we used locally preserving projection (LPP) to address that. After applying LPP, we get reduced dimension data. The LPP-reduced dimension emotion vector takes the input image and applies the cosine similarity to find the best mating emotion cluster. This is performed using a k-means clustering algorithm. Once the cluster is decided, it matches the pose of the face in the same cluster using a regressor face pose[2]. This considers the parameters such as roll pitch and yaw angles for the input image frame. Now, we have two sets of emotional attributes and matching pose features; one set is real facial features, and the other is a proxy set. To get the non-biometric facial attributes ResNet

---

[2] https://github.com/arnaldog12/Deep-Learning/tree/master/problems.

model pre-trained on the VGG Face dataset is used [10]. We used the GAN[3] network used in [1] for the final step, where the generator and discriminator use emotional attributes, pose, and non-biometric features using three loss functions to generate a new image.

3   Results and Discussion

We measure the performance of the used StyleGAN for the following standard parameters: Gender, emotion, eye, mouth, age, smiling, head pose, and blurring and obtained 37%, 49%, 61%, 46%, 24 (Mean (M)) 31 (standard deviation (SD)), 52 (M) 61 (SD), 3.3 (M) 6 (SD), 5 (M) 7(SD), respectively. The performance shows that the generated face has preserved sufficient emotional attributes with a de-identified face.

To test the methodology, we collected data from 10 engineering graduate and undergraduate students (2 females, 8 males) aged between 25–35 years. Before the study, participants read and signed an informed consent, which was reviewed and approved by the Institute Review Board of Indian Institute of Technology Bombay. The learners interacted with PyGuru [12] for ~90 min to learn python programming in a computer-based learning environment. The lab set-up consisted of a laptop-based system with a screen size of 59.8 cm × 33.6 cm and a screen resolution of 1920 × 1080 pixels. The learning environment was hosted on a web browser and collected the facial expressions from the laptop web camera. The laptop has an inclined support bar that increases the visibility of the entire face in an upright position. The laptop's camera captured the frontal video recording with image frames of 30 frames per second. Though the number of students is less, the number of image frames is 13,60,800. We also used the DAiSEE dataset [6], which consists of 112 users, while mapping the facial pose and using it to generate face images.

The collected facial expression data did not have the ground truth, and it plays a vital part in understanding the working of the deep learning classification method. The collected data is annotated (annotator's inter-rater reliability is Cohen's kappa = 0.88) following the same procedure mentioned in [3]. We performed learner-independent 10-fold cross-validation. The overall results show an accuracy, precision, recall, and F1 score of 89.81%, 87.23%, 91.11%, and 85.89%, respectively. We used the generated images, tested the same set of real face images, and compared the results. The overall results show an accuracy, precision, recall, and F1 score of 68.11%, 65.39%, 64.53%, and 66.48%, respectively. Also, we conducted the ablation study and it is observed that the lower performance is due to the small set of students considered in the study, not enough variations in non biometric features, and also we did not use data augmentation. Further, the data distribution of facial features may differ more from the training set. Optimizing the deep learning method and using a huge number of students in the training set can make the generative model more robust and perform better in emotion classification.

---

[3] https://github.com/NVlabs/stylegan.

4 Conclusion

StyleGAN was used with emotional, pose, and non-biometric feature attributes to de-identify faces while preserving facial expression features. Data was created and the DAiSEE database was used to generate the emotion cluster repository. Deep learning was used to train and test the learning-centered emotions on the created data. Results were promising and compared to using generated images as training data. Performance metrics showed that the methodology could generate de-identified face images while preserving facial expression, allowing for robust and generalized data in education without privacy issues. However, factors such as context, cultural background, and specific emotions must be considered when working with facial expression data. Future work will address these issues in a learning environment setting.

References

1. Agarwal, A., Chattopadhyay, P., Wang, L.: Privacy preservation through facial de-identification with simultaneous emotion preservation. SIViP **15**(5), 951–958 (2021)
2. Akgun, S., Greenhow, C.: Artificial intelligence in education: addressing ethical challenges in K-12 settings. AI Ethics **2**(3), 431–440 (2022)
3. Ashwin, T., Guddeti, R.M.R.: Affective database for e-learning and classroom environments using Indian students' faces, hand gestures and body postures. Futur. Gener. Comput. Syst. **108**, 334–348 (2020)
4. Chaudhry, M.A., Kazim, E.: Artificial intelligence in education (AIEd): a high-level academic and industry note 2021. AI Ethics **2**(1), 157–165 (2022)
5. D'Mello, S.: Monitoring affective trajectories during complex learning. In: Seel, N.M. (eds.) Encyclopedia of the Sciences of Learning, pp. 2325–2328. Springer, Boston (2012). https://doi.org/10.1007/978-1-4419-1428-6_849
6. Gupta, A., D'Cunha, A., Awasthi, K., Balasubramanian, V.: DAiSEE: towards user engagement recognition in the wild. arXiv preprint arXiv:1609.01885 (2016)
7. Huang, G.B., Ramesh, M., Berg, T., Learned-Miller, E.: Labeled faces in the wild: a database for studying face recognition in unconstrained environments. Technical report, Technical Report 07–49, University of Massachusetts, Amherst (2007)
8. McMahan, H.B., et al.: A general approach to adding differential privacy to iterative training procedures. arXiv preprint arXiv:1812.06210 (2018)
9. Pardo, A., Siemens, G.: Ethical and privacy principles for learning analytics. Br. J. Edu. Technol. **45**(3), 438–450 (2014)
10. Parkhi, O.M., Vedaldi, A., Zisserman, A.: Deep face recognition (2015)
11. Schroff, F., Kalenichenko, D., Philbin, J.: FaceNet: a unified embedding for face recognition and clustering. In: Proceedings of the IEEE Conference on Computer Vision and Pattern Recognition, pp. 815–823 (2015)
12. Singh, D., Subramaniam, H., Rajendran, R.: PyGuru: a programming environment to facilitate measurement of cognitive engagement. In: Proceedings of the International Conference on Computers in Education, pp. 363–368. IEEE (2022)
13. Szegedy, C., Vanhoucke, V., Ioffe, S., Shlens, J., Wojna, Z.: Rethinking the inception architecture for computer vision. In: Proceedings of the IEEE Conference on Computer Vision and Pattern Recognition, pp. 2818–2826 (2016)

14. Ashwin, T.S., Guddeti, R.M.R.: Automatic detection of students' affective states in classroom environment using hybrid convolutional neural networks. Educ. Inf. Technol. **25**(2), 1387–1415 (2020)
15. Wolf, L., Hassner, T., Maoz, I.: Face recognition in unconstrained videos with matched background similarity. In: 2011 IEEE Conference on Computer Vision and Pattern Recognition (CVPR), pp. 529–534. IEEE (2011)

Item Difficulty Constrained Uniform Adaptive Testing

Wakaba Kishida[1(✉)], Kazuma Fuchimoto[1], Yoshimitsu Miyazawa[2], and Maomi Ueno[1]

[1] The University of Electro-Communications, Tokyo, Japan
{kishida,fuchimoto}@ai.lab.uec.ac.jp, ueno@ai.is.uec.ac.jp
[2] The National Center for University Entrance Examinations, Tokyo, Japan
miyazawa@rd.dnc.ac.jp

Abstract. Computerized adaptive testing tends to select and present items frequently with high discrimination parameters because these items can discriminate examinees' abilities in a wide range. Unfortunately, that tendency leads to bias of item exposure. To address this shortcoming, we propose item difficulty constrained uniform adaptive testing. During the initial stage, an optimal item is selected and presented from a uniform item group generated by a modern uniform test assembly method. The method switches to the secondary stage when the examinee's ability converges. It selects and presents the optimal item with a difficulty parameter value near the examinee's ability estimate from the whole item pool. Empirical experiments demonstrate that the proposed method mitigates the bias of item exposure while maintaining low measurement error by reducing the number of presented items with high discrimination parameters which are likely to be presented frequently by earlier CAT methods.

Keywords: computerized adaptive testing · item response theory · uniform test assembly

1 Introduction

Computerized adaptive testing (CAT) selects and presents an optimal item from an item pool. That process, which is based on item response theory (IRT), maximizes the test information (Fisher information measure) at the examinee's current estimated ability [6,9]. However, conventional CAT often presents identical items from an item pool to examinees with similar abilities. That extremely increases bias of item exposure distribution. The bias leads to decreasing the reliability of tests because overexposed items are likely to be shared among examinees [6,9,11].

To resolve this shortcoming, various countermeasures and alternatives are proposed (e.g. [1,5,9,10]). Kingsbury and Zara [5] proposed a method, of dividing item pools into item groups. Thereafter, from the item group with the smallest value of item exposure among all of them, it selects and presents the optimal item (designated as KZ). Moreover, van der Linden [9] proposed a method selecting the optimal item from a shadow-test assembled by solving an integer programming problem with several constraints (designated as IP). Choi and Lim [1] proposed another shadow-test approach that minimizes the distance between a test information of a shadow-test and target information (designated as TI). As another approach, van der Linden and Choi [10] proposed a method controlling the item selection probabilistically (designated as Prob). However, these methods increase the bias of measurement accuracies among examinees.

Therefore, Ueno and Miyazawa [7] proposed a method that separates an item pool into several item groups using the test assembly method presented by Ishii et al. [3] in advance. This method was designated as uniform adaptive testing (UAT). The method selects and presents an item from a uniform item group assigned to each examinee. Their results demonstrated that UAT reduced the bias arose for measurement accuracies. However, UAT increases the measurement error through reduction of the item group size.

To overcome this difficulty, Ueno and Miyazawa [8] proposed two-stage uniform adaptive testing (TUAT). Initially, this method selects and presents the optimal item from a uniform item group generated by the method presented by Ishii and Ueno [4]. After the examinee's ability converges, the method switches to the secondary stage to select and present the optimal item from the whole item pool. They demonstrated that item exposure can be reduced by TUAT without any increase in the measurement error. Unfortunately, TUAT shows a marked tendency for frequent selection and presentation of items with high discrimination parameters because these items can discriminate examinee's abilities in a wide range. Consequently, reduction of bias of the item exposure by TUAT can be done only to a limited degree.

Therefore, we propose item difficulty constrained uniform adaptive testing. The proposed method generates numerous item groups in advance using the Hybrid Maximum Clique Algorithm with Parallel Integer Programming presented by Fuchimoto et al. [2], which assembles the greatest quantity of uniform tests. Similarly to TUAT, the algorithm initially selects and presents an optimal item from a uniform item group. When the examinee's ability converges, the proposed method subsequently selects and presents an optimal item with a difficulty parameter value near the examinee's ability estimate from the whole item pool. Empirical experimentation elucidate that the proposed method can mitigate the bias of item exposure while maintaining low measurement error.

2   Item Difficulty Constrained Uniform Adaptive Testing

To resolve the shortcomings presented by a state-of-the-art CAT, TUAT [8], this study proposes a new CAT method, item difficulty constrained uniform adaptive testing, which can reduce bias of item exposure.

2.1   Initial Procedure

The method proposed herein generates a large number of uniform item groups using Hybrid Maximum Clique Algorithm with Parallel Integer Programming, which was demonstrated by Fuchimoto et al. [2]. The uniform item group assembly method differs from that of TUAT [8].

The algorithm of the initial stage is similar to TUAT [8]. At the beginning of the initial stage, the method assigns a different uniform item group to each examinee. During this stage, an optimal item from a uniform item group is selected and presented to maximize Fisher information. This stage provides a rough ability estimate with keeping item exposure distribution as uniform as possible (see [8] for details.).

2.2 Secondary Procedure

The secondary procedure provides a more accurate ability estimate with preventing bias of item exposure from increasing. Similarly to TUAT, the method finishes the initial stage when the update difference of the estimate of an examinee's ability is less than a criterion value, which is the Switching Stage Criterion (SSC) [8]. Subsequently, the proposed method starts the secondary procedure. From the whole item pool, the method selects and presents an optimal item with a difficulty parameter value within the neighborhood of the examinee's ability estimate. The neighborhood interval of the examinee's ability estimate $\hat{\theta}$ is defined as

$$\hat{\theta} - \alpha SE(\hat{\theta}) < b < \hat{\theta} + \alpha SE(\hat{\theta}), \tag{1}$$

where $SE(\hat{\theta})$ represents the standard error of the examinee's ability estimate $\hat{\theta}$, and α denotes a hyperparameter. The SSC and the hyperparameter α are optimized to balance low measurement error and low bias of item exposure. More specifically, the selection procedure in this stage is as follows:

1. The difficulty interval is estimated from the current ability estimate $\hat{\theta}$ and its standard error.
2. From items with difficulty parameter values within the estimated difficulty interval, an optimal item that maximizes Fisher information is selected.
3. Based on the examinee's earlier response history, the current ability estimate is updated.
4. Procedures 1–3 are iterated until the update difference of the ability estimate falls to or below a constant value of ϵ.

The proposed method is expected to reduce the quantity of presented items with high discrimination parameters while maintaining low measurement error.

3 Empirical Evaluation

This section evaluates the effectiveness of the proposed method (Proposed) through comparison with earlier computerized adaptive testing methods: conventional CAT (designated as CAT), IP [9], TI [1], KZ [5], and TUAT [8]. We set the total test length as 30. The item group sizes used in KZ, TUAT and Proposed are equal to the test length.

Table 1. Experiment results

Item pool	Method	SD. exposure item	Max. No. exposure items	Number of non-presented items	Measurement error (RMSE)
simulated	CAT	1055.5	10000	832	**0.25**
	IP	984.0	5000	812	**0.25**
	Prob	987.8	5105	819	**0.25**
	TI	1000.4	10000	**0**	0.26
	KZ	918.0	6565	779	0.26
	TUAT (0.100)	864.7	6409	188	0.26
	Proposed (0.100, 0.8)	**682.3**	**4520**	68	0.26
real	CAT	1150.3	10000	836	**0.25**
	IP	1026.4	5000	809	0.26
	Prob	1034.8	5107	812	0.26
	TI	1047.6	10000	**7**	0.26
	KZ	1032.0	7364	792	0.26
	TUAT (0.075)	937.6	7381	274	0.26
	Proposed (0.100,0.6)	**672.7**	**5031**	263	0.27

A simulated item pool including 1000 items and a real item pool including 978 items were used to conduct experiments. For each item included in the simulated item pool, true parameters were generated from $\log a_i \sim N(-0.5, 0.2)$ and $b_i \sim N(0, 1)$, where a_i and b_i respectively signify the discrimination parameter of item i and the difficulty parameter of item i. The examinees' actual abilities are sampled from $\theta \sim N(0, 1)$ 10,000 times. For convergence to the same upper bound exposure counts, we performed our experiments with 5000 as IP upper bound exposure counts and with 0.5 as Prob upper bound exposure rate. A presented in Table 1, the results shown as the values in parentheses for TUAT represent the SSC value. Those for the Proposed represent the SSC value and hyperparameter α. Also, "SD. exposure item" stands for the standard deviation of the numbers of exposure items; "Max. No. exposure item" represents the maximum quantity of exposure items. The quantity of items which have not been presented is signified by the "Number of non-presented items".

Table 1 shows that TI provides the lowest values of "Number of non-presented items". However, the values of "SD. exposure item" are as large as those of CAT. Moreover, CAT and TI produce equal values of "Max. No. exposure item" as the number of examinees. These findings indicate that one or more items are exposed to all the examinees. Actually, IP, Prob, and KZ all provide lower values of "SD. exposure item" and "Max. No. exposure item" than those of CAT, but "Number of non-presented items" is still large. By contrast, Proposed provides the lowest values of "SD. exposure item" and "Max. No. exposure items" without increasing the measurement error considerably. Furthermore, Proposed has the second lowest values of "No. non-presented items". Next, we analyze the difference between TUAT and Proposed.

Figures 1a and 1b portray scatter plots of the number of exposure items and items' discrimination parameters for TUAT and Proposed. These figures indicate the important tendency of TUAT as able to select and present items with high discrimination parameters because these items can discriminate examinees'

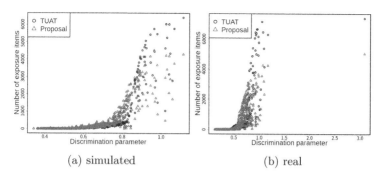

Fig. 1. Scatter plots presenting the numbers of presented items and discrimination parameters.

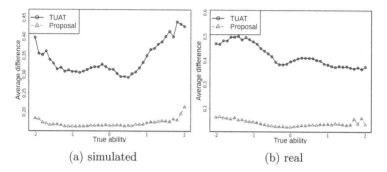

Fig. 2. Average difference between a difficulty parameter and an ability estimate in the secondary procedure

abilities in a wide range. A point of marked contrast is that the proposed method yields far fewer presented items with high discrimination parameters. Next, we analyze the reasons underlying this phenomenon.

Figures 2a and 2b depict the average differences between difficulty parameters and ability estimates in the secondary procedure. The difference between a difficulty parameter of the item presented to an examinee in the secondary procedure and the ability estimate is determined as

$$\sqrt{\frac{1}{n-l+1}\sum_{k=l}^{n}(\hat{\theta}_{k-1}-b_k)^2},\qquad(2)$$

where b_k denotes the difficulty parameter of the k-th presented item, and $\hat{\theta}_k$ represents the ability estimate after the k-th item is presented. Items from l-th to n-th are presented in the secondary procedure. Figures 2a and 2b portray an important benefit of TUAT: it often selects items with difficulty parameter values that differ greatly from the ability estimates. In contrast, the proposed method selects items with difficulty parameter values that are approximately equal to the ability estimates. As described previously, a marked tendency of

TUAT is the selection items with high discrimination parameters, even when the difficulty parameter values differ greatly from the ability estimates. This tendency consequently leads to bias of the item exposure. The proposed method avoids selection of items with difficulty parameter values that differ greatly from the ability estimates. As a result, the proposed method relaxes the item exposure bias that is a problem in TUAT.

4   Conclusion

First, the findings presented herein indicate that TUAT [8], which is a state-of-the-art CAT, has a tendency for the selection and presentation of items with high discrimination parameters frequently. To resolve this shortcoming, this study proposed item difficulty constrained uniform adaptive testing. Results of the empirical experiments showed that the proposed method provides a lower bias of item exposure than all comparison methods while maintaining low measurement error. That performance was achieved by reducing the number of presented items with high discrimination parameters, which are presented frequently by the earlier TUAT. Application of the proposed method is limited to IRT models with difficulty parameters. Relaxing the constraints of the proposed method is a goal for our future work.

References

1. Choi, S.W., Lim, S.: Adaptive test assembly with a mix of set-based and discrete items. Behaviormetrika **49**, 231–254 (2022)
2. Fuchimoto, K., Ishii, T., Ueno, M.: Hybrid maximum clique algorithm using parallel integer programming for uniform test assembly. IEEE Trans. Learn. Technol. **15**(2), 252–264 (2022)
3. Ishii, T., Songmuang, P., Ueno, M.: Maximum clique algorithm and its approximation for uniform test form assembly. IEEE Trans. Learn. Technol. **7**(1), 83–95 (2014)
4. Ishii, T., Ueno, M.: Algorithm for uniform test assembly using a maximum clique problem and integer programming. In: Artificial Intelligence in Education, pp. 102–112 (2017)
5. Kingsbury, G.G., Zara, A.R.: Procedures for selecting items for computerized adaptive tests. Appl. Measur. Educ. **2**(4), 359–375 (1989)
6. Ueno, M., Fuchimoto, K., Tsutsumi, E.: e-testing from artificial intelligence approach. Behaviormetrika **48**(2), 409–424 (2021)
7. Ueno, M., Miyazawa, Y.: Uniform adaptive testing using maximum clique algorithm. In: Artificial Intelligence in Education, pp. 482–493 (2019)
8. Ueno, M., Miyazawa, Y.: Two-Stage uniform adaptive testing to balance measurement accuracy and item exposure. In: Artificial Intelligence in Education, pp. 626–632 (2022)
9. van der Linden, W.J.: Review of the shadow-test approach to adaptive testing. Behaviormetrika, 1–22 (2021)
10. van der Linden, W.J., Choi, S.W.: Improving item-exposure control in adaptive testing. J. Educ. Meas. **57**(3), 405–422 (2020)
11. Way, W.D.: Protecting the integrity of computerized testing item pools. Educ. Meas. Issues Pract. **17**, 17–27 (1998)

"A Fresh Squeeze on Data": Exploring Gender Differences in Self-efficacy and Career Interest in Computing Science and Artificial Intelligence Among Elementary Students

Shuhan Li[1]([⊠]) [iD], Annabel Hasty[2] [iD], Eryka Wilson[3] [iD], and Roozbeh Aliabadi[4]

[1] Columbia University, New York, NY 10025, USA
shuhan.li.2000@gmail.com
[2] Quest Academy, Palatine, IL 60067, USA
[3] Keiser University, Fort Lauderdale, FL 33309, USA
[4] ReadyAI, Pittsburgh, PA 15206, USA

Abstract. Artificial intelligence (AI) is a growing field in both global job markets and educational spaces. This non-experimental quantitative study aims to explore how the educational program *A Fresh Squeeze on Data* affects students' self-efficacy and career choices and whether gender will differentiate the learning outcomes. Under the social cognitive career theory framework, this study designs questionnaires as data collection instrument. The results suggest that the program significantly improves students' comfortability with AI-related subjects but not for career interest or other measurements in self-efficacy. Unexpectedly, the program's effect is not divided by gender. Nevertheless, this study opens up conversations about assisting students from underrepresented backgrounds to envision success in AI courses and career pathways through an activity-driven curriculum. The paper also informs educators and researchers to devise culturally responsive pedagogy in teaching AI that empowers young girls before they develop a gendered career view.

Keywords: K-12 Education · AI Education · Gender Equity · Self-Efficacy · Career Interest

1 Introduction

The increasing importance of AI in major industries calls for relevant education among young learners. To prepare for their digitized future, students must cultivate data literacy earlier in their education. Educational institutes, in response, are gradually integrating relevant curricula into the classroom so that they can build up students' computing skills and familiarize them with emerging AI technologies.

However, gender inequality in AI-related fields still lingers. Women's underrepresentation can be explained by the dearth of female students in advanced computing science or AI-related courses. Despite many female students' exposure to data-driven

© The Author(s), under exclusive license to Springer Nature Switzerland AG 2023
N. Wang et al. (Eds.): AIED 2023, CCIS 1831, pp. 574–579, 2023.
https://doi.org/10.1007/978-3-031-36336-8_89

coursework in middle or high school, a majority discontinue their pursuit in these fields later in their life [4].

This non-experimental quantitative study delves into the effectiveness of a culturally responsive curriculum. From a socio-cognitive and pedagogical standpoint, this study will advocate for the active participation in school-led learning programs and the awareness of gender inclusivity in the design of AI educational materials.

2 Research Purposes and Questions

It is against the background of AI's rising importance and women's underrepresentation in the industry that this study is initiated. This paper explores the meaningfulness of a culturally responsive AI curriculum. It informs teachers of an innovative pedagogy that can both enhance AI content knowledge and achieve girls' empowerment. This paper asks the following research questions to navigate the subsequent analysis:

- RQ1: Does the participation in AI education programs impact students' self-efficacy and career interest in related fields?
- RQ2: To what extent are self-efficacy and career interest changed?
- RQ3: Does gender differentiate the changes in self-efficacy and career interest?

3 Social Cognitive Career Theory (SCCT)

SCCT provides the theoretical framework for this research. It assumes that individuals, their behaviors, and contexts jointly and bi-directionally affect each other [1, 8]. Self-efficacy is an important aspect in the SCCT circuit. It measures an individual's perceived competency in goals achievements [7]. This research also adopts the more unified model that takes demographics such as gender into consideration [11]. Social norms are part and parcel to shaping one's efficacy level. Hypothetically, women's low participation rate in AI-related fields can be attributed to inadequate learning opportunities and non-inclusive learning environment. These factors lower female learners' self-efficacy, thus preventing them from further career pursuits [5, 6].

4 Program Description

This study involves *A Fresh Squeeze on Data*, a student-centered, collaborative intervention program about data collection, data bias, and AI. Named after a picture book about problem-solving with data, this educational program encourages young students to be game changers and problem-solvers using the emerging technology of AI [10]. The children's book introduces a young girl who learns data collection from her mother, a data scientist, and reduces biases to solve problems for her community.

A Fresh Squeeze on Data lesson plan combines concepts with practices, leading students to effectively translate what they had learned to the real world. Each lesson lasts 90 min, consisting of two major hands-on activities. Through real-world simulations like running a candy stand and selecting an adopted pet, students are introduced to data collection, analysis, and bias identification [11]. These activities are combined with small group discussions to facilitate peer cooperations. The learning materials are accessible to Grade 3 to 5 students.

The program adopts a constructivist pedagogy. A wide range of activities hone students' teamwork, problem-solving, and inquiry skillsets [3]. Students directly interact with basic AI products such as Teachable Machine to train an AI model for categorization. This exercise familiarizes students with machine learning. Simple as the activity seems, it breaks down the dense technical knowledge into straightforward examples. In this way, students explore AI with greater ease and thus develop more interest in the subject.

Contrary to a conventional program with only a final project, this course places multiple hands-on activities in varying difficulties throughout the entire learning period [3]. Accordingly, students dedicate more time engaging with the learning materials. Each activity involves open-ended discussion questions [11]. This enables students to explore the technology through inquisitive trial and error. Compared to the lecture-style classrooms, *A Fresh Squeeze on Data* encourages students to demonstrate their agencies to learn and think critically about AI applications.

5   Method

5.1   Survey Design

Inspired by the survey design of previous literature [2, 9], this study utilizes similar instruments. Students are given the same questionnaire before and after the training program. The difference in their answers reflects the impact of the program on student outcomes. The survey contains five constructs, three of which measure self-efficacy and two measure career interest. Each construct contains two prompts that ask students to rate their opinion on a scale of 1 (strongly disagree) to 5 (strongly agree). The self-efficacy constructs gauge students' interest in, comfort with, and attitudes towards learning about AI and computing science. The career interest constructs investigate students' knowledge of and propensity towards occupations related to AI and data science. The last section asks for students' gender identification, age, and previous AI and data science experience.

5.2   Sample and Hypotheses

This study involves a sample size of 48 students (n = 48) who took part in the training program over the summer. Among them, 28 are male and 20 are female. Their ages range from 7 to 11. They come from diverse ethnic backgrounds. All sampled students have a certain degree of previous exposure in AI-related areas in either on-campus or extracurricular contexts. On average, the sampled students take AI- or STEM-related classes twice a week.

A repeated measures MANOVA is run on SPSS software to generate descriptive data on the survey results. Gender is the between-subject factor, while time is the within-subject factor. The dependent variables are the overall self-efficacy and career interest level for each student. Cumulative self-efficacy is calculated by adding the scores of the first five prompts, subtracting that of the sixth. Cumulative career interest is the total score of the last four prompts. The following hypotheses are tested at a confidence level of 0.95:

- H1: *A Fresh Squeeze on Data* impacts students' self-efficacy level and career interest in AI or computing science.
- H2: Gender differentiates students' change in self-efficacy level and career interest in AI or computing science.

6 Data Analysis

6.1 Impact on Self-efficacy

The mean for the overall self-efficacy level increases from 15.96 (sd = 4.156) to 16.48 (sd = 4.443) after the program. However, this increase is not statistically significant (p = 0.090 > 0.050). The overall effect of the program on students' self-efficacy level remains unclear.

Nonetheless, the results yield a significant change in students' answers to Prompt 4 (comfortability with learning AI and computing science), with a p-value of 0.010 < 0.050. The overall score increases from 3.52 (sd = 1.091) to 3.85 (sd = 1.148). This indicates that *A Fresh Squeeze on Data* reduces students' stress about learning AI-related knowledge. Part of the reason is that the course is conducted in an interactive environment using a child-friendly story book. The increased comfortability, in return, helps boost students' confidence in their learning ability.

6.2 Impact on Career Interest

The mean for students' career interest increases slightly from 14.21 (sd = 3.730) to 14.81 (sd = 3.745). However, the change is not significant (p = 0.203 > 0.050). The four items in the career interest constructs remain statistically insignificant as well. These unexpected results imply that there is little evidence to prove the correlation between the program and students' career choices in AI or computing science.

The unclear correlation can be partially attributed to the short timeframe. Students might not fully grasp the insights about AI or computing science, let alone consider these areas as future jobs. Even if they do, simply learning about a subject does not necessarily translate to job preference. More variables such as mentorship, duration of programming, and age should be considered. Additionally, a longitudinal study is needed to track the effects of similar learning programs on career interests over time.

6.3 Gender Disparity

Based on the between-subject effects no statistically significant gender difference is detected for self-efficacy (p = 0.315 > 0.050) or career interest (p = 0.087 > 0.050). Student outcomes in general are not differentiated by gender.

The undetected gender disparity in the sample can be attributed to the grade level of the sampled students. Students in Grades 3 to 5 might not have developed a gendered view of occupations, which research suggests may develop during Grades 6 to 8 [5]. Although it remains questionable whether gender will make a significant difference later in their lives, gender-inclusive programs are still crucial for underrepresented students at an early age before gendered views on career fields take root.

7 Limitations

Admittedly, the current study comes with shortcomings. The sample size is limited. A sample of 48 students with an unbalanced gender ratio may not accurately represent the demographics of Grades 3 to 5 students. A larger sample size with a more balanced gender ratio will provide a better representation of gender differences in the dependent variables.

The short time frame of the program also misrepresents students' actual learning outcomes. This project only took place over two class periods. However, elementary school students' career interests may take a much longer time to develop. Also, the scope of the knowledge in the curriculum is limited. More extensive work on the topic could be taught to elementary school students via long-term continuous programs and then evaluated.

Another limitation of this study is that the teacher is a strong female role model in computer science and AI. Thus, the survey may be vulnerable to observer bias. Students would speculate a "correct" response catering to the instructor's expectations. During the class periods, students might also tend to behave more extrovertedly than usual in front of their peers and instructors. These influences might cause students to report higher scores in their post-program surveys, resulting in overestimating the actual effects of the lesson.

Given the non-experimental nature of the study, there is no control groups for comparison. The lack of benchmarks may render questionable the actual effectiveness of the program. Nevertheless, whether to include a control group in this setting contains ethical concerns. Because the educational program pertains to students' prospects, it would be unfair to leave out a group of students for the sake of research observations.

8 Conclusions

The role of computing science and AI in society has increased dramatically, and educators need to prepare our young people for the digitalized future. *A Fresh Squeeze on Data* allows teachers to engage in interactive lessons with their students to discover concepts of bias in AI, machine learning, and other computing science fields. This study demonstrates how targeted interventions can affect students' self-efficacy and interest. However, it also demonstrates the need for longer term interventions and interventions conducted at earlier ages.

Future research can refine its methods in these four ways. First, longitudinal studies are needed to track the persistent effectiveness of the training program. Second, *A Fresh Squeeze on Data* can be integrated into other disciplines instead of being taught as a stand-alone subject. Thirdly, more demographic variables shall be considered, including race, socioeconomic class, school curricula, grade level, and ableness. Finally, similar studies should extend to areas where school affluence is differentiated.

Additional studies must also address questions such as the teachers' roles, parental support, and modeling [5]. Overall school support for the AI-related fields must be explored in greater precision. Nevertheless, by designing and implementing interventions such as *A Fresh Squeeze on Data*, teachers and schools may begin to close many of the learning gaps in AI, computer science, and data science that have persisted for far too long.

References

1. Bandura, A.: The explanatory and predictive scope of self-efficacy theory. J. Soc. Clin. Psychol. **4**(3), 359–373 (1986). https://doi.org/10.1521/jscp.1986.4.3.359
2. Chai, C.S., Lin, P.-Y., Jong, M.S.-Y., Dai, Y., Chiu, T.K.F., Qin, J.: Perceptions of and behavioral intentions towards learning artificial intelligence in primary school students. Educ. Technol. Soc. **24**(3), 89–101 (2021)
3. Condliffe, B., et al.: Project-based learning. A literature review. MDRC, 1–84 (2017)
4. Corbett, C., Hill, C.: Solving the equation: the variables for women's success in engineering and computing. American Association of University Women, 1111 Sixteenth Street NW, Washington, DC 20036 (2015)
5. Farland-Smith, D.: Exploring middle school girls' science identities: examining attitudes and perceptions of scientists when working "side-by-side" with scientists. Sch. Sci. Math. **109**(7), 415–427 (2009). https://doi.org/10.1111/j.1949-8594.2009.tb17872.x
6. Fouad, N.A., Santana, M.C.: SCCT and underrepresented populations in STEM fields. J. Career Assess. **25**(1), 24–39 (2016). https://doi.org/10.1177/1069072716658324
7. Lent, R.W., Brown, S.D., Hackett, G.: Toward a unifying social cognitive theory of career and academic interest, choice, and performance. J. Vocat. Behav. **45**(1), 79–122 (1994). https://doi.org/10.1006/jvbe.1994.1027
8. Lent, R.W., Brown, S.D.: Social cognitive approach to career development: an overview. Career Dev. Q. **44**(4), 310–321 (1996). https://doi.org/10.1002/j.2161-0045.1996.tb00448.x
9. Lin, P., Chai, C., Jong, M.S., Dai, Y., Guo, Y., Qin, J.: Modeling the structural relationship among primary students' motivation to learn artificial intelligence. Comput. Educ. Artif. Intell. **2**, 1–7 (2021). https://doi.org/10.1016/j.caeai.2020.100006
10. ReadyAI: Data bias lesson plan. Cloudera (2021). https://freshsqueezekids.com/. August 18, 2022
11. ReadyAI: A Fresh Squeeze on Data: Problem-Solving with Data. ReadyAI LLC (2021)

Simulating Learning from Language and Examples

Daniel Weitekamp[1(✉)], Napol Rachatasumrit[1], Rachael Wei[2], Erik Harpstead[1], and Kenneth Koedinger[1]

[1] Carnegie Mellon University, Pittsburgh, PA 15289, USA
weitekamp@cmu.edu
[2] University of Illinois Urbana-Champaign, Champaign, IL 61820, USA

Abstract. Simulations of human learning can be used as computational models for evaluating theories of learning. They can also be taught interactively to author intelligent tutoring systems. Prior simulated learner systems have learned inductively from worked examples and correctness feedback. This work introduces a mechanism where simulated learners can also learn from natural language. Using a neural grammar parser with additional symbolic processing steps, we simulate the production of loose interpretations of verbal instructions. These interpretations can be combined with worked examples to resolve the ambiguities of either form of instruction alone. We find that our system has practical benefits over an alternative method using github Copilot and slightly better accuracy.

Keywords: Learning Simulations · Computational Models of Learning · Natural Language Processing · Code Synthesis · Authoring Tools

1 Introduction

A computational model of learning is a student model that learns directly from instructional material, as humans do, and produces particular incorrect and correct responses as a result. By contrast statistical models of learning are fit to student data and typically reduce the cognitive complexities of learning to numerical predictions of performance [2]. Computational models of learning embody executable theories of learning that can be applied to learning materials to evaluate instruction *a priori* without fitting to student data [8]. While a statistical model may characterize learning as a change in the probability of correct responses, a computational model produces particular responses to question items and has interpretable underlying reasons for them.

Prior simulated student technologies used as computational models of learning have predominantly learned from worked examples and correctness feedback [5,6]. This work expands the set of instructional modalities that simulated students can learn from to include natural language instruction.

Worked examples alone can be ambiguous. The particular content of a worked example often lends itself to many different explanations. For instance for the problem:

```
Find the slope of the line that passes through (5, 4) and (7, 8)
```

The worked example solution is 2, and the correct explanation in this case is $(8 - 4)/(7 - 5) = 2$. However, there are also several incorrect explanations like $8/4 = 2$ or $7 - 5 = 2$. Prior simulated learners have used an error-prone brute-force guess-and-check style method called *how-learning* to induce these sets of operations. This work aims to allow simulated learners to learn more robustly by interpreting natural language instruction to disambiguate the operations used to produce worked examples.

Simulated students can also be used to rapidly author intelligent tutoring systems (ITS). ITS authoring can be a time consuming process. For instance, programming an hour of cognitive tutor based instruction takes about 200–300 hours of development time [1]. ITS authoring with simulated learners has been shown to be considerably faster even than GUI-based authoring methods [7]. Yet prior simulated learner based authoring systems have suffered from induction errors that put domain-general authoring out of reach. Typically simulated-learner's *how-learning* mechanisms help authors to program formulae that grade and producing bottom-out hint answers for problem steps. Enabling this functionality for domain-general authoring presents a challenge: when a large set of functions are made available to *how-learning*—which might include many kinds of functions beyond just arithmetic operations—*how-learning* may search through an intractably large space of function compositions, making it prone to stopping short on incorrect formulae that reproduce worked examples but that are incorrect in general (like in the example above). The method we present in this work allows authors to verbally clarify the composition of operations used in their worked examples. Our system interprets natural language and examples together to overcome the intractability of performing *how-learning* on demonstrated examples alone.

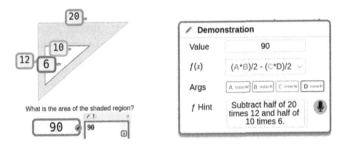

Fig. 1. (left) A worked example "90" for a triangle area composition problem. (right) the user's recorded spoken instruction "f Hint" and the induced formuale "f(x)".

2   Related Work

The majority of prior work on natural language processing of mathematical language pertains to the translation of word problems directly into equations [9]. By contrast, our system interprets operational langauge-based instruction, and outputs executable knowledge structures that can perform steps in mathematical procedures. Large language models (LLMs) like OpenAI's Codex [3] are closer to our method in terms of functionality, in the sense that, like our system, they can synthesize executable formulae. We compare our system to Codex via the Github CoPilot plugin.

3   Better Models of Novices Than LLMs with Neuro-Symbolic AI

Simulations of learning must make theoretical commitments about both learner's prior knowledge and about the content of the target knowledge being taught. The goal of a computational model of learning is to explain how knowledge is constructed and refined through learning. LLMs don't hold much promise of helping with these sorts of simulations because insofar as they exhibit mastery of capabilities that students typically learn, they have acquired those capabilities from many domain-specific experiences greater in number and diversity than any human would ever encounter in a lifetime (like millions of github repositories for Codex). While LLMs boast impressive generative capabilities, their learning process is considerably less data-efficient than human learning. Additionally their knowledge is largely encoded in unexplainable "blackbox" weights, acquired from often proprietary datasets. Most of all, LLMs simply know too much to be useful for modeling learning. A pre-trained model that already possess the domain-specific capabilities that one intends to simulate the acquisition of is useless for modeling a novices' learning trajectory from first experiences to mastery.

By contrast, our approach restricts itself to only the use of a pre-trained neural grammar parser [4] and coreference resolver[1], but uses no text generation models. Our use of a pre-trained grammer parser assumes that our simulated students can, as prior knowledge, parse the structure of English sentences, but does not assume an ability to translate mathematical language to executable operations or written equations. Our system transforms grammatically parsed sentences in several hard-coded yet domain-general processing steps to produce search policies for guiding a typical simulated learner's *how-learning* mechanism. These policies embody loose interpretations of sentences, which enclose small spaces of possible function compositions intended by the input sentence. Combining these policies with the typical search process used in *how-learning* disambiguates the formulae an instructor or ITS author intended to teach with their combined worked examples and natural language instruction (Fig. 2).

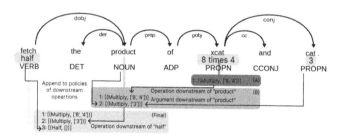

Fig. 2. Parsing and policy for "half the product of 8 times 4 and 3". Boxes (A), (B), and (Final) show the recursive process of a policy being constructed by traversing the grammatical parse of a sentence.

For authoring purposes, we don't need to constrain prior knowledge of simulated learners. Thus, if an LLM like Codex proves better than our approach at producing

---

[1] https://spacy.io/universe/project/coreferee.

target formulae—even if only because somewhere in its vast training set there is an example analogous to the target task—then it may ultimately be the preferable tool. Yet, authoring still requires domain-general generative capabilities, so if Codex is relying on instances from its training set, in lieu of broader generative capabilities, then it may fail when tasked with aiding authors at building one-of-a-kind materials. We evaluate this possibility by comparing our system with Codex on descriptions of made-up formulae.

4   Methods

We recruited 10 crowd workers through Prolific to generate natural language instructions (i.e. hints) for problem steps. Participants solved 14 unique math problem steps, and provided conceptual hints (i.e. "describe the concept in broad terms") and grounded operational hints, which we requested include all operations and values used to produce the answer, without using mathematical notation—they should be written as if spoken aloud. Conceptual hints were requested simply to highlight the requirements of operational hints and are not used in our evaluations. Participants were given a five question *check your understanding* survey to ensure they understood these distinctions.

Two of the authors independently coded each participants' grounded operational hints to mark if they were indeed both grounded (i.e. "mentions all required arguments and constants"), and operational (i.e. "mentions all required operations"). An inter-rater reliability of 97.2% was achieved, and the discrepancies were resolved through discussion. Hints marked as both grounded and operational, we refer to as *good* hints.

For each response where participants produced the correct answer, we ran the participant's hints and worked examples through our system, and also with worked examples only. We did the same with Codex via Github Copilot. In this case worked examples, which were used to eliminate functions based on return value. We also ran Copilot with hints only. Our system had 7 functions "Add, Multiply, Subtract, Divide, Half, Ones-Digit, and Square" available to it, which were sufficient for building function compositions for all of the target formulae, plus 8 additional functions not needed for any of the target tasks "TensDigit, Power, Double, Increment, Decrement, Log2, Sin, and Cos".

Typically Github Copilot produces a function implementation from a function header and doc-string. We filled our participants' grounded operational hints into Github Copilot as the doc-string of an empty Python function with the header foo():, and recorded the extended set of suggestions. Similar to our system's output this constituted a small variable set of candidate solutions. Typically Copilot uses arguments specified in the function header. This information is not present in the participant's hints, so we omit arguments and just evaluate whether Copilot suggested implementations that were functionally equivalent to the target formulae, expressed with constants instead of arguments. When examples were included, we also executed each of these functions to see if they reproduced the worked example.

For both systems we measured whether or not a correct formula was produced for each grounded operational hint, and counted the number of incorrect formulae produced. We considered any algebraic rearrangements (e.g. $A * B = B * A$) of the target formula to be correct. Thus, we use the average number of incorrect formulae as the principle measure of error magnitude, instead of proportion correct, which would be sensitive to returning several isomorphic formulae. The principal measures of the rate of correctness are (1) the percentage of the participant provided hints where the system produces at least one correct formula (i.e. *has correct*), and (2) the percentage where only the correct formula is produced (i.e. *only correct*).

To evaluate Codex's potential performance on one-of-a-kind formulae unlikely to be present in its training set, we repeated these evaluations for a set of 10 made-up formulae with accompanying grounded operational hints that we wrote ourselves.

5   Results

Each of our 10 participants finished 14 problems for a total of 140 responses. We removed 26 responses where participants produced incorrect answers and used the remaining 114 grounded operational hints for evaluation. Our system produced sets of formulae containing at least one correct formula 82.4% of the time, and 69.2% of the time only correct formulae were returned. Of the 114 grounded operational hints 87 were coded as *good*. On average our system performed better for the *good* hints, 86.2% had correct formulae and 73.6% had only correct formulae. For the set of all 114 responses our system produced an average of 1.54 unique incorrect formulae, whereas for the *good* hints it produced an average of 0.54 incorrect formulae.

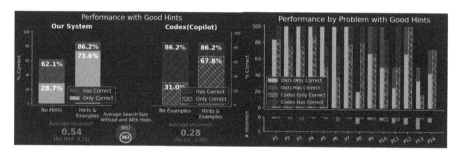

Fig. 3. Overall performance and performance by problem on *good* hints for our system and Github Copilot. Reduction in *how-learning* search size from hints is shown.

Overall our system's performance (Fig. 3) was improved considerably by parsing hints in addition to worked examples. When using only worked examples without hints, as in prior work, our system produced only correct solutions 28.7% of the time. Succeeding on just 4 of the 14 problems. Hints also reduced the average number of incorrect formulae from 8.33 to 0.54, and the average search space size of *how-learning* from 3952 function compositions to 162.

Github Copilot achieved the same *has correct* performance as our system on *good* hints with a slightly worse *only correct* performance of 67.4%. Our system performed nearly perfectly on problems 1–7 which had shorter formulae and participant hints—an average of 16 words, versus 24 words in problems 8–14. Most of our systems' errors came from difficulties parsing the verbose hints from problems 8–14. Copilot's performance was more varied, but was much higher than our system for some problems (e.g. P8, P11). This is likely partially due to mimicking of similar problems in Codex's training set, since in some cases it produced code with suspiciously domain-specific inline comments and variable names. However, Copilot also performed perfectly on our corpus of 10 made-up hints and worked examples, verifying that it is indeed strong at novel code generation. By contrast, our system also produced correct formulae for all 10 made-up hints, and only a few incorrect formulae for 4 of them.

6   Discussion

Both systems showed improvements consistent with the assertion that the combination of langauage and worked examples benefits instruction comprehension over either taken alone. Our system generally performed best on more concise hints—a pattern of performance that bodes well for computational modeling purposes. Humans often learn better from concise directed learning experiences. Our system performed most poorly on problems like problem P8 (Fig. 1), that are often scaffolded into multiple steps in ITS interfaces. Copilot by contrast had no consistent performance pattern, and showed signs of leveraging prior domain-specific knowledge.

7   Conclusion

Future computational modeling work may investigate how these language comprehension abilities compare to human capabilities. In this work we've demonstrated two means of generating knowledge from grounded operational instruction. This is however only a first step. Our system takes a relatively structured approach, compared to the method using Copilot, making it conducive to many future refinements and investigations. This opens opportunities to investigate questions of how learners interpret tutorial instruction, and how instruction and ITS hints may be improved as a result.

For ITS authoring purposes our crowd-worker results may understate the efficacy of our approach. For instance, authors may get better at generating hints over time, or use the experience as a starting point for authoring their own formulae directly—something that is not obvious, especially if non-arithmetic functions are involved. Additionally, the fairly small average number of incorrect formulae produced by both systems means correct formulae can be easily selected from among a small set of candidates. Overall, our method bodes well for incorperating natural language processing into ITS authoring and computational models of student learning.

References

1. Aleven, V., et al.: Example-tracing tutors: Intelligent tutor development for non-programmers. Int. J. Artif. Intell. Educ. **26**(1), 224–269 (2016)
2. Cen, H., Koedinger, K., Junker, B.: Learning factors analysis – a general method for cognitive model evaluation and improvement. In: Ikeda, M., Ashley, K.D., Chan, T.-W. (eds.) ITS 2006. LNCS, vol. 4053, pp. 164–175. Springer, Heidelberg (2006). https://doi.org/10.1007/11774303_17

3. Chen, M., et al.: Evaluating large language models trained on code. arXiv preprint arXiv:2107.03374 (2021)

4. Honnibal, M., Montani, I.: spaCy 2: Natural language understanding with Bloom embeddings, convolutional neural networks and incremental parsing (2017)

5. MacLellan, C.J., Harpstead, E., Marinier, R.P., III., Koedinger, K.R.: A framework for natural cognitive system training interactions. Adv. Cogn. Syst. **6**, 1–16 (2018)

6. Matsuda, N., Cohen, W.W., Koedinger, K.R.: Teaching the teacher: tutoring simstudent leads to more effective cognitive tutor authoring. Int. J. Artif. Intell. Educ. **25**(1), 1–34 (2015)

7. Weitekamp, D., Harpstead, E., Koedinger, K.: An interaction design for machine teaching to develop ai tutors. CHI (2020)

8. Weitekamp, D., Harpstead, E., MacLellan, C.J., Rachatasumrit, N., Koedinger, K.R.: Toward near zero-parameter prediction using a computational model of student learning. In: International Educational Data Mining Society (2019)

9. Zou, Y., Lu, W.: Text2math: End-to-end parsing text into math expressions. arXiv preprint arXiv:1910.06571 (2019)

Learner Perception of Pedagogical Agents

Marei Beukman$^{(\boxtimes)}$ and Xiaobin Chen

University of Tübingen, Tübingen, Germany
`marei.beukman@student.uni-tuebingen.de`

Abstract. Hyper-realistic human video generation (also referred to as "deepfake") is a recent development in artificial intelligence that may be applied to create pedagogical agents (PAs) for education purposes. Traditionally, PA research has been using 2D or 3D virtual figures to examine how their design may affect student learning or perceptions. In this study, we employed the latest human video generation technology to create PAs and investigated how students' perceived stereotypes of competent teachers would affect their preferences of a PA. It was found that students prefer to learn Japanese with Asian looking PAs over Caucasian or Black agents, supporting the hypothesis that real-world stereotype ideas can persist in the virtual world. Findings of the study offer references for how to better design PAs with generated human videos to boost learning motivation and enhance student perceptions.

Keywords: pedagogical agents · AI avatars · learner perception · stereotypes · language learning

1 Introduction

Pedagogical agents (PAs) are lifelike characters presented on a screen to guide students in digital learning environments [1]. PAs can help to increase learner motivation and attention which is beneficial not only in the learning process but also for learning success [2]. According to the social agency theory, this is due to the social presence PAs convey which leads learners to engage more deeply with the learning material [3].

PA designs vary from 2D or 3D cartoon figures to more realistic human-like figures. The latter are preferred, as higher levels of anthropomorphism have been found to lead to higher motivation and better learning outcomes [4]. State-of-the-art technology allows the design of hyper-realistic human PAs by using Artificial Intelligence (AI) to create avatars, i.e. virtual twins of real humans (e.g., Synthesia[1]). With machine learning techniques, videos of human-looking AI avatars can be synthesized using photos, text-to-speech, and lip synthesis; a process also called "deepfake". In the current study, we try to explore a traditional topic in PA research on how stereotypes affect students' preferences to learn with PAs with certain characteristics by using synthesized human PAs. As far as we know, this is the first study using human video synthesis technology in PA research.

---

[1] https://www.synthesia.io/.

N. Wang et al. (Eds.): AIED 2023, CCIS 1831, pp. 587–592, 2023.
https://doi.org/10.1007/978-3-031-36336-8_91

An often-cited theory for PA research is the Similarity Attraction Hypothesis which stipulates that humans are attracted more to other humans that match their own characteristics [5]. According to the theory, students prefer to learn from instructors that share their characteristics, such as age, gender, ethnicity, clothing, etc. [4]. This helps the student to identify with the instructor, which can result in higher attention and better learning outcomes. Empirical findings show that students indeed reported to prefer young, realistic looking PAs whose characteristics matched their own to those that did not match them in terms of gender or age [2]. Other studies have reported a celebrity endorsement effect where a "likeable person" (i.e. a famous person chosen by the student) is likely to increase motivation and learning outcomes [3,6].

In reality, students do not only prefer teachers with certain characteristics, but they also judge the teacher's competence on the subject matter. That is, students make stereotypical judgements about the teacher's abilities to do a good job in teaching a specific subject based on their feelings about how a good teacher for the subject should be like [7]. Despite the somewhat negative connotation of the term "stereotype" in common use, which refers to over-generalized beliefs about members of a social group, stereotypes are very important from a cognitive perspective as they help us to simplify our social environment and reduce the amount of processing, to make a fast reaction to new situations possible. As studies have found that stereotypes about teachers in the real world do persist in the virtual world [7], they should be considered when designing virtual PAs. However, despite a plethora of research on PA stereotypes (see, for example, the review in [8]), answers to how stereotypes affect student perception and learning outcomes are still unclear—conflicting results have been reported in previous studies. It is thus important to further investigate how PA characteristics affect students' preferences hence also motivation to learn with the PA. The use of synthesized human PAs makes it easy and cheap to vary the PA's characteristics while providing a hyper-realistic interactive experience to the students. Consequently, we aim at answering the following research questions in the current study:

1. Do learners have clear preferences in the choice of PAs with whom they will learn a new language (Japanese)? We vary the PAs' gender (male *vs* female), and ethnicity (Asian *vs* Caucasian *vs* Black).
2. If yes, what are the main causes of these preferences: The similarity between the PA and the students, personal preference, or perceived stereotypical competence?

We hypothesize that (1) learners will choose a PA that matches their gender, as predicted by the Similarity Attraction Hypothesis, and (2) learners will prefer an Asian looking PA for learning Japanese due to competence stereotypes.

Answers to these questions will provide valuable references for PA design and potentially also contribute to counteract stereotypical thinking. Well-designed PAs meeting the students' expectations will enable smooth interactions with the PA, allowing them to focus on the material instead of being distracted by unexpected features of the PA. On the other hand, non-stereotypical PAs can also be designed to extend the possibilities for social identification and role modeling.

2 Methods

Participants. 40 participants (20 males and 20 females) aged from 19 to 48 years (M = 27.93, SD = 6.35) that were fluent in English but did not speak Japanese took part in this study. Participants were recruited via crowd sourcing from all over the world. The sample included 17 white/Caucasian, 16 black, 3 Asian, and 4 participants of other ethnicities (Middle Eastern, Latino/Hispanic).

Stimuli. We created two sets of PAs, each set consisting of six AI avatars varying in the combination of gender (male/female) and ethnicity (Asian/Caucasian/Black). The commercial service Synthesia was used to create two AI-synthesized videos of each PA, a short one (9 s) in which the avatar introduces him-/herself and a one-minute learning video in which he/she additionally teaches three Japanese greetings: "Ohayō gozaimasu" (good morning), "Konnichiwa" (hello), and "Konbanwa" (good evening). The same script was used for each video and PAs of the same gender had the same voice to avoid a voice effect. Figure 1 shows screenshots of one set of PAs. The second set consists of avatars with the same characteristics but different appearances for countering appearance effect.

Fig. 1. One set of PAs that were used to generate the learning videos.

Procedure. The data was collected through an online questionnaire platform (SoSci Survey[2]). After a demographic questionnaire, six of the short videos were presented in random order, one for each condition (e.g. male+Asian, female+black). The PA for each condition was randomly chosen from one of the two sets. The participants were asked to choose their preferred PA to learn Japanese with, to give a reason for their choice, and to answer some questions about the choice. Then the one-minute learning video of the preferred PA was presented. Afterwards, the participants reported their experience (e.g. whether the teacher met the expectations) and answered some questions about their motivation, general preferences for virtual teachers, and stereotypes. In the end, the participants were tested on the meaning of the Japanese words introduced in the videos.

---

[2] https://www.soscisurvey.de.

3   Results

3.1   Preferences

Gender. The results show that in general the female PAs were preferred. As can be seen in Fig. 2, 60% of female and 55% of male participants preferred the female PAs. There was no significant correlation between participant gender and preferred PA gender (Chi-squared test of independence: $\chi^2(1, 40) = 0$, p $= 1$). Hypothesis (1) which predicts that learners would prefer PAs that match their gender can thus not be supported: Gender preference is not due to student-teacher similarity as predicted by the Similarity Attraction Hypothesis.

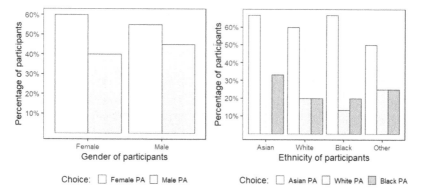

Fig. 2. Left: Percentage of female/male participants that chose a female/male PA. Right: Percentage of participants that chose an Asian/Caucasian/Black PA, grouped by the participants ethnicity.

Ethnicity. Overall, participants preferred the Asian looking PAs over the others: 62.16% of the participants chose an Asian looking PA (66.67% of all Asian participants, 60% of all white/Caucasian participants, 66.67% of all black participants, and 50% of the participants of other ethnicity). A majority of participants from all different ethnicities preferred the Asian looking PA, as predicted by Hypothesis (2). There was no significant correlation between the participants ethnicity and the ethnicity of the chosen PA (Chi-squared test of independence:: $\chi^2 (8, N = 40) = 6.83$, p-value $= .555$) which shows that also in terms of ethnicity participants did not prefer PAs with characteristics similar to themselves as predicted by the Similarity Attraction Hypothesis.

3.2   Perception, Experience, and Learning Outcome

After indicating which PA they preferred, participants were asked to write one or two short sentences to explain why they made this choice. The responses showed that a majority of people that chose an Asian looking PA (66.66%) did so because they thought the Asian PAs might be native Japanese speakers, which would be beneficial for learning. Overall, 45% of participants indicated that they made a choice according to whether they think the PA is a native speaker or not, 40% mentioned the appearance of the PA, 35% indicated that they liked the voice (which is surprising, as all PAs of the same gender had the same voice). Only one participant each indicated that his or her choice had to do with gender or age.

The results from the questionnaire about the perception of PAs showed that people were more convinced that the Asian looking PAs were proficient in Japanese (Statement: "This character is probably proficient in Japanese", Answer: Median = 5, on a five-point Likert scale with 1 = strongly disagree, 5 = strongly agree). PAs of other ethnicities got lower ratings in perceived competences (White: Median = 4, Black: Median = 3.5), despite the fact that all PAs presented the same text with the same voice (depending on gender). Furthermore, all PAs were virtual avatars, which theoretically enabled them to speak a large number of different languages.

After watching the learning video, participants were asked again to write down one or two short sentences, to report whether they were satisfied with the teacher and whether the teacher had turned out to be the way they had imagined him or her. Most participants (95%) reported that they were happy and satisfied with the teacher, that they liked the teacher, that the teacher met their expectations, and was helpful and competent. Only two participants reported that they were not satisfied, because the teacher looked too artificial. Two participants reported that the teacher was different than they expected (less empathetic, less expressive).

No significant learning outcome differences were observed among participants choosing different PA ethnicities.

4   Conclusions

We hypothesized that (1) learners would choose a PA that matched their gender, as predicted by the Similarity Attraction Hypothesis, and (2) learners would prefer an Asian looking PA for learning Japanese due to stereotyped competence. The results did not support hypothesis (1): Participants did not prefer PAs that matched their gender: both female and male participants showed preferences for the female PAs. Participants perceived the female PAs as more engaging, empathetic, and liked the female voice. This shows that the participants' gender did not have much influence on their preferences, rejecting the Similarity Attraction Hypothesis. Hypothesis (2) was supported by the results: A majority of participants from all different ethnicities preferred an Asian looking PA to teach them Japanese, because they expected the Asian looking PAs to be more proficient in Japanese. This result highlights the importance of matching a PAs design to its purpose, which can be facilitated by the use of synthesized AI avatars. This finding can potentially be applied to other domains, not only to language learning contexts.

However, even if it seems to be beneficial to match a PAs design to its purpose and to the learner's expectations, AI avatars can also be used to counteract stereotypical thinking. The field would profit from more research to clarify whether meeting a learner's expectations indeed helps him or her to focus on the topic, and how fast a learner can adapt to an "untypical" PA.

We acknowledge that the study has some limitations, namely that the age of the participants as well as the perceived age of the PAs might have had an effect on the results but was not evaluated. We are also aware that stereotypes, especially those connected to ethnicity, are delicate topics. Participants might not state the true reasons for their choice because they are uncomfortable writing them down or might not even be aware of them. The answers might thus be influenced by social acceptability. Further, the learning content was very limited and did not allow for a detailed analysis of learning outcome. Future work could implement synthesized human AI avatars into learning environments that offer more learning material and focus on learning outcome to address the question, whether the design of the PA can be a predictor for learning outcome. This would provide further implications for improved digital learning environments.

References

1. Clarebout, G., Heidig, S.: Pedagogical Agents. In: Seel, N. M. (ed.) Encyclopedia of the Sciences of Learning, pp. 2567–2571. Springer, US, Boston, MA (2012). https://doi.org/10.1007/978-1-4419-1428-6_942
2. Ozogul, G., Johnson, A.M., Atkinson, R.K., Reisslein, M.: Investigating the impact of pedagogical agent gender matching and learner choice on learning outcomes and perceptions. Comput. Educ. **67**, 36–50 (2013). https://doi.org/10.1016/j.compedu.2013.02.006
3. Domagk, S.: Do pedagogical agents facilitate learner motivation and learning outcomes?: the role of the appeal of agent's appearance and voice. J. Media Psychol. **22**(2), 84–97 (2010). https://doi.org/10.1027/1864-1105/a000011
4. Moreno, R., Flowerday, T.: Students' choice of animated pedagogical agents in science learning: a test of the similarity-attraction hypothesis on gender and ethnicity. Contemp. Educ. Psychol. **31**(2), 186–207 (2006). https://doi.org/10.1016/j.cedpsych.2005.05.002
5. Byrne, D., Nelson, D.A.: Attraction as a linear function of proportion of positive reinforcements. J. Pers. Soc. Psychol. **1**(6), 659–663 (1965). https://doi.org/10.1037/h0022073
6. Pi, Z., Deng, L., Wang, X., Guo, P., Xu, T., Zhou, Y.: The influences of a virtual instructor's voice and appearance on learning from video lectures. J. Comput. Assist. Learn. (November 2021), pp. 1–11 (2022). https://doi.org/10.1111/jcal.12704
7. Armando, M., Ochs, M., Régner, I.: The impact of pedagogical agents' gender on academic learning: a systematic review. Front. Artif. Intell. **5**(June), 1–23 (2022). https://doi.org/10.3389/frai.2022.862997
8. Dai, L., Jung, M.M., Postma, M., Louwerse, M.M.: A systematic review of pedagogical agent research: similarities, differences and unexplored aspects. Comput. Educ. **190**(December), 104607 (2022). https://doi.org/10.1016/j.compedu.2022.104607

Using Intelligent Tutoring on the First Steps of Learning to Program: Affective and Learning Outcomes

Maciej Pankiewicz[1]($\boxtimes$), Ryan Baker[2], and Jaclyn Ocumpaugh[2]

[1] Warsaw University of Life Sciences, Warsaw, Poland
maciej_pankiewicz@sggw.edu.pl
[2] University of Pennsylvania, Philadelphia, USA

Abstract. There exist several online applications for automated testing of the computer programs that students write in computer science education. Use of such systems enables self-paced learning with automated feedback delivered by the application. However, due to the complexity of programming languages, even the easiest tasks made available through such systems require understanding of several programming concepts and formatting. Therefore, a student's initial work in an introductory computer science course may be highly challenging, especially for students with no previous programming background.

To address this challenge, a highly-decomposed micro-task module has been developed and made available on an automated assessment platform with programming assignments. Impact of its introduction has been examined within an introductory programming university course with 239 participants. We investigated the micro-task module's impact on student affect, student performance on the platform, and student learning outcomes. Results of the experiment show that students in the experimental group (with micro-tasks enabled) significantly less frequently reported frustration, confusion and boredom, needed less time to solve tasks on the platform and achieved significantly better results on the final test.

Keywords: Computer science education · Programming micro-tasks · Affect

1 Introduction

Despite decades of research on helping students learn how to program [1, 2], many introductory students do not advance to more complicated topics and coursework [3]. Even when supported by educational tools and scaffolds, students with no programming background struggle in the first weeks of the course [4], where the complexity requires students to self-regulate through the acquisition of several cognitive skills that are introduced simultaneously [5–7]. These courses are also difficult to plan for because of their heterogeneous student populations. While some students have never programmed before or have only used graphical programming languages such as Scratch, other students have more experience. *Intelligent tutoring systems* (ITS) could be of help to these problems. However, developing ITS for this domain seems to be a challenge. A review of intelligent tutoring systems for programming [8] finds only one example of an intelligent tutoring system for a modern programming language in the years since 2010 [9].

N. Wang et al. (Eds.): AIED 2023, CCIS 1831, pp. 593–598, 2023.
https://doi.org/10.1007/978-3-031-36336-8_92

By contrast, the number of automated assessment platforms used in computer science education constantly increases [10]. The purpose of these platforms is to actually execute the code created by the student against a set of several test cases to provide immediate feedback on its correctness. Educational benefits often result from the fact that they may be used as a guide that helps students track if they are achieving their learning goals, simultaneously providing teachers more insight into student progress [10].

However, even when immediate feedback is available, students who do not understand why things have been marked as incorrect, may experience confusion, frustration or anxiousness and these negative affect states can negatively impact student outcomes [11–13]. If key aspects of intelligent tutors could be embedded into these platforms, it might be a feasible way to improve student outcomes. In this paper, we investigate a tool that breaks down the earliest steps of learning to program using micro-tasks (much like the earliest ITS for programming [1]), embedded within a test suite platform. We conducted a controlled experiment to investigate whether adding the micro-tasks to the test suite platform improves student learning and affect.

2   Methods

The Online Platform with the Micro-tasks Module. The online application *RunCode* [14] is a platform for automated execution and testing of programming code. It has been used since 2017 by students of computer science at the Warsaw University of Life Sciences. Usage of the platform is not mandatory, but students willingly use it (91% in Winter Semester 2021–2022). The *RunCode* app provides students with 146 programming tasks covering the following programming topics: types and variables (33), conditional statement (25), recursion (28), loops (17) and arrays (43). When a student submits a solution (programming code) to the task, the code is compiled, and tested. The student is then provided with several types of feedback, including the total score, the information from the compiler (if the code didn't compile) and detailed results for each test case executed on the submitted code (if the code compiled).

Most programming assignments in introductory courses require students to write several lines of code so that a program is ready to be executed against a defined set of test cases. With every line, statement, operator and punctuation, the complexity of the code increases. When faced with so many programming language components, students may find it hard to keep track of each detail of the created code. To prevent this problem, the platform provides students with a *micro-tasks* module designed to improve their understanding of a particular (single) element of programming code in more detail. Micro-tasks are coding tasks that focus on one particular aspect of the code and usually require the user to enter only one line of the code. They are designed to help students understand how different programming components work one-by-one. To achieve that, the line of code entered by the student is combined with a larger code structure. If the line of code is not correct, the student receives feedback on that specific element, in order to help them relate the errors that occurred to that one element being tested at the time. In contrast, when writing larger pieces of code, even a small syntax mistake may lead to a large number of compiler errors, which makes understanding the dependencies in the code much more difficult.

A total of 103 micro-tasks were created, spanning the topics of: types and variables (46), conditional statement (7), recursion (23), loops (9) and arrays (18). In the following sections, we use *micro-task* only when referring to a programming task that is made available in the introduced micro-task module. An example of such a task could be: "use the = operator and assign an appropriate value to the variable age." We refer to all the regular coding tasks available on the platform as a *task*. An example of such a task could be: "create a function that returns true if a given number is odd." To evaluate the impact of the micro-tasks module introduced to the system, students were randomly assigned to a control or experimental group. Sets of tasks for each topic were published later in the day after the corresponding classes. Students in the control group could access tasks on the platform immediately after they were published. For the experimental group, access to the main tasks was unlocked after solving a set of micro-tasks. Micro-tasks were not available to students in the control group.

Participants. This study's participants consisted of first-semester computer science students ($N = 276$) taking the *Introduction to Programming* course in Winter Semester 2022–2023 at the Warsaw University of Life Sciences. This course is required for computer science students at this university. The programming language used in this course is C#. A total of 239 students (28% female) consented to participate, with 174 submitting at least 1 solution on the platform during the study period (86 students in the control and 88 in the experimental group). In total, 31,011 submissions were collected (7,284 micro-tasks, and 23,727 regular programming tasks). In addition to these tasks, data consists of a single-item self-assessment of skills, results of the pre-test conducted during the first class session and the post-test conducted during the 10th class session.

Self-assessment of Skills: Novice and Experienced Programmers. During the first class session, students rated their level of knowledge on basic programming topics. Students were prompted, "Please rate on the scale 1 to 5 (where 1 – no knowledge at all and 5 – very good knowledge) your knowledge level for these topics taught in this course: types, variables, conditional statement, recursion, loops and arrays." Based on the self-assessment data, students were split into two groups: a "novice" group of students who reported their skill level as 1 and 2 ($N = 121$; 59 experimental, 62 control) and an "experienced" group of students who self-reported a skill level of 3 and above ($N = 118$; 54 experimental, 64 control). These values were selected to split students roughly evenly between the two groups. In the following sections we will refer to these groups as novice programmers and experienced programmers. Directly after the self-assessment, students were administered a pre-test to assess their knowledge level in a more objective manner. The test contained 8 multi-choice questions referring to the basic concepts taught during the course. The test questions were designed in a programming language-agnostic way, to take into account different programming languages that students may have previously learnt. The post-test was administered during the 10th class session, after all the relevant topics had previously been introduced during classes. The post-test contained 9 multi-choice questions.

Affective State While Using the Platform. Students self-reported their affective states while solving tasks on the platform through a dynamic HTML element. They were prompted to do so after they received submission results with the question: "Choose the option that best describes how you feel at the moment" with the following response options: *Focused, Anxious, Bored, Confused, Frustrated, Other* (in this order) and appropriate emoticons to visually highlight each of the available responses. This set of affective states was chosen based on their importance for learning and history of past research in AIED [12]. To avoid frustration that could arise from being required to fill out the survey too frequently, it was not presented after every submission, but randomly, with a probability of 1/3.

3   Results

Affective States. The most frequently reported state during regular programming tasks in both conditions was *focused* (more than half of collected responses in both conditions). A total of 6,051 survey responses were collected (experimental: 3,063, control: 2,988) from 128 students (experimental: 64, control: 64). Non-parametric Mann-Whitney U tests were calculated to compare differences in the frequencies of the affective states reported by each student (during regular programming tasks), due violation of normality assumptions. Due to multiple comparisons, the Benjamini-Hochberg alpha correction was used. Three affective states showed significant differences between conditions. For *boredom*, students in the experimental group reported marginally lower rates ($Mdn = 0$) than those in the control group ($Mdn = 0.0139$), ($W = 1566.5$, adjusted $\alpha = 0.01, p = 0.0125$, for a non-parametric Mann-Whitney U test). The same was true for *confusion*, where the experimental group ($Mdn = 0$) was marginally significantly lower than for the control group ($Mdn = 0.007$), ($W = 1641.5$, adjusted $\alpha = 0.02, p = 0.029$, for a non-parametric Mann-Whitney U test). *Frustration* followed the same pattern with the experimental group ($Mdn = 0$) reporting marginally significantly lower rates than the control group ($Mdn = 0.03$), ($W = 1674$, adjusted $\alpha = 0.03, p = 0.053$, for a non-parametric Mann-Whitney U test). Focus and anxiety, however, did not show significant differences between conditions. The experimental group reported a median that was not statistically significantly different from the control group ($Mdn = 0.72$ vs $Mdn = 0.52$), ($W = 2236$, adjusted $\alpha = 0.05, p = 0.369$, for a non-parametric Mann-Whitney U test). The same was true for *anxiety* ($Mdn = 0$ vs. $Mdn = 0.02$), ($W = 2236$, adjusted $\alpha = 0.04, p = 0.149$, for a non-parametric Mann-Whitney U test).

Learning Outcomes. To evaluate the impact of the introduced micro-task module on the post-test outcomes, we used a rank-based regression [15] (a non-parametric alternative to traditional likelihood or least squares estimators) used to test whether the pre-test and group significantly predicted the post-test results. The pre-test was a statistically significant predictor of the post-test, $t(236) = 14.53, p < 0.001$. The group was a statistically significant predictor of the post-test as well, $t(236) = 2.35, p = 0.02$; on average, students in the experimental condition performed 5.08% better on the post-test after controlling for pre-test. For novice programmers, the pre-test was a statistically significant predictor of the post-test, $t(118) = 3.50, p = 0.001$. The group was also a statistically significant predictor of the post-test, $t(118) = 2.07, p = 0.04$. In the group of novices, students in experimental condition performed on average 5.67% better on the post-test after controlling for pre-test. For experienced programmers, pre-test was again a statistically significant predictor of the post-test, $t(115) = 6.07$, $p < 0.001$, but group variable was not a statistically significant predictor of the post-test, $t(115) = 0.89$, $p = 0.373$.

Platform Usage. Within the experimental group, the time spent by novice programmers on micro-tasks ($Mdn = 190.47$ min.) was significantly higher than experienced programmers ($Mdn = 103.27$ min.), for a non-parametric Mann-Whitney U test ($W = 1406$, $p = 0.003$). To evaluate differences between students in the control and the experimental group in terms of time spent to solve tasks on the platform, we analyzed 7391 successful student attempts on tasks during the study period and evaluated the time students in both conditions needed to solve tasks. In doing so, we omitted tasks which fewer than 3 students successfully completed in each condition. Some students ceased work on a task without logging out; due to the fact that the platform does not monitor the keystroke-level data, a cutoff has been defined for tasks that have not been solved within the time of 2000 s (~33 min.). The vast majority of student attempts on tasks ended with a successful submission within that time (but several successful attempts were still in the 30–33 min range). The time needed to completely solve each task was significantly lower for the experimental group ($Mdn = 3.17$ min.) than the time for the control group ($Mdn = 4.7$ min.), $W = 1765.5$, $p < 0.001$, for a non-parametric Mann-Whitney U test. The time needed by novice programmers to completely solve each task was significantly lower for the experimental group ($Mdn = 3.45$ min.) than for the control group ($Mdn = 4.98$ min.), for a non-parametric Mann-Whitney U test ($W = 468.5$, $p < 0.001$). The time needed by experienced programmers to completely solve each task was marginally significantly lower for the experimental group ($Mdn = 3.3$ min.) than for the control group ($Mdn = 3.8$ min.), for a non-parametric Mann-Whitney U test ($W = 404$, $p = 0.053$).

4   Discussion

The results of the experiment show that students in the experimental group (micro-tasks enabled) were less likely to report frustration, confusion and boredom, needed less time than students in the control group to submit a correct solution to a programming task, and achieved better results on the post-test overall. However, the benefits of the micro-tasks for learning were clearer for students declaring no previous programming experience than for students with past experience. This finding indicates that not all students need the micro-tasks, suggesting that it may be best to provide them on the basis of self-report of expertise or, better yet, based on automatically inferring which students are likely to need them from initial within-system performance.

Limitation of this study is that we only studied impacts during the first ten classes of the semester; future work should study the impacts of micro-tasks over a longer time period. Finally, future work should study interventions of this nature in a broader selection of universities and introductory programming language, to establish generalizability. Overall, this initial evidence suggests that incorporating micro-tasks into introductory computer science platforms may have benefits both for learning and affect, potentially increasing the number of students who succeed in introductory courses and continue further into CS programs.

References

1. Anderson, J.R., Reiser, B.J.: The LISP tutor. Byte **10**(4), 159–175 (1985)
2. Brusilovsky, P.L.: Intelligent tutor, environment and manual for introductory programming. Educ. Train. Technol. Int. **29**(1), 26–34 (1992)
3. Robins, A.V.: Novice programmers and introductory programming. In: The Cambridge Handbook of Computing Education Research, pp. 327–376 (2019)
4. Prather, J., Pettit, R., McMurry, K., Peters, A., Homer, J., Cohen, M.: Metacognitive difficulties faced by novice programmers in automated assessment tools. In: Proceedings of the ACM Conference on International Computing Education Research, pp. 41–50 (2018)
5. Falkner, K., Vivian, R., Falkner, N.J.: Identifying computer science self-regulated learning strategies. In: Proceedings of the Conference on Innovation & Technology in Computer Science Education, pp. 291–296 (2014)
6. Renumol, V.G., Janakiram, D., Jayaprakash, S.: Identification of cognitive processes of effective and ineffective students during computer programming. ACM Trans. Comput. Educ. **10**(3), 1–21 (2010)
7. Alaoutinen, S.: Evaluating the effect of learning style and student background on self-assessment accuracy. Comput. Sci. Educ. **22**(2), 175–198 (2012)
8. Crow, T., Luxton-Reilly, A., Wuensche, B.: Intelligent tutoring systems for programming education: a systematic review. In: Proceedings of the Australasian Computing Education Conference, pp. 53–62 (2018)
9. Rivers, K., Koedinger, K.R.: Data-driven hint generation in vast solution spaces: a self-improving python programming tutor. Int. J. Artif. Intell. Educ. **27**(1), 37–64 (2017)
10. Paiva, J.C., Leal, J.P., Figueira, Á.: Automated assessment in computer science education: a state-of-the-art review. ACM Trans. on Comp. Education **22**(3), 1–40 (2022)
11. Rodrigo, M.M.T., et al.: Affective and behavioral predictors of novice programmer achievement. In: Proceedings of ACM SIGCSE, pp. 156–160 (2009)
12. Karumbaiah, S., Baker, R.S., Tao, Y., Liu, Z.: How does students' affect in virtual learning relate to their outcomes? A systematic review challenging the positive-negative dichotomy. In: Proceedings of the International Learning Analytics and Knowledge Conference, pp. 24–33 (2022)
13. Bosch, N., D'Mello, S.: The affective experience of novice computer programmers. Int. J. Artif. Intell. Educ. **27**(1), 181–206 (2017)
14. Pankiewicz, M.: Move in the right direction: impacting students' engagement with gamification in a programming course. In: EdMedia+ Innovate Learning, pp. 1180–1185. Association for the Advancement of Computing in Education (AACE) (2020)
15. Kloke, J.D., McKean, J.W.: Rfit: rank-based estimation for linear models. The R J. **4**(2), 57–64 (2012)

A Unified Batch Hierarchical Reinforcement Learning Framework for Pedagogical Policy Induction with Deep Bisimulation Metrics

Markel Sanz Ausin$^{(\boxtimes)}$ ⓘ, Mark Abdelshiheed ⓘ, Tiffany Barnes ⓘ, and Min Chi ⓘ

North Carolina State University, Raleigh, NC 27606, USA
{msanzau,mnabdels,tmbarnes,mchi}@ncsu.edu

Abstract. Intelligent Tutoring Systems (ITSs) leverage AI to adapt to individual students, and employ *pedagogical policies* to decide what instructional action to take next. A number of researchers applied Reinforcement Learning (RL) and Deep RL (DRL) to induce effective pedagogical policies. Most prior work, however, has been developed *independently* for a specific ITS and *cannot directly be applied to another*. In this work, we propose a Multi-**T**ask Learning framework that combines Deep **BI**simulation Metrics and DRL, named **MTL-BIM**, to induce a unified pedagogical policy for two different ITSs across different domains: logic and probability. Based on empirical classroom results, our unified RL policy performed significantly better than the expert-crafted policies and independently induced DQN policies on both ITSs.

Keywords: Deep Reinforcement Learning · Pedagogical Policy

1 Introduction

Intelligent Tutoring Systems (ITSs) are interactive e-learning environments that support students' learning by providing individualized instruction and on-demand help; and have shown to be highly effective [8,19]. In order to design an effective ITS, developers must determine *how* to teach the desired content. *Pedagogical policies* are the decision-making policies inside an ITS that decide what action to take next in the face of alternatives. While many ITSs exist for STEM domains, the pedagogical policies are often purposely built for *a single ITS in a single domain, and cannot work across ITSs.*

A number of researchers have studied Reinforcement Learning (RL) and Deep RL (DRL) for pedagogical policy induction [1,9,18]. The sequential decision-making nature of RL and DRL, combined with its ability to learn from a reward function, makes it a perfect fit to induce pedagogical policies for ITSs. In particular, Batch RL methods can be used to train an RL agent on historical logs of student-tutor interactions, rather than on simulated versions of students.

N. Wang et al. (Eds.): AIED 2023, CCIS 1831, pp. 599–605, 2023.
https://doi.org/10.1007/978-3-031-36336-8_93

In this work, we propose a general *Multi-Task Learning* framework using Deep **BI**simulation Metrics and DRL (**MTL-BIM**) to induce unified pedagogical policies across multiple ITSs. *MTL-BIM allows us to combine different training datasets so the RL agent can train a more robust pedagogical policy through a greater range of experience.* ITSs can differ widely in the level of granularity of their interventions: problem-level, step-level, or even micro-step. Such differences make it very challenging for RL-induced policies to be effective across multiple levels of granularity.

Our MTL-BIM combines batch DRL with deep bisimulation metrics to learn a shared *latent state representation* across multiple ITSs. It can combine multiple learning systems and their respective datasets and granularities to *train a single, unified pedagogical policy.* To evaluate MTL-BIM, we utilize two ITSs, named Deep Thought and Pyrenees, that teach how to solve logic proofs and probability, respectively. Deep Thought has a problem-level (high level) granularity, where the agent needs to take a decision for the entire problem. However, Pyrenees has problem-level (high level) and step-level (low level) granularities. The empirical results demonstrate the ability of our single policy to improve student learning across both tutors, showing that the students who used our pedagogical policy learned more effectively and efficiently than those who used an expert-designed policy tailored to each tutor separately.

2   MTL-BIM

Definition 1 (Bisimulation Metrics [2]). *From Theorem 2.6 in [11] with discount factor $c \in [0,1]$, and the Wasserstein metric W_1, the bisimulation metric is defined as: $d(s_i, s_j) = \max_{a \in \mathcal{A}}(1 - c) \cdot |\mathcal{R}_{s_i}^a - \mathcal{R}_{s_j}^a| + c \cdot W_1(\mathcal{P}_{s_i}^a, \mathcal{P}_{s_j}^a; d)$.*

Ferns et al. (2011) [11] introduced bisimulation metrics to relax the strict criterion for state aggregation used in Bismulation [17]. Bisimulation metrics define a pseudo-metric space $(\mathcal{S}, d)$, where a distance function $d : \mathcal{S} \times \mathcal{S} \mapsto \mathbb{R}_{\geq 0}$ is used to measure how behaviorally similar two states are. Our goal, inspired by [2], is to unify the states from the two different ITS domains ($\mathcal{D}_1$ and $\mathcal{D}_2$) into a shared latent space, and make their distance in latent space be the same as their bisimulation metric distance, thus making behaviorally similar states be closer to one another in latent space. We incorporate deep bisimulation metrics into a model-based DRL algorithms to build a unified framework for pedagogical policy induction.

In this work, we represent our two ITSs as domains: $\mathcal{D}_1$ and $\mathcal{D}_2$. The interactions in each domain can be modeled by a Markov Decision Process (MDP), described by the 4-tuple $(\mathcal{S}, \mathcal{A}, \mathcal{P}, \mathcal{R})$ where $\mathcal{S}$ is the state space, $\mathcal{A}$ is the action space, $\mathcal{P}(s'|s, a)$ is the transition function, and $\mathcal{R}(s)$ is the reward function.

Figure 1 (left) shows the general architecture of MTL-BIM, whereas Fig. 1 (right) shows all the neural networks in the model. It can be divided into two main parts: 1) Two Variational Auto-Encoders (VAEs) [16] (one per domain) combined with bisimulation metrics, which learns a shared latent representation across the two tasks. At each time step t, the encoder converts the observation space into a latent state. During training, we sample a batch of trajectories

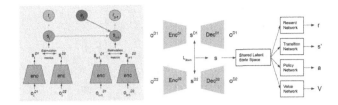

Fig. 1. MTL-BIM Framework and architecture. Left: general architecture. Right: Architecture of our framework for unifying two policies into a single system.

from each dataset, and we calculate the bisimulation metric between each pair of examples. We define the loss function used to train the encoder using bisimulation metrics as $\mathcal{L}_{bisim} = ||s^{\mathcal{D}_1} - s^{\mathcal{D}_2}||_1 - d(s^{\mathcal{D}_1}, s^{\mathcal{D}_2})$. To train our encoders to learn the desired representation, we make the L1 distance between every two latent states be the same as their bisimulation metric distance. 2) A model-based RL algorithm with a Recurrent State Space Model (RSSM) architecture [4], which gets the latent state from part 1 as input, and takes an action in the environment. We trained neural networks to learn a reward function ($r_t \sim p(r_t|h_t, s_t)$), a deterministic state ($h_{t+1} = f(h_t, s_t, a_t)$), a stochastic state ($s_t \sim p(s_t|h_t)$), a value function ($v_t = V(s_t)$), and a policy function ($a \sim \pi(a_t|s_t)$). The reward loss ($\mathcal{L}_R = \frac{1}{2}||r(s_t) - R(s_t)||^2$), transition loss ($\mathcal{L}_P = \frac{1}{2}||s_{t+1} - P(s_{t+1}|s_t, a_t)||^2$), value loss ($\mathcal{L}_V = \frac{1}{2}||v(s_t) - V(s_t)||^2$), and policy loss ($\mathcal{L}_\pi = -\mathbb{E}(\sum_{t=t}^{t+H} V(s_t))$) in the RSSM are trained jointly with the bisimulation loss and the VAE loss.

3   Domains & Empirical Evaluation

Domain $\mathcal{D}_1$: Deep Thought (UI shown in Fig. 2 (left)) is a data-driven, graph-based ITS for multi-step propositional logic problems, that uses pedagogical policies induced via DRL [9,13]. Deep Thought decides whether to present each problem as a Problem Solving (PS) or Worked Examples (WE). In PS, students *construct* logic proofs until the conclusion is reached. In WE, students simply *observe* how the tutor solves the problem. Students go through a pretest, training (20 problems), posttest procedure. Our state is comprised of the 142 features, including 10 student autonomy features, 29 temporal, 35 problem-solving, 57 general performance, and 11 hint usage. We use the learning efficiency as reward, which is calculated as the difference between the posttest and the pretest scores, divided by the training time.

Domain $\mathcal{D}_2$: Pyrenees (UI shown in Fig. 2 (right)), is a text-based ITS that teaches students how to solve probability problems. Pyrenees makes decisions in two different granularity levels: problem-level (high level) and step-level (low level). The former dictates whether the student sees the problem as a PS, a WE, or a Faded WE (FWE). During FWE, the student and the agent will collaborate when solving the problem (step-level). Students go through a standard *pretest-training-posttest*. Our state is represented by 142 features, which are different from Deep Thought. The reward function is the Normalized Learning Gain

Fig. 2. Left: UI for Deep Thought. Right: UI for Pyrenees.

(NLG), $NLG = \frac{posttest - pretest}{\sqrt{100 - pretest}}$. NLG measures student improvement from the pretest to the posttest, and is normalized to a range of $[-1, +1]$.

Empirical Evaluation: We conducted an empirical study using Deep Thought in Fall of 2020 with students at North Carolina State University, to evaluate MTL-BIM (55 students) against a baseline *PS-Only* policy (59 students). MTL-BIM is also compared against two policies that were carried out in the Spring of 2019 (S19): a human-crafted *Expert* policy (23 students), and an RL-based *DQN* [15] policy (30 students). For Pyrenees, we empirically compared our MTL-BIM policy (49 students) against a hierarchical control *Expert*-designed policy (49 students) in Fall of 2020. We also compared our MTL-BIM policy against an RL-based DQN policy (45 students) in Fall of 2018.

4 Results

Deep Thought: We have four groups: *MTL-BIM* and *PS-only, Expert* and *DQN*. Figure 3 (a) shows the mean and standard error of the students' posttest scores. A one-way ANCOVA test using the condition as a factor and the pretest score as a covariate showed a significant difference: $F(3, 162) = 22.32, p < 0.001$. Subsequent contrasts analyses showed *MTL-BIM* scored significantly higher than *Expert* ($t(75) = 5.235, p < 0.0001$) and *DQN* ($t(82) = 4.528, p < 0.001$). Similarly, *PS-Only* scored significantly higher than *Expert* ($t(79) = 6.852, p < 0.001$) and DQN ($t(86) = 6.389, p < 0.0001$). No significant difference was found between *MTL-BIM* and *PS-Only*, nor between *Expert* and *DQN*. Figure 3 (b) shows the mean and standard error of the four conditions' learning efficiency. A one-way ANCOVA test using condition as a factor and the pretest efficiency as a covariate showed no significant difference in the learning efficiency: $F(3, 162) = 1.857, p = 0.139$. Subsequent contrasts analyses showed a significant difference in the learning efficiency scores between the *MTL-BIM* and the *PS-Only* conditions ($t(111) = 2.537, p = 0.012$) in that the former scored significantly higher than the latter. In short, our MTL-BIM framework can induce pedagogical policies that are more efficient than the PS-Only policy (learning efficiency) and more effective than the Expert and DQN policies (posttest scores).

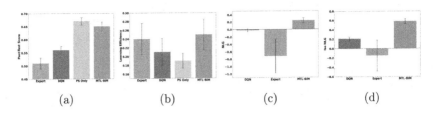

(a)   (b)   (c)   (d)

Fig. 3. Deep Thought: posttest (a): MTL-BIM=PS-Only > DQN=Expert. Learning Efficiency (b): MTL-BIM > PS-Only. **Pyrenees:** NLG (c): MTL-BIM > DQN= Expert. Isomorphic NLG (d): MTL-BIM > DQN=Expert.

Pyrenees: We have three groups: *MTL-BIM, Expert* and *DQN*. Figure 3 shows the NLG (c), and the Isomorphic NLG (d) across the three conditions. A one-way ANOVA test using the condition as a factor showed a significant difference in the NLG: $F(2, 140) = 3.319, p = 0.039$. Subsequent contrasts analyses showed that MTL-BIM scored significantly higher than both Expert $(t(96) = 2.073, p = 0.040)$ and *DQN* $(t(92) = 3.002, p = 0.003)$. However, no significant difference was found between the *Expert* and *DQN* conditions. Similarly, a one-way ANOVA test using the condition as a factor showed a significant difference in the isomorphic NLG: $F(2, 140) = 3.47, p = 0.034$. Subsequent contrasts analyses showed *MTL-BIM* significantly outperformed both *Expert* $(t(96) = 2.190, p = 0.031)$ and *DQN* $(t(92) = 5.480, p < 0.001)$ and no significant difference was found between the *Expert* and *DQN* conditions. In summary, after spending similar amounts of time on the Pyrenees, our *MTL-BIM* policy is more effective than *Expert* and the *individually trained DQN* policies, as shown by the NLG and Isomorphic NLG scores.

5 Related Work

Prior work has investigated using RL techniques to learn pedagogical policies for ITSs. For instance, [18] applied a Partially Observable Markov Decision Process (POMDP) to train online policies for faster training. Lately, DRL algorithms have also been used with similar purpose. [9] trained an offline DRL algorithm to learn a pedagogical policy that improves student learning when compared to an expert-crafted policy. The attempts to unify multiple ITSs have been very limited. The work by [10] managed to combine different sub-tasks in an educational game using MTL, and managed to more accurately predict posttest scores. Improving generalization in Machine Learning algorithms is a very active area of research, using Transfer Learning, Domain Adaptation or Multi-Task Learning. In RL, the generalization ability of agents has been widely studied through Multi-Task RL approaches [6,12]. [7] studied the generalization ability of RL agents and proposed a method that improves generalization by training the agent on a mixture of observations. The work by [14] presents an approach to Multi-Task Deep RL using attention to automatically learn the relationships between tasks. Finally, the work by [5] learns a shared latent representation of the state-action space across all the tasks. [3] also learned a shared representation across tasks that allow them to outperform the single-task versions of the algorithms in standard RL benchmarks.

6   Conclusion and Discussion

We developed MTL-BIM, a framework to learn a shared latent state representation using deep bisimulation metrics with a model-based DRL algorithm. We showed that MTL-BIM can learn a unified policy by combining both training datasets, and the students who used our MTL-BIM policy on Deep Thought learned more than the students who used both the Expert and the DQN policies, while learning more efficiently than the students assigned to the PS-Only policy. In Pyrenees, our results show that the MTL-BIM students improved significantly more than the students in both the Expert and the DQN policies, both in terms of general NLG and isomorphic NLG. This work provides a step forward in learning shared pedagogical policies across multiple ITSs.

Acknowledgements. This research was supported by NSF Grants: #1726550, #1651909, and #2013502.

References

1. Singla et al., A.: Reinforcement learning for education: Opportunities and challenges. arXiv preprint arXiv:2107.08828 (2021)
2. Zhang et al., A.: Learning invariant representations for reinforcement learning without reconstruction. arXiv preprint arXiv:2006.10742 (2020)
3. D'Eramo et al., C.: Sharing knowledge in multi-task deep reinforcement learning. In: ICLR (2019)
4. Hafner et al., D.: Dream to control: Learning behaviors by latent imagination. arXiv preprint arXiv:1912.01603 (2019)
5. Borsa et al., D.: Learning shared representations in multi-task reinforcement learning. arXiv preprint arXiv:1603.02041 (2016)
6. Higgins et al., I.: Darla: Improving zero-shot transfer in reinforcement learning. In: ICML, pp. 1480–1490. PMLR (2017)
7. Wang et al., K.: Improving generalization in reinforcement learning with mixture regularization. arXiv preprint arXiv:2010.10814 (2020)
8. Koedinger et al., K.R.: Intelligent tutoring goes to school in the big city. IJAIED 8, 30–43 (1997)
9. Sanz Ausin, M., Maniktala, M., Barnes, T., Chi, M.: Exploring the impact of simple explanations and agency on batch deep reinforcement learning induced pedagogical policies. In: Bittencourt, I.I., Cukurova, M., Muldner, K., Luckin, R., Millán, E. (eds.) AIED 2020. LNCS (LNAI), vol. 12163, pp. 472–485. Springer, Cham (2020). https://doi.org/10.1007/978-3-030-52237-7_38
10. Geden et al., M.: Predictive student modeling in educational games with multi-task learning. In: AAAI. vol. 34, pp. 654–661 (2020)
11. Ferns et al., N.: Bisimulation metrics for continuous mdps. SIAM **40**(6), 1662–1714 (2011)

12. Sodhani et al., S.: Multi-task reinforcement learning with context-based representations. arXiv preprint arXiv:2102.06177 (2021)
13. Shen, S., Mostafavi, B., Lynch, C., Barnes, T., Chi, M.: Empirically evaluating the effectiveness of pomdp vs. mdp towards the pedagogical strategies induction. In: Penstein Rosé, C., et al. (eds.) AIED 2018. LNCS (LNAI), vol. 10948, pp. 327–331. Springer, Cham (2018). https://doi.org/10.1007/978-3-319-93846-2_61
14. Bram et al., T.: Attentive multi-task deep reinforcement learning. Preprint arXiv:1907.02874 (2019)
15. Mnih et al., V.: Playing atari with deep reinforcement learning. NIPS (2013)
16. Kingma, D.P., Welling, M.: Auto-encoding variational bayes. Preprint arXiv:1312.6114 (2013)
17. Larsen, K.G., Skou, A.: Bisimulation through probabilistic testing. Inf. Comput. **94**(1), 1–28 (1991)
18. Rafferty, A.N., Brunskill, E., et al.: Faster teaching via pomdp planning. Cogn. Sci. **40**(6), 1290–1332 (2016)
19. Vanlehn, K.: The behavior of tutoring systems. Int. J. Artif. Intell. Educ. **16**(3), 227–265 (2006)

Impact of Experiencing Misrecognition by Teachable Agents on Learning and Rapport

Yuya Asano[1(✉)], Diane Litman[1], Mingzhi Yu[1], Nikki Lobczowski[2], Timothy Nokes-Malach[1], Adriana Kovashka[1], and Erin Walker[1]

[1] University of Pittsburgh, Pennsylvania, USA
{yua17,dlitman,miy39,nokes,aik85,eawalker}@pitt.edu
[2] McGill University, Quebec, Canada
nikki.lobczowski@mcgill.ca

Abstract. While speech-enabled teachable agents have some advantages over typing-based ones, they are vulnerable to errors stemming from misrecognition by automatic speech recognition (ASR). These errors may propagate, resulting in unexpected changes in the flow of conversation. We analyzed how such changes are linked with learning gains and learners' rapport with the agents. Our results show they are not related to learning gains or rapport, regardless of the types of responses the agents should have returned given the correct input from learners without ASR errors. We also discuss the implications for optimal error-recovery policies for teachable agents that can be drawn from these findings.

Keywords: Teachable agents · Automatic speech recognition · Rapport

1 Introduction and Related Work

Students benefit from teaching others more than being tutored [1]. This effect of learning by teaching also holds when they teach a virtual agent [7] or an embodied robot [9] (called a teachable agent or robot). Although interaction with teachable agents can be done through typing [6] or speech [2], the literature suggests speech-based interaction is powerful in tutorial dialogues in general because speech enables learners to complete tasks faster than typing due to ease of production [3]. However, speech-based teachable agents are susceptible to misrecognition made by automatic speech recognition (ASR) when converting speech input to text. It may change the flow of the dialogue between students and agents and thus affect students' learning and perception of the agents.

Past work has explored how misrecognition in different stages of speech-enabled tutorial dialogue systems (Fig. 1 shows the structure of our system) is related to students' learning gain and evaluation of the systems. D'Mello et al. [3] have found word error rates of ASR were not associated with students'

N. Wang et al. (Eds.): AIED 2023, CCIS 1831, pp. 606–611, 2023.
https://doi.org/10.1007/978-3-031-36336-8_94

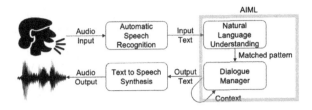

Fig. 1. The structure of our dialogue system. When it receives audio input, it converts audio to text. Then, Artificial Intelligence Markup Language (AIML) finds a pattern in the input text (NLU), selects an output (Dialogue Manager), and remembers it as a context. Finally, the system converts the text to speech.

learning gain but were related to their satisfaction with systems. Litman and Forbes-Riley [8] have found no correlations between errors made by a natural language understanding (NLU) module and students' learning. However, errors in earlier stages may not propagate to the outputs of a system. For example, mistakes in singular versus plural forms or "a" versus "the" made by ASR have little effect on NLU. Moreover, even if NLU fails to map "massive" to "big" because "massive" is out of vocabulary, the output may not change if it does not care whether a user says "big". Thus, errors in ASR or NLU may not represent misrecognition felt by end users. In this paper, we focus on mistakes that are retained until the final outputs of systems. Also, Dzikovska et al. [4] have shown that the frequency of a system not understanding user inputs and thus replying with a neutral response is negatively correlated with user satisfaction but not with learning gain. We instead distinguish the case where it is fine for dialogue systems to return a neutral response because learners say something irrelevant from the case where the systems should respond with something more specific.

We fill a literature gap on the effect of errors by dialogue systems on learners. First, we extend the work on tutoring systems that appear more knowledgeable than learners to teachable agents that are at most the same level. Second, we analyze errors observable by learners, instead of errors internal to systems. We define *dialogue misrecognition* as the situation where an agent responds differently to raw ASR inputs and "true" inputs. Finally, we evaluate an agent with students' sense of rapport with it rather than user satisfaction. Rapport is a predictor of learning for both human and agent tutees [9] and leaves a positive impression that helps agents establish a long-term relationship with learners [5]. Thus, it is likely a more direct metric than satisfaction with the effectiveness of human-agent collaborations. We have found dialogue misrecognition is not linked with learning or rapport. Our results are in line with the literature but contribute to it by measuring only misrecognition that impacts the flow of conversation and by using an outcome more suitable for human-agent collaborations.

2 Method

2.1 Dataset

We used the dataset from an experiment where 40 undergraduate students (35 female, 5 male; 17 White, 13 Asian, 5 Black, 1 Latino, 4 unknown; mean age

= 19.64, SD = 1.25) in a US city taught ratio word problems to a robot named Emma using spoken dialogue for 30 min. The study was over Zoom due to COVID-19. To talk to Emma, they pushed and held a button on a web application. She is designed to follow up on student explanations with questions or her own explanations. For example, she says *"So I multiply because I have three times as many people?"* after a student tells *"You have to multiply cups of seltzer by three."* She also guides students to a correct solution when they are unsure about the answer. For instance, if a student's utterance contains *"I don't know"*, she suggests the next step, saying *"Me either. Maybe I start by dividing?"*. We had two experimental conditions: students teaching in pairs (n=28) or alone (n=12). Our design was quasi-experimental; if only one of the students showed up out of a pair, we had them teach Emma alone. Students in pairs taught her collaboratively by alternating between discussing the problems with their partners and talking to her. We excluded one pair from our analysis because one of the students did not speak to Emma during the session.

The students individually took a pre-test before the session and a post-test and a survey after that to assess learning and rapport with Emma. We prepared two versions of pre- and post-tests each of which had 13 ratio problems similar to what they saw while teaching her. We removed five isomorphic problems from each test that proved to have different difficulties between versions, leaving eight problems for analysis [10]. On average, students scored 5.58 (SD = 1.90) in the pre-test and 6.82 (SD = 1.35) in the post-test. Rapport was measured on a six-point Likert scale devised by [9] (mean = 4.49, SD = .679) and represented the average of items asking about mutual positivity, attention, and coordination between the student and the robot [2,9].

2.2   Quantification of Dialogue Misrecognition

We simulated Emma's responses to "true" inputs as follows. First, we manually transcribed the students' utterances directed to her. Let U_t^H be the human transcribed utterance of a student directed to Emma in turn t and U_t^{ASR} be the utterance transcribed by ASR. Next, we sent U_t^H to Emma to get a simulated "true" response R_t^H. She selects R_t^H from a set of pre-authored responses written in Artificial Intelligence Markup Language (AIML), based on patterns in U_t^H and context C_t. We set C_1 to none and C_{t+1} to R_t^{ASR}, the response to U_t^{ASR}, because U_{t+1}^H is a reply to R_t^{ASR}. If there is no matching pattern in U_t^H given C_t, Emma returns an utterance randomly selected from a set of *generic* responses G that can make sense in any contexts such as *"I think I get it. What do I do next?"*. The size of G was 28. The responses in G do not change Emma's state, meaning that, unlike non-generic responses, they do not let students move to the next step of the problem. Finally, we calculated the proportions of the turns in which R_t^H differ from R_t^{ASR} to the total number of turns for each student (i.e. $P(R_t^H \neq R_t^{ASR})$, the proportion of **Overall** dialogue misrecognition). We did not use raw numbers of errors to normalize the differences in the number of turns for each student. In the case of students in pairs, we did not use the turns where their partner spoke to calculate the proportions but did use those turns

Table 1. Example interaction between students in a pair and Emma. Her responses in italics come from the set of generic responses G. In this example, each student had one turn, and Emma misrecognized student B's input. Therefore, $P(R_t^H \neq R_t^{ASR}) = 0$ for student A and $P(R_t^H \neq R_t^{ASR}) = 1$ for student B.

Speaker	Label	Utterance	Results
A	U_1^{ASR}	what is the ratio between the volume of paint and the surface area	AIML pattern: no match
	U_1^H	Emma, what is the ratio between the volume of paint and the surface area?	AIML pattern: no match
Emma	R_1^{ASR}	*I'm still learning this. I don't get it.*	$R_1^{ASR} = R_1^H$
	R_1^H	*I'm not sure actually.*	
B	U_2^{ASR}	the ratio between surface area and volume is 622	AIML pattern: no match
	U_2^H	The ratio between surface area and volume is 6 to 2.	AIML pattern: "ratio ... 6 ... 2"
Emma	R_2^{ASR}	*Can you explain a little more?*	$R_2^{ASR} \neq R_2^H$
	R_2^H	6 to 2 is the same ratio I used for step one. But the ratio of 1 to 3 seems an easier place to start?	

to define C_{t+1} and update Emma's dialogue manager. Note that we treated R_t^H and R_t^{ASR} as the same when both are from G, even if their surface forms are different. An example of the calculation of the proportions is shown in Table 1.

We further categorized dialogue misrecognition into three cases:

- **Prevented**: $P(R_t^H \notin G \wedge R_t^{ASR} \in G)$. This means the student could not move to the next step because Emma misrecognized their input.
- **Different**: $P(R_t^H \neq R_t^{ASR} \wedge R_t^H \notin G \wedge R_t^{ASR} \notin G)$. This often implies Emma suggested a different way to solve a problem from what the student said.
- **Proceeded**: $P(R_t^H \in G \wedge R_t^{ASR} \notin G)$. This represents the case where Emma went to the next step by accident due to misrecognition.

3   Results and Discussion

We examined how dialogue misrecognition is related to students' rapport with Emma and learning gain by running correlation analyses. We used Pearson's correlations for rapport and partial correlation for post-test scores controlled by pre-test scores because pre-test scores were positively correlated with post-test scores ($r = .443$, $p = .005$). Table 2 summarizes our analysis of the 38 students from both conditions. **Overall** misrecognition is not correlated with either rapport ($r = -.016$, $p = .922$) or learning ($r = .246$, $p = .142$). Of the three types of dialogue misrecognition, **Different** was the highest. None of these types was significantly correlated with learning or rapport. However, **Prevented** is marginally negatively correlated with rapport ($r = -.319$, $p = .051$).

Our results provide another piece of evidence that misrecognition by teachable agents is not necessarily relevant to learning gains or learners' perception of agents. Furthermore, this implies that an optimal error-recovery strategy for

Table 2. Descriptive statistics of the proportions of dialogue misrecognition and its correlations with rapport with Emma (Pearson's r, $df = 36$) and post-test scores (partial correlation controlled by pre-test scores, $df = 35$). **Different** type was the most likely. No correlation was significant.

Misrecognition types	Mean (SD)	Rapport (p-value)	Learning (p-value)
Overall	.146 (.080)	-.016 (.922)	.246 (.142)
Prevented	.026 (.031)	-.319 (.051)	.188 (.488)
Different	.103 (.070)	.141 (.399)	.271 (.104)
Proceeded	.017 (.025)	-.056 (.737)	-.131 (.440)

teachable agents should prefer moving to the next step, assuming that learners give reasonable inputs, rather than expressing they do not understand the inputs. This may sound counterintuitive because it may deprive learners of opportunities to realize their misunderstanding. Still, this disadvantage may be canceled out by exposure to correct solutions and more problems because the correlation between learning and the proportion that Emma returns a non-generic response when she is supposed to return a generic one is not significant. This policy may also aid inclusion because it can avoid generic responses stemming from ASR's poor performance in accented speech and minoritized dialects.

One limitation of this study is that many participants were at the ceiling (6 students scored 100% on the pre-test) and thus did not learn as part of the study, reducing our ability to examine correlations between dialogue misrecognition and learning. Another limitation is that our dialogue system is not a state-of-the-art end-to-end neural model. We used the Web Speech API for speech recognition off the shelf, which yielded a .226 word error rate on average $(SD = .102)$[1], and performed only pattern matching to decide Emma's response. Yet, our dialogue misrecognition measures can be used for end-to-end models that do not have internal components such as NLU. Also, due to the small sample size $(n = 38)$, we lack statistical power and could not include demographic variables or the experimental conditions as covariates. This stopped us from analyzing the effect of witnessing dialogue misrecognition encountered by a partner.

We proposed new measures of dialogue misrecognition to explore how changes in a conversation flow caused by errors that propagate through a dialogue system are related to rapport with teachable agents and learning gain. Our results indicate these changes are not linked to learning or rapport. This implies we do not need a sophisticated dialogue system with little misrecognition for teachable agents and that an optimal error-recovery policy can be as simple as presuming inputs from learners are reasonable. Future research can test how this policy affects rapport, learning, and other outcomes such as engagement.

---

[1] Word error rates were not correlated with rapport $(r = .196, p = .239)$, learning $(\rho = .246, p = .142)$, or overall dialogue misrecognition $(r = -.155, p = .354)$.

Acknowledgments. We would like to thank anonymous reviewers for their thoughtful comments on this paper. This work was supported by Grant No. 2024645 from the National Science Foundation, Grant No. 220020483 from the James S. McDonnell Foundation, and a University of Pittsburgh Learning Research and Development Center internal award.

References

1. Annis, L.F.: The processes and effects of peer tutoring. Human Learn.: J. Pract. Res. Appl. **2**(1), 39–47 (1983)

2. Asano, Y., et al.: Comparison of lexical alignment with a teachable robot in human-robot and human-human-robot interactions. In: Proc. of the 23rd Annual Meeting of the Special Interest Group on Discourse and Dialogue, pp. 615–622. Association for Computational Linguistics, Edinburgh, UK (Sep 2022)

3. D'Mello, S.K., Graesser, A., King, B.: Toward spoken human-computer tutorial dialogues. Human-Comput. Interact. **25**(4), 289–323 (2010)

4. Dzikovska, M.O., Moore, J.D., Steinhauser, N., Campbell, G.: The impact of interpretation problems on tutorial dialogue. In: Proceedings of the ACL 2010 Conference Short Papers, pp. 43–48. Association for Computational Linguistics, Uppsala, Sweden (Jul 2010)

5. Gulz, A., Haake, M., Silvervarg, A.: Extending a teachable agent with a social conversation module: Effects on student experiences and learning. In: Proceedings of the 15th International Conference on Artificial Intelligence in Education, p. 106–114. AIED'11, Springer-Verlag, Berlin, Heidelberg (2011)

6. Lee, K.J., Chauhan, A., Goh, J., Nilsen, E., Law, E.: Curiosity notebook: the design of a research platform for learning by teaching. In: Proceedings of the ACM on Human-Computer Interaction. vol. 5, pp. 1–26. ACM New York, NY, USA (2021)

7. Leelawong, K., Biswas, G.: Designing learning by teaching agents: The betty's brain system. Int. J. Artif. Intell. Ed. **18**(3), 181–208 (aug 2008)

8. Litman, D.J., Forbes-Riley, K.: Speech recognition performance and learning in spoken dialogue tutoring. In: INTERSPEECH, pp. 161–164 (2005)

9. Lubold, N.: Producing Acoustic-Prosodic Entrainment in a Robotic Learning Companion to Build Learner Rapport. Ph.D. thesis, Arizona State University (2018)

10. Steele, C., et al.: It takes two: Examining the effects of collaborative teaching of a robot learner. In: Rodrigo, M.M., Matsuda, N., Cristea, A.I., Dimitrova, V. (eds.) Artificial Intelligence in Education. Posters and Late Breaking Results, Workshops and Tutorials, Industry and Innovation Tracks, Practitioners' and Doctoral Consortium, pp. 604–607. Springer International Publishing, Cham (2022)

Nuanced Growth Patterns of Students with Disability

Sadia Nawaz[1,2](✉) ⓘ, Toshiko Kamei[1] ⓘ, and Namrata Srivastava[2] ⓘ

[1] Melbourne Assessment, University of Melbourne, Parkville, VIC 3010, Australia
sadia.nawaz@monash.edu
[2] Faculty of IT, Monash University, Clayton, VIC 3168, Australia

Abstract. Effective educational planning requires an accurate understanding of student growth. Few researchers have examined the growth trajectories of students' complex competencies, particularly of students with disability (SWD). Our study aims to fill this knowledge gap by analyzing individualized growth of SWD in such skills. We used Markov chain modeling on progress data from 7760 SWD. It was found that students who start out low have more difficulty progressing and that growth patterns vary among SWD and students with autism spectrum disorder (ASD). Students' individualized patterns of growth may provide nuanced expectations of growth rather than setting lower expectations based on a pre-defined group to which a student belongs. Such work can promote inclusion to provide all students with equitable access to learning.

Keywords: Disability · Markov chain · Learning progressions · 21st century skills

1 Introduction

Profiles of learning growth can represent the quantification of learning trajectories to analyze different growth patterns and set learning expectations [1]. Policy (e.g., [2]) has put forth the importance of an inclusive education system that meaningfully includes students with disabilities (SWD) to ensure their access to the same learning opportunities as all students [3]. However, there is a lack of data and information regarding the learning behaviors of SWD [4]. Consequently, they are often left out of standardized tests and a nuanced picture of their growth trajectories is often missing [4]. Individualized measures of success can ensure that all students are provided with proper expectations of growth [5].

Existing studies have indicated that SWD has different growth trajectories, with a tendency to fall behind their typically developing peers [4, 6]. These studies tend to group students according to their disability type, despite research showing that students within a diagnostic category such as autism spectrum disorder (ASD) can vary widely in how they progress and learn [7]. Therefore, the tendency to simply expect less of SWD [8] can have profound long-lasting impacts on students' learning success [9].

N. Wang et al. (Eds.): AIED 2023, CCIS 1831, pp. 612–618, 2023.
https://doi.org/10.1007/978-3-031-36336-8_95

With changing educational goals from traditional methods of pedagogical evaluation towards building competencies [10], more emphasis is now placed on complex competencies in education [11]. However, studies have tended to focus on reading and mathematics, for which assessment data is readily available. Therefore, for SWD, a nuanced picture of the growth trajectories in complex competencies is often missing [4]. SWD can have difficulties with complex competencies such as thinking [12] which limits their access to the general curriculum and learning. For instance, the Australian curriculum begins at Foundation level targeting the abilities of children 5 years of age [13]. Many students, including SWD may not be working within age-level expectations. Thus, we investigate the patterns of growth in foundational skills in one such competency – thinking skills (mental processes used to resolve a state of inquiry [14]).

2   The Current Study

Predominantly, longitudinal data has been analyzed using models that assume growth trajectories are smooth (when this is often not the case) [15]. We explored how analytics can analyze what learning growth in thinking skills looks like for SWD. Through analytics-based models (such as Markov chain (MC) modelling [16]), we attempted to reveal patterns of growth to understand what learning trajectories look like for SWD. MC modelling allows investigation of growth patterns and the patterns of students' regression or learning loss –when students revert to an earlier or more basic growth level [17]. Such understanding can pave ways for individualized expectations of growth for students, rather than relying on assumptions made due to disability type. To achieve this goal, the following research questions were explored in this study: (1) RQ1: What do generalized growth patterns look like for SWD in terms of their progression level transitions and, (2) RQ2: Are the transition patterns different between students with a diagnosis of ASD versus students without an ASD diagnosis?

3   Research Method

3.1   Assessment Context

This study analyzed 2017–2021 assessment data from a thinking skills assessment targeting school-age students (aged 5 to 20) with disabilities. The assessment was developed as one of a suite of instruments targeting foundational learning skills [18] as part of a larger program of work through two Australian Research Council (ARC) Linkage grants[1]. The thinking skills assessment was developed and validated through a process of expert review and empirical analysis [14]. It utilized a criterion-referenced framework

---

[1] The Victorian Department of Education and Training is the development sponsor and owner of all rights in the ABLES Tools. The ABLES Tools were derived from the Students with Additional Needs (SWANs) assessment and reporting materials which are owned by the University of Melbourne. The SWANs materials were developed with the support of the ARC as part of a Linkage partnership with the Centre for Advanced Assessment and Therapy Services and the University's foundation research partner, the Victorian Department of Education and Training.

to positively capture what SWD can do. It consisted of 18 indicators and 59 quality criteria for acquiring teacher observational data regarding competence in thinking skills. The maximum achievable score per indicator ranged from 2 to 4. The derivation of the scale was done through item response theory (IRT) modelling to convert observation-based scores (ranging from 0–59) to assignment of qualitative progression levels (1 to 5).

3.2 Data Pre-processing

The assessment data consisted of 18,112 observations. After keeping the entries where teachers' observations were complete across all indicators, it resulted in a set of 17,891 observations. The data was then converted from long to a wide format so all observations per student were in one row, finally resulting in a dataset consisting of 7760 students (2136 females and 5516 males). Of these students, 3620 had a diagnosis of ASD and 4032 did not have a diagnosis. No information was available for the remaining 108 students about their ASD status and gender.

3.3 Markov Modelling

Sequential growth data was analyzed using MC modelling [16]. This approach is used where data order matters and allows us to model observable sequences of events. As a stochastic process, MCs move in a sequence of steps through a set of states where the probability of entering a certain state depends on the state in the previous step and not on the earlier steps. Using MCs, students' growth patterns over time can be traced and compared. For comparison, we used t-tests and reported the results in terms of *p-value* statistics and *t-value* statistics. Next, Benjamini-Hochberg (BH) post-hoc correction was used to control for false positives as analysis involved multiple comparisons [19].

4   Results and Discussion

4.1 RQ1: Students' Generalized Patterns of Growth

Table 1 shows that students tended to remain at a given progression level, with over 60% staying at the same level (e.g., self-transitions for students at levels 1, 3, 4 and 5). The remaining transitions show that 25% of students progressed from level 1 → 2, while almost twice as many students (47%) progressed from level 2 → 3 (indicating slower growth from level 1). Similarly, when 25% of students progressed from level 3 → 4, only half transitioned from level 4 → 5. These patterns suggest fluctuations in students' growth where some stay at a given level – showing less growth while others progress.

Further, some students regressed or exhibited a learning loss [17]. For example, 8%, 19% and 31% students regressed from levels 3 → 2, 4 → 3 and 5 → 4 respectively. While these dips in students' growth are known from prior literature [17], it is the relative probabilities of students from a given level to another level that this study highlights and suggests making further investigations about. As students approach higher progression levels, the chances of regressing also increase where students may have difficulty consistently demonstrating higher levels of learning.

Table 1. Transition probabilities of students' generalized growth patterns (2017–2021).

Transition matrix		Next progression level				
		1	2	3	4	5
Previous progression level	1	0.60	0.25	0.14	0.01	0.00
	2	0.16	0.32	0.47	0.05	0.00
	3	0.03	0.08	0.64	0.25	0.01
	4	0.00	0.01	0.19	0.68	0.12
	5	0.00	0.00	0.01	0.31	0.67

Lastly, it appears that whether learning or regressing, some students can skip one or more levels altogether. For example, when 14% of students exhibited growth from level $1 \rightarrow 3$, 3% regressed from level $3 \rightarrow 1$, similarly, when 5% exhibited growth from level $2 \rightarrow 4$, 1% regressed from level $4 \rightarrow 2$; indicating a promising growth trend where there is a higher tendency to show a growth spurt than a learning loss.

4.2 RQ2: Comparing Growth Patterns – Known ASD Students Versus Their Peers

Next, growth patterns were compared for students *with* and *without* a diagnosis of ASD. First, the self-transitions were observed (see Table 2). The two groups had no significant difference for transitions between levels $2 \rightarrow 2$ and $5 \rightarrow 5$. However, the ASD group was significantly more likely to stay at level $3 \rightarrow 3$ and marginally more likely to stay at level $4 \rightarrow 4$ and this group was also significantly less likely to stay at level $1 \rightarrow 1$.

In terms of growth patterns, in some instances the ASD group was more likely than their peers to show growth transitions such as between levels $2 \rightarrow 3$, $3 \rightarrow 4$ and $4 \rightarrow 5$. This group is also more likely to exhibit growth spurts (skipping certain progression levels altogether) than their peers, e.g., from levels $1 \rightarrow 3$ and $2 \rightarrow 4$.

By contrast, in some instances, the ASD group was more likely to exhibit learning loss, e.g., from levels $3 \rightarrow 1$, $3 \rightarrow 2$, and $4 \rightarrow 3$. However, there were fewer instances of learning loss than of growth transitions. Then there were several transition sequences where the two groups did not seem to differ, suggesting that grouping of students based on disability type may not capture the growth patterns accurately. Individualized transitions, however, may shed light on these nuanced growth patterns.

Table 2. Growth pattern comparisons between the students *with* a diagnosis of ASD and students *without* a diagnosis of ASD using the t-tests. After the BH correction, results are significant (*) if p-value < 0.05 and marginally significant (.) if p-value < 0.10.

Transition Sequences	N_{ASD}	N_{NOT_ASD}	T (df)	P	Sig. after BH correction
1 → 1	257	331	−4.78 (1982)	<0.01	*
1 → 2	146	94	0.61 (2488)	0.54	
1 → 3	98	38	3.25 (2809)	<0.01	*
1 → 4	2	5	−1.48 (1615)	0.14	
1 → 5	0	1	−1.00 (1160)	0.32	
2 → 1	83	46	1.27 (2645)	0.21	
2 → 2	157	100	0.57 (2513)	0.57	
2 → 3	255	124	3.46 (2712)	<0.01	*
2 → 4	29	9	2.36 (2816)	<0.05	*
3 → 1	42	16	2.23 (2810)	<0.05	
3 → 2	128	46	4.16 (2820)	<0.01	*
3 → 3	913	499	3.58 (2669)	<0.01	*
3 → 4	348	203	2.15 (2639)	<0.05	
3 → 5	9	5	0.42 (2658)	0.67	
4 → 1	2	0	1.41 (1661)	0.16	
4 → 2	17	6	1.56 (2819)	0.12	
4 → 3	210	97	3.55 (2776)	<0.01	*
4 → 4	691	408	2.09 (2676)	<0.05	
4 → 5	130	63	2.34 (2771)	<0.05	*
5 → 3	4	1	1.04 (2761)	0.30	
5 → 4	62	43	0.04 (2550)	0.97	
5 → 5	139	86	0.56 (2374)	0.58	

5   Discussion

This study explores growth patterns in thinking skills for SWD, revealing varied growth patterns. For instance, students who are at lower levels of learning may face more difficulty progressing. This may be due to the nature of their disability. However, it also highlights potential issues with accessibility to foundational skills such as thinking which are fundamental for students to progress and learn [20]. These patterns may indicate that SWD lack the establishment of these skills, and this can compound the inaccessibility to learning that may contribute to expanding the learning gap. It would be beneficial to explore this further by examining the access SWD have to meaningful instruction in complex competencies as well as to investigate other complex competencies such as communication. Notably, SWD exhibit diverse growth patterns, and a comparison between students with and without ASD diagnoses shows that there are nuanced differences in the patterns observed. This indicates that a one-size-fits-all approach may not work.

6   Conclusion

Advanced analytic approaches to assessment can reveal learning patterns in foundational skills for all students, including those with complex needs. This study uses a criterion-referenced approach to assessment, which provides a rich data source to track student learning in complex competencies. By setting individualized expectations based on growth patterns, rather than group membership, this work addresses historically lower expectations for SWD. Such efforts can provide nuanced expectations of growth for SWD to support the inclusive practice of systems, schools, and teachers. This can contribute to Sustainable Development Growth (SDG) targets for inclusive quality education [3]. Ultimately, this can help to prevent the "Matthew effect" [21], facilitating all students to receive the necessary instruction to achieve appropriate expectations of success.

References

1. Kaffenberger, M.: A typology of learning profiles: tools for analysing the dynamics of learning. RISE Insight Series, vol. 15 (2019)
2. United Nations: Convention on the Rights of Persons with Disabilities (2022). https://www.un.org/development/desa/disabilities/convention-on-the-rights-of-persons-with-disabilities.html
3. United Nations: Sustainable Development Goals 4 Quality Education (2018). https://www.un.org/sustainabledevelopment/education/
4. Schulte, A.C., et al.: Achievement gaps for students with disabilities: stable, widening, or narrowing on a state-wide reading comprehension test? J. Ed. Psych. **108**(7), 925 (2016)
5. Dawkins, P.: Australian perspectives on benchmarking: the case of school education, in Australian Perspectives on Benchmarking. Australian Government Productivity Commission (2010)
6. Mattison, R.E., et al.: Longitudinal trajectories of reading and mathematics achievement for students with learning disabilities. J. Learn. Disabil. **56**(2), 132–144 (2023)
7. Baron-Cohen, S.: Editorial perspective: neurodiversity–a revolutionary concept for autism and psychiatry. J. Child Psychol. Psychiatry **58**(6), 744–747 (2017)
8. Thurlow, M.L., Quenemoen, R.F.: Revisiting expectations for students with disabilities. National Center on Educational Outcomes (NCEO), p. 8 (2019)
9. Foley, K.A.: Is 'Just More than Trivial' the best we can do? J. Leadersh. Ins. **19**(2), 48–49 (2020)
10. Segers, M., Dochy, F., Cascallar, E. (eds.): Optimising new modes of assessment: in search of quality and standards. Innovation and Change in Professional Education, p. 299. Kluwer, Dordrecht (2003)
11. Klieme, E., Hartig, J., Rauch, D.: The concept of competence in educational contexts. In: Assessment of Competencies in Educational Contexts, pp. 3–22 (2008)
12. Korinek, L., de Fur, S.H.: Supporting student self-regulation to access the general education curriculum. Teach. Except. Child. **48**(5), 232–242 (2016)
13. ACARA: The Australian curriculum. https://www.australiancurriculum.edu.au/

14. Kamei, T., Pavlovic, M.: Investigating differential item functioning to validate a thinking skills learning progression for students with intellectual disability and autism spectrum disorder. Int. J. Educ. Res. **106**, 101726 (2021)

15. Schwartz, A.E., et al.: The effects of special education on the academic performance of students with learning disabilities. J. Policy Anal. Manag. **40**(2), 480–520 (2021)

16. Ching, W.K., et al.: Markov chains. In: Hillier, F.S., Price, C.C. (eds.) Models, Algorithms and Applications, 2nd edn, vol. 189. Springer, New York (2006)

17. Gershkoff-Stowe, L., Thelen, E.: U-shaped changes in behavior: a dynamic systems perspective. J. Cogn. Dev. **5**(1), 11–36 (2004)

18. Abilities based learning and education support (ABLES) (2019). http://www.education.vic.gov.au/school/teachers/learningneeds/Pages/ables.aspx

19. Nawaz, S., Srivastava, N., Yu, J.H., Baker, R.S., Kennedy, G., Bailey, J.: Analysis of task difficulty sequences in a simulation-based POE environment. In: Bittencourt, I.I., Cukurova, M., Muldner, K., Luckin, R., Millán, E. (eds.) AIED 2020. LNCS (LNAI), vol. 12163, pp. 423–436. Springer, Cham (2020). https://doi.org/10.1007/978-3-030-52237-7_34

20. The future of education and skills: Education 2030, in OECD Publishing, O.f.E. Co-operation and Development, Editors (2018)

21. Huang, F.L., Moon, T.R., Boren, R.: Are the reading rich getting richer? Testing for the presence of the Matthew effect. Read. Writ. Q. **30**(2), 95–115 (2014)

Visualizing Self-Regulated Learner Profiles in Dashboards: Design Insights from Teachers

Paola Mejia-Domenzain[✉][iD], Eva Laini, Seyed Parsa Neshaei[iD], Thiemo Wambsganss, and Tanja Käser[iD]

EPFL, Lausanne, Switzerland
{paola.mejia,seyed.neshaei,tanja.kaeser}@epfl.ch

Abstract. Flipped Classrooms (FC) are a promising teaching strategy, where students engage with the learning material before attending face-to-face sessions. While pre-class activities are critical for course success, many students struggle to engage effectively in them due to inadequate of self-regulated learning (SRL) skills. Thus, tools enabling teachers to monitor students' SRL and provide personalized guidance have the potential to improve learning outcomes. However, existing dashboards mostly focus on aggregated information, disregarding recent work leveraging machine learning (ML) approaches that have identified comprehensive, multi-dimensional SRL behaviors. Unfortunately, the complexity of such findings makes them difficult to communicate and act on. In this paper, we follow a teacher-centered approach to study how to make thorough findings accessible to teachers. We design and implement `FlippED`, a dashboard for monitoring students' SRL behavior. We evaluate the usability and action-ability of the tool in semi-structured interviews with ten university teachers. We find that communicating ML-based profiles spark a range of potential interventions for students and course modifications.

Keywords: Flipped Classrooms · Clustering · Teacher Dashboard

1 Introduction

Over the past years, blended learning (BL), which combines in-person sessions with online learning, has gained increasing popularity. Students in *Flipped Classroom (FC)* courses, a variation of BL, complete pre-class activities before participating in face-to-face sessions. While FCs have the potential to enhance student learning (e.g.,[4]), they are not effective *per se* [8,12]. Independently regulating their learning (e.g., managing their time) is a challenging task for many learners. While there are multiple existing solutions to monitor students' self-regulated learning (SRL) skills [5,16], most of them overlook teachers' role in FCs and their ability to promote SRL skills and support students' learning experience [11].

Recently, tools designed for teachers to support SRL (e.g., [11,14]) have been valued positively. For example, in MetaDash [14], teachers perceived the visualization of students' emotions as a valuable tool for lesson design. Furthermore, teachers appreciated the possibility to monitor student progress and engagement using a Moodle Plugin [11]. Most tools [11,14] visualize aggregated statistics of student behavior only, whereas recent work has demonstrated that students exhibit complex SRL patterns [8,12]. In particular, [8] identified SRL profiles differing in levels of time management (regularity, effort, consistency, and proactivity) and metacognition (video control). These differences can get lost when aggregating them. Indeed,[11] found that providing aggregated information was not enough for supporting students. Therefore, machine learning (ML) approaches able to identify and represent comprehensive student behavior [8,12] build a promising basis for classroom orchestration.

Unfortunately, the findings of the aforementioned ML approaches are complex and therefore challenging to communicate and make accessible to teachers. The lack of trust in ML [10,13] or unclear visualizations [7,9] can hinder the adoption and use of ML-based teacher support tools.

In this paper, we therefore study how to best make comprehensive ML-based findings accessible to teachers in the context of SRL in FC settings. We identify students' multi-dimensional SRL profiles through clustering [8] and design multiple visualizations based on information visualization findings [1,3]. We then implement our findings into FlippED, a teacher dashboard for monitoring students' SRL behavior. We investigate teachers' responses to the tool by conducting semi-structured interviews with ten university teachers. With our mixed-method approach, we aim to answer the following research questions: How do teachers interact with and respond to an ML-based tool for FC (**RQ1**)? How actionable is the information provided in the dashboard (**RQ2**)?

2 Teacher-Centered Design and Evaluation Framework

To study how to communicate effectively ML-based findings to teachers, we followed a teacher-centered mixed-method approach. Requirements were compiled both from the existing literature and from 10 user interviews. Then, we identified multi-dimensional SRL profiles of students in an introductory mathematics FC course [8]. Next, we designed visualizations of the identified patterns according to the nature of the data as well as prior work on visual designs [1,3,6]. Finally, we iterated the design with seven different teachers and evaluated the final dashboard with ten university teachers.

2.1 Multi-dimensional SRL Profiles

We used the clustering framework suggested by [8] to obtain multi-dimensional profiles of students' SRL behavior. The framework analyzes five dimensions of SRL identified as important for online learning in higher education [2]: *Effort* (intensity of student engagement), *Consistency* (variation of effort over time), *Regularity* (patterns of working days and hours), *Proactivity* (anticipation or delay in course schedule), and *Control* (control of cognitive load). The pipeline consists of two main steps. First, behavioral patterns for each dimension are obtained using Spectral Clustering. In a second step, the resulting labels per dimension are clustered using K-Modes.

We applied the pipeline to log data collected from 292 students (29% identifying as female and 71% as male) of an undergraduate mathematics FC course [4]. After the first clustering step, we obtained different patterns for the five dimensions: for *Effort* and *Control*, difference in intensity (e.g., higher and lower intensity); for *Consistency*, constant intensity and increase of intensity during exam preparation; for *Proactivity*, up-to-date and delayed behavior; and for *Regularity*, students with high or low regularity patterns. We then integrated the obtained patterns using the second clustering step and obtained five different profiles. For example, the best-performing profile had higher *Control*, lower intensity *Effort*, constant intensity (*Consistency*), and up-to-date *Proactivity*. The worst-performing profile had similar characteristics except the students were not up-to-date (*Proactivity*).

A side effect of the richness of the analysis is that the profiles can be complex to understand. In the following, we explore how to best communicate the findings to empower teachers by providing easy-to-interpret and actionable information about their course, while maintaining a comprehensive analysis.

2.2   Teacher Dashboard

To address **RQ1** and **RQ2**, we designed a teacher dashboard displaying SRL profiles and behaviors. We evaluated the tool with ten teachers, analyzing their clickstream, think-aloud process, and semi-structured interview answers.

Design. We designed and implemented `FlippED`, a teacher dashboard for SRL. We then iterated the design with seven pilot teachers. The final design includes a navigation menu (see F1 in Fig. 1) with two parts: an overview displaying a *Summary* and *Student Profiles* and a detailed view illustrating students' behavioral patterns in the different dimensions. The user can select the course (F2) and the desired weeks (e.g., week $5 - 9$) (F3). In addition, there are help buttons (F4) throughout the dashboard providing further explanations. In the *Summary* page (Fig. 1-left), weekly statistics per dimension are displayed as well as the trend in comparison to the previous week. The description of the profiles and the associated grades are shown on the *Profiles* page (Fig. 1-center). Moreover,

Fig. 1. Example pages: *Summary*, *Student Profiles*, and *Groups*.

in the student behavior pages (Fig. 1-right), users can choose (F5) between the aggregated view and a view per group.

Accessibility. In order to make `FlippED` accessible to people with disabilities, we followed the guidelines from the *Web Accessibility Initiative* (e.g., using a colorblind palette, providing detailed alt-texts and captions for all graphs, supporting dark contrast mode, and having a clear voice-navigable interface) and achieved the conformance level of AAA (highest level).

Participants. We recruited ten university-level teachers (50% identified as female, 50% as male) with experience in BL and FC through a faculty email.

Procedure. We conducted semi-structured interviews asking participants to assume they were teaching a large FC course. The first task was to follow a think-aloud protocol and explore the dashboard. In the second task, participants were told that their students had just taken the midterm exam, and some students had surprisingly bad results. Thus, participants had to use the dashboard to give possible explanations. Lastly, we asked participants to assess whether and how they would use the dashboard in their classroom.

3   Results

3.1   Dashboard Usability (RQ1)

To understand the usability of `FlippED`, in a first analysis, we investigated participants' exploration (task 1) clickstream together with their think-aloud comments. Participants started in the *Summary* page and then either went into the *Profiles* page or the *Proactivity* page (the first menu item in *Student Behaviors*, see Fig. 1). Then participants went back and forth to understand the layout and the profiles. Once they understood the structure of the dashboard, they followed mostly an ordered exploration strategy, accessing the pages following the sequential order from the menu. Moreover, participants spent the longest time on the *Profiles* page (on average 10 minutes). Half of the participants were at first confused about the profiles but then said they would go back to the *Profiles* page regularly and use them to advise students on the best learning strategies.

In task 2, when asked to use the dashboard to identify possible causes for poor midterm performance, 80% of the participants went straight to the *Profiles* page and the remaining 20% visited the *Summary* page before the *Profiles* page. Then, participants described the properties of the poor-performing profiles.

In summary, in the beginning, during the exploration task, participants viewed the student behavior pages (Proactivity, Effort, Consistency, Control, and Regularity) to understand each dimension and the observed behaviors. Then, during the second task, they were more drawn to the overview pages (Summary and Profiles) to get summarized information as a basis for suggesting interventions.

3.2   Actionability of Information (RQ2)

In a second analysis, we examined the actionability of the provided information. Regarding the potential benefits of the dashboard, 80% of the participants said they would show the students the dashboard in class. In particular, they mentioned adding the *Proactivity* graphs to the course slides, showing the relationship between profile and grade to encourage proactive behaviors. As one participant illustrated, *I would also use this information as feedback to students on their working habits. I would [show them this information] and say: 'Why don't you try to change this?'.*

In addition, 70% of the interviewed teachers mentioned adapting the course in some way; some examples related to the *Proactivity* dimension were asking students to come up with questions before the interactive sessions, adding activities such as continuous evaluation (quizzes), or proposing additional materials. Based on the *Effort, Consistency* and *Regularity* pages, other possible actions teachers mentioned were sending motivational messages to all or some specific students, adapting the workload, sending automatic reminders to students who are not working regularly, and advising students in the lowest-grade profile to change their learning strategies. In addition, they proposed giving extra credit to promote watching videos and making the pre-class activities more entertaining. *In summary, the visualizations helped teachers come up with a wide range of actionable items like adapting the course and communicating with the students.*

4   Implications and Conclusion

In this work, we studied the usability (**RQ1**) and actionability (**RQ2**) of ML-based teacher dashboards in the context of SRL in FC. We identified student behavioral profiles in FC using clickstream and communicated the findings in a teacher dashboard. Then, we evaluated its usability and actionability in semi-structured interviews with ten university teachers.

In contrast to existing teacher dashboards that focused on communicating aggregated statistics (e.g.,[5,11,14]), we visualized and communicated intricate ML-based insights. Similar to [9,15], we found that most teachers judged the dashboard and visualizations as useful and actionable. In particular, participants appreciated the hierarchical design of the dashboard. They mentioned that the use of the overview pages (*Summary* and *Profiles*) displaying students' SRL profiles would be sufficient to design interventions on a weekly basis and used the detailed information on behavioral patterns mostly as a mean to understand the different SRL dimensions of the profiles. The hierarchical design of our dashboard solves two problems mentioned in previous work: not providing enough information for supporting student learning [11] and providing too much information [13]. Furthermore, teachers perceived some dimensions (e.g., *Proactivity*) as much more useful and actionable than others (e.g., *Control*).

Our results are mostly consistent with prior work on teacher dashboards. In the following, we emphasize design guidelines for complex dashboards that go beyond those emphasized in earlier work:

1. Structure the dashboard in a hierarchical way: include (1) overview and summary pages for daily use and (2) detailed information to gain a good understanding of the provided information.

2. Allow a flexible dashboard design that adjusts to the specific needs of the target teacher population. For example, omit dimensions that do not provide actionable items.

Our work sheds insights into the interpretability, usability, and actionability of ML-based teacher dashboards. In the future, we plan to study the generalizability and validity of our findings in different contexts, regions, and cultures and to evaluate the usage of `FlippED` in diverse classrooms for longer periods.

References

1. Albers, D., Correll, M., Gleicher, M.: Task-driven evaluation of aggregation in time series visualization. In: CHI (2014)
2. Broadbent, J., Poon, W.L.: Self-regulated learning strategies & academic achievement in online higher education learning environments: A systematic review. Internet High. Educ. **27**, 1–13 (2015)
3. Gleicher, M., Albers, D., Walker, R., Jusufi, I., Hansen, C.D., Roberts, J.C.: Visual comparison for information visualization. Inf. Vis. **10**(4), 289–309 (2011)
4. Hardebolle, C., Verma, H., Tormey, R., Deparis, S.: Gender, prior knowledge, and the impact of a flipped linear algebra course for engineers over multiple years. J. Eng. Educ. **111**(3), 554–574 (2022)
5. Jivet, I., Scheffel, M., Drachsler, H., Specht, M.: Awareness is not enough: Pitfalls of learning analytics dashboards in the educational practice. In: EC-TEL (2017)
6. Lee, S., Kim, S.H., Kwon, B.C.: VLAT: Development of a visualization literacy assessment test. IEEE Trans. Vis. Comput. Graph. **23**(1), 551–560 (2016)
7. Martinez-Maldonado, R., Pardo, A., Mirriahi, N., Yacef, K., Kay, J., Clayphan, A.: Latux: An iterative workflow for designing, validating, and deploying learning analytics visualizations. J. Learn. Anal. **2**(3), 9–39 (2015)
8. Mejia-Domenzain, P., Marras, M., Giang, C., Käser, T.: Identifying and comparing multi-dimensional student profiles across flipped classrooms. In: AIED (2022)
9. Nazaretsky, T., Bar, C., Walter, M., Alexandron, G.: Empowering teachers with AI: co-designing a learning analytics tool for personalized instruction in the science classroom. In: LAK (2022)
10. Nazaretsky, T., Cukurova, M., Alexandron, G.: An instrument for measuring teachers' trust in AI-based educational technology. In: LAK (2022)
11. Pérez-Sanagustín, M., Pérez-Álvarez, R., Maldonado-Mahauad, J., Villalobos, E., Sanza, C.: Designing a moodle plugin for promoting learners' self-regulated learning in blended learning. In: EC-TEL (2022)
12. Saint, J., Whitelock-Wainwright, A., Gašević, D., Pardo, A.: Trace-SRL: a framework for analysis of microlevel processes of self-regulated learning from trace data. IEEE TLT **13**(4), 861–877 (2020)
13. Verbert, K., Ochoa, X., Croon, R.D., Dourado, R.A., Laet, T.D.: Learning analytics dashboards: the past, the present and the future. In: LAK (2020)
14. Wiedbusch, M.D., et al.: A theoretical and evidence-based conceptual design of metadash: an intelligent teacher dashboard to support teachers' decision making and students' self-regulated learning. Front. Educ. **6**, 570229 (2021)
15. Wise, A.F., Jung, Y.: Teaching with analytics: Towards a situated model of instructional decision-making. J. Learn. Anal. **6**(2), 53–69 (2019)
16. Álvarez, R.P., Jivet, I., Pérez-Sanagustín, M., Scheffel, M., Verbert, K.: Tools designed to support self-regulated learning in online learning environments: A systematic review. IEEE TLT **15**(4), 508–522 (2022)

Classification of Brain Signals Collected During a Rule Learning Paradigm

Alicia Howell-Munson[1]([✉]) [ID], Deniz Sonmez Unal[2] [ID], Theresa Mowad[3],
Catherine Arrington[3] [ID], Erin Walker[2] [ID], and Erin Solovey[1] [ID]

[1] Worcester Polytechnic Institute, Worcester, MA 01609, USA
ahowell4565@gmail.com
[2] University of Pittsburgh, Pittsburgh, PA 15260, USA
[3] Lehigh University, Bethlehem, PA 18015, USA

Abstract. We propose incorporating biophysical data with behavioral data to inform digital learning environments on an individual's current cognitive state and how it relates to their learning. We used a rule learning paradigm drawn from cognitive psychology to define phases of rule learning across multiple domains. This paradigm can simulate an inductive reasoning framework seen during mathematics education while reducing the number of covariates compared to real-world settings. We combined the time series brain data with behavioral and contextual data in machine learning models for prediction of rule learning phases with the aim of developing approaches to incorporate a mixture of behavioral and neural data into digital learning designs.

Keywords: rule learning · inductive reasoning · functional near-infrared spectroscopy · brain-computer interfaces

1 Introduction

In realistic learning environments, students encounter multiple domains where they need to learn rules through inductive reasoning [4]. The mechanisms for inductive reasoning involve gathering information, generalizing it, classifying it, and chunking it, then recalling it. These mechanisms are collectively labeled as induction and refinement, a class of processes that lead to robust learning [6]. We aim to use brain data to improve knowledge modeling in intelligent tutoring systems (ITSs). While knowledge modeling from log data is well-studied in ITSs [8], it is still not possible to detect the precise moment a student has learned a rule [3]. Brain data may supplement behavioral data to provide a fine grained indication of when and to what degree a student has mastered a rule. Our first step is to demonstrate that brain signals can be used to recognize three stages of inductive reasoning as described in a cognitive science rule learning paradigm [5]. These stages form the fundamental mechanisms of induction and refinement processes which are the basis of important features of ITSs. Our goal is to expand upon this work and develop machine learning models that can classify a rule learning state of students and that transfer across task domains.

The National Science Foundation under Grant Nos. (1835307 and 1912474).

2   Related Literature

2.1 Neural Processes During Rule Learning Tasks. During rule learning, individuals move through phases of rule search, rule discovery, and rule following [5]. These phases can be studied in controlled laboratory tasks that measure behavior across a series of trials in standard rule learning tasks [4]. Stimuli such as numbers or spatial locations appear in a continuous stream and alternate between random non rule steps and rule sequences that follow a set rule or pattern. Participants must indicate with a key press whether the current stimulus matches a rule that they have discovered through inductive reasoning. Rule search is defined as when the participant is responding that they do not think the stimuli are following a pattern. Rule discovery is the first response where the participant indicates the stimuli are following a pattern. Finally, rule following is the continued response that the stimuli are following a rule. Prior work on rule learning has shown bilateral activation in the prefrontal cortex (PFC) with shifts between the medial, inferior, and posterior PFC regions between rule acquisition and rule following phases [4,5]. Similar brain patterns should be observed in learning tasks at points where students are just learning a rule and at points where they have mastered the rule, paralleling the concepts of induction, refinement, and fluency in cognitive states related to learning [4].

2.2 Neuroimaging and Usage of Neural Data with ITSs. Neuroimaging tools provide measures of brain activity while participants engage, in many tasks. Functional near-infrared spectroscopy (fNIRS), unlike more common tools like fMRI, is portable, easy to use, and quick to set up [9]. The fNIRS optodes are arranged on a mesh cap that emits near-infrared light into the head, then measures how much of the light was absorbed by the blood and tissue in the head with a detector. fNIRS typically has a range of 1–3cm into the cortex and the response time for peak signal after an event occurs is 4-7 s [9].

Researchers have used brain imaging data to study the neural processes underlying learning and to inform the design and development of intelligent tutoring systems. For example, Anderson and colleagues showed functional magnetic resonance imaging (fMRI) can be used to identify deep processing behaviors during problem-solving [1]. Others used the electroencephalogram (EEG) for understanding student engagement [10] during use of an ITS.

3   Data Collection and Curation

We conducted a study to build a dataset that can be used to explore whether fNIRS brain signals can be used to recognize stages of inductive reasoning. To do this, we expand on prior fMRI studies [5]. We aim to demonstrate to what extent it is possible to develop machine learning models that can classify the rule learning state of students, that transfer across task domains, and that do not require the student to provide individual data prior to using the model.

3.1 Rule Learning Task. We created isomorphic tasks in *numeric* and *spatial* domains for cross domain comparisons. Stimuli appeared sequentially on the screen with a fixation, '+', between stimuli. The *numeric* task involved the numbers 0 to 99; the *spatial* task involved a filled circle in one of 10 locations positioned in a circle. Stimuli shifted position alternating between sequences of 5-11 trials that followed a rule and 2-4 trials that occurred randomly within displacement between -4 and 4. Easy rules involved one step while hard rules followed two steps. A list of all 16 rules and their features are in Table 1. In the spatial task, subtraction steps moved counterclockwise, while addition steps moved clockwise. The order of the domain tasks was counterbalanced.

Table 1. The selected hard and easy rules all participants saw with examples.

Easy	Rule	+1	+2	+3	+4	- -1	−2	−3	−4
	Example	5, 6, 7	5, 7, 9	5, 8, 11	5, 9, 13	5, 4, 3	5, 3, 1	5, 2, 99	5, 1, 96
Hard	Rule	+1, +2	+2, +3	-2, -1	-3, -2	-2, +1	-2, +3	+2, -1	+2, −3
	Example	5, 6, 8	5, 7, 10	5, 3, 2	5, 2, 99	5, 3, 4	5, 3, 6	5, 7, 6	5, 7, 4

Our data processing approach followed the procedure outlined in [5]. Non-rule trials were removed from the analysis. Whenever a participant responded "f", the trial was coded as *rule search*. For rules that were successfully discovered, the first trial the participant responded "j" was coded as *rule discovery* and all subsequent trials were *rule following* until the rule ended. Rule sequences were coded as discovered if they responded "j" in the last two trials of the rule.

3.2 Equipment. The fNIRS signals were recorded using a NIRx NIRSport2 fNIRS device with a sampling rate of 10.17Hz. The device was configured with a fifty-two channel design using sixteen sensors and fifteen detectors (Fig. 1). A note was left in a visible area of the workstation that stated to press "j" if they thought it was a rule sequence and to press "f" if they thought it was not.

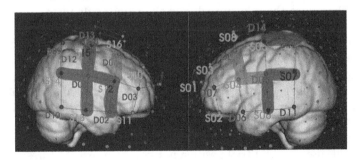

Fig. 1. Sensors (in red) emit light and detectors (in blue) receive light. Highlighted are the channels that correspond to regions of interest for rule acquisition (yellow) and rule following (green) based on previous work [5]. (Color figure online)

3.3 Participants. We recruited 22 university students as participants (5 male, 13 female, and 4 other) between 18 and 23 years old (M = 20.09, SD = 1.6). 71% of participants identified as white, 14% as Asian, 14% as mixed ethnicity, and 6% as black or African American. Participants were compensated with either coursework credit or monetary payment of $15.00. All participants signed an informed consent that described the procedure, compensation, and risks of the study. The informed consent was approved by the university's IRB. Seven participants were excluded from analysis due to errors in data collection, excessively noisy brain data, or failure to complete the task correctly, leaving 15 participants in the final analysis.

3.4 fNIRS Preprocessing. Channels were removed due to noise if they exceeded a 15% coefficient of variance threshold. A bandpass filter was applied to the raw data with a high cutoff frequency of 0.2 Hz and a low cutoff frequency of 0.01 Hz. We used the modified Beer-Lambert Law to convert the raw data into change in micromolar oxygenation with a differential pathlength factor of 7.25 for oxygenated data and 6.38 for deoxygenated data.

3.5 Dataset Overview for Machine Learning. We separated the data from the numeric and spatial domains into separate datasets to compare the performance of each machine learning model on the individual domain, combined domains, and cross-domain training and testing. The goal was to determine if the information domain affected a model's performance Each dataset contained two classes: rule acquisition and rule following. The datasets were marginally unbalanced with the acquisition class containing approximately 6% more trials. We selected fNIRS channels that corresponded to regions of activation for rule acquisition and rule following, as shown in Fig. 1 based upon previous results from the spatial task [5]. This resulted in 3 channels corresponding with rule acquisition and eight channels corresponding with rule following. Each time-series was approximately 5 s long, ensuring time for the hemodynamic response to start to peak, but reducing the influence from multiple trials within a time-series.

3.6 Machine Learning Feature and Classifier Selection. Models included a combination of a behavioral feature (response time), contextual features (rule difficulty, time since the rule started, and the unique rule identifier), and neural data (time series subsequences for each fNIRS channel). We used a logistic regression (LR) on the behavioral and contextual data and we used the time-series forest (TSF) classifier from sk-time on the fNIRS data. In addition, we used an ensemble method where each fNIRS channel was fitted to a model, then the channels would vote to determine the final label. For models that include computer logged features, the label from the LR would be included in the voting.

We fitted five separate models which were: behavioral model (BM) that had response times as its only feature, contextual model (CM) that had rule identity, difficulty, rule domain, and trial position in rule sequence as features; behavioral and contextual model (BCM); neural model (NM) that contained only fNIRS channels as features; neural and behavioral model (NBM); and neural, behavioral, and contextual features (NBCM).

4   Results

4.1 Behavioral Results. To confirm the behavioral results matched previous research, we analyzed the frequency of rule discovery in a 2 (Domain) x 2 (Difficulty) repeated measures ANOVA and the response time (RT) in a 3 (Rule learning phase) by 2 (Domain) x 2 (Difficulty) repeated measures ANOVA. We considered the number of rules discovered out of a maximum of 8 per condition. Rule discovery occurred in 78% of all rule sequences. There was a main effect of difficulty, with participants discovering significantly more easy rules, $F(1,21)$ = 29.105, $p < 0.001$, $n_p^2 = 0.581$. RTs sped up as participants moved from rule search to discovery to following, $F(2, 42) = 35.3$, $p < .001$, $n_p^2 = 0.627$. The other significant effect was of domain, $F(1, 21) = 15.8$, $p < .001$, $n_p^2 = 0.429$, with participants responding more quickly on spatial trials. These results aligned with previous research, giving validation to perform neural analysis [4,5].

4.2 Computer Logged Machine Learning Results. The computer logged data result in robust F1 scores for all three behavioral and contextual models. Across both numeric and spatial datasets, as well as the combined data and crossed domain training and testing, behavioral data resulted in the worse performance while the combination of behavioral and context data resulted in the best performance. The results from the mixed domain dataset mirrored those of the spatial dataset. Interestingly, the cross domain testing and training data sets resulted in F1 scores that were in line with models trained and tested within domains, suggesting that rule learning phases may share features across different domains.

Table 2. F1 scores on the models trained with combinations of RT, contextual, and neural features. All models performed better than the dummy classifier.

Dataset	BM	CM	BCM	NM	NBM	NBCM	Dummy
Numeric	0.69	0.76	0.78	0.50	0.57	0.69	0.39
Spatial	0.74	0.79	0.82	0.50	0.61	0.74	0.27
Numeric + Spatial	0.74	0.79	0.82	0.51	0.65	0.76	0.35
Numeric Train, Spatial Test	0.69	0.77	0.81	0.46	0.56	0.71	0.35
Spatial Train, Numeric Test	0.73	0.76	0.79	0.50	0.63	0.74	0.39

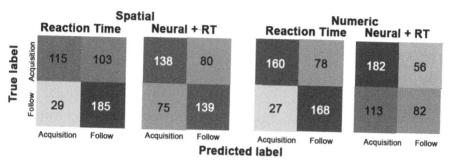

Fig. 2. Confusion matrices of true and predicted labels from the spatial (left) and numeric (right) datasets for BM and NBM. For both, the addition of neural features causes more accurate predictions of the acquisition class.

4.3 fNIRS Machine Learning Results. The classifiers including neural features consistently had a higher F1 score than the dummy classifier (Table 2). Classifiers using brain data are consistently better than random chance. In addition, the neural data improves the misclassification of the rule acquisition phase from the RT features in the spatial dataset (Fig. 2). The neural data similarly correctly predicts more acquisition labels on the numeric dataset; however, in exchange it misclassifies the majority of follow labels. Consistent with the computer logged features, the F1 scores of the cross domain training and testing remained consistent to the datasets of a singular information domain. This continues to support the logic that rule learning phases may share characteristics across task domains, even when incorporating the brain data. These results lead us to predict that brain data, RT, and possible context features would be ideal features for a classifier that works in real-time with an ITS.

5   Discussion

In this paper, we have shown the viability of predicting rule acquisition and rule following phases from fNIRS and behavioral data in a controlled psychological task in both numeric and spatial domains. Computer logged data is the most successful predictor of rule learning phase and should be relied upon when the data is available. However, we have shown that there is promise in using brain data to supplement computer logged data. This supplementation can be applied when computer logged data is either sparse or non-existent to maintain an informed state of the student's learning phase. In this study, we used a group model to predict the rule learning phase of students who were not included in the training data and showed the predictions carried across numeric and spatial domains. We believe that brain data can help fill the gaps during long pauses during ITS use and determine if a long pause is related to engagement [2] or disengagement [7] from the problem. Future work should transition models into real-world applications.

Acknowledgements. This material is based upon work supported by the National Science Foundation under Grant Nos. (1835307 and 1912474).

References

1. Anderson, J.R., Betts, S., Ferris, J.L., Fincham, J.M.: Cognitive and metacognitive activity in mathematical problem solving: prefrontal and parietal patterns. Cognit. Affect. Behav. Neurosci. **11**(1), 52–67 (2011)
2. Arroyo, I., Mehranian, H., Woolf, B.P.: Effort-based tutoring: An empirical approach to intelligent tutoring. In: Educational Data Mining 2010. Citeseer (2010)
3. Baker, Ryan S. J. D.., Gowda, Sujith M.., Corbett, Albert T.., Ocumpaugh, Jaclyn: Towards automatically detecting whether student learning is shallow. In: Cerri, Stefano A.., Clancey, William J.., Papadourakis, Giorgos, Panourgia, Kitty (eds.) ITS 2012. LNCS, vol. 7315, pp. 444–453. Springer, Heidelberg (2012). https://doi.org/10.1007/978-3-642-30950-2_57
4. Cao, B., Li, W., Li, F., Li, H.: Dissociable roles of medial and lateral pfc in rule learning. Brain Behav. **6**(11), e00551 (2016)
5. Crescentini, C., Seyed-Allaei, S., De Pisapia, N., Jovicich, J., Amati, D., Shallice, T.: Mechanisms of rule acquisition and rule following in inductive reasoning. J. Neurosci. **31**(21), 7763–7774 (2011)
6. Koedinger, K.R., Corbett, A.T., Perfetti, C.: The knowledge-learning-instruction (kli) framework: Toward bridging the science-practice chasm to enhance robust student learning. Cognitive Sci. (2010)
7. Muldner, K., Burleson, W., Van de Sande, B., VanLehn, K.: An analysis of students' gaming behaviors in an intelligent tutoring system: Predictors and impacts. User Model. User-Adap. Inter. **21**(1), 99–135 (2011)
8. Pelánek, R.: Bayesian knowledge tracing, logistic models, and beyond: an overview of learner modeling techniques. User Model. User-Adapted Inter., 313–350 (2017). https://doi.org/10.1007/s11257-017-9193-2
9. Solovey, E.T., et al.: Using fnirs brain sensing in realistic hci settings: experiments and guidelines. In: Proceedings of the 22nd Annual ACM Symposium on User Interface Software and Technology, pp. 157–166 (2009)
10. Stevens, Ronald H.., Galloway, Trysha, Berka, Chris: EEG-related changes in cognitive workload, engagement and distraction as students acquire problem solving skills. In: Conati, Cristina, McCoy, Kathleen, Paliouras, Georgios (eds.) UM 2007. LNCS (LNAI), vol. 4511, pp. 187–196. Springer, Heidelberg (2007). https://doi.org/10.1007/978-3-540-73078-1_22

Q-GENius: A GPT Based Modified MCQ Generator for Identifying Learner Deficiency

Vijay Prakash⬤ⓘ, Kartikay Agrawal⬤ⓘ, and Syaamantak Das$^{(\boxtimes)}$ⓘ

IDP in Educational Technology, Indian Institute of Technology Bombay,
Mumbai, India
`syaamantak.das@iitb.ac.in`

Abstract. This research proposes a novel methodology to address the challenge of identifying learner deficiencies from multiple choice questions (MCQs). The proposed approach involves manipulating the standard MCQ into a modified MCQ by generating distractors based on three different aspects: 'Fact', 'Process', and 'Accuracy'. 'Fact' refers to the fundamental information related to the answer of the question, 'Process' refers to the concepts and application methods required to solve the problem, and 'Accuracy' refers to the degree of closeness of the answer to its true value. To achieve this, a Generative Pre-trained Transformer (GPT) model was used to develop a system called 'Q-GENius', that can take a question stem and some other optional additional information as input, and generate a modified MCQ with key and three distractors, each having a deficiency either in the 'Fact,' 'Process,' or 'Accuracy' aspect of the answer. The study was validated using inter-annotator agreement of subject matter experts on the accuracy and reproducibility of the model. The validation was performed on four different subjects. The results demonstrated that the proposed approach of modified MCQ was effective in identifying learner deficiencies when compared to standard MCQs. Overall, the proposed methodology has the potential to improve the accuracy of identifying learner deficiencies in MCQs.

Keywords: MCQ generator · Learner deficiency · Generative Pre-trained Transformer

1 Introduction

Assessing the knowledge of learners is a crucial aspect of education. Multiple-choice questions (MCQs) have been widely used as an assessment tool due to their efficiency in evaluating a large number of students in a short period. However, identifying learner deficiency from MCQs is difficult [1]. Distractors play a critical role in identifying the understanding of a student regarding a particular concept or topic [2]. In this study, a novel methodology is proposed to create

© The Author(s), under exclusive license to Springer Nature Switzerland AG 2023
N. Wang et al. (Eds.): AIED 2023, CCIS 1831, pp. 632–638, 2023.
https://doi.org/10.1007/978-3-031-36336-8_98

distractors for three different aspects, which are 'Fact', 'Process', and 'Accuracy'. *'Fact'* refers to the fundamental information related to the answer of the question; *'Process'* refers to the concepts and application methods required to solve the problem; and *'Accuracy'* refers to the degree of closeness of the answer to its true value. A Generative Pre-trained Transformer (GPT) model [3] was used to develop a system called 'Q-GENius'. This name incorporates the idea of generating questions, while also implying that the model can identify areas where the learner may need improvement. The use of "genius" also adds a positive connotation, suggesting that using the software will lead to smarter, more knowledgeable learners. This name plays on the idea of boosting learning by identifying knowledge gaps and generating questions to fill them. The question dataset used for this research was from standard 10th grade of Indian NCERT secondary school system.

2 Literature on Problem Background

2.1 Problem of Learner Deficiency Identification in MCQs

MCQs have been criticized for their inability to identify learners' deficiencies [4]. According to research [5], MCQs may only measure students' ability to recognize information, rather than their ability to apply knowledge or analyze information. Additionally, MCQs may not capture the depth and complexity of learning, as they often test surface-level knowledge rather than deep understanding [1,5]. Furthermore, MCQs may not provide sufficient feedback to students on their areas of weakness or strength. Although MCQs may not be the best tool to identify learner deficiencies, they can still be useful in providing some insight into students' knowledge gaps. By analyzing students' responses to MCQs, teachers can identify areas where students are struggling and adjust their teaching accordingly. One approach to identifying learner deficiencies with MCQs is the use of diagnostic testing [6]. By analyzing the results of diagnostic tests, educators can gain a better understanding of students' knowledge gap and provide remedial solutions accordingly.

2.2 Challenges in Developing MCQs Automatically

One of the most challenging aspects of generating MCQs is generating plausible distractors [7], which are the incorrect options. The distractors should be similar in format and style to the correct answer to make the question more challenging for the student. However, the distractors should also be significantly different from the correct answer to ensure that the question tests the student's knowledge and understanding, rather than just their ability to identify patterns. Generating automatic plausible distractors [8] requires a deep understanding of the subject matter and the ability to create multiple variations of the same question, each with different distractors. This process can be time-consuming, and generating high-quality distractors automatically remains a significant challenge.

2.3   Research Gap

Based on the existing literature, the following questions were considered for this research:
RQ1. Is it possible to develop a modified MCQ that can identify learner deficiency?
RQ2. Can an automated system be developed that can be generalized across subject domains to generate such modified MCQs?

3   Methodology for the Proposed Model

3.1   Question Dataset Details

For this research, 120 questions was selected from 10th grade school level Science subjects (Physics, Chemistry, Mathematics and Biology). The questions were classified into two categories namely: numerical or non-numerical (conceptual/theory based). The following table (Table 1) shows the distribution of the questions type and subject. Initially all the questions were solved to find the correct answer and was validated by subject matter expert about the correctness of the answer.

Table 1. Distribution of question types and Subject.

Sl. No.	Subject	Question Type	No. of Questions
1.	Mathematics	Numerical	20
		Non-Numerical	20
2.	Physics	Numerical	20
		Non-Numerical	20
3.	Chemistry	Non-Numerical	20
4.	Biology	Non-Numerical	20
Total			120

3.2   Generative Pre-trained Transformer (GPT) as Core Model

GPT is a neural network architecture that uses unsupervised learning to pre-train a language model on a huge corpus of text and then fine-tunes the model on specific tasks such as language translation, question answering, or text completion. The transformer architecture of GPT allows it to capture long-term dependencies and generate coherent, context-appropriate text. The propose research opted to use reinforcement learning (RL) based language transformers as core model [9] instead of using regular supervised learning (SL) models. The question that arises is why not use the labels directly with SL to finetune the model? There are multiple reasons behind this. First, in SL, the objective is to minimize the difference between true label and model outputs. The labels are ranking scores of responses to specific prompts. Thus, a regular SL would tune a model to predict the rank but not to generate response. An option could have been reformulating the task into a constrained optimization problem such that there is a combined loss consisting of an 'output text loss' and a 'reward score' term that can be optimized jointly with SL. But, to generate distractors, as the model need additional coherent information, there is a need for additional cumulative rewards. Second, empirically RL tends to perform better than SL. This is because SL uses a token level loss (that can be summed or averaged over the text passage) while RL takes the entire passage as a whole into account.

3.3   Description of the Model Architecture

The following figure (Fig. 1) shows the overall model architecture.

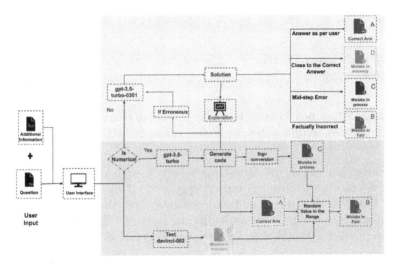

Fig. 1. Model architecture for Key and Distractor Generation for Numerical and Non-Numerical Problems.

Input to the System: The user has to provide the system with a required question stem and some optional additional information (keywords, formulae etc.) associated with the question concepts. This is used by the system to generate the required correct answer and corresponding plausible distractors. The user also selects the type of the question i.e. numerical or non-numerical. Here is an example: **Stem**: *A current of 10A flows through a conductor for two minutes. Calculate the amount of charge passed through any area of cross section of the conductor.* **Additional Information:** None. **Question Type:** Numerical

Numerical Type of Questions: For numerical type of questions given by the user, gpt-3.5-turbo system[1] produces the corresponding code in string format, based on which the key is generated. In cases of erroneous code being generated, the model switch back to the Theoretical Model. The signs in the string are manipulated to find the 'Process' distractor. Text-Davinci 002 model[2] are known to give low accuracy for Numerical problems. Hence it is used for creating the 'Accuracy' distractor. For generating the 'Factual' distractor, the closest value in the range of the other 3 distractors, was considered, so that the values don't look much different from each other. Here are the key and distractors generated for the above question:

Key: *The amount of charge which passed through any area of cross section of the conductor will be 1200 C.*

Factual Deficiency Distractor: *The current flow in the conductor is calculated for two minutes, but the area of cross-section is not known.*

Process Deficiency Distractor: *The total number of electrons flowing will be 7.5×10^{21} electrons.*

Accuracy Deficiency Distractor: *The amount of charge which passed through any area of cross section of the conductor will be 2400 C.*

Non-Numerical Type of Questions: For theoretical/concept based questions, gpt-3.5-turbo-0301 model is used to generate the key and three distractor options based on the given conditions (prompts) from the solution of the question. To generate the prompt, an additional layer of GPT (System) was used based on pre-defined conditions for distractor generation.

4   The Experiment

120 questions which were manually solved by the subject matter experts, were given input into the system sequentially. The proposed model generated the key and the three distractors for each question. The key to each question and the plausible distractors were validated using inter-annotator agreement (using Fleiss Kappa) by three annotators. The overall inter-annotator agreement was 0.816 making it '*almost perfect agreement*'.

---

[1] https://platform.openai.com/docs/models/gpt-3-5. Last accessed 5 March 2023.
[2] https://platform.openai.com/docs/models/gpt-3. Last accessed 5 March 2023.

5   Result and Observation

The detailed result is shown in Table 2. For some questions (35 out of 120), some additional information was required apart from the stem. These type of questions were mostly non-numerical questions from Biology and Physics. The result was analyzed based on the generation of the correct key. If a key generated was incorrect, the remaining distractors of the question was considered invalid*. It was observed that 17 questions out of 120 were absolutely wrong in terms of generating the correct key. The break-up is as follows: 10 (Numerical, Maths), 3 (Numerical, Physics), 1(Non-numerical Maths), 1 (Non-numerical, Physics), 1 (Non-Numerical, Chemistry) and 1 (Non-Numerical, Biology).

For numerical questions in Mathematics which involves multiple equations to generate the correct answer, the proposed model is unable to achieve the requisite accuracy in most of the cases (8/10 questions). Example: *A box contains 12 balls out of which x are black. If one ball is drawn at random from the box, what is the probability that it will be a black ball? If 6 more black balls are put in the box, the probability of drawing a black ball is now double of what it was before. Find x.*

For non-numerical questions (e.g. complex knowledge based questions) the proposed model is not generating all the possible keys. Example:*A person is able to see objects clearly only when these are lying at distance between 50 cm and 300 cm from his eye. What kind of defect of vision he is suffering from?'* The correct answer is Myopia and Hypermetropia. The model generated only Myopia.

Table 2. Distribution of Key and Distractor generated by the proposed model.

Score	Quality of distractor	Key	Fact	Process	Accuracy	Count	Percentage
1–3	Error/Missing/No output Not even suitable for standard MCQs	Incorrect	Invalid*	Invalid*	Invalid*	17	14.16%
3–4	As per standard MCQs	Average	Average	Average	Average	6	5%
4–5	Good (better than standard MCQs)	**Good**	**Good**	**Good**	**Good**	**34**	**28.3%**
5	Very Good as per proposed model	**Very Good**	**Very Good**	**Very Good**	**Very Good**	**63**	**52.5%**

6   Future Work and Conclusion

For certain Maths problem, especially problems related to graph generation is not feasible through the current GPT model as it is primarily a language learning model. Thus the future work is currently focused on generating distractors for graph based questions which can identify the learning deficiency.

References

1. Bridgeman, B., Rock, D.A.: Relationships among multiple- choice and open-ended analytical questions. J. Educ. Meas. **30**(4), 313–329 (1993)

2. Shakurnia, A., Ghafourian, M., Khodadadi, A., Ghadiri, A., Amari, A., Shariffat, M.: Evaluating functional and non-functional distractors and their relationship with difficulty and discrimination indices in four-option multiple-choice questions. Educ. Med. J. **14**(4), 1–8 (2022)
3. Radford, A., Narasimhan, K., Salimans, T., Sutskever, I.: Improving language understanding by generative pre-training (2018)
4. Schuwirth, L.W., Van Der Vleuten, C.P.: Different written assessment methods: what can be said about their strengths and weaknesses? Med. Educ. **38**(9), 974–979 (2004)
5. Palmer, E.J., Devitt, P.G.: Assessment of higher order cognitive skills in undergraduate education: modified essay or multiple choice questions? research paper. BMC Med. Educ. **7**(1), 1–7 (2007)
6. Guo, R., Palmer-Brown, D., Lee, S.W., Cai, F.F.: Intelligent diagnostic feedback for online multiple-choice questions. Artif. Intell. Rev. **42**, 369–383 (2014)
7. Alsubait, T., Parsia, B., Sattler, U.: Ontology-based multiple choice question generation. KI-Künstliche Intelligenz **30**, 183–188 (2016)
8. Ch, D.R., Saha, S.K.: Automatic multiple choice question generation from text: a survey. IEEE Trans. Learn. Technol. **13**(1), 14–25 (2018)
9. Rodriguez-Torrealba, R., Garcia-Lopez, E., Garcia-Cabot, A.: End-to-end generation of multiple-choice questions using text-to-text transfer transformer models. Expert Syst. Appl. **208**, 118258 (2022)

Towards Automatic Tutoring of Custom Student-Stated Math Word Problems

Pablo Arnau-González[1](✉) , Ana Serrano-Mamolar[2] ,
Stamos Katsigiannis[3] , and Miguel Arevalillo-Herráez[1]

[1] Departament d'Informàtica, Universitat de València, Valencia, Spain
{pablo.arnau,miguel.arevalillo}@uv.es
[2] Departamento de Ingenieria Informatica, Universidad de Burgos, Burgos, Spain
asmamolar@ubu.es
[3] Department of Computer Science, Durham University, Durham, UK
stamos.katsigiannis@durham.ac.uk

Abstract. Math Word Problem (MWP) solving for teaching math with Intelligent Tutoring Systems (ITSs) faces a major limitation: ITSs only supervise pre-registered problems, requiring substantial manual effort to add new ones. ITSs cannot assist with student-generated problems. To address this, we propose an automated approach to translate MWPs to an ITS's internal representation using pre-trained language models to convert MWP to Python code, which can then be imported easily. Experimental evaluation using various code models demonstrates our approach's accuracy and potential for improvement.

Keywords: Math Word Problems · Algebra Tutoring · Intelligent Tutoring Systems · Automatic Code Generation

1 Introduction

Math Word Problem (MWP) solving is the task of providing a numerical solution to a mathematical problem expressed in natural language [2]. The computation of the numerical answer of the MWP requires the correct identification of the quantities expressed in the problem statement, together with the relationships between these quantities [4]. MWPs are widely used in Intelligent Tutoring Systems (ITSs) to teach math and arithmetic by emulating human tutor tasks such as interpreting problem statements, validating processes, and adapting problems to individual learners [1]. However, ITSs are limited to supervising registered problems and cannot assist with self-introduced problems, as they have to be registered in the system's own knowledge representation schema.

In this work, we present a two-stage process to automate the encoding of problem solutions from natural language statements. A large code model, i.e. a Large Language Model (LLM) trained on source code, generates an intermediate representation of the problem solution as Python code, which can then be used to create a bipartite graph solution specification using compiler technology.

N. Wang et al. (Eds.): AIED 2023, CCIS 1831, pp. 639–644, 2023.
https://doi.org/10.1007/978-3-031-36336-8_99

This automation has benefits for learners and experts, enabling problem-specific supervision and easy addition of new exercises to the ITS. In our evaluation on a dataset of 1,000 MWP, our method correctly solved 39% of the problems, consistent with other state-of-the-art MWP solving methods, while at the same time facilitating the automated encoding of MWPs for ITS.

2   Proposed Methodology

Given a natural language problem description S that consists of n words w_i, $S = \{w_0, w_1, ..., w_{n-1}\}$, the proposed approach will generate Python source code that first defines a list Q_S of m quantities that appear in S, $Q_S = \{q_0, q_1, ..., q_{m-1}\}$, and then computes and returns the numerical answer y_S to the problem S. The main benefit of proposing a solution expressed in source code is that it allows to construct a graph using the mathematical relations between the quantities, and simultaneously allows for the automatic naming of the identified quantities so that the system is capable of semantically matching the user input to the correct quantity. The generated Python source code is essentially an intermediate representation of the MWP that can be subsequently converted to the internal representation of an ITS in an automated manner. To generate Python code for a MWP, we propose initialising a LLM's prompt with an example problem statement and the expected output code, followed by the target problem statement. The example problem is introduced as a code comment followed by a function definition (sol()), with known quantities defined in a Map-like structure using Python's dictionary data structure, as shown in Fig. 1a. One operation is defined per line until the solution is defined and returned. The model output is the source code of the sol() function for the target problem, as shown in Fig. 1b. The code can then be compiled into the required representation for any ITS.

The pre-trained LLMs examined in this work are the 350M, 2B, 6B, and 16B parameter variations of Saleforce's CodeGen [9] model ("mono" version), a transformer-based model trained on a general text corpus and fine-tuned first on a corpus with source code from multiple languages and then on a Python corpus, and the 1B and 6B parameters variations of Facebook's InCoder [3] Casual Language model that has been trained on a corpus that contained source code from Github and StackOverflow.

3   Results

Our proposed approach is evaluated against a common benchmark dataset for MWP solving, SVAMP [10]. SVAMP is a collection of 1,000 MWP, expressed in natural language (English), along with the numerical answer and an algebraic expression to solve the problem. The performance of the examined models on the SVAMP dataset was evaluated according to the accuracy metric, defined in this case as the percentage of problems that were solved with the first solution returned by the model.

```
""" A book has 3 chapters. The first chapter is 91 pages long
the second chapter is 23 pages long and the third chapter is
25 pages long. How many more pages does the first chapter have
than the second chapter? """
def sol():
    context = dict()
    context['number of chapters'] = 3
    context['number of pages first chapter'] = 91
    context['number of pages sencond chapter'] = 23
    context['number of pages third chapter'] = 25
    context['pages more first chapter'] = (
        context['number of pages first chapter']
        - context['number of pages second chapter']
    )
    return context['pages more first chapter']
""" Each pack of dvds costs 76 dollars.
If there is a discount of 25 dollars on each pack.
How much do you have to pay to buy each pack? """
def sol():
```

```
context = dict()
context['price of dvds'] = 76
context['discount'] = 25
context['price of dvds with discount'] = (
    context['price of dvds'] - context['discount']
)
return context['price of dvds with discount']
```

(a) Prompt   (b) Output

Fig. 1. Example input problem statement with one example provided to the model and expected solution for a MWP.

Table 1. Accuracy for the best-performing temperature (t) for each of the examined models

Model	# Params	Temperature	Accuracy
InCoder [3]	1B	0.1	0.061
	6B	0.3	0.174
CodeGen [9]	350M	0.1	0.088
	2B	0.1	0.272
	6B	0.3	0.335
	16B	0.3	**0.391**

Note: M: Million, B: Billion. Best performance in bold.

For the performance evaluation procedure, a random problem from SVAMP was selected and manually solved by implementing the respective sol() function, according to the specifications described in Sect. 2. This problem was then used as the initialisation of all the language models' prompts, in order to ensure a fair evaluation across the different examined models. Then, each model was queried with the randomly chosen example problem and the statement for the problem to be solved, as shown in Fig. 1a. Each model was queried to generate $n = 10$ solutions for each of the 1,000 problems of the SVAMP dataset, which were then tested by running the generated Python source code and evaluating whether the result was equal to the expected solution or not.

Both examined models (CodeGen, InCoder) were evaluated for all the publicly available parameter number variations. CodeGen was tested for all the aforementioned model variations, while InCoder was tested for the 1B and 6B parameter versions. The inference was carried out by tuning Softmax's temperature (t) parameter to values 0.1, 0.3, and 0.5. The temperature is used to control the randomness of a model's predictions, with higher values of temperature resulting in a model becoming more random and less certain of its predictions, whereas lower values result in more certain predictions. To this end, each parameter number variation of the CodeGen and InCoder models was evaluated three times using the three aforementioned temperature values, respectively.

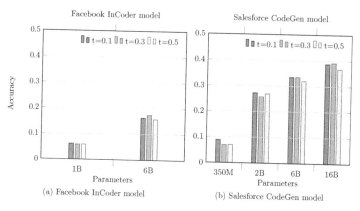

(a) Facebook InCoder model   (b) Salesforce CodeGen model

Fig. 2. Accuracy obtained for different temperatures (t) and number of parameters for each model. (a) Facebook InCoder; (b) Salesforce CodeGen.

Results in terms of accuracy, are reported in Table 1 for the temperature that provided the best accuracy for each examined model variation. From this table, it is evident that the 16B parameter CodeGen model achieved the best performance on the SVAMP dataset in terms of the examined metric, reaching an accuracy of 39.1%. In addition, one of the main takeaways from Table 1 is that there is a clear relationship between the number of parameters and the performance, with variants of a model with more parameters achieving higher accuracy (Pearson's $\rho = 0.77$), as shown in Fig. 2a and 2b, for the InCoder and CodeGen models respectively. Figure 2 depicts the accuracy for each examined model variant and temperature value. The figure shows clearly the relationship between the number of model parameters and the accuracy. However, it also makes evident that this relationship is not purely linear, but rather it appears to follow a logarithmic trend. In addition, it is evident that the temperature parameter has minimal effects on the achieved accuracy.

The presented results compete directly with the latest state-of-the-art MWP solving methods, with CodeGen-16B's maximum accuracy of 39.1% outperforming 6 out of the 9 current top performing methods, as shown in Table 2, without being explicitly designed to solve MWPs. In addition and in contrast to other available MWP solving methods, the proposed approach enables the automation of the task of translating MWP from natural language to the internal representation of ITSs, thus addressing one of their major limitations. Consequently, other state-of-the-art MWP solving methods cannot replace the proposed approach for the task at hand, as they cannot provide the required source code representation of the MWP. It is expected that as large language models get even bigger in terms of the number of parameters, the proposed approach will be capable of surpassing the state-of-the-art MWP solving methods without requiring specific domain knowledge.

Table 2. Current top performing MWP solving methods on the SVAMP dataset.

Model	Year	Problem encoding	Accuracy
DeductReasoner [5]	2022	No	0.473
Roberta-Graph2Tree [10]	2021	No	0.438
Roberta-GTS [10]	2021	No	0.410
CodeGen-16B (Ours)	2023	**Yes**	0.391
Graph2Tree [12]	2020	No	0.365
BERT-Tree [8]	2021	No	0.324
GTS [11]	2019	No	0.308
Roberta-Roberta [6]	2022	No	0.303
BERT-BERT [6]	2022	No	0.248
GroupAttn [7]	2022	No	0.215

4   Conclusions

In this work, we presented a method for solving MWPs, expressed in natural language, by using Large Language Models to produce Python source code that can solve the problem and can automatically convert it to the internal representation of Intelligent Tutoring Systems. Apart from the automated solving of MWPs, the proposed approach is also capable of naming the quantities in the MWP. Together, they allow the translation of problem statements into the ITS' internal knowledge representation schema, thus allowing learners to add new MWPs in an ITS and tutors to add new MWPs at scale. Our experimental evaluation showed that Saleforce's CodeGen model with 16B parameters achieved the highest accuracy (39.1%) for the proposed approach on the SVAMP MWP dataset. In addition, despite some state-of-the-art MWP solving methods achieving higher accuracy, they are not able to encode the MWP in a form that would enable automated import to an ITS, thus they cannot replace the proposed method.

The success of the proposed approach relies heavily on the performance of code generation models, as shown by the significant performance differences between the examined CodeGen and InCoder models. However, the observed correlation between the size of the model and performance suggests that the quality of the generated solutions will improve with the future development of more advanced models with a larger number of parameters.

Nevertheless, the proposed solution has some limitations. Since it uses plain Python, it is not capable of proposing solutions to problems that cannot be arithmetically solved. In addition, it only generates one solution graph per problem. Although this is in general the most obvious solution, it is not necessarily the only one. Future work will seek to find strategies to deal with these weaknesses. Additionally, it would be interesting to study the different generated source code snippets in order to synthesise all the possible resolution paths to a given problem.

Acknowledgement. This research has received support from project TED2021-129485B-[C42/C43], funded by the Ministry of Science and Innovation (Strategic Projects Focused on the Green and Digital Transition); project PGC2018-096463-B-I00, funded by MCIN/AEI/10.13039/501100011033 and "ERDF A way of making Europe"; grant CIAPOS/2022/163, project AICO/2021/019 and Grupos de Investigación Emergentes, funded by Generalitat Valenciana; Margarita Salas 2022–2024 Grant, awarded by University of Valencia [MS21-29], and by University of Burgos funded through NextGenerationEU funds; and project "AGENCY", funded by the Engineering and Physical Sciences Research Council [EP/W032481/1], United Kingdom.

References

1. Arnau, D., Arevalillo-Herráez, M., González-Calero, J.A.: Emulating human supervision in an intelligent tutoring system for arithmetical problem solving. IEEE Trans. Learn. Technol. **7**(2), 155–164 (2014). https://doi.org/10.1109/TLT.2014.2307306
2. Bobrow, D.G.: Natural language input for a computer problem solving system. Technical Report. AIM-066, Massachusetts Institute of Technology (1964)
3. Fried, D., et al.: Incoder: a generative model for code infilling and synthesis. arXiv:2204.05999 (2022)
4. Jie, Z., Li, J., Lu, W.: Learning to reason deductively: math word problem solving as complex relation extraction. In: Annual Meeting of the Association for Computational Linguistics (2022)
5. Jie, Z., Li, J., Lu, W.: Learning to reason deductively: math word problem solving as complex relation extraction. In: Proceedings of the 60th Annual Meeting of the Association for Computational Linguistics, pp. 5944–5955 (2022)
6. Lan, Y., et al.: Mwptoolkit: an open-source framework for deep learning-based math word problem solvers. In: Proceedings of the AAAI Conference on Artificial Intelligence, vol. 36, pp. 13188–13190 (2022). https://doi.org/10.1609/aaai.v36i11.21723
7. Li, J., Wang, L., Zhang, J., Wang, Y., Dai, B.T., Zhang, D.: Modeling intra-relation in math word problems with different functional multi-head attentions. In: Proceedings of the 57th Annual Meeting of the Association for Computational Linguistics, Florence, Italy, pp. 6162–6167 (2019). https://doi.org/10.18653/v1/P19-1619
8. Li, Z., et al.: Seeking patterns, not just memorizing procedures: contrastive learning for solving math word problems. arXiv preprint arXiv:2110.08464 (2021)
9. Nijkamp, E., et al.: Codegen: an open large language model for code with multi-turn program synthesis. ArXiv preprint, abs/2203.13474 (2022)
10. Patel, A., Bhattamishra, S., Goyal, N.: Are NLP models really able to solve simple math word problems? In: Proceedings of the 2021 Conference of the North American Chapter of the Association for Computational Linguistics: Human Language Technologies, pp. 2080–2094 (2021)
11. Xie, Z., Sun, S.: A goal-driven tree-structured neural model for math word problems. In: Proceedings of the 28th International Joint Conference on Artificial Intelligence, pp. 5299–5305 (2019)
12. Zhang, J., et al.: Graph-to-tree learning for solving math word problems. In: Proceedings of the 58th Annual Meeting of the Association for Computational Linguistics, pp. 3928–3937 (2021)

A Software Platform for Evaluating Student Essays in Interdisciplinary Learning with Topic Classification Techniques

Bryan Lim Cheng Yee, Chenyu Hou, Gaoxia Zhu, Fun Siong Lim, Shengfei Lyu, and Xiuyi Fan[(✉)]

Nanyang Technological University, Singapore, Singapore
xyfan@ntu.edu.sg

Abstract. Interdisciplinary learning aims to address the growing need to solve complex problems that go beyond the boundaries of a single discipline. To better facilitate interdisciplinary learning, it is crucial to perform robust evaluations of students' works. In this paper, we presented TopicWise, an interdisciplinary learning evaluation tool for student essays, as an initial step to address this research gap. TopicWise is developed in the context of a digital literacy course, which was part of the interdisciplinary collaborative core curriculum used in an Asian university. TopicWise reads student essays and detects the number of disciplines presented and then estimates the degree of disciplinary integration. TopicWise delivers evaluation results similar to human grader.

Keywords: Interdisciplinary Learning · Essay Evaluation

1 Introduction

Interdisciplinary learning is concerned with integrating knowledge, methods, and insights from multiple disciplines to solve complex problems in challenging situations [4]. Increasing life complexities and societal issues require higher education institutions to support learners' interdisciplinary learning. However, how to evaluate students' interdisciplinary learning remains challenging [7]. Previous literature mainly relies on surveys to measure students' self-reported interdisciplinary learning [2], which is subjected to validity related issues as such the intrusiveness nature of surveys, respondent bias and social desirability [3].

In this work, we approach interdisciplinary learning evaluation as an instance of *Automated Essay Evaluation (AEE)* [13]. As one of the most important educational applications of natural language processing (NLP), AEE is "the process of evaluating and scoring written prose via computer programs" [14]. Research in AEE began with Page's landmark work on the Project Essay Grader system [11] and remained active ever since. Klebanov & Madnani [6] present a

N. Wang et al. (Eds.): AIED 2023, CCIS 1831, pp. 645–651, 2023.
https://doi.org/10.1007/978-3-031-36336-8_100

recent overview of AEE systems. We see that AEE research has been predominately focused on generating holistic scores from essays. Such scoring systems have been successfully used in standardized tests such as TOEFL, PTE or GRE. Although researchers have started to investigate AEE using other evaluation criteria, such as *coherence* (e.g. [15]), *clarity* (e.g. [12]) and *persuasiveness* (e.g. [10]), there is little work in evaluating the interdisciplinarity of essays.

To fill this gap, we proposed three indices, *Disciplinary Grounding Index (DGI)*, *Disciplinary Integration Index (DII)* and *Disciplinary Evenness (DEI)* to quantitatively evaluate essay interdisciplinarity. All three indices are modelled after Mansilla et.al.'s work [9] on interdisciplinary writing rubrics and adapted to our local course context. We focus on these indices as interdisciplinary learning is not only about contributing knowledge from different disciplinary perspectives but more importantly, integrating these knowledge to achieve coherence in writing [8]. Jointly, these indices capture the number of disciplines in an essay, as well as their degree of integration.

The contribution of this work is threefold. Firstly, from an evaluation and feedback perspective, we have explored three indices for quantifying the interdisciplinarity of student essays. Secondly, from an AI/NLP perspective, we have adopted and applied topic classification techniques in a novel application in AEE. Lastly, from a software engineering perspective, we have developed a software tool to support students and teachers in interdisciplinary learning and teaching.

2   Interdisciplinary Essay Evaluation Indices

Inspired by the interdisciplinary writing rubric work by Mansilla et.al. [9], we proposed three indices for evaluating students' interdisciplinary essays with topic classification techniques. Specifically, we focused on assessing an essay's *disciplinary grounding* and *integration*. Considering in our course, we do not require students to critically reflect on the contributions and limitations of their solutions. We introduce our indices as follows.

The Disciplinary Grounding Index (DGI) is proposed to evaluate *"the extent to which do students use disciplinary knowledge from multiple disciplines in their writing."* DGI assesses the number of disciplines used in an essay. Let us consider an essay E composed of n paragraphs $\{s_1, \ldots, s_n\}$ in which each paragraph s_i includes a set of disciplines $d_i = \{d_1^i, \ldots, d_{mi}^i\}$. The first index, DGI, is:

$$\text{DGI}(E) = \frac{|\bigcup_{i=1}^{n} d_i|}{n}. \tag{1}$$

Intuitively, DGI is a normalized measure that counts the number of disciplines in an essay. For each essay, DGI examines the disciplines for each paragraph and joins the findings for the entire essay. The overall count is then normalized by the number of paragraphs in the essay. This normalization is useful as it provides a more standardized comparison between essays of different lengths. Intuitively, DGI is always greater than 0. The more disciplines an essay covers relative to its length, the higher its DGI index is.

The Disciplinary Integration Index (DII) and the Disciplinary Evenness Index (DEI) jointly evaluate *"the extent to which do the students properly integrate knowledge, methods, and insights from multiple disciplines in their writing."* Disciplinary integration is jointly represented by these two indices as integration needs to be assessed by the amount or percentage of co-occurrence of knowledge from multiple disciplines in an essay (DII) as well as the balance between different disciplinary in the essay (DEI).

For an essay E with n paragraphs $s_1, \ldots, s_n$, let $D = \{d_1, \ldots, d_n\}$ be the set of disciplines exist for all paragraphs in E. In other words, the paragraph s_i covers the set of disciplines d_i. The second index, DII, is such that:

$$\text{DII}(E) = \frac{|\{d|d \in D, |d| > 1\}|}{n}. \tag{2}$$

DII is the ratio between the number of paragraphs that cover more than one discipline and the number of paragraphs. It is a normalized measure with respect to the number of paragraphs. Intuitively, we consider an essay to have a higher integration score if a bigger portion of it includes more than one discipline. Note that by restricting the measurement to paragraphs, we enforce a "tight intergration" between different topics. Thus making DII a distinct measure than DGI, as which counts the total number of covered disciplines. DII is between 0 and 1; and higher scores mean greater integration.

DEI uses the *Gini-Simpson Index* developed in biodiversity research [5] to measure the evenness of discipline distribution. With our notation, for an essay E that covers k topics, let $c_1, \ldots, c_k$ be the number of paragraphs labelled by discipline $1, \ldots, k$, respectively. Let $l = \sum_{i=1}^{k} c_i$. The third index, DEI, is:

$$\text{DEI}(E) = 1 - \sum_{i=0}^{k} \left(\frac{c_i}{l}\right)^2. \tag{3}$$

This definition indicates that DEI is a measure between 0 and 1. The higher its value, the greater the evenness.

Intuitively, suppose an essay that covers three disciplines A, B and C with 10 paragraphs, such that 2 paragraphs covering discipline A solely, 7 paragraphs covering 2 disciplines A, B, and 1 paragraph covering all 3 disciplines A, B, C. Then its indices are such that: DGI 0.3, as the essay covers three disciplines with 10 paragraphs; DII 0.8, as 8 out of the 10 paragraphs are interdisciplinary; and DEI 0.54, which is computed as $l = 10 + 8 + 1 = 19$, $c_1 = 10/19 = 0.53$, $c_2 = 8/19 = 0.42$, and $c_3 = 1/19 = 0.053$. The resulting 0.54 is read as a passing score as the essay achieves a good evenness between disciplines A and B, but out of balance with discipline C.

The key to achieving a successful computation for DGI, DII, and DEI is accurately identifying disciplines from essays. Specifically, for each paragraph s_i in an essay, developing a robust mechanism for calculating d_i is core to this work. This is a typical use case for topic classification techniques. In our implementation, an Multi-label classification model is chosen as it can produce several labels (disciplines) for a piece of input text. This is suitable for our application as each paragraph can include multiple disciplines, especially when disciplinary integration occurs.

3   Software Realization

To make our indices more accessible for teachers to understand students' interdisciplinary learning status and highlight areas that need more work as well as for students to self-evaluate and improve their interdisciplinary work, we developed TopicWise. With TopicWise, students upload their essay assignments to the platform. Upon submission, they can view the disciplines present in their own assignment, together with the interdisciplinary (DGI) and integration (DII & DEI) scores. In contrast, teachers have an overview of the discipline analysis of the submitted assignments of all of their students in various classes and can compare the results at a glance. They will also be able to create and manage different assignments for the students.

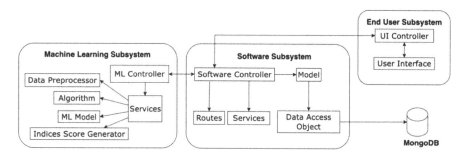

Fig. 1. Overview of Software Architecture for TopicWise.

An overview of our software architecture is shown in Fig. 1. There are four main subsystems in our design: (1) Software Subsystem, (2) End User Subsystem, (3) MongoDB, and (4) Machine Learning Subsystem. The *Software Subsystem* uses the *Model-View-Controller* architectural pattern [1] that separates the subsystem into separate logical components. The *Model* component encapsulates all data-related logic that transfers data between different components. The *View* component processes interface requests via the *End User Subsystems*. The *Controller* component functions as an interface between the View, Model, and the *Machine Learning Subsystem* and handles incoming requests and business logic transactions. The *Routes* and *Services* components complement the Controller by routing requests to the corresponding data objects. The *Data Access Object* provides an abstract interface to access the database.

The End User subsystem comprises the *User Interface* (UI) display and the *UI controller*. The User Interface functions allow end users to directly interact with the system and view the desired output that they have requested. The

UI controller consolidates the requests made by the end user and sends them to the software controller. The main database used for this project is MongoDB, an open-source NoSQL database management program that can manage document-oriented information, and store or retrieve information.

4   Evaluation

We evaluated the performance of our software on student essays collected from a digital literacy course. In the August semester of 2022, we redesigned weekly in-class activities and final essay requirements to further support students' interdisciplinary learning and collected 12 group essays from 71 voluntary participants. The participants came from four classes, with about 50 students in each class who formed nine student teams. Each team was composed of students from different schools such as engineering, science, business, and social science. At the end of the semester, each team was asked to submit one 3,000 words essay based on a topic negotiated by themselves relevant to any content discussed in the class. To promote interdisciplinary learning, the students were asked to approach their writing from multiple perspectives, such as technological, scientific, economic, and societal perspectives. It is worth noting as the students in this course were first-year undergraduate students, they were yet to be equipped with deep domain knowledge. Thus, "interdisciplinary learning" in this context was grounded to actively joining knowledge from large domains such as "social science" and "engineering", rather than, e.g., applying "convex optimization" techniques for the purpose of "probabilistic logic" inferences.

Amongst the 12 essays, two were submitted as image files with illegible OCR results. Thus, 10 group essays containing 236 paragraphs were successfully collected. Each paragraph of the essays was read by human raters and manually labelled with discipline labels. Two human raters independently labelled (coded) the paragraphs and achieved agreement on 165 of them. These 165 paragraphs are used in our study with their commonly agreed labels as the ground truth. We also applied transfer learning in this step to employ an one-hidden layer classifiers trained from student essays as a post-processor to fine-tune predictions.

Table 1. Topic Classification Performance on Student Essays.

Discipline	Accuracy	Precision	Recall	F1-score
Technology	0.79	0.85	0.77	0.80
Sciences	0.74	0.60	0.68	0.63
Business	0.79	0.69	0.78	0.72
Humanities	0.73	0.43	0.61	0.49

With a 80/20 training/testing split on the essays, the classification performance of the final model on student essays is shown in Table 1. To verify that our software tool performs interdisciplinary evaluation comparable to humans, we compute DGI, DII, and DEI scores computed from manual coding and machine predictions. The results are shown in Fig. 2. On all three scores, strong correlations between manual coding and ML predictions have been found, with correlations on DGI 0.75, DII 0.80 and DEI 0.80, respectively. These results how that our interdisciplinary AEE tool performs interdisciplinary evaluation in the same way as a human grader in the vast majority of the cases.

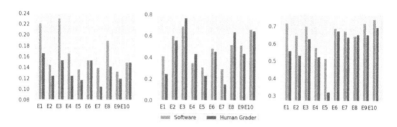

Fig. 2. DGI, DII and DEI Scores Computed from Software Tool and Manual Coding.

5   Conclusion

The evaluation of interdisciplinary learning remains a practical and research challenge, despite its increasing attention. This study proposes a set of indices to quantify interdisciplinarity and TopicWise, a software realisation for automatically measuring the indices. TopicWise is based on topic classification techniques developed in NLP. The proposed indices are inspired by rubrics for evaluating interdisciplinarity in student essays, and the software has been validated on student essays collected from a digital literacy course. The results show that the software-generated evaluation results are comparable to human raters. Limitations of this work include the focus on only three indices for measuring interdisciplinarity and a broad classification of disciplines into four categories. Future work could explore other evaluation dimensions, refine topic models, and explore large language model-based approaches.

Acknowledgement. This project is supported by the Centre for Teaching, Learning and Pedagogy, Nanyang Technological University under its EdeX Programme.

References

1. Bass, L., et al.: Software Architecture in Practice. Addison-Wesley, Boston (2003)
2. Berasategi, N., et al.: Interdisciplinary learning at university: assessment of an interdisciplinary experience based on the case study methodology. Sustainability **12**(18), 7732 (2020)

3. Gonyea, R.M.: Self-reported data in institutional research: review and recommendations. New Direct. Inst. Res. **2005**(127), 73–89 (2005)
4. Ivanitskaya, L., Clark, D., Montgomery, G., Primeau, R.: Interdisciplinary learning: process and outcomes. Innov. High. Educ. **27**(2), 95–111 (2002)
5. Jost, L.: Entropy and diversity. Oikos **113**(2), 363–375 (2006)
6. Klebanov, B.B., Madnani, N.: Automated evaluation of writing-50 years and counting. In: Proceedings of ACL, pp. 7796–7810 (2020)
7. Klein, J.T.: Evaluation of interdisciplinary and transdisciplinary research: a literature review. Am. J. Prevent. Med. **35**(2), S116–S123 (2008)
8. MacLeod, M., van der Veen, J.T.: Scaffolding interdisciplinary project-based learning: a case study. Eur. J. Eng. Educ. **45**(3), 363–377 (2020)
9. Mansilla, V.B., et al.: Targeted assessment rubric: an empirically grounded rubric for interdisciplinary writing. J. High. Educ. **80**(3), 334–353 (2009)
10. Nguyen, H.V., Litman, D.J.: Argument mining for improving the automated scoring of persuasive essays. In: Proceedings of AAAI, pp. 5892–5899. AAAI Press (2018)
11. Page, E.B.: The imminence of... grading essays by computer. Phi Delta Kappan **47**(5), 238–243 (1966)
12. Persing, I., Ng, V.: Modeling thesis clarity in student essays. In: Proceedings of ACL, pp. 260–269. The Association for Computer Linguistics (2013)
13. Shermis, M.D., Burstein, J.: Handbook of Automated Essay Evaluation. Routledge, Abingdon (2013)
14. Shermis, M.D., Burstein, J.C.: Automated Essay Scoring: A Cross-Disciplinary Perspective. Routledge, Abingdon (2003)
15. Somasundaran, S., et al.: Lexical chaining for measuring discourse coherence quality in test-taker essays. In: Proceedings of COLING, pp. 950–961. ACL (2014)

Automated Scoring of Logical Consistency of Japanese Essays

Sayaka Nakamoto[✉] and Kazutaka Shimada

Kyushu Institute of Technology, Iizuka, Japan
nakamoto.sayaka478@mail.kyutech.jp, shimada@ai.kyutech.ac.jp

Abstract. Automated essay scoring (AES) is to estimate the scores of essays automatically. Two types of AES models are commonly used: handcrafted feature-based and neural-based models. In this paper, we introduce AES systems based on the two types for evaluating the logical consistency of Japanese essays. In addition, to enhance the performance of models, we integrate the neural-based model with the handcrafted features: a hybrid AES system. In the experiment, we show the effectiveness of our hybrid AES system. Besides, most of our AES models obtained higher QWK scores than human evaluators.

Keywords: Automated Essay Scoring · Machine Learning · Education Technology

1 Introduction

In recent years, short essays have been increasingly used in entrance exams to evaluate thinking and expression skills. However, manually scoring these essays is costly in terms of work time, personnel costs, errors, and fairness issues. To solve the problems, automated essay scoring (AES) has gained attention to provide cost-effective and reliable evaluations.

Our study aims to develop a high-quality AES system for Japanese short essays. In this paper, the target dataset was developed by Ohno et al. [7]. It contains essays with scores based on four criteria: (1) comprehensiveness, (2) validity, (3) logical consistency, and (4) spelling and grammar. While previous studies have mainly focused on comprehensiveness and validity, we focus on logical consistency. As one of the aims of general essay exams is to evaluate logical thinking skills, scoring logical consistency is one of the most important parts of the AES.

In the history of the AES research, effective word-level and sentence-level features have been found. On the other hand, large language models with neural network have been applied to the AES. To enhaunse the performance, hybrid models have been getting attracted. We introduce three approaches: handcrafted feature-based, neural-based, and hybrid approaches.

N. Wang et al. (Eds.): AIED 2023, CCIS 1831, pp. 652–658, 2023.
https://doi.org/10.1007/978-3-031-36336-8_101

2   Related Work

This section briefly introduces existing handcrafted feature-based and neural-based AES systems.

One of the earliest practical English AES systems, e-rater [2], was developed by Burstein et al. The improved e-rater V2.0 [1] utilized features such as grammar, vocabulary level, relevance to the prompt, and sentence length. It assigned feature weights empirically to score essays on a variety of prompts. E-rater is now widely used in entrance exams in the US. Ishioka et al. developed a Japanese AES system called Jess [5], inspired by e-rater. Jess evaluated essays on the basis of rhetorical relations, logical structure, and content. Since there was no Japanese essay dataset at the time, they utilized newspaper columns as the training data. Despite not using essay data in the training phase, it demonstrated reasonable results compared to e-rater. In this study, we propose a handcrafted feature-based system using similar features as Jess for a pre-scored Japanese essay dataset.

Kiyono et al. [6] have proposed two Japanese AES systems with neural-based models: LSTM and LSTM with attention mechanisms. In their experiment about comprehensiveness scores, LSTM with attention mechanisms obtained higher accuracy. Uto et al. [8] have proposed an English AES hybrid system based on BERT and handcrafted features. A lot of handcrafted features, such as the numbers of words and the numbers of nouns, were utilized. Their results indicated that the accuracy was improved by adding handcrafted features. In contrast to Kiyono et al., we evaluate our systems for logical consistency scores. In this paper, we combine BERT and other handcrafted features based on Jess.

3   Dataset

The Japanese short essay dataset developed by Ohno et al. [7] consists of 4,815 short essays covering 9 themes such as global, science, east asia, and so on[1]. Each theme consists of several questions, and the size of the essay, namely character limits, is different in each theme. There are 9 themes such as global, science, east asia, and so on. Size of essays in each questions ranges 105–328, and character limit ranges 100–800. As a result, the dataset contains a wide variety of essays. When writing the essays, reference documents related to the theme were provided to the respondents. The documents are also included in the dataset. Test participants are undergraduate students and graduated students.

Each essay contains five-grade scores on four criteria: (1) comprehensiveness, (2) validity, (3) logical consistency, and (4) spelling and grammar. In addition, the dataset includes evaluation results by human evaluators. The target of our study is logical consistency. Table 1 shows an example of rubrics for logical consistency evaluation. According to it, logical structures and essay contents are evaluated in logical consistency.

---

[1] https://gsk.or.jp/catalog/gsk2021-b (in Japanese).

Table 1. Rubric for logical consistency evaluation (Theme: Global).

Grade	Evaluation criteria
5	The writer has been able to logically explain why certain phenomena are occurring through the economy and society framework The document should contain numerical values as evidence.
4	The writer has been able to logically provide suitable arguments with numerical evidences from lecture documents.
3	The description is reasonable and objective There is no leap in logic in the argument.
2	The description is not always reasonable and objective There are some leaps in logic in the argument.
1	The document has not been written in a logical structure. Unconvincing.

4   Proposed Method

Figure 1 shows the overview of our method. The handcrafted feature-based method only utilizes handcrafted features as input. The neural-based method utilizes BERT output, and the hybrid method utilizes both of them.

4.1   Handcrafted Feature-Based Method

We apply twelve features and Bag-of-Words features to a regression-based machine learning approach because of a five-grade score for logical consistency. In this paper, we utilize Random Forest Regressor as the regression model[2]. The twelve handcrafted features are as follows:

- Median and maximum sentence length
- Median and maximum clause length
- Median and maximum number of phrases in clauses
- Kanji ratio
- Number of attributives declined or conjugated words
- Diversity of vocabulary (K characteristic value [9])
- Number of big words
- Percentage of passive sentences
- Number of common words between an essay and a reference document

Most of them are inspired by Jess. Bag-of-Words features are based on the word frequency in each document.

---

[2] Although we evaluated other regression models such as SVR, Random Forest Regressor tended to be better than them.

4.2 Neural-Based Method

We apply BERT [4] as the neural model. BERT is a pre-trained language model. In this paper, we employ Japanese BERT pre-trained by Tohoku University[3]. In BERT, the special token [CLS] is utilized for classification tasks. The logical consistency score is predicted via the linear layer with dropout as a pre-process. We fine-tune the BERT model with the dataset described in Sect. 3, on the basis of the Mean Squared Error (MSE) loss function.

4.3 Hybrid Method

We concatenate the [CLS] token from BERT and handcrafted features, and then the model learns the linear layer with dropout. BERT is also fine-tuned by the training data.

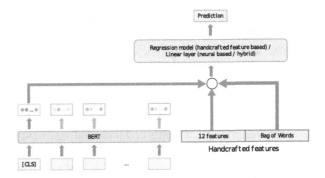

Fig. 1. The overview of our method.

5 Experiments

5.1 The Evaluation Criterion: QWK

In the experiment, we evaluate our models with QWK [3]. QWK is calculated as follows:

$$\kappa = 1 - \frac{\sum_{i,j} w_{i,j} O_{i,j}}{\sum_{i,j} w_{i,j} E_{i,j}} \tag{1}$$

where $O_{i,j}$ is the number of records, and $E_{i,j}$ is the expected number of records. i is the true label, and j is the predicted label. $w_{i,j}$ is a penalty calculated using $(i - j)^2$. Here, assume that the distributions of each true and predicted label are independent. $E_{i,j}$ is calculated as follows:

$$E_{i,j} = \frac{Num_i}{Num_{all}} \times \frac{Num_j}{Num_{all}} \times Num_{all} \tag{2}$$

---

[3] https://github.com/cl-tohoku/bert-japanese.

where Num_i is the number of records labeled i, Num_j is the number of records labeled j, Num_{all} is the number of records.

5.2   Experimental and Parameter Settings

The experimental data setting follows the previous studies of this dataset. The true scores are set to the average of hand-scored results from two evaluators. Essays excluded in the previous studies are also excluded. The number of essays finally used is 4803 out of 4815.

For the handcrafted feature-based method, the output of the regressor is a float value. Therefore, it is rounded to an integer when QWK is calculated. For the neural-based and hybrid methods, the logical consistency scores are converted from $1\tilde{\;}5$ range to $-1.0\tilde{\;}+1.0$ for the BERT training. Then, the predicted value from BERT ($-1.0\tilde{\;}+1.0$) is converted to an integer between $1\tilde{\;}5$ when QWK is calculated.

The loss function utilized in neural and hybrid methods is MSE, and the backpropagation is applied to the linear layer and the final layer of BERT. The optimizer is AdamW, and the learning rate is $5e-5$.

5.3   Results and Analysis

Table 2 shows the experimental result. Hybrid methods tended to produce high QWK scores, as compared with BERT-only and handcrafted feature-based methods. In particular, the hybrid method with all combinations, namely BERT with 12 features and Bag of Words (BoW), produced the best performance. It indicates that both handcrafted features have an impact on improving the prediction. In addition, neural-based and hybrid methods are mostly better than handcrafted feature-based methods. In other words, neural-based approaches are helpful in predicting the logical consistency score. For handcrafted features, BoW is better than the twelve features. Therefore, the essay's contents, namely words, are one of the most important factors for the prediction.

The QWK score between human evaluators is mentioned in the manual book of the dataset, and it is 0.561. Most of our systems achieved higher QWK than human evaluators. This result shows the effectiveness of our AES methods.

Table 2. Result of logical consistency prediction. [†] denotes methods that outperformed the score of human evaluators.

Type	Method		QWK
Handcrafted feature-based	Random Forest Regressor	12 features only	0.440
		BoW only	0.603[†]
		12 features+BoW	0.591[†]
Neural-based	BERT	–	0.596[†]
Hybrid	BERT	+12 features	0.607[†]
		+BoW	0.632[†]
		+12 features+BoW	**0.647[†]**
Human evaluators			0.561

6   Conclusions

We developed three types of Japanese AES systems: handcrafted feature-based, neural-based, and hybrid methods. Our experimental results indicated that hybrid methods tended to yield superior QWK scores. In addition, more than half of the variants by combinations of features outperformed the score from human evaluators. Thus, our proposed methods are effective for the AES task.

Since the feature combination is effective, feature engineering is one important future work for the improvement of QWK. For the logical consistency score, the structure of the essay is the most important factor. Discourse makers such as "because" and "therefore" have important roles for logical consistency. Incorporating these features into our method is interesting future work. The simultaneous/parallel learning with other criteria, such as comprehensiveness and validity, is another future work, although we focused on one criterion, namely the logical consistency.

References

1. Attali, Y., Burstein, J.: Automated essay scoring with e-rater vol 2. J. Technol. Learn. Assess. **4**(3) (2006)
2. Burstein, J., et al.: Automated scoring using a hybrid feature identification technique. In: 36th Annual Meeting of the Association for Computational Linguistics and 17th International Conference on Computational Linguistics, vol, 1, pp. 206–210 (1998)
3. Cohen, J.: Weighted kappa: nominal scale agreement provision for scaled disagreement or partial credit. Psychol. Bull. **70**(4), 213–220 (1968)
4. Devlin, J., Chang, M.W., Lee, K., Toutanova, K.: BERT: pre-training of deep bidirectional transformers for language understanding. In: Proceedings of the 2019 Conference of the North American Chapter of the Association for Computational Linguistics: Human Language Technologies, vol. 1 (Long and Short Papers), pp. 4171–4186 (2019). https://doi.org/10.18653/v1/N19-1423

5. Ishioka, T., Kameda, M.: Automated Japanese essay scoring system based on articles written by experts. In: Proceedings of the 21st International Conference on Computational Linguistics and 44th Annual Meeting of the Association for Computational Linguistics, pp. 233–240 (2006)
6. Kiyono, M., Koichi, T.: Automatic scoring of Japanese essay using neural networks. Forum Inf. Technol. **18**(2), 239–240 (2019). (In Japanese)
7. Ohno, M., et al.: Construction of open essay writing data and automatic essay scoring system for Japanese. In: Proceedings of The 15th International Conference of the Pacific Association for Computational Linguistics, vol. 17, pp. 215–220 (2017)
8. Uto, M., Xie, Y., Ueno, M.: Neural automated essay scoring incorporating handcrafted features. In: Proceedings of the 28th International Conference on Computational Linguistics, pp. 6077–6088 (2020)
9. Yule, C.U.: The Statistical Study of Literary Vocabulary. Cambridge University Press, Cambridge (2014)

Exercise Generation Supporting Adaptivity in Intelligent Tutoring Systems

Tanja Heck[(✉)] [iD] and Detmar Meurers[iD]

University of Tübingen, Keplerstraße 2, 72074 Tübingen, Germany
{tanja.heck,detmar.meurers}@uni-tuebingen.de

Abstract. Exercise generation makes it possible to systematically provide the range and number of activities needed to support macro-adaptive activity sequencing. We propose an approach to adaptivity that balances the pedagogical need for variable contexts and activity types and their inherent complexity with the goal to incrementally complexify activities along the overall learning trajectory. To accommodate such a macro-adaptivity less strictly based on activity complexity, we integrate more micro-adaptive, scaffolding features offering within-exercise variability and refine the mastery criterion.

Keywords: exercise generation · adaptivity · ITS · language learning

1 Introduction

Intelligent Tutoring Systems (ITS) in principle offer the means to macro-adaptively select learning material that is tailored to an individual learner's needs. Given the substantial heterogeneity in many education contexts, macro-adaptive systems require learning material that is systematically varied to adapt to the individual differences. This is hard to specify manually, so we see automatic exercise generation as a necessary prerequisite for large-scale user-adaptive systems.

Exercise generation has received substantial attention, especially in the domain of language learning [8]. A particular challenge in this domain is the integration of the exercises into the overall classroom [18], which requires the exercise vocabulary and semantic topic to be appropriate. In addition, different learners may struggle with different linguistic material in the syntactic context (e.g., interrogative, negated or embedded structures) of the forms being practiced, as well as with different skills targeted by different exercise types [2, p. 5]. Generating a range of exercises from the same text, introducing variants of the syntactic context, and generating different exercise types makes it possible to cater to individual differences while keeping vocabulary and semantic topic under control.

© The Author(s), under exclusive license to Springer Nature Switzerland AG 2023
N. Wang et al. (Eds.): AIED 2023, CCIS 1831, pp. 659–665, 2023.
https://doi.org/10.1007/978-3-031-36336-8_102

Different from other setups, our adaptive sequencing approach is not solely based on the complexity of an exercise. We argue for pedagogically motivated sequences incrementally introducing the learner to the learning targets in varied contexts, noting that the role of practice variability is well-established in other areas of skill acquisition [7]. Some of the burden of ensuring that the learner can successfully tackle a given activity is thereby shifted to the micro-adaptivity dynamically providing the scaffolding needed by this learner. We thus try to strike a balance between supporting practice that is varied (also in complexity) and incrementally increasing the complexity along the individual learning trajectory.

Realizing our approach for the FeedBook system targeting English as a foreign language in secondary school [19], we tailor generation of form-based grammar exercises towards supporting different adaptivity strategies. Seen through the lens of the three-step approach of [8], the pre-processing step accounts for variability of syntactic contexts, question construction for micro-adaptive variants, and post-processing for variability of skill contexts.

In what follows, Sect. 2 discusses related work on adaptivity and exercise generation. Section 3 then describes the exercise generation setup grounding our macro- and micro-adaptive approach presented in Sect. 4.

2   Related Work

Aiming to find the best exercise sequence, **macro-adaptive** systems pursue either global or local path planning strategies [11]. Global approaches determine the entire learning path through methods ranging from expert knowledge based decision trees [21] and fuzzy logic with a focus on remedial practice to iteratively operating evolutionary algorithms and data driven artificial neural networks [17], and learning graphs [6]. In contrast to these global methods not adjusting to changing learner properties, local approaches determine the next practice material dynamically after completing the previous one. They comprise Elo ratings to jointly determine and update learner skills and exercise complexities, item response theory for static assessments, as well as mastery learning based knowledge tracing [10]. Most of these methods rely on complexity estimates for the exercises or a large amount of learner data to train the models and face performance issues [14] and the typical cold-start problem of data-enabled methods. An alternative we are therefore considering is to redefine mastery criteria based on exposure to varied contexts and general exercise performance to support moving away from purely complexity-driven exercise assignment.

Micro-adaptive scaffolds have mostly been integrated in systems for self-regulated learning. Implemented as prompts, tools, pedagogical agents, strategies, or feedback [9], they help learners to identify learning objects relevant to them based on their learner models. In macro-adaptive systems, where the learning objects are automatically and adaptively assigned, micro-adaptivity is usually limited to adaptive feedback [1].

Exercise generation can functionally be considered in the context of question generation. Language exercises dominate the recent review of question generation [8] showing that many systems focus on question complexity during automatic generation but not contextual diversity. This makes sense considering that the goal of the systems is predominantly to generate questions for adaptive testing, where it is important to know and control the complexity of the test items, not to offer varied material to support effective practice.

In contrast, our approach focuses on exercises for practice in an ITS designed to foster learning, therefore prioritizing contextual diversity over a singular focus on exercise complexity [12]. In the domain of grammar exercises for language learning, accounting for variable contexts has received strikingly little attention. In contrast to the approach in [3] at least allowing users to apply transformations to the sentences then used for exercise generation, other systems generally only require the extracted sentences to contain the linguistic constructions targeted by the exercises, without considering the syntactic context. While approaches exist to filter sentences based on linguistic constructions [4,5], these are not fine-grained enough to allow filtering for specific contextual properties. To fill the gap, we present a system for exercise generation that offers multiple ways of introducing varied syntactic contexts into language exercises.

3   Our Approach

We build on [8] by combining pre-processing, question construction and post-processing as illustrated in Fig. 1.

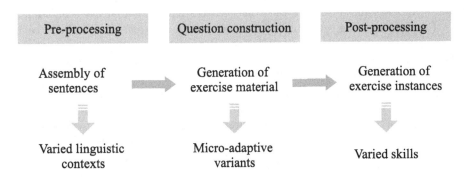

Fig. 1. Three step question generation for high-variability language exercise generation

Pre-processing targets the provision of sentences from which exercises can be generated. Sentences can either be extracted from a corpus or generated by means of Natural Language Generation (NLG) tools [15]. Corpus-based sentences have the advantage of offering authentic language material and being contextualized, which has been argued to have positive effects on motivation

[13]. Our current system implementation supports two corpora, the BookCorpus [16] and the web. The initial user interface allows users to restrict the selected sentences based on properties of the targeted constructions and the linguistic context. In order to identify suitable sentences, the texts of the selected corpus are processed with Natural Language Processing (NLP) tools. In addition, we also support AI-generated sentences for a topic specified by the user. This allows us to enable users to control for linguistic properties and the vocabulary of the generated sentences [15]. By relying on the GPT-3[1] Python interface instead of pursuing a rule-based approach, we avoid the cumbersome task of implementing generation rules and facilitate scaling to other linguistic phenomena. Besides extracting or generating sentences, users can also upload their own texts. In this case, they are responsible themselves for providing appropriate linguistic contexts. An additional means of accounting for context variability consists in manipulating the selected sentences using NLP and NLG based transformations. Whether and which property variations are generated can be configured by the user and also depends on the learning target. While it typically makes sense to vary contextual properties either through sentence selection or through sentence manipulations, both strategies can also be applied jointly.

Question construction generates all material relevant to the exercises. This includes exercise elements such as distractors for Single Choice (SC) exercises, hints in parentheses for Fill-in-the-Blanks (FiB) exercises, or chunks for Jumbled Sentences exercises. Generating exercise elements mostly relies on linguistic annotations, with distractor generation in addition requiring NLG functionalities. The NLP and NLG pipelines of the pre-processing step are used for this purpose. For most exercise elements, multiple variants are equally suitable for the exercises, differing merely in complexity, such as instructions of varying specificity. We store these variants with the exercises for micro-adaptive scaffolding.

Post-processing generates concrete exercise instantiations where each instantiation practices a distinct skill. We differentiate between *form, use, semantic parsing, word order,* and *metalinguistic knowledge.* Since different learners may struggle in different areas of the skill dimension, it makes sense to generate all possible exercises so that additional practice can target the problematic skill. The exercise data compiled in the previous step supports multiple instantiations for each skill with slightly different characteristics. Where different exercises do not target a different skill, they can be merged into a single exercise. The element variations, for example the number of distractors, the presentation of base words for FiB exercises in parentheses or as bags of words above the exercise, or the specificity of instructions are stored in the exercise instantiation with indications of their nature so as to allow the micro-adaptive algorithm to dynamically select the appropriate one. This results, for example, in a joint exercise instantiation for SC and FiB exercises.

---

[1] https://github.com/openai/openai-python.

4 Outlook

Adaptivity-specific requirements must be considered when generating exercises.

The envisioned **macro-adaptive** algorithm performs on four levels. (1) The learning target and skill to be practiced are based on the learner's progress within the domain model. Learners get exercises for a target until they have mastered it. The mastery criterion consider the ratio of correctly solved exercises for the learning target in all required skill dimensions. (2) The linguistic material which is given to the learner consists of the sentences from which an exercise is generated. The algorithm restricts the selection of exercises to those whose sentences align with the learner's curriculum and general language proficiency. (3) Since different linguistic properties have inherently different complexities, the order in which learners are exposed to them is pre-set. As the learner interacts with the system, the algorithm can learn which properties are most challenging for a learner and put special focus on them in the selected practice material. While strong learners should be exposed to as many properties in an exercise as possible, weaker learners need to be given exercises where all sentences have similar properties. (4) The linguistic constructions that the exercises target determine the concrete exercise which a learner is given. Similarly to the selection of properties, the algorithm prefers exercises with larger numbers of targeted linguistic constructions for stronger learners, while prioritizing exercises with less constructions for weaker learners. These considerations result in a ranking of the exercises that have passed the filters of the preceding three levels. The highest-ranked exercise is then presented to the learner.

We envision **micro-adaptive** scaffolding to be available in three places inside exercises. (1) Feedback is provided to learners after each interaction with the system. It is available in English as well as in the learner's native language, using either domain-specific or simple language. Learners always first receive feedback in domain-specific English. As they struggle to correct their errors, these settings are adjusted. (2) Similarly, instructions are available in multiple languages and providing more or less detail. For example, Mark-the-Words exercises can be made more specific by indicating the number of words to find in the text. (3) Exercise elements provide considerable potential for scaffolding. Gap-filling exercises, for instance, start with input fields for the target answers and their lemmas and additional semantic distractors as a bag of words. Scaffolding first eliminates the semantic distractors, then places the lemmas in parentheses behind the correct gap, and finally transforms the input fields into drop-downs to choose from a number of distractors which can then step by step be reduced. Determining the most appropriate scaffolding steps within a place should be based on the learner's global proficiency or on data from the learner model [20].

5 Conclusion

We presented an implementation of language exercise generation for user-adaptive systems. We elaborated on the theoretical framing of our approach and outlined important features of exercises used in adaptive ITS. By also illustrating the macro- and micro-adaptive mechanisms envisioned in the implementation of the ITS, we showed the relevance to real-live usage of the requirements we impose on exercise generation. Future research will determine the concrete realization of mappings of learner proficiency to macro-adaptive characteristics, as well as the selection of appropriate micro-adaptive interventions.

References

1. Bimba, A.T., Idris, N., Al-Hunaiyyan, A., Mahmud, R.B., Shuib, N.L.B.M.: Adaptive feedback in computer-based learning environments: a review. Adapt. Behav. **25**(5), 217–234 (2017)
2. Grellet, F., Francoise, G.: Developing Reading Skills: A Practical Guide to Reading Comprehension Exercises. Cambridge University Press, Cambridge (1981)
3. Heck, T., Meurers, D.: Generating and authoring high-variability exercises from authentic texts. In: Proceedings of NLP4CALL 2022, pp. 61–71 (2022)
4. Heck, T., Meurers, D.: Parametrizable exercise generation from authentic texts: effectively targeting the language means on the curriculum. In: Proceedings of BEA 2022, pp. 154–166. ACL (2022)
5. Hoshino, A., Nakagawa, H.: A Cloze test authoring system and its automation. In: Leung, H., Li, F., Lau, R., Li, Q. (eds.) ICWL 2007. LNCS, vol. 4823, pp. 252–263. Springer, Heidelberg (2008). https://doi.org/10.1007/978-3-540-78139-4_23
6. Karampiperis, P., Sampson, D.: Adaptive learning resources sequencing in educational hypermedia systems. Educ. Technol. Soc. **8**(4), 128–147 (2005)
7. Kim, T., Wright, D.L., Feng, W.: Commentary: variability of practice, information processing, and decision making-how much do we know? Front. Psychol. **12** (2021)
8. Kurdi, G., Leo, J., Parsia, B., Sattler, U., Al-Emari, S.: A systematic review of automatic question generation for educational purposes. Int. J. Artif. Intell. Educ. **30**(1), 121–204 (2020)
9. Lim, L., et al.: Effects of real-time analytics-based personalized scaffolds on students' self-regulated learning. Comput. Hum. Behav. **139**, 107547 (2023)
10. Liu, Q., Shen, S., Huang, Z., Chen, E., Zheng, Y.: A survey of knowledge tracing. CoRR (2021)
11. Mac, T.T., Copot, C., Tran, D.T., De Keyser, R.: Heuristic approaches in robot path planning: a survey. Robot. Auton. Syst. **86**, 13–28 (2016)
12. Melton, A.W.: The situation with respect to the spacing of repetitions and memory. J. Verb. Learn. Verb. Behav. **9**(5), 596–606 (1970)
13. Peacock, M.: The effect of authentic materials on the motivation of EFL learners. ELT J. **51**(2), 144–156 (1997)
14. Pelánek, R., Řihák, J.: Experimental analysis of mastery learning criteria. In: Proceedings of UMAP 2017, pp. 156–163, UMAP 2017. ACM, New York, USA (2017)
15. Perez-Beltrachini, L., Gardent, C., Kruszewski, G.: Generating grammar exercises. In: Proceedings of BEA 2012, pp. 147–156. ACL, Montréal, Canada, June 2012
16. Presser, S.: Homemade BookCorpus (2020). github.com/soskek/bookcorpus/issues/27. Accessed 8 Nov 2022
17. Rasheed, F., Wahid, A.: Sequence generation for learning: a transformation from past to future. Int. J. Inf. Educ. Technol. **36**(5), 434–452 (2019)
18. Rosell-Aguilar, F.: Top of the pods - in search of a podcasting "Podagogy" for language learning. Comput. Assist. Lang. Learn. **20** (2007)
19. Rudzewitz, B., Ziai, R., De Kuthy, K., Meurers, D.: Developing a web-based workbook for English supporting the interaction of students and teachers. In: Proceedings of NLP4CALL & LA, pp. 36–46 (2017)

20. Vogel, F., Kollar, I., Fischer, F., Reiss, K., Ufer, S.: Adaptable scaffolding of mathematical argumentation skills: the role of self-regulation when scaffolded with CSCL scripts and heuristic worked examples. Int. J. Comput.-Supported Collaborative Learn. 1–26 (2022). https://doi.org/10.1007/s11412-022-09363-z
21. Wang, Y.h., Tseng, M.H., Liao, H.C.: Data mining for adaptive learning sequence in English language instruction. Expert Syst. Appl. **36**(4), 7681–7686 (2009)

Context Matters: A Strategy to Pre-train Language Model for Science Education

Zhengliang Liu[1], Xinyu He[1], Lei Liu[2], Tianming Liu[1(✉)],
and Xiaoming Zhai[1(✉)]

[1] University of Georgia, Athens, GA 30666, USA
tliu@uga.edu, xiaoming.zhai@uga.edu
[2] Educational Testing Service, Princeton, NJ, USA

Abstract. This study aims at improving the performance of scoring student responses in science education automatically. BERT-based language models have shown significant superiority over traditional NLP models in various language-related tasks. However, science writing of students, including argumentation and explanation, is domain-specific. In addition, the language used by students is different from the language in journals and Wikipedia, which are training sources of BERT and its existing variants. All these suggest that a domain-specific model pre-trained using science education data may improve model performance. However, the ideal type of data to contextualize pre-trained language model and improve the performance in automatically scoring student written responses remains unclear. Therefore, we employ different data in this study to contextualize both BERT and SciBERT models and compare their performance on automatic scoring of assessment tasks for scientific argumentation. We use three datasets to pre-train the model: 1) journal articles in science education, 2) a large dataset of students' written responses (sample size over 50,000), and 3) a small dataset of students' written responses of scientific argumentation tasks. Our experimental results show that in-domain training corpora constructed from science questions and responses improve language model performance on a wide variety of downstream tasks. Our study confirms the effectiveness of continual pre-training on domain-specific data in the education domain and demonstrates a generalizable strategy for automating science education tasks with high accuracy. We plan to release our data and SciEdBERT models for public use and community engagement.

1 Introduction

Writing is critical in science learning because it is the medium for students to express their thought processes. In classroom settings, educators have engaged students in writing explanations of phenomena, design solutions, arguments, etc. [10,15], with which students develop scientific knowledge and competence. However, it is time-consuming for teachers to review and evaluate natural language

---

Z. Liu and X. He—Co-First Author.

N. Wang et al. (Eds.): AIED 2023, CCIS 1831, pp. 666–674, 2023.
https://doi.org/10.1007/978-3-031-36336-8_103

writing, thus preventing the timely understanding of students' thought processes and academic progress. Recent development in machine learning (ML), especially natural language processing (NLP), has proved to be a promising approach to promoting the use of writing in science teaching and learning [17]. For example, various NLP methods have been employed in science assessment practices that involve constructed responses, essays, simulation, or educational games [14]. In this rapidly developing domain, the state-of-the-art Bidirectional Encoder Representations from Transformers (BERT) model [4], a transformer-based machine learning architecture developed by Google, demonstrates superiority over other machine learning methods in scoring student responses to science assessment tasks [1].

Studies have shown that the performance on NLP tasks can be improved by using domain-specific data to contextualize language models [5]. Several BERT-based language models, such as SciBERT [3], AgriBERT [12], BioBERT [8], and ClinicalRadioBERT [11], have demonstrated significant success on domain-specific tasks. Therefore, it is reasonable to speculate that ML-based scoring of students' scientific writing can be improved if we have a domain-specific language model for scientific education. In this case, we need to find the proper domain-specific data that are directly relevant to student writing. It is important to note that student responses are preliminary expressions of general science knowledge and lack the rigor of academic journal publications. In addition, their writing is also influenced by the developmental progress of writing skills and the length of the required tasks. These characteristics of student writing are challenges for using NLP tools to score students' writing [6,9]. Therefore, to further improve the application of large pre-trained language models to automatically score students' scientific writing, we use different datasets to improve BERT and compare their performance on various downstream tasks. We propose a general framework, SciEdBERT (Fig. 1), where we continually pre-train BERT-based language models on education-specific data to guide foundational models to acquire knowledge and vocabulary relevant to science education. In this work, we make the following contributions:

1. We provide a method to improve model performance on the downstream tasks by contextualizing BERT with the downstream context in advance.
2. We prove the effectiveness of domain-specific data in improving BERT-based model performance.
3. We will release our language models, which can be further tested and used in other science education tasks.

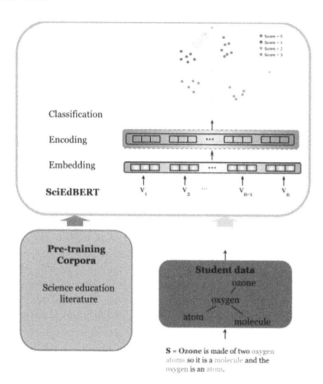

Fig. 1. The SciEdBERT framework. A student response instance is classified based on the latent representation of word vectors.

2 Methodology

2.1 Architecture/Background

The BERT (Bidirectional Encoder Representations from Transformers) language model [4] is based on the transformer architecture [13]. It is trained using the masked language modeling (MLM) objective, which requires the model to predict missing words in a sentence given the context. This training process is called pre-training. The pre-training of BERT is unsupervised and only requires unlabeled text data. During pre-training, word embedding vectors are multiplied with three sets of weights (query, key and value) to obtain three matrices $\mathbf{Q}$, $\mathbf{K}$, and $\mathbf{V}$, respectively. These matrices are then used to calculate attention scores, which are weights that measure the importance among input words. For example, in the example "I love my cats.", the word "I" should (ideally) be strongly associated with the word "my", since they refer to the same subject (Fig. 2).

Fig. 2. An example of BERT's attention mechanism

For each word, the attention scores are then used to weigh intermediate outputs that sum up to the final vector representation of this word.

$$Attention(Q, K, V) = softmax(\frac{QK}{\sqrt{d_k}})V \tag{1}$$

where d_k refers to the dimension of the **K** matrix.

BERT takes a sequence of words as the input, and outputs a latent representation of input tokens in the form of word vectors. This latent representation, or embedding, captures the semantics, positional information, and contextual information of the input sentence. It can be further used for downstream NLP tasks. To use BERT for practical natural language understanding applications, it is necessary to fine-tune the model on the target task. BERT can be fine-tuned on a wide variety of tasks, such as topic classification and question answering, by adding task-specific layers on top of this pre-trained transformer. Fine-tuning is a supervised learning process. During this process, BERT is trained on a labeled dataset and the parameters of the model are updated in training to minimize the task-specific loss function.

2.2 Domain-Specific Training

BERT is a fundamental building block for language models. In practice, it has many variants that are tailored to the purposes and peculiarities of specific domains [2,3,8,11,12]. For example, BioBERT [8] is a large language model trained on biomedical publications (PubMed) and delivers superior performance on biomedical and chemical named entity recognition (NER), since it has a large and contextualized vocabulary of biomedical and chemical terms.

Substantial evidence indicates that language models perform better when then target and source domains are aligned [5,8]. In other words, continual pre-training BERT-based models with in-domain corpora could significantly improve their performance on downstream tasks [5]. In addition, there is much correlation between model performance and the extent of in-domain training. Specifically, training with more relevant in-domain text and training-from-scratch can further improve pre-trained language models [5].

In this work, we incorporate prior experience in NLP, specifically that of domain-specific training, to train our SciEdBERT models designed specifically for science education tasks.

2.3 Training Design

We follow a pyramid-shaped training scheme to maximize our models' utilization of domain-relevant data.

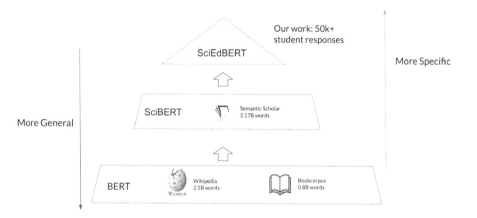

Fig. 3. The pyramid training scheme

In Fig. 3, we can see that SciBERT [3] is a science-oriented version of BERT developed through in-domain pre-training. As shown in Table 2, some of the models we developed for this experiment in this study are further extensions of SciBERT through continual pre-training on various science education data.

The primary benefit of following the pyramid training scheme is to avoid diluting the relatively scarce in-domain data with the vastly more abundant general text data. If instead a language model is trained on a combined corpus of general text and domain-specific data, the effects of in-domain training will be insignificant.

3 Experiment

3.1 Dataset

We employ several datasets to train the language models, including the Academic Journal Dataset for Science Education (SciEdJ), a large dataset of students' written Responses (SR1), and a small dataset of students' responses to four argumentation tasks (SR2). Then, we use seven tasks from the large dataset (7T) and the four argumentation tasks (4T) as two datasets to fine-tune the trained language model. Below we briefly introduce these datasets.

Training Dataset. We use three datasets to train the language model. The SciEdJ is a collection of 2,000 journal articles from journals in science education. We select ten journals in science education with the highest impact factors according to Web of Science, including *Journal of Research in Science Teaching, International Journal of Science Education, Science Education*, etc. We only include the most recent and the most relevant publications in the dataset. The SR1 dataset is a collection of over 50,000 middle school student short responses to 49 constructed responses of NGSA assessment tasks in the United States.[1] Students are anonymous to researchers and not traceable. The SR2 dataset is a collection of 2,940 student responses from a set of argumentation assessment tasks [7].

---

[1] http://nextgenscienceassessment.org.

Fine-Tuning Dataset. We employ two datasets to evaluate the model performance. The 7T dataset includes seven tasks selected from the SR1 dataset, including short-constructed student responses and human expert-assigned labels. Overall, the 7T dataset includes 5,874 labeled student responses (915 for task H4-2, 915 for task H4-3, 834 for task H5-2, 883 for task J2-2, 743 for task J6-2, 739 for tasks J6-3, and 845 for task R1-2). The 4T dataset includes 2940 student responses and their labels from SR2 dataset (e.g., 770 for item G4, 642 for item G6, 765 for item S2, and 763 for item S3). All the samples in the two datasets are written responses from middle school students to explain science phenomena. Trained experts are employed to assign scores to student responses according to scoring rubrics developed by science education researchers, and the inter-rater reliability is at or above satisfactory level (details see [15, 16]).

3.2   Baselines

Our study aims to examine how the context of training data matters to pre-trained models' (e.g., BERT) performance and explore strategies to further improve model performance. To achieve this goal, we employ various datasets to train and fine-tune the models. First, we use the original BERT as the pre-trained model and 7T as the downstream task. This is the baseline model. We then train a BERT model from SR1 and use 7T as the downstream task. Given that the 7T is grounded in the context of SR1, a comparison between the two fine-tuned models (based on BERT vs. SR1-BERT) can address our goals.

Second, we repeat this training and fine-tuning process using BERT with SR2 and 4T datasets. To examine the generalization of the findings, we also employ 4T as the downstream task in other pre-trained models, including SciBERT [3], a BERT model trained on SciEdJ (i.e., SciEJ-BERT), a SciBERT model trained on SciEdJ (i.e., SciEdJ-SciBERT), a BERT model trained on SR2 (i.e., SR2-BERT), and a SciBERT model trained on SR2 (i.e., SR2-SciBERT), with increasingly closer contextualization between the pre-trained models and the downstream tasks.

3.3   Results

As Table 1 presents, the average accuracy of SR1-BERT (0.912) is slightly higher than the accuracy of BERT (0.904). Among the seven tasks, SR1-BERT achieves higher accuracy than BERT on four tasks and are on par with BERT on the remaining three tasks. This indicates that the accuracy of automatic scoring can be improved to a certain extent by training the model with in-domain training data.

This indication is clearer in our second experiment with the 4T dataset. As Table 2 presents, overall, SR2-SciBERT has the highest average accuracy (0.866), which indicates training the model with the contexts of the downstream tasks can improve the accuracy of automatic scoring.

Table 1. Comparing different model performance on 7T task

Item	Accuracy	
	BERT	SR1-BERT
H4-2	0.913	0.929
H4-3	0.831	0.831
H5-2	0.958	0.970
J2-2	0.920	0.926
J6-2	0.959	0.973
J6-3	0.845	0.845
R1-2	0.864	0.864
Average	0.904	0.912

The model with the second highest accuracy (0.852) is SR2-BERT. SR2-BERT has the same performance as SR2-SciBERT on S3 and even higher accuracy (0.821) than SR2-SciBERT (0.815) on G4. On S2, SR2-BERT's performance (0.915) is only second to SR2-BERT. Only on G6, SR2-BERT has a lower accuracy (0.719) than comparison models. Therefore, although the two models share the same average accuracy, based on the accuracy results on each individual task, SR2-BERT performs better than BERT. This is also in line with our previous findings that context matters in improving model performance.

SciEdJ-SciBERT and SciEdJ-BERT have the lowest average accuracy scores (0.842) among the models. Only on G4 do these two models perform better than BERT. This indicates that the context of science education publications cannot help BERT learn the language of student responses better. In fact, on the contrary, such context may introduce confusion to the machine learning process.

In summary, SR2-SciBERT and SR2-BERT achieve the best results among the models, which indicates that contextualizing the language models with the same language of the downstream tasks can improve the model's performance.

Table 2. Comparing model performance on the 4T tasks

Item	Accuracy				
	BERT	SciEdJ-BERT	SciEdJ-SciBERT	SR2-BERT	SR2-SciBERT
G4	0.792	0.804	0.815	0.821	0.815
G6	0.766	0.727	0.742	0.719	0.766
S2	0.895	0.882	0.889	0.915	0.928
S3	0.934	0.954	0.921	0.954	0.954
Average	0.847	0.842	0.842	0.852	0.866

4   Conclusions

This study investigates training language models with different contextual data and compares their performance on eleven constructed response tasks. The results indicate that using the in-domain data directly related to downstream tasks to contextualize the language model can improve a pre-trained language model's performance. In automatic scoring of students' constructed responses, this means continual pre-training the language model on student responses and then fine-tuning the model with the scoring tasks. In science education, using SciEdBERT can further improve model performance as SciEdBERT is well-versed in scientific vocabulary. Our study confirms the effectiveness of using domain-specific data to pre-train models to improve their performance on downstream tasks and validate a strategy to adapt language models to science education.

Acknowledgement. The study was funded by National Science Foundation(NSF) (Award # 2101104, 2138854, PI: Zhai). Any opinions, findings, conclusions, or recommendations expressed in this material are those of the author(s) and do not necessarily reflect the views of the NSF.

References

1. Amerman, H., Zhai, X., Latif, E., He, P., Krajcik, J.: Does transformer deep learning yield more accurate sores on student written explanations than traditional machine learning? In: Paper submitted to the Annual Meeting of the American Educational Research Association, Chicago (2023)
2. Araci, D.: FinBERT: financial sentiment analysis with pre-trained language models. arXiv preprint arXiv:1908.10063 (2019)
3. Beltagy, I., Lo, K., Cohan, A.: SciBERT: a pretrained language model for scientific text. arXiv preprint arXiv:1903.10676 (2019)
4. Devlin, J., Chang, M.W., Lee, K., Toutanova, K.: BERT: pre-training of deep bidirectional transformers for language understanding. arXiv preprint arXiv:1810.04805 (2018)
5. Gu, Y., et al.: Domain-specific language model pretraining for biomedical natural language processing. ACM Trans. Comput. Healthcare (HEALTH) 3(1), 1–23 (2021)
6. Ha, M., Nehm, R.H.: The impact of misspelled words on automated computer scoring: a case study of scientific explanations. J. Sci. Educ. Technol. 25(3), 358–374 (2016)
7. Haudek, K.C., Zhai, X.: Exploring the effect of assessment construct complexity on machine learning scoring of argumentation (2021)
8. Lee, J., et al.: BioBERT: a pre-trained biomedical language representation model for biomedical text mining. Bioinformatics 36(4), 1234–1240 (2020)
9. Litman, D.: Natural language processing for enhancing teaching and learning. In: Thirtieth AAAI Conference on Artificial Intelligence (2016)
10. Novak, A.M., McNeill, K.L., Krajcik, J.S.: Helping students write scientific explanations. Sci. Scope 33(1), 54 (2009)

11. Rezayi, S., et al.: ClinicalRadioBERT: knowledge-infused few shot learning for clinical notes named entity recognition. In: Lian, C., Cao, X., Rekik, I., Xu, X., Cui, Z. (eds.) International Workshop on Machine Learning in Medical Imaging, MLMI 2022. LNCS, vol. 13583, pp. 269–278. Springer, Cham (2022). https://doi.org/10.1007/978-3-031-21014-3_28

12. Rezayi, S., et al.: AgriBERT: knowledge-infused agricultural language models for matching food and nutrition. In: IJCAI. IJCAI (2022)

13. Vaswani, A., et al.: Attention is all you need. In: Advances in Neural Information Processing Systems, vol. 30 (2017)

14. Zhai, X., Haudek, K.C., Shi, L., Nehm, R.H., Urban-Lurain, M.: From substitution to redefinition: a framework of machine learning-based science assessment. J. Res. Sci. Teach. **57**(9), 1430–1459 (2020)

15. Zhai, X., Haudek, K.C., Ma, W.: Assessing argumentation using machine learning and cognitive diagnostic modeling. Res. Sci. Educ. 1–20 (2022)

16. Zhai, X., He, P., Krajcik, J.: Applying machine learning to automatically assess scientific models. J. Res. Sci. Teach. **59**(10), 1765–1794 (2022)

17. Zhai, X., Yin, Y., Pellegrino, J.W., Haudek, K.C., Shi, L.: Applying machine learning in science assessment: a systematic review. Stud. Sci. Educ. **56**(1), 111–151 (2020)

Identifying Usability Challenges in AI-Based Essay Grading Tools

Erin Hall[(⊠)], Mohammed Seyam[(⊠)], and Daniel Dunlap[(⊠)]

Virginia Polytechnic Institute and State University, Blacksburg, VA 24061, USA
{erinehall,seyam,dunlapd}@vt.edu

Abstract. Automated Essay Scoring (AES) efforts have recently made it possible for platforms to provide real-time feedback and grades for student essays. With the growing importance of addressing usability issues that arise from integrating artificial intelligence (AI) into educational-based platforms, there have been significant efforts to improve the visual elements of User Interfaces (UI) for these types of platforms. However, little research has been done on how AI explainability and algorithm transparency affect the usability of AES platforms. To address this gap, a qualitative study was conducted using an AI-driven essay writing and grading platform. The study involved participants of students and instructors, and utilized surveys, semi-structured interviews, and a focus group to collect data on users' experiences and perspectives. Results show that user understanding of the system, quality of feedback, error handling, and creating trust are the main usability concerns related to explainability and transparency. Understanding these challenges can help guide the development of effective grading tools that prioritize explainability and transparency, ultimately improving their usability.

Keywords: Usability · Algorithmic Transparency · Explainability · Artificial Intelligence · Writing · Feedback

1 Introduction and Related Works

As more platforms integrate AI into their functionality, there is an additional layer of usability that needs to be addressed, specific to how end-users interact with complex systems of this nature [3]. While a large part of explainable AI (XAI) efforts focus on making AI algorithms more explainable and transparent to its developers, there has been an increasing emphasis on exploring how AI explainability and algorithm transparency affect the usability of AI-driven platforms for non-technical users [3].

In the educational sector, Automated Essay Scoring (AES) techniques have made it possible to utilize AI to provide feedback to students, thus promoting equitable education and creating a level playing field by ensuring all students conform to a set of foundational writing conventions [9]. Similarly, AES can play a role in the essay grading process, however, the practice of using AI for

© The Author(s), under exclusive license to Springer Nature Switzerland AG 2023
N. Wang et al. (Eds.): AIED 2023, CCIS 1831, pp. 675–680, 2023.
https://doi.org/10.1007/978-3-031-36336-8_104

summative assessment is not an approach that is widely adopted, due to the fact that no existing AI algorithms can grade the content of writing with 100% accuracy and essay evaluation is largely subjective [9]. Thus, most tools of this nature use a human-machine teaming approach where the system serves as a form of guidance to graders [6]. This paper is concerned with tools that adopt such an approach. That being said, there are concerns in using AI in essay grading, including lack of AI explainability due to the use of black box models and the potential introduction of undetected bias [7].

Due to variations across different bodies of research, it is necessary to provide clear definitions for a number of terms used in this paper.

Explainable AI (XAI) is an emerging research area that includes techniques that provide a level of transparency to machine learning models [4]. These techniques attempt to explain how the output of a model was generated based on the input provided. Attempts to open black box models are included in this research area [4].

AI literacy is defined to be "a set of competencies that enables individuals to critically evaluate AI technologies; communicate and collaborate effectively with AI; and use AI as a tool online, at home, and in the workplace" [5]. It involves a general understanding of how artificial intelligence works and what an algorithm is doing, which helps bridge the gap between the mathematical principles that drive an AI system and the user's mental model of how the system actually works [5].

Algorithmic transparency is a practice that can increase users' AI literacy by providing feedback that indicates how the algorithm is producing its output [8]. Algorithmic transparency can "empower users to make informed choices" and "judge [the algorithm's] potential consequences" [8].

This research aims to explore how graders interact with AI-grading assistants. This will be investigated by collecting data on their experiences with using a tool of this nature and providing a synthesis of how explainability and transparency relate to usability. The following research questions guide this study:

1. How does explainability and transparency relate to usability?
2. How do explainability and transparency efforts affect how graders use a system?

2   Methodology

To address the above research questions, a review of relevant literature was conducted to synthesize the relationship between explainability, algorithm transparency, and usability. Following this, a qualitative study was conducted on an AI essay writing and grading platform. The essay grading tool has two main goals: to provide students with feedback before they turn in their essay to promote a higher quality floor, and to automate more tedious aspects of essay grading to leave graders with more time to focus on evaluating the content and ideas presented in an essay.

The study employed three surveys, individual interviews, and a focus group to collect data on the experiences and perspectives of the users of the tool. The participants were told that each survey would take 10–15 min to complete and the interviews and focus groups would last no longer than one hour. All interviews were conducted virtually over Zoom and were recorded and transcribed using Zoom's audio transcription feature.

While the essay grading tool is targeted to both students and graders and offers unique features for both, this paper is primarily concerned with its role as a grading assistant. However, understanding the perspective of students is necessary to understanding how instructors use the platform effectively, thus, both graders and students were recruited as participants in this study. The participants were recruited to provide a broad range of perspectives across different roles within the academic context. Due to limited access to the essay grading tool at the institution, the study included nine participants, consisting of five teaching assistants of a CS course, one instructor of a non-CS course, and three CS students who had used the tool that semester. While all 9 participants were individually interviewed, only the 6 graders participated in the surveys, and one instructor, one TA, and one student participated in the focus group. All participants consented to take part in the study, no compensation was provided for their participation, and efforts were taken to keep their comments anonymous.

The first survey involved gathering background data on the participants and their relevant experiences. The second survey aimed to collect the participants' initial impressions of using the tool. The third survey, interviews, and focus group all aimed to obtain a more in depth understanding of the participants' experiences using the tool. The interviews and focus group were semi-structured, using a loose guide of questions to cover certain topics with follow-up questions asked as needed. The focus group was designed to encourage open discussion between participants. The surveys, interviews, and focus group covered topics such as overall platform experience, ease of finding automated grade explanations, understanding of automated grade explanations, understanding of the algorithm for calculating different rubric grades, and trust in the system.

The audio recordings and transcripts of the individual interviews and focus group were evaluated using reflexive thematic analysis [1, 2]. The transcripts were first thoroughly studied to gain familiarity with the data. Then, initial codes were generated for each piece of relevant data in the transcripts. Codes were iteratively reviewed, condensed, reworded, and refined. Finally, each of these codes were organized into broader themes.

3   Results

Following the completion of the study, 224 initial codes were extracted from the survey responses, interview, and focus group transcripts. These codes were then analyzed for similarities and condensed into a smaller list. 10 broader themes emerged that captured the main findings of the data. Each coded piece of data was organized into one of these 10 themes. To provide a more condensed overview, these themes were grouped into four categories as described below:

Understanding of System: this category includes concerns related to poor understanding of how the system works (e.g. having a perception of how the algorithm works that does not align with what it is actually doing), understanding how the system works through assumptions (e.g. assumptions that are inaccurate and thus lead to an incorrect mental model of how the system works), inconsistent mental models of the system across different types of users (e.g. students think grades are calculated one way and graders think a different way), and changes to their grading workflow that are not ideal (e.g. spending more time assessing the validity of automated grades)

Quality of Feedback: this category includes concerns that explanations are not detailed enough (e.g. graders are unable to expand on feedback when questioned by students) and difficulties finding explanations

Error Handling: this category includes concerns that the algorithm has errors (e.g. explanations are presented as facts that are actually inaccurate) and that the system grades in a way that does not align with the goals of the course instructor (e.g. technical writers are not always as concerned with the flow of an essay)

Creating Trust: this category includes concerns that users do not trust the system (e.g. they grade everything manually due to lack of trust) and that there are inconsistencies in grade overriding across different graders (e.g. differences in trust lead to differences in overriding grades)

4   Discussion

The study revealed several explainability and transparency issues that present usability concerns in AI-based writing and grading systems. Participants reported a steep learning curve, and some had misconceptions about how the algorithm worked, even after using the tool for a whole semester. Several participants mentioned they were not sure what students could see, but assumed they could see the number of automated points before turning in their papers, which was not the case. Lack of algorithmic transparency in several areas caused users to make assumptions like this, causing those participants to be more hesitant to override grades over fairness concerns. On top of this, when different information was presented to different users, they had varying mental models of the system which resulted in misunderstandings and disconnects. For example, students described getting specific feedback on what was wrong with their essays, while graders were only able to see general explanations that were not student-specific. Having a poor understanding of the system made several participants feel the need to double check the system's work, which took just as much time as manual grading. These insights all highlight the need to consider the user's understanding of the algorithm and to take efforts to bridge the gap between the user's mental model of the algorithm and the tool's conceptual model.

The quality of feedback and explanations also contributed to how well users understood the system. Several participants described the tool's feedback as being vague and lacking detail. Most participants mentioned wanting more information on how grades were calculated to help them better understand the system, make more informed decisions on when to override grades, and to be able to better expand on this feedback when questioned by students. Additionally, a few participants mentioned having difficulties finding explanations for certain parts of the rubric, which left them with an incomplete understanding of students' automated grades. Greater transparency and higher quality explanations can help build trust between the users and the system, and improve grading and feedback quality.

Effective error handling is key to helping build user trust in a system. Knowing where the algorithm performs well and poorly can enable users to make more informed decisions about when to override or put more effort into verifying automated grades. Bugs in the algorithm, such as flagging the "can" in "American" as a hedge word, led several participants to distrust the algorithm and feel the need to spend more time double checking its work. The algorithm's influence on students' writing style was another common concern that several participants would have liked the system to address in some capacity. While perfect accuracy is difficult to achieve, addressing the system's accuracy to its users and indicating which feedback is factual or suggestive can help build trust between users and the system.

Finally, user trust in a system is a key component to understanding how they use these types of tools. Participants who reported having low trust in the system often questioned automated grades more and spent just as much time verifying their validity than they would have grading entirely manually. This detracted from the overall goal of the system, which is to help graders spend less time on more tedious parts of the grading process. On top of this, the participants had varying levels of trust in the system which resulted in drastic differences in automated grade overriding habits, with some saying they barely overrode while others overrode every paper. While some users may come in with preconceptions and an inherent distrust of AI, improving users' understanding of a system by providing high quality explanations and acknowledgement of the system's shortcomings can help build trust between the users and the system, allowing them to spend less time on evaluating the shortcomings of the system and more time on the core task of evaluating students' ideas and content quality.

Although black boxed models can make AI explainability difficult, adopting an AI-based grading tool requires a certain level of explainability to provide sufficient explanations to graders and feedback to students. To minimize usability challenges brought on by the use of AI, utilizing explainable AI to improve algorithm transparency can help users interact with a system more effectively.

5   Conclusion and Future Works

In conclusion, this paper has identified a set of usability challenges related to AI explainability and algorithm transparency that are commonly found in AI-driven essay grading and feedback platforms. These usability concerns are important to consider in order to improve the usability of AI-based essay writing and grading tools, and include user understanding of the system, the quality of feedback provided to users, how errors are handled, and how much users trust the system.

Moving forward, there are several avenues for future research in this area. Further analysis will be applied to this data to evaluate the overall usability of the tool. Additionally, it would be valuable to collect data from more users in non-technical fields to determine how these findings apply to different contexts. Such research could inform the development of more effective and equitable educational technology solutions in the future.

References

1. Braun, V., Clarke, V.: Using thematic analysis in psychology. Qual. Res. Psychol. **3**, 77–101 (2006). https://doi.org/10.1191/1478088706qp063oa
2. Braun, V., Clarke, V.: Reflecting on reflexive thematic analysis. Qual. Res. Sport Exerc. Health **11**(4), 589–597 (2019)
3. Haque, A.K.M.B., Islam, A.K.M.N., Mikalef, P.: Explainable artificial intelligence (XAI) from a user perspective: a synthesis of prior literature and problematizing avenues for future research. Technol. Forecasting Soc. Change **186**, 122120 (2023)
4. Kumar, V., Boulanger, D.: Explainable automated essay scoring: deep learning really has pedagogical value. Front. Educ. **5** (2020). https://doi.org/10.3389/feduc.2020.572367, https://www.frontiersin.org/article/10.3389/feduc.2020.572367
5. Long, D., Magerko, B.: What is AI literacy? Competencies and design considerations, pp. 1–16. Association for Computing Machinery (2020). https://doi.org/10.1145/3313831.3376727
6. Lyons, J.B., Wynne, K.T., Mahoney, S., Roebke, M.A.: Chapter 6 - Trust and Human-Machine Teaming: A Qualitative Study, pp. 101–116. Academic Press (2019). https://doi.org/10.1016/B978-0-12-817636-8.00006-5, https://www.sciencedirect.com/science/article/pii/B9780128176368000065
7. Petch, J., Di, S., Nelson, W.: Opening the black box: the promise and limitations of explainable machine learning in cardiology. Can. J. Cardiol. **38**(2), 204–213 (2022)
8. Rader, E., Cotter, K., Cho, J.: Explanations as mechanisms for supporting algorithmic transparency (2018)
9. Semire, D.: An overview of automated scoring of essays. J. Technol. Learn. Assess. **5**(1) (2006). https://ejournals.bc.edu/index.php/jtla/article/view/1640

Enhancing Engagement Modeling in Game-Based Learning Environments with Student-Agent Discourse Analysis

Alex Goslen[1]([✉]), Nathan Henderson[1], Jonathan Rowe[1], Jiayi Zhang[4], Stephen Hutt[2], Jaclyn Ocumpaugh[4], Eric Wiebe[1], Kristy Elizabeth Boyer[3], Bradford Mott[1], and James Lester[1]

[1] North Carolina State University, Raleigh, NC, USA
amgoslen@ncsu.edu
[2] University of Denver, Denver, CO, USA
[3] University of Florida, Gainesville, FL, USA
[4] University of Pennsylvania, Philadelphia, PA, USA

Abstract. Pedagogical agents offer significant promise for engaging students in learning. In this paper, we investigate students' conversational interactions with a pedagogical agent in a game-based learning environment for middle school science education. We utilize word embeddings of student-agent conversations along with features distilled from students' in-game actions to induce predictive models of student engagement. An evaluation of the models' accuracy and early prediction performance indicates that features derived from students' conversations with the pedagogical agent yield the highest accuracy for predicting student engagement. Results also show that combining student problem-solving features and conversation features yields higher performance than a problem solving-only feature set. Overall, the findings suggest that student-agent conversations can greatly enhance student models for game-based learning environments.

Keywords: Student engagement · Game-based learning · Discourse analysis

1 Introduction

Student engagement plays a central role in effective learning across a wide range of educational settings [7]. Students who disengage often develop a superficial understanding of the material [5]. Pedagogical agents have shown potential in enhancing student engagement through discursive interaction with students [8] and improving student learning [18]. Additionally, positive interactions with pedagogical agents have been shown to help learners feel more engaged [1].

In game-based learning environments, understanding student engagement is multi-faceted and may involve measuring students' degree of attention to stimuli in the game (involvement) and their affective response to those stimuli (situational interest) [4]. Pedagogical agents offer the potential to provide such insights for measuring and modeling

© The Author(s), under exclusive license to Springer Nature Switzerland AG 2023
N. Wang et al. (Eds.): AIED 2023, CCIS 1831, pp. 681–687, 2023.
https://doi.org/10.1007/978-3-031-36336-8_105

student engagement [9]. Although prior work has investigated how dialogue with conversational pedagogical agents impacts learners [17], little is known about the relationship between students' dialogue and their overall engagement.

This paper analyzes student discourse with a conversational pedagogical agent in a game-based learning environment. Specifically, we examine six machine learning techniques using neural word embeddings of student-agent discourse and in-game problem solving behavior as input features to predict student engagement. We evaluate the models in terms of accuracy and early prediction performance, and we examine implications of the results for the design of game-based learning environments.

2 Related Work

There is growing literature on utilizing discourse for learning analytics and predicting student engagement in learning environments. Modeling student engagement is an important step in developing adaptive learning environments that can mitigate issues like disengagement or mind-wandering [2]. Emerson et al. [4] investigated features related to student interactions with non-player characters (NPC) in a game-based learning environment and observed that students who interacted more frequently with NPCs showed lower overall interest in the game-based learning environment. This suggests that using students' problem-solving actions and conversational behavior to predict student engagement may help inform adaptive responses for pedagogical agents.

Advances in natural language processing allow for improvements in analyzing unstructured text in dialogue-based learning environments [10]. Analysis of such has included manual annotation and bag-of-words methods to derive meaning from students' text-based utterances [11] and demonstrated the ability of pre-trained neural embeddings to detect student engagement in reflection tasks [6]. More recently, BERT [3] has proven useful for modeling different types of conversational strategies [16].

3 Game-Based Learning Environment

CRYSTAL ISLAND is a game-based learning environment designed to support middle school students learning microbiology through the narrative of an illness outbreak on a remote island research station. A text-based conversational pedagogical agent, Alisha, was integrated into the game [15]. Alisha provides students with an opportunity to share updates about their problem-solving progress and receive pedagogical support by prompting in-game actions like exploring the island or talking to an NPC. Alisha's dialogue was controlled by a finite state machine based on students' dialogue acts [13].

4 Method

4.1 Study Procedure

We use data from 77 middle school students who played CRYSTAL ISLAND over a 3-day study at a public, urban middle school in North Carolina. These students completed both pre- and post-game surveys (32 male, 38 female, and 7 who did not report gender).

Following gameplay, students were asked to complete three surveys that were used as a proxy for engagement-related constructs. These were drawn from the original version of the User Engagement Scale (UES) [14] and a revised version (UESz) [19].

4.2 Feature Generation for Predictive Models of Student Engagement

This study utilizes three subscales most relevant to understanding engagement in game-based learning environments: the novelty subscale (NO) and felt involvement (FI) subscale from the original UES and the focused attention (FAz) subscale from the UESz [20]. Novelty (NO) captures a basic measure of situational interest. Focused attention (FAz) captures a retrospective rating of flow-like experience. Felt involvement (FI) can be described as the enjoyment at the intersection of the two previous engagement constructs. We binarized this data using a median split to indicate low and high levels of FAz ($med = 24$, $SD = 7.36$), FI ($med = 11$, $SD = 3.05$), and NO ($med = 10$, $SD = 3.01$).

Student-Agent Conversation Features. There was a total of 2,634 chat messages sent throughout gameplay, with 1,523 messages originating from Alisha and 1,111 messages originating from students. On average, students sent 14.4 messages ($SD = 15.8$) and Alisha sent 14.5 messages ($SD = 14.7$) across the entirety of gameplay. To generate representations of the students' utterances for use by the pre-trained BERT model [3], tokens were generated for the separate words for each utterance, with stop words being retained due to their contextual relevance. Utterance-level feature vectors were produced by averaging across each token sequence. We constructed both sequential and non-sequential embedding input representations. Sequential models utilized a two-dimensional vector representation, where each row represented a single utterance and each column represented a conversation feature. By summing across sequences, a one-dimensional feature representation was generated for the non-sequential models.

In-Game Problem-Solving Features. To create predictive models of student engagement, we derived problem-solving features from students' gameplay interactions with CRYSTAL ISLAND captured by trace logs: action type, action argument, and location [12]. Action type features represent students' actions during gameplay, action arguments provide details about students' problem-solving actions, and location features denote where the gameplay action took place. We extracted these features from students' gameplay data and converted each action into a binary one-hot vector. Sequential and non-sequential features were generated similarly to the conversation features.

Combined In-Game Problem-Solving and Student-Agent Conversation Features. Finally, we combined in-game features and student-agent conversation features into a single combined representation via sequence-level concatenation.

4.3 Models and Evaluation

The six supervised machine learning techniques used were support vector machines (SVMs), random forest (RF), logistic regression (LR), Naïve Bayes (NB), multilayer perceptron (MLP), and long short-term memory networks (LSTMs). LSTM models utilized sequential input representations and the other models utilized non-sequential input representations. We used student-level cross-validation splits to eliminate data leakage. Given variability in students' total gameplay time ($M = 75.41$ min, $SD = 29.41$ min), and a limited number of conversations per student, we split the data cumulatively using five-minute increments and stopped early prediction models after 10 increments.

The models were evaluated across gameplay time intervals in terms of accuracy as well as two early prediction metrics: convergence rate and standardized convergence point. Convergence rate measures the percentage of early predictions that have an accurate final prediction value. Standardized convergence point measures the point at which a model only makes correct predictions from there [19]. Higher convergence rates and lower convergence points correspond to improved predictive performance.

5   Results

Table 1 shows all results from the three representations. All six machine learning models yielded engagement models that improved on a majority-class baseline in terms of accuracy for FAz using conversation-only features. SVM, NB and LR achieved the highest predictive performance in terms of convergence rate and standardized convergence point for FAz, with MLP and LSTM performing well for FI. LSTMs had the lowest convergence rate and highest convergence point of all methods tested, likely due to the small size of the student-agent conversation dataset. Most models failed to converge for NO, implying that novelty might be more difficult to predict in this case.

Results for early prediction models constructed with in-game problem-solving features showed lower accuracy for focused attention and felt involvement and higher accuracy for novelty. For all three engagement measures, most models performed worse than the majority baseline. Overall, models constructed with only in-game problem-solving features performed worse than models using only student-agent conversation features in terms of convergence rate and convergence point with the exclusion of LSTM. These results show promise for using information about student-agent conversational behavior for predicting engagement in game-based learning environments.

All models follow similar trends to the gameplay-only results shown in part G of the table. Notably, the accuracy values for the combined feature representation models (C + G) were not higher than the conversation-only feature models (C). However, the combined representation yields much better predictive performance for FAz and NO, meaning the combined representation predictions converge at a higher rate and generate more accurate predictions. This finding implies that the combined feature representation might be optimal for real-time predictions. All three evaluation metrics appear to improve from the combined feature representation relative to the in-game problem-solving feature representation for focused attention prediction, which suggests that the addition of student-agent conversation features improved predictive model performance.

Table 1. Model results for all experiments. C represents student-agent conversation input, G represents gameplay only input, and C + G represent the combined input representation.

	Majority	FAz			FI			NO		
		Acc	CR	CP	Acc	CR	CP	Acc	CR	CP
		55.84	N/A	N/A	57.14	N/A	N/A	55.84	N/A	N/A
C	SVM	**63.90**	74.00	87.92	55.90	64.00	**94.14**	52.57	58.00	101.50
	RF	61.72	62.00	94.10	58.55	58.00	101.20	**58.16**	60.00	99.12
	NB	59.95	78.00	**86.96**	48.79	26.00	104.60	49.90	28.00	106.16
	LR	61.21	**84.00**	88.98	58.69	**72.00**	96.94	57.71	**80.00**	**98.96**
	MLP	58.62	72.00	96.04	**62.70**	70.00	94.32	55.81	58.00	101.82
	LSTM	58.59	62.00	95.92	62.16	54.00	99.16	59.50	58.00	96.40
G	SVM	52.69	**74.00**	**94.86**	56.01	66.00	**86.30**	51.91	60.00	93.70
	RF	54.39	72.00	96.74	**58.53**	**70.00**	92.10	**59.53**	58.00	97.54
	NB	**56.64**	56.00	97.94	51.25	44.00	99.78	56.50	42.00	97.30
	LR	52.15	68.00	97.42	47.66	54.00	96.28	53.00	68.00	94.24
	MLP	53.63	62.00	97.20	49.87	58.00	95.34	52.98	56.00	97.38
`	LSTM	53.97	92.00	85.50	55.37	90.00	79.24	53.84	**88.00**	**82.10**
C + G	SVM	53.06	82.00	88.44	53.80	**60.00**	**87.62**	53.58	58.00	94.22
	RF	54.61	80.00	94.92	**57.01**	**60.00**	91.96	**57.51**	66.00	92.86
	NB	**56.83**	**92.00**	**82.64**	51.28	50.00	93.94	55.36	**86.00**	**83.94**
	LR	48.68	66.00	97.46	48.70	46.00	99.42	51.92	56.00	97.08
	MLP	50.05	56.00	98.10	47.09	52.00	97.96	52.57	54.00	96.74
	LSTM	52.75	90.00	83.48	54.29	92.00	90.70	55.66	86.00	81.04

6 Discussion and Conclusion

Student conversations with pedagogical agents in game-based learning environments can inform predictive models of engagement. Analyses of student-agent conversational behavior in a game-based learning environment for science problem-solving reveal that the conversation features performed best in terms of predictive accuracy for focused attention and felt involvement and improved on a majority baseline. We also found that accuracy, convergence rate, and standardized convergence point improved when adding student-agent conversation features to in-game problem-solving features, suggesting student-agent conversation is an important indicator for engagement modeling.

A limitation of the study is that the surveys were collected retrospectively, immediately after students' completion of the game. For a better understanding of student engagement, intermittent engagement reporting data could be introduced throughout gameplay. Understanding changes in engagement over the course of the student's interaction with the game-based learning environment might impact the observed relationship between student discourse patterns and engagement. Additional measurements of engagement could be also explored since novelty and felt involvement are more summative in nature. Analyzing student-agent conversations as a means of predicting student engagement shows significant promise and could contribute to cultivating student interest and engagement during science problem solving in game-based learning environments. Future work should focus on investigating additional text analytic techniques, such as sentiment analysis, to enhance automated analysis of student discourse. Furthermore, mapping trajectories of student engagement over time could prove useful to instructors for guiding pedagogical interventions. Further analysis could also be done to see how patterns of student conversational engagement relate to student learning gains, and real-time tracking of these features could inform adaptive scaffolding for supporting engagement.

Acknowledgements. This research was supported by funding from the National Science Foundation under grants IIS 2016943, IIS 2016993, and IIS 1409639. Any opinions, findings, and conclusions expressed in this material are those of the authors and do not necessarily reflect the views of the NSF.

References

1. Burgoon, J., et al.: Application of expectancy violations theory to communication with and judgments about embodied agents during a decision-making task. Int. J. Hum.-Comput. Stud. **91**, 24–36 (2016)
2. Dermouche, S., Pelachaud, C.: Engagement modeling in dyadic interaction. In: Proceedings of the 2019 International Conference on Multimodal Interaction, pp. 440–445 (2019)
3. Devlin, J., et al.: BERT: pre-training of deep bidirectional transformers for language understanding. North American Association for Computational Linguistics (NAACL) (2018)
4. Emerson, A., et al.: Multimodal learning analytics for game-based learning. Br. J. Educ. Technol. **51**(5), 1505–1526 (2020)
5. Forbes-Riley, K., Litman, D.: Benefits and challenges of real-time uncertainty detection and adaptation in a spoken dialogue computer tutor. Speech Commun. **53**(9–10), 1115–1136 (2011)
6. Geden, M., et al.: Predictive student modeling in game-based learning environments with word embedding representations of reflection. Int. J. Artif. Intell. Educ. **31**(1), 1–23 (2021)
7. Gobert, J., Baker, R., Wixon, M.: Operationalizing and detecting disengagement within online science microworlds. Educ. Psychol. **50**(1), 43–57 (2015)
8. Graesser, A.: Conversations with AutoTutor help students learn. Int. J. Artif. Intell. Educ. **26**(1), 124–132 (2016)
9. Graesser, A., et al.: Advancing the science of collaborative problem solving. Psychol. Sci. Public Interest **19**(2), 59–92 (2018)
10. Hirschberg, J., Manning, C.: Advances in natural language processing. Science **349**(6245), 261–266 (2015)
11. Lin, Z., et al.: MinTL: minimalist transfer learning for task-oriented dialogue systems. arXiv preprint arXiv:2009.12005 (2020)

12. Min, W., et al.: Multimodal goal recognition in open-world digital games. In: 13th Artificial Intelligence and Interactive Digital Entertainment Conference, pp. 80–86 (2017)
13. Min, W., et al.: Predicting dialogue acts for intelligent virtual agents with multimodal student interaction data. International Educational Data Mining Society (2016)
14. O'Brien, H., Toms, E.: The development and evaluation of a survey to measure user engagement. J. Am. Soc. Inf. Sci. Technol. **61**(1), 50–69 (2010)
15. Pezzullo, L.G., et al.: "Thanks Alisha, Keep in Touch": gender effects and engagement with virtual learning companions. In: André, E., Baker, R., Hu, X., Rodrigo, M.M.T., du Boulay, B. (eds.) AIED 2017. LNCS (LNAI), vol. 10331, pp. 299–310. Springer, Cham (2017). https://doi.org/10.1007/978-3-319-61425-0_25
16. Pugh, S., et al.: Do speech-based collaboration analytics generalize across task contexts?. In: 12th International LAK Conference, pp. 208–218 (2022)
17. Sikström, P., et al.: How pedagogical agents communicate with students: a two-phase systematic review. Comput. Educ. **188**, 104564 (2022)
18. Tegos, S., et al.: Conversational agents for academically productive talk: a comparison of directed and undirected agent interventions. Int. J. Comput.-Support. Collab. Learn. **11**(4), 417–440 (2016)
19. Wiebe, E., et al.: Measuring engagement in video game-based environments: investigation of the User Engagement Scale. Comput. Hum. Behav. **32**, 123–132 (2014)
20. Zhang, J., et al.: Investigating student interest and engagement in game-based learning environments. In: Rodrigo, M.M., Matsuda, N., Cristea, A.I., Dimitrova, V. (eds.) AIED 2022. LNCS, vol. 13355, pp. 711–716. Springer, Cham (2022). https://doi.org/10.1007/978-3-031-11644-5

Understanding the Impact of Reinforcement Learning Personalization on Subgroups of Students in Math Tutoring

Allen Nie$^{(\boxtimes)}$ ⓘ, Ann-Katrin Reuel ⓘ, and Emma Brunskill ⓘ

Stanford University, Stanford, CA, USA
{anie,anka,ebrun}@cs.stanford.edu

Abstract. Reinforcement learning has the promise to help reduce the cost of creating effective educational software through automatically adapting the experience to each individual. Most reinforcement learning algorithms aim to learn an automated pedagogical strategy that optimizes performance on average across the population and outputs a decision policy that may rely on complex representations, like deep neural networks, that are largely opaque. Yet, in most educational contexts, we would like a deeper understanding of educational interventions, such as if the machine-learned pedagogical strategy differs in its benefits to different students, how it differentiates instruction across individuals or situations, and if the personalized strategy learned has benefits over alternative personalizations or automated strategies. Here we explore such analyses for a reinforcement learning decision policy for educational software teaching students about the concept of volume. While some related work covers part of these analyses, we suggest that conducting all three such analyses can help enhance our understanding of the impact of a reinforcement learning decision policy in education and help inform stakeholders' decisions around the use of a particular learned decision policy.

Keywords: Reinforcement Learning · Conditional Average Treatment Effect · Offline Policy Evaluation

1 Introduction

Reinforcement learning algorithms can learn adaptive decision policies that map from a context to an intervention in order to optimize an expected outcome. Such algorithms hold great promise for optimizing educational software to best support the learning experience of individual students. However, most reinforcement learning algorithms optimize for outcomes on average, often employ complex, hard-to-interpret models like deep neural networks, and frequently lack

---

A.-K. Reuel—Equal contribution.

© The Author(s), under exclusive license to Springer Nature Switzerland AG 2023
N. Wang et al. (Eds.): AIED 2023, CCIS 1831, pp. 688–694, 2023.
https://doi.org/10.1007/978-3-031-36336-8_106

formal guarantees of convergence to a globally optimal solution. Therefore, an important area of inquiry is to better understand the impact of a particular reinforcement learned pedagogical policy on different students.

While some studies have examined the outcomes and implications of reinforcement learning policies in education, the amount of research conducted in this area remains limited. Prior work on college students using a logic tutoring system suggested that some students may be relatively insensitive to different automated pedagogical policies, but some other students benefited significantly from the RL policy personalization [3]. Concurrent to our paper, Abdelshiheed et al. [1] found that an adaptive deep RL policy yielded substantial gains for students who initially were unlikely to try new metacognitive strategies for a logic tutor, but seemed to have little effect for students who already employed such strategies. In the context of a machine learning method for video recommendations for algebra learning, Leite et al. [4] used causal decision trees to identify if subgroups of students significantly varied in their treatment effects. To our knowledge, a more holistic set of analyses to understand the personalization done by a learned reinforcement learning policy, and its impact on student subgroups, has not been proposed.

To gain insight into the impact and effects of a personalization policy (highlighted in Fig. 1), we suggest three useful analyses:

(R1) **Subgroup Identification**: Identifying subgroups of students according to their differential treatment effect under the RL adaptive policy vs a standard control.

(R2) **Analysis of Personalization**: Employing insights from the model interpretation to analyze the difference in RL automated tutoring strategies for different subgroups of students.

(R3) **Impact of the Specific RL Personalization**: Constructing alternative policies and using offline policy evaluation methods to estimate the impact of the specific RL policy personalization on subgroups of students.

We present a case study that uses these analyses to advance our understanding. To do so, we use data from a study in which RL was used to personalize an educational tool to help elementary school students learn about the concept of volume.

2   Study Description

In the study on student learning of volume concepts, some students used a narrative-based, artificial intelligent educational software tool. While these students work through a series of volume-related problems, the software can provide adaptive support in response to student questions. The reinforcement learning policy selects among four pedagogical strategies: providing direct hints, encouragement, Socratic questioning or prompting the student to reframe, or a simple acknowledgment.

In this study, 270 participants in grades 3–5 were recruited from across the United States. Children were randomly assigned to one of two conditions: a standard interface that provided students with a volume practice task without hints or a narrative and a condition with a storyline and a reinforcement-learning augmented agent. 67 children used the control system, and 203 children used the system with RL agent-mediated guidance. Gender and grade were balanced between the two conditions.

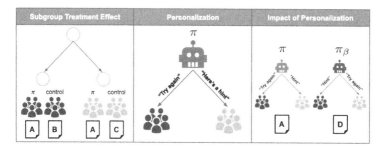

Fig. 1. Three useful Analyses of RL Personalization: (left) Understanding differential treatment effects of the learned RL policy, vs a standard benchmark approach. (middle) Features that describe differences in the learned decision policy. Blue and yellow icons represent subgroups of students with different covariates (for example, low and high pre-test scores). (right) Exploring if different personalization policies yield significantly more or less effective teaching. (Color figure online)

3   Analysis and Results

3.1   Student Subgroup Identification

We first examine if there are significant differences among subgroups in terms of the impact of reinforcement learning vs. a control condition. Similar to Leite et al. [4], we employ a subgroup treatment effect analysis using a two-stage cluster-robust causal forest [10] to estimate the individual treatment effect and identify subgroups. A causal forest is an ensemble of causal trees that have been grown on a random subsample of the data during training to predict individual treatment effect [2]. Causal forests are more robust to nuances of the data-splitting procedure, but are a bit more limited for identifying consistent subgroups, which may vary substantially across trees in the forest. Leite et al. [4] addressed this by using a best linear analysis. However we are particularly interested in the non-linear but interpretable benefits of decision trees. Therefore we proposed an alternate heuristic method to identify the student subgroups and estimate the subgroup-level treatment effect, which preserves a tree representation flexibility. For our data, the student features we include are "gender" (0/1), "math anxiety" (9–45), and "pre-test score" (0–8).

We first subsample 43% of our data to build the causal forest. Then we use the R-loss [5], which calculates the expected difference between the estimated and the true treatment effect, to select the best tree out of the ensemble. We follow this tree's decision rules and allocate students in each subgroup. Then we use the holdout 50% of our data to calculate the conditional average treatment effects (CATE) for each subgroup. This constitutes an honest estimation and mimics the procedures in honest forests. See Wager et al. [10] for a formal discussion.

We built a causal forest with 500 trees and a minimum node size of 7. Due to the small size of the dataset, we further set the sample fraction that is used to grow an individual tree to 0.8. The 'grf' R package was used to fit the causal forest [9]. We use a difference-in-means estimator to calculate the CATE for each subgroup and construct the 95% confidence interval.

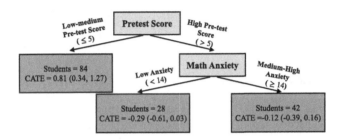

Fig. 2. The best tree selected from the causal forest. It shows three identified subgroups with the respective CATE for each group and the 95% confidence interval.

Three subgroups have been identified, with splits occurring based on the features 'math anxiety' and 'pre-test score.' We show them in Fig. 2. Finally, the holdout students ($n = 154$) are divided into subgroups determined by the best causal tree before the respective CATE is calculated. Students with a low pre-test score had the highest average group average treatment effect in our tree: these students particularly benefited from the RL adaptive policy condition compared to a simple control condition. The impact on other students overlaps with zero, suggesting a slight negative or null result. This is consistent with past work [3] that founds benefits in a logic tutor with RL most benefited students who initially appeared to be struggling.

3.2 Analysis of Personalization

The learned RL policy uses a neural network to produce a probability distribution over the four pedagogical strategies given an input learning context. In previous work on this dataset, Ruan et al. [8] found that integrated gradients, a method in explainable machine learning, suggested math anxiety and student pre-test scores are the most influential student features to the policy's decision-making. Here, we compute the average probability of taking each action for the student subgroups, created by taking the top and bottom 25%-quantile of both features. We note that the two groups that exhibit the largest difference in terms of probability between all actions are the students with high pre-test scores and high math anxiety and students with low pre-test scores and low math anxiety. However, a chi-squared analysis did not show a difference between the groups.

Table 1. We show the average probability of policy taking each action in each student subgroup. Top/bottom means the top or bottom 25% quantile of the pre-test score. High/low means the top or bottom 25% quantile of the math anxiety.

RL Policy Action	Top × High	Bottom × Low	Top × Low	Bottom × High
Pr(Direct Hint)	57.6%	35.7%	50.7%	43.2%
Pr(Acknowledgment)	3.35%	18.3%	7.36%	8.14%
Pr(Encouragement)	20.6%	25.9%	27.9%	16.8%
Pr(Guided Prompt)	18.4%	20.1%	14.1%	31.8%

3.3   Impact of Personalization

In the previous section, we identified two subgroups of students that induce the most difference in the action probabilities of the policy. We choose the two subgroups to be the Top × High and Bottom × Low groups defined in Table 1. We now explore the following counterfactual: if the policy had not assigned the learned personalized interventions to these two groups of students, would we have seen a difference in the learning outcomes of these two groups?

To explore this, we built an alternative anti-policy π_β, in contrast to our original policy π. We construct π_β with the following procedure: First, we compute the maximum for both the pre-test and the math anxiety scores across all students in group A (Top × High). We then replace the pre-test and math anxiety scores for all students in group B (Bottom × Low) with these two maxima. All other features in group B (negative/positive sentiment in text, failed attempts, etc.) remain unchanged. Similarly, we take the original data in group B and compute the minimum across all students for the pre-test and the math anxiety scores in this group. Subsequently, we replace the corresponding feature values in group A with these two minima. Again, all other features in group A remain the same. This feature swap allows us to obtain a counterfactual policy without re-training by rerunning the original policy on the alternated dataset.

An alternative to adaptive, differentiated policies is a static, non-personalized decision policy. This might be particularly useful if sometimes student features are noisy or mismeasured. Therefore, we also estimated the performance of a non-personalized static policy $\bar\pi$, constructed by computing the average probability of each action over all states: $\bar\pi(a) = \mathbb{E}_s[\pi(a|s)]$. To estimate the performance of these alternative policies, we use a popular offline policy evaluation technique: weighted importance sampling [7], which allows us to use historical data collected by policy π to compute the expected student improvement (Y) of π_β.

Let $w_i = \prod_{j=1}^{L} \frac{\pi_\beta(a|s)}{\pi(a|s)}$ for the i-th student: $\mathbb{E}_{\tau \sim \rho_{\pi_\beta}}[Y] = \mathbb{E}_{\tau \sim \rho_\pi}\left[\frac{w_i}{\sum_{j=0}^{n_1} w_j}Y\right]$ and $\mathrm{Var}_{\pi_\beta}(Y) = \mathbb{E}_{\tau \sim \rho_\pi}\left[\left(\frac{w_i}{\sum_{j=0}^{n_1} w_j}\right)^2 (Y - \mathbb{E}_{\pi_\beta}[Y])^2\right]$. Suppose there is a significant difference between $\mathbb{E}_\pi[Y]$ and $\mathbb{E}_{\pi_\beta}[Y]$. In that case, we can conclude that the policy's choice of personalization impacts these subgroups of students. We can derive the variance of the weighted importance sampling estimator, with details in Owen [6].

Table 2. Mean & variance of expected student performance improvement (WIS)

Sub-group	$\mathbb{E}_\pi[Y]$ (Original)	$\mathbb{E}_{\pi_\beta}[Y]$ (Anti)	$\mathbb{E}_{\tilde{\pi}}[Y]$ (Static)
Top Pre-test × High Anxiety	0.42 (0.06)	0.26 (0.04)	0.33 (0.06)
Bottom Pre-test × Low Anxiety	3.55 (0.23)	1.71 (1.27)	3.21 (0.54)

We report our findings in Table 2. There is a significant benefit for both student subgroups from using the original policy (see $\mathbb{E}_\pi[Y]$) over our estimate of the alternate anti-optimized π_β (see $\mathbb{E}_{\pi_\beta}[Y]$). The potential change in performance vs using the static policy is small, indicating there exists a static policy that is robust to potential mismeasurement of student features. This provides stakeholders options on which policy to implement.

4 Conclusion

Here we proposed analyses to help understand the personalization done by an RL policy, and the impact outcomes. We apply our framework to data from a real-life study on math tutoring, showing RL personalizes in an impactful way.

Acknowledgements. This work was supported by a Stanford Hoffman-Yee & a NSF #2112926 grant.

References

1. Abdelshiheed, M., Hostetter, J.W., Barnes, T., Chi, M.: Leveraging deep reinforcement learning for metacognitive interventions across intelligent tutoring systems. In: Artificial Intelligence in Education: AIED 2023 (2023)
2. Athey, S., Tibshirani, J., Wager, S., et al.: Generalized random forests. Ann. Stat. **47**(2) (2018)
3. Ausin, M.S.: Leveraging deep reinforcement learning for pedagogical policy induction in an intelligent tutoring system. In: Proceedings of EDM (2019)
4. Leite, W.L., et al.: Heterogeneity of treatment effects of a video recommendation system for algebra. In: Learning@Scale, pp. 12–23 (2022)
5. Nie, X., Wager, S.: Quasi-oracle estimation of heterogeneous treatment effects. Biometrika **108**(2), 299–319 (2021)

6. Owen, A.B.: Monte Carlo theory, methods and examples (2013)
7. Precup, D.: Eligibility traces for off-policy policy evaluation. Computer Science Department Faculty Publication Series, p. 80 (2000)
8. Ruan, S., et al.: Reinforcement learning tutor better supported lower performers in a math task. arXiv preprint arXiv:2304.04933 (2023)
9. Tibshirani, J., Athey, S., Wager, S.: grf: Generalized random forests. R package version 120 (2020)
10. Wager, S., Athey, S.: Estimation and inference of heterogeneous treatment effects using random forests. JASA **113**(523), 1228–1242 (2018)

Automated Assessment of Comprehension Strategies from Self-explanations Using Transformers and Multi-task Learning

Bogdan Nicula[1,2], Marilena Panaite[1], Tracy Arner[3], Renu Balyan[4], Mihai Dascalu[1,5(✉)], and Danielle S. McNamara[3]

[1] University Politehnica of Bucharest, 313 Splaiul Independentei, 060042 Bucharest, Romania
{bogdan.nicula,marilena.panaite,mihai.dascalu}@upb.ro
[2] Fotonation SRL, Blvd. Timisoara, Nr. 4A, 061328 Bucharest, Romania
[3] Department of Psychology, Arizona State University, PO Box 871104, Tempe, AZ 85287, USA
{tarner,dsmcnama}@asu.edu
[4] Math/CIS Department, SUNY at Old Westbury, Old Westbury, NY 11568, USA
balyanr@oldwestbury.edu
[5] Academy of Romanian Scientists, Str. Ilfov, Nr. 3, 050044 Bucharest, Romania

Abstract. Self-explanation practice is an effective method to support students in better understanding complex texts. This study focuses on automatically assessing the comprehension strategies employed by readers while understanding STEM texts. Data from 3 datasets (N = 11,833) with self-explanations annotated on different comprehension strategies (i.e., bridging, elaboration, and paraphrasing) and an overall quality score was used to train various machine learning models in both single-task and multi-task setups. Our end-to-end neural architecture considers RoBERTa as an encoder applied to the target and self-explanation texts, combined with handcrafted features for assessing text cohesion and filtering out low-quality examples. The best configuration obtained a .699 weighted F1-score for the overall self-explanation quality.

Keywords: Self-explanations · Language models · Multi-task learning

1 Introduction

The ability to read and comprehend text is a critical skill for learners to be successful in their educational and future career goals. However, recent assessment data suggest that only 33% of fourth-graders and 31% of eighth-graders in the United States are at or above proficiency when reading and comprehending grade-level texts [11]. More importantly, students' level of proficiency has decreased significantly since the global pandemic forced instruction to occur in online environments [1].

© The Author(s), under exclusive license to Springer Nature Switzerland AG 2023
N. Wang et al. (Eds.): AIED 2023, CCIS 1831, pp. 695–700, 2023.
https://doi.org/10.1007/978-3-031-36336-8_107

Reading comprehension is the ability to obtain meaning from text. Prior to deriving meaning from a text, readers must first decode letters and corresponding sounds to create words, access the meaning of the words, and then combine the words to create meaning from the text. As such, reading comprehension is a complex phenomenon that has been described with many theoretical frameworks [10] that vary in their explanation of these processes.

Explaining text to oneself can serve to scaffold inference generation and improve the coherence of the reader's mental representation of the text. As such, considerable evidence indicates that explicit self-explanation reading training (SERT) improves comprehension, particularly for low-knowledge and less skilled readers [8]. SERT includes instruction on five sub-strategies; comprehension monitoring, paraphrasing, bridging, prediction, and elaboration. Comprehension monitoring is a critical skill for students to recognize where they have gaps in knowledge. The use of prediction has been shown to be less common in the self-explanation of science texts. Therefore, in this study, we focus on paraphrasing, bridging, and elaboration.

Paraphrasing involves restating text in one's own words. This process is important in developing the reader's textbase level understanding by prompting them to think about the words and select words from their own lexicon that are more familiar and, thus, easier to understand and remember. *Bridging* is a critical strategy in the development of the reader's mental model of the text because authors cannot include every piece of relevant information. Therefore, readers must link ideas in the text to fill in the missing information. *Elaboration* is similar to bridging in that the reader is linking ideas from the text. However, elaborative inferences involve the reader connecting information from the text to their own knowledge base.

The aim of this work is to develop an automated model capable of evaluating the overall quality of self-explanations as well as the comprehension strategies employed by the readers.

2   Method

2.1   Corpus

The present study includes three different datasets collected at a large university in the Southwest United States. Participants in each of the studies were provided with training that was either direct instruction or worked examples of different methods of generating self-explanations (i.e., paraphrasing, bridging, elaboration). The direct instruction training consisted of short vignettes with descriptions and worked examples that demonstrated each method of generating self-explanations. The worked examples for training included a target sentence and an example response that highlighted each method participants could use to produce a self-explanation. In some studies, the training was presented in the context of a larger study using the iSTART Intelligent Tutoring System [9]. Following the training phase, participants were instructed to generate self-explanations of target sentences in relatively complex science texts.

In total, the three datasets contain 11,833 entries. Each datapoint consists of the participant's self-explanation, the target sentence, and scores for the presence of a paraphrase, bridge, or elaboration, as well as overall quality. These datasets have been split into train/dev/test by a ratio of 54.5%/27.5%/18%.

The 4 dimensions are represented as categorical variables having either 3 or 4 classes. The 0 class contains poor SEs, 1 contains acceptable SEs, whereas 2 and 3 contain good SEs. In the case of bridging and elaboration presence, the final 2 classes (marked with bold) were merged to minimize class imbalance (see Table 1).

Table 1. Class distribution per task.

Dimension	Num classes	Class 0	Class 1	Class 2	Class 3
Paraphrase presence	3	1487	1992	8354	
Bridge presence	4	4869	981	**4569**	**1414**
Elaboration presence	3	9382	**777**	**1674**	
Overall	4	799	4207	5093	1734

2.2   Neural Architecture

Our model consists of a combination of deep learning Natural Language Processing (NLP) [5] techniques and feature engineering. The core component of the model is the Transformer [13] block which processes the SE and source text pairs, thus generating a set of contextualized embeddings. Transformer-based architectures achieve state-of-the-art results in most NLP tasks. We consider both BERT [4] and RoBERTa [6] models as building blocks for our neural end-to-end architecture. Pretrained versions were loaded using the Huggingface library [14].

The embeddings generated by the Transformer block are combined with a set of handcrafted features that extract lexical, syntactical, and semantic information from both the SE and the source text. The aggregated features are processed via a fully connected layer. A smaller set of specific filtering features is generated separately to function as a mask to filter out part of the aggregated features and, in some cases, label the entire example as a 0. The resulting set of filtered features is used as input for the task classifiers in the multi-task learning setup, each having a single fully-connected layer. In the case of single-task training, the model has only one classifier.

A set of handcrafted features was generated to complement the representations generated from the Transformer block. These features were computed using the ReaderBench framework [3] for both the source text and the self-explanation and consisted of: lexical and semantic distance metrics, lexical and part-of-speech n-gram overlap, as well as surface, lexical, syntactic, semantic complexity indices.

The previously developed workflows for automatic self-explanation scoring [12] included a separate filtering step in the pipeline for poorly scored predictions (label 0). Based on the zero prediction rules from this filtering step, a smaller set of handcrafted features were generated to filter out irrelevant SEs. The filtering rules penalize SEs that are short, or contain copy-pasted text or frozen expressions.

RoBERTa models consistently outperformed the BERT alternatives. As such, our final configuration considered RoBERTa. Out of the 124.8M parameters, 180,912 parameters were trained at a max learning rate of $2e-4$, while the rest of the 124.6M parameters (all related to the pretrained Transformer model) were trained with a smaller LR of $1e-5$. A linear learning rate scheduler with warmup was used. The learning rate was gradually increased in the first five epochs up to the maximum value of 2e-4 and then gradually decreased to 0. The model was trained for 25 epochs using an AdamW optimizer [7] and a CrossEntropy loss for each task. A weighted CrossEntropy loss is used in order to account for the class imbalances present in all 4 tasks, as seen in Table 1.

Multi-task learning [2] is an approach by which a neural network can be trained simultaneously on multiple related tasks while using a shared representation. This type of training can improve generalization and is more efficient resource-wise than training one model per task. For the multi-task setup, multiple weighting schemes were considered for the individual task losses. Only the best-performing weighting scheme was used in the final experiment. This scheme considered the overall dimension to be as important as the sum of the other 3, and the paraphrase and bridging presence dimensions being twice as important as elaboration presence.

3   Results

The experiments done as part of this study were meant to determine the best model for assessing the 4 tasks. A second objective was that of determining whether a multi-task model could obtain similar or better performances than the 4 single-task models while training with similar resource constraints. For this reason, both the multi-task and the single-task models were trained for the same amount of time (25 epochs), using the same batch size, learning rate scheduler, optimizer, and regularization settings. The models are trained using a CrossEntropy loss, or a weighted sum of CrossEntropy losses, in the case of the multi-task model. They are evaluated throughout the run on the test set by computing the F1-score (see Table 2).

One aspect that is made clear when looking at the results is that the weighted Cross-Entropy loss does not completely address the issue of class imbalance. In the case of elaboration, we see that there is a 1-to-13 ratio between class 1 and class 0 on the test set. This offers an explanation for the large discrepancy in F1- scores for both models on this task, with the F1-score for class 0 being close to 0.9, while the F1-score for class 0 is barely 0.2. A similar but less extreme phenomenon can be observed for other tasks (e.g., class 1 has considerably fewer samples than classes 0 and 2 for bridging presence).

Table 2. Classification results for single-task, multi-task, and legacy models.

Strategy	Support	Model	F1-score				
			Weighted	Class 0	Class 1	Class 2	Class 3
Paraphrase presence	(90/141/1911)	STL	0.843	0.32	0.14	0.92	
		MTL	0.818	0.21	0.23	0.89	
Bridge presence	(859/24/1259)	STL	0.785	0.71	0.08	0.85	
		MTL	0.765	0.66	0.03	0.85	
Elaboration presence	(1996/146)	STL	0.899	0.95	0.21		
		MTL	0.882	0.93	0.72		
Overall	(61/688/926/467)	STL	0.694	0.24	0.73	0.69	0.71
		MTL	**0.699**	0.25	0.73	0.7	0.71

* STL - single-task learning; MTL - multi-task learning

When looking at both per-class and weighted F1-scores, we observe that there are very few differences between the multi-task and single-task models (i.e., the difference between the 2 models is smaller than 0.025 F1-score in all 4 cases). The single-task model seems to perform slightly better on the 3 tasks involving comprehension strategies, while the multi-task model has a better performance on the overall task.

4   Conclusions and Future Work

The aim of this paper was to develop a machine learning model to assess self-explanations in terms of an overall quality score and three individual scores that denote the presence of specific comprehension strategies. The model receives the raw text of the SE and of the source sentence and computes the four scores. Furthermore, the model is designed to automatically classify empty, illegible, or bad examples as class zero without requiring a separate filtering stage in its pipeline. Two separate approaches for meeting these criteria were developed. Firstly, 4 single task models were trained, each for predicting one of the scores. Secondly, a multi-task model was trained to solve the four tasks simultaneously, using an analogous architecture and training regimen.

The best result when predicting the overall quality score was a 0.699 F1-score obtained by the multi-task model, which was only slightly better than the 0.694 F1-score obtained by the single-task model. On the other three tasks, the single-task models obtained slightly better results, but the difference between single-task and multi-task training was never larger than 0.025.

Both single-task and multi-task models consisted of a similar number of parameters (i.e., 124.8M) and were trained using the same optimizer, the same learning rate scheduler, the same batch size, and an identical number of epochs. This underlines the value of the multi-task model, which managed to obtain similar or even better performance as the 4 single-task models while requiring only a quarter of the resources.

The current study is a significant advance in the ability to provide feedback to short responses to multiple science texts (i.e., any text, any time), both in terms of the specific strategy used by the reader, as well as the overall quality of the self-explanation. This was achieved by leveraging a larger dataset than available previously combined with recent advances in the field of AI.

Acknowledgements. This work was supported by the Ministry of Research, Innovation, and Digitalization, project CloudPrecis, Contract Number 344/390020/ 06.09.2021, MySMIS code: 124812, within POC, the Ministry of European Investments and Projects, POCU 2014–2020 project, Contract Number 62461/03.06.2022, MySMIS code: 153735, the IES (NSF R305A130124, R305A190063), the U.S. Department of Education, and the NSF (NSF REC0241144; IIS-0735682).

References

1. Adedoyin, O.B., Soykan, E.: Covid-19 pandemic and online learning: the challenges and opportunities. In: Interactive Learning Environments, pp. 1–13 (2020)
2. Caruana, R.: Multitask learning. Mach. Learn. **28**, 41–75 (1997)
3. Dascalu, M., Crossley, S., McNamara, D., Dessus, P., Trausan-Matu, S.: Please ReaderBench This Text: A Multi-dimensional Textual Complexity Assessment Framework, pp. 251–271. Nova Science Publishers Inc., New York (2018)
4. Devlin, J., Chang, M.W., Lee, K., Toutanova, K.: BERT: pre-training of deep bidirectional transformers for language understanding. In: Proceedings of the 2019 Conference of the NAACL, vol. 1, pp. 4171–4186 (2019)
5. Jurafsky, D., Martin, J.H.: Speech and Language Processing: An Introduction to Natural Language Processing, Computational Linguistics, and Speech Recognition. Pearson Prentice Hall, Upper Saddle River (2009)
6. Liu, Y., et al.: RoBERTa: a robustly optimized BERT pretraining approach. CoRR abs/1907.11692 (2019), http://arxiv.org/abs/1907.11692
7. Loshchilov, I., Hutter, F.: Decoupled weight decay regularization. In: International Conference on Learning Representations (2018)
8. McNamara, D.S.: SERT: self-explanation reading training. Discourse Process. **38**(1), 1–30 (2004)
9. McNamara, D.S., Levinstein, I.B., Boonthum, C.: iSTART: interactive strategy training for active reading and thinking. Behav. Res. Methods Instrum. Comput. **36**(2), 222–233 (2004)
10. McNamara, D.S., Magliano, J.: Toward a comprehensive model of comprehension. Psychol. Learn. Motiv. **51**, 297–384 (2009)
11. National Assessment of Educational Progress: The nation's report card: mathematics and reading at U.S. department of education, institute of education sciences. National Center for Education Statistics (2022). https://www.nationsreportcard. gov/reading/nation/scores/
12. Panaite, M., et al.: Bring It on! Challenges encountered while building a comprehensive tutoring system using *ReaderBench*. In: Penstein Rosé, C., et al. (eds.) AIED 2018. LNCS (LNAI), vol. 10947, pp. 409–419. Springer, Cham (2018). https://doi.org/10.1007/978-3-319-93843-1_30
13. Vaswani, A., et al.: Attention is all you need. CoRR abs/1706.03762 (2017). http:// arxiv.org/abs/1706.03762
14. Wolf, T., et al.: Transformers: State-of-the-art natural language processing. In: EMNLP (2020)

Ensuring Fairness of Human- and AI-Generated Test Items

William C. M. Belzak(✉) ⓘD, Ben Naismith ⓘD, and Jill Burstein ⓘD

Duolingo, Pittsburgh, PA 15209, USA
wbelzak@duolingo.com

Abstract. Large language models (LLMs) have been a catalyst for the increased use of AI for automatic item generation on high-stakes assessments. Standard human review processes applied to human-generated content are also important for AI-generated content because AI-generated content can reflect human biases. However, human reviewers have implicit biases and gaps in cultural knowledge which may emerge where the test population is diverse. Quantitative analyses of item responses via differential item functioning (DIF) can help to identify these unknown biases. In this paper, we present DIF results based on item responses from a high-stakes English language assessment (Duolingo English Test - DET). We find that human- and AI-generated content, both of which were reviewed for fairness and bias by humans, show similar amounts of DIF overall but varying amounts by certain test-taker groups. This finding suggests that humans are unable to identify all biases beforehand, regardless of how item content is generated. To mitigate this problem, we recommend that assessment developers employ human reviewers which represent the diversity of the test-taking population. This practice may lead to more equitable use of AI in high-stakes educational assessment.

Keywords: Assessment · Fairness and Bias · Differential Item Functioning

1 Introduction

Automated item generation (AIG) has been used in the past decade as an efficient way to generate item content for high-stakes assessments (e.g., question prompts, response options) [1]. Advances in large language models (LLMs), such as Generative Pre-trained Transformer models (e.g., GPT-4) have accelerated the capabilities of AIG, as these methods can produce wider varieties of item content with less human input. On one hand, this has drastically increased the efficiency of content development, allowing for cheaper, more accessible assessments. On the other hand, LLMs are not immune to generating the same types of biased content that humans may produce (e.g., gender and ethnic stereotypes [2]). Thus, humans remain integral to reviewing items according to fairness guidelines [e.g., 3, 4].

Theoretical fairness guidelines such as those by Zieky [3] provide assessment developers with a sensible approach for evaluating whether item content is offensive, inappropriate, or may give unfair advantages to certain groups of test takers. For instance,

N. Wang et al. (Eds.): AIED 2023, CCIS 1831, pp. 701–707, 2023.
https://doi.org/10.1007/978-3-031-36336-8_108

Zieky [3] recommends avoiding topics that may elicit strong, negative emotions in test takers such as religion, death, and slavery, unless they are required for valid measurement. However, there is little empirical research demonstrating that human reviewers who use these guidelines can sufficiently filter out all such "construct-irrelevant" content, especially given that humans are prone to implicit biases [4]. For assessments with a diverse population of test takers, this problem may be exacerbated if reviewers are not representative of the test-taking population.

The first aim of this paper is to evaluate whether human reviewers can sufficiently identify all construct-irrelevant content in human- and AI-generated item content. We use statistical analyses (i.e., differential item functioning [DIF]) on a high-stakes assessment of English proficiency, the Duolingo English Test (henceforth, the DET), to answer the following research question: Do human- and AI-generated items exhibit DIF despite being human-reviewed? The second aim is to compare human- and AI-generated item content with respect to bias; in other words: do AI-generated items exhibit more or less DIF compared to human-generated items?

2   Background

Currently, GPT-4 is the state-of-the-art LLM driving the most widely used text generation systems (such as ChatGPT). AI researchers openly acknowledge that while there are many benefits of LLMs, Large language models (LLMs) can contribute to fairness and bias (FAB) issues [2, 5]. Although there has been a considerable amount of research on fairness in machine learning and LLMs, the scope of this paper is on FAB reviews with regard to AI-generated content for high-stakes assessment. We focus on this below.

LLMs are trained on human-created texts. Human biases – for example, stereotypes and prejudicies – may be embedded in these texts and then reflected in automatically generated text. Findings from FAB word co-occurrence experiments in Brown et al. [6] suggested stereotypes related to gender, race, and religion. For instance, words describing women were more often related to their appearance than those describing men. From an assessment perspective, language reflecting stereotypes may introduce construct-irrelevant factors which do not assess the intended skill and could unfairly impact test-taker performance, such as distracting the test taker from the task at hand.

Theoretical FAB review guidelines have been used to review human content creation for high-stakes assessments [3, 4]. These guidelines were developed to mitigate the generation of construct-irrelevant test item content which may unfairly impact test-taker performance. Unfairness due to content knowledge may be accentuated when an assessment is intended for diverse populations, where test takers represent varied linguistic and cultural contexts. We examine a high-stakes test of English language proficiency to investigate the sufficiency of human reviewers and FAB guidelines in this context, and to compare human- and AI-generated content with respect to the presence and magnitude of bias.

3   Duolingo English Test

The Duolingo English Test (DET) is an online English proficiency test that measures a test-taker's ability to communicate in English-medium settings (e.g., universities). DET test items are developed by experts in language testing, primarily by leveraging automatic item generation based on authentic sources of English language content [7]. From 'seed' items, a large item bank is generated with items appropriate for test takers of all proficiency levels from A1–C2 on the Common European Framework of Reference [8]; for example, reading comprehension passages and questions were created using GPT-3 [9]. Items are then reviewed by humans using established FAB guidelines [3].

The DET is taken by a highly diverse testing population, making it an ideal case for investigating a wide range of potential human biases. During 2022 and 2023, there were test takers from 218 countries/dependent territories (most commonly India and China), with an approximately even distribution of self-reported male and female identities (52.4% vs. 47.5%), a median age of 22 (80.9% between 16–30), and a larger proportion of Windows operating system users compared to Mac (74.3% vs. 25.2%).[1]

4   Methods and Data

We analyze the DET for Differential Item Functioning (DIF), a standard statistical analysis used in assessment research. DIF is defined as systematic test-taker differences in item correctness, controlling for true differences in English language proficiency.[2] Test-taker differences refer to differences in background characteristics, including gender, age, computer operating system, and nationality. All items on the DET are reviewed by humans using theoretical FAB guidelines [3]. Our first goal is to determine whether statistical analyses identify biases in item content that human reviewers are unaware of, and which the FAB guidelines do not account for. Our second goal is to compare DIF results between human- and AI-generated items.

We focus on a single item type from the DET – namely, a C-test task of reading comprehension – because this task contains more content per item compared to other item types, and thus may be more liable to exhibiting DIF due to content familiarity. A C-test task is a passage that contains a number of 'damaged' words (the last part of the word is missing) which the test taker must complete (e.g., see p. 6). Partial credit is given for correctly completing the missing text of a single word.[3] DIF analysis is done at the C-test task (or passage) level, rather than at the damaged word level.

---

[1] Test-taker operating system is a proxy for socioeconomic status (SES), with Mac correlating to higher SES.

[2] This definition describes "intercept DIF" but not "slope DIF". We focus on evaluating intercept DIF here to simplify the presentation of results. See Millsap [10] for a more general definition of DIF.

[3] A score of .5 means that half of the damaged words in the passage were completed correctly.

We collected DET response data from May 2022 to May 2023 for 1,967 C-test tasks. Humans generated 62% of the tasks, and GPT-3 generated the rest. After excluding C-test tasks with fewer than 250 responses, the mean number of responses per human-generated task was N = 566 and per GPT-3-generated task was N = 540.

We evaluate 23 background variables for DIF in each C-test task (see Table 1). All background variables except age are dummy coded (e.g., gender is 1 if female, 0 if male/other). Age is coded in years (e.g., 20 years old). We use linear regression and Wald tests of statistical significance [11] to test for DIF, such that item correctness is predicted by the overall DET score and each background variable separately. The DET score controls for true differences in English language proficiency, whereas the effect of each background variable on item correctness provides a direct test of DIF.

Although there are a multitude of DIF methods, we use the regression approach because it is particularly flexible and powerful [11]. It can handle large amounts of missing data, and it has high sensitivity in detecting DIF (low Type II error). This is critical in cases where the failure to identify bias is highly detrimental. A downside of this method is that it has lower specificity (high Type 1 error) in larger samples. To adjust for this lack of specificity, we use effect sizes to determine the severity of DIF.

5   Findings

In Table 1, we show percentages of items that exhibited DIF for each background variable and each method of item generation at varying effect sizes (ES). The ESs of .0025, .01, .05, and .1 can be thought about in terms of the proportion of damaged words completed. For instance, if a DIF effect for gender (female) equals .05, this is equivalent to females correctly completing 5% more or less of the damaged C-test passage compared to males, controlling for gender differences in English proficiency.

Notably, DIF effects were small in absolute magnitude for both human- and GPT-3-generated items despite a large percentage of C-test tasks exhibiting DIF for certain groups of test takers (e.g., China, India). It is important to point out, however, that intersections of test-taker backgrounds, such as female Indian test takers may receive an item that has many (dis-)advantageous effects of DIF, leading to more bias.

We find that human- and GPT-3-generated items exhibited similar proportions of DIF across all items. However, some background variables appeared to be (dis-)advantaged more often by human-generated items compared to GPT-3, and vice versa. For instance, human-generated items tended to show slightly higher rates of DIF for countries in the eastern hemisphere (e.g., China, India), whereas GPT-3 items tended to show slightly higher rates of DIF for countries in the western hemisphere (e.g., Canada, United States). This pattern was not observed perfectly, however.

The presence of DIF does not indicate the direction of bias, that is, whether DIF advantages or disadvantages a particular background variable. This is also shown in Table 1 as "Average Effect Size". Most average effect sizes of DIF are in the same direction for both GPT-3 and human-generated items, although the absolute magnitude of DIF is larger or smaller for some background variables. This finding is consistent with varying proportions of items with DIF across background variables.

Table 1. Percentage of Items with DIF and Average Effect Size of DIF (GPT-3 vs. Human)

	GPT-3					Human				
	Percentage of Items[a] with Effect Size >=				Average Effect	Percentage of Items[a] with Effect Size >=				Average Effect
Background variable	.0025	.01	.05	.1	Size[b]	.0025	.01	.05	.1	Size[b]
Gender	19	19	0	0	0.002	**21**	**21**	**1**	0	**0.021**
Age	9	0	0	0	0.000	7	0	0	0	-0.001
Operating System	**19**	**19**	1	0	**0.025**	14	14	**1**	0	0.006
China	41	41	12	0	0.024	**46**	**46**	**21**	**1**	**0.034**
India	37	37	13	0	0.012	**43**	**43**	**24**	**3**	-0.041
Canada	**9**	**9**	**7**	**2**	-0.073	7	7	6	1	-0.051
Brazil	**11**	**11**	**10**	**2**	-0.037	10	10	10	2	-0.017
South Korea	**10**	**10**	**9**	1	-0.050	8	8	7	**1**	-0.049
Indonesia	**11**	**11**	**8**	0	-0.016	6	6	5	**0**	-0.009
United States	**9**	**9**	**8**	1	-0.075	5	5	5	**1**	-0.054
Iran	**9**	**9**	8	1	0.023	**9**	**9**	**8**	**2**	**0.041**
Mexico	**9**	**9**	**8**	**3**	-0.049	7	7	7	2	0.002
Pakistan	**7**	**7**	**7**	1	-0.037	6	6	6	**2**	**-0.080**
Bangladesh	3	3	3	1	-0.029	**4**	**4**	**4**	1	**-0.048**
Columbia	3	3	3	1	-0.001	**7**	**7**	**7**	**2**	**0.035**
France	4	4	4	1	0.020	**6**	**6**	**6**	**2**	**0.030**
Ukraine	0	0	0	0	**0.086**	5	5	5	2	0.071
Japan	4	4	4	1	0.024	4	4	4	1	**0.058**
Saudi Arabia	3	3	3	**2**	-0.120	**5**	**5**	**5**	**3**	-0.096
Nigeria	3	3	3	1	0.015	**4**	**4**	**3**	1	**0.029**
Turkey	**2**	**2**	**2**	0	-0.078	2	2	2	**1**	-0.061
Vietnam	1	1	1	0	0.046	**2**	**2**	**2**	**1**	**0.067**
Russia	2	2	2	1	**0.081**	2	2	2	1	0.075

a. The values indicate the percentage of items where the absolute DIF effect (for each background variable) is greater than or equal (>=) to .0025, .01, .05, and .1. Bold values indicate the effect is larger for either GPT-3 or human generated items.
b. The average effect size is computed using all non-zero DIF effect sizes for each background variable. Bold values indicate the average effect size is larger *in absolute value* for either GPT-3 or human generated items.

6 Discussion

This paper evaluated whether human review of item content (according to established FAB guidelines) is sufficient for identifying biased test content. In particular, we wanted to know the following: (1) Do human- and AI-generated items exhibit DIF despite being human-reviewed? (2) Do AI-generated items exhibit more or less DIF compared to human-generated items?

First, our DIF analyses suggest that human-reviewed items exhibit DIF despite being human-reviewed, as nearly 40% of the items analyzed showed some degree of DIF (ES $\geq$ .01) regardless of the method of item generation. Although the effects for each background variable were often small in isolation, test takers with particular combinations may be affected more severely due to a multitude of DIF effects. For instance, the following is an example of an item with larger DIF effects favoring younger, female, French, Mac-using test takers (in total, ES $\cong$ .25) (Fig. 1):

This item demonstrates that seemingly innocuous content (e.g., "charity event") may (dis-)advantage certain test takers (e.g., French vs. non-French). Additionally, there are no clear guidelines from Zieky [4] about "charity events" or similar kinds of content

Since my company was experiencing success, I wanted to fine-tune my processes by finding a more efficient way to promote our services. After researching what others have done, I decided to hold a charity event supporting local children. I contacted a local children's hospital and worked out a deal with them to host our annual fundraising dinner. The event was a huge success, and I hope to make it even bigger next year. I'm sure that the exposure will help us reach new clients and get more jobs.

Note: The underlined parts of the passage indicate damaged words.

Fig. 1. Example of AI-generated C-test passage demonstrating DIF effects

which would lead human reviewers to reject this item. Other variables may also explain the source of DIF, e.g., test takers' socioeconomic status (as suggested by DIF in favor of Mac operating system users). Nevertheless, the fact that no cultural explanation is evident highlights the potentially 'invisible' nature of bias. This task was retired from the DET operational item bank based on this analysis.

Second, we did not find strong evidence that GPT-3-generated items exhibited more or less DIF compared to human-generated items. This finding suggests that human reviewers evaluate item content in similar ways, regardless of the item generation method.

Of course, statistical analyses of DIF are used frequently in high-stakes assessments. Zieky [4] recommends that developers use DIF analyses to investigate whether test items show unfair disadvantages. With the rise of LLMs, however, reviewing content rather than creating it has become the primary job of humans. The findings here thus emphasize previous recommendations of using multi-pronged FAB reviews which integrate both theoretically-motivated human reviews and empirically-motivated quantitative analyses [4].

Based on our findings, we recommend that, as much as possible, human reviewers represent the diversity of the test-taking population. This practice that more diverse perspectives are considered during the human review process, which may reduce the number of items exhibiting DIF due to unknown reviewer biases.

References

1. Gierl, M.J., Haladyna, T.M. (eds.): Automatic Item Generation: Theory and Practice. Routledge, Abingdon (2012)
2. Mehrabi, N., Morstatter, F., Saxena, N., Lerman, K., Galstyan, A.: A survey on bias and fairness in machine learning. ACM Comput. Surv. (CSUR) **54**(6), 1–35 (2021)
3. Zieky, M.J.: Fairness in test design and development. In: Dorans, N.J., Cook, L.L. (eds.) Fairness in Educational Assessment and Measurement, pp. 9–31. Routledge (2016)
4. Zieky, M.J.: Developing fair tests. In: Downing, S.M., Haladyna, T.M. (eds.) Handbook of Test Development, pp. 97–115. Routledge (2015)
5. Sherman, J.E.: Multiple levels of cultural bias in TESOL course books. RELC J. **41**(3), 267–281 (2010)
6. Brown, T., et al.: Language models are few-shot learners. Adv. Neural Inf. Process. Syst. **33**, 1877–1901 (2020)
7. Cardwell, R., Naismith, B., LaFlair, G.T., Nydick, S.W.: Duolingo English Test: Technical Manual [Duolingo Research Report]. Duolingo (2023). https://go.duolingo.com/dettechnical manual

8. Council of Europe: Common European Framework of Reference for Languages: Learning, Teaching, Assessment. Press Syndicate of the University of Cambridge, UK (2001)
9. Attali, Y., et al.: The interactive reading task: transformer-based automatic item generation. Front. Artif. Intell. **5**, 903077 (2022)
10. Millsap, R.E.: Statistical Approaches to Measurement Invariance. Routledge, Abingdon (2011)
11. Swaminathan, H., Rogers, H.J.: Detecting differential item functioning using logistic regression procedures. J. Educ. Meas. **27**(4), 361–370 (1990)

Deidentifying Student Writing with Rules and Transformers

Langdon Holmes[1]([✉]) [iD], Scott A. Crossley[1] [iD], Wesley Morris[1] [iD],
Harshvardhan Sikka[2] [iD], and Anne Trumbore[3]

[1] Vanderbilt University, Nashville, USA
langdon.holmes@vanderbilt.edu
[2] Georgia Institute of Technology, Atlanta, USA
[3] University of Virginia, Charlottesville, USA

Abstract. As education increasingly takes place in technologically mediated settings, it has become easier to collect student data that would be valuable to researchers. However, much of this data is not available due to concerns surrounding the protection of student privacy. Deidentification of student data is a partial solution to this problem, but student-generated text, a form of unstructured data, is a major challenge for deidentification strategies. In response to this problem, we develop and evaluate two approaches for the automatic detection of student names. We develop one system using a rule-based approach and one using a transformer-based approach that relies on finetuning a pretrained large language model. Our findings indicate that the transformer-based approach to student name detection shows more promise, especially when there is a high degree of variation between texts in a dataset.

Keywords: deidentification · anonymization · large language models · massively open online course · named entity recognition

1 Introduction

Educational artificial intelligence systems rely on large amounts of student data. However, as learning technologies make it easier to collect and utilize this data, there are growing concerns surrounding student privacy. Anonymization of student data is an important strategy for developing educational tools while protecting student privacy.

The first step in automated deidentification is to label personally identifiable information (PII). Rule-based approaches to labeling work by applying a set of rules or labeling functions to a text. For instance, Lison et al. developed a Python library to facilitate the creation of rule-based labeling systems, Skweak [3]. In an application, Skweak was used to achieve an F1 of 81% on per-token named entity labeling in a corpus of Wall Street journal articles, which demonstrates the potential for rule-based approaches to detect PII in unstructured text. In the educational domain, Bosch et al. used a rule-based approach that aggregated a

N. Wang et al. (Eds.): AIED 2023, CCIS 1831, pp. 708–713, 2023.
https://doi.org/10.1007/978-3-031-36336-8_109

set of text-level features using a machine learning model. They achieved as high as 95% recall of student names in a university classroom's discussion forum [1]. Deep-learning-based approaches to PII labeling rely on transformers, which are a neural network architecture that is widely used in natural language processing. Previous work has shown success in the medical domain with this approach, achieving PII recall as high as 99% on some medical datasets [5].

In the current study we build and assess a transformer-based student name labeling system alongside a rule-based system. We focus on student names because they are the most prevalent form of PII in student-generated text [2] and because they are challenging to detect. Our rule-based name labeling system works by aggregating a set of rules that detect student names. Our transformer-based name labeling systems were developed by finetuning a pretrained transformer model on domain-specific labeled training data. We evaluate these systems to assess their potential as part of an automated deidentification system for student-generated text.

2 Methods

Student writing samples were collected by Coursera from students enrolled in a publicly available, online course. Course completion required students to submit a written essay in which they reflected on how the course content could be applied to a problem with which they are familiar. Submissions were required to be in PDF format, and the files were retained on a third-party hosting platform. In total, 221,043 submission events were recorded. 29,142 of these PDF files were converted to plain text. We selected 5,797 of the documents in the corpus for hand-coding of PII. Each document was annotated by two undergraduate students during an internship at a research university. The annotators were instructed to label any student names, including names that could refer to a classmate, and any names associated with the student (such as a colleague or a family member). The authors of cited texts, the instructors of the course, and public figures were not considered student names. Each document was seen by at least one annotator. All documents that included any student name annotations were reviewed by the first author to ensure accuracy. Student names were included in 845 submissions, and there were 1,155 student name annotations in total.

This dataset was split into training, validation, and test partitions, which comprised 60%, 20%, and 20% of the data, respectively. The training set was used exclusively to finetune the models used in our transformer-based system. The validation set was used for validating the transformer-based system during finetuning and for developing the rule-based name annotation system. The test set was used to evaluate the performance of both our rule-based and transformer-based systems. We report results only on the testing set for both systems.

2.1   Rule-Based System

We divide the task of labeling student names into two parts. The first part of the task is to identify which spans correspond to person names. The second part of the task is to identify which person names belong to students or might otherwise constitute PII. The system is illustrated in Fig. 1 and discussed below.

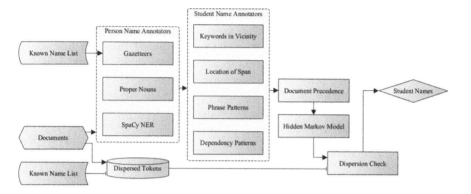

Fig. 1. A flowchart illustrating our rule-based student name labeling system.

We use three labeling functions to extract a set of person names from each document. These functions rely on part-of-speech tagging, a list of known names (called a gazetteer), and a general purpose NER model. The sole purpose of these functions is to supply a list of person names to the next group of labeling functions, which will determine which person names belong to students. We use a part-of-speech (POS) tagger to identify all proper nouns and label them as person names. We also search the document using a large list of known names and label any matches as person names [6]. We aggregate the output of these three labeling functions into a list of person name spans.

The next step is to identify which of these person names are student names. Vicinity annotators take a candidate span and search its context on the left and right hand side for a keyword token that might indicate it is a student name (e.g., the word "reflection" which often appears in the title). Location annotators consider the position of a span in the document. We created location annotators for spans that appear on a line by itself, for spans that appear in repeated lines, for spans that appear in the first lines of the document, and for the final span of the document. Phrase matcher annotators look for specific sequences of text, such as "Author: [proper noun]." We also developed syntactic patterns, such as "I am [proper noun]." Lastly, we use a precedence annotator to extend student name annotations to repeated mentions. If the person name "Brian" is labeled as a student name by any function, the student name label is extended to any repeated mentions of "Brian."

In order to aggregate the output of all labeling functions into a single set of labels, we implement a hidden Markov model using Skweak. After the hidden

Markov model, we apply a post-processing step that checks for the dispersion of the labeled tokens throughout the full corpus of texts. Any token that appears in at least five documents is considered well dispersed, and therefore cannot uniquely and independently identify the author of the text. However, a common first name or other well-dispersed token may be used to identify a student in combination with other so-called quasi-identifiers. As a result, we reduce the list of named tokens using a set of known names extracted from Wikidata [7]. Any labeled spans which pass the dispersion check labeling function are the final student name labels.

2.2   Transformer-Based System

We developed two transformer-based models from Roberta, a pretrained large language model [4]. Each model used different weights for precision and recall during training. Sequence labeling tasks typically weight precision and recall equally. For the purposes of deidentification, we reason that recall should be weighted more highly because any false negatives could constitute a breach of student privacy. On the other hand, false positives are not as harmful. As a result, we trained the first model (Finetuned 1) with precision weighted at .25 and recall at .75. A second model (Finetuned 2) was trained with precision weighted at .10 and recall at .90.

We evaluate the performance of all systems in terms of recall, precision, and F1 score. The F1 score is the harmonic mean between precision and recall. Since the performance of the transformer-based system can only be evaluated on the testing set (N = 1,160), we use the same set to evaluate all our methods. There were 556 tokens that were part of student names in the labeled test set.

3   Results

We first evaluated the set of person name labeling functions on the held-out test set. Since documents were labeled for student names and not all person names, non-student person names are treated as false positives, resulting in lower-than-expected precision. Since all student names are person names, recall can be interpreted in a straightforward manner. The combined set of all person name labeling functions recalled 95% of student names with a precision of .03%. The best performing person name annotator was the general purpose NER, which recalled 81% of student names, with a precision of 33%. The full name gazetteer had a precision of 15%, which was higher than the other person name gazetteers, while also exhibiting a recall of 19%. The part of speech annotator, which labeled all proper nouns as persons, had the highest recall of any single labeling function. It recalled 95% of student names, which was the same as the combined results of all person name labeling functions. The part of speech annotator also had higher precision, .5%, than the combined system (.03%).

We then evaluated the student name labeling functions on the set of spans labeled as person names by the previous group of functions. The precision, recall, and F1 score of each labeling function are reported in Table 1. The hidden Markov model, which aggregates the outputs of these functions, achieved a recall of .59 and a precision of .30. The dispersion filter, applied to the output of the hidden Markov model, resulted in an approximately unchanged recall of .59 and a precision of .41. The overall F1 score of this system was .48.

Table 1. Performance of student name labeling functions on the test set ($N = 1{,}160$).

	Precision	Recall	F1
Dispersion filter	.41	.59	.48
Hidden Markov model	.30	.59	.40
Precedence annotator	.19	.15	.17
Phrase Patterns	.35	.17	.23
Dependency Patterns	.16	.01	.03
Repeated line (header/footer)	.27	.14	.18
Last span in document	.78	.06	.12
Line by itself	.39	.21	.27
Keyword Vicinity	.50	.01	.01

We tested two finetuned transformer-based models. The first model, Finetuned 1, was trained to weight recall at .75 and independently achieved a per-token recall of .64 and a precision of .90. The second model, Finetuned 2, was trained to weight recall at .90 and independently achieved a per-token recall of .84 and a precision of .68. Both of these models had approximately the same F1 score of .75, despite the different precision/recall trade off.

Finally, we created a third system that aggregates the labeling functions and both finetuned transformer-based models. This configuration resulted in a recall of .85 and a precision of .34. After applying the dispersion filter, recall was reduced to .83 and precision increased to .43. Overall, the combined system performs worse (F1 = .56) than the transformer-based systems alone.

4   Discussion

We evaluated the performance of rule-based and transformer-based approaches to the automatic detection of student names in student-generated text. We focused on student names as an important and challenging form of PII that has important implications in learning technologies. We developed two systems, specifically designed for our dataset, and evaluated their performance. Our results indicate that a rule-based system does not perform as well as a transformer-based system for student-generated text collected from a publicly available online course. Neither does the rule-based system contribute to the performance of the transformer-based system. We conclude that a rule-based system is not well-suited to the complex task of student name labeling in highly variable student-generated text. Rule-based systems have proven to be effective for some text types, but they are brittle to formatting errors and stylistic variation that are to be expected in larger datasets collected from open courses. These challenges are particularly evident for the case of student names, which appear in a variety of linguistic contexts. An effective student name labeling system must also distinguish student names from other, non-student names such as referenced authors. Our transformer-based system was a more effective student name detector, suggesting that a generalizable deidentification system for student-generated text should adopt this approach for the detection of student names.

While the problem of automatic text deidentification is a challenging one, it does not appear intractable. There are few fields that would stand to benefit as much as learning analytics if an effective, automatic system were developed. Such a system, paired with ethical practices such as informed consent, would promote open science in while also protecting student privacy.

This material is based upon work supported by the National Science Foundation under Grant 2112532 Any opinions, findings, and conclusions or recommendations expressed in this material are those of the author(s) and do not necessarily reflect the views of the National Science Foundation.

Acknowledgement. This material is based upon work supported by the National Science Foundation under Grant 2112532 Any opinions, findings, and conclusions or recommendations expressed in this material are those of the author(s) and do not necessarily reflect the views of the National Science Foundation.

References

1. Bosch, N., Crues, R.W., Shaik, N.: "Hello, [REDACTED]": protecting student privacy in analyses of online discussion forums. In: Proceedings of The 13th International Conference on Educational Data Mining, p. 11 (2020)
2. Holmes, L., Crossley, S.A., Haynes, R., Kuehl, D., Trumbore, A., Gutu, G.: Deidentification of student writing in technologically mediated educational settings. In: Proceedings of the 7th Conference on Smart Learning Ecosystems and Regional Development (SLERD), Bucharest, Romania (2023)
3. Lison, P., Barnes, J., Hubin, A.: Skweak: weak supervision made easy for NLP. In: Proceedings of the 59th Annual Meeting of the Association for Computational Linguistics and the 11th International Joint Conference on Natural Language Processing: System Demonstrations, pp. 337–346. Association for Computational Linguistics, Online, August 2021. https://doi.org/10.18653/v1/2021.acl-demo.40
4. Liu, Y., et al.: RoBERTa: a robustly optimized BERT pretraining approach. arXiv:1907.11692 [cs], July 2019
5. Murugadoss, K., et al.: Building a best-in-class automated de-identification tool for electronic health records through ensemble learning. Patterns **2**(6), 100255 (2021). https://doi.org/10.1016/j.patter.2021.100255
6. Remy, P.: Name dataset. GitHub (2021)
7. Vrandečić, D., Krötzsch, M.: Wikidata: a free collaborative knowledgebase. Commun. ACM **57**(10), 78–85 (2014). https://doi.org/10.1145/2629489

Comparative Analysis of Learnersourced Human-Graded and AI-Generated Responses for Autograding Online Tutor Lessons

Danielle R. Thomas$^{(\boxtimes)}$ [ID], Shivang Gupta [ID], and Kenneth R. Koedinger [ID]

Carnegie Mellon University, Pittsburgh, PA 15213, USA
{drthomas,shivang,koedinger}@cmu.edu

Abstract. Machine learning and artificial intelligence (AI) are ubiquitous, although accessibility and application are often misunderstood and obscure. Automatic short answer grading (ASAG), leveraging natural language processing (NLP) and machine learning, has received notable attention as a method of providing instantaneous, corrective feedback to learners without the time and energy demands of human graders. However, ASAG systems are only as valid as the reference answers, or training sets, they are compared against. We introduce an AI-based, machine learning method of autograding online tutor lessons that is easily accessible and user friendly. We present two methods of training set creation using: a subset of learnersourced, human-graded tutor responses from the lessons; and a surrogate model using the recently released AI-chatbot, *ChatGPT*. Findings indicate human-created training sets perform considerably better than AI-generated training sets (F1 = 0.84 and 0.67, respectively). Our straightforward approach, although not accurate enough for wide use, demonstrates application of directly available machine learning based NLP methods and highlights a constructive use of *ChatGPT* for pedagogical purposes that is not without limitations.

Keywords: Machine learning · Automated short answer grading · Chatbots · Tutor training · Natural language processing · Constructed response

1 Introduction

Machine learning and artificial intelligence (AI) are rapidly growing fields within computer science, capturing human interest, as evidenced by over a million users accessing AI-chatbot *ChatGPT* within days after its release [1]. Current global revenue from AI software reportedly exceeds over $50 billion dollars and is predicted to grow by over 13 times within the next decade [3]. However, despite the ubiquitous nature of machine learning and recent advances in AI, there is much obscurity and misunderstanding surrounding practicality of application. In other words, what exactly can AI-generated content do and not do? Concurrently, automatic short answer grading (ASAG) has received considerable attention recently due to advances in machine learning and AI-based technologies [2, 9]. ASAG is defined as the process of grading constructed-response questions, also called *short answer* tasks, using an automatic grading method

N. Wang et al. (Eds.): AIED 2023, CCIS 1831, pp. 714–719, 2023.
https://doi.org/10.1007/978-3-031-36336-8_110

[2]. Often possessing more than one correct answer and necessitating a deeper depth in knowledge than multiple-choice questions, short answer tasks promote critical thinking and enhance knowledge development [4, 7]. Despite the benefits of participating in short answer tasks for the learner, there are several drawbacks for the human grader. Assessing textual answers from short answer tasks is time consuming, frequently includes nuances allowing for diverse variation in correct and incorrect responses and often contains fairness and bias issues [4].

We present a machine learning method of automating short answer tasks within online tutor lessons on how to effectively give praise to students titled, *Giving Effective Praise* and how to respond effectively to students who have recently made an error named, *Reacting to Errors*. We also report the results of two methods of creating a training dataset: by using a learnersourced subset of human-graded tutor responses; and by creating a surrogate model using ChatGPT. The primary research questions include:
RQ1. Can we effectively automate the scoring process for short answer tasks through the use of *off-the-shelf* machine learning based natural language processing (NLP) methods?
RQ2. How does the surrogate training set created by ChatGPT compare in performance to the training set of human-graded responses? In other words, how do AI-generated datasets compare to learnersourced datasets?

2   Related Work

Automated Short Answer Grading. Automated short answer grading (ASAG) Methods are an increasingly popular area of research with *short-answer grading*, analyzing semantic content, or the meaning behind a short piece of writing as a common approach [5]. Numerous short-answer systems are reported, such as the well-known, C-rater system used in the National Assessment for Educational Progress (NAEP) research [5]. However, C-rater and similar methods work well for specific content-level questioning, such as science or math concepts, and are *not* designed to assess questions prompting opinions and personal experiences [5]. In addition, despite advancing performance, there still exists cumbersome challenges, including the mismatch of using pre-trained language models not well suited for math-related domains and the need to train one model per question being assessed [9]. Furthermore, nuanced responses from real-life scenarios may host a wide range of appropriate answers. One-to-one training of models by question removes connections and variabilities across questions and increases model storage tremendously [9]. Other limitations to ASAG include: difficulty in evaluating complex or nuanced responses; extensive instructor input [2]; bias within the reference answers; and need to constantly "retrain" the model. If the reference responses are formulated by respondents not representative of the population being tested, the scores generated by the ASAG system will not be accurate.

3   Method

The two training sets analyzed, providing insight into the comparisons between learnersourced and AI-generated datasets, are: *HUMAN*, the subset of human-graded tutor responses; and *AI-Bot*, the datasets created using the open source, AI-chatbot, ChatGPT. The short answer tasks within *Giving Effective Praise* require tutors to predict how to best respond to a scenario challenging a student's motivation. In the second tutor lesson, *Reacting to Errors*, tutors practice responding to a student who has recently made an error, in a way that increases motivation and engagement. Figure 1 displays one of two analogous scenarios (used interchangeably for pretest and posttest) and the corresponding short answer tasks (see [8]).

Giving Effective Praise: You're tutoring a student named Kevin. He is struggling to understand a math problem. When he doesn't get the answer correct the first time, he wants to quit. After trying several different approaches, Kevin gets the problem correct. As Kevin's tutor, you want him to continue working on his math assignment.
1. What exactly would you say to Kevin to provide effective praise that will increase his motivation to continue working and increase his engagement?

Reacting to Errors: You're tutoring a student named Lucy. She is having trouble solving a math problem. She just finished adding a 3-digit and 2-digit number and has made a common mistake (shown right).
1. What exactly would you say to Lucy, regarding her mistake, to effectively respond in a way that increases her motivation to learn?

Fig. 1. One of the two analogous scenarios in *Giving Effective Praise* and *Reacting to Errors*.

Learnersourced Datasets. The *HUMAN* training sets created from randomly removing a subset of tutors' correct and incorrect responses are named *HUMAN_Praise* and *HUMAN_Errors*. Tutors were unpaid volunteers (51% male, 52% White, 21% Asian, 56% older than 50 years old, 94% reporting experience) from a national, online tutoring organization [8]. The *Giving Effective Praise* training set consisted of 61 correct and 40 incorrect responses for a total of 101 manually coded responses. The *Reacting to Errors* training set was made up of 51 correct and 53 incorrect responses, with training sets manually coded according to the schema described previously in [8]. Examples of correct tutor responses for *Giving Effective Praise* include: *"I appreciate your effort and willingness to work hard to get to the answer"; "You can do this! You are learning and developing skills, and making great progress as we work through this."* Less-desired, or incorrect, responses were: *"You are doing a great job so far. Let's see how we can improve"; "Great job! Why don't we try solving [a] few more problems so that you feel confident about these problems?"* Examples of correct tutor responses for *Reacting to Errors* include: *"Almost there. Take a look at where the ones, tens, and hundreds are lined up. You can do it!"; "Hi, Kanye! Thank you for showing me how you did the problem. Can you explain your thinking process?"* Incorrect responses were: *"Nice try, Kanye, but you made a very common mistake. Let's try lining up the numbers again"; "You have the correct idea, but the numbers are not aligned correctly."*

AI-Generated Datasets. The *AI-bot* training sets were created by inputting scenarios and short answer tasks (see Fig. 1) into the ChatGPT dialogue box and recording the AI-bot's responses[1]. For each pretest and posttest scenario *ChatGPT* generated 25 correct and 25 incorrect responses, for a total of 100 AI-generated short answer responses.

---

[1] ChatGPT Dec 15 version. Retrieved December 22, 2022. https://openai.com

Subsequent prompts were added to shorten the length of chatbot output by requesting responses be 20 or less words in length, as the average length of output was 20 words for *HUMAN* datasets— typical human responses were shorter in length compared to AI-generated responses. To gather AI-classified correct responses, we input the short answer task (see Fig. 1) for each scenario. To collect incorrect responses, we explicitly asked *ChatGPT* to provide incorrect, or *"wrong,"* responses. Examples of correct tutor responses for *Giving Effective Praise*, include: *"Great job, Kevin! You were really persistent in trying to solve that problem"; "Keep it up, you're making great progress!"* Less-desired, or incorrect, responses include: *"Great job! You're so smart"; "I'm glad you finally got the answer correct. I was getting frustrated with you."; "Good job! You are better than the other students."* A correct response example for *Reacting to Errors* includes: *"Hi Lucy! Keep practicing math problems and make sure to line up place values correctly when adding. Ask for help if you need it. We're here to support you!"* AI-generated incorrect responses were: *"It looks like there might be an error in your calculation. Can you take another look and see if you can find it?"; "You're wasting my time."; "This is so simple, I can't believe you can't do this."*

Model Training. *HUMAN* and *AI-bot* datasets were trained using *LightSide*, a free open text mining toolbench for automated text analysis related to classification and feature extraction [6]. Feature extractions were similar across datasets consisting of: *unigrams, bigrams, parts of speech bigrams*, line length, and the inclusion of punctuation. Additionally, feature extracted *stem n-grams* were added which includes root words (e.g., *"work", "working", "reworked"*), sacrificing inflection but gaining generality [6]. Using Naïve Bayes as a classifier, we performed a 10-fold cross-validation on all datasets. For the *Giving Effective Praise* lesson, *HUMAN* obtained an average accuracy of 81% and Kappa score of 0.62; *AI-bot* obtained an average accuracy of 95% and Kappa score of 0.90. For the *Reacting to Errors* lesson, *HUMAN* obtained an average accuracy of 82% and Kappa score of 0.63; *AI-bot* obtained an average accuracy of 94% and Kappa score of 0.88.

4 Results

RQ1: Absolute Performance of Models. The absolute performance of the *HUMAN* and *AI-bot* datasets are shown in Table 1 using a testing set, separate from the training set, of tutor responses for *Giving Effective Praise* ($n = 88$) and *Reacting to Errors* ($n = 160$). Classifiers of model performance include the simple probabilistic model Naïve Bayes (NB), logistic regression (LR), and support vector machine (SVM). Absolute performance was adequate for *HUMAN* training sets, with accuracy ranging from 67% to 80% and F1 score ranging 0.62 to 0.84, depending on the measure used and lesson. *AI-bot* datasets did not perform as well with accuracy ranging from 52% to 70% and F1 score ranging from 0.48 to 0.67. Kappa score was considerably lower than F1 score most likely due to learners performing relatively well on the lessons with 85% and 68% of tutors getting the short answer task correct for *Giving Effective Praise* and *Reacting to Errors,* respectively.

Table 1. *HUMAN* and *AI-bot* absolute performance for both lessons.

	Giving Effective Praise						Reacting to Errors					
	HUMAN			AI-bot			HUMAN			AI-bot		
	Acc	κ	F1	Acc	K	F1	Acc	K	F1	Acc	κ	F1
NB	0.69	0.41	0.72	0.66	0.37	0.67	0.71	0.40	0.63	0.65	0.22	0.48
LR	0.78	0.50	0.77	0.60	0.23	0.64	0.68	0.38	0.64	0.70	0.30	0.51
SVM	0.80	0.55	0.84	0.52	0.17	0.49	0.67	0.35	0.62	0.68	0.32	0.57

RQ2: Comparison Between Human-Created and AI-Generated Models. Comparing human-created to AI-generated datasets, the results indicate *HUMAN* performed considerably better than *AI-bot* datasets, with performance ranges for HUMAN higher than *AI-bot* datasets (see RQ1 above). For *Giving Effective Praise*, the *HUMAN* dataset achieved an F1 score as high as 0.84 (using SVM) compared to 0.67 (using NB) for *Giving Effective Praise*. Overall performance was lower for *Reacting to Errors*, although the *HUMAN* dataset still performed better with an F1 score of 0.64 (using LR) compared to the *AI-bot*, which scored 0. 57 (using SVM).

5 Discussion and Conclusion

Overall, the performance of ready-made machine learning based NLP is not quite ready for use by a wide audience, The AI-generated training sets, *AI-bot*, underperformed compared to learnersourced training sets, *HUMAN*, evidenced by lower absolute performance measures by *AI-bot* (Table 1). We do not mean to imply that more sophisticated methods will not work, just that straightforward approaches like those directly available and easily accessible in websites like *LightSIDE* and ChatGPT do not produce high enough levels of accuracy. However, these user-friendly approaches may be useful, with appropriate framing. For example, for tutor responses classified as incorrect, lesson agnostic, corrective feedback responses to learners could be: *"An automated comparison of your response to others indicates that it may be [just right/a bit off]. Would you like to try to modify your response and get feedback on your change?"; "Would you like to see an example of a similar [correct/incorrect] response?"* Improved model performance may be possible by gaining additional learnersourced data and by conducting more specific training on parts of responses that are right or wrong.

The biggest limitation of AI-generated datasets is overfitting evidenced by very high training performance scores (>94%). This is most likely due to lack of variation among responses as ChatGPT had a difficult time providing authentic and nuanced responses. Many of the correct responses were synonymous and contained similar words such as, *"Great job on your math work, Kevin. Keep going"* and *"Nice work, Kevin! Keep it up!"* To generate incorrect answers, we had to explicitly ask it to provide "wrong" responses to students for the provided scenarios, of which ChatGPT responded: *"Well, I guess you're just not cut out for math"; "I can't believe you made such a dumb error"; "That's a stupid mistake."* The use of prompt engineering could increase diversity among AI-generated responses. In addition, blatantly absurd and ill-mannered responses highlight that although ChatGPT may be impressive, it is not without limitations, such as often generating inappropriate and harmful feedback. Collaborative and focused efforts among the AIED community are needed to ensure constructive application and responsible use of ChatGPT and similar systems for educational purposes.

Acknowledgements. This work is supported with funding from the Chan Zuckerberg Initiative (Grant #2018-193694), Richard King Mellon Foundation (Grant #10851), Bill and Melinda Gates Foundation, and the Heinz Endowments (E6291).

References

1. Altman, S.: [@sama]. ChatGPT launched on Wednesday. Today it crossed 1 million users! Twitter, 5 December 2022. https://twitter.com/sama/status/1599668808285028353
2. Burrows, S., Gurevych, I., Stein, B.: The eras and trends of automatic short answer grading. Int. J. Artif. Intell. Educ. **25**(1), 60–117 (2015)
3. Grand View Research: Artificial Intelligence Market Size, Share & Trends Analysis Report By Solution, By Technology (Deep Learning, Machine Learning, Natural Language Processing, Machine Vision), by End Use, By Region, and Segment Forecasts, 2022 – 2030. Market Analysis Report (2022)
4. Lan, A. S., Vats, D., Waters, A.E., Baraniuk, R.G.: Mathematical language processing: automatic grading and feedback for open response mathematical questions. In: Proceedings of the Second (2015) ACM Conference on Learning@ Scale, pp. 167–176 (2015)
5. Leacock, C., Chodorow, M.: C-rater: automated scoring of short-answer questions. Comput. Hum. **37**(4), 389–405 (2003)
6. Mayfield, E., Rosé, C.P.: LightSIDE: open source machine learning for text. In Handbook of Automated Essay Evaluation, pp. 146–157. Routledge (2013)
7. Mello, R.F., et al.: Enhancing instructors' capability to assess open-response using natural language processing. In: Hilliger, I., Muñoz-Merino, P.J., De Laet, T., Ortega-Arranz, A., Farrell, T. (eds.) Educating for a New Future: Making Sense of Technology-Enhanced Learning Adoption. EC-TEL 2022. Lecture Notes in Computer Science, vol. 13450, pp. 102–115. Springer, Cham (2022). https://doi.org/10.1007/978-3-031-16290-9_8
8. Thomas, D.R., Yang, X., Gupta, S., Adeniran, A., McLaughlin, E.A., Koedinger, K.R.: When the tutor becomes the student: design and evaluation of efficient scenario-based lessons for tutors. In LAK23: 13th International Learning Analytics and Knowledge Conference (LAK 2023), 13–17 March 2023, Arlington, TX, USA. ACM, USA (2023)
9. Zhang, M., Baral, S., Heffernan, N., Lan, A.: Automatic short math answer grading via in-context meta-learning. arXiv preprint arXiv:2205.15219 (2022)

Using Similarity Learning with SBERT to Optimize Teacher Report Embeddings for Academic Performance Prediction

Menna Fateen[(✉)][iD] and Tsunenori Mine[iD]

Kyushu University, Fukuoka, Japan
menna.fateen@m.ait.kyushu-u.ac.jp

Abstract. Student performance prediction continues to be a focus of research in educational data mining due to its many potential benefits. While teachers' assessment reports are a crucial part of the educational process, they have not been commonly used in performance prediction. We propose a model that uses similarity learning as an embedding-enhancing technique. Results outperform earlier research with an average accuracy of 73% for detecting strong performance.

Keywords: similarity learning · sentiment analysis · teacher reports · performance prediction · text mining · machine learning

1 Introduction

Academic performance prediction has been an active area of research in educational data mining [6]. By obtaining early warning signs of potential struggles, performance prediction models can aid in the identification of students who may need additional support before they become significant problems. However, while various factors have been investigated, teachers' observation reports have rarely been used. Teachers play an important role in traditional classes and are responsible for assessing their students' performance. Such assessment routines have been shown to be affected by the types of mindsets that teachers have [5].

With this in mind, we aim to investigate if teachers' reports can be used to enhance performance prediction models. We build upon [3] 's work in which sentiment and topic features were extracted to build an interpretable input vector. In their work, the initial step was topic modeling. While topic modeling can give an in-depth understanding of the reports, it requires extensive data pre-processing and is also an unsupervised problem which can lead to ambiguity in results. Therefore, we investigate of sentiment in teacher reports and how it can help to improve prediction with a more straightforward approach. We further pre-train a siamese Sentence-BERT [7] network on the similarity of the reports according to the sentiment. After finetuning SBERT, we can use the model to obtain optimized embeddings for the prediction task which acts as a simpler method than manually extracting features. Moreover, since it is common for multiple

© The Author(s), under exclusive license to Springer Nature Switzerland AG 2023
N. Wang et al. (Eds.): AIED 2023, CCIS 1831, pp. 720–726, 2023.
https://doi.org/10.1007/978-3-031-36336-8_111

Table 1. Number of reports and students and date ranges in experiments

	Finetuning Dataset	Testing Dataset
Reports	26,931	14,313
Range	March 2021–August 2021	May 2020–October 2020
Students	433	180

teachers with different styles to contribute to a single student's reports, it could be challenging to effectively compare and analyze the reports. Similarity representation learning can help in overcoming this problem and to the best of our knowledge, has not been applied in this context. Our research aims to answer two main questions. **RQ1:** Does a correlation exist between sentiment scores extracted from teacher reports and the students' final performance? **RQ2:** Can similarity learning utilizing sentiment scores enhance teacher report embeddings for performance prediction? We address these questions through our experiments and verify the results with a dataset covering reports on different subjects. We compare our results with previous research [4] that has used teacher reports for performance prediction and show how our method can contribute to the field.

2   Data Description

The dataset used in this study consists of teacher observation reports provided by a cram school in Japan for middle school students. The reports contain brief remarks for each student, an understanding score, and information about the material covered in the lessons which cover subjects: Japanese, Math, English, Science, and Social Studies. The feedback is mainly written for parents and consists of around 4 short sentences. The dataset also includes regular test scores of a subset of the attending students from their respective regular schools, which were used as features to predict their performance. Additionally, the provided results of simulation exams taken by the students before the actual entrance exam were used as target variables. For the experiments, the reports were split into two subsets, a fine-tuning set and a testing set and are described in Table 1. The finetuning dataset of 2021 was used for finetuning SBERT on similarity learning only. To evaluate the performance of the model, we used the unseen reports from the 2020 testing dataset. The order of using 2020 for testing and 2021 for finetuning is irrelevant because the students and reports from both years differ and therefore the datasets are considered independent.

The experimental setup employs three main feature sets for comparison. The first feature set is derived from the teacher reports and includes comment embeddings, students' average sentiment scores, and average understanding scores. The quartiles of both scores are also appended to the vectfor. The second feature set includes only the regular test scores, while the third feature set adds the regular scores to the teacher comment embeddings. The study compares the performance of these feature sets in predicting the students' performance in the simulation exams taken in September 2020 and September 2021.

3 Methodology

3.1 Preprocessing

The teachers at the cram school are required to write down reports for the students after each class to encourage them and record their progress. However, since teachers write these reports daily for each student, some parts of the reports are often repetitive and could be considered redundant. To eliminate redundancy, we first divided the reports into sentences. Frequency per sentence was then calculated for the entire dataset. The threshold for sentence removal was determined according to the frequency of repetition, sentences repeated more than 300 times were eliminated. This number was determined through experimental settings to optimize for sentiment score and final score correlation. It was challenging to determine this threshold due to the nature of the dataset where repetition is common. In prior work [2–4], such sentences were not eliminated.

3.2 Sentiment Score Extraction

In order to determine the sentiment of the teachers' comments which are not initially accompanied by sentiment scores, we utilized Twitter-XLM-roBERTa-base, a multilingual model trained on 198M tweets and finetuned for sentiment analysis [1]. The model generates either a positive, negative or neutral label and a confidence score. After extracting the labels for each sentence, the whole report was assigned a label based on the 'NPN' rule, or a precedence rule where if one sentence in a report was labeled 'Negative' then the whole report is assigned 'Negative'. If no sentences were 'Negative' but there existed one sentence with a 'Positive' label, then the report is assigned 'Positive' and finally if no sentences are either 'Negative' or 'Positive' then the report is assigned a 'Neutral' label.

Many of the redundant sentences removed previously had positive sentiments which affected the label assigned to the whole report. After the elimination of these sentences, the percentage of positive reports dropped from 66% to 52% and neutral reports increased from accounting to 16.2% of the reports to 28.2%. Moreover, the number of total 2020 reports dropped by 11.5% meaning that 1,651 reports consisted of redundant sentences only.

To address **RQ1**, a bivariate correlation analysis was conducted on the fine-tuning and testing datasets to examine the relationship between the ratio of positive sentiment reports to the total reports of each student and their simulation exam score. The results showed a moderate correlation between sentiment ratio and final score, with a correlation coefficient value of 0.2 for the fine-tuning dataset and an average correlation coefficient value of 0.15 for the testing dataset. Neutral-labeled reports were removed to increase the correlation coefficient to 0.19 for the testing dataset. Overall, the relationship between sentiment ratio and final score, although moderate, has been proven to exist.

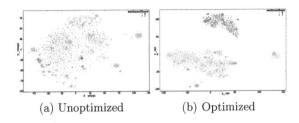

(a) Unoptimized   (b) Optimized

Fig. 1. Math report embeddings labeled according to sentiment score

3.3 SBERT Finetuning

To generate more optimized embeddings than those used in previous work [2–4], we fine-tune Sentence-BERT [7] with similarity learning. The goal of SBERT is to bring sentence embeddings with similar meanings closer in the vector space and separate those with different meanings. To fine-tune SBERT for semantic similarity, we build pairs of reports that were labeled positive and assign these pairs a similarity score of 1 and pairs of positive and negative reports with a score of 0. We randomly built 34,914 pairs of positive and positive reports and 17,457 pairs of positive and negative reports. The number of positive pairs was built to be twice the number of negative and positive ones to account for the imbalance in distribution. The siamese network is then trained with a mean square error loss function between the cosine similarity of two input embeddings, $rprt_x$ and $rprt_y$ and the assigned similarity score $sim_{xy} = \{0, 1\}$ as defined in Eq. 1.

$$\mathcal{L} = \frac{1}{n} \sum^{n} (sim_{xy} - cos(embed(rprt_x), embed(rprt_y)))^2 \qquad (1)$$

After finetuning SBERT, optimized sentence embeddings were obtained for the testing dataset from the model by averaging the word vectors in the report. Following [3,4], we summed the embeddings for each student. To visualize the optimized embeddings, we additionally extracted sentiments for the testing set. Using t-SNE, we plot the Math report embeddings obtained from the finetuned model (Fig. 1b) and those obtained from the same SBERT base network model but with no finetuning (Fig. 1a). Each point refers to a report and is labeled with the assigned sentiment. As seen in Fig. 1, the network was able to cluster the embeddings according to the assigned sentiments.

3.4 Performance Prediction with Optimized Embeddings

We trained five separate regression and classification XGBoost models for each subject. We then evaluate the model when predicting the final subject score with mean absolute error (MAE) and with accuracy when predicting both grade and strong/weak. Strong/weak is defined as obtaining a score exceeding 50% while the 4-class grades are predicted according to the converted letter grades: (A: 100-75, B: 75-50, C: 50-25, D: 25-0). Figure 2 shows an illustration of the steps.

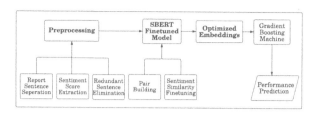

Fig. 2. Proposed Methodology Steps

4   Experimental Results

Experiments on the testing dataset were evaluated using 10-fold cross-validation. Table 2 shows the average MAE and accuracy of 4-class and strong/weak when using optimized embeddings with our method compared to using embeddings extracted from the same-base SBERT network with no finetuning. On average the best-performing embedding was our optimized embeddings. Due to unbalance in the 4-grade class distribution in the testing datasets, we compare the accuracy of the model with the baseline of majority class classification. While FS_1 using both embeddings was unable to outperform the 4-class baseline, adding the regular scores to the embeddings showed an increase of 9% in accuracy when using the optimized embeddings vs an increase in 4% with the unoptimized ones. On average, FS_3 was able to outperform FS_2 with the optimized embeddings showing the advantage of finetuning. FS_2 shows the same performance in both embedding types because the feature set does not rely on report embeddings.

To compare with [4], we similarily build vectors with no redundancy elimination and with unoptimized embeddings. The average of the subject results in each method is shown in Table 3. Even though the 'Students Model' approach [4] outperformed the similarity learning approach in terms of MAE with FS_1, our approach was able to outperform in all other cases.

Table 2. MAE, 4-Score and strong/weak classification accuracies. Bold show the best embedding method results of each subject per feature set.

		Optimized			Unoptimized			4 Class Baseline
		MAE	4 Class	S/W	MAE	4 Class	S/W	
FS1	Japanese	**9.55**	0.53	**0.59**	9.66	0.55	0.55	**0.58**
	Math	**12.08**	**0.46**	**0.61**	12.25	0.43	0.61	0.46
	Science	**17.32**	**0.42**	**0.59**	18.52	0.39	0.48	0.41
	Social Studies	**13.93**	0.41	**0.66**	14.13	0.43	0.59	**0.44**
	English	14.21	0.44	0.56	**13.92**	0.42	**0.62**	**0.45**
	Average	**13.42**	0.45	**0.60**	13.70	0.45	0.57	**0.47**
FS2	**Japanese**	9.58	0.56	0.66	9.58	0.56	0.66	0.61
	Math	9.50	0.46	0.67	9.50	0.46	0.67	0.46
	Science	12.28	0.56	0.78	12.28	0.56	0.78	0.41
	Social Studies	11.87	0.46	0.69	11.87	0.46	0.69	0.44
	English	11.44	0.56	0.76	11.44	0.56	0.76	0.42
	Average	10.94	0.52	0.71	10.94	0.52	0.71	0.47
FS3	Japanese	9.17	0.58	**0.62**	10.13	0.54	0.62	**0.61**
	Math	**9.82**	**0.58**	**0.70**	10.07	0.55	0.68	0.46
	Science	12.36	**0.52**	**0.78**	**11.65**	0.48	0.74	0.41
	Social Studies	11.28	**0.50**	**0.74**	**11.21**	0.43	0.64	0.44
	English	11.61	**0.50**	0.78	**11.01**	0.47	**0.79**	0.42
	Average	10.85	**0.54**	**0.73**	**10.81**	0.49	0.69	0.47

Table 3. Similarity learning results compared with 'Students Model' approach

		Similarity Learning			Students Model		
		MAE	4-Class	S/W	MAE	4-Class	S/W
FS1	Average	**13.42**	**0.45**	**0.60**	14.09	0.43	0.56
FS3	Average	10.85	**0.54**	**0.73**	**10.78**	0.51	0.70

5   Conclusion

In this study, we investigated the use of teacher reports and sentiment scores in performance prediction. The results showed a correlation between sentiment scores and performance and that similarity learning can enhance prediction models. Our study encourages the exploration of other factors correlated with performance and the use of such attributes in similarity learning for more accuracy.

Acknowledgements. This work was supported in part by JST, the establishment of university fellowships towards the creation of science technology innovation (JPMJFS2132), KAKENHI (JP21H00907) and e-sia Corporation.

References

1. Barbieri, F., Espinosa Anke, L., Camacho-Collados, J.: Xlm-t: multilingual language models in twitter for sentiment analysis and beyond. In: Proceedings of the Language Resources and Evaluation Conference, pp. 258–266. European Language Resources Association, Marseille (2022). https://aclanthology.org/2022.lrec-1.27
2. Fateen, M., Mine, T.: Predicting student performance using teacher observation reports. In: International Educational Data Mining Society (2021)
3. Fateen, M., Mine, T.: Extraction of useful observational features from teacher reports for student performance prediction. In: International Conference on Artificial Intelligence in Education, pp. 620–625. Springer, Heidelberg (2022). https://doi.org/10.1007/978-3-031-11644-5_58
4. Fateen, M., Ueno, K., Mine, T.: An improved model to predict student performance using teacher observation reports. In: Proceedings of the 29th International Conference on Computers in Education Conference, ICCE (2021)
5. Murphy, L., Thomas, L.: Dangers of a fixed mindset: implications of self-theories research for computer science education. In: Proceedings of the 13th Annual Conference on Innovation and Technology in Computer Science Education, pp. 271–275 (2008)
6. Namoun, A., Alshanqiti, A.: Predicting student performance using data mining and learning analytics techniques: a systematic literature review. Appl. Sci. 11(1), 237 (2020)
7. Reimers, N., Gurevych, I.: Sentence-bert: sentence embeddings using siamese bert-networks. arXiv preprint arXiv:1908.10084 (2019)

Using Simple Text Mining Tools to Power an Intelligent Learning System for Lengthy, Domain Specific Texts

John Sabatini and John Hollander[(✉)]

University of Memphis, Memphis, USA
jmhllndr@memphis.edu

Abstract. This paper describes the development of a system for tracking the opportunities for vocabulary and conceptual learning within lengthy subject area texts. Specifically, we leveraged text mining tools to observe the cumulative frequency of words throughout a novel and chapters from two college-level textbooks. We describe how cumulative lexical occurrence and frequency information may be made accessible and useful to researchers and educators without the need for them to have deep knowledge of NLP or other computational linguistic skills. We applied these tools to three different types of texts, a narrative novel, and two textbooks and provide some preliminary results. Finally, we describe potential methods of using this information to structure learning activities utilizing topical analytics and automatic item generation techniques. By leveraging cumulative lexical frequency throughout a text, researchers and educators may be able to isolate words that may be both difficult and important in comprehending any given text, as well as building domain-specific knowledge.

Keywords: Text mining · lexical frequency · automatic item generation

1 Introduction

Students in secondary education and beyond gain a significant portion of their knowledge in various subject areas through reading, and reading outside of class is required and utilized as the foundation for lectures, class activities, and homework. Throughout, students encounter words that they have rarely, if ever, seen or heard before. Other times, difficult words may be colloquially familiar to students, but their technical or domain-specific meanings are not yet clear. However, the most difficult words in a text are often also critically important to deeply comprehending and learning from it [1]. In fiction, particularly in narrative texts such as novels, these difficult-but-important words may include proper nouns, such as character or place names, or unique concepts created within the universe of a given book. In non-fiction works, such as textbooks, these words are often specialized terms with highly specific definitions that do not frequently occur in other contexts.

Previous research illustrates the effectiveness of scaffolding learning around vocabulary, such as simple definitions for essential concepts and components, can contribute

© The Author(s), under exclusive license to Springer Nature Switzerland AG 2023
N. Wang et al. (Eds.): AIED 2023, CCIS 1831, pp. 727–733, 2023.
https://doi.org/10.1007/978-3-031-36336-8_112

to the development of academic vocabulary and content knowledge [2]. Using elaboration or another scaffolding technique, such as linguistic simplification, can also enhance students' comprehension of text content [3]. In practice, most text modifications use a combination of simplification and elaboration techniques aligned with curriculum and individual learner needs.

As a result of this research, educational technologies have been developed to use alongside text-based activities to improve learning. For example, the Language Muse system [4] is an online tool designed to aid teachers in creating linguistically-focused instructional materials for content-area curriculum, including the development of lesson plans, activities, and assessments. In the Language Muse application, the developers created a text analysis tool, coupled with an automated item generation and scoring engine based using NLP features. The system took as input any brief (~1000 word), subject area text (e.g., biology). NLP analytic features would identify affordances for generating automatic items aligned to a knowledge framework for language development. This system generated up to 17 different item types, most of which were selected response items that could be automatically scored. Short answer and summary constructed response items were scored using machine learning algorithms based on rubrics.

In a related project explored how to automatically generate vocabulary items as they relate to knowledge domains [5]. For breadth, the team used topical analysis algorithms to generate co-occurrence parameters to cluster knowledge corpora into topical vocabulary/knowledge lists. These statistics were merged with general linguistic item statistics. The results were automatically generated items that predicted background knowledge in the domain. Vocabulary depth has multiple dimensions. Polysemy and multi-word phrase detectors and classifications are critical for distinguishing common from technical meanings of terms (e.g., prime number vs. prime rib), an important distinction when learning in technical domains. In this project, the team used semantic network maps derived to generate conceptual depth of knowledge items. For example, students had to identify relationships among *warrant, claim,* and *evidence* in demonstrating knowledge of *argument structures.* Using such techniques, the team was able to assess depth of knowledge of technical terms and their interrelationships, efficiently and reliably.

These systems created the grist for learning-embedded assessment using the testing effect to stimulate learning [6]. However, each had several shortcomings with respect to creating a more intelligent learning system for advanced learning in a subject area. Language Muse only operated on short texts (~1000 words) and therefore was limited with respect to how one might design learning to build long-term acquisition of knowledge in a domain. By utilizing a more robust content area corpora consisting of standards documents and subject area texts, the vocabulary project team was able to create richer topical vocabulary lists for probing student knowledge, but the lists were not aligned to specific text content. Here, we take the next step in conceptualizing how to use text mining and a landscape map of where, when, and how frequently topical vocabulary occur and recur in longer texts, as a prerequisite step to building algorithms that systematically probe students' learning as they process content area textbook chapters or novels. We demonstrate how these analytics can be operated across longer texts using simple tools that can be made available to educators and researchers for exploring how content is represented via general and technical vocabulary. In the conclusion, we provide a brief discussion of how these tools could be incorporated into an intelligent learning system.

2 Methods

We utilize the tidytext R package [7] to break down three texts: the first chapters of a college-level history textbook (Ch.1 "Indigenous America" [8]), a biology textbook (Ch. 1 "Introduction to Biology" [9]), and Mary Shelley's *The Modern Prometheus* (Franken-stein [10]). We unnested the texts into matrices such that each word is represented by a row, removing stop words. We appended to this matrix a few count-based measures: the total number of words in the text up to that point (i.e., each word's index), the number of times each word has appeared in the text so far (i.e., cumulative frequency), and the number of times each word appears in the text as a whole (i.e., total frequency). We also appended linguistic feature variables such as each word's log frequency in the SUBTLEX-US corpus [11] as well as each word's character length. In this poster, we present results based solely on word frequencies.

3 Results

Snapshots of results are organized here to demonstrate the educational relevance of assembling this basic textual information in a format that is accessible to educators and as drivers of an automated learning system. To this end, we present sample tables, figures, and analyses that may convey this utility. Table 1 displays a list of the most frequently used words in each text. Note that for the History and Biology texts, the content focus is immediately apparent from the repetition rate. For the novel, the top recurring words seem to be aligned with thematic content. This table also displays difficult words (defined by their rarity in the SUBTLEX-US corpus) in each text and how often they appear. In the History and Biology texts, these terms are indicative of the upper bound of vocabulary knowledge and learning in each domain. In the novel, terms like 'wretchedness' and 'endeavor' speak to the theme and content. Note: if we include variants (wretch, wretched, wretchedly, wretchedness), we observe 65 occurrences.

In Table 2, we clustered the total words in the text by bands of SUBTLEX frequency. This tiering of the words helps isolate the volume of highly frequent words in the text (1000 most frequent words); moderately frequent words (1001 to 10000); and rarer content words (greater than 10000). Of these rarer words, we would apply co-occurance filters (such as we developed in the Vocabulary project) to distinguish content-relevant vs. -irrelevant terms, as well as distinguishing proper nouns from dictionary terms.

As a final step, Fig. 1 shows how we might graphically map the positional occurrence of terms across a longer text. For example, we see that the term Native (Americans) occurs frequently starting in the first 1000 words of the Chapter. The first occurrence of the word Spanish, on the other hand, is not until the 2000th word in the chapter. Figure 2 illustrates a more complex example by displaying the 5 most frequent words in the biology text. By building item probes in relative proximity to the density of occurrences (before, during, after), we can more selectively help students to identify relevant content and reinforce it throughout learning.

Table 1. Top 10 words most frequent and most difficult (rarest in SUBTLEX corpus) within each text and their relative frequency.

Most frequent					
History		Biology		Frankenstein	
word	n	word	n	word	n
native	61	science	99	life	115
spanish	58	biology	62	father	112
world	47	scientific	52	eyes	104
america	39	organisms	49	time	98
Most difficult					
mestizos	5	inductive	9	wretchedness	14
indies	4	binomial	3	endeavouring	9
benin	2	genera	3	traversed	6
benignly	1	physiologists	3	darted	5
cassava	1	generalizations	2	cheerfulness	4

Table 2. Word count by frequency tier.

SUBTLEX tier	num. words	% Total Words	Unique Words	% Unique Words	Sample of Unique Words in Text
History					
top 1,000	585	11.94%	170	2.95%	world, american, people, city, power history, day, land
1,001–10,000	2849	58.14%	1139	19.79%	north, Atlantic, press, university, social, population, peoples, empire
10,001 +	1466	29.92%	823	14.30%	festered, seasonally, stockades, tributaries, undergrowth, appointees, dissemination, earthen
Biology					

(*continued*)

Table 2. (*continued*)

SUBTLEX tier	num. words	% Total Words	Unique Words	% Unique Words	Sample of Unique Words in Text
top 1,000	455	16.09%	135	10.97%	life, world, questions, change, called, earth, human, system
1,001–10,000	2218	78.38%	724	58.81%	method, scientists, species, study, applied, exist, specific, results
10,001 +	894	40.33%	372	30.22%	observations, tissues, inductive, phylogenetic, biosphere, populations, molecule, disciplines
Frankenstein					
top 1,000	3028	22.84%	464	7.70%	death, dear, friend, heard, love, human, words, country
1,001–10,000	7234	54.56%	2934	48.68%	cottage, despair, happiness, chapter, horror, scene, creature, joy
10,001 +	2995	22.59%	2629	43.62%	proceeded, labours, misfortunes, reflections, wept, awoke, reflected, solitude

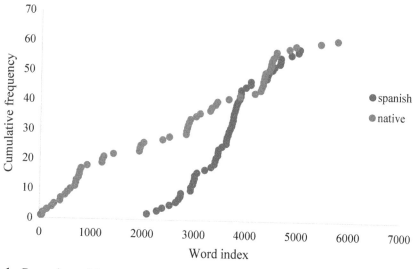

Fig. 1. Comparison of the terms "spanish" and "native" as they appear throughout the history text.

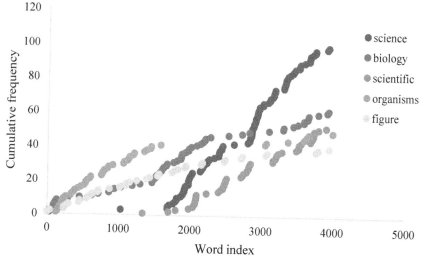

Fig. 2. Comparison of the 5 most frequent terms as they appear throughout the biology text.

4   Conclusion

In this paper, we have provided a glimpse into the components of an intelligent learning system that utilizes simple text mining techniques with automatic item generation to enhance content and vocabulary learning. We described previous systems that we have developed to identify developmental, topical vocabularies which can be used to generate test items that are indicators of breadth and depth of content knowledge. We also described previous research into a system that can generate a variety of item types from texts. Here we have demonstrated how using simple frequency counts of where and when topical vocabulary occur in the text, we could clue in educators to affordances of longer texts for vocabulary and concept learning, and how we could power an automated system that previews key content, tests for learning as terms occur, and delivers post learning quizzes as follow up to evaluate long term retention. Next text analytic steps include sentence/syntactic level adjustments for polysemy (which sense of a word is used in sentence), multi-word phrases (Native American), morphology (lemmas, affixes), and semantic relations (knowledge mapping).

References

1. Marks, C.B., Doctorow, M.J., Wittrock, M.C.: Word frequency and reading comprehension. J. Educ. Res. **67**(6), 259 (1974)
2. Nagy, W.E., Herman, P.A.: Breadth and depth of vocabulary knowledge: Implications for acquisition and instruction. Nat. Vocab. Acquis. 19–35 (2014)
3. Sharma, P., Hannafin, M.J.: Scaffolding in technology-enhanced learning environments. Interact. Learn. Environ. **15**(1), 27–46 (2007)
4. Burstein, J., Madnani, N., Sabatini, J., McCaffrey, D., Biggers, K., Dreier, K.: Generating language activities in real-time for English learners using language muse. In: Proceedings of the Fourth ACM Conference on Learning@ Scale, pp. 213–215 (2017)
5. Deane, P., Lawless, R.R., Li, C., Sabatini, J., Bejar, I.I., O'Reilly, T.: Creating vocabulary item types that measure students' depth of semantic knowledge. ETS Res. Rep. Ser. **2014**(1), 1–19 (2014)
6. Rowland, C.A.: The effect of testing versus restudy on retention: a meta-analytic review of the testing effect. Psychol. Bull. **140**(6), 1432 (2014)
7. Silge, J., Robinson, D.: tidytext: Text mining and analysis using tidy data principles in R. J. Open Source Softw. **1**(3), 37 (2016)
8. Locke, J.L., Wright, B.: The American Yawp: A Massively Collaborative Open US History Textbook, vol. 1–1877. Stanford University Press (2019)
9. Fowler, S., Roush, R., Wise, J.: Concepts of Biology (OpenStax). OpenStax (2013)
10. Shelley, M.: The Modern Prometheus. U. K. Lackington Hughes Harding Mavor Jones, 1818
11. Brysbaert, M., New, B., Keuleers, E.: Adding part-of-speech information to the SUBTLEX-US word frequencies. Behav. Res. Methods **44**(4), 991–997 (2012). https://doi.org/10.3758/s13428-012-0190-4

Question Classification with Constrained Resources: A Study with Coding Exercises

Luiz Rodrigues[1]([✉]) [ID], Filipe Pereira[2] [ID], Jario Santos[1] [ID], Elaine Oliveira[3] [ID], Isabela Gasparini[4] [ID], Rafael Mello[1,5] [ID], Leonardo Marques[1] [ID], Diego Dermeval[1] [ID], Ig Ibert Bittencourt[1,7] [ID], and Seiji Isotani[6,7] [ID]

[1] Center for Excellence in Social Technologies (NEES), Federal University of Alagoas, Maceió, Brazil
luiz.rodrigues@nees.ufal.br
[2] Department of Computer Science, Federal University of Roraima, Boa Vista, Roraima, Brazil
[3] Institute of Computing, Federal University of Amazonas, Manaus, Brazil
[4] Department of Computer Science, Santa Catarina State University, Joinville, Brazil
[5] Computing Department, Federal Rural University of Recife, Recife, Brazil
[6] Institute of Mathematics and Computer Science, University of São Paulo, São Carlos, Brazil
[7] Graduate School of Education, Harvard University, Cambridge, MA, USA

Abstract. Evidence-based learning strategies, such as the testing effect, might help address the achievement gap. However, exploiting the testing effect depends on having a set of instructional activities with fine-grained tagging. While instructors might find questions in textbooks, they often lack fine-grained tagging, and data labeling is laborious. Despite much research on text classification, to our best knowledge, state-of-the-art question classifiers are mostly based on extensive models (i.e., BERT) and English text. Respectively, those are incompatible with the resource-constrained devices (e.g., mobile) and languages (e.g., Portuguese) of many underprivileged countries in the global south. Therefore, we developed a question classifier on top of DistilBERT, a version of BERT compatible with resource-constrained applications, using grid search and hold-out. Based on a corpus of 1045 coding questions written in Brazilian Portuguese, we found a model that achieved a near-perfect performance on unseen data, similar to last-generation results using BERT for English text. Thus, we present a step towards equitable education by i) providing underprivileged Portuguese-speaking countries with the support that enables opportunities already available for first-world countries and ii) demonstrating the feasibility of creating resource-constrained applications compatible with state-of-the-art AIED systems.

Keywords: Natural Language Processing · Retrieval Practice · Inclusive and Equitable Learning · Resource-limited

N. Wang et al. (Eds.): AIED 2023, CCIS 1831, pp. 734–740, 2023.
https://doi.org/10.1007/978-3-031-36336-8_113

1 Introduction

The Covid-19 pandemic has led to an increased achievement gap in several countries [9]. Instructors and policy-makers could use evidence-based learning strategies, such as the testing effect, to address this gap. The testing effect has shown benefits for learning achievement compared to standard activities, and it could help instructors recommend personalized learning activities based on students' knowledge gaps [1]. However, finding annotated questions with fine-grained tags is challenging. For instance, they might find relevant questions in textbooks and other educational resources, which often lack fine-grained annotations.

Helping instructors with this issue is important because data tagging is laborious. To address that issue, one might use Question Classification (QC) [2] to automatically tag questions from a textbook's PDF file or printed version, which might be converted to text through a mobile Optical Character Recognizer (e.g., [3]). However, the state-of-the-art on QC is mainly based on extensive models, such as BERT, and is often limited to English texts [6,7]. Consequently, these models are incompatible with the *resource-constrained* devices (i.e., low-cost and low-consuming devices with restricted memory, storage, and internet access) commonly available in global south's countries and their common languages (e.g., Portuguese and Spanish) [4]. This context presents equity issues, as people in these countries do not have the same opportunities as those who speak English and have access to technology. Accordingly, we need question classifiers that adequately consider such languages and resource constraints.

To start addressing that need, we conducted an experimental study to develop a resource-constrained model for fine-grained QC. While prior research primarily uses BERT and is limited to English texts [6,7], we explore DistilBERT, a compressed version of BERT that retains 97% of its language understanding capabilities, has 40% fewer parameters and runs inferences 60% faster [8]. Thereby, we considered DistilBERT a compelling alternative for resource-constrained applications targeting global south countries. Based on a corpus of 1045 Portuguese-written coding questions, we validated and tested our classifier, which achieved a near-perfect performance on unseen data (F-measure and Kappa > 0.84 [10]). This result is comparable to related work [6,7], although our classifier is compatible with resource-constrained applications.

Thus, we provide empirical evidence on the feasibility of achieving state-of-the-art performance with a resource-constrained classifier for Portuguese-written questions. This model facilitates creating resource-constrained educational technologies (e.g., for mobile applications), which more closely align with the resources available in Portuguese-speaking regions and schools of the global south. Additionally, DistilBERT's ability to handle Spanish words facilitates expanding our classifier to help instructors of most American Latin and Caribbean countries. Hence, our classifier (available in t.ly/POMs) provides a step towards ensuring inclusive and equitable education for all by helping the deployment of test-based learning in the global south's Portuguese-speaking countries without the burden of manually annotating questions' fine-grained topics. Furthermore, this paper proves the possibility of achieving state-of-the-art performance while targeting resource-constrained devices and will serve as the foundation of AIED systems underdevelopment in projects of Brazil's Ministry of Education.

2   Method

This paper developed and tested a neural network for fine-grained QC on questions written in Brazilian Portuguese using the multilingual DistilBERT. For this, we used a dataset of coding exercises extracted from CodeBench[1], which encompasses 1045 exercises classified into nine fine-grained topics (see Table 1). Because topics might overlap (e.g., *if-else* is within *nested if-else*), each exercise only features the most complex topic required to solve it.

Table 1. Description of the dataset used for question classification in this study.

Topic	Description	Count (%)
Nested if-else	Handling nested conditional structure	161 (0.154)
Arrays	Operations on arrays/lists	160 (0.153)
Sequential Structure	Handling variables and arithmetic operations	157 (0.150)
If-else	Handling conditionals	136 (0.130)
Matrices	Handling bidimensional arrays/lists	134 (0.128)
While-loop	Handling repetitions using *while*	117 (0.112)
For-loop	Handling repetitions using *for*	114 (0.109)
Strings	Handling strings	47 (0.045)
Print and Input	Printing information and handling a single variable	19 (0.018)

For preprocessing, we conducted two steps. Following similar research (e.g., [7]), we first cleaned the whole dataset by i) removing stopwords; ii) lowercasing all characters; iii) removing Unicode and additional spaces, tabs, and newline symbols; and iv) replacing all punctuation with a blank space. Next, we used DistilBERT and its respective encoder [8], as we describe next.

We used artificial neural networks to develop the classifier, considering they have achieved state-of-the-art performance in similar tasks (e.g., [6]). Our network topology featured five layers. First, the input, which receives the prepared texts. Second, DistilBERT's encoder. Third, DistilBERT, which receives the encoded texts and generates their embeddings. Fourth, a hidden layer aimed to learn our problem's specific patterns, which we chose based on a grid search that combined a layer type (simple RNN, GRU, Bidirectional LSTM, or Convolutional) and a number of units (8, 16, 32, or 64). Note that we opted for a single

---

[1] https://codebench.icomp.ufam.edu.br/.

hidden layer to prevent exponentially increasing the model's complexity in light of the resource-constrained context. Lastly, the output layer is dense with nine units and softmax activation to support our problem's nine classes [5].

Furthermore, Sparse Categorical Cross Entropy was the loss function and Adam was the optimizer. Whereas the number of epochs was 1000000 to ensure the model had time to converge, we set a callback to stop training once the model failed to improve validation loss for 10 epochs. We also used a checkpoint callback to recover the model with the smallest validation loss once the train stopped. Those are standard practices for training neural networks that enable parameter optimization while preventing overfitting and unnecessary computation [5].

We used grid search, based on the search space described before, to find the best-hidden layer for our classifier and validate its performance. This choice is inspired by prior research [7], which also informed our convolutional layer setting: kernel size was eight, followed by a Global Max Pooling. To mitigate overfitting, we ran each of the 16 possible combinations 10 times. In each one, we split the training set using a different seed and saved 10% of it for validation. This split followed the procedure described before, and we evaluated each combination based on its overall performance on validation data. Finally, we relied on Macro F-measure (F1) and Cohen's Kappa to evaluate the grid search results because accuracy might be misleading for unbalanced data like ours [5].

To further mitigate overfitting, we used a stratified hold-out strategy based on labels' distribution (see Table 1). We used 70% (n = 731) of our data for training (i.e., grid search's input that was split while validating each setting) and the remainder (n = 314) for testing the grid search's best settings on unseen data. To select the best settings, we first found the one that achieved the highest F1/Kappa on grid search. Second, we selected all settings where the 95% confidence intervals (CI) of their F1/Kappa overlapped with that of the best model. Those were then trained using the whole training set. Notably, this training differs from that of grid search in a single aspect: Because no validation data was available in this step, we opted for stopping training after a single epoch without improvement in the training loss to prevent overfitting. Lastly, the best models were evaluated on the testing set to check their generalization to unseen data.

3 Results and Discussion

Table 2 presents the grid search results. It shows the convolutional layer with 16, 32, and 64 units and the bidirectional layer with 32 units achieved the best performances, considering their CIs overlap with that of the best setting (F1: 0.852, Kappa: 0.835). Accordingly, Table 2 also presents the results of evaluating these models on the testing set (n = 314). For settings using a convolutional layer, the results confirm those of the grid search: all of them yielded both F1 and Kappa values above 0.80 and within the grid search's CI. In contrast, for the setting based on the bidirectional LSTM, the results for F1 (0.794) and Kappa (0.762) were close but below the grid search's lower CIs (0.802 and 0.775, respectively). Thus, we considered the setting using a 64-unit convolutional layer as our best model given it performed best in both grid search and testing phases. This model (see t.ly/POMs) is ready to use in Python applications and can be converted and used in mobile and web-based applications using Tensorflow.

Table 2. Results shown as Mean (Lower CI - Upper CI). To validate each setting, we calculated metrics based on a validation set (n = 73; 10% of the training set) randomly extracted from the training set for 10 trials. For testing, we only selected settings whose CIs overlapped with the mean performance of the highest mean on validation.

Setting		Grid Search		Test	
Layer	Units	F1	Kappa	F1	Kappa
RNN	8	0.542 (0.503–0.581)	0.491 (0.448–0.535)		
RNN	16	0.659 (0.633–0.686)	0.617 (0.588–0.646)		
RNN	32	0.684 (0.659–0.710)	0.644 (0.614–0.673)		
RNN	64	0.680 (0.654–0.706)	0.639 (0.607–0.671)		
GRU	8	0.678 (0.657–0.698)	0.640 (0.618–0.662)		
GRU	16	0.750 (0.720–0.779)	0.718 (0.685–0.752)		
GRU	32	0.765 (0.735–0.796)	0.738 (0.704–0.773)		
GRU	64	0.767 (0.746–0.787)	0.739 (0.716–0.761)		
Bidirectional LSTM	8	0.735 (0.695–0.776)	0.702 (0.657–0.746)		
Bidirectional LSTM	16	0.784 (0.753–0.814)	0.756 (0.723–0.789)		
Bidirectional LSTM	32	**0.820** (0.802–0.838)	**0.796** (0.775–0.818)	0.794	0.762
Bidirectional LSTM	64	0.800 (0.782–0.817)	0.776 (0.754–0.798)		
Convolutional	8	0.787 (0.754–0.820)	0.763 (0.728–0.799)		
Convolutional	16	**0.812** (0.789–0.836)	**0.789** (0.764–0.813)	0.818	0.791
Convolutional	32	**0.837** (0.819–0.856)	**0.815** (0.793–0.837)	0.842	0.817
Convolutional	64	**0.852** (0.828–0.877)	**0.835** (0.808–0.863)	0.862	0.842

Based on grid search and hold-out, our classifier achieved a near-perfect performance (kappa = 0.842) on unseen data. This performance is comparable to that of prior research [6,7], but based on a resource-constrained model able to encode some languages of global south countries. Accordingly, our contribution enables the development of AIED systems to mitigate the achievement gap based on test-based strategies, as well as helps promote the inclusion of Portuguese-speaking underprivileged countries in the global south by using a model able to handle some of their languages. For developers, we provide a ready-to-use question classifier for coding problems, which can be embedded into resource-constrained applications to i) reduce the burden of manually annotating questions' fine-grained topics, ii) enable the automatic generation of quizzes tailored to student's needs, and iii) drive the generation of topic-based questions. For researchers, we demonstrate the feasibility of creating resource-constrained-compatible question classifiers with state-of-the-art performance.

Furthermore, this paper presents the foundation for research funded by our country's Ministry of Education that aims at developing AIED systems for addressing the achievement gap in underprivileged schools, which may lead to the design and implementation of public education policies for the whole country. Based on that context, we call for future research to expand our approach and aid underprivileged learners and instructors with other subjects and from different countries. Notice, however, that our findings are based on a corpus of 1045 introductory programming questions written in Brazilian Portuguese. While that limits our contribution's applicability and generalization, our results were comparable to those of related work, despite not fine-tuning DistilBERT, using a single hidden layer, and the relatively small sample. Additionally, although our dataset is established in the higher education context, it is still applicable to introductory programming learning in primary and secondary education. In Brazil, for instance, the computing curriculum in primary and secondary education was approved in 2022 and needs to consider the resources and connectivity constraints throughout the country to be implemented.

Thus, this paper readily contributes to educational technologies for introductory programming, demonstrates the feasibility of helping underprivileged populations with AIED for resource-constrained devices, and encourages studies to develop and deploy resource-constrained AIED applications to help address the achievement gap. Ultimately, this study is a step towards attending to the United Nations' goal of ensuring inclusive and equitable quality education for all - at least for Portuguese-speaking countries in the global south.

Acknowledgments. This study was financed in part by Brazilian National Council for Scientific and Technological Development (CNPq - 163932/2020-4, 308458/2020-6, 308395/2020-4, and 308513/2020-7); Coordination for the Improvement of Higher Education Personnel (CAPES - Finance Code 001); São Paulo State Research Support Foundation (FAPESP - 2013/07375-0); the Acuity Insights under the Alo Grant program; the Samsung-UFAM Project for Education and Research (SUPER).

References

1. Carpenter, S.K., Pan, S.C., Butler, A.C.: The science of effective learning with spacing and retrieval practice. Nat. Rev. Psychol. 1(9), 496–511 (2022)
2. Ferreira-Mello, R., André, M., Pinheiro, A., Costa, E., Romero, C.: Text mining in education. Wiley Interdisc. Rev. Data Min. Knowl. Disc. 9(6), e1332 (2019)
3. Freitas, E., et al.: Learning analytics desconectada: Um estudo de caso em análise de produçoes textuais. In: Anais do I Workshop de Aplicações Práticas de Learning Analytics em Instituições de Ensino no Brasil, pp. 40–49. SBC (2022)
4. Gašević, D.: Include us all! directions for adoption of learning analytics in the global south. In: Learning Analyfics for the Global South, pp. 1–22 (2018)
5. Géron, A.: Hands-on machine learning with Scikit-Learn, Keras, and TensorFlow (2022)
6. Mohammed, M., Omar, N.: Question classification based on bloom's taxonomy cognitive domain using modified tf-idf and word2vec. PloS One 15(3) (2020)
7. Pereira, F.D., et al.: Towards supporting cs1 instructors and learners with fine-grained topic detection in online judges (2022)
8. Sanh, V., Debut, L., Chaumond, J., Wolf, T.: Distilbert, a distilled version of bert: smaller, faster, cheaper and lighter. arXiv preprint arXiv:1910.01108 (2019)

9. UNESCO: Reimagining our futures together: a new social contract for education. UN (2022)
10. Viera, A.J., Garrett, J.M., et al.: Understanding interobserver agreement: the kappa statistic. Fam. Med. **37**(5), 360–363 (2005)

Even Boosting Stereotypes Increase the Gender Gap in Gamified Tutoring Systems: An Analysis of Self-efficacy, Flow and Learning

Maria Takeshita[1]([⊠]) [ID], Geiser Chalco Challco[1] [ID], Marcelo Reis[1] [ID], Jário Santos[2] [ID], Seiji Isotani[2,3] [ID], and Ig Ibert Bittencourt[1,3] [ID]

[1] NEES: Center for Excellence in Social Technologies, Federal University of Alagoas, Maceio, AL 57072-970, Brazil
mjst@ic.ufal.br
[2] University of Sao Paulo, Sao Paulo, SP 13566-590, Brazil
[3] Harvard Graduate School of Education, Cambridge, MA 02138, USA

Abstract. The stereotype threat, either racial or gender-related, has been proven as one of the factors influencing negatively students worldwide, causing from declines in academic performance and decrease in engagement to complete dropout from schools and universities. Gamification, or the use of game elements in context of non-games, has been widely used increase engagement and improve performance of students, with overall positive effects. Nonetheless, gamified environments are often stereotyped and cause opposite effects, preventing desired results from happening. Students can present inherent factors capable of opposing these threats, such id the case of self-efficacy, i.e., the ability to believe that one is capable of taking the necessary actions to complete a given task. In this study, we applied an experimental design with gamified environments with neutral and gender-stereotyped boosts and threats to male and female students of two technical schools. Results indicated that boost messages presented higher influence over male participants through increase expectations of self-efficacy. Moreover, female participants presented relatively higher performance results under neutral or threat conditions while compared to gender-oriented boosts. Based on these results, it is possible to conclude that boost messages were less effective, even causing negative effects on the performance female participants.

Keywords: Gamification · Self-efficacy · Stereotype Threat · Logic Teaching

1 Introduction

Gamification is a concept that involves the use of game elements in non-gaming contexts to enhance engagement and motivation [1]. It has been studied extensively in education due to its potential to make activities less tedious and improve performance, activating psychological factors [1–4]. One psychological factor that gamification can activate is self-efficacy, which is associated with self-awareness, perseverance, and autonomy, all of which are crucial for success. However, gamification is not without its flaws, one of which is the potential for non-inclusive design elements that perpetuate exclusive situations and lead to negative outcomes, creating a case of stereotype threat.

© The Author(s), under exclusive license to Springer Nature Switzerland AG 2023
N. Wang et al. (Eds.): AIED 2023, CCIS 1831, pp. 741–746, 2023.
https://doi.org/10.1007/978-3-031-36336-8_114

Stereotype threat occurs when an individual is concerned about being negatively evaluated as a member of a stigmatized group, which can negatively impact academic performance. On the other hand, positive motivational processes, known as a stereotype boost, can result from unconscious stimuli that reinforce positive group stereotypes [5, 6]. These often act over beliefs in the ability to perform required actions to fulfill tasks, which can be associated with self-efficacy [7, 8]. Stereotype threat and boost have been studied through the Stereotype Threat Theory (STT) and Stereotype Boost Theory (SBT), respectively, and evidence suggests that they can also influence levels of personal confidence, such as self-efficacy [5, 6, 9–11]. Moreover, stereotype boosts or threats can also influence levels of personal confidence (i.e., self-efficacy), and evidence of this relationship can be obtained through academic performance [7, 12–15].

In this study, we designed an experiment that used gender-stereotyped boost and threat messages in gamified environments to evaluate their effects on engagement, performance, and self-efficacy of students. We aim to examine the characteristics of these messages and determine whether gender boosts or threats are occurring, influencing results. We assume that there statistically significant differences the three metrics exist on participants of different genders in neutral and gender-stereotyped gamified environments. Identifying and activating positive stereotype triggers is crucial for optimizing learning, and this study seeks to contribute to this area of research.

2 Methods

2.1 Participants Profile and Sampling Procedure, Measures and Covariables

Participants were students from two public schools located in the countryside of Northeastern Brazil. These students were enrolled in years one to three of Systems Development courses. The experiment was conducted at Informatics and Languages laboratories, between April and May of 2021. In order to ensure inclusion of students in remote attendance, the experiment was performed via Google Classrooms.

In order to account for potential covariate factors, an ANCOVA was used in this study, because allows for adjustment of response variables effects influenced by uncontrolled variation. Such as status, preferences and biases in behavior. The data was analyzed in terms of flow experience, self-efficacy, and academic performance in the environments with gender stereotypes and conditions of boost, threat, and control.

2.2 Research Design, Flow State, Self-efficacy and Performance

We explored the relationship between self-efficacy and performance in gender-stereotyped gamified environments using positive and negative gender-stereotyped messages with logic puzzles. Participants completed pre- and post-tests to assess changes in engagement, self-efficacy, and performance. Consent from participants was obtained using TCLE #466/12 CNS before conducting the experiment. Participants completed pre-tests of flow disposition and self-efficacy, followed by message display and logic puzzle, and then post-tests of flow state and self-efficacy. We used adapted versions of the DFS and FSS questionnaires in Brazilian Portuguese [16, 17]. Participants were randomly assigned to either the control or experimental gamified environments and chose avatars to answer the puzzle. A total of 142 students participated (47 females and 97 males). The gamified platform included boost messages and points system, where correct answers added 5 points to the score, and achieving 25 or 50 points granted badges.

3 Results

3.1 Participants Profile

Data collected during the experiment was organized with entries for each participant, including: id, test results, environment of participation (stereotyped/control), gender and condition (boost/threat). Participants were unaware of the assigned group and the messages in the gamified platforms. This setting was designed based on previous studies [18] and modified to insert phrases directed to both genders.

3.2 Flow Experience Per Condition and Gender & Environment

Statistical tests produced mean values for the dispositional flow (DFS) and flow (FSS) of participants in the three conditions. Results showed higher DFS values for participants under stereotype threat but no significant differences. On the other hand, the boost and default/control conditions had higher average values for FSS. Statistically significant differences were identified for DFS ($p < 0.001$) but not for DFS and environments.

Results grouped according to gender and environments did not indicate statistically significant differences between variables, but females obtained higher average in gender-oriented environments. Results of the analysis of covariance of DFS for these variables detected differences ($p < 0.001$) and pairwise comparisons of flow test indicated that the control group (adj M = 3.924, SD = 0.489) and female (adj M = 3.938, SD = 0.534) differed from male stereotype environments (adj M = 3.562, SD = 0,475) with $p < 0.05$.

3.3 Self-efficacy by Condition and Gender & Environment

Results for self-efficacy indicated higher means in boost conditions, except for the pre-test of the control group. Pairwise comparisons indicated that boost (adj M = 6.671, SD = 0.803) differ from control condition (adj M = 6.047, SD = 1.356). Covariance analyses results for self-efficacy and for condition presented significance of $p < 0.001$.

Results for the self-efficacy of gender and environments indicated higher average values for males in control environments and lower for females in male-stereotyped environments and significant differences for these variables in post-tests of self-efficacy (Table 1). Values of control environment (adj M = 6.215, SD = 1.144) and female (adj M = 6.271, SD = 1.027) differed significantly from male (adj M = 6.98, SD = 0.401).

Table 1. Covariance analysis test results regarding participants' self-efficacy according to stereotype message (stType) and gender. *** (0 to 0.001), ** (0.001 to 0.01).

Effect	DFn	DFd	SSn	SSd	F	p	ges	p.sign
self-efficacy	1	110	36.825	89.755	45.131	< 0.001	0.291	***
stType	2	110	7.885	89.755	4.832	0.01	0.081	**
gender	1	110	7.119	89.755	8.725	0.004	0.073	**
stType:gender	2	110	1.753	89.755	1.074	0.345	0.019	ns

3.4   Learning Performance by Condition and Per Gender and Environment

Results for learning performance presented higher values in boost and lowest for threat conditions. Significant effects were identified for condition ($F(2.139) = 3.833$, $p = 0.024$, ges $= 0.052$) and for points earned. The boost (adj M $= 12.855$, SD $= 4.786$) differed significantly from threat condition (adj M $= 10.955$, SD $= 4.356$). Significant differences in points per environment and gender were identified respectively with $p < 0.01$ and $p < 0.05$. Points in female stereotyped (adj M $= 11$ and SD $= 4.243$) differed significantly from the male environments (adj M $= 13.558$ and SD $= 4.697$) and the factor gender also presented significant differences (female: adj M $= 10.9$ and SD $= 4.644$; male: adj M $= 13.558$ and SD $= 4.697$). Results of descriptive statistics indicate significant differences between stereotype tests and genders (Table 6), but not between engagement (flow) in environment or in comparisons between groups. Results also indicated statistically significant differences in self-efficacy post-tests and in performance between boost and control (Table 2).

Table 2. Results of analysis of variance (ANOVA) for estimated marginal means (EMMs) of engagement (FSS), self-efficacy (self) and performance (points) between boost, threat, and control (Default) groups with p-values adjusted through the Bonferroni method. * (0.01 to 0.05), ** (0.001 to 0.01).

var	group 1	group 2	estim	se	df	stats	p	p.adj	sign
FSS	Boost	Threat	0.176	0.089	135	1.979	0.050	0.150	ns
FSS	Boost	Default	-0.012	0.095	135	-0.121	0.904	1.000	ns
FSS	Threat	Default	-0.188	0.097	135	-1.932	0.055	0.166	ns
self	Boost	Threat	-0.090	0.056	137	-1.620	0.108	0.323	ns
self	Boost	Default	-0.202	0.059	137	-3.443	0.001	0.002	**
self	Threat	Default	-0.112	0.062	137	-1.813	0.072	0.216	ns
points	Boost	Threat	-0.372	0.144	139	-2.589	0.011	0.032	*
points	Boost	Default	-0.312	0.154	139	-2.019	0.045	0.136	ns
points	Threat	Default	0.061	0.158	139	0.383	0.703	1.000	ns

4 Discussion

The study aimed to investigate how stereotype boosts impact self-efficacy, flow, and performance in a gamified environment. Results showed that females experienced a stereotype boost through increased self-efficacy and had higher performance scores than males in gender-aligned settings. Self-efficacy levels fluctuated in females but remained stable in males. Both genders had higher flow experiences in gender-aligned settings, possibly due to feeling threatened by stereotyped messages in opposite gender settings. Self-efficacy was highest in neutral settings, indicating that negative messages were not preferred. Positive stereotype boosts did not have the expected effect, and males may have felt threatened by female boosts due to normative conditions [19–21].

Results also showed that stereotype boosts led to better performance, which aligns with previous literature. Males had higher performance scores than females, possibly due to unconsciously activated mental constructs when exposed to gender-aligned information. While stereotypes may be activated, they can also be refuted, and self-efficacy is linked to personal competence. It is important to identify factors that influence self-efficacy, boost activation, and performance to improve education, motivation, and confidence in students [5, 6, 9, 18, 22]. Such findings may be related to the exposure to the information contained aligned to one's gender, unconsciously activating mental constructs via external stimuli [23–26].

In summary, the study found that stereotype boosts can impact self-efficacy, flow, and performance in a gamified environment, with females experiencing increased self-efficacy and higher performance scores in gender-aligned settings. Self-efficacy levels fluctuated in females, and both genders had higher flow experiences in gender-aligned settings. Males had higher performance scores, possibly due to unconsciously activated mental constructs. It is crucial to understand factors that influence self-efficacy, boost activation, and performance to improve education, motivation, and confidence.

References

1. Kapp, K.M.: The Gamification of Learning and Instruction: Game-Based Methods and Strategies for Training and Education. John Wiley & Sons (2012)
2. de Sousa Borges, S., et al.: A systematic mapping on gamification applied to education. In: Proceedings of the 29th Annual ACM Symposium on Applied Computing (2014)
3. Dicheva, D., et al.: Gamification in education: a systematic mapping study. J. Educ. Technol. Soc. 18(3), 75–88 (2015)
4. Hervás, R., et al. Gamification mechanics for behavioral change: a systematic review and proposed taxonomy. In: Proceedings of the 11th EAI International Conference on Pervasive Computing Technologies for Healthcare (2017)
5. Steele, C.M., Aronson, J.: Stereotype threat and the intellectual test performance of African Americans. J. Pers. Soc. Psychol. 69(5), 797 (1995)
6. Steele, C.M., Spencer, S.J. Aronson, J.: Contending with group image: the psychology of stereotype and social identity threat. In: Advances in experimental social psychology. Elsevier, pp. 379-440 (2002)
7. Bandura, A.: Self-efficacy: toward a unifying theory of behavioral change. Psychol. Rev. 84(2), 191 (1977)

8. Bong, M., Clark, R.E.: Comparison between self-concept and self-efficacy in academic motivation research. Educ. Psychol. **34**(3), 139–153 (1999)
9. Davies, P.G., et al.: Consuming images: How television commercials that elicit stereotype threat can restrain women academically and professionally. Pers. Soc. Psychol. Bull. **28**(12), 1615–1628 (2002)
10. Levy, B.: Improving memory in old age through implicit self-stereotyping. J. Pers. Soc. Psychol. **71**(6), 1092 (1996)
11. Shih, M.J., Pittinsky, T.L., Ho, G.C.: Stereotype boost: Positive outcomes from the activation of positive stereotypes (2012)
12. Bandura, A., *Fearful expectations and avoidant actions as coeffects of perceived self-inefficacy*. 1986
13. Hackett, G.: Role of mathematics self-efficacy in the choice of math-related majors of college women and men: a path analysis. J. Couns. Psychol. **32**(1), 47 (1985)
14. Pajares, F., Miller, M.D.: Role of self-efficacy and self-concept beliefs in mathematical problem solving: a path analysis. J. Educ. Psychol. **86**(2), 193 (1994)
15. Siegle, D., McCoach, D.B.: Increasing student mathematics self-efficacy through teacher training. J. Adv. Acad. **18**(2), 278–312 (2007)
16. Jackson, S.A., Martin, A.J., Eklund, R.C.: Long and short measures of flow: the construct validity of the FSS-2, DFS-2, and new brief counterparts. J. Sport Exerc. Psychol. **30**(5), 561–587 (2008)
17. Bittencourt, I.I., et al.: Validation and psychometric properties of the Brazilian-Portuguese dispositional flow scale 2 (DFS-BR). PLoS ONE **16**(7), e0253044 (2021)
18. Albuquerque, J., et al.: Does gender stereotype threat in gamified educational environments cause anxiety? Experimental study. Comput. Educ. **115**, 161–170 (2017)
19. Fennema, E.H., Sherman, J.A.: Sex-related differences in mathematics achievement and related factors: a further study. J. Res. Math. Educ. **9**(3), 189–203 (1978)
20. Meece, J.L., et al.: Sex differences in math achievement: toward a model of academic choice. Psychol. Bull. **91**(2), 324 (1982)
21. Levine, D.U., Ornstein, A.C.: Sex differences in ability and achievement. J. Res. Dev. Educ. (1983)
22. Kaye, L.K., Pennington, C.R.: "Girls can't play": the effects of stereotype threat on females' gaming performance. Comput. Hum. Behav. **59**, 202–209 (2016)
23. Bargh, J.A., Chartrand, T.L.: Studying the mind in the middle: a practical guide to priming and automaticity research. In: Handbook of Research Methods in Social Psychology. Handbook of Research Methods in Social and Personality Psychology, pp. 253–285 (2000)
24. Chartrand, T.L., Bargh, J.A.: Nonconscious motivations: their activation, operation, and consequences (2002)
25. Kawada, C.L., et al.: The projection of implicit and explicit goals. J. Pers. Soc. Psychol. **86**(4), 545 (2004)
26. Fiske, S.T., Gilbert, D.T., Lindzey, G.: Handbook of Social Psychology, vol. 2. John Wiley & Sons (2010)

DancÆR: Efficient and Accurate Dance Choreography Learning by Feedback Through Pose Classification

İremsu Baş, Demir Alp, Lara Ceren Ergenç, Andy Emre Koçak, and Sedat Yalçın[✉]

Hisar School Computer Science ideaLab, Istanbul, Turkey
{iremsu.bas,demir.alp,lara.ergenc,emre.kocak,
sedat.yalcin}@hisarschool.k12.tr

Abstract. Educational philosophies have slowly focused on differentiated and self-learning systems that respectively emphasize tailoring instruction to meet individual needs and gathering, processing, and retaining knowledge without the help of another person in recent years. In this regard, creating opportunities to further self-learning resources has become increasingly important. Some people started to prefer these over traditional learning practices for various reasons, such as the difficulty of transportation in metropolises or fitting their timetables to in-person lessons. The creation of such platforms or opportunities for physical education, however, proves to be more difficult as individuals require continuous and precise feedback regarding the usage of their bodies. Accordingly, we have developed an augmented reality application that presents a platform for dance that focuses on differentiated and self-learning principles with accurate feedback. We built the augmented reality (AR) app prototype using the Swift programming language and used the MoveNet pose detection model along with our own neural network to capture the body position. Our proposal could prove a valuable addition to learning physical activities assisted by AI systems since the application of AI technologies to dance and physical education could be improved by further investigation and research.

Keywords: Machine Learning · AI-Assisted Tutoring · Motivational Diagnosis and Feedback · AR

1 Introduction

Dance is an incredibly influential medium in which many cultures throughout the span of history have expressed themselves. It has been a significant part of many religions and various cultural identities. One of the main aspects behind this substantial presence of dance within history is its efficiency as a tool of expression and social bonding. This is still accurate today, as research indicates that dance can significantly increase feelings of closeness among groups [1]. As a result, dance stands out as a tool that, in many ways, is inherent to how humans socialize and, in turn, express themselves. This makes a focus on dance in education much more valuable. On top of this, dance is closely associated with

N. Wang et al. (Eds.): AIED 2023, CCIS 1831, pp. 747–755, 2023.
https://doi.org/10.1007/978-3-031-36336-8_115

physiological and psychological benefits to individuals, specifically adolescents as well. Dance allows adolescents' physical fitness to be positively impacted while improving their sleep cycle and motor skills, as well as improving their overall health due to its medical benefits [2]. It also facilitates the improvement of mental state, well-being, confidence, and perceived competence [2]. As a result, a focus on dance in physical education can become highly beneficial, improving the students in various ways, both physically and emotionally. Thus it becomes crucial to capitalize on these benefits of dance and investigate how it can be further integrated into the lives of individuals. Specifically, systems that innovate dance education and make it more accessible become increasingly valuable. To do so, however, employing accurate feedback systems becomes vital as feedback constitutes an integral part of dance.

Feedback is a crucial component of effective learning in general because it helps learners reflect on their own performance and identify knowledge gaps. Feedback helps to bridge the gap between the current level of understanding and the desired level of understanding. It should provide guidance or assistance to the learner to help them progress towards their learning goals. Feedback can be delivered in various ways, such as by positively influencing learners' attitudes and behaviors or by using cognitive processes like modifying their current understanding, validating their responses, highlighting the need for additional information, suggesting alternative approaches, or providing guidance on how to progress [4].

In the context of feedback in teaching environments, a formative assessment is when the learner revises their work more than once according to continuous feedback, and a summative assessment is when the work done by the learner is evaluated at the end of the learning process. The most effective way of learning occurs when both feedback types work together. Formative and summative evaluations are inseparable. These feedback systems work like diagnostic measurements; they conclude information about the initial state of the student, the distance between the start and the targeted level, and the learner's ability to apply the present acquired knowledge to different situations and where difficulties are faced [3, 5].

On this note, AI is highly beneficial in the process of implementing and automating feedback systems. As a result, implementing AI can garner significant importance in educational fields. In the last few years, many intelligent systems, such as Chat-GPT, Khan Academy, Duolingo, Writing Pal, and ALEKS, have actively taken roles in education. These intelligent tutoring systems (ITS) are critical applications of artificial intelligence (AI) in education, centering differentiated learning and self-learning. ITS provides personalized, adaptive instruction to students, leveraging AI to analyze student data and provide feedback and support. The system's ultimate goal is to provide each student with an optimal pathway through the learning materials tailored to their individual needs and abilities, leading to better learning outcomes and more effective use of time and resources [6].

Besides academic applications, there are also physical training aspects of AI, such as remote sports and physiotherapy training. As a result, the development of such tools provides the necessary fundamentals for applying AI technologies to dance education due to the established foundation on interactive whole-body experiences, motion capture (optical and inertial), and virtual and augmented reality. [7] Augmented reality provides unique opportunities for dance education as it furthers whole-body interaction systems, giving real-time feedback in a manner that develops movement capture technologies and allows users to directly visualize ideal movements in their environments or on their bodies [7].

Consequently, this study will focus on a new physical education system for dance via AR and AI systems in DancÆR and its feedback algorithm, creating an interactive space for our users that proves an optimal learning environment.

2 Related Works

Dance video games such as Just Dance and Dance Central are famously known for spreading dance exercises for entertainment. The user follows a teacher displayed on the screen and tries to imitate their movements. Another alternative is dance instruction videos, where a tutor explains the choreography in a step-by-step manner and then combines the choreography with music. However, one shortcoming of both instructional videos and most video games is that they don't provide sufficient feedback to users, which impedes the users' ability to learn and improve their dance moves. To address this problem, Chan et al. proposed a virtual reality dance training system providing feedback utilizing a motion capture system [12]. Another approach with feedback is proposed by Tian et al. using a joint tracking algorithm and Unity, focusing more on cultural dances [11]. In Choi et al.'s proposal, the feedback algorithm for joint tracking using pose estimation models is thoroughly investigated, concluding that the best feedback is achieved by comparing the teacher and student's joint position and angular similarity [8, 9]. A comparable application designed by Shen et al. for a yoga self-exercising system checks the synchronization of the instructor and the learner [10]. Building on these foundations while focusing on summative and formative feedback, our proposal approaches dance education through a novel perspective that essentially imitates private instruction through an intelligent system that combines summative and formative feedback systems. On top of gradually teaching choreographies, our algorithm optimizes how instruction is conducted per the user's needs, the way a traditional dance instructor might approach their practice. In that sense, DancÆR provides a unique and easily accessible alternative to dance education.

3 AR Dance Instruction

The project, DancÆR, will function as a step-by-step choreography learning application with built-in functions that guide users in learning dance routines. The application will first introduce users to dance routines sourced from various online sources and dance databases, such as AIST Dance Video Database [13]. Once our algorithm analyzes the chosen routine, it will break down each choreography into step-by-step movements following rhythm-based segmentation of the backing track. After that, the application will present the movements to the user through a generated AR instructor that shows the progressive movements to the sound of metronomic counting that starts with a much slower tempo than the song itself. Afterward, the user will begin to mimic the movements of the instructor while the app analyzes the movements of the user, comparing them to the original data provided, and then correct the movements by projecting the accurate positioning onto the screen of the user in a low opacity avatar that we call movement markers which the user will try to fill completely when dancing. While slowly teaching the dance, the instructor will note all the points in which the user struggles and place such movement markers onto the screen when the movements need to be carried out. While applying this system, the user will continue learning the choreographies at their own speed, ensuring that no instruction is too fast or slow-paced for our users. On top of this, the instructor will provide the user with general advice, such as suggestions regarding the fluidity of movement. The algorithm prepared will gradually increase the tempo of the dance and then introduce the backing track until the choreography can be executed accurately at the song's original tempo. DancÆR will adapt to the user's learning experience by progressing through the steps delineated above in accordance with the user's success per our move estimation algorithm, further expanded on in section four. The instructor will also provide the user cues on the backing track, signaling when to switch movements. After a routine is learned and carried out at 100% of its original speed, the instructor will ask the user to perform it. Afterward, a score and general feedback will be provided. This will allow the users a significantly more enjoyable experience as the score provided will function as a form of gamification. Each time a user gets to 100% accuracy on a new choreography, they will gain a set amount of points corresponding to the difficulty level of the routine. The users will also have the option of sharing their performance on the DancÆR platform and other social media platforms, further emphasizing the social aspect of dance and allowing users to engage with their peers. After a routine is fully learned, the users can self-evaluate by taking notes and implementing markers on the choreography to respective bars on top of the summative feedback and recording provided at the end of the dance session. With these constructive criticisms, the user will be aware of the parts they need or want to improve; they will be able to change certain parts of the choreography to add their own spin to it and upload it to our platform as an alternative version.

DancÆR will provide its users with the opportunity to replace their bodies with a 2D avatar if they want to keep a certain level of anonymity within our platform, allowing them to have the option of continuing to engage with our platform and its user base even if they don't feel comfortable with posting their videos online.

4 Technical Specifications of Artificial Intelligence Algorithm

Our application has three main interfaces: the home page that allows the user to choose the dance session, the AR dance session page, and the self-evaluation page. On the AR dance session page, the users will be constantly tracked and given feedback to via a convolutional neural network that computes user accuracy through vector feature comparisons and timing comparisons. Additionally, the user's session will be recorded, which will appear on the self-evaluation page. Our ML model will provide the user with the mistakes in their dance movements, both positionally and rhythmically, and the user will be able to improve their performance with this feedback. To create this feedback system, we used a weighted approach with two considerations when determining the user's accuracy: live feedback from joint movement data and timing of the user, determined through pose classification, though we plan to include a measure for the fluidity

Fig. 1. User-interface (left) and flowchart (right) of DancÆR

of movement in future updates. Our algorithm compares the teacher's (vector features extracted from professional video) joint position ratios and angles to the students' joints for live feedback during the AR dance session. We were inspired by the proposal of Choi et al., a dance pose evaluation that simultaneously performs an affine transformation and an evaluation method to compare the joint position and joint angle information [9]. The results of this comparison will be used later in Eq. 2, found in Fig. 2 to represent features extracted from the user and professional data (f_l^i and f_t^i), respectively. The secondary component of user assessment is timing. Dance choreographies are generally counted as 8 bars, and while learning the dance accurately, it is important to know which beat corresponds to which movement. Our algorithm compares the user's timing with the sample video's timing so that the user's movement is within the correct beats. The prototype application was built for IOS devices using Swift and Apple's development kit SDK. Therefore when choosing which pose estimation model to use, we had to pick a model optimized for embedded systems while retaining a high accuracy during joint tracking and classification. That's why we used the TensorFlow Lite implementation of the MoveNet pose estimation model in tandem with our own Neural Network to achieve our desired effect of pose classification. This choice allowed us to implement high-end pose estimation technology like MoveNet (which can track 17 joints including arms, legs, hips, etc.) onto mobile platforms without encountering many performance issues, as most other models would prove too processor intensive. The program starts by retrieving the frame data, preprocessing it for MoveNet, and inputting it into the MoveNet model. After running inference through MoveNet, coordinates of the 17 joints (feature vectors) are extracted from the frame. This data is then processed through a sequential model that is comprised of dense layers to predict the pose classes. Additionally, this data is also fed to the algorithm that compares joint angles and joint positions to that of the teachers. Finally, all of this data is evaluated to get a percentage score of the user's accuracy in the dance. To compare the dancer's movements to the sample and evaluate the accuracy of the similarity of the choreography calculations based on an equation that incorporates both pose and timing accuracies was utilized. This approach was inspired by the work of Kim et al., who proposed a method to calculate the dance similarity between a dancer's movements and a sample video [14]. The equation for calculating similarity is as follows (Fig. 1):

$$S = \frac{1}{N}\sum_{j=1}^{N} S_t^j S_p^j \qquad (1)$$

In this equation, N represents the number of sequences of the dancer's performance, S_t is the timing accuracy, whereas S_p is the position accuracy. Kim et al.'s calculations for Sp and St are shown below [14].

$$S_p = \frac{1}{T_t^e - T_t^s + 1} \sum_{i=T_t^s}^{T_t^e} exp(-\frac{||f_l^i - f_t^i||}{\beta}) \qquad (2)$$

$$S_t = 1 - min(1, exp(\frac{|T_t - T_l| - \tau}{\alpha\tau})) \qquad (3)$$

In equation numbered two to express position accuracy, f_l^i and f_t^i express the vector features of the dancer (referred to as learner in Kim et al.'s proposal) and the sample video (referred to as a teacher in Kim et al.'s proposal) in which the upper and lower torso vector features are represented. The deviation of the dancer's movements from the sample video is shown as β. The s and e superscripts for T, respectively, indicate the start and end points of the dance sequence [14]. The result of said equation gives an approximate of the difference of the vector coordinates of the user (learner) to that of the teacher (vector data from sample video). An increase in the deviation of the vectors causes a lower score as an output. On the other hand, in the equation numbered three to calculate timing accuracy, T_t and T_l are the middle time of the video sample and the dancer, respectively, calculated using the following equations: $T_t = \frac{T_t^s + T_t^e}{2}$ and $T_l = \frac{T_l^s + T_l^e}{2}$. The maximum deviation of timing is governed by the parameter τ and the slope of Eq. (1) is controlled by the parameter α [14]. By using Eq. 3, the algorithm can determine how offbeat the user is during their choreography. The timing of which is taken as timestamps throughout the song with a countdown that extracts features (vector data from video taken at 60 fps) based on the time signature and the beats per minute (bpm) of the song. Using these equations, it is possible to quantify the similarity between the dancer's movements and the sample video, considering both the position and timing accuracies.

To gauge the performance of this system, we created an example dataset for the choreography of the song Gangnam Style. We chose this dance as it was popular enough to easily collect a large dataset and simple enough to classify since it featured two moves. The two dance movements were separated into classes for supervised learning. We recorded professional/hobbyist dancers performing Gangnam Style. The images were slightly blurred to keep the subject in the foreground, and image noise was increased to decrease the chances of overfitting since there were many similar frames in the data. In total, we used 10000 images for training and 1000 for testing. The training was conducted with this dataset and 50 epochs, resulting in a training accuracy of 0.9583 and a validation accuracy of 0.9345 (see Fig. 2).

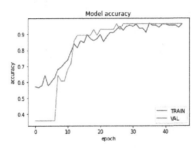

Fig. 2. Pose Classification Model Accuracy Metrics

5 Expected Research Plan

DancÆR is expected to function as a vital technology that would not only hasten the speed at which a dance choreography can be learned but also increase the accuracy with which said choreography could be performed. In this context, we believe that the constant feedback provided through the platform that fixes and guides every movement will prove to function as a form of private tutelage with extreme attention to detail, allowing the user to feel more confident in their performance as well. Thus we believe that DancÆR will allow the user to become more proficient in dance and have a markedly positive emotional experience. After these claims are tested through experimentation, DancÆR will be further optimized to provide a smoother user experience. Its feedback methods will be improved, and the necessary alterations will be made to its AI feedback algorithm.

The testing procedure will be executed under one experimental group that tests DancÆR compared to two control groups. One will utilize forms of non-interactive instruction without feedback, those seen in online dance videos, etc. In contrast, the other will be in the form of private lessons with in-person instruction. The importance of utilizing two control groups in this experiment is that it allows the experimental design to evaluate whether DancÆR demonstrates any statistically significant difference from existing alternatives in the market while also assessing how well DancÆR compares to traditional dance education. With this comprehensive evaluation, the experiment will reveal whether or not DancÆR is a viable alternative if results reveal a significant difference from pre-existing alternatives and show that the prototype can come close to the effectiveness of private instruction.

The overall sample will be 200 people assorted into groups via random assignment. The sample will be collected within the Hisar High School population and other affiliated high schools through random selection to reflect the project's target audience. With this in mind, a selected choreography of intermediate difficulty will be presented to all groups (control and experimental), and participants will be given a time frame of two weeks to learn the choreography and be evaluated on the accuracy, timing as well as flow (reduced rigidness) of movement. Within the experimental group, participants will practice the dance routine for two weeks with the guidance of DancÆR's feedback. In contrast, the first control group will practice independently through mimicry of the provided instruction video, and the second will practice guided by a private instructor.

Afterward, the evaluation will be carried out through our AI algorithm, trained with videos and images of professional dancers performing the routine and a subjective grading system via scoring by ten third-party dance instructors. Though the latter mode of scoring is subjective, it will provide insight for further correlational analysis between the performance of our algorithm compared to the feedback from professionals in the field.

The experiment will also evaluate participants' emotional experience while considering the effects of certain confounding variables, such as previous experience in dance. This is why participants will fill out a detailed questionnaire where they are required to express their experience in dance and other forms of sports as well as experience in music and art to provide us with insight into how history in music, art, and physical exercise can influence the results of their performance. Moreover, the questionnaire will also acquire detailed information regarding how the participants felt throughout the process, such as their confidence, engagement, or enjoyment levels. Through analyzing both qualitative and quantitative data, the testing process will prove a vital part of DancÆR's future and give direction to attempts to further the prototype.

6 Future Plans

We recognize the importance of social interaction in learning, such as its benefits to motivation or peer-to-peer support. Thus, we believe a beneficial contribution to our proposal would be to design the app to allow users to connect with their friends globally and dance together. This could be achieved by converting the captured poses of various users from different locations during a live session and instantaneously converting them into AR avatars while synchronizing their movements with music and eliminating the video and audio delay. This way, the users can dance with their friends live and synchronously, where collaboration and prosociality heighten [1], emphasizing the undeniably social component of dance and learning while going through the learning process at their own pace.

7 Conclusion

Dance is a form of art that is enjoyable to watch and has several physical and emotional benefits. It is a great exercise that improves flexibility, strength, balance, and coordination. In recent years, digital environments and computer vision technologies have been utilized in various ways in dance. Our project, DancÆR, is an example of how artificial intelligence (AI) and augmented reality (AR) can be used to enhance the learning experience of dancers. Our system uses pose classification technology to enhance students' learning experience. It provides immediate, informative feedback through one-by-one movement training, formative feedback, and summative feedback for a self-reflection opportunity, encouraging development, creativity, and exploration. DancÆR is a promising project for exploring ML education, and it presents potential new directions in Physical Education and sustainability in educational paths.

Acknowledgment. We would like to thank our colleague Melis Alsan who invested in lending assistance for the UI/UX programming of our research's prototype. Their contributions were critical to the success of this research, and we are deeply grateful for their hard work and dedication.

References

1. Reddish, P., Fischer, R., Bulbulia, J.: Let's dance together: synchrony, shared intentionality and cooperation. PLoS ONE **8**(8), e71182 (2013)
2. Tao, D., et al.: The physiological and psychological benefits of dance and its effects on children and adolescents: a systematic review. Front. Physiol. **13** (2022)
3. Shute, V.J.: Focus on Formative Feedback. Rev. Educ. Res. **78**(1), 153–189 (2008)
4. Hattie, J., Timperley, H.: The power of feedback. Rev. Educ. Res. **27**, 50–51 (2016)
5. Baht, B.A., Bhat, G.J.: Formative and summative evaluation techniques for improvement of learning process. Eur. J. Bus. Social Sci. **7**(5), 776–785 (2019)
6. Miao, F., Holmes W., Huang, R., Zhang, H.: Understanding AI and education: emerging practices and benefit-risk assessment, AI and education: a guidance for policy makers. In: UNESCO, pp. 13–15 (2021)
7. Raheb, K.E., Stergiou, M., Katifori, A., Ioannidis, Y.: Dance interactive learning systems: a study on interaction workflow and teaching approaches. ACM Comput. Surv. **52**(3), 50 (2019)
8. Lee, J., Choi, J., Chuluunsaikhan, T., Nasridinov, A.: Pose evaluation for dance learning application using joint position and angular similarity. In: Adjunct Proceedings of the 2020 ACM International Joint Conference on Pervasive and Ubiquitous Computing and Proceedings of the 2020 ACM International Symposium on Wearable Computers (UbiComp-ISWC 2020), pp. 67–70. Association for Computing Machinery, New York (2020)
9. Choi, J., Lee, J., Nasridinov, A.: Dance self-learning application and its dance pose evaluations. In: Proceedings of the 36th Annual ACM Symposium on Applied Computing (SAC 2021), pp. 1037–1045. Association for Computing Machinery, New York (2021)
10. Shen, S., Huang, W., Anggraini, I.T., Funabiki, N., Fan, C.: Design of OpenPose-based of exercise assistant system with instructor-user synchronization for self-practice dynamic yoga. In: Proceedings of the 10th International Conference on Computer and Communications Management (ICCCM 2022), pp. 246–251. Association for Computing Machinery, New York (2022)
11. Tian, F., Zhu, Y., Li, Y.: Design and implementation of dance teaching system based on Unity3D. In: 6th International Conference on Intelligent Computing and Signal Processing (ICSP), pp. 1316–1320. IEEE, Xi'an (2021)
12. Chan, J.C.P., Leung, H., Tang, J.K.T., Komura, T.: A virtual reality dance training system using motion capture technology. IEEE Trans. Learn. Technol. **4**(2), 187–195 (2011)
13. Tsuchida, S., Fukayama, S., Hamasaki, M., Goto. M.: AIST dance video database: multi-genre, multi-dancer, and multi-camera database for dance information processing. In Proceedings of the 20th International Society for Music Information Retrieval Conference (ISMIR 2019) (2019)
14. Kim, Y., Kim, D.: Interactive dance performance evaluation using timing and accuracy similarity. In: ACM SIGGRAPH 2018 Posters (SIGGRAPH 2018), vol. 67, pp. 1–2. Association for Computing Machinery, New York (2018)

Intelligence Augmentation in Early Childhood Education: A Multimodal Creative Inquiry Approach

Ilene R. Berson[1]([✉]) [iD], Michael J. Berson[1] [iD], Wenwei Luo[2] [iD], and Huihua He[2] [iD]

[1] University of South Florida, Tampa, FL 33620, USA
iberson@usf.edu
[2] Shanghai Normal University, Shanghai, China

Abstract. This paper explores the intersection of early childhood education and AI, with a focus on promoting play, imagination, and creativity. The post-digital era has made technology a constant presence in our lives, and AI has enabled new and innovative ways of interacting with technology. However, it's important to ensure that the use of AI in early childhood education aligns with the core principles of early childhood pedagogy. Intelligence augmentation (IA) can play a significant role in promoting multimodal creative inquiry among young students, supporting critical thinking, problem-solving, and creativity. Drawing from practitioner research in preschool settings, we explore the potential of an AI-powered music-making tool to promote creativity and self-expression. The human-in-the-loop approach is crucial in aligning AI integration with early childhood pedagogy, ensuring that technology is used to support, rather than replace, teachers' role in the classroom. The paper highlights the importance of maintaining a human presence in the learning process to create educational experiences that are developmentally appropriate and transformative for young children.

Keywords: Early Childhood · Intelligence Augmentation · Music Exploration · Multimodal Play

1 Introduction

1.1 Multimodal Creative Inquiry

Play, imagination, and creativity are essential components of early childhood development, and as AI-enabled technologies become increasingly ubiquitous, young children have begun appropriating these digital technologies into their play [1–3] While AI has the potential to revolutionize the way we learn, early childhood education, which focuses on the education of children from birth to age 8 years old [4], is a unique field that requires a specialized approach. Although the vast majority of early childhood AI applications have focused on teaching or tutoring systems, AI social agents, and curriculum to promote AI literacy [5–12], our work is positioned at the intersection of multimodal creative inquiry and digital innovations that align with early childhood pedagogy.

Catalyzing Creativity with AI. In early childhood education, intelligence augmentation (IA) [13] can play a significant role in promoting multimodal creative inquiry

© The Author(s), under exclusive license to Springer Nature Switzerland AG 2023
N. Wang et al. (Eds.): AIED 2023, CCIS 1831, pp. 756–763, 2023.
https://doi.org/10.1007/978-3-031-36336-8_116

among young students. By providing new and innovative tools and resources, IA can help students engage with complex information and ideas in a more meaningful and creative way. Multimodal creative inquiry is a critical component of early childhood education, as it allows students to explore and understand a topic using multiple modes of representation, such as visual, auditory, and kinesthetic. AI-powered tools can support this type of inquiry by helping young children identify patterns and relationships that might not be apparent using traditional methods, leading to new insights and discoveries.

In the context of early childhood education, this approach goes beyond the use of digital technologies as tools to develop skills and cognition. It considers how digital innovations address various dimensions of social practices, including operational, cultural, and critical aspects of communication, across time and space. This type of research can help to ensure that AI-enhanced innovations are used in a way that supports and enhances the relationship between the teacher and the child, rather than undermining or disrupting it. This can be especially important in early childhood education, where strong relationships between teachers and children are key to ensuring positive learning outcomes.

Citizen DJ

In early childhood, there is an incredible range of opportunities for children to explore and engage with sound. From listening to music to inventing their own unique sounds, music and movement provide children with a diverse array of formal and improvised experiences that allow them to express themselves and find joy in their identities [14–16]. Like other forms of play, musical play is centered on variation, improvisation, and human interaction [17], and our work has reflected on how AI-enhanced innovations can catalyze children's creativity by presenting complex information in accessible formats that may inspire innovation.

Brian Foo, the 2020 Innovator in Residence at the United States Library of Congress, created Citizen DJ [18] as a free tool for experimenting with sounds and rhythms to foster self-expression and creativity. The platform supplements the original metadata by highlighting features that are important to creators, putting aesthetics at the forefront. Pre-processing methods segment the audio into small pieces, and machine learning analyzes the tempo and key of each audio file as well as generates supplemental metadata tags for each audio file, including information about the genre, instrumentation, and date of the recording. The resulting interface groups each piece by its sonic quality (i.e., opera vs. piano vs. comedic script) while allowing users to reference and know the provenance of the source material.

This feature streamlines the process of selecting and remixing audio files and provides a simple way for users to create rhythmically coherent remixes without needing to manually adjust the timing of each drum hit. Users can select and isolate notes and rhythms from the extensive sound collection of the Library of Congress and mix them with drumbeats to create unique compositions.

Citizen DJ demonstrates how technology can be used to promote cultural heritage and democratize access to archives while enabling creative self-expression. Inspired by the innovative practices of the hip-hop movement, Citizen DJ encourages exploration of found sounds and the integration of disparate and obscure sounds from different cultures into new creations. The focus is on awe and wonder as users engage with the raw materials of history.

As a result of the Music Modernization Act of 2018, the recent expansion of the sound recording collection at the Library of Congress has increased the variety of sounds available for use and reuse, with over 10,000 sound recordings entering the public domain. This presents a wealth of material for early childhood educators to engage their students in musical play and to encourage them to create and share music, using the tool to weave an intertextual tapestry that reflects their own unique expression. By engaging with Citizen DJ, educators and children can discover interpersonal connections and explore how these historic samples can help them tell their own stories, all while keeping the human element of creativity and expression in the loop.

2 Theory

2.1 Papert's Constructionism

This work is based on the constructionist approach developed by Papert [19, 20], which involves examining how children interact with objects and how these interactions promote self-directed learning, leading to the acquisition of new knowledge. Papert's theory emphasizes the importance of learning through hands-on experience rather than innate cognitive abilities [19]. It helps us understand how ideas are generated and transformed when expressed through different media, contexts, and individual minds.

2.2 Multimodality of Digital Play

Evidence from field-based research around the globe has demonstrated the potential benefits for integration of digital play into early childhood curricula and practice [21–23]. The process of multimodal meaning-making expands the scope of learning and teaching beyond language by encompassing various forms of communication that occur during classroom activities within a shared social context. Multimodality centers on how children interact with new technologies in their everyday lives, incorporating various dimensions such as language, visuals, audio, and movement in space [24]. Such interactions offer novel avenues for children to derive meaning from their experiences, as they transition between digital and physical modalities.

The data that emerges from everyday classroom moments, where children interact with the material environment and teachers facilitate play with digital objects, is valuable for analysis [21]. There is a growing understanding that children's multimodal 'lifeworlds' consist of an array of digital resources that enrich their learning experiences [24, 25]. As children navigate complex multimodal engagements in the classroom, AI-enhanced technologies offer an additional modality that extends children's "plurality of identities (people, places, activities, literacies), possibility awareness (of what might be invented, of access options, of learning by doing and of active engagement), playfulness of engagement (the exploratory drive) and participation (all welcome through democratic, dialogic voice)" [24].

3 Method

3.1 Participants and Procedures

The quality of education heavily relies on the interactions between people, specifically the role of teachers. In early childhood education, there is often a gap between the expected success of new educational technology and the actual outcomes, which can be due to a lack of understanding of the classroom environment and the essential role that teachers play. People may overestimate the capabilities of technology to replace traditional teaching methods, resulting in an incomplete view of the classroom environment and its key elements. When the technology does not meet expectations, educators may either revert to their original teaching methods or become skeptical of future technology altogether. This pattern of anticipated failures highlights the necessity for a more comprehensive understanding of the classroom environment and the critical role of teachers in implementing educational technology.

We have previously used a Design-based Implementation Research (DBIR) approach [26, 27] to address these limitations and explore the dynamics of teacher-child interactions while engaged with digital innovations, such as robotics, tangible technologies, and extended reality [23, 28–31]. DBIR emphasizes collaboration and sustainability when implementing technology in education, recognizing the unique needs of all stakeholders involved [32, 33]. In the case of early childhood education, this requires a deep understanding of the social dynamics of the classroom, including the ways in which children and teachers interact. Observing and analyzing these interactions can provide valuable insights into the types of interactions that are most supportive of positive educational outcomes for children.

We analyzed qualitative data collected from twenty preschool classrooms serving children ages 4–5 years old, including video observations of children playing with Citizen DJ, as well as reflective journals from the participating early childhood educators. Over the course of a month, 22 play sessions with Citizen DJ were recorded, and we analyzed focal excerpts of video that captured small group guided play with the AI-enabled platform.

4 Findings

In one of the play sessions, the teacher introduced Citizen DJ after reading the children's book "When the Beat Was Born" by Laban Carrick Hill, which told the history of rhythm, rap, and hip hop. The teacher had set up a music station in the classroom where small groups of 3–4 children could work with Citizen DJ to create their own musical compositions using "Carnival of Venice". This famous song inspired other musical compositions such as "(How Much Is) That Doggie in the Window?" and "My Hat, It Has Three Corners," which connected to children's existing funds of knowledge.

After a whole class introduction to the features of Citizen DJ, the teacher provided scaffolded support to a small group of children who gathered around the music station and began experimenting with the tools and features of Citizen DJ. They listened to different versions of "Carnival of Venice" and tried to identify specific instruments and sounds they heard. The children began to randomly play with a variety of samples and beats and gradually became more intentional in the combinations of sounds they preferred.

One child suggested that they could blend drums with the song to make it sound even better. She experimented with several options. As another child listened, he suggested that they could use some of the electronic sounds available in Citizen DJ to make the song sound like it was made by aliens in outer space. The teacher encouraged the children to be creative and try out different ideas.

As the children worked, they began to develop their own unique musical compositions based on "Carnival of Venice." The children took turns playing their compositions for each other and danced along to the music.

The integration of AI-enabled digital play through Citizen DJ showed promise in advancing pedagogy. The platform promoted inquiry-based dialogue as children independently and collaboratively explored, improvised, and created with sound. Through collaborative problem-solving, the children became co-facilitators of knowledge. The role of the teacher was less prominent, acting more as a facilitator, observer, or guide. Overall, the AI-enabled platform put children at the center of their learning experience, empowering them to drive and determine their own outcomes even with the use of technology. Despite the complexity of interacting with Citizen DJ, this approach prioritized play-based and child-initiated learning, emphasizing experiential, collaborative, and interactive approaches to learning and development. It aligned with the constructionist approach, by exploring new avenues for learning that highlight children's creative expression through music [34].

5   Implications for Design and Implementation

Our research highlights the importance of teachers in mediating children's interactions with AI-enabled technologies to support playful learning. AI innovations that engage children in generative and creative exploration appear to be an important precursor to introducing children to generative AI. By automating some of the more tedious and time-consuming aspects of music production, machine learning can help to make music creation more accessible to children and young learners, allowing them to focus on exploring and experimenting with different sounds and rhythms. However, it is important to note that while machine learning can be a valuable tool for enhancing creative music exploration, it is not a substitute for human creativity and expression. As Brian Foo has emphasized, the focus of Citizen DJ is on promoting experimentation and self-expression through the use of digital tools, rather than relying on automated or algorithmic processes.

The growing prevalence of AI in daily life has significant ethical and social implications for both designers and users. Human-in-the-loop [35] approaches are critical for aligning AI integration with early childhood pedagogy, as they ensure that technology is used to augment and support, rather than replace, the role of teachers in the classroom. In this approach, the technology is designed to work alongside the teacher and the children. Collaborative efforts among multiple stakeholders, including children and educators, are needed to address these complex issues. This helps to ensure that the technology is used in a way that is consistent with the educational goals and pedagogical approach of the teacher, and that it is used in a manner that supports and enhances the learning experience of the children.

References

1. Shani, C., Libov, A., Tolmach, S., Lewin-Eytan, L., Maarek, Y., Shahaf, D.: Alexa, do you want to build a snowman?" Characterizing playful requests to conversational agents. In: Conference on Human Factors in Computing Systems - Proceedings. Association for Computing Machinery (2022). https://doi.org/10.1145/3491101.3519870
2. Kucirkova, N., Hiniker, A.: Parents' ontological beliefs regarding the use of conversational agents at home: resisting the neoliberal discourse. Learn Media Technol. (2023). https://doi.org/10.1080/17439884.2023.2166529
3. Hoffman, A., Owen, D., Calvert, S.L.: Parent reports of children's parasocial relationships with conversational agents: trusted voices in children's lives. Hum. Behav. Emerg. Technol. **3**, 606–617 (2021). https://doi.org/10.1002/hbe2.271
4. UNESCO: Early Childhood Care and Education (2023). https://www.unesco.org/en/early-childhood-education/need-know
5. Prentzas, J.: Artificial intelligence methods in early childhood education. Stud. Comput. Intell. **427** (2013). https://doi.org/10.1007/978-3-642-29694-9_8
6. Su, J., Yang, W.: Artificial intelligence in early childhood education: a scoping review. Comput. Educ.: Artif. Intell. **3** (2022). https://doi.org/10.1016/j.caeai.2022.100049
7. Williams, R., Park, H.W., Oh, L., Breazeal, C.: PopBots: designing an artificial intelligence curriculum for early childhood education. In: 33rd AAAI Conference on Artificial Intelligence, AAAI 2019, 31st Innovative Applications of Artificial Intelligence Conference, IAAI 2019 and the 9th AAAI Symposium on Educational Advances in Artificial Intelligence, EAAI 2019 (2019)
8. Kim, D.-H.: An analysis of early childhood teachers' current status and awareness of using artificial intelligence. J. Korea Open Assoc. Early Childhood Educ. **27** (2022). https://doi.org/10.20437/koaece27-1-07
9. Su, J., Ng, D.T.K.: Artificial intelligence (AI) literacy in early childhood education: the challenges and opportunities. Comput. Educ.: Artif. Intell. (2023). https://doi.org/10.1016/j.caeai.2023.100124
10. Su, J., Zhong, Y.: Artificial Intelligence (AI) in early childhood education: Curriculum design and future directions. Comput. Educ.: Artif. Intell. **3** (2022). https://doi.org/10.1016/j.caeai.2022.100072
11. Sadam, K.: Implementation of AI pop bots and its allied applications for designing efficient curriculum in early childhood education. Art. Int. J. Early Childhood Spec. Educ. (2022). https://doi.org/10.9756/INT-JECSE/V14I3.271
12. Kewalramani, S., Kidman, G., Palaiologou, I.: Using Artificial Intelligence (AI)-interfaced robotic toys in early childhood settings: a case for children's inquiry literacy. Eur. Early Child. Educ. Res. J. **29**, 652–668 (2021). https://doi.org/10.1080/1350293X.2021.1968458
13. Ruiz, P., Fusco, J.: Teachers partnering with artificial intelligence: augmentation and automation. (2022). https://digitalpromise.org/2022/07/06/teachers-partnering-with-artificial-intelligence-augmentation-and-automation/
14. Welch, G.F.: The challenge of ensuring effective early years music education by non-specialists. Early Child Dev Care. **12**, 1972–1984 (2021). https://doi.org/10.1080/03004430.2020.1792895
15. Kirby, A.L., Dahbi, M., Surrain, S., Rowe, M.L., Luk, G.: Music uses in preschool classrooms in the U.S.: a multiple-methods study. Early Child Educ. J. **51**, 515–529 (2023). https://doi.org/10.1007/s10643-022-01309-2
16. Young, S.: Early childhood music education research: an overview. Res. Stud. Music Educ. **38**, 9–21 (2016)

17. Nieuwmeijer, C., Marshall, N., van Oers, B.: Musical play in the early years: the impact of a professional development programme on teacher efficacy of early years generalist teachers. Res. Pap. Educ. (2021). https://doi.org/10.1080/02671522.2021.1998207

18. Foo, B.: Citizen DJ. (2020). https://citizen-dj.labs.loc.gov/

19. Ackermann, E.: Piaget's constructivism, Papert's constructionism: what's the difference? Future of learning group publication **5**, (2001). http://learning.media.mit.edu/content/public ations/EA.Piaget_Papert.pdf

20. Lodi, M., Martini, S.: Computational thinking, between papert and wing. Sci. Educ. (Dordr). **30** (2021). https://doi.org/10.1007/s11191-021-00202-5

21. Marsh, J., Plowman, L., Yamada-Rice, D., Bishop, J., Scott, F.: Digital play: a new classification. Early Years **36**, 242–253 (2016). https://doi.org/10.1080/09575146.2016.116 7675

22. Disney, L., Geng, G.: Investigating young children's social interactions during digital play. Early Child Educ. J. **50**, 1449–1459 (2022). https://doi.org/10.1007/s10643-021-01275-1

23. Berson, I.R., Murcia, K., Berson, M.J., Damjanovic, V., McSporran, V.: Tangible digital play in Australian and U.S. preschools. Kappa Delta Pi record **55**, 78–84 (2019). https://doi.org/ 10.1080/00228958.2019.1580986

24. Arnott, L., Yelland, N.J.: Multimodal lifeworlds: pedagogies for play inquiries and explorations. J. Early Childhood Educ. Res. **9**, 124–146 (2020)

25. Arnott, L., Palaiologou, I., Gray, C.: Digital and multimodal childhoods: exploration of spaces and places from pedagogy and practice. Global Stud. Childhood **9**, 271–274 (2019). https:// doi.org/10.1177/2043610619885464

26. Penuel, W.R., Fishman, B.J., Cheng, B., Sabelli, N.: Developing the area of design-based implementation research. SRI International, Menlo Park (2011)

27. McKay, S.: Quality improvement approaches: Design-based implementation research. Carnegie Foundation Blog (2017). https://www.carnegiefoundation.org/blog/quality-improv ement-approaches-design-based-implementation-research/

28. Berson, I., Cross, M., Ward, J., Berson, M.: People, places, and pandas: engaging preschoolers with interactive whiteboards. Soc. Stud. Young Learn. **26**, 18–22 (2014)

29. Berson, I.R., Berson, M.J., Carnes, A.M., Wiedeman, C.R.: Excursion into empathy: exploring prejudice with virtual reality. Soc. Educ. **82**, 96–100 (2018)

30. Berson, I.R., Berson, M.J., McKinnon, C., Aradhya, D., Luo, W., Shapiro, B.R.: An exploration of robot programming as a foundation for spatial reasoning and computational thinking in preschoolers' guided play. Early Child Res. Q. **65**, 57–67 (2023). https://doi.org/10.1016/ j.ecresq.2023.05.015

31. Berson, I.R., Berson, M.J., Connors, B.C., Reed, L.E., Almuthibi, F.H., Alahmdi, O.A.: Using mixed reality to create multimodal learning experiences for early childhood. In: Cherner, T. and Fegely, A. (eds.) Bridging the XR Technology-to-Practice Gap. pp. 151–162. Association for the Advancement of Computing in Education (2023). https://www.learntechlib.org/p/222 293/

32. Holstein, K., McLaren, B.M., Aleven, V.: Co-designing a real-time classroom orchestration tool to support teacher–AI complementarity. J. Learn. Anal. **6**, 27–52 (2019). https://doi.org/ 10.18608/jla.2019.62.3

33. Chubb, J., Missaoui, S., Concannon, S., Maloney, L., Walker, J.A.: Interactive storytelling for children: a case-study of design and development considerations for ethical conversational AI. Int. J. Child Comput. Interact. **32** (2022). https://doi.org/10.1016/j.ijcci.2021.100403

34. Powell, S., Somerville, M.: Drumming in excess and chaos: music, literacy and sustainability in early years learning. J. Early Child. Lit. **20**, 839–861 (2020). https://doi.org/10.1177/146 8798418792603

35. Mosqueira-Rey, E., Hernández-Pereira, E., Alonso-Ríos, D., Bobes-Bascarán, J., Fernández-Leal, Á.: Human-in-the-loop machine learning: a state of the art. Artif Intell Rev. **56**, 3005–3054 (2023). https://doi.org/10.1007/s10462-022-10246-w

Promoting Students' Pre-class Preparation in Flipped Classroom with Kit-Build Concept Map

Yusuke Hayashi[1]([✉]), Yuta Aikawa[1], Yuki Kawaguchi[1], Huazhe Sha[2], Mayumi Sugiura[2], Katsusuke Shigeta[2], and Tsukasa Hirashima[1]

[1] Hiroshima University, Higashi-Hiroshima 7390046, Hiroshima, Japan
hayashi@lel.hiroshima-u.ac.jp
[2] Hokkaido University, Sapporo 08600808, Hokkaido, Japan

Abstract. This paper provides a comprehensive overview of using the Kit-build concept map in a flipped classroom setting to aid in the learning process. The Kit-build concept map is an effective tool for helping learners create concept maps during pre-class preparation, which enhances their understanding of basic concepts and helps them acquire fundamental knowledge to be applied during lectures. Additionally, this paper explores the various achievements of using the Kit-build concept map, including improved student engagement and a deeper understanding of the material covered. There are still several challenges that need to be addressed in the future, such as ensuring that the tool is accessible to all learners and used effectively to achieve learning outcomes. This paper argues that the Kit-build concept map is a valuable tool for both students and educators and that its continued use and refinement can lead to significant improvements in the learning process.

Keywords: flipped classroom · pre-class preparation · concept map · kit-build

1 Introduction

The flipped classroom approach allows for more interactive and collaborative learning activities during class time [2]. In a flipped classroom, learners first acquire basic knowledge through self-paced learning before engaging in in-person instruction. This approach is made possible through the use of technology to deliver content outside of the classroom.

Pre-class preparation can be useful in enhancing students' understanding during face-to-face learning in flipped classrooms [11]. However, it has been reported that if it is not done properly, the desired effects may not be achieved. Proper preparation for pre-class activities by students requires not only preparing teaching materials, but also analyzing how students are learning during pre-class activities and providing effective guidance.

The aim of this research is to suggest a technique to assist in the adequate preparation of pre-class work by using a kit-building concept map. For conceptual development in

N. Wang et al. (Eds.): AIED 2023, CCIS 1831, pp. 764–771, 2023.
https://doi.org/10.1007/978-3-031-36336-8_117

flipped classrooms, the goal of a video lecture is for learners to understand core concepts [5, 10]. Quizzes are a typical method of engaging learners in learning and checking their understanding [1, 12]. However, many learners only learn enough to pass the quizzes and course assessments [3]. Writing essays can encourage learners to understand the content at a deeper level and more accurately measure their understanding than multiple-choice tests. However, it is a costly approach, and automatic assessment is still an emerging technology [8]. In this study, we propose the use of Kit-build concept map in pre-class preparation of flipped classroom that is less burdensome for both students and teachers than essays, and allows students to gain a broader understanding of fundamental knowledge compared to quizzes.

2 Description of the AIEd Implementation

2.1 Kit-Build Concept Map

This study employs Kit-Build Concept Maps [7, 13], which are reconstructive concept maps, as AIEd technology. Concept map is a kind of the graphical representation of information in which each node represents a concept, and each link identifies the relationship between the two concepts it connects. Kit-build concept map is a method for creating concept maps by reconstructing existing ones. For example, when using this method in a classroom, the teacher first organizes the information they want to convey to learners as a concept map. Learners will use the deconstructed concept map to reconstruct the information acquired in class as a concept map (Fig. 1).

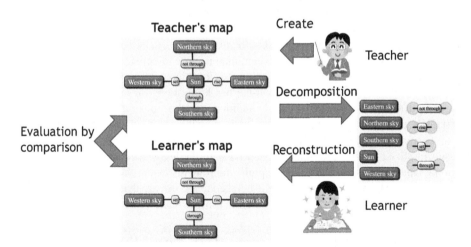

Fig. 1. An overview of Kit-build concept mapping

There are two benefits of using concept maps in Kit-Building to organize learning materials in class: learners can be supported to organize learning materials as concept maps, and it is easy to identify the gaps in understanding between teachers and learners. Creating a concept map requires learners to have a variety of cognitive skills. For instance, they need to identify important concepts from a large amount of information, name the relationships between them and structure them accordingly, and evaluate and modify the accuracy and completeness of the content. Kit-build concept map supports the creation of concept maps by providing components for them. Research on note-taking suggests that learners engage in two processes: selection and connection. It is particularly important to make effective selections during the selection process, as failure to do so may lead to an inability to recover during the connection process. It has been pointed out that it is important to reduce the load on the learner's selection process and focus on the connection process. Accordingly, Kit-building concept map provides learners with parts to prevent failures in the selection process, allowing them to focus on the connection process.

As an AI technology, Kit-Build Concept Map can work as a domain model and an open learner model [3] and can easily provide feedback learners. The teacher's map works as a domain model representing what learners to focus on in class. Learner model can be easily made as the difference between learner's map and teacher's map because learners' maps are composed with components of teacher's map. Displaying the difference of maps enables to be open learner model to learners. The difference between the teacher. One simple form of feedback is to compare a student's concept map to a teacher's one. Showing the differences to learners can help them recognize the gap between their understanding and that of the teacher regarding the course content. Moreover, for video lectures, it is possible to suggest re-watching particular parts of the lecture videos to fill the gaps [6]. This can be achieved by correlating each proposition in the concept map with the corresponding explanation in the lecture video. This is because the teacher's concept map used in the Kit-build concept map represents what the teacher wants to convey in class, and it is composed of what was shown during the class. Thanks to this feedback, learners can focus their review of lecture videos only on the parts where they have a gap in understanding compared to the teacher.

There is a report that Kit-build concept map is more effective than fill-in-the-blank questions [8]. In this study there was two classes in which one conducted Kit-build concept map and the other answer to fill-in-the-blank questions for clarification of understanding after each lecture in the unit. In addition to it, they took written comprehension test on the unit. The test results show the test score in the class with the Kit-build concept map is significantly higher than one in the class with fill-in-the-blank questions. This result implies that fill-in-the-blank questions help the learners review isolated propositions whereas Kit-build concept map also help them review the association among propositions. The basic limitation of Kit-build concept map are that teachers must make a concept map about information they want to provide learners with by themselves.

2.2 Implementation of Kit-Build Concept Map in Flipped Classroom

The educational goal in this study is to promote the understanding of learners' basic knowledge in pre-class preparation for flipped classroom in universities. In a flipped classroom, fundamental knowledge learned in pre-class preparation becomes familiar to learners, and activities to deepen that understanding can be conducted in class. To achieve this, it is necessary for each learner to commit to pre-class preparation in order to engage with fundamental knowledge.

In this study, we used kit-building concept maps as a learning tool for pre-class preparation in three flipped classroom sessions in a Japanese national university. These sessions focused on creating learning materials related to digital literacy education and students learn instructional design. This class has been implemented as a flipped classroom for several years. In 2022, we introduced the Kit-build concept map to enhance the acquisition of basic knowledge about instructional design during pre-class preparation. Video materials for pre-class preparation are prepared by one of the co-authors, who is the responsible teacher. Students are required to watch them before class. During the lecture, the teacher first reflected on the comments made by the students on the previous lecture's comment sheet. After that, the students conducted exercises using the basic knowledge about instructional design gained from pre-class preparation. After each lecture, students write about what they learned and noticed during the class on a comment sheet. In the end, the students applied the knowledge they gained in class to create a learning-task analysis diagram for the teaching materials they were developing. We consider the effectiveness of Kit-build concept map for pre-class preparation in three flipped classroom sessions by comparison between the classes in 2021 and 2022. The numbers of valid data are 23 and 29 respectively.

3 Reflection of the Challenges and Opportunities Associated with the Implementation

The challenge of this study is to enhance learners' comprehension of fundamental knowledge that will be utilized in in-person lectures through the use of the kit-build concept map during pre-class preparation for the flipped classroom. The instructor created a concept map in accordance with the content of the lecture videos used for pre-class preparation. Figure 2 shows one of the concept maps created by teacher. During the classes in 2022, students organized their learning from lecture videos by creating concept maps using Kit-build concept map in addition to watching the videos for pre-class preparation. Figure 3 shows the kit provided students to reconstruct concept map. Figure 4 shows the interface of feedback and video review. Table 1 show the average score of Kit-build concept map (Full marks are 100 points). This shows most of students can almost have completed Kit-build concept map. During the lecture, as in previous years, the teacher conducted a reflection based on the comment sheets written by the students after the previous class. In addition, the teacher also conducted a reflection on pre-class preparation using concept maps.

To evaluate the effectiveness of introducing a kit-built concept map, we compared the quality of comment sheets and learning-task analysis diagrams created by students in this year's and last year's lectures. For the evaluation of learning-task analysis diagrams, we established a rubric and graded each year's submissions based on the same criteria. Rubrics are composed of eight items, and each item is scored on a scale of 0 to 2, resulting in a total of 18 points. As a result, the quality of the task analysis chart is significantly higher this year than last year. The basic approach for describing the contents of the comment sheet is to fragment them by sentence and code them based on their "content" and "type". The "types" used for coding include what has been learned, impressions of what has been learned, and considerations regarding what has been learned. Regarding

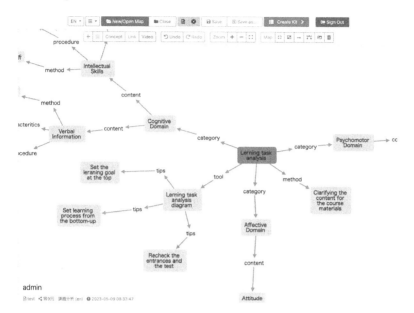

Fig. 2. A concept map created by the teacher (translated into English)

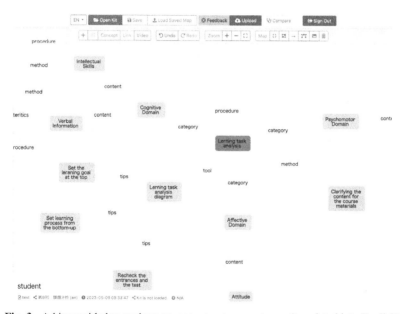

Fig. 3. A kit provided to students to reconstruct concept map (translated into English)

the "content," it is classified into two axes: "declarative knowledge" and "procedural knowledge," and "general content" and "specific content related to specific theories or models." According to the aggregated results, the total amount of description in

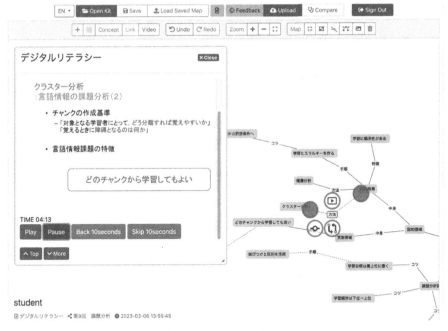

Fig. 4. Feedback and video review

Table 1. Averages of map score

	map score	
	first	last
1st lesson	88.7	92.6
2nd lesson	85.5	95.1
3rd lesson	95.1	95.8

the comment sheet has increased in 2022 compared to 2021, with the majority of the increase being comments related to declarative knowledge. Furthermore, it was found that comments on declarative knowledge tend to focus also on "general content". We consider that this result is influenced by the ease of writing declarative knowledge by concept maps (Tables 2, 3 and 4).

The teacher's impression is that Kit-build concept map is useful to easily check learners' comprehension of pre-class preparation. Especially, learners' concept maps reflect how the leaners relate what they have learned in pre-class preparation. Although it takes cost to prepare teacher's concept map and to kindly show learners how to use Kit-build concept map in classroom, he is satisfied with the use of Kit-build concept map in the class.

Table 2. Numbers of comment categorized by type of knowledge.

	N	Procedural knowledge	Declarative knowledge	Total	Fisher's exact test
2021	23	87 (3.78)	24 (1.04)	111 (4.83)	0.000
2022	29	104 (3.59)	126 (4.34)	230 (7.93)	

Number inside parentheses indicates the quantity per individual.

Table 3. Numbers of comment categorized related to procedural knowledge.

	N	general content	specific content	Total	Fisher's exact test
2021	23	66 (2.87)	21 (0.91)	87 (3.78)	0.48
2022	29	84 (2.90)	20 (0.69)	104 (3.59)	

Number inside parentheses indicates the quantity per individual.

Table 4. Numbers of comment categorized related to declarative knowledge.

	N	general content	specific content	Total	Fisher's exact test
2021	23	2 (0.09)	22 (0.96)	24 (1.04)	0.002
2022	29	52 (1.79)	74 (2.55)	126 (4.34)	

Number inside parentheses indicates the quantity per individual.

4   Description of Future Steps

Future challenges include conducting more detailed analysis and confirming repro-ducibility. We are planning to analyze how students' interests and understanding have changed through a qualitative approach to analyzing the contents of the comment sheet. Furthermore, in the future, it is expected that by recording the dialogue among students during exercises in class, in addition to the comment sheets, a more detailed analysis of the learning process can be conducted. To evaluate the effectiveness of the Kit-build concept map in pre-class preparation for a flipped classroom, we will continue to use and refine it, while also collecting additional data.

References

1. Admiral, W., Huisman, B., Pilli, O.: Assessment in massive open online courses. Electron. J. e-Learning **13**, 207–216 (2015)
2. Bergmann, J., Sams, A.: Flip Your Classroom: Reach Every Student in Every Class Every Day. International Society for Technology in Education (2012)
3. Bull, S., Kay, J.: Open learner models. Advances in Intelligent Tutoring Systems, pp. 301–322 (2010). https://doi.org/10.1007/978-3-642-14363-2_15
4. Deb, S., Pal, A., Bhattacharya, P.: Design considerations for self paced interactive notes on video lectures - a learner's perspective and enhancements of learning outcome. Proc. of IHCI **2017**, 109–121 (2017)

5. Drake, J.R., O'Hara, M., Seeman, E.: Five principles for MOOC design: with a case study. J. Inf. Technol. Educ. Innovations Pract. **14**, 125–143 (2015)

6. Hayashi, Y., Maeda, K., Honda, T., Hirashima, T.: Sectional review recommendations based on learner's comprehension in video-based learning. In: Proceedings of the ICCE 2018, pp. 328–333 (2018)

7. Hirashima, T., Yamasaki, K., Fukuda, H., Funaoi, H.: Framework of Kit-Build concept map for automatic diagnosis and its preliminary use. Res. Pract. Technol. Enhanc. Learn. **10**(1), 1–21 (2015)

8. Kitamura, T., Hayashi, Y., Hirashima, T.: Generation of fill-in-the-blank questions from concept map and preliminary comparison between multiple-choice task and Kit-Build task. J. Inf. Syst. Educ. **18**(1), 11–15 (2019)

9. Reilly, E.D., Stafford, R.E., Williams, K.M., Corliss, S.B.: Evaluating the validity and applicability of automated essay scoring in two massive open online courses. Int. Rev. Res. Open Distrib. Learn. **15**(5), 83–98 (2014)

10. Seery, M.K., Donnelly, R.: The implementation of pre-lecture resources to reduce in-class cognitive load: a case study for higher education chemistry. Br. J. Edu. Technol. **43**, 667–677 (2012)

11. Shibukawa, S., Taguchi, M.: Exploring the difficulty on students' preparation and the effective instruction in the flipped classroom: a case study in a physiology class. J. Comput. High. Educ. (2019)

12. Wachtler, J., Hubmann, M., Zöhrer, H., Ebner, M.: An analysis of the use and effect of questions in interactive learning-videos. Smart Learn. Environ. **3**(1), 1–16 (2016). https://doi.org/10.1186/s40561-016-0033-3

13. Yamasaki, K, Fukuda, H, Hirashima, T, Funaoi, H.: Kit-build concept map and its preliminary evaluation. In: Proceedings of 18th International Conference on Computers in Education, pp. 290–294 (2010)

AIED Unplugged: Leapfrogging the Digital Divide to Reach the Underserved

Seiji Isotani[1,2(✉)] , Ig Ibert Bittencourt[1,2] , Geiser C. Challco[2,3] ,
Diego Dermeval[2] , and Rafael F. Mello[2,4]

[1] Harvard Graduate School of Education, Cambridge, MA 02138, USA
{seiji_isotani,ig_bittencourt}@gse.harvard.edu
[2] NEES: Center for Excellence in Social Technologies, Federal University of Alagoas,
Maceio, AL 57072-970, Brazil
[3] Federal Rural University of the Semi-arid Region, Pau dos Ferros, RN 59900, Brazil
[4] Federal Rural University of Pernambuco, Recife, PE 52171-900, Brazil

Abstract. Artificial Intelligence in Education (AIED) is a driving force
to improve education. Nevertheless, policymakers from the Global South
fear that AI will increase the digital divide and reduce the opportunities for students in these regions to thrive. To address this problem, we
analyzed the past 30 years of data on four aspects of the digital divide.
Then, based on these findings and a series of discussions with stakeholders (e.g., policymakers), we proposed the concept of *AIED Unplugged*.
An approach to creating AI-based educational technologies that do not
require changes in current school settings (e.g., infrastructure), do not
rely on stable internet access, and do not ask for digital skills to use
them. We applied this concept to redesign an education policy in Brazil
to help students improve their writing skills. Our results show a reduction in time, cost and complexity to running the policy, and a positive
impact on more than 500,000 students in 7,000 schools in the country.

Keywords: Global South · public policy · educational technology

1 The Global Movement Toward AI

Artificial Intelligence (AI) has gained considerable attention in the past years
from the market, governments, and civil society. AI is revolutionizing our lives
to the point that moving from point A to point B has become difficult without
using an intelligent digital resource. According to Ng [3], AI can be considered
the new electricity. He argues that intelligent technologies will transform every
industry in the next several years, just as electricity did in 100 years.

In the context of Education, AI has been considered by several educational
stakeholders as a driving force of transformation to build back better after the
worldwide decrease in students' performances due to the extreme measures taken
during the covid-19 pandemic [2]. In May 2019, ministers of state, hundreds of

government representatives, academic institutions, highly-regarded members of civil society, and the private sector met in Beijing. The result of this meeting was the Beijing Consensus on Artificial Intelligence in Education (AIED) which reaffirmed the commitment to the 2030 Agenda for SDG and provided recommendations to governments and other stakeholders in UNESCO's Members States [9]. Following the Beijing Consensus, in 2021, UNESCO released the AI and education guidance for policy-makers [10]. This guidebook highlighted three key policy questions that need system-wide responses to fully unleash the opportunities and mitigate the potential risks: How can AI be leveraged to enhance education? How can we ensure AI's ethical, inclusive, and equitable use in education? And How can education prepare humans to live and work with AI?

Although this movement has significantly promoted more AI innovations in Education, it also highlights the clear need for a more inclusive AIED. Since AI strongly depends on digital devices and internet connectivity, **AIED has become the new source of educational inequality**. Vinuesa and colleagues [12] show that although AI has positively impacted the world's 17 most pressing needs (also known, as Sustainable Development Goals - SDGs), it also has *negatively* impacted some SDGs, such as SDG 1 (End Poverty), SDG 4 (Quality Education), and SDG 10 (Reduce Inequality).

In this context, many policymakers from Global South countries are concerned that introducing AI in Education may bring potentially unequal benefits to students. High-income students with access to the internet and other digital resources will benefit the most from using AI. Meanwhile, low-income students lacking basic infrastructure (such as electricity) at home and school may observe a decrease in the opportunities to thrive, since most jobs they currently seek and are able to get will disappear because of AI and other digital technologies.

To face the challenge of offering the benefits of AI to underserved students without the need for high amounts of infrastructure investments, we propose a novel concept referred to as *AIED Unplugged*. In this work, we will discuss the fundamentals of this concept with an example developed together with policymakers to bypass the digital divide and use of AI to impact millions of low-income students in Brazil that have been struggling to learn Portuguese.

2   AIED Divide

AI is the "new electricity"! And this can be considered terrible news for people living in the Global South, where low-income countries are mostly located. The fear that AI will increase the gap between high- and low-income countries comes from the fact that the current digital divide has already accelerated this process [1]. The digital divide can be defined as the readiness (in terms of attitudes, access, skills, usage, and culture) of people and society to benefit from technology [8]. High-income countries not only have the needed infrastructure to benefit from technology, but most of their citizens are motivated and have the digital skills to use technology, positively impacting their personal and professional activities.

To better understand the digital divide between high-, middle- and low-income countries across the years and identify potential opportunities to support public policies, in this work, we analyzed data from international organizations such as the World Bank[1] and the US National Center for Education Statistics[2]. We collected 30 years of data and conducted a regression analysis to create a model considering two key elements of the digital divide: infrastructure (access to electricity, internet, and device: mobile phones) and digital skills. The results of our analysis are shown in Fig. 1. The model curve based on 30 years of collected data is shown as continuous lines and the expected results for the following years (i.e., prediction) are shown as dashed lines.

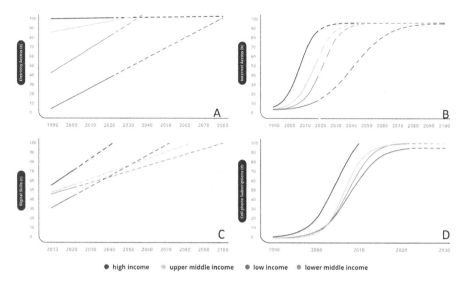

Fig. 1. Digital Divide among high-, middle- and low-income countries: (A) Electricity access; (B) Internet access/usage; (C) Digital Skills; (D) Cellphone access/subscription.

According to Fig. 1, great disparities exist between high-, middle- and low-income countries. For example, Fig. 1A shows that Only 40% of the population in low-income countries have access to electricity while high-income countries are very close to 100%. Furthermore, it will take over 50 years for low-income countries to reach the same level of access to electricity currently available in high- and upper-middle-income countries. The same pattern can be observed regarding internet access/usage and digital literacy as the regression models presented in Figs. 1B and 1C indicate that low-income countries will reach the same levels of high-income countries only after the year of 2100.

Furthermore, in the context of education, according to a policy brief from UNICEF published in 2021, it is estimated that 1.4 trillion dollars will be

---

[1] https://data.worldbank.org/.
[2] https://nces.ed.gov/surveys/icils/.

required in the coming decade to reduce the digital divide and enable students to benefit from technology during the learning process [13].

Although policymakers have responded to this threat, there are several digital divides, and they are continually changing. In this context, the challenge of benefiting from AI in Education is even higher since most research and applications of AI require internet connectivity and physical devices for all students (i.e., affordability). Affordability determines individual access to AI and abilities to participate fully in the AI revolution. The required infrastructure and skills to develop and use AIED applications restrict its potential benefits to locations with sufficient computing power, access to relevant internet bandwidth, and high levels of digital skill personnel. This is what we refer to as the **AIED Divide**.

The AIED Divide transcends geographic, socioeconomic, gender, and race boundaries. It demands temporal, material, mental, social, and cultural resources to enable the meaningful use of AIED solutions and promote better educational outcomes. Nevertheless, there is an important lesson the covid-19 pandemic brought to many fields, including AI and Education: we can build back better and restore opportunities for children and youth to gain the skills to build a more inclusive and sustainable society [7]. To do this, we need to reimagine our technology and its appropriation to promote a pedagogy of cooperation and solidarity, bringing to the discussion different stakeholders with complementary and even contradictory views to improve the world [10].

3   AIED Unplugged: Fundamentals

Although there is a nearly 100-year gap (see Fig. 1) hard to bridge the global north and south, it is possible to reimagine the future and bypass the various digital divide we face in the world. As a movement to build back better, we can promote more creative and innovative ways to create public policies that help the development and use of AI technologies to tackle the challenges of education and, thereby, reach minorities and marginalized communities such as migrants, underserved students, internally displaced people, refugees, imprisoned people, indigenous peoples, students with special rights (instead of special needs), people affected by natural disasters or wars, and so on. By facing the challenge of ensuring opportunities for quality learning for all, different approaches to thinking of AI in Education can be observed as potential ways of leapfrogging the AIED divide, for example, through the use of mobile solutions that are already accessible worldwide, including in low-income countries as shown in Fig. 1D).

Despite the opportunities for innovations to address the problems of the AIED divide, a literature review carried out by Nye [4] shows that very few Intelligent tutoring systems and other AI technologies for education have been developed to address the challenges of the developing world. According to Nye, most AIED technologies require digital skills, do not allow hardware sharing, and cannot be used on low-cost mobile devices. These problems persist throughout the years, as indicated by a more recent report from UNESCO [5].

Inspired to innovate in the field, since 2017 we have been conducting a series of workshops, interviews, meetings and policy design activities together with policymakers, teachers, students and other stakeholders in the field of education in Brazil. We have been defining, testing, and refining key elements that are fundamental to creating education policies with the support of AI technologies and reaching communities that are usually left behind due to the limitations described in the section *AIED Divide*.

As a result, we proposed the concept of *AIED Unplugged*. The term "unplugged" refers to the fact that in our context: (i) AI solutions should not constantly access the internet; and (ii) the target users (e.g., students) lack digital skills and access to resources (hardware, internet, etc.) being disconnected from the digital world [4]. AIED Unplugged extends and operationalizes the ideas of Jugaad innovations [6] in the field of AI in Education. Juggad innovation is a concept that originated in India and focused on finding creative and accessible solutions to problems. Two important elements of Juggad innovation are *resourcefulness*, the ability to find creative solutions using the resources that are already available in a particular setting; and *simplicity*, the focus on creating solutions that are easy to use (not requiring extensive training of users) and maintain. Considering that, the key elements of the AIED Unplugged are:

- **Conformity**. Rather than disrupting the educational environment, requiring extensive training and changes in infrastructure, the AI-based solution should be developed considering the available infrastructure, resources and pedagogical practices.
- **Disconnect**. The AI-based solution should not require internet access to work. Conversely, it should use the internet whenever available to update AI models, collect data and provide user feedback.
- **Proxy**. We cannot assume that target users (e.g., students) own hardware to access an AI-based solution or have the skills to create a login account in a system. Thus, The AI-based solution may consider a proxy between the target user and the AI solution.
- **Multi-user**. AI-based solutions should be created considering that hardware and software are constantly shared among users and proxies. Thus, any solution that requires users to log in or need to record individual interactions to update the AI models (e.g., the user model in an intelligent tutoring system) will most likely not work in our context.
- **Unskillfulness**. AI-based solutions should be created to be simple enough that do not require additional digital skills other than what most people with access to a cellphone already possess (such as clicking an icon, taking a picture, sending/writing a message, making calls, etc.).

The objective of AIED Unplugged is to transform the way we think, design, build, and use intelligent educational technologies to overcome the lack of infrastructure, digital skills and other aspects of the AIED divide that are prevalent in lower-middle- and low-income countries. The ultimate goal of AIED Unplugged is to leapfrog the divide and use AI technology to benefit underserved students and communities. Technology leapfrogging, following the AIED Unplugged approach, is seen as a way to rapidly increase the pace of a country's economic development and thereby reduce the gap between developed and developing nations. Indeed, Leapfrogging experiences have been successfully applied in several countries of the Global South for different purposes, including education [11].

4   Redesigning an Education Policy with AIED Unplugged

To carry out a policy for national assessment and identify the level of writing skills of K-12 students in Brazil, the federal government requests students in public schools to write essays. These essays are written on a piece of paper and are physically mailed to a center that is responsible for digitalizing and distributing the essays for peer review. This center is also responsible for mailing back the results of these assessments back to schools and students. The whole process takes about four months to complete and cannot be used for formative assessments (that would help students to learn). According to policymakers involved in this policy, the status-quo solution to carry out the national assessment for writing skills is high-cost, time-consuming and complex due to: (i) the need for expensive industrial scanners; (ii) the use of mail services that need to reach places with limited access (e.g., schools in the middle of the Amazon forest or in semi-arid regions with no roads); (iii) the necessity of paying people to evaluate the essays; (iv) the requirement of training to evaluate the essays using a specific rubric; and (v) the need for controlling the peer review process.

Using the concept of AIED Unplugged in this work we aimed at improving the writing skills of K-12 students without increasing the burden on teachers and considering the social inequalities of the country, which means a high number of schools without internet access and digital devices, a high rate of students and teachers with low or intermediate levels of digital skills, lack of qualified/trained teachers in many vulnerable schools (especially in North and Northeast of Brazil), and lack of pedagogical practices that focus on provide formative feedback on writing essays.

To follow the key elements of the AIED Unplugged (vide Sect. 3) we did not require any change in the school settings and infrastructure nor asked teachers and students to learn new digital skills. We create an AI application for mobile devices that allows the digitalization, correction, and pedagogical diagnosis of handwritten texts in Portuguese. The overall flow of the technology is shown in Fig. 2. The students' essays are obtained from sheets of paper with QR codes and markings (Fig. 2-1). The teacher acts as a proxy of students using the mobile application, which has a simple menu automatically populated with students' names in the class (Fig. 2-2), and takes a photo of their essays. The photos remain on the mobile device and are uploaded to a server when an internet connection is available. When the photo is uploaded, the server uses Computer Vision to make the segmentation of the image into Portuguese words (Fig. 2-3 to Fig. 2-5). The current accuracy of this process is between 92% to 95%. Then, we use Natural Language Processing to automatically assess the essays according to a specific rubric (Fig. 2-6). To provide feedback for students and support for teachers we developed paper-based dashboards that can be printed to work with students (Fig. 2-7). Finally, we are developing an unplugged intelligent tutoring system to provide paper-based feedback to students and recommend following educational resources to improve writing skills Fig. 2-8).

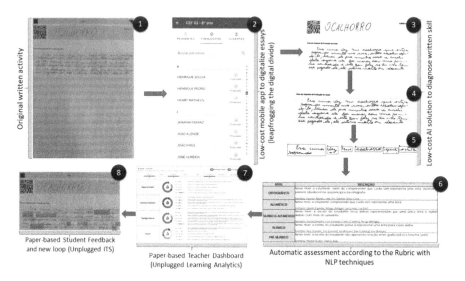

Fig. 2. An AI technology created using the concept of AIED Unplugged

We have been using this technology since 2022. To date, more than 1.5 million essays produced by over 500,000 k-12 students in 7,000 different schools in Brazil have been assessed. The maximum time from the beginning of the process to the end of it is 72 h. This means that students receive formative feedback on their essays in less than three days. This is a radical change compared to the previous solution that required four months to complete the same task as discussed at the beginning of this section. The costs and complexity of assessing essays nationwide were also reduced. Furthermore, as a result of this work, we have created the largest dataset[3] of essays written in Brazilian Portuguese with their respective transcripts.

Due to the success of our work in completely redesigning the implementation of a policy, the Brazilian Ministry of Education and the World Bank worked together to pass a decree creating the Brazilian Policy of Learning Recovery in Basic Education[4] which aims to encourage the development and use of AIED Unplugged-based technologies to help improve current practices that guarantee quality and equity in education.

Based on this experience, we intend to further explore and better conceptualize the definition of AIED Unplugged. We want to enable the design of policies

---

[3] This dataset will be released freely to the public in 2023–2024.

[4] http://www.planalto.gov.br/ccivil_03/_ato2019-2022/2022/decreto/D11079.htm.

that use AI technologies to bypass the problem of the AIED Divide and extend the benefits of AI to those who do not have access to technology. Particularly for this policy, our future work will extend the concept of AIED Unplugged to create intelligent tutoring systems that can work in multi-user settings, with multiple learner models running simultaneously, and using a paper-based interface where hints and feedback are given to a proxy (e.g., a teacher) who will then adapt its practices to improve students' learning experiences.

Acknowledgements. We wish to express our gratitude to all participants in this national project, including researchers, policymakers, teachers, educators and students. This work was supported by the Brazilian Ministry of Education.

References

1. Bastion, G.D., Mukku, S.: Data and the global south: Key issues for inclusive digital development (2020). https://us.boell.org/en/2020/10/20/data-and-global-south-key-issues-inclusive-digital-development
2. Chanduvi, J.S., et al.: Where are we on education recovery? UNICEF Report (2022). https://www.unicef.org/reports/where-are-we-education-recovery
3. Lynch, S.: Andrew ng: why ai is the new electricity. Insights by Stanford Business, vol. 11 (2017). https://www.gsb.stanford.edu/insights/andrew-ng-why-ai-new-electricity
4. Nye, B.D.: Intelligent tutoring systems by and for the developing world: A review of trends and approaches for educational technology in a global context. Int. J. Artif. Intell. Educ. **25**, 177–203 (2015)
5. Pedro, F., Subosa, M., Rivas, A., Valverde, P.: Artificial intelligence in education: Challenges and opportunities for sustainable development. UNESCO Report (2019). https://unesdoc.unesco.org/ark:/48223/pf0000366994
6. Radjou, N., Prabhu, J., Ahuja, S.: Jugaad Innovation: Think Frugal, be Flexible, Generate Breakthrough Growth. John Wiley & Sons, Hoboken (2012)
7. Reimers, F.M. (ed.): Primary and Secondary Education During Covid-19. Springer, Heidelberg (2022). https://doi.org/10.1007/978-3-030-81500-4
8. Tsatsou, P.: Digital divides revisited: what is new about divides and their research? Media Cult. Soc. **33**(2), 317–331 (2011)
9. UNESCO: Beijing consensus on artificial intelligence and education (2019). https://unesdoc.unesco.org/ark:/48223/pf0000368303
10. UNESCO: AI and education: Guidance for policy-makers. UNESCO (2021). https://unesdoc.unesco.org/ark:/48223/pf0000376709
11. Vegas, E., Ziegler, L., Zerbino, N.: How ed-tech can help leapfrog progress in education. Center for Universal Education (2019). https://files.eric.ed.gov/fulltext/ED602936.pdf
12. Vinuesa, R., et al.: The role of artificial intelligence in achieving the sustainable development goals. Nat. Commun. **11**(1) (2020). https://doi.org/10.1038/s41467-019-14108-y
13. Yao, H., et al.: How much does universal digital learning cost? policy brief. UNICEF Office of Research-Innocenti (2021). https://www.unicef-irc.org/publications/pdf/How-Much-Does-Universal-Digital-Learning-Cost.pdf

Can A/B Testing at Scale Accelerate Learning Outcomes in Low- and Middle-Income Environments?

Aidan Friedberg[✉]

EIDU, Berlin, Germany
`aidan.friedberg@eidu.com`

Abstract. On current trends the world will fail to reach the objectives set in the UN's Sustainable Development Goals for Education by 2030 or even within the 21st century. Changing this trend will require a significant acceleration in learning outcomes. Digital personalised learning (DPL) tools are a potentially cost-effective intervention that can contribute to this acceleration. In particular, the continuous experimentation afforded by these tools through software A/B testing, has considerable potential to create compounding improvements in learning outcomes. This paper provides an overview of EIDU, an educational platform combining student focused DPL content with digital structured pedagogy programmes in public pre-primary schools in Kenya. Collection of student's longitudinal unsupervised assessment data at scale creates the possibility of learning outcome focused A/B testing. This is a novel contribution to the development and research field as up until now this type of capability has largely been confined to students in high-income environments.

1 Description of the AIED Implementation or Its Intended Use

While digital educational tools are becoming more prevalent in high-income countries, they are still almost non-existent at scale in low- and middle-income countries (LMIC). As a result, most of the insight into the impact of digital tools in these environments is largely left to the snapshots garnered by infrequent studies. Recent reviews of the available evidence have found digital interventions to potentially be a cost-effective way to improve learning outcomes in the LMIC environment [1, 2]. This is especially pertinent given the stagnation in learning outcomes experienced by many LMIC countries over the last decade despite high levels of school enrolment [3].

EIDU is a digitally supported learning platform for schools in the LMIC context with the aim of ensuring inclusive and equitable quality education, in line with UN Sustainable Development Goal 4 (Quality Education). In addition to accelerating learning outcomes for students on the platform, EIDU partners with the educational development and research community to generate generalisable insights on how digital interventions can be optimised for the LMIC context.

1.1 Current Learner Population

EIDU works predominantly in public pre-primary schools in Kenya as well as with students in low-cost private primary schools up to Grade 2. As of January 2023, EIDU has 100,000 monthly active learners from 1,600 schools engaging with the EIDU platform in Kenya. In co-ordination with the Kenyan government, EIDU will scale to all public pre-primary schools in Kenya over the next two years – reaching 2 million pre-primary learners by the end of 2025. Additionally, EIDU will pilot the digital learning platform in public primary schools in Kenya 2023 and has already piloted the platform in Nigeria and Ghana.

1.2 Implementation Model

EIDU provides classrooms with one to two low-cost smartphones with the EIDU application pre-installed. There are two elements to the application: a student-facing Digital Personalised Learning (DPL) tool and a teacher-facing area for supporting structured pedagogy. The environments we operate mean it can be challenging to standardise an implementation model. Generally, a dedicated 'EIDU Corner' is created in the classroom with a table and chair. Individual learners sit at this corner while engaging with our DPL software. Teachers choose the subjects and specific competencies for students to focus on. Teachers also use the software to view student progress and to access digital lesson plans as part of a structured pedagogy programme. Teachers are regularly provided training and are supported by our coaching staff and by county government officers in using the EIDU platform.

1.3 Digital Personalised Learning

Our DPL Tool is content agnostic, allowing for any content creator to integrate their content with the platform using a software development kit (https://dev.eidu.com). We target content with proven impact in the LMIC context such as onebillion [4]. Content needs to be provided in the form of Work Units. Work Units are a self-contained collection of 3 to 5 small exercises targeting a specific skill.

By creating skill mappings of all work units we can automatically group them to match the relevant curriculum structure for a given country. Taking Kenya as our example, content is structured in strands and substrands as per the Kenyan curriculum ensuring teachers can easily integrate the tool into their teaching. As of January 2023, we cover all Kenyan strands and substrands in Mathematics and Language up to Grade 2. Any new content is first sent to the Kenyan Institute for Curriculum Development (KICD) for review. Once approved as appropriate and relevant for students it is made available on our platform.

After choosing the substrand they want their learners to practice on, teachers select the first learner to engage with the device. A session for an individual learner lasts approximately 5 min. With learners averaging 40 min of usage per week as of January 2023. Learners can self-manage handover of devices as following the end of a session the profile picture of the next learner appears. All identifying and sensitive learner data is encrypted locally on the device so that EIDU only has access to anonymised data.

1.4 Personalisation

To optimally integrate work units from different content providers an effective personalisation algorithm is key. Our current implementation takes an approach similar to the Deep Knowledge Tracing (DKT) algorithm described by Piech et al., 2015 [5]. This is a Long Short-Term Memory (LSTM) neural network which takes a learner's performance history as a binary input sequence (solved/not solved) and outputs a vector of probabilities for solving a set of work units from a chosen substrand. This vector of success probabilities can be viewed as the knowledge profile of a learner within a substrand given the available content. We then infer performance multiple time-steps into the future to evaluate how this knowledge profile would change depending on which work unit was played next. The work unit which generates the highest expected gain in a learner's knowledge profile, taking into account the probability of success, is chosen as the next unit for the learner. We have A/B tested numerous personalisation implementations and will continuously do so to ensure we provide the most meaningful and impactful experiences for learners.

1.5 Structured Pedagogy Programme

EIDU also provides a teacher area where digitised structured pedagogy programmes can be accessed. In Kenya we have digitised the Tayari programme. Tayari was an early childhood development and education (ECDE) paper-based intervention in Kenya. The pilot, which ran from November 2014 to July 2018, aimed to develop a cost-effective and scalable model of ECDE that would ensure that children who join Grade 1 are cognitively, physically, socially, and emotionally ready to start, and succeed in primary school. Upon evaluation, Tayari was found to significantly improve school readiness outcomes for pre-primary children with a standardised effect size of 0.5 for public pre-primary schools [6]. EIDU has adapted the original Tayari intervention so it can be delivered to schools digitally. Teachers receive trainings in the use of the programme along with digital lesson guides.

1.6 Learning Data

The World Bank and UNICEF in their 2021 review of the global education crisis concluded "To tackle the learning crisis, countries must first address the learning data crisis, by assessing students' learning levels" [7]. Being able to measure and evaluate learning robustly is also key to tracking the evolution of learning outcomes for learners on our platform. We have digitized established digital assessment batteries like EGMA, EGRA and MELQO. We then run concurrent validity tests on these digital items to ensure we are still measuring the same outcomes as the original paper test (minimum Pearson's $r > 0.65$). Learners are tested on random units during their regular usage sessions and are then re-tested 6 weeks later, creating large amounts of longitudinal learning data. As of January 2023, we already gather over 300,000 unique test results in Numeracy and Literacy from 60,000 learners per month. We also collect simple demographic data such as gender and age to measure group effects.

1.7 A/B Testing Framework

All significant changes to the platform are A/B tested using the learning data described in 1.6 as well usage data to evaluate the impact of any changes. The design process for these A/B tests is described in more detail in 2.1. By continuously improving the platform through A/B testing we aim to generate compounding value in terms of learning gains [8]. Although a well-established practice in the software industry, continuous experimentation and A/B testing is still in its infancy within academia as a whole [9]. A small number of empirical studies have demonstrated the potential A/B testing has for education science [10, 11]. EIDU is now partnering with the EdTech Hub to expand this area of research. The project, funded by the Bill and Melinda Gates Foundation, will explore how A/B testing can improve learning outcomes for students on the platform while generating generalisable knowledge on how digital interventions can be optimised for the LMIC environment.

2 Reflection of the Challenges and Opportunities Associated with the Implementation

2.1 Development Process

EIDU designed and rolled-out its first pilot with low-cost private schools in Kibera, Nairobi – the largest informal settlement in Africa – in 2016. This was done to ensure EIDU's implementation model could be proven to work even in challenging environments and so that teachers and learners from within marginalised communities could meaningfully contribute to designing a solution that works for them. Since then, we have tried to ensure the teachers and learners we serve are at the heart of our development process. We do this through dedicated User Experience researchers who themselves were teachers in Kenya and so have a strong level of sensitivity and understanding to teacher needs. Additionally in partnership with county governments we organise cluster meetings for each ward – a group of 20 to 50 geographically proximate schools. Here teachers from the same community come together and can share their experience and challenges with each other, with EIDU and with their county representative. This ensures both teachers and county officials are part of our development process. Finally, we also partner with organisations like Women's Educational Researchers of Kenya (WERK) and EdTech Hub to conduct in-depth qualitative and quantitative research, providing a qualified external perspective on our implementation model and how we can improve this.

Insights generated from these processes are worked on by our development and product teams to create new features or improvements. These are then released to a small alpha group of highly engaged low-cost private schools who have been using EIDU for several years. Teachers are actively asked for their feedback on any changes while classroom observations and data analysis allow us to understand the learner experience.

After passing alpha testing a feature can be released as an A/B test. Depending on the feature, partitions can be randomised on a learner, class, or school level. If randomised on a learner level and released to our whole population as of January 2023 this would mean an A/B test would have 50,000 learners in each partition, leading to considerable test sensitivity.

Tests are run for a pre-determined period usually one school term (3 calendar months). Tests are continuously monitored and can be ended prematurely if found to be unhelpful. Where evaluation through assessment outcomes is inconclusive, we explore alternative metrics such as progress of learners in the curriculum or other usage metrics. Figure 1 shows an example of a personalisation A/B test on our learning analytics platform.

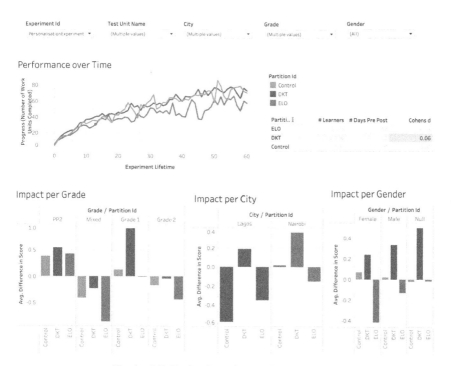

Fig. 1. A/B Testing Learning Analytics Platform

Once the partition generating the largest improvement in learning outcomes has been identified, this becomes our default implementation and the next improvement or feature to iterate on is identified and tested. This agile iterative process opens up the possibility of drastically reducing the duration of the design – implement – evaluation loop within education in the LMIC context.

2.2 Challenges

2.2.1 Content

There are several challenges inherent to our implementation context. One is that we are generally curators of content rather than creators, meaning we are limited by what others in the world create. The majority of digital content that is created is targeted at high-income country environments, this can make it difficult to find cultural or context appropriate content for our schools. Moreover, content in local languages can be difficult to obtain, although we do have Swahili language content from onebillion. Finally, our personalisation algorithm can only work with content atomised as work units. AI capable of providing hints or changing content within work units is not possible in our current implementation.

2.2.2 Connectivity

Although EIDU was designed to work offline, to download new content, new versions or to upload usage data an internet connection is required. We provide all devices with sponsored mobile data, allowing them to upsync and downsync EIDU data for free. We are seeing increasing challenges as we expand into more rural areas regarding connectivity. Although this does not stop schools having access to content it does mean this content may become outdated creating unequal access to newer benefits. Additionally, if rural learners are underrepresented in our A/B testing population we risk marginalising them further. We have prototyped peer-to-peer solutions which we will continue to test with the aim of mitigating this.

2.2.3 Unsupervised Assessments

To a large degree we rely on learners engaging with assessments in an unsupervised environment in order to track longitudinal learning data. This leads to a significant source of uncertainty. Learners can play on accounts that are not theirs which can mean on an individual level we cannot be sure of the fidelity of a single learner's assessment history. However, the evolution of learning for a class or a school as a whole still has high validity.

More critical is that learners are currently less engaged by assessment units than by our regular educational work units. This can be seen through classroom observations as well as distributions of scores in our data when compared to supervised tests as shown in Fig. 2. Our digital units perform very similar to the equivalent paper assessment under the supervision of an enumerator during validation. However, once a learner engages with this unit in an unsupervised environment there is a high skew to 0 scorers. We have made some progress in improving engagement through A/B testing. In Fig. 3 we see how the simple introduction of coins (gained for correct answers) and hearts (lost from incorrect answers), improves engagement by reducing the percentage of learners scoring 0. This also demonstrates how A/B testing can be used to measure and improve the engagement of learners on the platform.

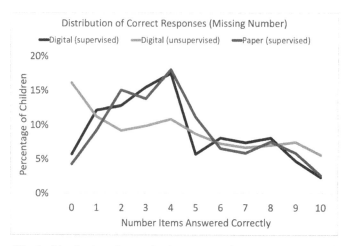

Fig. 2. Distribution of supervised vs. unsupervised assessment scores

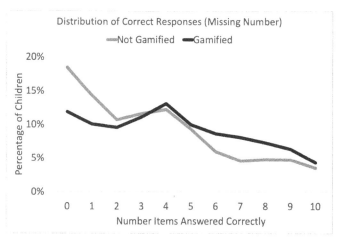

Fig. 3. Feedback Gamification Feature and A/B test results

3 Description of Future Steps

3.1 Next Steps

We are focused on overcoming the challenges mentioned in 2.2 to scale our learning platform sustainably and effectively. After proving this implementation can scale nationwide and can accelerate learning outcomes through compounding value, we hope it can be used as a model for other countries in sub-Saharan Africa and beyond. The EdTech Hub are currently in the middle of conducting a large scale external randomised control trial with ~2,800 learners exploring the impact the EIDU platform has in Kenyan public pre-primary schools. Baseline data was collected in October 2022 and endline data will be collected in November 2023.

3.2 Partnerships

A key motivation for this paper is EIDU's belief that collaboration from across the educational spectrum is crucial to meaningfully combat global learning poverty. EIDU has partnered with research and philanthropic institutions to try to strengthen the ties between practitioners, industry, and researchers. We have partnered with the EdTech Hub to explore how our learning data can create generalisable knowledge on how learning outcomes can be improved in the LMIC context throughout 2023 and 2024 using A/B testing. We have also now released the personalisation plugin first described at AIED22 [12]. Details can be found at https://dev.eidu.com/category/personalization-plugin-dev elopment. We are actively seeking to partner with researchers in the field of AIED and learning personalisation, using this plugin to help improve learner outcomes while generating valuable insight on how edtech can help learners in low and middle income environments.

References

1. Rodriguez-Segura, D.: Educational technology in developing countries: a systematic review. University of Virginia EdPolicy Works Working Papers, p. 2021 (2020). Accessed 17 Dec 2020
2. Major, L., Francis, G.A., Tsapali, M.: The effectiveness of technology-supported personalised learning in low-and middle-income countries: a meta-analysis. Br. J. Edu. Technol. **52**(5), 1935–1964 (2021)
3. Angrist, N.: Mapping the global learning crisis. Educ. Next 22(2). (2022)
4. Levesque, K., Bardack, S., Chigeda, A., Bahlibi, A., Winiko, S.: Two-year RCT of EdTech in Malawi (2022)
5. Piech, C., et al.: Deep knowledge tracing. In: Advances in Neural Information Processing Systems, p. 28 (2015)
6. Ngware, M., et al.: Impact evaluation of Tayari school readiness program in Kenya (2018)
7. UNICEF: The state of the global education crisis: a path to recovery (2021)
8. Vaquero, L.M., Twomey, N., Dias, M.P., Camplani, M., Hardman, R.: Towards continuous compounding effects and agile practices in educational experimentation. arXiv preprint arXiv:2112.01243 (2021)
9. Ros, R., Runeson, P.: Continuous experimentation and A/B testing: a mapping study. In: Proceedings of the 4th International Workshop on Rapid Continuous Software Engineering, pp. 35–41, May 2018
10. Savi, A.O., Ruijs, N.M., Maris, G.K., van der Maas, H.L.: Delaying access to a problem-skipping option increases effortful practice: application of an A/B test in large-scale online learning. Comput. Educ. **119**, 84–94 (2018)
11. Ritter, S., Murphy, A., Fancsali, S.: Curriculum-embedded experimentation. In: Proceedings of the Third Workshop on A/B Testing and Platform-Enabled Research (at Learning@ Scale 2022) (2022)
12. Friedberg, A.: Introducing EIDU's solver platform: facilitating open collaboration in AI to help solve the global learning crisis. In: Proceedings of the 23rd International Conference on Artificial Intelligence in Education, AIED 2022, Part II, Durham, UK, 27–31 July 2022, pp. 104–108. Springer International Publishing, Cham, July 2022. https://doi.org/10.1007/978-3-031-11647-6_18

A Case Study on AIED Unplugged Applied to Public Policy for Learning Recovery Post-pandemic in Brazil

Carlos Portela[1,2(✉)] ⓘ, Rodrigo Lisbôa[1,3] ⓘ, Koiti Yasojima[1,3] ⓘ,
Thiago Cordeiro[1,4] ⓘ, Alan Silva[1,4] ⓘ, Diego Dermeval[1,4] ⓘ, Leonardo Marques[1,4] ⓘ,
Jário Santos[1,5] ⓘ, Rafael Mello[1,6] ⓘ, Valmir Macário[1,6] ⓘ, Ig Ibert Bittencourt[1,4,7] ⓘ,
and Seiji Isotani[1,5,7] ⓘ

[1] Center for Excellence in Social Technologies (NEES), Maceió, Brazil
csp@ufpa.br, {rodrigo.lisboa,koiti.yasojima}@ufra.edu.br,
{thiago,alanpedro,ig.ibert}@ic.ufal.br,
diego.matos@famed.ufal.br, leonardo.marques@cedu.ufal.br,
{rafael.mello,valmir.macario}@ufrpe.br, sisotani@imc.usp.br
[2] Federal University of Pará, Cametá, Brazil
[3] Federal Rural University of Amazônia, Paragominas and Capitão Poço, Brazil
[4] Federal University of Alagoas, Maceió, Brazil
[5] University of São Paulo, São Paulo, Brazil
[6] Federal Rural University of Pernambuco, Recife, Brazil
[7] Harvard Graduate School of Education, Cambridge, MA, United States

Abstract. This article discusses the opportunities and challenges of applying Artificial Intelligence in Education (AIED) unplugged in Brazil's post-pandemic context from a Public Policy that considers the national reality. In this context, an intelligent platform is being adopted to support learning recovery, focusing on developing writing skills at elementary school. To meet this objective, AI algorithms were developed with computer vision for transcribing hand-written texts and natural language processing techniques for correcting texts in Portuguese and giving feedback to students to improve their writing skills.

Keywords: Artificial Intelligence · Public Policy · Learning Recovery

1 AIED Implementation in Learning Recovery

1.1 Learning Poverty in a Post-pandemic World

According to World Bank [1], learning poverty was very high even before the Covid-19 pandemic hit because, in 2019, the average global learning poverty rate in low- and middle-income countries was 57 percent. The pandemic has closed schools, leading to an unprecedented disruption of learning worldwide. A simulation based on the latest available data and evidence indicates that the pandemic caused a risen from 57 to 70 percent in Global learning poverty (Fig. 1).

N. Wang et al. (Eds.): AIED 2023, CCIS 1831, pp. 788–796, 2023.
https://doi.org/10.1007/978-3-031-36336-8_120

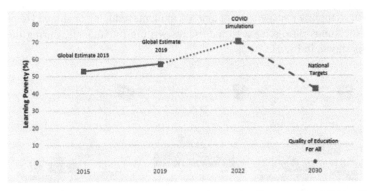

Fig. 1. Data, simulation results, and target – Global Learning Poverty from 2015 to 2030 [1].

The low- and middle-income countries are aligned with a collective goal to halve learning poverty, and their plan implies reducing learning poverty to 42 percent by 2030. To achieve this target, governments, educators, civil society, and stakeholders' partners must collectively commit to learning recovery, since now. In the specific case of Brazil, the Covid-19 pandemic caused an increase in students with Portuguese language problems, which scaled from 15.5% in 2019 to 33.8% in 2021 [2].

1.2 Learning Recovery in Brazil

To reduce these problems, the Brazilian Ministry of Education and World Bank defined a Brazilian Policy of Learning Recovery in Basic Education (decree N. 11.079, May 2022), which aims to encourage implementing teaching practices that promote improving the quality of education with equity.

From this policy, efforts focused on diagnosing the scenario of learning Portuguese language in the public education. According to the Basic Education Assessment System (SAEB) [3], there was a drop in learning levels in the Portuguese language from 2017 to 2021. The students with the worst performances were those in the first years of elementary school, the age group most affected by the schools closed during the pandemic, as shown in the graphs in Fig. 2.

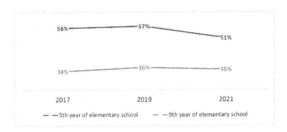

Fig. 2. Learning levels in the Portuguese language - Brazil from 2017 to 2021 [3].

With the temporary suspension of classes, students reducing or stopped to practicing writing. The status quo, related to the essays writing by students, during the remote teaching is shown in Fig. 3.

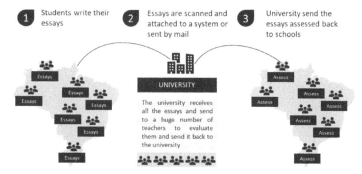

Fig. 3. Status quo during remote teaching in Brazil.

The problems identified in this approach were the high cost with scanners or mail services, high cost with evaluation of the essays, high cost and high complexity or teacher training, time-consuming, and no further support provided.

This approach was not scalable due to the increasing the burden on teachers and considering the social inequalities of the country, which means a high number of schools without internet access and digital devices, high rate of teacher with low or intermediate levels of digital skills, lack of teachers and many vulnerable schools (especially in North and Northeast of Brazil), lack of qualified teachers to provide feedback on writing essays.

Given a post-pandemic scenario and the classes return, how to improve the writing skills of students without increasing the burden on teachers?

Research points out that technologies that assess text production can help to improve public school students' writing skills. However, most of research focus is on textual correction techniques and accuracy. In this context, we develop a solution that can evaluate and correct textual production, focusing on providing personalized feedback for students using Artificial Intelligence in Education (AIED), grouping them according to their learning level [4] and tackle the problem at scale in Brazil.

1.3 Intelligent Platform to Support Developing Writing Skills

An Intelligent Platform (http://plataforma-integrada.nees.ufal.br/) was developed to support developing writing skills for elementary school students in Brazil. It groups several associated technologies that, when executed sequentially, allow the digitalization, correction, and pedagogical diagnosis of handwritten texts in Portuguese, applying Artificial Intelligence in Education (AIED) unplugged.

Initially, a paper-based and human-AI approach is used to analyze handwritten essays (Fig. 4, steps 1 to 6). The students' texts are obtained from Answer Sheets, with QR codes and markings, where handwritten articles are written. A mobile application was developed to capture the digitized texts and data on the student, class, school, and the like. The captured images are accessed by AI algorithms, which perform binarization to improve image quality, recognize relevant areas (Fig. 5-A) and eliminate elements from the manuscript sheet. Then, segmentation of the image into words is performed (Fig. 5-B), reaching an accuracy percentage between 92.5%–95%.

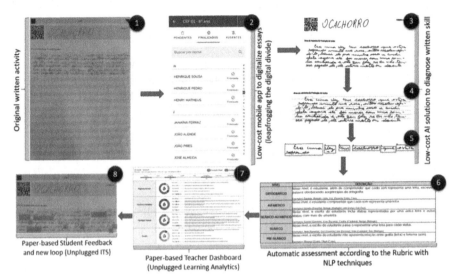

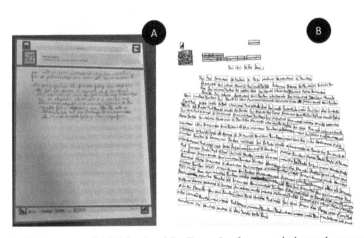

Fig. 4. Example of AIED unplugged applied in the developing writing skills.

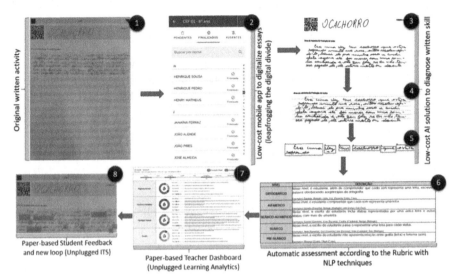

Fig. 5. A - Relevant areas highlighted and B - Example of manuscript's word segmentation.

From this step, the model provides a paper-based dashboard in PDF format (downloaded) using Unplugged Learning Analytics (ULA) (Fig. 4, step 7). After giving feedback, the following educational resource is an Unplugged Intelligent Tutoring System (Fig. 4, step 8). The redesign considered the use of Computer Vision and Natural Language Processing techniques [5] to automatize the assessment of essays and considered two applications of AIED Unplugged: Unplugged Educational Data Mining (Computer Vision and Natural Language Processing to assess handwritten essays), Unplugged Learning Analytics (paper-based dashboard), and unplugged Intelligent Tutoring Systems (to provide paper-based students feedback and recommend next educational resource).

1.4 Impacted Participants

By the end of 2022, more than 1 million digitalization of texts produced by basic education students in Brazil were carried out. Thus, the solution impacted more than 500,000 students, 20,000 teachers, and approximately 7,000 schools, dispersed in about 1,350 cities, reaching all 27 Brazilian states.

2 Challenges and Opportunities

2.1 Implementation and Evaluation

The implementation of AIED in Brazil learning recovery followed four stages based on Teaching at the Right Level (TaRL) [4], as shown in Fig. 6.

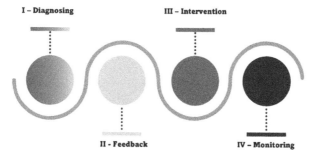

Fig. 6. AIED unplugged applied in Brazilian Policy on Learning Recovery.

In stage I, it is necessary to diagnose students' writing level. The teacher must submit a written proposal. Then the students write and hand in the text for correction. The teacher corrects and asks the student to rewrite, making adjustments. Using a mobile application, teachers can scan texts through an application. Then, the written text will be analyzed by a Human-AI approach, and feedback will be elaborated for the teachers. At this time, a reduction in the burden on teachers is expected.

450 essays from the elementary school were considered to assess the entire process. Each competency was evaluated with scores between 1 and 5. Two evaluators with experience correcting essays did the correction independently (Evaluator A and Evaluator B). In this way, two comparisons were made: 1st – Evaluation between the two human evaluators (blue bar); 2nd – Evaluation of the score recommended by the AI and human evaluators (red bar). Figure 7 presents the results of the evaluation per competence. To

analyze equity in education between urban and rural areas, Fig. 8 presents the final score of evaluation.

Then, in stage II, the feedback is disponible to the teachers. Based on text correction feedback from writing skills, teachers can understand feedback results and group students according to type and level. An intelligent platform suggests grouping students by level and writing competence. One of the main expected results of this policy is reducing teachers' overload. Therefore, technology can be a pedagogical ally, shortening the time of correcting texts but mainly providing systematized data to support teachers' decision-making.

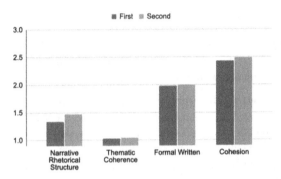

Fig. 7. Evaluation of Human-IA Approach (Score per Competence).

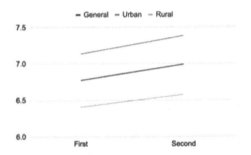

Fig. 8. Final Score (Urban x Rural Areas).

The teacher creates and views students' performance statistics based on the analysis and feedback given on the written and digitized texts. It facilitates the work and expands the faculty's vision of the result and the impact of their actions.

In stage III, it is necessary to carry out the necessary interventions. Teachers must intervene in the recovery of students according to their difficulties with textual writing. Based on AIED feedback, students can be grouped according to the type and level of performance in writing texts. Subsequently, recommended strategies for assisting students according to the difficulties of textual writing begin to be used. In addition, technology can also be used as a learning guide, allowing educators to identify writing styles and select support materials to help students improve their writing skills.

With these students attending, there is an improvement in the quality of the written texts and a consequent improvement in the general performance of the students. The platform provides support material so that the teacher can follow pedagogical paths that serve the students according to the grouping.

Finally, in stage IV, occurs the monitoring of learning recovery. After a recovery cycle, which involves the diagnosis, feedback, and intervention, the teacher must conduct another assessment of the student's writing level. Thus, you can see if there was an improvement in the writing ability of this student. In this way, it is possible to increase student interest in reading, engagement in classes, and school self-efficacy.

2.2 Opportunities

For school managers, this solution is an accountability opportunity. They can use the generated data to assess the effectiveness of teacher training, student progress, and performance policies and apply improvements to the pedagogical environment, such as:

Opportunity 1: Encourage the practice of writing and producing texts. The production of texts is considered a fundamental skill for evaluating students' comprehension and knowledge articulation. In assessing school learning, writing is a source of evaluating the level of understanding and articulation of knowledge. To improve this skill, it is necessary to practice writing texts.

Opportunity 2: Provide adequate feedback for writing improvement. Feedback should be made available to teachers and students according to writing competence: formal record; thematic coherence; textual typology; and cohesion. From this feedback, students can learn better the content worked and produce higher-quality texts. In Brazil case, it was possible to compare the statistical significance about the first and last text written, digitized, and corrected. The AIED can identify the viable texts and generate adequate feedback. Until feb/2023, was generated 184,180 (1st to 4th year of elementary school) and 177,692 (5th to 9th year of elementary school) dashboard with texts feedback, representing around 36% of processing done by AIED with viable texts to correction.

Opportunity 3: Expand the solution to other countries. Currently, the solution only corrects Brazilian Portuguese text production, but it can be expanded to other languages and countries according to your particularities. The expansion into other languages may occur as the flow for digitizing, recognizing, and extracting text is already well established. The application of the policy to other countries is feasible, as its application in Brazil allowed for identifying and overcoming several challenges in improving text writing and learning recovery on scales in a country of continental proportions and social plurality.

2.3 Challenges

Challenge 1: Group students for recovery learning according to their difficulties. The presented policy only suggests of grouping students by level and writing competence. Each competency can be divided into five levels, according to the correction (e.g., level 3 of thematic coherence). In this way, it is recommended to apply the TaRL approach in learning recovery.

Challenge 2: Conduct pedagogical courses by groups of students. A mediator (teacher or tutor) must carry out pedagogical courses that serve the students according to the grouping. Students should participate and complete the proposed pedagogical courses. Only at the final of this process, its will be possible to evaluation the polity impact in the reduce learning loss caused by the Covid-19 pandemic and to measure the increased equity in learning.

3 Future Steps

3.1 Next Steps

Step 1: Develop a community of practices module in Intelligent Platform. To support the learning recovery, is being developing a community of practices to produce didactic materials to support learning recovery. This module allows teachers create, in a collaborative way, materials to help students improve their writing skills. Additionally, it provides curation functionality for evaluating and approving material.

Step 2: Establish an education equity journey. To replicate the approach proposed in this policy, it is intended to promote a journey to recover learning equally to students. Thus, it will be possible to manage programs to have a customization of each policy knowing that they follow the same flow for creation and use AIED solutions.

3.2 Improvement Points

Point 1: Improve the feedback for students' text correction. This improvement point depends on the consolidation of a reference matrix for text correction according to writing competence: formal record; thematic coherence; textual typology; and cohesion. It depends on the consensus of Portuguese language teachers for the algorithm training. A standard matrix is being validated with the Ministry of Education.

Point 2: Improve the grouping of students for recovery learning. The grouping students is not yet being realized by the intelligent platform. At this moment, the grouping is being carried out by the teachers according to the feedback obtained. The grouping student's functionality is under development on the platform.

Point 3: Improve the impact analysis of the policy. Despite the promising results, the Policy of Learning Recovery in Basic Education (decree N. 11.079, May 2022) does not have a single year. Therefore, it is still not possible to assess its real impact on the reduce learning poverty caused by the Covid-19 pandemic. Data continues to be collected on the platform so that the impact of the policy can be measured.

References

1. World Bank Homepage. https://www.worldbank.org/en/topic/education/publication/state-of-global-learning-poverty. Accessed 01 Mar 2023
2. Basic Education Census of Brazil. https://censobasico.inep.gov.br. Accessed 03 Mar 2023
3. Basic Education Assessment System. https://www.gov.br/inep/pt-br/areas-de-atuacao/avaliacao-e-exames-educacionais/saeb. Accessed 03 Mar 2023
4. Teaching at the Right Level Homepage. https://www.teachingattherightlevel.org/. Accessed 05 Mar 2023
5. Bulut, O., MacIntosh, A., Walsh, C.: Leveraging natural language processing for quality assurance of a situational judgement test. In: 23rd International Proceedings on AIED 2022. Springer, Durham (2022). https://doi.org/10.1007/978-3-031-11647-6_14

Enabling Individualized and Adaptive Learning – The Value of an AI-Based Recommender System for Users of Adult and Continuing Education Platforms

Sabine Digel[1(✉)], Thorsten Krause[2], and Carmen Biel[3]

[1] Universität Tübingen, Münzgasse 11, 72070 Tübingen, Germany
sabine.digel@uni-tuebingen.de
[2] German Research Center for Artificial Intelligence, Parkstr. 40, 49080 Osnabrück, Germany
thorsten.krause@dfki.de
[3] German Institute for Adult Education, Heinemannstr. 12–14, 53175 Bonn, Germany
biel@die-bonn.de

Abstract. The extent to which individualized and adaptive learning can be supported by recommender systems is increasingly being discussed in the field of adult and continuing education (ACE). Aspects of accessibility and customization of learning platforms play just as much a role as the added value of AI from a pedagogical perspective. This paper addresses the question of how recommender systems can be used to support self-directed learning of adult learners with heterogenous prerequisites and learning needs. Building on the initial situation of the target group as well as assumptions of learning opportunity-use models, an idea of AI and humans acting in partnership is designed and at the same time made to the object of investigation.

Keywords: Recommender System · Individualized and Adaptive Learning · Education · Adult and Continuing Education (ACE)

1 Introduction

Artificial intelligence is advancing to become a central topic in the debates on the quality of adult and continuing education (ACE) conducted in science and the public. There is tension between the high discursive expectations and the low diffusion of AI in practice. On the one hand, the low use has to do with the fact that the necessary technical infrastructures as a basis for the use of AI are not available in large parts of the educational landscape [6]. On the other hand, there is a lack of concepts for using AI-based technologies in a didactically meaningful way [9].

In this context, a learning goal-oriented integration of support systems as well as transparency of the integration for the users should be considered. With the image of an equal coexistence of humans and AI technology [4], the potential of recommender systems for individualized and adaptive learning unfolds. Learners can be supported

N. Wang et al. (Eds.): AIED 2023, CCIS 1831, pp. 797–803, 2023.
https://doi.org/10.1007/978-3-031-36336-8_121

in their learning decisions and learning processes by recommender systems that refer to learners' learning prerequisites, learning interactions, and learning outcomes. Such tailored accessibility of learning offers, in turn, enables a low-effort and flexible use of further education of practitioners with especially time-limited resources.

This is the starting point of the project KUPPEL for the development of an AI-supported learning architecture using a recommender system for demand-oriented and sustainable competence development of the heterogeneous target group of teachers in adult and continuing education.

2 The Recommender System of the KUPPEL Cloud to Support Learners' Decisions and Learning Processes

2.1 Initial Situation

The existing ACE platforms are often based on step-by-step models of expertise development, are therefore linear in structure, and do not take sufficient account of the different profiles and learning needs of teachers due to their different approaches to ACE. Furthermore, the range of learning platforms offered is designed in an institution- and provider-specific manner, a phenomenon that applies to the entire adult and continuing education landscape [12]. There is a lack of cross-topic and cross-provider offerings or solutions that help integrate offerings of existing ACE platforms and enable individualized and adaptive learning.

If – which is rare – recommender systems are implemented in digital education offerings, mostly recommendation types based on content criteria or the usage behavior of the mass of learners are used [3]. Especially in the field of ACE personal profiles of learners and personal learning behavior have hardly been considered to date. Furthermore, it is not known what effects are associated with the different recommendation types and what combinations are conducive to enabling learners towards self-active handling of recommendations and are effective for learning processes and outcomes. Thus, an ideal-typical merging of different algorithm logics with pedagogical approaches is still pending [9].

2.2 Requirements for Recommender Systems for ACE Platforms

One of the main goals of the project KUPPEL is to develop a recommender system prototype for ACE platforms and to address some of these topics. This prototype should support the professional development of ACE teachers while considering the heterogeneity of the ACE field. ACE teachers have several different backgrounds and expertise levels [2] as well as experience in learning online [1] that a recommender system should consider while adapting learning arrangements and making suggestions for the next learning steps.

When identifying requirements for a recommender system, a constructivist approach regarding adult learning, which organizes the learning process in a self-directed and problem-oriented way, and which places the adult learner at the center of a learning process [8] seems to be promising under the described conditions. Nonetheless, and consistent with moderate constructivism, instructional support is also needed as digital learning environments and self-directed learning have their own prerequisites [8].

Therefore, the recommender system should support adaptivity in learning, while providing assistance and reasoning about the given recommendations (explainable AI) so that ACE teachers are more aware of their potential needs, and can choose accordingly to their current demands and in line with their level of expertise and experience.

When asking for relevant parameters to suggest recommendations in a pedagogically meaningful way learning opportunity-use models can provide a heuristic. The models are based on the assumption that characteristics of learning offers, learners' learning prerequisites, their learning activities, and their learning output are interdependent components for the quality of teaching and learning [13]. Defining quality by satisfaction and learning success of the self-directed learner, the question arises as to how a recommender system can combine learning object data, personal profile data, learning behavior data, and learning result data in a way to promote an attractive and profitable individualized and adaptive learning for users of ACE platforms. This question is the focus of the submitted paper.

2.3 Trial Architecture

Testing the recommender prototype, a viable basis in terms of content is needed. For this purpose, we developed a competence-based framework curriculum for digitalization training for ACE teachers which can be based on different ACE platforms. The curriculum: DTrain is divided into a total of six modules which, when completed in its entirety, should have a learning scope of approx. 130 h. For two of these modules, we will use already existing learning units (comprising contents and tasks) as well as adapt and create new learning units and assessments for two learner types during the project period. Each learning unit is provided with a comprehensive set of metadata (e.g., on competencies regarding the curriculum, module and thematic affiliation, expected learning duration, addressed learner type, modus of learning, etc.) which the recommender system utilizes to generate learning sequences according to the mentioned requirements.

In the testing phase during the project, learning units located on two different ACE platforms. The two learning management systems are connected to each other through a virtual cloud via interfaces, are used. These interfaces send necessary information (metadata and learner data) from the corresponding platforms to the virtual cloud, where the recommender system is located.

The developed recommender system adaptively recombines existing learning sequences from multiple learning management systems into new ones. To adapt recommendations as much to the user as possible, we employ several data sources:

- Personal profile data including demographic data, learning interests, and requirements
- Personal learning behavior data including navigational patterns and status data
- Personal learning results including completed learning units and assessments
- Other learners' profile data and learning behavior data

We use different recommendation mechanisms to process the different data types: Knowledge-based systems use concrete rules derived from expert knowledge and provide recommendations based on personal data. They are best for providing science- and experience-based recommendations but cannot incorporate information about other learners' data as such rules would be too complex to define. Content-based systems provide recommendations if navigational patterns may be used, i.e., last visited learning units. They can recommend sequences similar to previously visited content but are naturally unable to incorporate other learners' data. Collaborative filtering systems provide recommendations based on other users' learning behavior and personal data [11]. Provided sufficiently much data, they can accurately determine which content a user will likely interact with next. Hence, they are most flexible and do not require hand-written rules. But it is also harder to incorporate domain knowledge into their recommendations. If multiple systems are eligible for use, we average the corresponding outputs with a hybrid recommendation scheme.

Sequence generation always starts with an empty sequence. The system iteratively chooses a learning unit and appends it to the sequence until the user's preferred sequence length is reached. Knowledge-based modules inform the systems in each iteration which learning units are eligible to append without violating human-defined rules.

3   Challenges and Opportunities

3.1   Filter Technologies and Criteria for the Recommendations

Adaptively recombining learning sequences from different platforms poses several challenges, especially to collaborative filtering methods.

First, collaborative filtering methods aim to minimize a mathematical objective that the researcher must specify. In the context of learning recommendations, such objectives could be to maximize competency or satisfaction. These data are, however, scarce as users only perform few assessments and provide few ratings. Consequentially, recommendations become inaccurate. For this reason, modern practice and science across several domains focus on implicit feedback, data about who consumes what, to generate recommendations. Users generate implicit feedback automatically and in large numbers, enabling more accurate recommendations. The corresponding objective is to maximize the likelihood of a user interacting with a recommendation and may be less beneficial to the learning process. Choosing an objective, therefore, implies a tradeoff between accuracy and relevance.

Second, not all platforms can provide all types of data, i.e., about how much time a user spent on a particular task due to technical limitations. Our recommendation system must therefore be able to handle sparse user and item data.

Third, collaborative filtering requires a fully connected network of user-item interactions. Initially, however, users of one learning platform have not yet interacted with the content of another. Equivalently, items have only been interacted with by users from their source platform. The orthogonality between the platform's interaction graphs prevents collaborative filtering models to predict how useful the content of one platform is to users of the other. This problem is similar to the cold start problem in recommender systems. We seek to alleviate it by considering the competency framework and curriculum-related information associated with each learning unit.

Fourth and contrarily to classical recommender systems, recombining learning sequences is not an item-to-item recommendation task. Instead, the task is to recommend a sequence of learning units based on the previous sequences the user interacted with. Recommending sequences of learning units, or – in modular curricula – sets of learning units is a basket recommendation task [14], requires advanced inference methods. Dealing with sequence-wise instead of item-wise feedback poses another challenge: The individual learning unit's contribution to a consumed learning sequence's quality cannot directly be estimated. The sequence may have been interacted with by the user because of or despite the individual learning unit. This problem has been dealt with before, for example, in the context of natural language processing [15], and as well requires advanced inference methods. We will iteratively extend our prototype to tackle these constraints.

Last, exploring the solution space, which recommender systems rely on, can generate many infeasible learning sequences. For example, we found that 57 learning sequences each containing 4 learning units can be recombined into 190 billion new sequences of which only 0.18 percent are feasible. To shield learners from low-quality output, we reduce the number of infeasible learning sequences during exploration by constraining the solution space with a knowledge-based approach that filters out most infeasible learning sequences.

3.2 Integration of Recommendations in the ACE Platform

Places and Times to Provide Recommendations. Due to the well-explored heterogenic learning prerequisites of teachers in ACE [2] there are varying demands when it comes to when and where recommendations should be made. A low level of affinity for technology on the one hand, and the fact that teachers like to make decisions autonomously on the other hand, the frequency and abundance of recommendations should be carefully dosed. Furthermore – as the critical analysis of learning paths in adaptive and AI learning programs from Kerres et al. [10] shows - small-step diagnosis of learning levels can significantly disrupt the flow of the learning process.

Referring to these challenges we designed our recommendation API for maximum flexibility so that LMS providers can decide when and where their users should receive recommendations. In the testing phase during the project we plan to provide two places where learners will receive recommendations: 1) on their dashboard after login and 2) at the end of each learning sequence. The dashboard helps learners discover the learning offer. Its recommendations include a variety of topics, segmented into typical teaching-learning situations and competence facets. At the end of each learning sequence, we recommend similar learning sequences to allow learners to directly continue their path.

Autonomy of the System Versus the Learners. Due to the support needed by learners related to their prerequisites to cope with self-directed learning, tension arises between automatically generated recommendations of the recommender system and autonomous decisions regarding the recommendations on the part of the users. Neither the machine nor the human is perfect. A recommender system not constrained by rules can identify patterns in data, apply them to new cases and in the process learn implicit rules of which people themselves are not aware [7]. When AI decides, it should choose the best alternative not only from an educational but also from an ethical perspective [7]. In any case, the transparency of the procedure towards adult learners appears to be crucial. They should be able to understand why they receive which recommendations. On the basis of explanations that point out the reasons for the personalized recommendations and open up several possible paths, it may even be possible to make the decision itself the responsibility of the learners. In this context, AI is not seen as reduced to information transfer in the background, but as an expert supporting self-directed learning.

Referring to this challenge we plan to give learners a choice whether they allow personal profile data, personal learning data, personal learning results as well as the profile and learning behavior data of the other users to be included or excluded from the recommendations they receive. By (de)activating checkboxes the learners are able to change their choice at any time.

3.3 Developmental Research to Meet the Requirements and Needs of the Target Group and Explore Concrete Implementation Ideas

In order to really adapt the recommender system to the needs of our target group of teachers in adult education, our developments are gradually accompanied by evaluation studies which aim to collect requirements and needs as well as to analyze the acceptance and the usability of the designed solutions from the user's point of view. In doing so, we follow the basic idea of design-based research and regard the practitioners as development partners with whom we would like to find feasible ways of dealing with the challenges outlined above [5]. In this way, we hope to ensure a high level of customization and sustainability of our developments.

Likewise, the interoperable nature of the developed architecture serves to enable other adult and continuing education providers to join, thus creating larger networks of AI-enabled ACE platforms, which in turn can strengthen the reach and accessibility of the offering in practice. As a basis for such an upscaling, technical and content-related success factors for connecting third parties to the Recommender API are also identified.

4 Conclusion and Outlook

Since the added value of recommender systems for educational platforms depends in any case on the readiness of the addressed target groups to accept the technical support, it is worth investing time in a careful exploration of possible scenarios. Already during the development, it seems to be essential to include the users' perspective in order to promote their acceptance.

From a didactic point of view, the way we have chosen to create an integrated recommender system that meets as many different needs as possible and lets the adult learners decide for themselves whether and when they receive which recommendations and in which form seems promising. Further research following the project period will have to show to what extent the learners make use of the individual controllability of recommendations and what competence growth adult learners can achieve on the basis of which adaptive learning paths.

Acknowledgments. This contribution originates from the research project KUPPEL (KI-unterstützte plattformübergreifende Professionalisierung erwachsenen-pädagogischer Lehrkräfte), funded by the German Federal Ministry of Education and Research (BMBF), ref. no. 21INVI0802, 21INVI0803, 21INVI0805.

References

1. Bildungsberichterstattung, A. (ed.) Bildung in Deutschland 2020. wbv Media (2020). https://doi.org/10.3278/6001820gw
2. Autorengruppe wb-personalmonitor (ed.) Das Personal in der Weiterbildung. Arbeits- und Beschäftigungsbedingungen, Qualifikationen, Einstellungen zu Arbeit und Beruf. DIE Survey. wbv Media (2016)
3. Biel, C., Brandt, P., Hellmich, C., Kilian, L., Schöb, S.: Lern-Empfehlungen von der Maschine - Adaptives Recommending am Beispiel von EULE. weiter bilden **26**, 22–25 (2019)
4. Borgwardt, A.: Bit für Bit in die Zukunft - Künstliche Intelligenz in Wissenschaft und Forschung. Friedrich-Ebert-Stiftung, Berlin (2020)
5. Edelson, D.C.: Design research – what we learn when we engage in design. J. Learn. Sci. **11**, 105–112 (2002)
6. Fischer, F., Hartmann, P., Mattern, D., Mayer, P., Paul, L. (eds.) Künstliche Intelligenz in der Bildung der Zukunft – individuell, flexibel, vernetzt und lebenslang? Whitepaper. DHBW Karlsruhe (2020)
7. Herrmann, A.: Autonome KI als Partner des Menschen – Ethische Perspektiven im Spannungsfeld von Kommerzialisierung und Kollaboration. In: R. Fürst (ed.) Digitale Bildung und Künstliche Intelligenz in Deutschland. Nachhaltige Wettbewerbsfähigkeit und Zukunftsagenda, pp. 257–270. Springer, Wiesbaden (2020)
8. Kerres, M.: Mediendidaktik: Konzeption und Entwicklung digitaler Lernangebote, 5. Aufl., De Gruyter Studium, Oldenbourg (2018). https://doi.org/10.1515/9783110456837
9. Kerres, M., Buntins, K.: Recommender in AI-enhanced learning: an assessment from the perspective of instructional design. Open Educ. Stud. **2**, 101–111 (2020). https://doi.org/10.1515/edu-2020-0119
10. Kerres, M., Buntins, K., Buchner, J., Drachsler, H., Zawacki-Richter, O.: Lernpfade in adaptiven und künstlich-intelligenten Lernprogrammen: Eine kritische Analyse aus Sicht der Mediendidaktik. In: de Witt, C., Gloerfeld, C., Wrede, S.E. (eds.) Künstliche Intelligenz in der Bildung. Springer, Wiesbaden (in press)
11. Krause, T., Stattkus, D., Deriyeva, A., Beinke, J.H., Thomas, O.: Beyond the rating matrix: debiasing implicit feedback loops in collaborative filtering. In: Wirtschaftsinformatik Proceedings, p. 12 (2022). https://aisel.aisnet.org/wi2022/ai/ai/12
12. Schrader, J.: Fortbildung von Lehrenden der Erwachsenenbildung: Notwendig? Sinnvoll? Möglich? In: Schrader, J., Hohmann, R., Hartz, S. (eds.) Mediengestützte Fallarbeit. Konzepte, Erfahrungen und Befunde zur Kompetenzentwicklung von Erwachsenenbildnern, pp. 25–68. Bertelsmann, Bielefeld (2010)
13. Seidel, T.: Angebots-Nutzungs-Modelle in der Unterrichtspsychologie. Integration von Struktur- und Prozessparadigma. Zeitschrift für Pädagogik **60**, 850–866 (2014). https://doi.org/10.25656/01:14686
14. Vančura, V: Neural basket embedding for sequential recommendation. In: Proceedings of the 15th ACM Conference on Recommender Systems (2021)
15. Williams, J.D., Raux, R.: Henderson, M: The dialog state tracking challenge series: a review. Dialogue Discourse **7**, 4–33 (2016)

"Learning Recorder" that Helps Lesson Study of Collaborative Learning

Hajime Shirouzu[1]([envelope]) [ORCID], Moegi Saito[1], Shinya Iikubo[1], and Kumiko Menda[2]

[1] Nahomi Institute for the Learning Sciences, Kanagawa, Japan
{shirouzu,saitomoegi,iikubo}@ni-coref.or.jp
[2] Akiota Town Tsutsuga Elementary School, Hiroshima, Japan
tsutsugashokoucho@gakko.akiota.jp

Abstract. Renovation of education requires students to engage in a newer style of learning, or *collaborative learning* and teachers to implement *lesson study* of that learning. Such learning, however, creates a situation wherein multiple student groups simultaneously engage in dialogues in a noisy classroom. Thus, we developed a "Learning Recorder" system, which has a 360-degree video camera to record students when set in the center of the group, and sends the voice data to an Automatic Speech Recognition system for immediate transcription. This paper explains how this system has evolved from our practical experiences with AI-powered lesson study and how it helped even a novice teacher look back and learn from the student dialogue. We propose that an information appliance like the Learning Recorder draws teachers' attention to student learning, solicits multiple interpretations, and brings about collaborative learning among teachers.

Keywords: Collaborative Learning · Lesson Study · Knowledge Constructive Jigsaw · Automatic Speech Recognition · Learning Recorder

1 Introduction

Renovation of education requires students to engage in a newer style of learning, or *collaborative learning* [1] and teachers to implement *lesson study* [2] of that learning. The newly revised Japanese curriculum guidelines, the Courses of Study, for example, introduced "proactive, interactive, and deep learning", emphasizing teachers' collaborative lesson study to ensure such learning in each and every classroom. However, teachers lack appropriate tools to do this. We thus developed a series of learning-in-class monitoring systems and tested them in hundreds of educational settings from elementary school to high school over a period of 10 years. In this paper, we report how our practical experiences produced, or crystalized into a "Learning Recorder" system, and how the system helped teachers by showing an example of lesson study by a novice teacher and his supportive, professional community.

In this section, we first outline an image of student collaborative learning, then explain the teachers' lesson study, and finally explore the technology they required.

1.1 Student Collaborative Learning

A fourth-grade science class is just beginning. Students are settling into their seats as the teacher poses a question: Why do we use ceiling fans as well as heaters if we want to quickly warm up the whole room in the winter?

Instead of listening to a lecture, students are assigned one of three pieces of information. One piece refers to the direction of the air flow from an air conditioner depending on the season. Another refers to the different weights of cold and warm air while the third explains the mechanism of how hot-air balloons fly. The classroom buzzes with energy as the students find clues to the main question. Next, students discuss their findings in small groups, integrating the ideas of their classmates. They question, compare, explore, speculate and reflect. Their voices fill the room, challenging the preconceived notion that classrooms should be quiet. Natural curiosity drives them towards an answer and helps them retain the knowledge for longer.

What I just described is not a typical Japanese classroom. It is part of a growing network of schools across Japan adopting a new method–one that encourages creativity, collaboration and problem-solving rather than uniformity and memorization.

At the Consortium for Renovating Education of the Future (CoREF)–launched at the University of Tokyo in 2008 and taken over by the Nahomi Institute for Learning Sciences from 2021–we have helped to transform Japan's educational system by applying the latest research in the cognitive and learning sciences.

What we have found is that an active, collaborative approach to learning can improve the quality of the education. The method we promote–the Knowledge Constructive Jigsaw method [3]–encourages students to piece together a deeper understanding of a topic by considering it from multiple angles, working as part of a group.

1.2 Teachers' Lesson Study

It is a big departure from traditional education in Japan, which is why we must consider every aspect of implementation–not only the effect on the students, but also the impact on the teachers, lesson plans, resources and technology.

To use the Knowledge Constructive Jigsaw method in their classrooms, teachers must adapt. Instead of delivering a static curriculum to passive students, they should follow an active "Plan-Do-Check-Act (PDCA)" cycle–designing student-centric lessons, monitoring and providing feedback on student activity, and collaborating with colleagues so as to continuously improve. This cycle, originating from Japan and recognized as an effective form of professional development [2], in which a group of teachers plan lessons, observe live classroom lessons, collect data, and perform an analysis, is already known as *lesson study*. Renovation calls for revitalization of lesson study.

In theory, this is a great idea. In practice, it is more difficult. Inside classrooms, teachers simply cannot monitor all of the student dialogues, as there are likely to be multiple small groups working simultaneously. Outside the classroom, they rarely find colleagues with the same subject matter or grade in their schools; about 40% of elementary schools and 20% of junior high schools have less than one class per grade in Japan. Simply compiling a grounded theory of the instructional method and the community established around such theory and method does not help tackle this challenge.

1.3 Assistive Technology for Lesson Study

Given the constraints of the traditional classrooms and schools in Japan, we see technology as the best option in overcoming the above challenge.

Here is how the technology should work in the cycle of lesson study:

1. When teachers create their lesson plans, they look for tested plans and recommendations in a *digital teaching platform* [4] that includes the database of lessons.
2. The teachers hold discussions with their colleagues and researchers across Japan via the platform in order to simulate how their students will learn. Then, they tell a *learning-in-class monitoring system* which key words or phrases they expect will come up during the lesson.
3. In the actual lesson, the monitoring system records and transcribes the student dialogues, scanning for expected key words and phrases.
4. After the lesson, the teachers review the output of the system along with audio- and video-recording to assess student learning and refine the lesson plans while creating a rule of thumb of how students learn.
5. Lessons are archived in the database of the platform for other teachers to use, with principles of lesson design being gradually extracted.

We developed two types of technology: the *"Gakufu* (Learning Note) system" [5] as the digital teaching platform and the *"Gakkan* (Learning Viewer) system" as the monitoring system. In this paper, we focus on the latter system due to space constraints, especially on how the Learning Recorder expanded from the Learning Viewer to embed the functions of recording, transcribing and reviewing in an all-in-one device.

2 System Development Along with Community Growth

2.1 Community Growth for Lesson Study

This study builds on the community of CoREF, focusing on the "New Learning Project", which began in 2010 mainly for teachers of elementary and junior high schools. As of the end of 2022, 1,162 teachers from 28 organizations, including the boards of education (hereafter BOE), in 19 prefectures have participated.

CoREF builds on a theory of *constructive interaction* [6], which explains that when two persons, engaging in solving a shared problem, exchange the roles of a task-doer who proposes solutions and a monitor who reflects upon them, such role exchange potentially deepens each participating individual's understanding.

The Knowledge Constructive Jigsaw is an instructional framework for constructive interaction to take place in the class. It consists of five learning steps for students: (1) writing an individual answer to the day's given problem, (2) an expert-group activity which allows each student to accumulate three pieces of knowledge relevant in solving the problem, (3) a jigsaw-type activity where students from different expert groups get together to exchange and integrate the pieces of relevant knowledge and form an answer, (4) a crosstalk activity to exchange their ideas for solutions across groups in the entire class, and (5) writing down an individual answer again to the same problem. Each student works independently (Step 1), takes responsibility for the role division (Step 2) and the role exchange within groups (Step 3) and across groups (Step 4), and compiles, in the end, what he or she has learned (Step 5).

The method serves as an instructional framework as well as the stage for the collaborative lesson study of teachers, since it requires teachers to design the learning content, that is, the "problem (main task)" and "learning materials (for the expert activity)".

For this purpose, the project used a mailing list (hereafter ML) for online discussion on the planning and reporting of the lessons in each subject group across the schools and the BOEs from the initial year 2010. To date, 7,099 contributions have been made to the ML. In addition, CoREF devised a common format of a "teaching plan" and "reflection sheet", accompanied by "teaching materials (e.g., worksheets, readings)", which makes discussions smoother. At the end of every fiscal year, CoREF publishes an annual report with a DVD which includes sets of a plan, materials and reflection sheets of tested practices. As of the end of FY2022, there are a total of 1,002 teaching materials, covering all subjects in all grades of elementary and junior high schools. This resource is not intended to guarantee the success of the lesson, but to provide other teachers with chances to modify it to learn from the dialogues of their *own* students.

In this way, the community as a whole helped the participating teachers work together to take turns teaching the lesson as the main teacher (task-doer) and observing it (monitor), in order to ensure constructive interaction also took place among the teachers. Such interaction potentially let each teacher turn the PDCA cycle in a timely manner while enriching their knowledge about student learning and lesson design.

The teachers, however, needed additional supports for assessment of student learning, not in the sense of summative assessment for ranking individuals but in the sense of formative assessment for improvement of lessons and student learning. Although the Knowledge Constructive Jigsaw gives information on learning, since it lets each student write down their answers to the same question twice (Steps 1 and 5), the teachers want to find out why the answers changed and what kind of dialogue took place during the two points in time. This is why an AI-powered lesson study offers advantages.

2.2 System Development: From Learning Viewer to Learning Recorder

CoREF aims to support dialogue analysis, by auto-transcribing the students' dialogues by means of the Learning Viewer and through the provision of transcripts, which are electronically searchable using keywords by means of a subordinate component of the Learning Viewer, the "Conversation Analyzer" system. The initial version of the Conversation Analyzer had two functions (Fig. 1): first, highlighting utterances that included the keywords of the lesson, and second, changing the scope of analysis between one group (zoom-in) and all groups or the whole class (zoom-out).

So as to be able to provide transcription data, the Learning Viewer was equipped with an automatic speech recognition function (IBM Watson Speech to Text). In order to accumulate dialogue data and create an acoustic model of the children's voices, CoREF used a set of a unidirectional headset microphone and IC-recorder per student, through which accuracy was improved [7]. The two systems were then integrated so that the teachers reviewed the transcribed dialogues while listening to the actual voices of the students. Yet, it took CoREF a month to receive data from the teachers, have it recognized by the Learning Viewer and visualize it on the Conversation Analyzer (due to the underlying difficulty in synchronizing the voice data of multiple students).

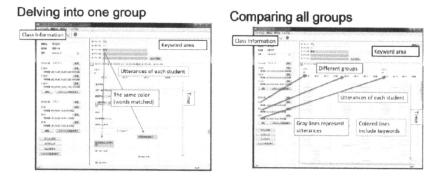

Fig. 1. Conversation Analyzer (left: Zoom-in window; right: Zoom-out window).

a. Shooting group students.   b. Video and transcription windows.

Fig. 2. Learning Recorder.

CoREF thus developed the Learning Recorder, which has a 360-degree video camera to record students when set in the center of the group (Fig. 2a: black square box), and sends the voice data to the Learning Viewer or broadcasts it live. Directly after the lesson, the teachers are able to use the transcription of jigsaw dialogue synchronized with video data. They scan the transcription, hold a green, triangular indicator to jump to anywhere they like, and watch the video while listening to the dialogue (Fig. 2b).

Although a similar video-recorder of group activities exists in Japan [8], it lacks the transcribing function. Such a function is built into many smart speakers like Alexa, but they do not specialize in visualizing group dialogues. With the transcribing and visualizing functions, the teachers have control over how and to what extent they reflect upon student learning through using AI, without being replaced by AI.

To date, the Learning Viewer and Recorder have been used in 189 lessons for 2,526 students ranging from elementary to high school across almost all subjects. The systems were usable in the standard internet environment of the schools without any technical supports. Although the speech recognition accuracy remained at around 60% to 70% for the student group conversations, the teachers were able to interpret them because of their rich understanding of the context for the conversations. They even registered keywords into the Learning Viewer to raise the level of recognition accuracy.

Those systems, especially when coupled with a "hypothesis-testing lesson study" [5], made a big difference. In this style of lesson study, participating teachers 1) listen to the hypothesis of the main teacher prior to the lesson, 2) observe the same student or group together with other teachers during the lesson, 3) share and discuss observations on the learning processes of target students compared with the hypothesis of the main teacher, 4) make recommendations on the lesson design and the teacher's moves, 5) listen to the self-reflection of the main teacher, and 6) write down their own findings from the lesson study. The Learning Recorder was used in the second activity, the data of which was used in the third activity *after* sharing and discussion without the system. The teachers often found many *missed utterances* of the students and newer aspect of student learning processes by *reinterpreting* how they deepened their understanding.

3 The Case of a Novice Teacher

3.1 Method: Data Source

We report here how such reinterpretation took place even within a novice teacher with the help of other teachers. The lesson study was held in an elementary school of a small town, a partner BOE of the New Learning Project, in January 2023. This teacher YO had just graduated from university to work for this town from April 2022. Two other teachers (supervisors from the BOE) attended the pre- and post-lesson studies.

The lesson was a Grade 4 mathematics class on the unit of "composite figures". YO found appropriate materials from the Learning Note, which asked "How do you find the area of this figure?" (Fig. 3a). YO solved this problem by himself, only to think that it was too difficult for his students, even though they had mastered how to calculate the areas of rectangles and squares. Then YO studied the Courses of Study and several other lesson plans to learn that the learning objective of this unit was "to be able to find the area of a composite figure by detecting rectangles and squares through decomposing, synthesizing and deforming". He further located the email in the Learning Note that said, "It is *after* confirming the 'equation and answer' that students turn to focus on 'ways of thinking' such as justification".

YO devised three expert materials: "decomposing the whole figure into the square and rectangle (and add)" (expert material A), "filling in the top-left square to make the bigger rectangle (and subtract)" (material B), and "making two of the same figures to synthesize them to make the biggest rectangle (and divide)" (material C; Fig. 3b). All the materials omitted numerical information on the length, since the teacher expected his students to focus not on the calculation but on the "ways of thinking".

3.2 Results

Ten students in the class tackled the problem. Each group of three or four took responsibility for each expert material; three students who took charge of different materials gathered as a jigsaw group (three groups in total; one group had four members). They reached the correct answer mainly by decomposition and synthesis in the crosstalk.

One supervisor KM observed the expert group using material A; the other one YY did the same for the expert group using material C. They also observed different groups in the jigsaw activity. All the group dialogues were recorded by the Learning Recorder.

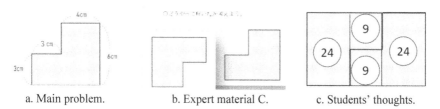

a. Main problem.   b. Expert material C.   c. Students' thoughts.

Fig. 3. Problem, Materials and Students' Thoughts for a Grade 4 Math Lesson.

Directly after the lesson, YO expressed the unexpected success of his lesson. In the post-lesson study, the two supervisors shared their observations with the teacher. Table 1 shows YY's notes on the student dialogue around expert material C. Yet, she missed the utterances from Lines 2 to 5 (the part shaded in gray in Table 1). As a result, the children *appeared* to understand the material correctly, since it required understanding of the concept of *doubling the area*, which would be later necessary in learning about the area of triangles. However, when three teachers went on to discuss the student learning in the jigsaw activity, KM wondered why the students did not use the concept of doubling, if the students from expert C group had successfully understood it.

Then they looked back at the dialogue data on the Conversation Analyzer, only to find that the murmur-like utterances from Lines 2 to 5 had been missed. The students broke the composite figure into two parts, brought numerical information of the lengths from the main problem (Fig. 3a) into the expert material (Fig. 3b), added the areas of all parts (Fig. 3c; Line 6 in Table 1), and failed in understanding the concept of doubling. The three teachers, novice as well as veteran, were all surprised and excited that they reinterpreted the learning process of the students to make sense of it.

Teacher YO reported, "I had my hands full with proceeding with my plan during the lesson, but now understand what I thought I observed was different from what the students thought by using these systems. Besides, my instruction at Line 5 might have diverted students from their natural courses of thinking which goes back and forth between the abstract and concrete, and hurried them up towards reaching the solution".

Table 1. Student dialogue on expert material C.

Line	Speaker	Utterances
1	C1	Ah, how about this? If we look at it in this way, we find a rectangle…
2	C1	But I do not know how to calculate this
3	C2	It is 3 cm long. These are of the same length as this (pointing to the main problem in Fig. 3a), 4 cm and 6 cm
4	C1	Then it is nine square centimeters, nine plus nine equals eighteen, and eighteen plus twenty-four plus twenty-four (Fig. 3c)
5	T	You do not have to talk about figures, but just about how you think
6	C2	If we put them (four rectangles in Fig. 3c) together, it becomes the (biggest) rectangle
7	C1	Yes. Then we add up all the areas, and then divide, right?
8	C3	Oh, that is what the material means? I've got it
9	C2	The sum is 66 square centimeters, and 66 divided by two equals 33
10	C1	Now we've solved it. (C3: Yes, we did.)

4 Benefits, Future Steps, and Challenges

As benefits, the Learning Recorder, coupled with pedagogy (the theory of constructive interaction, and the Knowledge Constructive Jigsaw), the community (the New Learning Project, and the BOE members), and the renovated method of lesson study (the hypothesis-testing lesson study), helped teachers look back and learn from the student dialogue. It was not only effective but also efficient, for example, taking only two hours for the pre-lesson study and seventy minutes for the post-lesson study in the case above.

As a future step, we propose that an *information appliance* like the Learning Recorder, when set in the center of the student group and in the center of the cycle of teachers' lesson study, draws teachers' attention to student learning, solicits multiple interpretations, brings about collaborative learning among the teachers, and leads them to become *collaborative, reflective practitioners.*

Challenges for that step are 1) how to establish and scale up a teacher community and 2) how to analyze and visualize the teacher development through the system use.

Acknowledgement. This work was supported by JSPS KAKENHI to the first (No. 20K20816), second (23K02770) and third authors (23K02727) and Akiota Town Board of Education.

References

1. Hmelo-Silver, C. E., Chinn, C. A., Chan, C., O'Donnell, A. M.: The international handbook of collaborative learning. Routledge, Oxon (2013). https://doi.org/10.4324/9780203837290
2. Lewis, C.: Lesson study. Research for Better Schools (2002)

3. Miyake, N.: Case report 5: Knowledge construction with technology in Japanese classrooms (CoREF). In: Kampylis, P., Law, N., Punie, Y. (eds.) ICT-Enabled Innovation for Learning in Europe and Asia, pp.78–90. European Commission, Joint Research Centre (2013)
4. Dede, C., Richards, J.: Digital Teaching Platforms: Customizing Classroom Learning for Each Student. Teachers College Press, New York (2012)
5. Iikubo, S., Shirouzu, H., Saito, M.: Design-based research to improve the lesson study project for co-evolution of research, development and practices. In Chinn, C., Tan, E., Chan, C., Kali, Y. (eds.) Proceedings of the 16th International Conference of the Learning Sciences, pp. 1669–1672. ISLS, Hiroshima (2022). https://doi.org/10.22318/icls2022.1669
6. Miyake, N.: Constructive interaction and the iterative process of understanding. Cogn. Sci. **10**(2), 151–177 (1986)
7. Shirouzu. H., Saito, M., Iikubo, S., Nakayama, T., Hori, K.: Renovating assessment for the future. In Kay, J., Luckin, R. (eds.) Proceedings of the 13th International Conference of the Learning Sciences, pp.1807–1814. ISLS, London (2018)
8. Hylable Homepage. https://www.hylable.com/. Accessed 5 Mar 2023

"Learning Note" that Helps Teachers' Lesson Study Across Time and Space

Shinya Iikubo[1], Hajime Shirouzu[1][(✉)] [ID], Moegi Saito[1], and Hideko Hagiwara[2]

[1] Nahomi Institute for the Learning Sciences, Kawasaki, Kanagawa, Japan
{iikubo,shirouzu,saitomoegi}@ni-coref.or.jp
[2] Akiota Town Kake Elementary School, Hiroshima, Japan
kakeshokoucho@gakko.akiota.jp

Abstract. Renovation of education requires students to engage in a newer style of learning, or *collaborative learning*. For the design of such a lesson, teachers need to develop a new type of knowledge: *pedagogical content knowledge* (PCK) and *horizon content knowledge*. Mere sharing of lesson materials makes little contribution to such knowledge development, since the materials are only end products, and do not include their development processes. Thus, we developed a *digital teaching platform*, the "Learning Note", which links discussions on lesson design with lesson materials, while laying out the materials on unit maps for teachers to acquire a horizontal view of long-term learning processes. This paper explains the overall results of how the teachers used this platform, showing the example of a novice teacher who managed to design a lesson in the course of a unit with the help of her colleagues. We propose that the digital teaching platform enables teachers to engage in continuous lesson study across time and space.

Keywords: Collaborative Learning · Lesson Study · Knowledge Constructive Jigsaw · Digital Teaching Platform · Learning Note

1 Introduction

Renovation of education requires students to engage in a newer style of learning, or *collaborative learning* [1] and teachers to implement *lesson study* [2]. The newly revised Japanese curriculum guidelines, the Courses of Study, for example, introduced "proactive, interactive, and deep learning", emphasizing teachers' lesson study to ensure such learning in each and every classroom. However, teachers lack appropriate tools to do this. We thus developed an online community for lesson study from 2010 and a *digital teaching platform* [3] from 2018, and used the platform together with a total of about 500 elementary and secondary teachers. It is also an important issue for AIED to design the *digital teaching platform* for collaborative learning other than personalized, e-learning systems such as Udemy and Khan Academy. This paper reports the overall results of how the teachers used this platform, showing the example of a novice teacher who managed to design a lesson in the course of a unit with the help of her colleagues. In this section, we explain the knowledge that teachers need to develop, the lesson study that they should engage in, and the technology that they required.

N. Wang et al. (Eds.): AIED 2023, CCIS 1831, pp. 813–820, 2023.
https://doi.org/10.1007/978-3-031-36336-8_123

1.1 Teachers' Knowledge and Lesson Study

Educational renovation calls for a new instructional method that encourages creativity, collaboration and problem-solving rather than uniformity and memorization. It is a big departure from traditional education in Japan, which is why we must consider every aspect of implementation–not only the effect on the students, but also the impact on the teachers, lesson plans, resources and technology.

First, teachers have to design lessons using both content knowledge and pedagogical knowledge, the integrated knowledge of which Shulman [4] referred to as *pedagogical content knowledge* (PCK).

Second, even when teachers succeed in designing lessons using pedagogical content knowledge, they also need to acquire *horizon content knowledge* (i.e., understanding of how topics are connected over the span of the curriculum across years) [5].

The teachers should follow an active "Plan-Do-Check-Act (PDCA)" cycle–designing student-centric lessons, monitoring and providing feedback on student activity, and collaborating with colleagues so as to continuously improve lessons and student learning for a long term. This cycle, originating from Japan and recognized as an effective form of professional development [2], in which a group of teachers plan lessons, observe live classroom lessons, collect data, and perform an analysis, is already known as *lesson study*. Renovation calls for revitalization of lesson study.

In theory, this is a great idea. In practice, it is more difficult. Japanese teachers rarely find colleagues with the same subject matter or grade in their schools; about 40% of elementary schools and 20% of junior high schools have less than one class per grade.

We thus developed a set of an online community and a digital teaching platform to support lesson study across time and space.

1.2 Technology for Lesson Study and Digital Teaching Platform

Given the constraints of the traditional classrooms and schools in Japan, we see technology as the best option in overcoming the above challenge.

Here is how the technology should work in the cycle of lesson study:

1. When teachers create their lesson plans, they look for tested plans and recommendations in the digital teaching platform that includes the database of lessons.
2. They hold discussions with their colleagues and researchers across Japan via the platform in order to simulate how their students will learn. Then, they tell a *monitoring system* which key words or phrases they expect will come up in the lesson.
3. In the actual lesson, the monitoring system records and transcribes the student dialogues, scanning for expected key words and phrases.
4. After the lesson, the teachers review the output of the system along with audio- and video-recording to assess student learning and refine the lesson plans while creating a rule of thumb of how students learn.
5. Lessons are archived in the database of the platform for other teachers to use, with principles of lesson design being gradually extracted.

We developed two types of technology: the "*Gakufu* (Learning Note) system" as the digital teaching platform and the "*Gakkan* (Learning Viewer) system" as the monitoring system. In this paper, we focus on the former system due to space constraints.

Before describing the system and its usage, we will briefly explain the background of the digital teaching platform in the learning sciences [6]. Online communication tools such as emails and bulletin boards were used from an early stage, all efforts of which led to confirm the importance of the *online community* for teacher learning such as in building a resource library by archiving the best interactions. In addition, the design-based research often required decisions and customization by each teacher for classroom learning without any direct support from researchers or master teachers. Such a situation called for a technology-supported environment for teaching, called digital teaching platforms that are designed to support teacher planning and reflection.

In Japan, however, websites that introduce teaching materials often only present lesson plans (instructional plans) and materials, and it is often unclear what kind of student learning takes place. This leads teachers to accept the materials as *best practice*, i.e., something that should be done without any changes being made in class. In this study, we aimed to prevent this by adding a *reflection sheet*, so that the teaching materials would be accepted as *just an example* of lesson study and become a resource for each teacher's own trial and error research on teaching materials.

2 Online Community: CoREF

This study builds on the community of the Consortium for Renovating Education of the Future (CoREF)–launched at the University of Tokyo in 2008 and taken over by the Nahomi Institute for Learning Sciences from 2021. CoREF helped to transform Japan's educational system by applying the cognitive and learning sciences. This paper focuses on the "New Learning Project", which began in 2010 mainly for teachers of elementary and junior high schools. As of the end of 2022, 1,162 teachers from 28 organizations, including the boards of education (hereafter BOE), in 19 prefectures have participated.

CoREF built on a theory of *constructive interaction* [7], and introduced an instructional method of "Knowledge Constructive Jigsaw" [8] for constructive interaction to take place in the class. The method also serves as the stage for the lesson study of teachers, since it requires teachers to design the learning content.

For this purpose, the project used a mailing list (hereafter ML) for online discussion on the planning and reporting of the lessons in each subject group across the schools and the BOEs from the initial year 2010. To date, 7,099 contributions have been made to the ML. In addition, CoREF devised a common format of a "teaching plan" and "reflection sheet", accompanied by "teaching materials (e.g., worksheets, readings)", which makes discussions smoother. At the end of every fiscal year, CoREF publishes an annual report with a DVD which includes sets of a plan, materials and reflection sheets of tested practices. As of the end of FY2022, there are a total of 1,002 teaching materials, covering all subjects in all grades of elementary and junior high school. This resource is not intended to guarantee the success of the lesson to other teachers, but to provide them with opportunities to modify it to learn from their *own* students.

In this way, the community helped the participating teachers, experienced as well as novice, work together to engage in the lesson study by taking turns teaching the lesson as the main teacher and observing it, in order to ensure constructive interaction took place. Such interaction potentially let each teacher turn the PDCA cycle in a timely manner while enriching their knowledge about student learning and lesson design.

3   Digital Teaching Platform: Learning Note

Various issues, however, emerged from the accumulated trials of lesson study supported by the mature technologies like the mailing list and DVD. The teachers, for example, needed a usable platform that linked the ML discussions with the lesson materials, while retaining the benefit of a convenient mailing list.

Thus, CoREF developed a website specifically for the project members, the Learning Note system, which collects and categorizes the emails posted on the ML by lesson topic, displays posts with attached files in chronological order, enables searches by keywords and automatically recommends related topics. It was named the "Topics" page and put into operation in October 2018. It stored 7,099 emails along with attached files such as plans and materials, and categorized them using a simple, n-gram algorithm into 1,949 topics. It also had an automatic recommendation function for topics with similar content using a Doc2Vec algorithm.

In May 2019, we added a "Developed Materials" page that incorporated data from the developed materials and allowed users to search aspects by subject, grade level, year, etc., and to browse through links to related topics (Fig. 1).

The "Topics" page and the "Developed Materials" page made it possible to link the flow of information from the ML discussions on lesson design with the stock of developed materials resulting from the cycle of lesson study. This revealed that the teaching materials ultimately taught in the lesson undergo discussions via the ML and repeated revisions before they are complete, and enable each teacher to learn from this process.

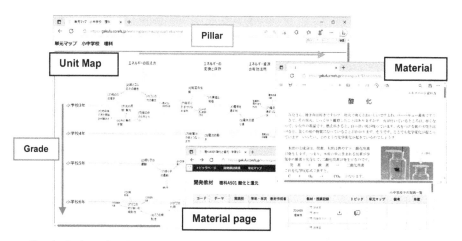

Fig. 1. Referencing developed teaching materials (in front) from the unit map (behind).

The "Topics" page and the "Developed Materials" page, however, only deal with one lesson at a time. In April 2021, we added a "Unit Map" page to enable visualization of the relationship between the lessons and units in order to support the creation of horizon content knowledge (Fig. 1). The Courses of Study provided a map-like visualization of the content structure. For instance, the science curriculum of the elementary and junior high school units is structured around the pillars of "energy", "particles", "life" and "earth", and are illustrated in the diagram horizontally through a pillar structure and vertically by grade level. The "Energy" pillar consists of three items: "How to understand energy", "Energy conversion and conservation" and "Effective use of energy resources", with units related to each being arranged. We linked this map to the teaching materials developed and accumulated in this project manually in science, mathematics and social science. This allows the user to understand how units are connected over the span of the curriculum across years and to design a single lesson in such a way that it links to the learning that comes before and after it.

4 Results: Real Use in the Field

4.1 Daily Operational Results

To date, 485 members such as teachers, administrators and researchers, or 41.7% of the community, have registered for this system, which has generated a total of 107,597 page views (PV) in four years and five months. Figure 2 shows the monthly status of account registrations and page views after the start of the Learning Note. As described above, the "Topics" page became operational in October 2018, the "Developed Materials" page was added in May 2019, and the "Unit Map" page was added in April 2021. At the beginning of the addition of each page, and about once a year for the later years, we introduced the system to all participants using a mailing list, explaining the addition and benefits of the pages and how to register.

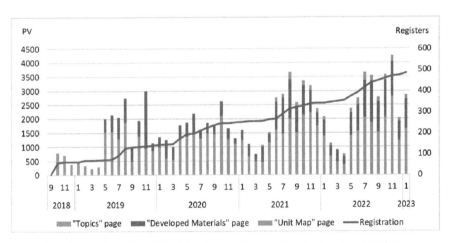

Fig. 2. Number of monthly page views and accumulated account registrations.

From the monthly trends of registered users and page views in Fig. 2, the number of registrations increases most after each announcement while gradually increasing in the other months. The number of PVs was counted separately for the "Topics" page, "Developed Materials" page, and "Unit Map" page. The number of PVs increases with the number of account registrations at the time of system-related announcements at the beginning of the fiscal year, further increases from summer to fall when there are many occasions requiring the creation of lessons such as open classes, and then decreases towards the end of the fiscal year (the Japanese fiscal year starts from April and ends in March). This periodicity in the number of PVs closely portrays the active and inactive periods of the teachers' lesson study, or the reality of their working style.

As shown in Fig. 2, the teachers tend to view the "Topics" page the most (about 60% of the whole PV), and then the "Developed Materials" page (about 30%), indicating that they do not use this platform as a resource to download materials and copy them for their lesson but as a resource to learn from it. Yet, they do not make full use of the "Unit Map" page (only about 3%), and the transition between these pages is also inactive, implying that an issue of flexible, dynamic search remains.

4.2 The Case of a Novice Teacher

In order to explore both possible benefits and future challenges of the system, we report how a novice teacher used the Learning Note with the help of other teachers. The lesson study was held in an elementary school of a small town, a partner BOE of the New Learning Project, in November 2022. The teacher, KM, had been working as an elementary school teacher for six years, and had moved to this school two years earlier.

Although it had been scheduled that Teacher KM would practice the open class of the sixth graders in November, she had not decided what subject and unit she would teach (Japanese elementary school teachers teach all subjects). She was also wondering whether she should design the lesson from scratch or not. Veteran teachers advised her to pick *any* materials appropriate for the period of the lesson from the Learning Note.

Figure 3 shows the pages that KM explored. She searched for materials from the unit on the "Edo period" in social science both from the "Developed Materials" and the "Unit Map" pages. In the former, she checked the boxes of "Social Science" and "Grade 6", and entered "Edo period" into the search box (top-left page in Fig. 3). In the latter, she looked for materials in the gray boxes linked to the white box of the unit "Early-Modern History" (bottom-right page in Fig. 3). Clicking the gray boxes led her to the corresponding materials pages. Two relevant materials (bottom-left columns in Fig. 3) came up, which dealt with "the townspeople culture of the Edo period". KM consulted the veteran teachers about comparing the two sets of materials. They dealt with the same content, had a similar main problem, and contained similar information. Yet, they differed in the complexity of the reading materials and the period when each lesson had been taught: one was taught at the beginning of the unit, but the other had been at the end. KM decided to combine the strengths of the two: making them easy to read and comprehensive enough to serve as an *advanced organizer* for the unit.

An in-person pre-lesson study took place in the school just two days before the lesson, with a review of the exchanges of emails and the reflection sheets on the two lessons. Then they simulated how their students would deepen their understanding, in what areas they might have difficulty, and what support they might require.

The pre-lesson study took only 20 min, but the teachers including KM realized the importance of thinking about students' prerequisites for understanding this lesson (e.g., industrial development in the Edo period had made the townspeople richer than before),

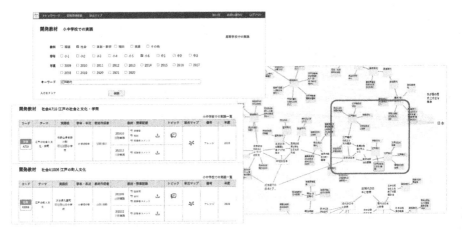

Fig. 3. Pages on the Learning Note explored by the teacher.

the role of this lesson on students' learning of the unit (i.e., having an outlook on the unit, the details of which would be elaborated in the following lessons) and the depth of understanding that KM should pursue as an instructional goal in this lesson.

In the evening, one veteran teacher posted an email on the ML on behalf of KM, explaining the content and materials of KM's lesson with reference to the two materials that she had built on, and the result of their lesson study. Within an hour, two researchers reacted to that email: one researcher simulated the students' learning processes in detail by reviewing the two original materials, while the other encouraged this style of *efficient* lesson study to other teachers and BOEs. Both comments gave KM confidence as well as making her point of view clearer on student learning. KM devised various things, for example, preparing pre-reading materials which were intended to enrich the students' image of townspeople and farmers before and after the Edo period.

In the lesson, the students reached her expectations for some points, for example, getting to know that the Edo period had fewer battles and the townspeople enjoyed peace, but not for other points, for example, not being able to imagine the "improvement of the people's standard of living" at that time.

Although the veteran teachers attended the lesson and observed it with the help of the Learning Viewer, the observations were made not only by them but also by KM. She further explained that she would ask the same question "Why did the townspeople culture prevail in the Edo period?" again at the end of the unit in order to examine to what extent her students would have deepened their understanding.

This implies that the pre-lesson study using the Learning Note enriches observations on student dialogues and brings insightful understanding of student learning and lesson design. Especially when modifying existing materials, even a novice teacher is able to focus on the differences between the original and the modified materials, and the relationship between the instructional goal and the actual learning processes of the students.

5 Benefits, Challenges, and Future Steps

As benefits, although the Learning Note only utilizes mature AI technologies (N-gram, Doc2Vec), it helps even a novice teacher acquire the PCK through letting her not simply copy the existing teaching material but retrace its development process, compare the instructional goal with the actual student learning, and compare her plan with the original one. In addition, the visualization of the long-term learning structure (the Unit Map) helps the teacher adjust the goal of each lesson and deal flexibly with the diverse ideas of the students, depending on the *horizon content knowledge*, or at least a longer-term view into the learning processes. Moreover, each teacher not only learns from the platform, but also contributes to it through posting their thoughts, materials and reflections on the ML. Such accumulation then serves as a resource for new comers. As a result, the teachers engage in continuous lesson study across time and space.

One of the challenges is that the Learning Note fulfills its potential when embedded in the living community of lesson study. Although KM initially used the system for practical purposes, she gradually moved towards *deeper lesson study* under the implicit guidance of the veteran teachers. They guided her around the system such as suggesting where to look and how to make full use of information. We have not been able to succeed in recreating such human guidance through AI. As future steps, we intend to clarify how each practitioner uses this digital platform and in what kind of interaction with others in a physical environment, in order to combine the strengths of both the digital and physical environments, towards creating more effective but efficient lesson study.

Acknowledgement. This work was supported by JSPS KAKENHI to the first (No. 23K02727), second (23K02770) and third authors (20K20816) and Akiota Town Board of Education.

References

1. Hmelo-Silver, C.E., Chinn, C.A., Chan, C., O'Donnell, A.M.: The International Handbook of Collaborative Learning. Routledge, Oxon (2013)
2. Lewis, C.: Lesson Study: A Handbook of Teacher-Led Instructional Change. Research for Better Schools, Philadelphia (2002)
3. Dede, C., Richards, J.: Digital Teaching Platforms. Teachers College Press, New York (2012)
4. Shulman, L.S.: Those who understand. Educ. Res. **15**(2), 4–14 (1986)
5. Ball, D.L., Thames, M.H., Phelps, G.: Content knowledge for teaching: what makes it special? J. Teach. Educ. **59**(5), 389–407 (2008)
6. Fishman, B.J., Davis, E.A., Carol, K.K.C.: A learning sciences perspective on teacher learning research. In: Sawyer, R.K. (ed.) The Cambridge Handbook of the Learning Sciences, 2nd edn., pp. 707–725. Cambridge University Press, New York (2014)
7. Miyake, N.: Constructive interaction and the iterative process of understanding. Cogn. Sci. **10**(2), 151–177 (1986). https://doi.org/10.1016/S0364-0213(86)80002-7
8. Miyake, N.: Case report 5: knowledge construction with technology in Japanese classrooms (CoREF). In: Kampylis, P., Law, N., Punie, Y. (eds.) ICT-Enabled Innovation for Learning in Europe and Asia, pp. 78–90. European Commission, Joint Research Centre (2013)

Author Index

Printed in the United States
by Baker & Taylor Publisher Services